T0140053

Lecture Notes in Computer Science 13579

More information about this series at https://link.springer.com/bookseries/558

Xiang Zhao · Shiyu Yang · Xin Wang ·
Jianxin Li (Eds.)

Web Information Systems and Applications

19th International Conference, WISA 2022
Dalian, China, September 16–18, 2022
Proceedings

Editors
Xiang Zhao
National University of Defense Technology
Changsha, China

Xin Wang
Tianjin University
Tianjin, China

Shiyu Yang
Guangzhou University
Guangzhou, China

Jianxin Li ⓘ
Deakin University
Melbourne, VIC, Australia

ISSN 0302-9743   ISSN 1611-3349 (electronic)
Lecture Notes in Computer Science
ISBN 978-3-031-20308-4   ISBN 978-3-031-20309-1 (eBook)
https://doi.org/10.1007/978-3-031-20309-1

This Springer imprint is published by the registered company Springer Nature Switzerland AG
The registered company address is: Gewerbestrasse 11, 6330 Cham, Switzerland

Preface

It is our great pleasure to present the proceedings of the 19th Web Information Systems and Applications Conference (WISA 2022). WISA 2022 was organized by the China Computer Federation Technical Committee on Information Systems (CCF TCIS) and Dalian Ocean University. WISA 2022 provided a premium forum for researchers, professionals, practitioners, and officers closely related to information systems and applications to discuss the theme of artificial intelligence and information systems, focusing on difficult and critical issues as well as the promotion of innovative technology for new application areas of information systems.

WISA 2022 was held in Dalian, Liaoning, China, during September 16–18, 2022. WISA 2022 focused on intelligent cities, government information systems, intelligent medical care, fintech, and network security, emphasizing the technology used to solve the difficult and critical problems in data sharing, data governance, knowledge graphs, and block chains.

This year we received 214 submissions, each of which was assigned to at least three Program Committee (PC) members to review. The peer review process was double blind. The thoughtful discussions on each paper by the PC resulted in the selection of 42 full research papers (an acceptance rate of 19.63%) and 22 short papers. The program of WISA 2022 included keynote speeches and topic-specific invited talks by famous experts in various areas of artificial intelligence and information systems to share their cutting-edge technologies and views about the state of the art in academia and industry. The other events included industrial forums, the CCF TCIS salon, and a PhD forum.

We are grateful to the general chairs, Ge Yu (Northeastern University) and Zhijun Li (Dalian Ocean University), as well as all the PC members and external reviewers who contributed their time and expertise to the paper reviewing process. We would like to thank all the members of the Organizing Committee, and the many volunteers, for their great support in the conference organization. Especially, we would like to thank publication chairs, Xiang Zhao (National University of Defense Technology), Shiyu Yang (Guangzhou University), Chunying Li (Guangdong Polytechnic Normal University), and Bin Xu (Northeastern University), for their efforts on the publication of the conference proceedings. Last but not least, many thanks to all the authors who submitted their papers to the conference.

August 2022   Xin Wang
                                                                                                     Jianxin Li

Organization

Steering Committee

Baowen Xu	Nanjing University, China
Ge Yu	Northeastern University, China
Chunxiao Xing	Tsinghua University, China
Ruixuan Li	Huazhong University of Science and Technology, China
Xin Wang	Tianjin University, China

General Chairs

Ge Yu	Northeastern University, China
Zhijun Li	Dalian Ocean University, China

Program Committee Chairs

Xin Wang	Tianjin University, China
Jianxin Li	Deakin University, Australia

Program Committee Vice-chairs

Zhuoming Xu	Hohai University, China
Haofen Wang	Tongji University, China

Workshop Co-chairs

Chunxiao Xing	Tsinghua University, China
Ruixuan Li	Huazhong University of Science and Technology, China

Publication Co-chairs

Xiang Zhao	National University of Defense Technology, China
Shiyu Yang	Guangzhou University, China
Chunying Li	Guangdong Polytechnic Normal University, China
Bin Xu	Northeastern University, China

Publicity Co-chairs

Zhenxing Li   Agile Century, China
Bohan Li   Nanjing University of Aeronautics and
                                Astronautics, China

Sponsor Chair

Bin Cao   Beijing Small and Medium Enterprises
                                Information Service Co., Ltd., China

Website Chair

Mingjian Liu   Dalian Ocean University, China

Organizing Committee Co-chairs

Hong Yu   Dalian Ocean University, China
Han Chen   Dalian Big Data Industry Development Research
                                Institute, China

Organizing Committee Vice-chair

Sijia Zhang   Dalian Ocean University, China

Program Committee

Luyi Bai   Northeastern University, China
Tiecheng Bai   Tarim University, China
Zhifeng Bao   RMIT University, Australia
Bin Cao   Beijing SME Online Co., Ltd., China
Yu Cao   University of Massachusetts Lowell, USA
Xingong Chang   Shanxi University of Finance and Economics,
                                China
Kong Chao   Anhui Polytechnic University, China
Lemen Chao   Renmin University of China, China
Ling Chen   Yangzhou University, China
Yanping Chen   Xi'an Jiaotong University, China
Yanhui Ding   Shandong Normal University, China
Zhiming Ding   Institute of Software, Chinese Academy of
                                Sciences, China
Yongquan Dong   Jiangsu Normal University, China
Yuefeng Du   Liaoning University, China

Yang-Geng Fu	Fuzhou University, China
Guoqiang Gai	Yunhe Enmo (Beijing) Technology Co., Ltd., China
Zaobin Gan	Huazhong University of Science and Technology, China
Ening Gao	Northeastern University, China
Hong Gao	Harbin Institute of Technology, China
Lina Gong	Nanjing University of Aeronautics and Astronautics, China
Qiuyu Guo	Guangzhou University, China
Wenzhong Guo	Fuzhou University, China
Yanbo Han	North China University of Technology, China
Qinming He	Zhejiang University, China
Tieke He	Nanjing University, China
Zhenting Hong	Guangxi University of Science and Technology, China
Mengxing Huang	Hainan University, China
Shi-Lin Huang	IBM, China
Fangjiao Jiang	Jiangsu Normal University, China
Shujuan Jiang	China University of Mining and Technology, China
Weijin Jiang	Xiangtan University, China
Cheqing Jin	East China Normal University, China
Shenggen Ju	Sichuan University, China
Xiangjie Kong	Zhejiang University of Technology, China
Huaizhong Kou	Yellow River Conservancy Commission, China
Yue Kou	Northeastern University, China
Wang-Chien Lee	Pennsylvania State University, USA
Bin Li	Yangzhou University, China
Bohan Li	Nanjing University of Aeronautics and Astronautics, China
Chunying Li	Guangdong Polytechnic Normal University, China
Dong Li	Liaoning University, China
Feifei Li	University of Utah, USA
Jianxin Li	Deakin University, Australia
Juanzi Li	Tsinghua University, China
Lin Li	Wuhan University of Technology, China
Qingzhong Li	Shandong University, China
Ruixuan Li	Huazhong University of Science and Technology, China
Wei Li	Harbin Engineering University, China
Weimin Li	Shanghai University, China
Ximing Li	Jilin University, China

Haofen Wang	Tongji University, China
Jun Wang	iWudao Technology, China
Lin Wang	Japan Advanced Institute of Science and Technology, Japan
Xiaoguang Wang	Virginia Tech University, USA
Xibo Wang	Shenyang University of Technology, China
Xin Wang	Tianjin University, China
Xingce Wang	Beijing Normal University, China
Yanlong Wen	Nankai University, China
Shengli Wu	Jiangsu University, China
Feng Xia	Dalian University of Technology, China
Chunxiao Xing	Tsinghua University, China
Li Xiong	Emery University, USA
Bin Xu	Northeastern University, China
Guandong Xu	University of Technology Sydney, Australia
Huarong Xu	Xiamen University of Technology, China
Lei Xu	Nanjing University, China
Lizhen Xu	Southeast University, China
Zhuoming Xu	Hohai University, China
Zhongmin Yan	Shandong University, China
Shiyu Yang	Guangzhou University, China
Yajun Yang	Tianjin University, China
Hua Yin	Guangdong University of Finance and Economics, China
Jianming Yong	University of Southern Queensland, Australia
Jinguo You	Kunming University of Science and Technology, China
Hong Yu	Dalian Ocean University, China
Jeffrey Xu Yu	Chinese University of Hong Kong, China
Mei Yu	Tianjin University, China
Minghe Yu	Northeastern University, China
Qiao Yu	Jiangsu Normal University, China
Shui Yu	University of Technology Sydney, Australia
Ziqiang Yu	Yantai University, China
Fang Yuan	Hebei University, China
Guan Yuan	China University of Mining and Technology, China
Long Yuan	Nanjing University of Science and Technology, China
Xiaojie Yuan	Nankai University, China
Chen Zhang	CreateLink Technology, China
Fan Zhang	Guangzhou University, China

Guigang Zhang	Institute of Automation, Chinese Academy of Sciences, China
Haiwei Zhang	Nankai University, China
Jing-An Zhang	Shanxi Datong University, China
Meihui Zhang	Beijing Institute of Technology, China
Rui Zhang	University of Melbourne, Australia
Rui-Ling Zhang	Luoyang Normal University, China
Sijia Zhang	Dalian Ocean University, China
Weifeng Zhang	Nanjing University of Posts and Telecommunications, China
Xiaolin Zhang	Inner Mongolia University of Science and Technology, China
Yan Zhang	Mudanjiang Normal University, China
Ying Zhang	Nankai University, China
Yong Zhang	Tsinghua University, China
Yongxin Zhang	Shandong Normal University, China
Zhiqiang Zhang	Zhejiang University of Finance and Economics, China
Erping Zhao	Tibet Nationalities Institute, China
Feng Zhao	Huazhong University of Science and Technology, China
Fengda Zhao	Yanshan University, China
Gansen Zhao	South China Normal University, China
Xiang Zhao	National University of Defense Technology, China
Xiangjun Zhao	Jiangsu Normal University, China
Jiping Zheng	Nanjing University of Aeronautics and Astronautics, China
Jiantao Zhou	Inner Mongolia University, China
Junwu Zhu	Yangzhou University, China
Mingdong Zhu	Henan Institute of Technology, China
Qiaoming Zhu	Soochow University, China
Qingsheng Zhu	Chongqing University, China
Fang Zuo	Henan University, China

Contents

Natural Language Processing

World Wide Web

Machine Learning

Query Processing and Algorithm

Recommendation

Data Privacy and Security

Blockchain

Knowledge Graph

Temporal Knowledge Graph Embedding for Link Prediction

Yi Zhang[1], Zhi Deng[2], Dan Meng[3], Liang Zhou[1], Mengfei Li[1], Qijie Liu[1], and Chao Kong[1(✉)]

[1] School of Computer and Information, Anhui Polytechnic University, Wuhu, China
zhangyi@ahpu.edu.cn, lzhou@ahpu.edu.cn, lmf@stu.ahpu.edu.cn,
lqj@stu.ahpu.edu.cn, kongchao@ahpu.edu.cn
[2] School of Computer Science, Northwestern Polytechnical University, Xi'an, China
dengcai@mail.nwpu.edu.cn
[3] OPPO Research Institute, Shenzhen, China
mengdan@oppo.com

Abstract. Link prediction aims to infer the behavior of the network evolution process by predicting missed or future relationships based on currently observed connections. It has become an attractive area of research since it allows us to understand how networks will evolve. Early studies cast the link prediction task as an entity identifying problem on graphs and adopt vertex representation strategies to perform predictive analysis. Although these methods are effective to some extent, they overlook the special properties of network evolution.

In this paper, we propose a new method named TKGE, short for _Temporal Knowledge Graph Embedding_, to learn the evolutional representations of temporal knowledge graph for link prediction task. Specifically, we employ the self-attention mechanism to incorporate the static structural information and dynamic temporal information by aggregating the context from related entities. By introducing the position embedding characterizing the dynamic information of temporal knowledge graph, TKGE can generate the evolutional embedding of entities and relations for downstream applications, such as link prediction, recommender system, and so on. We conduct experiments on several real datasets. Both quantitative results and qualitative analysis verify the effectiveness and rationality of our TKGE method.

Keywords: Temporal knowledge graph · Representation learning · Self-attention · Link prediction

1 Introduction

The knowledge graph (KG), also known as the semantic network, represents the network of real-world entities and illustrates the relations between them. It

---

Y. Zhang, Z. Deng and D. Meng—These authors contribute equally to this work.

X. Zhao et al. (Eds.): WISA 2022, LNCS 13579, pp. 3–14, 2022.
https://doi.org/10.1007/978-3-031-20309-1_1

has been widely used in various applications, such as language representation learning, question answering, recommender systems, and so on.

To date, existing KGs are usually constructed by machine learning algorithms automatically, but there are many hidden relations that have not been observed, so they are often incomplete. In view of this situation, the link prediction task in KG aims to predict the missing entities or relations, which is also called knowledge inference [1] and knowledge completion [2]. The real-world KGs are usually dynamic, in which more new entities and new types of relations are observed in the KGs over time. We call such KGs with dynamic changes over time as temporal knowledge graphs (TKGs). Figure 1 shows an example of a temporal knowledge graph that evolves over time from the year 1998 to 2013. As can be seen from the observed TKG facts in Fig. 1, Tencent was founded by Pony Ma in 1998 and had multiple relations with other entities from 2010 to 2013, resulting in the semantics of the entity of Tencent shifts over time. Hence, modeling temporal information in KGs is crucial to understanding how the knowledge evolves over time. In this paper, we study the problem of learning TKG embedding for link prediction.

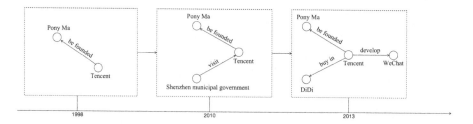

Fig. 1. An example of the evolution of temporal knowledge graph.

However, the TKG embedding for link prediction is often a challenging task due to the following reasons: (1) it is arduous to simulate the strong time dependency in TKG; (2) there are some potential factors that affect the network evolution.

In this paper, we propose an approach, called TKGE, to learn the evolutional embeddings of entities and relations in TKG. We formulate the link prediction task as a conditional probability problem covering entity prediction and relation prediction. For the provided approach, we would like to address the two challenges highlighted earlier. Specifically, we utilize self-attention mechanism to model TKG from both static structural information and dynamic evolutions of the graph to simulate the complex time dependency better. We employ the multi-head self-attention to model the TKG from many aspects. With multi-head self-attention, TKGE can perform very efficiently but still generate accurate evolutional embeddings. Extensive results on the six datasets have shown the effectiveness and rationality of the proposed model.

2   Related Work

The link prediction task on TKG aims to predict missing quadruples. It is further divided into two subtasks: entity prediction and relation prediction. Traditional research on KG link prediction tasks mainly focuses on static KGs, that is, facts do not change with time, and less research on TKGs. However, the time sequence information in the TKG is very helpful to capture the dynamic trend of facts, and the interactive data in the real world is not invariable.

In general, TKG link predictions are categorized in two settings: interpolation and extrapolation. Given a TKG with the time interval $[t_0, t_n]$, interpolation aims to predict missing facts at timestamp t such that $t_0 \leq t \leq t_n$ and extrapolation, which aims to predict new facts at timestamp t such that $t \geq t_n$. For interpolation, DE-SimplE [3] draws inspiration from diachronic word embeddings, developing a diachronic embedding function to generate a hidden representation for the entity at any given time. However, it is hard to help supplement the KG on future timestamps.

For extrapolation, existing methods are achieved through various techniques such as temporal point process framework [4,5], CNN-based methods [6] and deep recurrent models [7,8]. Mei et al., propose N-SM-MPP [4] model for link prediction using the Hawkes process to capture the dynamic temporal information. In addition, DyRep [5] considers representation learning as a latent mediation process. However, these two methods are more suitable for model TKG in continuous time. For the multi-relational, directed graph structure of each KG, REGCN [6] utilize GCN, which it's quite a powerful model for graph-structured data to characterize structural dependencies. Jin et al., develop an autoregressive framework [7], called RE-NET, utilizing recurrent event encoder and neighborhood aggregator to model the information of dynamic temporal and concurrent events in the same timestamp. Despite the effectiveness of the methods mentioned above, these methods have limitations to some degree, because they do not consider the special properties of TKG that is there exist some potential factors that affect the network evolution.

3   Problem Formulation

We first give notations used in this paper and then formalize the TKG representation learning problem to be addressed.

Notations. A TKG is defined as a sequence of static KGs with timestamps, $G = \{G_1, \cdots, G_n\}$ where n represents the number of timestamps. Each KG G_t is a multi-relational, directed graph which contains all the facts that co-occur at timestamp t and we define it as $G_t = \{V, R, \mathcal{E}_t\}$ where V denotes the set of entities, R denotes the set of relations, and $\mathcal{E}_t$ denotes the set of facts at timestamp t. Any fact in $\mathcal{E}_t$ can be represented as a quadruple (subject entity, relation, object entity, timestamp) and is denoted by (s, r, o, t), where $s, o \in V$ and $r \in R$.

Problem Definition. The TKG representation learning task aims to learn latent representations $\mathbf{h}_{o,t} \in \mathbb{R}^{F'}$ and $\mathbf{r}_t \in \mathbb{R}^{F'}$ for each node $o \in V$ and relation

$r \in R$ at time $t = \{1, 2, ..., n\}$, such that $\mathbf{h}_{o,t}$ preserves both the graph structures and dynamic temporal information. Formally, the problem can be defined as:

Input: A sequence of static KGs with timestamps, $G = \{G_1, \cdots, G_n\}$, entity embedding matrices $\mathbf{H}_t$ and relation embedding matrices $\mathbf{R}_t$.

Output: A map function $f : V \cup R \to \mathbb{R}^{F'}$, which maps each entity and relation in G to a F'-dimensional embedding vector.

4   Methodology

In this section, we first introduce our proposed method, TKGE, which consists of two components. The structural self-attention module is used to capture the static structural information of the KG at each timestamp t and the temporal self-attention module is used to capture the evolution of the dynamic KG over time. The overall framework of the model is shown in Fig. 2. First, the structural self-attention module uses self-attention [9,10] to aggregate the local static structural information of each entity in each timestamp t generating structural embeddings for entities. Then, by introducing position embedding, temporal self-attention module can capture the evolution of KG in different timestamps. Finally, based on the representations of the entities and relations learned above, we can use various scoring functions to perform prediction tasks in future timestamps. It is worth mentioning that, intuitively, some latent facts in the real world can affect the evolution of graphics. This is evident in the case of citation networks, papers of different research fields may expand their citation papers at significantly varying rates. Since we introduce multi-head self-attention mechanism [11,12] in the structural self-attention module and temporal self-attention module, which can not only characteristic the evolution of the KG from different aspects but also reduce the deviation of prediction and improve the stability of the model.

4.1   Structural Self-attention

To capture the static structural information in the KG at each timestamp t, this component is designed to realize the mapping from a series of TKGs $G = \{G_1, \cdots, G_n\}$ to a series of entity embedding matrices $\{\mathbf{H}_1, \cdots, \mathbf{H}_t\}$. The initial entity embedding matrix is obtained by random initialization.

Specifically, the structural self-attention mechanism calculates the structural representation of the entity by learning the importance of the relevant entity to the target entity and assigning different weights to each relevant entity: $\mathbf{h}_{o,t} = \sigma(\frac{1}{c_o} \sum \alpha_{s,o} W^s (\mathbf{h}_{s,t} + \mathbf{r}_t))$, where $\mathbf{h}_{s,t}$ and $\mathbf{r}_t$ are the embeddings of entity s and relation r at timestamp t, respectively. c_o is the in-degree of the object entity o. $\sigma(\cdot)$ is a nonlinear activation function. In particular, $\mathbf{h}_{s,t} + \mathbf{r}_t$ represents the translation attribute between the entity and the relation in the triplet. The calculation formula of the specific weight $\alpha_{s,o}$ is as follows:

$$\mathbf{e}_{s,o} = \sigma(A_{s,o} \cdot \alpha^T [W^s \mathbf{h}_{ot} || W^s \mathbf{h}_{s,t}]), \forall (s, r, o) \in \mathcal{E}_t, \tag{1}$$

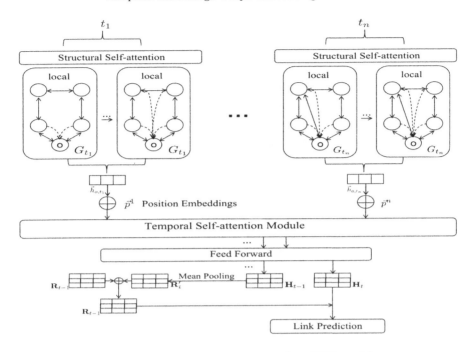

Fig. 2. The overall architecture of the TKGE model.

$$\alpha_{s,o} = \frac{exp(e_{s,o})}{\sum_{(s,r),\exists(s,r,o)\in\mathcal{E}_t}}, \tag{2}$$

where $A_{s,o}$ represents the weight matrix of the link between the subject entity and the object entity at the current timestamp t. In this paper, the number of edge occurrences is used as the weight. $\alpha \in \mathbb{R}^{2D}$ and $W^s \in \mathbb{R}^{F\times D}$ denote the weight vector and the parameter matrices, respectively. $\|$ is the concatenation operation and $\sigma(\cdot)$ denotes the non-linear activation function. Significantly, this module is composed of multiple stacked structural self-attention layers. By stacking multiple layers, our method can further consider the high-order relation between entities to generate the final structural embedding of the entity at each timestamp.

4.2 Temporal Self-attention

To capture the dynamic temporal information in TKG, we first utilize position embedding $\{\mathbf{p}^1, \cdots, \mathbf{p}^\top\}, \mathbf{p}^t \in \mathbb{R}^F$ to describe the dynamic temporal sequence information of each static KG. Then, temporal self-attention module takes a series of embeddings for a particular entity o at different timestamps calculated by $\mathbf{h}_{o,t} + \mathbf{p}^t$ as input and returns a sequence of temporal entity embeddings at different timestamps. Meanwhile, this component utilize GRU to realize the mapping from a series of TKGs $G = \{G_1, \cdots, G_n\}$ to a series of relation embedding matrices $\{\mathbf{R}_1, \cdots, \mathbf{R}_n\}$. The initial relation embedding matrix is obtained by random initialization. Specifically, the input and the output for each entity o is denoted by $\{\mathbf{x}_{o,t_1}, \cdots, \mathbf{x}_{o,t_n}\}, \mathbf{x}_{o,t} \in \mathbb{R}^{D'}$ and $\{\mathbf{h}_{o,t_1}, \cdots, \mathbf{h}_{o,t_n}\}, \mathbf{h}_{o,t} \in \mathbb{R}^{F'}$. Where n, D' and F' denote the number of timestamps and the dimension of the input and the output vector. In practice, we compute the input and the output embeddings for entity o at different timestamps, packed together into a matrix $\mathbf{X}_o \in \mathbb{R}^{T \times D'}$ and $\mathbf{H}_o \in \mathbb{R}^{T \times F'}$. Specifically, the output embedding of the entity is mainly calculated according to $\mathbf{H}_o = \beta_o(\mathbf{X}_o\mathbf{W}_v)$, where $\beta_o \in \mathbb{R}^{T \times T}$ is the weight matrix of attention. This calculation first converts the queries, keys, and values to different spaces through the learned matrix $\mathbf{W}_q \in \mathbb{R}^{D' \times F'}, \mathbf{W}_k \in \mathbb{R}^{D' \times F'}, \mathbf{W}_v \in \mathbb{R}^{D' \times F'}$, and then uses the scaled dot-product attention mechanism to calculate the attention score and weight:

$$e_o^{ij} = (\frac{((\mathbf{X}_o\mathbf{W}_q)(\mathbf{X}_o\mathbf{W}_k)^\top)_{ij}}{\sqrt{F'}} + M_{ij}). \tag{3}$$

$$B_o^{ij} = \frac{exp(e_o^{ij})}{\sum_{k=1}^n exp(e_o^{ik})}, \tag{4}$$

where $\mathbf{M} \in \mathbb{R}^{T \times T}$, is the mask matrix, which is used to encode autoregressive attributes, that is, to use information from a few previous timestamps to describe the information at a later timestamp. If $i \leq j$, $M_{ij} = 0$, otherwise $M_{ij} = -\infty$.

Intuitively, the relation embedding contains the information of the entity in the corresponding fact, so the relation's embedding at timestamp t will be affected by the entity information related to the relation r at timestamp t and its own information at timestamp $t - 1$, where the related entities is denoted by $E_{r,t} = \{i|(i, r, o, t) or (s, r, i, t) \in \mathcal{E}_t\}$. Inspired by [6], by applying the average pooling operation in the embedding matrix of the entities related to the relation r at timestamp $t - 1$, $H_{t-1,E_{r,t}}$, the calculation formula of relation embedding is: $\mathbf{r}'_t = [pooling(\mathbf{H}_{t-1,E_{r,t}})||\mathbf{r}]$, where $\mathbf{r}$ is the embedding vector of the relation r. Especially, when there is no relation at timestamp t, $\mathbf{r}'_t = 0$. Then, according to the relation embedding matrix $\mathbf{R}'$ and $\mathbf{R_{t-1}}$ at timestamp $t - 1$, this paper utilizes GRU to obtain the updated relationship matrix.

After obtaining the evolutional embeddings of entities and relations, we perform link prediction problems in a probabilistic way. Considering the good results shown by GCN in KG link prediction, we choose ConvTransE [13] as the decoder. Therefore, the specific calculation formula of the entity conditional probability vector is as follows:

$$\mathbf{p}(o|s, r, \mathbf{H}_t, \mathbf{R}_t) = \sigma(\mathbf{H}_t ConvTransE(\mathbf{s}_t, \mathbf{r}_t)), \tag{5}$$

where $\sigma(\cdot)$ denotes the sigmoid function. Similarly, we can get the relational conditional probability vector according to the following formula:

$$\mathbf{p}(r|s, o, \mathbf{H}_t, \mathbf{R}_t) = \sigma(\mathbf{R}_t ConvTransE(\mathbf{s}_t, \mathbf{o}_t)), \tag{6}$$

where $\mathbf{s}_t, \mathbf{r}_t$ and $\mathbf{o}_t$ are the corresponding elements in the entity and relation embedding matrices $\mathbf{H}_t$ and $\mathbf{R}_t$, respectively.

4.3 Parameter Learning

This paper takes the entity and relation prediction task as a multi-classification problem, where each category corresponds to an entity or relation, and $\mathbf{y}_{t+1}^o \in \mathbb{R}^{|V|}$ or $\mathbf{y}_{t+1}^r \in \mathbb{R}^{|R|}$ is used to represent the label vectors of the entity or relation prediction task under the $t+1$ timestamp. If a value is 1 indicates the corresponding fact happens. The corresponding loss functions are as follows:

$$L^o = \sum_{t=0}^{n-1} \sum_{(s,r,o,t+1)\in\mathcal{E}_{t+1}} \sum_{i=0}^{|V|-1} y_{t+1,i}^o \log p_i(o|s,r,\mathbf{H}_t,\mathbf{R_t}), \tag{7}$$

$$L^r = \sum_{t=0}^{n-1} \sum_{(s,r,o,t+1)\in\mathcal{E}_{t+1}} \sum_{i=0}^{|R|-1} y_{t+1,i}^r \log p_i(r|s,o,\mathbf{H}_t,\mathbf{R_t}), \tag{8}$$

where n represents the number of timestamps in the training set, $y_{t+1,i}^o$ and $y_{t+1,i}^r$ is the value of the i-th element of $\mathbf{y}_{t+1}^o$ and $\mathbf{y}_{t+1}^r$ respectively.

4.4 Discussion

To verify the efficiency of our algorithm, this section analyzes the time complexity of the model. First, the self-attention mechanism mainly includes three steps: similarity calculation, softmax, and weighted summation. After the analysis, the time complexity of this part is $O(n^2d)$, where n is the maximum length of the sequence, and d is the dimension of embedding. In addition, we use GRU component to calculate relation embedding, where the time complexity of pooling operation is $O(|R|D)$, where D is the maximum number of related entities with relation r at timestamp t, and R is the number of elements in the relation set. Therefore, the total time complexity is $O(n^2d + |R|D)$.

5 Experiments

To evaluate the entity and relation embeddings learned by TKGE, we employ them to address the link prediction problem in TKG. Link prediction in TKG can be regarded as a classification task, which is mainly divided into entity prediction and relation prediction. Through experiments, we aim to answer the following research questions:

RQ1: How does TKGE perform compared with state-of-the-art temporal knowledge graph embedding methods?

RQ2: Is the construction of structural and temporal self-attention component helpful to learn more desirable representations for temporal knowledge graph?

RQ3: How do the key hyper-parameters affect the performance of TKGE?

In what follows, we first introduce the experimental settings, and then answer the above research question in turn.

Table 1. Descriptive statistics of datasets.

Name	$|V|$	$|R|$	$|\varepsilon_{train}|$	$|\varepsilon_{valid}|$	$|\varepsilon_{test}|$	Time interval
ICEWS18	23,033	256	373,018	45,995	49545	24 h
ICEWS14	6,869	230	74,845	8,514	7,371	24 h
ICEWS05-15	10,094	251	368,868	46,302	46,159	24 h
WIKI	12,554	24	539,286	67,538	67,538	1 year
YAGO	10,623	10	161,540	19,523	20,026	1 year
GDELT	7691	240	1,734,399	238,765	305,241	15 mins

5.1   Experimental Settings

Dataset. We utilize six commonly used TKGs datasets: ICEWS18, ICEWS14, ICEWS05-15, WIKI, YAGO, and GDELT. The first three datasets come from the data captured and processed by the Integrated Crisis Early Warning System (ICEW). The data in the GDELT comes from various international news reports, such as the New York Times, Washington Post, etc. The statistics of the datasets are summarized in Table 1.

Evaluation Protocols. To evaluate the performance of link prediction task, we performed the same processing on the dataset as RE-GCN. For the ICEWS14 and ICEWS05-15 datasets, we randomly select 80% of the instances as the training set, 10% of the instances as the validation set, and the remaining 10% as the test set. The details of the dataset division are shown in Table 1. Based on previous research, this paper selects two evaluation indicators commonly used in link prediction, namely Mean Reciprocal Ranks (MRR) and Hits@K.

Baselines. We compare TKGE with two types of baselines: TKG link prediction models under the interpolation setting and TKG link prediction models under the extrapolation setting. For interpolation, we select HyTE [14], TTransE [15] and TA-DistMult [16] as comparison method. Similarly, for extrapolation, we select RGCRN [17], CyGNet [18], RE-NET [7] and RE-GCN [6] as comparison method. Especially, RGCRN is an extended model of GCRN, which originally for the homogeneous graphs. Specifically, RE-NET replace GCN with R-GCN.

Parameter Settings. We implement our method TKGE in TensorFlow and carry it out on Tesla V100. The training epoch is limited to 200. We adopt the Adam optimizer for parameter learning with a learning rate of 10^{-3}. For all models, we set the dimension of embedding as 200 and the batch size as 256 for a fair comparison.

5.2   Performance Comparison (RQ1)

Entity Prediction. In this task, the results of all the comparison methods are presented in Table 2. Firstly, the TKGE is superior to TKG link prediction models under the interpolation setting, such as HyTE, TTransE, and TA-DistMult, because TKGE additionally captures the static structural information of the

Table 2. Entity prediction performance on different datasets.

Model	ICEWS18			ICEWS14			ICEWS05-15		
	MRR	H@3	H@10	MRR	H@3	H@10	MRR	H@3	H@10
HyTE	7.48%	7.34%	16.04%	16.79%	24.85%	43.96%	16.05%	20.16%	34.73%
TTransE	8.46%	8.98%	22.35%	12.84%	15.75%	33.62%	16.52%	20.76%	39.27%
TA-DistMult	16.45%	18.12%	32.52%	26.23%	29.75%	45.25%	27.55%	31.46%	47.33%
RGCRN	23.49%	26.67%	41.97%	33.35%	36.65%	51.52%	35.94%	40.03%	54.62%
CyGNet	24.99%	28.57%	43.55%	34.62%	38.85%	53.12%	35.42%	40.23%	54.43%
RE-NET	26.18%	29.82%	44.39%	35.74%	40.15%	54.85%	36.85%	41.86%	57.63%
RE-GCN	27.50%	31.18%	46.56%	37.72%	42.51%	58.82%	38.22%	43.12%	59.95%
TKGE	**28.84%**	**32.46%**	**48.74%**	**39.73%**	**44.91%**	**62.79%**	**39.63%**	**44.33%**	**62.33%**

Model	WIKI			YAGO			GDELT		
	MRR	H@3	H@10	MRR	H@3	H@10	MRR	H@3	H@10
HyTE	25.48%	29.14%	37.74%	14.49%	39.85%	46.96%	6.65%	7.51%	19.03%
TTransE	20.46%	23.98%	33.05%	26.14%	36.25%	47.62%	5.42%	4.86%	15.27%
TA-DistMult	26.45%	31.37%	38.88%	44.89%	50.71%	61.21%	10.45%	10.46%	21.67%
RGCRN	28.67%	31.47%	38.59%	43.74%	48.54%	59.65%	18.64%	19.79%	32.43%
CyGNet	30.89%	33.87%	41.25%	46.62%	52.55%	61.12%	18.12%	19.13%	31.43%
RE-NET	30.88%	33.82%	41.39%	46.74%	52.75%	61.85%	**19.85%**	**20.86%**	**33.83%**
RE-GCN	39.83%	44.38%	53.85%	52.32%	65.61%	75.82%	19.22%	20.46%	33.28%
TKGE	**43.94%**	**49.67%**	**60.06%**	**57.95%**	**71.96%**	**82.71%**	19.77%	20.58%	32.67%

KG at each timestamp t and the temporal evolution information of the TKG. This improvement demonstrates that the accuracy of evolutional embedding generated from TKGE has significantly improved. Secondly, the temporal models for the extrapolation setting, such as CyGNet, RE-NET, and RE-GCN, have good performance. The result verifies the effect of the repetitive patterns and the direct neighbors on the entity prediction task. Among them, the RE-GCN is the most efficient method. Compared with CyGNet and RE-NET, RE-GCN considers the static structure information in each timestamp t and the dynamic temporal information over time and can obtain satisfactory evolutional representations. Thirdly, our proposed method TKGE performs very well on most datasets. It is noteworthy that compared with other datasets, the experiments result in GDELT are not as superior as others. After further analysis, the reason for this phenomenon may be that there are many abstract entities that don't specify concrete entities (e.g., teacher and school). Specifically, with the time interval increasing, the performance gap between the last two rows in the Table 2 is becoming larger. For WIKI and YAGO, because the long time interval determines the data at each timestamp t having more static structural information, it is helpful to improve the performance of the entity prediction task. Since the TKGE method simultaneously models the static structural and dynamic temporal information, it is efficient for link prediction tasks in TKG.

Relation Prediction. Due to the space limitation, we only compare TKGE with some typical TKG methods and show the performance on MRR. Specifically, we

Table 3. Relation prediction performance on different datasets.

Model	ICEWS18	ICEWS14	ICEWS05-15	WIKI	YAGO	GDELT
RGCRN	37.16%	38.07%	38.29%	88.89%	90.11%	18.57%
RE-GCN	39.49%	39.74%	38.55%	95.62%	95.15%	19.12%
TKGE	**41.52%**	**41.43%**	**38.74%**	**97.92%**	**97.26%**	**19.71%**

Table 4. TKGE with and without structural self-attention or temporal self-attention.

Model	ICEWS18	ICEWS14	ICEWS05-15	WIKI	YAGO	GDELT
TKGE	28.84%	39.73%	39.63%	43.94%	57.95%	19.77%
TKGE-NS	27.63%	38.22%	36.98%	39.81%	54.72%	19.02%
TKGE-NT	27.23%	37.57%	34.45%	39.59%	53.08%	18.87%

select the RGCRN and RE-GCN, because we can use them for relation prediction tasks directly. As illustrated in Table 3, we can observe that the performance of the proposed model TKGE is always better than other baselines in all the datasets. The superiority of the TKGE demonstrates that our designed structural and temporal self-attention module can generate more accurate evolutional entity and relation representations. Compare with the entity prediction task, the performance gap on the relation prediction task is smaller. This is probably because the number of relations was limited, making relation prediction tasks easier than entity prediction tasks. For example, due to the number of relation on WIKI and YAGO being 24 and 10, the experiment results on both dataset is superior to other datasets.

5.3   Utility of Structural and Temporal Self-attention (RQ2)

To demonstrate the effectiveness of our designed structural and temporal self-attention module we compare TKGE with its variants TKGE-NS and TKGE-NT. Two variants represent the model removing the structural self-attention module and the model removing temporal self-attention modules, respectively. As shown in Table 4, we can see that the performance of TKGE-NS and TKGE-NT are lower than TKGE in all datasets. This result verifies the validity of the structural and temporal self-attention module.

5.4   Hyper-Parameter Studies (RQ3)

Due to the space limitation, we only analyze the effect of the number of heads h in multi-head structural and temporal self-attention, since this number plays a crucial role to model TKG. Specifically, except for the measured parameters, we keep other parameters fixed for fairness. We analyze entity prediction tasks on datasets ICEWS14 and YAGO individually. As illustrated in Table 5, we find that multi-head self-attention can improve the performance of TKGE effectively.

Table 5. Impact of hyper-parameter h on entity prediction.

Dataset	$h = 1$	$h = 2$	$h = 4$	$h = 8$	$h = 16$
ICEWS14	38.03%	38.17%	38.29%	39.73%	38.65%
YAGO	56.24%	56.49%	57.74%	57.95%	57.23%

In addition, with the parameter h increasing, the performance of entity prediction tasks will increase firstly then reduce after the locally optimal value. In this paper, the best number of attention heads is 8. In general, the multi-head self-attention which can model TKG from many angles is an effective method for entity prediction tasks.

6 Conclusions

We have presented TKGE, a novel approach for embedding TKG. It jointly models both the static structural and temporal information in learning evolutional representations for entities and relations. Extensive experiments on six real-world datasets demonstrate the effectiveness and rationality of our TKGE method. In this work, we have regarded TKG as a series of static KG snapshots, thus we only explore the discrete-time approach. Since discrete-time approach characterizes temporal information in relatively coarse levels, the learned embeddings may lose some information between snapshots. To address this issue, we plan to extend our TKGE method to explore continuous-time approaches [19–21] to integrate more fine-grained temporal information.

Acknowledgment. This work was supported in part by the National Natural Science Foundation of China Youth Fund (No. 61902001), the Open Project of Shanghai Big Data Management System Engineering Research Center (No. 40500-21203-542500/021), the Industry Collaborative Innovation Fund of Anhui Polytechnic University-Jiujiang District (No. 2021cyxtb4), and the Science Research Project of Anhui Polytechnic University (No. Xjky072019C02, No. Xjky2020120). We would also thank the anonymous reviewers for their detailed comments, which have helped us to improve the quality of this work. All opinions, findings, conclusions and recommendations in this paper are those of the authors and do not necessarily reflect the views of the funding agencies.

References

1. Cheng, K., Yang, Z., Zhang, M., Sun, Y.: Uniker: a unified framework for combining embedding and definite horn rule reasoning for knowledge graph inference. In: EMNLP, pp. 9753–9771 (2021)
2. Che, F., Zhang, D., Tao, J., Niu, M., Zhao, B.: Parame: regarding neural network parameters as relation embeddings for knowledge graph completion. In: AAAI, pp. 2774–2781 (2020)

3. Goel, R., Kazemi, S.M., Brubaker, M., Poupart, P.: Diachronic embedding for temporal knowledge graph completion. In: AAAI, pp. 3988–3995 (2020)

4. Mei, H., Eisner, J.: The neural hawkes process: a neurally self-modulating multivariate point process. In: NIPS, pp. 6754–6764 (2017)

5. Trivedi, R., Farajtabar, M., Biswal, P., Zha, H.: Dyrep: learning representations over dynamic graphs. In: International Conference on Learning Representations (2019)

6. Li, Z., et al.: Temporal knowledge graph reasoning based on evolutional representation learning. In: SIGIR, pp. 408–417 (2021)

7. Jin, W., Qu, M., Jin, X., Ren, X.: Recurrent event network: autoregressive structure inferenceover temporal knowledge graphs. In: EMNLP, pp. 6669–6683 (2020)

8. Kong, C., Chen, B., Li, S., Chen, Y., Chen, J., Zhang, L.: GNE: generic heterogeneous information network embedding. In: WISA, pp. 120–127 (2020)

9. Vaswani, A., et al.: Attention is all you need. In: NIPS, pp. 5998–6008 (2017)

10. Cheng, S., Xie, M., Ma, Z., Li, S., Gu, S., Yang, F.: Spatio-temporal self-attention weighted VLAD neural network for action recognition. IEICE 104-D, pp. 220–224 (2021)

11. Liu, J., Chen, S., Wang, B., Zhang, J., Li, N., Xu, T.: Attention as relation: learning supervised multi-head self-attention for relation extraction. In: IJCAI, pp. 3787–3793 (2020)

12. Xu, Y., Huang, H., Feng, C., Hu, Y.: A supervised multi-head self-attention network for nested named entity recognition. In: AAAI, pp. 14185–14193 (2021)

13. Shang, C., Tang, Y., Huang, J., Bi, J., He, X., Zhou, B.: End-to-end structure-aware convolutional networks for knowledge base completion. In: AAAI, pp. 3060–3067 (2019)

14. Dasgupta, S.S., Ray, S.N., Talukdar, P.P.: Hyte: hyperplane-based temporally aware knowledge graph embedding. In: EMNLP, pp. 2001–2011 (2018)

15. Leblay, J., Chekol, M.W.: Deriving validity time in knowledge graph. In: WWW, pp. 1771–1776. ACM (2018)

16. García-Durán, A., Dumancic, S., Niepert, M.: Learning sequence encoders for temporal knowledge graph completion. In: EMNLP, pp. 4816–4821 (2018)

17. Schlichtkrull, M.S., Kipf, T.N., an Rianne van den Berg, P.B., Titov, I., Welling, M.: Modeling relational data with graph convolutional networks. In: ESWC, vol. 10843, pp. 593–607 (2018)

18. Zhu, C., Chen, M., Fan, C., Cheng, G., Zhang, Y.: Learning from history: modeling temporal knowledge graphs with sequential copy-generation networks. In: AAAI, pp. 4732–4740 (2021)

19. Garg, K., Panagou, D.: Fixed-time stable gradient flows: applications to continuous-time optimization. IEEE Trans. Autom. Control. **66**(5), 2002–2015 (2021)

20. Chien, J., Chen, Y.: Continuous-time attention for sequential learning. In: AAAI, pp. 7116–7124 (2021)

21. Zhang, L., Zhao, L., Qin, S., Pfoser, D., Ling, C.: TG-GAN: continuous-time temporal graph deep generative models with time-validity constraints. In: WWW, pp. 2104–2116 (2021)

A Multi-modal Knowledge Graph Platform Based on Medical Data Lake

Ruoyu Wang[✉]

School of Methematical Sciences, Beihang University, Beijing, China
18377446@buaa.edu.cn

Abstract. In recent years, the application of Knowledge Graphs (KGs) in the medical domain has developed rapidly. Embedding medical knowledge graphs into low-dimensional vectors to implement certain applications is the focus of recent research. However, current approaches focus on using structured data while ignoring external information from prior knowledge and multi-modal data, such as visual and textual data. In this paper, we propose a multi-modal knowledge graph platform based on medical data lake. It enhances Knowledge Graph Embedding (KGE) from structural data by using prior knowledge like doctors' advice. After the generation of embedding of knowledge graphs, we use a fully-connected layer to combine the structural vectors and multi-modal embedding from unstructured data for link prediction. Moreover, we provide recommendations to help doctors diagnose and research based on the knowledge graph embedding model. Our platform has been used in a clinical decision support system, which is capable of discovering unveiled relations between entities and giving suitable advice for doctors.

Keywords: Knowledge graph · Medical data lake · Multi-modal data

1 Introduction

Over the past few years, research on knowledge graph (KG) has thrived, which produced a significant number of remarkable achievements. A wide variety of applications are implemented, including recommendations system, Q&A and information retrieval [2,8]. Knowledge graph is a type of knowledge base that uses a graph-structured data model to store semantic information in the form of entities and the relationship between them. It is composed of many triplets (head entity-relation-tail entity) and each triplet represents one piece of knowledge.

In the medical domain, there are different types of entities, such as diseases, symptoms, treatments, etc and a variety of relationships between them. The information in the knowledge graph can be supplemented by prior knowledge from some physicians and data from other models, which assists providers by validating diagnoses and identifying treatment plans based on individual needs. When it comes to applications like Q&A and recommendation systems, the key point is that the accuracy of the answers and recommendations is tightly linked

X. Zhao et al. (Eds.): WISA 2022, LNCS 13579, pp. 15–27, 2022.
https://doi.org/10.1007/978-3-031-20309-1_2

with the correctness and completeness of the knowledge graphs. Most of the popular knowledge graphs (e.g., YAGO, DBPedia, or Wikidata) remain incomplete despite the great effort invested in their creation and maintenance, even though they have already stored large amounts of entities and relationships. To complete the knowledge graph, researchers proposed the knowledge graph embedding (KGE) method, which is to embed the entities and relations into low-dimensional vectors. In the KGE field, the most representative translational distance model is TransE [1]. Many researchers have proposed a significant number of improved models to increase its accuracy, such as TransH [17] and TransD [6]. Apart from that, researchers have also tried to use the information from multi-modal data to supplement the embedding [12]. They have tried to integrate textual information [19] and visual information [20] into knowledge graphs. However, it is rare to include multi-modal data in existing medical knowledge graphs.

With the advances of medical technology, enormous amounts of multi-modal medical data are generated, and traditional database is not able to store and manage the data efficiently. A new type of database called data lake is good at handling this problem [14,21]. In data lake, various types of data can be stored and processed, such as Electronic Medical Records (EMRs), Magnetic Resonance Imaging (MRI) scans, Computerized Tomography (CT) scans, X-ray and Positron Emission Tomography (PET) scans, etc. Moreover, doctors can give their advice and document their experience in it. All of the metadata can be saved for traceability and safety in data lake. For example, images like magnetic resonance imaging (MRI) can be embedded into embedding vectors, and the vectors will be stored to represent specific features. Due to the powerful storage capacity and convenient operations, data lake supports plenty of applications like data exploration. Since its advantages in data exploration and abundant multi-modal data, we can efficiently construct a knowledge graph based on it, and import the multi-modal data and documented experiences to improve its quality. Therefore, we propose a Multi-Modal Knowledge Graph Platform (MMKGP) based on medical data lake.

Then this knowledge graph can provide advice and recommendations for doctors. We address our contributions as follows:

1. A multi-modal knowledge graph platform that uses structured data, multi-modal data and doctors' experiences to construct a highly complete and accurate knowledge graph based on medical data lake.
2. A translation-based model of KGE which can make use of the constraints of pairs of relations according to prior knowledge.
3. A method to use external information in multi-modal medical data to find missing relations of knowledge graph.
4. The platform has been used in a clinical decision support system, which is capable of discovering unveiled relations between entities and giving suitable advice for doctors

In the following, we will first review the related work in Sect. 2, and then introduce the overview of our proposed framework in Sect. 3. How to use prior knowledge to get more accurate embedding is demonstrated in Sect. 4, and how to find missing relations with the help of unstructured data is presented in Sect. 5. After that, we will explain how to use the generated high-quality knowledge graph to help doctors in Sect. 6 followed by the conclusion in Sect. 7.

2   Related Work

In this chapter, we discuss the knowledge graph embedding and data lake.

There are many knowledge graph embedding models. In 2013, TransE model was proposed [1], which considers relations as translating operations between head and tail entities. A lot of translation-based models were proposed to achieve a better performance in representation learning tasks of knowledge graph after TransE, such as link prediction [9]. For instance, TransH [17], which considers relations as the vectors on a hyperplane, was proposed to overcome TransE's drawbacks in dealing with 1-to-N, N-to-1 and N-to-N relations. However, entities and relations are still in the same semantic space, which limits the modeling ability. TransR [9] appeared and assumed that entities and relationships belong to different semantic spaces. Furthermore, TransD [6]was proposed to simplify TransR by projection matrix. Besides translation-based models, semantic matching models (such as RESCAL [13], where the entire knowledge graph is encoded as a 3D tensor) and graph neural networks models (such as R-GCN [15], aggregating different nodes according to different relationships) are other good ways to represent knowledge graphs.

Later attempts focus on using background knowledge to constraint the representation learning [4,5,16] or integrating extra information beyond triplets [12]. In 2018, Ding et al. noticed the connection between the pairs of relations [4], and proposed proposes to use the background information to constrain the embeddings of relations in RESCAL. Xie et al. focused on using the extra information from multi-modal data to supplement the embedding of knowledge graphs. They proposed methods to make use of textual description [19] and image information [20] in the construction of knowledge graphs. In the medical domain, myDIG [7] and SemTK [3] proposed knowledge graph building tools which can extract triplets from website and texts. EMKN [22] constructed an EMR-based medical knowledge graph by extracting the medical entities. Liu et al. proposed to extract triplets from 1454 clinically pediatric cases, and then combined them with expert experiences and book knowledge in the generation of knowledge graphs [10].

However, it is not enough to combine the translation-based model just with the multi-modal information or just with prior knowledge. There is no model that combines structured, multi-modal information and prior knowledge together to build a medical knowledge graph (See Table 1).

Last decade has witnessed great development of the application of knowledge graphs. Bordes et al. proposed to embed knowledge graphs into vectors and use these vectors to do link prediction which can specify the unrevealed relations between the entities [1]. Apart from that, we can also break through the boundary of the input knowledge graph and expand the application into

Table 1. Comparison of the information contained in the model

Method	Triplets	Multi-modal	Prior knowledge
TransE [1]	+	–	–
IKRL [20]	+	+	–
CNA [4]	+	–	+
myDIG [7]	+	–	–
SemTK [3]	+	–	–
EMKN [22]	+	–	–
HKDP [10]	+	–	+
Ours	+	+	+

Note: (+) means this kind of information is contained in the model and (–) means this type of this kind of information is not contained in the model.

boarder domains. In 2013, Weston et al. proposed to use knowledge graphs to extract unseen relation between two entities [18] Another important application of knowledge graph is the wide use in the recommender systems [8]. We can use knowledge graph as a big database and give recommendation for the users according to the their preference and history operation. Moreover, knowledge graph can be also used in the Q&A systems [2], which is very helpful in the medical domain.

To store and manage multi-modal data in the medical domain, Zhang et al. proposed to use medical data lake to deal with unstructured data [21]. As for unstructured data, data lake embeds it into vectors by state-of-the-art models automatically, and then it can be regarded as a sub-tree that contains the nodes representing its path information, feature vectors, and others. Data lake provides a convenient way for us to query data and explore data efficiently in the construction of knowledge graph.

3   Architecture of MMKGP

MMKGP is established on the medical data lake, in which three sources of information are available: prior knowledge provided by doctors, structured data or primary knowledge graph, unstructured data such as linguistic and visual information. The architecture of MMKGP is presented in Fig. 1.

The experience from medical experts can be understood as prior knowledge, and we can learn the constraints of pairs of relations and their weights from it. The structured data is put into the TransE model, then we can get the structured embedding of entities and relations. After that, the structured embedding and the weights of pairs of constrained relations are imported into the first model to generate the updated structured embedding. Next, based on the unstructured embedding, we construct a fully-connected layer to complement the missing relations between pairs of entities. Finally, a highly-quality knowledge graph can be generated, which includes newly found relations, and can help doctors make diagnoses and research.

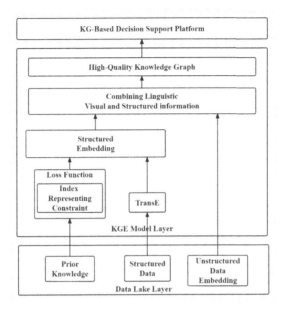

Fig. 1. Overview

4 Translation-Based Model Enhanced by Prior Knowledge

We take doctors' professional experience as prior knowledge in the construction of knowledge graph. At the same time we mine the constraints of the relations from the knowledge graph automatically [4]. For the two types of prior knowledge, we use the same method to represent it, which is the constraint weight of pairs of relations. The weight of relation A and relation B represents the certainty that we can infer B from A. For example, if knowledge graph has triplet (h, r_1, t) and the constraint weight of r_1 and r_2 is -0.95, we can conclude that triplet (h, r_2, t) is correct with 95% certainty.

In the traditional translation-based model, the constraints of relations is usually ignored. In fact, the constraints can provide a lot of information. For example, there are two diseases called diabetes mellitus and coronary heart disease, and we know that diabetes mellitus is "cause of" Mike (See Fig. 2. So we have one triplet $(diabetes\_mellitus, cause\_of, coronary\_heart\_disease)$ in the knowledge graph. Due to the relationship between "cause_of" and "complication_of", we need to control the embedding vector of "complication_of" is nearly the reverse of the embedding vector of "cause_of". And then, based on the principle of TransE, we can easily conclude that coronary heart disease is "complication_of" diabetes mellitus so that we can add up a new triplet $(coronary\_heart_disease, complication\_of, diabetes\_mellitus$ into the knowledge graph. It is just a simple example, but we can understand the importance of constraints in KGE and why these constraints can significantly improve the accuracy in link prediction.

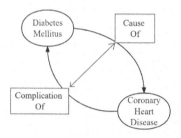

Fig. 2. Constrained relations

4.1   TransE Model

In traditional translation-based model, given an existing triplet (h, r, t) where h represents the head entity, r represents the relation and t represents the tail entity, the relation is embedded into a translation vector $\mathbf{r}$ so that the embedded head entity $\mathbf{h}$ and the embedded tail entity $\mathbf{f}$ can be connected by $\mathbf{r}$ in low error, e.g., $\mathbf{h}+\mathbf{r} \approx \mathbf{t}$. For example, if triplet $(COVID-19, type\_of, pneumonia)$ holds, after the embedding in model TransE, we are looking forward to the following equation satisfied in low error:

$$\mathbf{V}_{\mathbf{COVID-19}} + \mathbf{V}_{\mathbf{type\_of}} \approx \mathbf{V}_{\mathbf{pneumonia}},$$

where $V_{r/e}$ means the embedding vector of the relation r or the entity e. The scoring function of TransE is defined as the distance between $\mathbf{h} + \mathbf{r}$ and $\mathbf{t}$:

$$f_r(h, t) = \|\mathbf{h} + \mathbf{r} - \mathbf{t}\|_{\frac{1}{2}},$$

The score is expected to be small if the triplet (h, r, t) is correct.

4.2   Constraint for Relations

Based on TransE, we propose to add the constraints for relations. In the medical domain, on the one hand, we can get the entailment of relations from doctors' experience. On the other hand, the constraints can be automatically calculated by modern rule mining system [11]. We use a weight to represent the degree that the former approximately constraint the latter, e.g., PARENT_OF and CHILD_OF. The maximum of the absolute value of the weight is 1, and the minimum is 0. The larger the value is, the more restrictive the latter is by the former and the more similar this pair of the relations' embedding vectors should be. The sign means the direction of the latter relation. If the sign is negative, we should calculate the constraint of the reversed latter vector with the former vector.

We first explore the strict constraint, e.g., weight $\lambda = 1$. This strict constraint $r_1 \rightarrow r_2$ means that if relation r_1 holds, then relation r_2 holds. And the constraint can be roughly represented in the following equation:

$$\mathbf{r_1} = \mathbf{r_2}$$

where $\mathbf{r_i}$ means the embedding vector of the relation r_i.

For non-strict constraints, e.g., weight $\lambda = 0.95$, we use a power function to measure the entailment of the vectors:

$$f_\lambda(\mathbf{r_1}, \mathbf{r_2}) = \|\mathbf{r_1} - \mathbf{r_2}\|_{\frac{1}{2}}^{\|\lambda\|},$$

where $\mathbf{r_i}$ means the embedding vector of the relation r_i and λ is the value of constraint weight. When weight $\lambda = \pm 1$, which means the constraint is strict, this function degenerated into the distance between $\mathbf{r_1}$ and $\mathbf{r_2}$. Due to the extension to the model, we should also add up the scoring function of constrained relations. The scoring function is defined in the following equation:

$$[f_\lambda(\mathbf{r_1}, \mathbf{r_2}) = \|\mathbf{r_1} - \mathbf{r_2}\|_{\frac{1}{2}}^{\|\lambda\|} - \gamma]_+,$$

where γ is the margin value can be given by users and $[x]_+ = max(0, x)$. During the training, we not only calculate the scoring function of triplets, but also calculate the scoring function of constrained relations, and add them up into one combined scoring function. And then, we use the stochastic gradient descent (SGD) to adjust parameters to lower the function.

5   Knowledge Graph Completion with Multi-modal Data

Our method is based on the implementation of data lake. The data in the data Lake includes structured data, semi-structured data, unstructured data and binary data, so as to form a centralized data storage containing all forms of data. We add a fully connected layer into this model so that it can consider the external information from entities when the model needs to make link prediction or triplet classification. There is much external information in the multi-modal data which can help us improve our knowledge graph embedding. Picture or linguistic information stored in data lake can correct the wrong relationship or add some new relationships.

For example, a person who suffers from influenza has symptoms of cough and runny nose (See Fig. 3). The triplet is (cough, is_symptom_of, influenza) and (runny_nose, is_symptom_of, influenza). A coughing person needs cough syrup(cough,need_medicine, cough syrup). TransE may predict that a person with a runny nose only will also need cough syrup (runny_nose, need medicine, cough_syrup). But we get a picture of a prescription showing that doctors use nasal spray instead of cough syrup to treat runny nose patients. After the visual embedding, we can amend the relation among runny nose, cough syrup and nasal spray(runny_nose, need_medicine, nasal_spray).

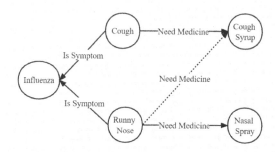

Fig. 3. An example using multi-modal embedding.

5.1 Dataset

Our data is collected from three hospitals, including more than 500 diseases and medical records of 3 million patients. This dataset includes 8293284 concepts, 83591932 entities and 295848293 relationships, of which 32256360 are the relationships between concepts.

5.2 Evaluation Criterion

Triple Energy: The energy includes three parts:

- Structural Energy: we set the energy function in terms of the rules of TransE as $E_s = \|h_s + r_s - t_s\|$.
- Multi-modal Energies: First, we define the multi-modal representations hm and tm of the head and the tail entities. There are two parts of multi-modal energy functions: $E_{m1} = \|h_m + r_s - t_m\|$ and $E_{m2} = \|(h_m + h_s) + r_s - (t_m + t_s)\|$.
- Structural-multi-modal Energies: We need to make sure structural and the multi-modal representations are in the same space, so we define the energy functions as: $E_{sm} = \|h_s + r_s - t_m\|$ and $E_{ms} = \|h_m + r_s - t_s\|$.

Finally, we add all of them to define our triple energy.

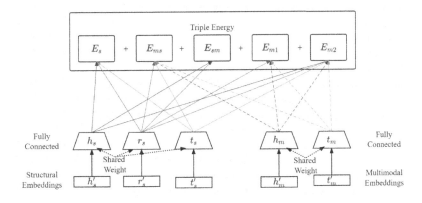

Fig. 4. Overview of the fully connect layer and triple energy

Objective Function: First, we need two negative triple sets. One is to replace head entities named T'_{head}, and another is to replace the tail entities named T'_{tail}. Then we set $L = L_h + L_t$ to represent the loss.

$$L_h = \sum_T \sum_{T'_{tail}} MAX(E(h,r,t) - E'(h,r,t'))$$

$$L_t = \sum_T \sum_{T'_{head}} MAX(E(h,r,t) - E'(h',r,t)$$

Finally, in order to consider various information during knowledge graph embedding, we set fully connected layer that can map both structural and multi-modal representations into the same place (See Fig. 4). Weights are also shared between those inputs.

5.3   Model Training and Result

Model Training: After completing these preparations, we can start the model training.

Our method is divided into two parts. The first is to train TransE+AER. We set batch size to 100, embedded size to 100 dimensions, learning rate to 1.0 and use PyTorch to carry out the random gradient descent method. With the help of OpenKE, we take part of the positive and negative triples in each iteration. Then update the parameters to enter the next iteration. The loss of each iteration is displayed on the terminal panel. The model saves the real-time model parameters and optimal model parameters. Before the next part training, we test this embedding result to make a control group.

In the second part, we use the structural embeddings to continue our training. We set batch size to 100, initial learning rate to 0.001, and use the Adam optimizer to get the best results. We combine linguistic embeddings and visual embeddings to get the final KG.

Table 2. Link prediction results

Method	Raw		Filter	
	MR	Hits@10	MR	Hits@10
TransE	160.0	0.641	152.1	0.643
TransE+AER	148.3	0.720	145.6	0.720
TransE+Multi	25.2	0.666	25.02	0.672
Ours	18.7	0.766	18.6	0.772

Comparison of Various Methods: We compared our result with several methods, including TransE, TransE+AER, TransE+multi-modal. Table 2 shows the results. Our idea combining the advantages of both those methods can do better in knowledge graph embedding and Link Prediction. We get a lower MR and higher Hits@10. Prior knowledge helps us add more information of relations so that knowledge graph becomes more comprehensive. Multi-modal data can help correct the unsuitable relations between entities. Therefore, we get a more detailed and accurate knowledge graph.

6 Knowledge Graph-Based Clinical Decision Support System

To help doctors to diagnose and do research, we have generated a knowledge graph and tried to improve its quality by using our platform. During the quality improvement, new triplets considering as pieces of knowledge will be generated. They can clarify the relations between pairs of specific entities. The newly derived knowledge can help doctors to find new connections between entities. After the improvement, we get a high-quality knowledge graph which can efficiently and accurately perform tasks like recommendation and Q&A. According to the above, we can divide the applications of the knowledge graph into applications during improvement and applications after improvement.

6.1 Link Prediction & Correction

While constructing a high-quality knowledge graph, we have done link prediction to complete the knowledge graph, which can help doctors find out the missing relationship between two entities. The system stores these newly derived triplets into a list to save them. After the completion, this list will be printed out and shown to the doctors and researchers. They can check it carefully and try to point out the potential reason for these pieces of knowledge, which may give researchers clues and ideas.

Apart from the completion, this system can also check the accuracy of existing knowledge. Doctors can put forward suggestions to modify the existing triples according to their own knowledge. After checking every triplet, the list will be printed out to the doctors so that they can check whether any mistakes happen or the data is wrong. Then the staff or doctors can correct the wrong data to reduce potential risks. Some existing knowledge may be wrong. For example, this system may hint to doctors that some of the treatments for a specific disease are inappropriate. Especially if doctors have corresponding conjecture, this hint may lead to a retest to the treatment and may help patients get proper curing.

6.2 Recommendation and Q&A System

One application is to use this system to recommend treatments or medicines according to the patients' basic information like age, sex, weight, etc. and their symptoms. For example, in the meeting with a patient, we import his name and symptoms. After checking such as CT, this system can extract the information from the CT image. By comprehensive analysis of symptoms and the checking reports, this system can clearly understand the condition of the patient. Then it can give diagnosis to the disease suffered by the patient and its recommendations to the corresponding treatment. The system also considers the age and the weight of the patient and gives its recommendation to the dose of medicine.

Apart from recommendation, knowledge graph supports Q&A system [2] as well. During patient meetings, diagnose and research, doctors can type their questions into this system. Then state-of-the-art models are used to get the potential answer of the question on the knowledge graph. This is especially helpful for the new graduate or intern doctors because they don't have many experiences. It can also be applied to the online consultation services. People with some simple symptoms can query online and get the answer quickly so that they don't need to spend a lot of time going to a hospital and wait for the meeting with doctors.

7 Conclusion

We present a multi-modal platform to construct high-quality knowledge graph-based on medical data lake. Based on the fact that experts' experience can give essential guidance in the medical domain, we propose to make use of this experience to calculate the constraint weights for the pairs of relation. These weights are used to enhance the embedding vector of the relations. Considering the neglected external information from unstructured data, we construct a fully-connected layer to combine both structured and unstructured embedding of data to find the missing relations between entities.

Acknowledgement. This work was supported by National Key R&D Program of China(2020AAA0109603).

References

1. Bordes, A., Usunier, N., Garcia-Duran, A., Weston, J., Yakhnenko, O.: Translating embeddings for modeling multi-relational data. In: Burges, C.J.C., Bottou, L., Welling, M., Ghahramani, Z., Weinberger, K.Q. (eds.) Advances in Neural Information Processing Systems, vol. 26. Curran Associates, Inc. (2013)

2. Bordes, A., Weston, J., Usunier, N.: Open Question Answering with Weakly Supervised Embedding Models. Springer, New York, Inc (2014)
3. Cuddihy, P., Mchugh, J., Williams, J.W., Mulwad, V., Aggour, K.S.: Semtk: an ontology-first, open source semantic toolkit for managing and querying knowledge graphs (2017)
4. Ding, B., Wang, Q., Wang, B., Guo, L.: Improving knowledge graph embedding using simple constraints. In: Meeting of the Association for Computational Linguistics (2018)
5. Guo, S., Wang, Q., Wang, L., Wang, B., Guo, L.: Jointly embedding knowledge graphs and logical rules. In: Proceedings of the 2016 Conference on Empirical Methods in Natural Language Processing (2016)
6. Ji, G., He, S., Xu, L., Liu, K., Zhao, J.: Knowledge graph embedding via dynamic mapping matrix. In: Meeting of the Association for Computational Linguistics & the International Joint Conference on Natural Language Processing (2015)
7. Kejriwal, M., Szekely, P.: MYDIG: personalized illicit domain-specific knowledge discovery with no programming. Future Internet 11(3), 59 (2019)
8. Yehuda, K., Robert, B., Chris, V.: Matrix factorization techniques for recommender systems. Computer 42, 30–37 (2009)
9. Lin, H., Liu, Y., Wang, W., Yue, Y., Lin, Z.: Learning entity and relation embeddings for knowledge resolution. In: Procedia Computer Science international Conference on Computational Science, ICCS 2017, 12–14 June 2017, pp. 345–354. Zurich, Switzerland (2017)
10. Liu, P., et al.: HKDP: a hybrid knowledge graph based pediatric disease prediction system. In: International Conference on Smart Health (2016)
11. Galárraga, L., Teflioudi, C., Hose, K., Suchanek, F.M.: Fast rule mining in ontological knowledge bases with amie +. VLDB J. 24, 707–730 (2015)
12. Mousselly-Sergieh, H., Botschen, T., Gurevych, I., Roth, S.: A multimodal translation-based approach for knowledge graph representation learning. In: Proceedings of the Seventh Joint Conference on Lexical and Computational Semantics, pp. 225–234 (2018)
13. Nickel, M., Tresp, V., Kriegel, H.P.: A three-way model for collective learning on multi-relational data, pp. 809–816 (2011)
14. Ren, P., et al.: HMDFF: a heterogeneous medical data fusion framework supporting multimodal query. In: Siuly, Si., Wang, H., Chen, L., Guo, Y., Xing, C. (eds.) HIS 2021. LNCS, vol. 13079, pp. 254–266. Springer, Cham (2021). https://doi.org/10.1007/978-3-030-90885-0_23
15. Schlichtkrull, M., Kipf, T.N., Bloem, P., Berg, R.V., Welling, M.: Modeling Relational Data with Graph Convolutional Networks. The Semantic Web (2018)
16. Wang, Q., Wang, B., Guo, L.: Knowledge Base Completion Using Embeddings and Rules. AAAI Press (2015)
17. Wang, Z., Zhang, J., Feng, J., Chen, Z.: Knowledge graph embedding by translating on hyperplanes. Proc. AAAI Conf. Artif. Intell. 28(1) (2014)
18. Weston, J., Bordes, A., Yakhnenko, O., Usunier, N.: Connecting language and knowledge bases with embedding models for relation extraction. IEEE (2013)
19. Xie, R., Liu, Z., Jia, J., Luan, H., Sun, M.: Representation learning of knowledge graphs with entity descriptions. In: Proceedings of the AAAI Conference on Artificial Intelligence, vol. 30 (2016)
20. Xie, R., Liu, Z., Luan, H., Sun, M.: Image-embodied knowledge representation learning. In: AAAI Press (2016)

21. Zhang, Y., Sheng, M., Zhou, R., Wang, Y., Dong, J.: HKGB: an inclusive, extensible, intelligent, semi-auto-constructed knowledge graph framework for healthcare with clinicians' expertise incorporated. Information Processing & Management, p. 102324 (2020)
22. Zhao, C., Jiang, J., Xu, Z., Guan, Y.: A study of EMR-based medical knowledge network and its applications. Comput. Meth. Programs Biomed. **143**, 13–23 (2017)

Fusion of Natural Language and Knowledge Graph for Multi-hop Reasoning

Xun Lu, Feng Zhao[(⊠)], and Hai Jin

National Engineering Research Center for Big Data Technology and System,
Services Computing Technology and System Lab, Cluster and Grid Computing Lab,
School of Computer Science and Technology, Huazhong University of Science and
Technology, Wuhan, China
xlu@hust.edu.cn, zhaof@hust.edu.cn

Abstract. Multi-hop reasoning has been widely studied for its important
application values in the domain of intelligent search and question answering. Real-world applications are often dominated by natural language
input, and it is difficult to directly model the relation semantics in natural
language to fit the multi-hop reasoning model. In addition, the extremely
large scale and complex structure of knowledge graphs increase the challenge. We propose a *natural language and knowledge graph fusion* model
(NLKGF) for multi-hop reasoning. NLKGF embeds knowledge graph by
fusing natural language semantics during the graph propagation process
and adds a relation attention mechanism to enable entity representations
to perceive the contribution of different relations. A relation path encoder
is designed to encode the relation path by an improved recurrent neural
network, the reasoning entity is obtained by calculating the correlation
score with the natural language. We tested the performance of the NLKGF
model on two datasets requiring multi-hop reasoning. The experimental
results show that NLKGF beats advanced benchmark models in multi-hop
reasoning tasks, which proves superiority of our model.

Keywords: Knowledge graph · Muiti-hop reasoning · Graph neural
network

1 Introduction

As a highly structured heterogeneous graph, *knowledge graph* (KG) is widely
used in reasoning research. Traditional KG reasoning algorithms are mostly
based on simple relationships, it is difficult to recognize complex multi-hop relationships, reducing the reliability of reasoning. Multi-hop reasoning is a kind
of complex reasoning in KG, its task is to reason tail entities through KG by
given entity and multi-hop relation path [1], which is the key to build intelligent
applications such as *question answering* (QA). However, real applications are
mostly based on natural language input, these natural language questions contain complex multi-hop relation semantics that are difficult to extract directly.

X. Zhao et al. (Eds.): WISA 2022, LNCS 13579, pp. 28–39, 2022.
https://doi.org/10.1007/978-3-031-20309-1_3

Currently, information extraction methods [2–4] have become the mainstream direction of the research, this method embeds knowledge graph and natural language questions into the same semantic space to calculate the similarity measure. Although it performs well, it still faces several challenges as follows: (1) Difficulty in extracting subgraphs. The traditional approach tends to miss the real answer entities, which makes the low answer coverage and affects the upper limit of reasoning accuracy, and the scale of subgraphs remains large. (2) Difficulty in utilizing information. Existing methods are not able to fuse the semantic information of natural language with the structured features of knowledge graphs, only a single part of it is utilized. (3) Difficulty in selecting path. The multi-hop relation path of KG is very important to the reasoning process, a large number of candidate entities increase this difficulty.

In this paper, we design a fusion model of *natural language and knowledge graph* (NLKGF) to resolve the above challenges. NLKGF embeds KG through iteratively updating the representation of natural language and entity nodes. The embedding of entities and relations will be combined with paths for deep multi-hop reasoning. The main contributions of this paper are summarized as follows:

- We propose a method to extract subgraphs of the KG that can greatly decrease the scale of the KG and have a high answer coverage. The KG subgraphs will be used for the subsequent fusion model.
- To enable knowledge graphs to understand the relation semantics of natural language, we propose a model that can embed KG by fusing natural language representations and heterogeneously updating the entity node representations. We also consider relation path information for enhancing reasoning performance,

We tested NLKGF model on two datasets and the experimental results shows our model beats advanced benchmark models in multi-hop reasoning tasks.

2   Related Works

In this section, we will introduce the related work from the methods of semantic parsing and information extraction respectively.

Semantic Parsing. The method of semantic parsing hinges on how to convert natural language into a logical representation for reasoning in knowledge graph execution. This method follows a strict execution sequence and is the earliest method used by researchers due to the strong interpretability. Earlier approaches mainly relied on manual construction of rules and templates to parse natural language [5]. Some researchers try to model KG with different network structures, such as query graph [6], Graph2Seq [7]. Das et al. [8] proposed a memory network fusing textual information and knowledge graphs, and Kun et al. [9] improved the structure of the memory network to perform shallow multi-hop reasoning through multiple memory slots.

Information Extraction. The method of information extraction does not require defined features or templates. Bordes et al. [2] first used the knowledge representation learning approach to embed the natural language and the candidate subgraphs of the knowledge graph into the same semantic space to compute similarity reason entities. Dong et al. [10] focused on the information of multiple dimensions of the answer entity. Saxena et al. [11] used the ComplEx [12] to embed KG and viewed multi-hop reasoning as a link prediction problem, which achieved the best on incomplete KG. GNN can effectively aggregate the structural features of graphs and provide a new way of node representation. Facing the problem of incomplete knowledge graph, Sun et al. [3] used GNN to fuse text information based on the idea of early fusion. Then he further proposed an intelligent subgraph retrieval mechanism named PullNet [13] to avoid a large number of invalid reasoning. Feng et al. [4] improved the GCN [14] and fused multi-hop paths into the KG embedding process to provide better interpretability. In addition, Zhang et al. [15] realized multi-hop reasoning of knowledge graph based on probabilistic model, and He et al. [16] used the idea of knowledge distillation to learn the missing intermediate state supervision signal in the multi-hop reasoning process.

3   Model

3.1   Description

Formally we defined KG as $G = (E, R, T)$, where E, R, T respectively denotes entities, relations, and triples set. The triples is defined as $T = \{(h, r, t) | h, t \in E, r \in R\}$. Given a vocabulary V, the question is defined as a sequence of words $q = \{(w_1, w_2, \ldots, w_n) | w_i \in V\}$. NLKGF should correctly find the answer entity v by multi-hop reasoning when given a question and corresponding KG. To accomplish this task, we first need to extract a subgraph $G' \in G$ related to the question, the subgraph is defined as $G' = \{(E', R', T') | dist(e_i, u) \leq k, e_i \in E'\}$, where e_i is the entity in subgraph and $dist$ is the minimum distance between the topic entity u and e_i. The subgraph G' should have high answer coverage to enable NLKGF model reason the correct answer entity. The embedding of the subgraph is obtained through a process of fusion of the KG and natural language, while the relation path is extracted for enhancing the multi-hop reasoning performance.

3.2   Framework

The architecture of NLKGF is shown in Fig. 1, NLKGF extracts subgraphs related to natural language question by restricted subgraph algorithm, and then implements multi-hop reasoning through the structural fusion module and the relation reasoning module.

The subgraphs are extracted by a restricted subgraph algorithm based on relation linking. Then the embedding of KG and question are trained through the method of graph neural network, and the final correlation score is calculated between the relation path and the question to reason the answer entity. Our model is divided into three parts:

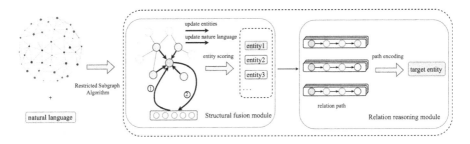

Fig. 1. NLKGF model architecture

Subgraph Retrieval. The subgraph retrieval process is implemented by a restricted subgraph algorithm based on relation linking. The relation linking model extracts the relation set related to the question, and the restricted subgraph algorithm filters out irrelevant entities to reduce the size of the KG, making our model scalable for large-scale KG.

Structural Fusion. The structural fusion module fuses the natural language question representation with the embedding of KG to obtain feature vectors by dynamically updating the node representations of the question and KG, and additionally introduces a mechanism of relation attention that enables nodes to learn the contribution of different relations in the subgraph.

Relation Reasoning. The relation reasoning module extracts the relation paths between all candidate entities and topic entities, and then extracts features through encoding layer to calculate the correlation score with the question.

3.3 Subgraph Retrieval

Relation Linking. Relation linking, whose task is to extract all the implicit relations associated with the set of relations in the KG from unstructured natural language, is the mapping process of relations from natural language to the KG. Fig. 2 shows the result of the relation linking for the question "Which country is the director of the Interstellar". The relation such as "has_nationality" related to question has higher score, meanwhile, we can find the 2-hop relation "born_in" and "belong_to" also have high score if the relation "has_nationality" in KG is missing, which is 0.61 and 0.53 respectively. By setting a threshold α, relations with score greater than α are added to the relation set, which will be used in the restricted subgraph algorithm.

We do not focus on the design of the relation linking model. Based on large-scale pre-training models, the question are directly input to the RoBERTa to get the word embeddings, the representation of the question h_q will be encoded by a LSTM and use sigmoid function to calculate the linking score.

Restricted Subgraph Algorithm. *Restricted subgraph algorithm* (RSA) uses a improved breadth-first search algorithm to extract subgraphs from KG. We restrict the subgraph scale by limiting the search depth, all the triples whose

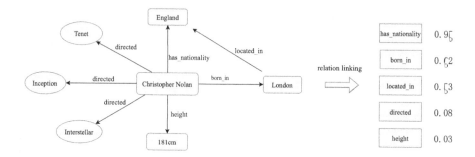

Fig. 2. Example of relation linking

relation is in the relation set are considered as valid triples and added into part of the subgraph. We use the random sampling method and set an upper limit n to avoid oversizing search space. Due to the superior performance of the relation linking model, RSA is able to extract subgraphs with high answer coverage and low subgraph scale.

3.4   Structural Fusion

The structural fusion module is based on the idea of graph neural networks to get node representation of G'. First we need to initialize the representation of entity node $h_i^{(0)}$ and question $h_q^{(0)}$, and then $h_q^{(0)}$ are fused during graph propagation process to enhance the encoding performance of GNN, the final entity node representations are used to calculate entity scores to filter candidate entities.

Node Initialization. The representation of entity nodes in the KG are embedded into fixed-size vector $h_i^{(0)} \in R^F$, where $h_i^{(0)}$ can be initialized from the pre-trained KG embedding or Gaussian distribution and F is the embedding size. We use RoBERTa to initialize the question representation $h_q^{(0)}$, and then use a linear layers to convert $h_q^{(0)}$ into the same dimension F:

$$h_q^{(0)} = W^T(RoBERTa(w_1, w_2, \ldots, w_n))$$

Fusion Propagation Strategy. A propagation process of graph neural network can generally be represented by two states: message propagation and node update. The message propagation process is used to capture the structural information of the knowledge graph context to obtain a distributed representation of the information, and the node update process is used to extract the semantics captured by the message propagation process to update the next layer of entity representation. Following this, we describe the message propagation process as:

$$m_i^{(l)} = \sum_r \sum_{j \in N_r(i)} \alpha_{ij} W^{(l-1)} h_j^{(l-1)} \tag{1}$$

where $m_i^{(l)}$ denotes the propagation message representation of entity i during the lth iteration, $N_r(i)$ denotes the neighbors of the entity i as the incoming edge, α_{ij} is the attention weight that is described in next section. This progress can aggregate the features of the neighbor nodes. After obtaining the message representation, we update entity nodes and question as:

$$h_i^{(l)} = \sigma \left(h_i^{(l-1)} \oplus h_q^{(l-1)} \oplus m_i^{(l)} \right) \tag{2}$$

$$h_q^{(l)} = \sigma \left(\sum_{i \in S_q} h_i^{(l-1)} \right) \tag{3}$$

where $\oplus$ denotes the cascade operation. We can find that the update of the entity node is represented by the previous propagation process of the entity node $h_i^{(l-1)}$ and the question $h_q^{(l-1)}$, and the propagation message representation $m_i^{(l)}$ obtained by cascade operations, σ is a feedforward neural network used to transform the dimension of the feature vector to achieve a stronger fit. After an update operation, $h_q^{(l)}$ is computed by topic entity nodes representation, here S_q denotes the set of topic entities.

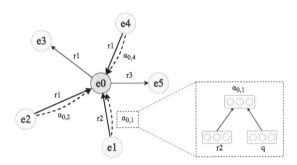

Fig. 3. The aggregation process of different relation

Relation Attention. Due to the heterogeneity of the subgraph, it is necessary to consider the attention weight of different relations of the entity neighbor nodes in the propagation process. By calculating the similarity between the questions and different relations, messages can be propagated in the direction of more relevant relations. Figure 3 describes this progress, the dashed arrows represent the edges that are participating in the propagation. Each propagation process takes into account the directionality of the triples, so only the nodes with the entity nodes as the incoming edges are selected as neighbor nodes to participate in the calculation. Neighbor nodes give different contribution values to entity nodes through the attention weights of different relations. We compute relation attention coefficients as:

$$e_{ij} = x_r^T h_q^{(l-1)} \tag{4}$$

$$\alpha_{ij} = \frac{\exp\left(e_{ij}\right)}{\sum_{k \in N_i} \exp\left(e_{ik}\right)} \tag{5}$$

where α_{ij} reprsents the attention coefficient of the relation r and e_{ij} represents the similarity score between the relationship r and the question q, which is directly calculated by dot product, x_r is the relation representation.

Entity Scoring. We use all entity node representations in the subgraph for the entity scoring process, after completing l times propagation, the entity representation is represented as $h_v(v \in E^{'})$. The representation effectively fuses structural features of KG and natural language semantics, and focuses more on the relations associated with the natural language question. The probability score s of each candidate entity is calculate by the sigmoid function:

$$s = \sigma\left(w^T h_v^l + b\right)$$

The entity scoring process uses L2 regularization to constrain the parameters. Here, the candidate entities with higher scores are fed to the next module to reason the target answer entities.

3.5   Relation Reasoning

We extract the relation paths of all topic entities and candidate entities and obtain the hidden feature representation through the relation path encoder. For a relation path, the embedding of entities and relations trained by the structural fusion module are used for the initialization of the path. Due to the strong directionality of the relation path, Bi-LSTM is used to encode the path.

The contribution of different nodes in the relation path to the final feature vector is different. We calculate node attention in the relationship path, which is similar to the calculation method of relationship attention.

$$\alpha_i = \frac{\exp\left(h_q \times h_i\right)}{\sum_{j=1}^{n} \exp\left(h_q \times h_j\right)} \tag{6}$$

where α_i denotes the attention coefficients of different nodes in the relation path, h_q is the embedding of the question. After obtaining the attention coefficients corresponding to each node in the relation path, the final representation of the relation path is calculated as:

$$h_r = \sum_{i=1}^{n} \alpha_i \times h_i \tag{7}$$

For all relation paths, the similarity score is calculate by the Euclidean measure between the relation path representation h_r and the natural language question representation h_q. In order to make the representation of the relation path and question as close as possible in the semantic space, contrastive loss is chosen to optimize the experimental results. The candidate entity that has the highest similarity score of the relation path is adopted as the reasoning result. The relation reasoning module effectively fuses the semantics of relation paths and improves the reliability of reason results.

4 Experiments

4.1 Experiment Setup

Dataset. We conduct experiments on two datasets: MetaQA is a dataset in the movie domain, which contains more than 30k natural language questions, all questions have been classified according to the hop of three different types. WebQuestionSP dataset is from the knowledge base FreeBase, whose natural language questions mostly contain two-hop relations. All answer entities can be found in the FreeBase knowledge base. All details are shown in Table 1.

Baselines. We compare our model with the following baselines: (1) Probability-based method: VRN [15]. (2) Graph neural network methods: GraftNet [3], Pull-Net [13], and RecHyperNet [17]. (3) Embedding-based method: EmbedQA [11].

Table 1. Statistics of datasets

Datasets	MetaQA			WebQuestionSP
	1-hop	2-hop	3-hop	
Train	96106	118980	114196	2882
Dev	9992	14872	14274	250
Test	9947	14872	14274	1530
Entities	43234			8239091
Relations	18			575
Triples	269482			499198889

Metrics. We use Hits@N and F1-score for evaluation. The Hits@N metric indicates the average percentage of answer entities among the top N candidate entities with the highest relevance score. The F1-score is calculated by a threshold α that we tune as the binary criterion for the test dataset.

4.2 Model Comparison

Comparison Result. Table 2 shows the Hits@1 metrics on different models, our model achieves almost the best result, indicating the superiority of our model.

In particular, Hits@1 for 1-hop in the MetaQA dataset are second only to RecHyperNet, while the experimental results of 2-hop and 3-hop both achieve the best performance, indicating that NLKGF performs better in handling natural language questions with multi-hop relations. On the WebQuestionSP dataset, the model still achieves the leading position, indicating that the model has well applicability on large-scale knowledge graphs.

Table 2. Comparison of Hits@1 performance on different models

Datasets	MetaQA			WebQuestionSP
	1-hop	2-hop	3-hop	
VRN(2018)	97.5	82.7	48.9	–
GraftNet(2018)	97	94.8	77.7	66.4
PullNet(2019)	97	**99.9**	91.4	68.1
EmbedQA(2020)	97.5	98.8	94.8	66.6
RecHyperNet(2021)	**99.1**	99.2	95	68.1
NLKGF(ours)	97.5	**99.9**	**99.9**	**69.5**

Performance on Relation Linking. Table 3 represents the accuracy of relation linking model. It can be seen that the performance of relation linking is excellent. The accuracy of the MetaQA all reaches 100%. This is due to the substantial quantities of training data and fewer types of relations on metaQA datasets. When the training data is significantly decreased and the relation composition is more complex, WebQuestionSP also achieves 73.8% accuracy on the test dataset, indicating the reliability of the linking results.

Table 3. Accuracy of the relation linking model on two datasets

Datasets	MetaQA			WebQuestionSP
	1-hop	2-hop	3-hop	
Train	100	100	100	80.5/98.5
Test	100	100	100	73.8/93

The result of relation linking will greatly affect the coverage of subgraphs, so we expand the selection range to get higher accuracy. As the selection range increases, the accuracy improves substantially, however it leads to the rapid expansion of the subgraph size and brings disturbance from noisy entities. In the tradeoff of high accuracy and low subgraph size, we select top5 as the relation set to extract subgraphs, which is shown in Table 4, the accuracy reaches 93% now on WebQuestionSP test dataset.

Table 4. Results of extracting subgraphs

Subgraph	MetaQA			WebQuestionSP
	1-hop	2-hop	3-hop	
Avg entities	3.2	9.92	64.11	21.9
Avg triples	2.3	10.13	93.68	25.4
Max entities	95	233	927	480
Max triples	94	361	1296	687

Ablation Experiment. We ablate three different components respectively to verify the effectiveness. Table 5 shows Hit@1 and F1-score results on WebQuestionSP dataset. When removing the relation path and directly using the structural fusion module for separate reasoning, the Hits@1 metrics decreases by 3.8%, and the F1-score decreases by 1.8%, proving that the relation path is able to improve the multi-hop reasoning performance. Relation attention enables NLKGF to perceive the relevance of different relations to propagate messages in a more relevant direction. When the relation attention mechanism is ablated, the results perform worst, demonstrating the necessary of the relation attention. Using pre-trained models is also helpful to improve the reasoning performance.

Table 5. Results of ablation experiments on the WebQuestionSP dataset

Model	Hits@1	F1-score
NLKGF	69.5	64.2
– Relation Path	65.7	62.4
– Relation Attention	64.6	60.2
– RoBERTa	66.2	63.1

5   Conclusion

In this paper, the model NLKGF is proposed for multi-hop reasoning tasks. NLKGF designs a structural fusion module to fuse natural language representations in the process of graph propagation, enabling entity representations to express richer semantics, and then enhance the performance of multi-hop reasoning by encoding relation paths. In addition, a restricted subgraph algorithm based on relation linking is designed, which can effectively extract knowledge graph subgraphs and restrict the range of candidate entities, enabling the model to be extended to large-scale knowledge graphs.

References

1. Qiu, L., et al.: Dynamically fused graph network for multi-hop reasoning. In: Proceedings of the 57th Conference of the Association for Computational Linguistics, pp. 6140–6150 (2019)
2. Bordes, A., Chopra, S., Weston, J.: Question answering with subgraph embeddings. In: Proceedings of the 2014 Conference on Empirical Methods in Natural Language Processing, pp. 615–620 (2014)
3. Sun, H., Dhingra, B., Zaheer, M., Mazaitis, K., Salakhutdinov, R., Cohen, W.W.: Open domain question answering using early fusion of knowledge bases and text. In: Proceedings of the 2018 Conference on Empirical Methods in Natural Language Processing, pp. 4231–4242 (2018)
4. Feng, Y., Chen, X., Lin, B.Y., Wang, P., Yan, J., Ren, X.: Scalable multi-hop relational reasoning for knowledge-aware question answering. In: Proceedings of the 2020 Conference on Empirical Methods in Natural Language Processing, pp. 1295–1309 (2020)
5. Liang, P.: Lambda dependency-based compositional semantics. arXiv preprint arXiv:1309.4408 (2013)
6. Yih, W., Chang, M., He, X., Gao, J.: Semantic parsing via staged query graph generation: question answering with knowledge base. In: Proceedings of the 53rd Annual Meeting of the Association for Computational Linguistics and the 7th International Joint Conference on Natural Language Processing of the Asian Federation of Natural Language Processing, pp. 1321–1331 (2015)
7. Xu, K., Wu, L., Wang, Z., Yu, M., Chen, L., Sheinin, V.: Exploiting rich syntactic information for semantic parsing with graph-to-sequence model. In: Proceedings of the 2018 Conference on Empirical Methods in Natural Language Processing, pp. 918–924 (2018)
8. Das, R., Zaheer, M., Reddy, S., McCallum, A.: Question answering on knowledge bases and text using universal schema and memory networks. In: Proceedings of the 55th Annual Meeting of the Association for Computational Linguistics, pp. 358–365 (2017)
9. Xu, K., Lai, Y., Feng, Y., Wang, Z.: Enhancing key-value memory neural networks for knowledge based question answering. In: Proceedings of the 2019 Conference of the North American Chapter of the Association for Computational Linguistics: Human Language Technologies, pp. 2937–2947 (2019)
10. Dong, L., Wei, F., Zhou, M., Xu, K.: Question answering over freebase with multi-column convolutional neural networks. In: Proceedings of the 53rd Annual Meeting of the Association for Computational Linguistics and the 7th International Joint Conference on Natural Language Processing of the Asian Federation of Natural Language Processing, pp. 260–269 (2015)
11. Saxena, A., Tripathi, A., Talukdar, P.P.: Improving multi-hop question answering over knowledge graphs using knowledge base embeddings. In: Proceedings of the 58th Annual Meeting of the Association for Computational Linguistics, pp. 4498–4507 (2020)
12. Trouillon, T., Welbl, J., Riedel, S., Gaussier, É., Bouchard, G.: Complex embeddings for simple link prediction. In: Proceedings of the 33rd International Conference on Machine Learning, pp. 2071–2080 (2016)
13. Sun, H., BedraxWeiss, T., Cohen, W.W.: Pullnet: open domain question answering with iterative retrieval on knowledge bases and text. In: Proceedings of the 2019 Conference on Empirical Methods in Natural Language Processing and the 9th

International Joint Conference on Natural Language Processing, pp. 2380–2390 (2019)

14. Zhao, B., Xu, Z., Tang, Y., Li, J., Liu, B., Tian, H.: Effective knowledge-aware recommendation via graph convolutional networks. In: Proceedings of the 17th International Conference on Web Information Systems and Applications, pp. 96–107 (2020)

15. Zhang, Y., Dai, H., Kozareva, Z., Smola, A.J., Song, L.: Variational reasoning for question answering with knowledge graph. In: Proceedings of the Thirty-Second AAAI Conference on Artificial Intelligence, pp. 6069–6076 (2018)

16. He, G., Lan, Y., Jiang, J., Zhao, W.X., Wen, J.: Improving multi-hop knowledge base question answering by learning intermediate supervision signals. In: Proceedings of the Fourteenth ACM International Conference on Web Search and Data Mining, pp. 553–561 (2021)

17. Yadati, N., Dayanidhi, R.S., Vaishnavi, S., Indira, K.M., Srinidhi, G.: Knowledge base question answering through recursive hypergraphs. In: Proceedings of the 16th Conference of the European Chapter of the Association for Computational Linguistics, pp. 448–454 (2021)

Commonsense Knowledge Construction with Concept and Pretrained Model

Hanjun Cai, Feng Zhao$^{(\boxtimes)}$, and Hai Jin$^{(\boxtimes)}$

National Engineering Research Center for Big Data Technology and System,
Services Computing Technology and System Lab, Cluster and Grid Computing Lab,
School of Computer Science and Technology, Huazhong University of Science and
Technology, Wuhan, China
zhao@hust.edu.cn, jin@hust.edu.cn

Abstract. *Commonsense knowledge* (CSK) is the information that people use in daily life but do not often mention. It summarizes the practical knowledge about how the world works. Existing machines have knowledge but lack commonsense because they do not understand and master commonsense knowledge in the same way that humans do. In the latest works, crowdsourcing-based method is costly and has low coverage, knowledge base completion method can highly fit samples, and methods extracted from unstructured data have the defects of low quality. CG&BF is commonsense knowledge construction with a concept-based generator and a BERT-based filter. We utilize semantic search for node matching and entropy encoder for filtering triples with high abstraction. Two algorithms based on concept aggregation and path credibility are proposed to obtain high-quality CSK triples. We subsequently finetuning a BERT to filter incorrect triples. We obtain 500,000 CSK triples based on ConceptNet, which is superior to other construction methods in novelty and quality. In the reading comprehension task, the three-way attention network is selected as the basic model and the knowledge we generate enables the base model to perform better, which proves that the output of CG&BF has higher quality and ease of use.

Keywords: Knowledge graph · Commonsense knowledge · Pretrained language model

1 Introduction

Commonsense knowledge (CSK) [1] is a consensus among human beings, which is rarely explicitly mentioned in the corpus. Commonsense does not focus on a specific instance but on the relationship between concepts. For example, "A house has rooms" is commonsense knowledge; "The Palace of Versailles has 700 rooms" is not commonsense knowledge but world knowledge. It's time-consuming and laborious to extract high-quality commonsense knowledge through handcrafting. In addition, some studies have shown that the effect of advanced completion models will decline rapidly with the decrease of density [2]. With the increase of human knowledge, the artificial construction of commonsense knowledge base is not a long-term solution.

© The Author(s), under exclusive license to Springer Nature Switzerland AG 2022
X. Zhao et al. (Eds.): WISA 2022, LNCS 13579, pp. 40–51, 2022.
https://doi.org/10.1007/978-3-031-20309-1_4

At the moment, manual construction [3,4] is time-consuming and low coverage, automatic extraction from raw data [5–8] and commonsense knowledge base completion [2,9–11] have the phenomenon that the quality is not high enough and fits the training data, which makes it difficult to perform well in other tasks. We propose a general method for building commonsense knowledge called CG&BF. CG&BF (commonsense knowledge construction with concept-based generator and BERT-based filter) selects high-quality commonsense knowledge as the seeds in the existing database and then expands knowledge by introducing additional conceptual knowledge. In this paper, we employ CG&BF to obtain about 500,000 CSK triples from ConceptNet, which show better performance in the reading comprehension task. We also conduct experiments to demonstrate their better novelty and quality.

Contributions. The contributions of this paper are summarized as follows:

- We propose a method to mine commonsense in the pretrained model by guiding the model with a binary classification task. The pretrained model is not allowed to complete the generation part, but to do the discriminant work.
- We innovatively propose a method to generate effective commonsense knowledge by combining concept aggregation and semantic search. Multiple concept co-decision can help select a more suitable concept and generate the same high-quality triples.
- Our model obtaines a total of about 500,000 pieces of CSK triple, which is more uncommon and serviceable than what other methods produce.

2   Related Works

In this section, we provide background information about commonsense knowledge generation and review related works.

Manual Construction. Traditional approaches to acquire commonsense knowledge, such as ConceptNet, inherited from the crowdsourcing project OMCS [3] in 1999, and ATOMIC [4], crowdsourced on MTurk in 2017, often require laborious and expensive manual annotation. Additionally, coverage is not satisfactory because the build job is to expand the seeds proposed by the researchers.

Knowledge Base Completion. In the era without pretrained models, Li et al. [9] used LSTM [12] to score the credibility of the language, generate additional new triples randomly, and then obtain triples with medium confidence quality of ConceptNet. Saito et al. [10] employ BiLSTM as an encoder to replace the original LSTM and use an additional LSTM as a decoder to complete the generation. After Transformer [13] are presented, COMET [11] takes seed knowledge from ATOMIC [4] and ConceptNet and utilizes them for GPT-2 pre-training. Besides they employ the model to predict t with the input h and r. Due to the characteristics of GPT-2, the same h and r can lead to similar or confusing t. Further work has ATOMIC$_{20}^{20}$ [14] which employs GPT-3 [15] and BART [16] to generate new tail. Feldman et al. [17] proposed an unsupervised method using a pretrained language model to get the point-wise mutual information of the candidate triples and took it as the basis of the credibility of the triples to get new commonsense knowledge. The knowledge completion of traditional knowledge graph [18] mostly relies on knowledge representation like TransE [19] series. Malaviya et al. [2] combined the graph structure information learned from GCN [20] with the node semantic information obtained from BERT to enrich the context and then employed ConvTransE [21] as a decoder to complete commonsense knowledge base.

Automatic Extraction from Raw Data. ASCENT [6] has focused on higher-order semantics from the outset, arguing that attributes of entities are associated with additional commonsense relationships. Accordingly, ASCENT builds a pipeline consisting of degree classifiers and text taggers to mine commonsense knowledge from the english corpus. For a particular kind of commonsense knowledge, Xu et al. [7] designed two tasks to extract *LocatedNear* commonsense. One is to find the sentence containing *LocatedNear* commonsense, and the other is to find the entities with *LocatedNear* relationship from the sentence. Xiao et al. [8] focused on the *CapableOf* commonsense relationship. They acquired commonsense knowledge through knowledge reasoning and web search engine. Xiao et al. [22] also mined the verb-oriented commonsense knowledge through conceptualization.

These methods more or less have the following problems:

1. Low coverage. Commonsense knowledge collected by crowdsourcing is often some seed concepts proposed by researchers and then expanded and added by researchers, which can only cover part of commonsense commonly used in daily life.
2. High fitting. The large-scale pretrained model can be used to generate a tail entity. However, the tail entities generated for the same commonsense triple are similar because of large-scale training.
3. Low quality. Few commonsense knowledge exists in the large-scale corpus, and high-quality knowledge can be obtained by summarizing and condensing to a certain extent.

3   Methodology

In this section, we first formalize the problem and then elaborate on the model for commonsense knowledge generation.

3.1   Framework of CG&BF

The core of the model in this paper is to employ a simple generator as the carrier of the varying ability of the pretrained model and then utilize a simple binary task to guide and mine the commonsense knowledge hidden in the pretrained model. Therefore, the model is divided into a concept-based generator and a BERT-based filter. The architecture of CG&BF is shown in Fig. 1.

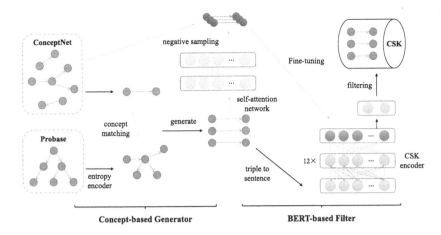

Fig. 1. CG&BF model architecture

Problem Definition. Given a set KG that contains commonsense knowledge (h, r, t), where $h, t \in E_1$ and r is a commonsense relation just like $CapableOf$ and $UsedFor$. Given a set $ConceptKG$ that contains concept triples (h, r, t), where $h, t \in E_2$ and r is a relation like IsA and $InstanceOf$. There are some nodes not only in E_1 but also in E_2. Our goal is to obtain commonsense knowledge based on KG with the conceptual information in $ConceptKG$.

Module 1: Concept-Based Generator. This module introduces three parts of the commonsense knowledge generation algorithm, namely, semantic similarity matching method based on Transformer used in concept matching, two generative algorithms based on concept aggregation and credibility greedy strategy to dig commonsense knowledge, and entropy encoder for filtering abstract concepts.

Module 2: BERT-Based Filter. This module includes natural language converters and filters fine-tuned based on BERT. The model consists of BERT, a fully connected layer, and a soft-max layer, and is trained with negative-sampled data.

3.2 Concept-Based Generator

There may be high credibility commonsense knowledge hidden between entities with the same concept. As shown in Fig. 2, hounds can help hunters, watchdogs can guard homes, and guide dogs can guide the blind, then we can get a hidden relationship between "dogs" and "helping humans" according to the conceptual relationship. We develop an algorithm based on concept aggregation to discover new triples like above.

As is shown in Fig. 3, the blue nodes represent commonsense nodes, the green nodes denote a concept, the orange edges denote conceptual relationships, such as IsA and $TypeOf$, and the other colored edges denote commonsense relationships, such as $CapableOf$ and $UsedFor$. Nodes 1, 3, and 5 have the

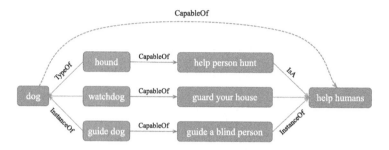

Fig. 2. Example of dogs in conceptual abstraction

same concept of node 7. Nodes 1 and 3 have blue relationships. Nodes 4 and 6 have a common concept of node 8, so it can be concluded that there may be a relationship between node 7 and node 8 with a blue edge. In the same way, node 8 and node 9 may have a relationship with a yellow edge. For the validity of commonsense knowledge, filters are eventually used to ensure the quality of the generated knowledge.

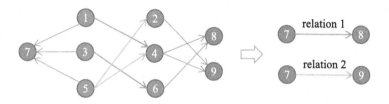

Fig. 3. Concept selection based on concept aggregation

The concept selection based on concept aggregation takes into account the effectiveness of generating knowledge, and removes a large number of concepts that are not relevant enough in the initial generation, but does not consider the concept relationship between multiple hops. We generate the confidence for this edge based on the frequency of the conceptual relationship. Then a greedy strategy is utilized to search on the incoming and outgoing edges for related concepts. This can mine commonsense knowledge between concepts at the same level and at a different level. As shown in Fig. 4, the number on the orange edge represents the credibility of the edge, and the depth of the search is represented by *depth*. When the *depth* is 1, the nearest matched nodes are 1 and 2, and the commonsense triplet from node 1 to node 2 is generated. When *depth* is 2, generate commonsense triples from node 6 to node 4.

In the process of concept selection, the degree of abstraction of the concept also has a great influence on the concept. For example, the "dog" may eventually be abstracted to the meaningless concept of "object". We cannot just rely on absolute depth to measure the abstraction of a concept. The top-level nodes

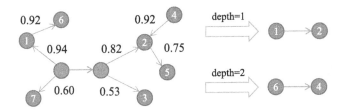

Fig. 4. Concept selection based on credibility greedy strategy

tend to have more hyponyms, while the lower-level nodes have fewer hyponyms. Therefore, we use entropy to encode the differences at different conceptual levels. The entropy can reflect the influence of the number of hyponyms on the abstraction while avoiding the effect of the level gap. Let t_i be the number of hyponyms of t at level i. t_{total} is the total number of hyponyms of t. We have

$$p_i(e) = \frac{t_i}{t_{total}} \tag{1}$$

$$Entropy(e) = -\sum_{i=2} p_i(e) \log p_i(e) \tag{2}$$

3.3 BERT-Based Filter

The information about commonsense knowledge triples is gradually blurred and lost in the process of concept replacement. For this reason, there is plenty of noise in the generated triples. We fine-tuning BERT to filter invalid triples and triples need to be converted to natural sentences before filtering.

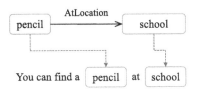

Fig. 5. Natural language conversion

The idea of the conversion algorithm from triples to natural language is effortless to grasp as shown in Fig. 5. Since the relationships of commonsense knowledge are predefined and are not much, a specific template is designed for each commonsense relationship. Afterwards the head entity and tail entity are combined with the template and entered into BERT. This paper describes the screening of commonsense knowledge as a binary classification task. Figure 6 is the model architecture consisting of a pretrained BERT and fully connected

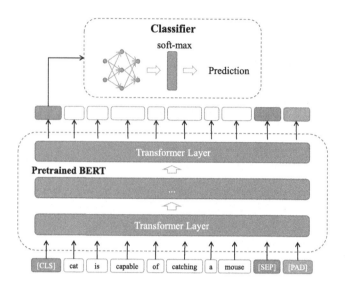

Fig. 6. BERT-based filter architecture

layer and a softmax layer. The model employs cross-entropy as the loss function of the model to optimize the training results.

In data preparation, a part of the CSK triples in ConceptNet is extracted as true and the triples generated by negative sampling are labeled as false. We make use of approximately 200,000 triples to fine-tune BERT to achieve 94% accuracy in model predictions.

4   Experiments

In this section, we conduct extensive experiments to evaluate the effectiveness of our approach.

4.1   Experiment Setup

The CG&BF model utilizes ConceptNet and Probase for commonsense knowledge generation. Additionally, to demonstrate the effectiveness and ease of use of the commonsense database, the effect comparison is made on the downstream task reading comprehension of commonsense knowledge.

Baselines. Commonsense construction methods for comparison include LAMA [23], COMET [11], TransOMCS [3], ASCENT [5], and commonsense knowledge resource obtained by them.

- TransOMCS. This is a dataset obtained by pattern matching with Concept-Net based on language graph ASER extracted from unstructured text.
- ASCENT. ASCENT is a pipeline that can automatically collect, extract, and integrate commonsense knowledge from any English text corpus.

– COMET. COMET uses the ConceptNet and ATOMIC seeds for pre-training on GPT-2 and utilizes the pretrained model to generate the tail entity.
– LAMA. The method utilizes BERT's complementary prediction ability to generate more knowledge.

Metrics. The evaluation of the model is divided into two parts. In the first part, the commonsense triples generated by the model are evaluated with the words, novelty, and perplexity (PPL). In the second part, the generated CSK triples of different models are compared in a downstream task. The effect of the model with different additional commonsense resources will be evaluated through three indicators: Precision, Recall, and F1-score.

$$Perplexity(s) = \sqrt[n]{\prod_{i=1}^{n} \frac{1}{p(w_i|w_1, ..., w_{i-1})}} \tag{3}$$

4.2 Model Comparison

CSK Generation. The seed set used to generate the experiment is about 600,000 commonsense triples, and the used concept knowledge is about 350,000. As shown in Table 1, a total of 994,072 commonsense triples were generated under the two concept selection algorithms, and a total of 499,376 triples were filtered. In the comparison experiments, the COMET model selects models with beam search values of 1 and 10 for comparison. TransOMCS selects the top 1% of triples for comparison.

After comparison, it is found that CG&BF is stronger than other models in generation ability, and can obtain more and higher quality CSK triples. It is 4% to 60% higher than other models in triple novelty, and about 12% higher than other models in PPL.

Table 1. Model generation comparison on ConceptNet

Model	Words	Turtle	Novelty	Novelty$_h$	Novelty$_c$	Novelty$_t$	PPL
OMCS	36954	207427	–	–	–	–	–
COMET@1	715	1200	33.96%	0	5.27%	33.96%	103.67
COMET@10	2232	1200	64.95%	0	27.15%	**64.95%**	112.15
TransOMCS	**37517**	184816	95.71%	**61.43%**	**75.65%**	46.19%	99.59
CG&BF	22295	**994072**	**99.90%**	42.59%	35.63%	46.01%	105.33
CG&BF(Filtered)	19453	**499376**	99.81%	37.61%	36.32%	40.66%	**88.26**

Reading Comprehension. To verify the quality and validity of the generated commonsense knowledge, relevant experiments are done on the commonsense downstream task to prove it. We select MCScript [24], a commonsense reading comprehension task for horizontal comparison in TransOMCS. We choose TriAN [25] as the base model and attach different commonsense resources to test the effects of different resources on model Precision, Recall, and F1-score. The evaluation results are shown in Table 2. Our model is 0.09%-0.54% higher than other methods and can improve the precision rate of the base model TriAN by 0.9%, the recall rate by 0.35%, and the F1 score by 0.52%.

Table 2. Reading comprehension task MCScript comparison

Commonsense knowledge resource	Precision(%)	Recall(%)	F1(%)
TriAN	83.30	48.41	61.24
+ConceptNet	83.89	48.75	61.67
+LAMA@1	83.47	48.33	61.22
+LAMA@10	83.52	48.34	61.24
+COMET@1	83.54	48.42	61.31
+COMET@10	83.58	48.44	61.33
+ACENT	83.75	48.37	61.33
+TransOMCS	83.75	48.25	61.23
+CG&BF	**84.20**	**48.76**	**61.76**

Effects of Semantic Search. The original intention of using Transformer embedding for concept matching based on semantic similarity is to match more concept nodes. The method can map the knowledge in two different knowledge bases to the same semantic space. Before and after the method is used, the matching coverage rate of the nodes in the commonsense base is tested. The coverage rate is defined as the ratio of the nodes of the KG that can match the concept in the $ConceptKG$ to the total number of KG nodes. When the semantic similarity is above the threshold p, it is defined as a successful match. It can be seen from Table 3 that when the semantic search method is not used, the concepts in the concept base can only cover 14.66% of the nodes in the KG. When using semantic search, the matching rate can be up to 36.69% by controlling the threshold p of concept matching.

Table 3. Effects of semantic search

Threshold (p)	1.0	0.9	0.8	0.7	0.6	0.5
Coverage	14.66%	16.87%	22.17%	31.15%	35.29%	36.69%

Effects of Natural Language Conversion. We use the same training set, adopt the same negative sampling method, and use different conversion methods to evaluate the difference between natural language conversion and direct input. As shown in Table 4, template conversion can better ensure the integrity of the original semantics of triples. Consequently, the model achieves significant improvements in precision, recall, and F1-score of 1.62%, 1.39%, and 1.54% respectively after BERT fine-tuning.

Table 4. Effects of natural language conversion

Conversion	Precision(%)	Recall(%)	F1(%)
Template	**97.27**	**59.86**	**74.11**
Direct	95.65	58.47	72.57

5   Conclusion

This paper proposes a method for building a commonsense knowledge base by combining concept-based generators and BERT-based filters. Experiments show that a large amount of high-quality knowledge can be obtained by selecting seeds from ConceptNet, which is of higher quality and novelty than commonsense knowledge produced by other advanced methods. In the reading comprehension task, we experiment on TriAN using various commonsense knowledge resources acquired by other advanced methods. The knowledge of CG&BF can make the model achieve better results, proving that CG&BF's commonsense knowledge has higher quality and usability.

References

1. Bhargava, P., Ng, V.: Commonsense knowledge reasoning and generation with pre-trained language models: a survey. In: Proceedings of the AAAI Conference on Artificial Intelligence (2022). Arxiv:2201.12438
2. Malaviya, C., Bhagavatula, C., Bosselut, A., Choi, Y.: Commonsense knowledge base completion with structural and semantic context. In: Proceedings of the AAAI Conference on Artificial Intelligence, pp. 2925–2933 (2020)
3. Zhang, H., Khashabi, D., Song, Y., Roth, D.: Transomcs: from linguistic graphs to commonsense knowledge. In: Proceedings of the Twenty-Ninth International Joint Conference on Artificial Intelligence, pp. 4004–4010 (2021)
4. Sap, M., et al.: Atomic: an atlas of machine commonsense for if-then reasoning. In: Proceedings of the AAAI Conference on Artificial Intelligence, pp. 3027–3035 (2019)
5. Zhang, H., Liu, X., Pan, H., Song, Y., Leung, C.W.K.: Aser: a large-scale eventuality knowledge graph. In: Proceedings of The Web Conference 2020, pp. 201–211 (2020)

6. Nguyen, T.P., Razniewski, S., Weikum, G.: Advanced semantics for commonsense knowledge extraction. In: Proceedings of the Web Conference 2021, pp. 2636–2647 (2021)

7. Xu, F.F., Lin, B.Y., Zhu, K.: Automatic extraction of commonsense locatednear knowledge. In: Proceedings of the 56th Annual Meeting of the Association for Computational Linguistics, pp. 96–101 (2018)

8. Liu, J., Xiao, Y., Wang, A., He, L., Shao, B.: Capableof reasoning: a step towards commonsense oracle. In: Proceedings of the 43rd International ACM SIGIR Conference on Research and Development in Information Retrieval, pp. 1797–1800. ACM (2020)

9. Li, X., Taheri, A., Tu, L., Gimpel, K.: Commonsense knowledge base completion. In: Proceedings of the 54th Annual Meeting of the Association for Computational Linguistics, pp. 1445–1455 (2016)

10. Saito, I., Nishida, K., Asano, H., Tomita, J.: Commonsense knowledge base completion and generation. In: Proceedings of the 22nd Conference on Computational Natural Language Learning, pp. 141–150 (2016)

11. Bosselut, A., Rashkin, H., Sap, M., Malaviya, C., Celikyilmaz, A., Choi, Y.: Comet: commonsense transformers for knowledge graph construction. In: Proceedings of the 57th Conference of the Association for Computational Linguistics, pp. 4762–4779 (2019)

12. Hochreiter, S., Schmidhuber, J.: Long short-term memory. Neural Comput. **9**(8), 1735–1780 (1997)

13. Vaswani, A., et al.: Attention is all you need. In: Proceedings of the Conference on Neural Information Processing Systems, pp. 6000–6010 (2017)

14. Hwang, J.D., et al.: Comet-atomic 2020: on symbolic and neural commonsense knowledge graphs. In: Proceedings of the AAAI Conference on Artificial Intelligence, pp. 6384–6392 (2021)

15. Brown, T., et al.: Language models are few-shot learners. In: Proceedings of the Conference on Neural Information Processing Systems, pp. 1877–1901 (2020)

16. Lewis, M., et al.: Bart: denoising sequence-to-sequence pre-training for natural language generation, translation, and comprehension. In: Proceedings of the 58th Annual Meeting of the Association for Computational Linguistics, pp. 7871–7880 (2020)

17. Davison, J., Feldman, J., Rush, A.M.: Commonsense knowledge mining from pre-trained models. In: Proceedings of the 2019 Conference on Empirical Methods in Natural Language Processing and the 9th International Joint Conference on Natural Language Processing, pp. 1173–1178 (2019)

18. Jiang, S., Nie, T., Shen, D., Kou, Y., Yu, G.: Entity alignment of knowledge graph by joint graph attention and translation representation. Web Information Systems and Applications, pp. 347–358 (2021)

19. Bordes, A., Usunier, N., Garcia-Duran, A., Weston, J., Yakhnenko, O.: Translating embeddings for modeling multi-relational data. In: Proceedings of the Conference on Neural Information Processing Systems, pp. 2787–2795 (2013)

20. Kipf, T.N., Welling, M.: Semi-supervised classification with graph convolutional networks. In: Proceedings of the 5th International Conference on Learning Representations (2017)

21. Shang, C., Tang, Y., Huang, J., Bi, J., He, X., Zhou, B.: End-to-end structure-aware convolutional networks for knowledge base completion. In: Proceedings of the AAAI Conference on Artificial Intelligence, pp. 3060–3067 (2019)

22. Liu, J., et al.: Mining verb-oriented commonsense knowledge. In: Proceedings of the 36th IEEE International Conference on Data Engineering, pp. 1830–1833 (2020)

23. Petroni, F., et al.: Language models as knowledge bases? In: Proceedings of the 2019 Conference on Empirical Methods in Natural Language Processing and the 9th International Joint Conference on Natural Language Processing, pp. 2463–2473 (2019)
24. Ostermann, S., Roth, M., Modi, A., Thater, S., Pinkal, M.: Semeval-2018 task 11: machine comprehension using commonsense knowledge. In: Proceedings of the 12th International Workshop on Semantic Evaluation, pp. 747–757 (2018)
25. Wang, L., Sun, M., Zhao, W., Shen, K., Liu, J.: Yuanfudao at semeval-2018 task 11: three-way attention and relational knowledge for commonsense machine comprehension. In: Proceedings of The 12th International Workshop on Semantic Evaluation, pp. 758–762 (2018)

Simplifying Knowledge-Aware Aggregation for Knowledge Graph Collaborative Filtering

Honghai Zhang, Yifan Chen, Xinyi Li, and Xiang Zhao[✉]

Laboratory for Big Data and Decision, National University of Defense Technology,
Changsha, China
xiangzhao@nudt.edu.cn

Abstract. Incorporating knowledge graph (KG) for recommendation has been well considered in recent researches since it can alleviate the sparsity and cold-start problem of collaborative filtering. To capture the rich semantics of knowledge graph, existing KG-based models utilize graph neural networks (GNNs). However, we empirically find that the feature transformation and nonlinear activation designs in GNN contribute little to the recommendation performance. We propose *simplified knowledge-aware attention network* (SKAN) that simplifies the knowledge-aware aggregation by removing the two designs. To ensure the personalization during propagation, we apply weighted aggregation with user-specific attentions. We further aggregate the interacted items of users to enhance the user representation learning. We apply the proposed model on three real-world datasets, and the empirical results suggest that simplified knowledge-aware attention network (SKAN) significantly outperforms several compelling state-of-the-art baselines.

Keywords: Recommender system · Knowledge graph · GNNs

1 Introduction

Recommender systems deal with information overload by filtering out irrelevant information and providing only relevant information to users. They have been widely used in various scenarios, such as music, movie and power domain [6,18].

In recent years, in order to alleviate the problems of cold start and sparse data, adding knowledge graph (KG) as side information (by aligning the items with entities in a KG [17]) to the recommender system has proven highly effective. KG is a semantic graph composed of nodes and edges that contain rich semantic knowledge. When applied to recommendation systems, the accuracy, diversity and interpretability can be vastly improved [2]. By integrating multi-source heterogeneous information, the KG represents rich entity relationships, and helps the recommender present accurate items to users.

X. Zhao et al. (Eds.): WISA 2022, LNCS 13579, pp. 52–63, 2022.
https://doi.org/10.1007/978-3-031-20309-1_5

The main challenge of existing KG-based models is how to learn effective user/item representations for recommendation. GNNs have become the dominant method to learn graph representations [5]. KGCN [10] is a representative work. To enhance the representation of users and items, graph convolutional neural network (GCN) is utilized to aggregate information from the neighbors in the KG. Another representative work is KGAT [11], which proposes collaborative knowledge graph (CKG) to combine user-item bipartite graph (UIG) with KG. In these KG-based works, feature transformation and nonlinear activation are two common designs in GNN. However, as suggested by a recent work of lightGCN [4], the two designs play a negative role in graph collaborative filtering (GCF). It is natural to believe that the two designs are also redundant for knowledge graph collaborative filtering (KGCF). For KGCF, we mainly care about the propagation of user personalized preference. We contend that the two designs may hinder the personalized propagation of user preference in KG.

In this light, we put forward *simplified knowledge-aware attention network* (SKAN) to simplify the aggregation of existing KG-based methods, which is to boost personalized preference propagation in KG. In specific, to propagate user preference, we follow the design of KGCN to incorporate user representations into the attention mechanism. Different from KGCN, however, inspired by lightGCN, we remove the feature transformation and nonlinear activation of GNN for knowledge graph collaborative filtering (KGCF). Furthermore, to better propagate user preference, we propose to aggregate user neighbors to form user representations. This is fundamentally different from existing works, since they generally perform aggregation over items [8,9].

Contributions. To summarize, We make the following contributions:

- We propose simplified knowledge-aware attention network, which is an end-to-end framework that learns the effective user/item representations from their interactions and KG by modeling users and items separately. SKAN uses a graph neural network aggregation approach that is more suitable for collaborative filtering.
- SKAN focuses on users' preference for relationships, while we verify the negativity of two common designs of GCNs in KG-based recommender systems—feature transformation and nonlinear activation on recommendation results.
- We perform empirical experiments on three real-world recommendation scenarios whose results demonstrate the superiority of SKAN over compelling state-of-the-art baselines, and we validate the effectiveness of modeling users and new information aggregation approaches in ablation experiments.

Organization. The rest of the paper is structured as follows: Section 2 discusses related works, and Sect. 3 formalizes the task. We present our method in Sect. 4, followed by experiments in Sect. 5. Section 6 concludes the paper.

2 Related Work

Recent research of KG-based recommender systems can be divided into three categories [3]: 1) path-based methods, 2) embedding-based methods and 3) unified-based methods. Unified-based methods combine the idea of path-based methods and embedding-based methods to realize recommendation, and unified-based methods avoid the disadvantage of the first two methods: the path needs to be set manually and the high-order semantic information of the graph cannot be obtained. Representative model of unified-based recommendation include RippleNet, KGCN, KGAT, CKAN [12] and LKGR [1], etc.

RippleNet is a memory-network model that propagates usersâĂŹ potential preference in the KG and explores their hierarchical interests. But note that the importance of relations is weakly characterized in RippleNet. In addition, the size of ripple set may go unpredictably with the increase of the size of KG, which incurs heavy computation and storage overhead.

KGCN captures local neighborhood information well and considers neighbor node weights for recommendations. It also pays attention to the importance of relationship. However, it isolates user from item's attribute graph, and ignores user modeling. In addition, with the further research on the application of graph-based neural network in recommendation, it shows that the way of node aggregation in KGCN does not improve the recommendation effect, and results in a waste of computation and storage costs.

Afterwards, KGAT introduces a user-item interaction matrix in the KG and weights the graph relationships to recommend them, while CKAN holds that users are represented by interacting items, and items are represented by those items that are interacted by the users. Distinctively, LKGR employs information propagation strategies in the hyperbolic space to encode heterogeneous information from historical interaction and KG, but fails to take into account the preference of each user for different relationships.

3 Task Formulation

In recommendation, users are defined as $\mathcal{U} = \{u_1, u_2 \ldots u_m\}$, and items as $\mathcal{I} = \{i_1, i_2 \ldots i_n\}$. We represent interaction data as a user-item bipartite graph $\mathcal{G}_1$, which is defined as $\{(u, y_u, i) \mid u \in \mathcal{U}, i \in \mathcal{I})\}$, When the user interacts with the item, $y_{ui} = 1$, otherwise $y_{ui} = 0$. Besides, we have side information to items, organized in the form of a knowledge graph $\mathcal{G}_2$, which is a directed graph composed of subject-property-object triple facts. Formally, it is $\{(h, r, t) \mid h, t \in \mathcal{E}, r \in \mathcal{R}\}$, where each triple describes that there is a relationship r from head entity h to tail entity t. For example, (*Jacky Chen, ActorOf, Police Story*) states the fact that *Jacky Chen* is an actor of the movie *Police Story*.

We adopt the notion of *CKG* [11], which encodes user behaviors and item knowledge as a unified relational graph. In CKG, user behavior is expressed by a triple $(u, \text{Interact}, i)$, where $y_{ui} = 1$ is represented as an additional relation *Interact* between user u and item i. Then, based on the *item-entity* alignment set, the user-item graph can be seamlessly integrated with the KG, producing a unified graph $\mathcal{G} = \{(h,r,t) \mid h,t \in \mathcal{E}', r \in \mathcal{R}'\}$, where $\mathcal{E}' = \mathcal{E} \cup \mathcal{U}$ and $\mathcal{R}' = \mathcal{R} \cup \{\text{Interact}\}$.

In summary, the task input is a collaborative knowledge graph $\mathcal{G}$ that combines the user-item bipartite graph $\mathcal{G}_1$ and knowledge graph $\mathcal{G}_2$, and the output is a function that predicts the probability $\hat{y}_{ui}$ that user u would adopt item i.

4 Methodology

We now present the simplified knowledge-aware attention network (SKAN). The core of this paper lies in the design of the knowledge-aware propagation. To facilitate the personalized propagation, nodes in SKAN obtains the neighbor information via linear aggregation. Without the feature transformation and nonlinear activation, during propagation the model can well concentrate on user preferences, rather than the KG semantics. Figure 1 depicts the overall framework.

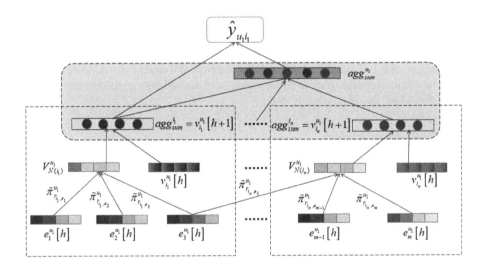

Fig. 1. The overall framework of SKAN.

4.1 Personalized Knowledge Aggregation

We propagate user preferences over the unified graph $\mathcal{G}$ through node aggregation. The high-order propagation is typically achieved through stacking multiple GNN layers. We then introduce the single GNN layer first. For every node except the users in $\mathcal{G}$, SKAN performs personalized aggregation of the information from node neighbors. Since the distribution of node degree is generally long-tail, for different entities in the graph, the number of neighbors varies largely. To address this degree imbalance, we follow [15] to sample fixed length of neighbors for every nodes. The sampled set of nodes are denoted by $\mathcal{N}$.

SKAN perform weighted aggregation with user-specific attentions:

$$e_{\mathcal{S}(e)} = \sum_{e' \in \mathcal{N}(e)} \pi^u_{r_{e,e'}} e', \tag{1}$$

where $\pi^u_{r_{e,e'}}$ is the personalized attention score. We use function $f(\boldsymbol{u}, \boldsymbol{r})$ to calculate the attention, e.g.,

$$\pi^u_{r_{e,e'}} = \frac{\exp\left(f(\boldsymbol{u}, \boldsymbol{r}_{e,e'})\right)}{\sum_{e' \in \mathcal{N}(e)} \exp(f(\boldsymbol{u}, \boldsymbol{r}_{e,e'}))}, \tag{2}$$

We further aggregate the neighbor representation with the entity representation, where we remove the feature transformation and the nonlinear activation.

$$agg^i_{\text{sum}} = e + e_{\mathcal{S}(e)}. \tag{3}$$

Note that Eq. (3) is slightly different from that in [4]. In LightGCN [4], the entity representation itself is not aggregated. We argue that this is necessary for KGCF. The reason is that LightGCN is working on the user-item bipartite graph, where the neighbor information of the item are users only. Items can be well defined by the aggregations of interacted users in the bipartite graph. In comparison, the KG has other entities to distract the node information. Therefore, the entity representation is necessary to balance the possible distraction, especially when the entity is an item.

4.2 User Aggregation

Many KG-based recommender systems pay too much attention to the modeling of the item-side and ignore the modeling of the user-side. Similar to the item-side, the user representation in SKAN is also composed of two parts of information, one is the user's own information, and the other is the historical item information that the user has interacted with. Different from the item side, when aggregating user neighbor nodes, only one layer of aggregation is done. Since users have different preference for interactive goods, we introduce the user's attention mechanism for items as follows:

$$\lambda_{u,e} = g(\boldsymbol{u}, \boldsymbol{e}), \tag{4}$$

Users' preference for items is normalized to

$$\tilde{\lambda}_{u,e} = \frac{\exp\left(\lambda_{u,e}\right)}{\sum_{e \in \mathcal{N}(u)} \exp\left(\lambda_{u,e}\right)}, \tag{5}$$

where u is the representation of user, e is the representation of item that the user has interacted with, $\mathcal{N}(u)$ denotes the set of items connected to u. After sampling and aggregating, the items interacted by users are represented as

$$U_{\mathcal{S}(u)} = \sum_{e \in \mathcal{N}(u)} \tilde{\lambda}_{u,e} e, \tag{6}$$

The final representation and calculation method of user is as follows:

$$agg_{\text{sum}}^{u} = \boldsymbol{u} + U_{S(u)}. \tag{7}$$

4.3   Prediction Layer

From the above, we can see that in single-layer SKAN, both entity representation and user representation are jointly represented by itself and its 1-hop neighbors. When the single-layer SKAN is extended to multi-layer, the model can deeply mine the high-order information of knowledge spectrum, so as to obtain more accurate entity representation and user representation. Specifically, the initial representation of each entity (order 0 representation) is propagated to its adjacent entities to obtain the first-order entity representation. Then we can repeat this process, that is, further propagate and aggregate the 1-hop representation to obtain the 2-hops representation, so recursively. Therefore, after obtaining the vector representation of the item and user, we input them into the function F to predict the probability together:

$$\hat{y}_{uv} = F(\mathbf{u}, \mathbf{v}), \tag{8}$$

In order to improve the computational efficiency, we use the negative sampling strategy in the training process. The complete loss function is as follows:

$$\mathcal{L} = \sum_{u \in \mathcal{U}} \left(\sum_{v: y_{uv}=1} \mathcal{J}(y_{uv}, \hat{y}_{uv}) - \sum_{i=1}^{T^u} \mathbb{E}_{v_i \sim P(v_i)} \mathcal{J}(y_{uv_i}, \hat{y}_{uv_i}) \right) + \lambda \|\mathcal{F}\|_2^2. \tag{9}$$

where $\mathcal{J}$, P and T^u are the representations of cross-entropy loss, a negative sampling distribution and the number of negative samples for user u, respectively. The last term is the L2-regularizer.

5   Experiments

In this section, we evaluate the proposed model under three real world scenarios, with the aim of answering the following research questions:

 – **RQ1**: How does SKAN perform compared with state-of-the-art methods?
 – **RQ2**: How do different components affect SKAN?

5.1   Datasets

In order to verify the effectiveness of our proposed method, we conducted comparative experiments on three public datasets, i.e., Movie-Lens20M, Book-Crossing and Dianping-Food. The statistics of the datasets are presented in Table 1.

Table 1. Statistics of the three datasets: Movie-Lens20M, Book-Crossing, and Dianping-Food. The inter-avg means the average interactions per user, the link-avg means the average links per item.

Dataset	Movie-Lens20M	Book-Crossing	Dianping-Food
# users	138,159	17,860	2,298,698
# items	16,954	14,967	1,362
# interactions	13,501,622	139,746	23,416,418
# inter-avg	98	8	10
# entities	102,569	77,903	28,115
# relations	32	25	7
# KG triples	499,474	151,500	160,519
# link-avg	29	10	118

- **Movie-Lens20M**[1] is a widely used benchmark dataset in movie recommendations, which consists of approximately 20 million explicit ratings (ranging from 1 to 5) on the MovieLens website.
- **Book-Crossing**[2] is collected from the book-crossing community, which consists of trenchant ratings (ranging from 0 to 10) from different readers about various books.
- **Dianping-Food**[3] is provided by Dianping.com, which contains over 10 million interactions (including clicking, buying, and adding to favorites) between approximately 2 million users and 1 thousand restaurants. The corresponding KG contains 28, 115 entities, 160, 519 edges and 7 relation-types.

5.2   Baselines

In order to demonstrate the effectiveness of SKAN, we compare it with the state-of-the-art methods, including CF-based method (BPRMF), embedding-based methods, path-based method (PER) unified-based methods (RippleNet, KGCN, KGNN-LS, KGAT, CKAN, LKGR):

- **BPRMF** [7] is a Bayesian personalized ranking model based on matrix factorization (MF), which realizes personalized recommendation by learning individual user preference.
- **CKE** [14] is a representative embedding-based method, which leverages KG embeddings of entities derived from TransR as ID embeddings of items under the MF framework. Where, KG relations are only used as the constraints in TransR to regularize the representations of endpoints.

---

[1] https://grouplens.org/datasets/movielens/.
[2] https://searchengineland.com/library/bing/bing-satori.
[3] https://github.com/hwwang55/KGNN-LS/raw/master/data/restaurant/Dianping-Food.zip.

– **PER** [13] treats the KG as HIN and extracts meta-path based features to represent the connectivity between users and items. In this paper, we use all item-attribute-item features for PER (e.g., âĂİmovie-director-movie").
– **RippleNet** [8] is an end-to-end framework that uses KG to realize recommendation system. It combines the embedding-based and path-based method into the recommendation system based on KG for the first time. Through the method of preference propagation in KG, users' potential hierarchical interests are continuously and automatically found.
– **KGCN** [10] is a state-of-the-art unified-based method which extends spatial GCN approaches to the KG domain. By aggregating high-order neighbor information, both structure information and semantic information of the KG can be learned to capture usersâĂŹ potential long-distance interests.
– **KGNN-LS** [16] is based on KGCN, which transforms KG into user-specific graphs, and then considers user preference on KG relations and label smoothness in the information aggregation phase, so as to generate personalized item representations.
– **KGAT** [11] is a propagation-based recommender model. It applies a unified relation-aware attentive aggregation mechanism in UKG to generate user and item representations.
– **CKAN** [12] is based on KGNN-LS, which utilizes different aggregation schemes on the user-item graph and KG respectively, to encode knowledge association and collaborative signals.
– **LKGR** [1] is a state-of-the-art hyperbolic GNN method with Lorentz model, which employs different information propagation strategies in the hyperbolic space to encode heterogeneous information from interactions and KG.

5.3   Experimental Settings

We implemented SKAN in TensorFlow, and we randomly divided the dataset into training set, evaluation set and test set, and their division ratio is $6 : 2 : 2$. In our experiments, click-through rate (CTR) prediction was the recommendation scenario, and we used AUC and F1 to evaluate the effectiveness.

The hyper-parameter settings for SKAN are as follows, where neighbor sampling size, dimension of embeddings, depth of receptive field, L2 regularizer weight, learning rate, training times are denoted as K, d, H, λ, η and n respectively: for Movie-Lens20M, $K = 8$, $d = 32$, $H = 2$, $\lambda = 1e - 7$, $\eta = 2e - 3$, $n = 20$; for Book-Crossing, $K = 8$, $d = 64$, $H = 3$, $\lambda = 2e - 5$, $\eta = 2e - 4$, $n = 10$; for Dianping-Food, $K = 4$, $d = 8$, $H = 2$, $\lambda = 1e - 7$, $\eta = 2e - 2$, $n = 10$. The best settings for hyper-parameters in all comparison methods are reached by either empirical study or following their original papers.

5.4   Performance Comparison (RQ1)

We first report the performance of all the methods, and then analyze the experimental results. The experimental results are reported in Table 2, where %Imp. denotes the relative improvement of the best performing method (bolded) over the strongest baselines (underlined). We find that:

Table 2. The results of AUC and F1 in CTR prediction.

Dataset	Movie-Lens20M		Book-Crossing		Dianping-Food	
Metric	AUC	F1	AUC	F1	AUC	F1
BPRMF	0.958	0.914	0.658	0.611	0.832	0.764
CKE	0.7321	0.7385	0.7323	0.6363	0.6462	0.6559
PER	0.838	0.792	0.605	0.572	0.766	0.697
RippleNet	0.976	0.927	0.721	0.647	0.863	0.783
KGCN	0.977	0.93	0.684	0.631	0.845	0.774
KGNN-LS	0.975	0.929	0.676	0.631	0.852	0.778
KGAT	0.976	0.928	0.731	0.654	0.846	0.785
CKAN	0.976	0.929	0.753	0.673	0.878	0.802
LKGR	0.979	0.9336	0.7056	0.6536	0.871	0.7932
SKAN(ours)	**0.9839**	**0.9411**	**0.7735**	**0.6998**	**0.9206**	**0.8444**
%Imp.	0.50%	0.80%	2.72%	3.98%	4.85%	5.29%

(1) SKAN consistently yields the best performance on all the datasets. In particular, it achieves significant improvement even over the strongest baselines w.r.t. AUC by 0.5%, 2.72% and 4.85% in Movie-Lens20M, Book-Crossing and Dianping-Food datasets, respectively. We believe that the following reasons have led to this good result: 1)modeling the user makes the embedded representation of the user more accurate; 2)Removing the nonlinear activation function and transformation matrix is indeed helpful to improve the recommendation effect. We will design ablation experiments later to verify the above views.

(2) From the results of three datasets, the ranking of the results of all models' performance on the three datasets from high to low is Movie-Lens20M, Dianping-Food, Book-Crossing. By looking at the data in Table 1, we believe that the reason for this may be that the average interaction per user and the average links per item are different. On average, Movie-Lens20M and Dianping-Food datasets have richer interactions and links than Book-Crossing datasets. Therefore, for datasets with poor interaction and links, the recommended model does not have enough information to understand the potential embedding.

(3) From the experimental results, it is also easy to see that KG is very obvious to improve the recommendation effect. KG-based methods is more effective than CF-based methods, especially the unified-based methods on the three datasets across two evaluation metrics.

(4) Through the comparison of results of all KG-based methods, The unified-based methods are better than embedding-based methods and the path-based methods on three datasets. This shows the importance of high-order information in KG-based recommender systems. Among all methods, the path-based method performs the worst, which may be because it is difficult to define

the optimal path in reality. The poor performance of the embedding-based method may be due to its static nature, which ignores the information connectivity in the knowledge graph, can not use the multi hop relationship between entities, can not mine the high-order relationship in the graph.

5.5   Ablation Studies (RQ2)

In this section, we examine the contributions of main components in our model to the final performance by comparing SKAN with the following two variants:

(1) SKAN$_{fn}$: Inspired by lightGCN, we remove the nonlinear activation function and transformation matrix from the aggregation function. The model without removing the nonlinear activation function and transformation matrix is recorded as SKAN$_{fn}$.
(2) SKAN$_{user}$: User modeling is an important module in SKAN. The model without user modeling is called SKAN$_{user}$.

Table 3. Ablation study on feature transformation and nonlinear activation.

Dataset	Movie-Lens20M		Book-Crossing		Dianping-Food	
Metric	AUC	F1	AUC	F1	AUC	F1
SKAN$_{fn}$	0.9836	**0.9413**	0.7605	0.6923	0.9173	0.8392
SKAN	**0.9839**	0.9411	**0.7735**	**0.6998**	**0.9206**	**0.8444**
%Imp.	0.03%	−0.028%	1.71%	1.08%	0.36%	0.62%

Table 4. Ablation study on user modeling.

Dataset	Movie-Lens20M		Book-Crossing		Dianping-Food	
Metric	AUC	F1	AUC	F1	AUC	F1
SKAN$_{user}$	0.974	0.9287	0.7049	0.6603	0.8409	0.7703
SKAN	**0.9839**	**0.9411**	**0.7735**	**0.6998**	**0.9206**	**0.8444**
%Imp.	1.02%	1.34%	9.73%	5.98%	9.48%	9.62%

Are Nonlinear Activation Functions and Transformation Matrices Useful? The KG-based recommender systems generally has nonlinear activation function and feature transformation matrix in the aggregation function part, but lightGCN points out that they not only do not improve the recommendation effect, but increase the difficulty of training. Because The premise that the nonlinear activation function and transformation matrix can work is that the nodes have rich feature representation. It can be seen from the results in Table 3 that this conclusion is still applicable to the KG-based recommender systems. Notice that, SKAN does not perform best on Movie-Lens20M in Table 3, since the average interactions per user are so abundant that the nodes have rich feature representation (Table 4).

Is User Modeling Working? Most KG-based recommender systems ignore the modeling of users. SKAN aggregates historical items that users have interacted with to obtain better user representations. From the experimental results, user modeling can improve the recommendation effect.

6 Conclusion

In this work, we propose a recommendation system model to make better use of item attribute information. First, we build a collaborative knowledge graph, then we model entities and users on CKG to get their embedded representations by simplified knowledge-aware attention network which simplifies the knowledge-aware aggregation by removing feature transformation and nonlinear activation. Specifically, we use neighbor sampling and aggregation with attention mechanism for entities and users, in which the attention mechanism takes into account the user's preference for relationships and items. Experiment results show that SKAN outperforms state-of-the-art baselines on three real-world recommendation scenarios. In future, it is an exciting research direction that using KG information to explore the interpretability of recommender systems.

Acknowledgement. This work was in part supported by NSFC under grant No. 61872446, and The Science and Technology Innovation Program of Hunan Province under grant No. 2020RC4046.

References

1. Chen, Y., et al.: Modeling scale-free graphs with hyperbolic geometry for knowledge-aware recommendation. In: Proceedings of the 15th ACM International Conference on Web Search and Data Mining, pp. 94–102 (2022)
2. Cheng, H.T., et al.: Wide & deep learning for recommender systems. In: Proceedings of the 1st Workshop on Deep Learning for Recommender Systems, pp. 7–10 (2016)
3. Guo, Q., et al.: A survey on knowledge graph-based recommender systems. IEEE Trans. Knowl. Data Eng. **34**, 3549–3568 (2020)
4. He, X., Deng, K., Wang, X., Li, Y., Zhang, Y., Wang, M.: LightGCN: simplifying and powering graph convolution network for recommendation. In: Proceedings of the 43rd International ACM SIGIR Conference on Research and Development in Information Retrieval, pp. 639–648 (2020)
5. Kipf, T.N., Welling, M.: Semi-supervised classification with graph convolutional networks. arXiv preprint arXiv:1609.02907 (2016)

6. Wang, Y., Gao, S., Li, W., Jiang, T., Yu, S.: Research and application of personalized recommendation based on knowledge graph. In: Xing, C., Fu, X., Zhang, Y., Zhang, G., Borjigin, C. (eds.) WISA 2021. LNCS, vol. 12999, pp. 383–390. Springer, Cham (2021). https://doi.org/10.1007/978-3-030-87571-8_33

7. Rendle, S., Freudenthaler, C., Gantner, Z., Schmidt-Thieme, L.: BPR: bayesian personalized ranking from implicit feedback. arXiv preprint arXiv:1205.2618 (2012)

8. Wang, H., et al.: RippleNet: propagating user preferences on the knowledge graph for recommender systems. In: Proceedings of the 27th ACM International Conference on Information and Knowledge Management, pp. 417–426 (2018)

9. Wang, H., Zhang, F., Xie, X., Guo, M.: DKN: deep knowledge-aware network for news recommendation. In: Proceedings of the 2018 World Wide Web Conference, pp. 1835–1844 (2018)

10. Wang, H., Zhao, M., Xie, X., Li, W., Guo, M.: Knowledge graph convolutional networks for recommender systems. In: The World Wide Web Conference, pp. 3307–3313 (2019)

11. Wang, X., He, X., Cao, Y., Liu, M., Chua, T.S.: KGAT: knowledge graph attention network for recommendation. In: Proceedings of the 25th ACM SIGKDD International Conference on Knowledge Discovery & Data Mining, pp. 950–958 (2019)

12. Wang, Z., Lin, G., Tan, H., Chen, Q., Liu, X.: CKAN: collaborative knowledge-aware attentive network for recommender systems. In: Proceedings of the 43rd International ACM SIGIR Conference on Research and Development in Information Retrieval, pp. 219–228 (2020)

13. Yu, X., et al.: Personalized entity recommendation: a heterogeneous information network approach. In: Proceedings of the 7th ACM International Conference on Web Search and Data Mining, pp. 283–292 (2014)

14. Zhang, F., Yuan, N.J., Lian, D., Xie, X., Ma, W.Y.: Collaborative knowledge base embedding for recommender systems. In: Proceedings of the 22nd ACM SIGKDD International Conference on Knowledge Discovery and Data Mining, pp. 353–362 (2016)

15. Hamilton, W., Ying, Z., Leskovec, J.: Inductive representation learning on large graphs. Adv. Neural Inform. Process. Syst. **30** 1024–1034 (2017)

16. Wang, H., et al.: Knowledge-aware graph neural networks with label smoothness regularization for recommender systems. In: Proceedings of the 25th ACM SIGKDD International Conference on Knowledge Discovery & Data Mining, pp. 968–977 (2019)

17. Zhao, X., Zeng, W., Tang, J., Wang, W., Suchanek, F.: An experimental study of state-of-the-art entity alignment approaches. IEEE Trans. Knowl. Data Eng. **34**, 2610–2625 (2020)

18. Chen, Y., Wang, Y., Zhao, X., Zou, J., Rijke, M.D.: Block-aware item similarity models for top-n recommendation. ACM Trans. Inform. Syst. **38**, 1–26 (2020)

Bi-Directional Neighborhood-Aware Network for Entity Alignment in Knowledge Graphs

Jingwen Bai, Tiezheng Nie[(⊠)], Derong Shen, Yue Kou, and Ge Yu

School of Computer Science and Engineering, Northeastern University, Shenyang 110004, China
baijingwen0219@163.com, {nietiezheng,shenderong,kouyue,
yuge}@cse.neu.edu.cn

Abstract. As an important research work in knowledge fusion, entity alignment can promote the sharing and integration of multi-source knowledge graphs. Recently, entity alignment based on graph neural networks has received a lot of attention for its ability to capture the topology of entities, but it ignores the noise in neighbor subgraphs and the impact of distant neighbors on central entities. In addition, the knowledge graph is a sparse structure, with the vast majority of entities obeying the long-tail effect.But existing works pay little attention to the alignment of long-tail entities. To address the above problems, this paper proposes an entity alignment approach, which aggregates bi-directional multi-hop neighbors to enrich the context of the central entity, and uses entity names to supply entities with less structural information. The feature fusion module can dynamically adjust weights for the significance of different features. Experimental results show that the overall performance of our model is superior than that of GNN-based methods.

Keywords: Entity alignment · Long-tail · Knowledge graph · Cross-lingual

1 Introduction

Knowledge graph is widely used in many knowledge-driven applications, such as recommendation systems and search engines. Existing knowledge graphs such as YAGO, and DBpedia are built by independent organizations through their own building rules and languages, so the same entity may have different meanings and distribution in different knowledge graphs. This makes it possible to design a technology that can integrate heterogeneous knowledge among knowledge graphs from different sources. Entity alignment aims to connect entities with the same meanings in different knowledge graphs and provide support for downstream applications.

The existing methods of entity alignment can be divided into two categories. (1) Translation-based. TransE [1], a method that assumes head entity is translated from its tail entity representations using relation vectors. It can capture the local semantics of relation triples (2) GNNs-based. This kind of method is encoded by using GNN to recursively aggregate the neighbors of entities. However, with the introduction of GNN

© The Author(s), under exclusive license to Springer Nature Switzerland AG 2022
X. Zhao et al. (Eds.): WISA 2022, LNCS 13579, pp. 64–76, 2022.
https://doi.org/10.1007/978-3-031-20309-1_6

into the entity alignment task, the recent models still face some challenges in dealing with heterogeneous knowledge graphs and long-tail entities.

Challenge1: How to Aggregate the Neighbors of Entity More Selectively?

Due to the different construction rules, there are differences in the topological structure of some entities. So it's easy to generate different representations for same entity pairs. AliNet [3] proposes distant neighborhood information to improve the heterogeneity of the knowledge graphs, but it did not consider the different importance of the one-hop neighbors. Besides, although GMNN [4], GTEA [17] and NMN [8] solve the one-hop noise neighbors of entities, the useful two-hop neighbors have not been fully considered, which may affect the integrity of knowledge graph embedding.

In addition, many methods based on graph neural networks regard a knowledge graph as an undirected graph, thus ignoring that entities have different contexts in different relational directions of neighbors. Directional neighbors can distinguish entities more effectively. So RDGCN [5] introduces the dual relation information of KGs through graph interaction.

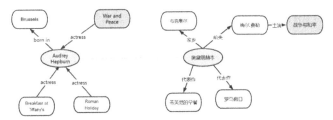

Fig. 1. The tough case of entity alignment. The blue and black arrows denote input and output relationships respectively.

For example, Fig. 1 shows a pair of equivalent entities from Chinese and English knowledge graphs. Here, both central entities refer to the same real-world identity, *Audrey Hepburn*, a famous actor. However, the two entities have the same neighborhoods with different relationship directions, and their common neighbor, *war and peace,* appears in different hop and direction. If only the one-hop output neighbors are considered, it is likely to ignore some more informative neighbors. So it is necessary to pay more attention to the relational directions and the two-hop neighbors.

Challenge 2: How to Improve the Representation of Long-tail Entities?

Recent research [20] indicates that current data sets are much denser than real-world KGs, hence the existing methods can't perform satisfactory results when dealing with long-tail entities. It is of great significance to propose a method that emphasizes the alignment of long-tail entities.

Even though DAT [14] improved the alignment of long-tail entities by introducing the degree of entities, it ignores the degree difference between the same entity pairs, while SEA [10] reduces a large degree difference between neighbors by conducting adversarial training. In addition, EVA [13] enhances the representation of entities by introducing the images of entity, but it relies on limited image data sets. AttrGNN [19] proposed that only using the name of the entity can achieve more than 60% alignment. Especially for long-tail entities, the role of entity names will be more important than their structural information, and it can better represent the embedding of entities.

To solve the proposed challenges, we summarize the main contributions of this paper as follows:

1) We propose to jointly leverage the entity's neighbors and name for entity alignment. And make full use of graph convolution network to encode the multi-hop neighbors of entities and aggregate bi-directional neighbors of entities through highway mechanism.
2) We design a **Bi**-directional **N**eighborhood-**A**ware(BNA) network based on the Bi-attention mechanism, which can dynamically compute the weight of entity's feature and make two features of the entity be able to effectively complement each other.
3) Our method is superior to the other GNN-based methods on three cross-lingual data sets. Experiments on sparse data set also further improve the alignment of long-tail entities.

2   Related Work

2.1   Embedding-Based Methods

In recent years, embedding-based methods have emerged as vital means for entity alignment. It can encode KGs into individual low-dimensional vector spaces. Early works utilize TransE [1] to embed the knowledge graph. Like MTransE [11] and Mutil-View [9]. BootEA [22] uses a basic bootstrapping approach to expand the data sets. KDCoE [18] and AtteNet [16] use description and attributes to enrich entity embedding. Since TransE [1] is only trained in a single triple, it may lack the ability to take advantage of the global view of the entity. The graph neural network GNN can obtain a comprehensive and powerful entity representation. GCN-Align [2] tries to use entities and their attributes as low-dimensional vectors.AliNet [3] introduces distant neighbors to overcome the ubiquitous heterogeneity in graphs. NMN [8] estimates the structure and neighborhood similarity between two entities. In addition, RDGCN [5] takes into account the dual semantic relationship between neighbors. When the pre-trained word vector of the entity name label is initialized, GMNN [4] and HGCN [6] bring excellent alignment results.

Although the existing studies have improved the results of entity alignment in different perspectives, GNN-based models still inevitably are affected by the heterogeneity and sparseness of the knowledge graph, and they ignore the importance of multi-hop neighbors and give poor consideration to the alignment of long-tail entities.

2.2   Phenomenon of Long-Tail

Most entities in the knowledge graph are sparse and follow the long-tail distribution. The long-tail entity are rarely connected with other entities, so it has less structural information. As shown in Fig. 2, we investigate the degree distributions of entities on

EN-FR-15K (V1), which is a data set closer to realistic KGs. Then we calculate the average degrees by counting the number of relational triples in which the entity appears. The number of entities with degrees in [1,6) accounts for nearly two-thirds. Although RDGCN [5] achieves the satisfactory result by dual relationship, for long-tail entities with few paths involve. The calculation accuracy of semantic relations of these entities is often low, and the alignment effect is significantly reduced. DAT [14] enriches the structural information of long-tail entities by iterative self-training. EVA [13] enhances the representation of long-tail entities by introducing images and descriptions of entities. However, these methods usually require pre-aligned attributes and data preprocessing. The long-tail is still a thorny problem in entity alignment.

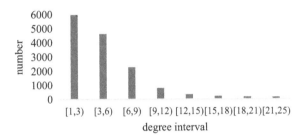

Fig. 2. Average degrees of EN-FR-15K (V1)

3 Problem Formalization

Knowledge graphs represent structural information about entities in real world as triples. The relational triples T in the KG can be represented as $T = (h, r, t)$, where h and t represent the head entity and the tail entity, respectively, and r denotes multiple relationships between two entities. Our goal is to design a framework to automatically find out more unaligned entities through the existing pre-aligned entity pairs.

Definition 1 Knowledge Graph. A knowledge graph is defined as a directed graph $G = (E, R, T)$, where $E = \{e_1, e_2 \ldots e_n\}$ is the set of entities and $R = \{r_1, r_2 \ldots r_n\}$ is the relation set. Relational triples T is represented as $T = \{E \times R \times E\}$.

Definition 2 Entity Alignment. For two knowledge graphs $G_1 = (E_1, R_1, T_1)$, $G_2 = (E_2, R_2, T_2)$, Given a pre-aligned entity pairs $A = \{(e_1, e_2) | e_1 = e_2, e_1 \in E_1, e_2 \in E_2\}$. The task of entity alignment is to find the remaining entity pairs based on A.

4 Methodology

This section introduces our proposed model BNA. The model is divided into two stages. In the stage of entity embedding, firstly, we make full use of GCN to encode neighbors for each entity, then represent the entity's name by the pre-trained model Bert. Finally, two features are forwarded to the feature fusion module. In the alignment stage, under the guidance of Bi-attention, the similarity weights are dynamically allocated for each entity's feature and help the alignment model to determine the importance of each feature. At last, the similarity function is used to learn and predict the equal entity pairs in the two knowledge graphs. To provide an overview, we show the framework of BNA(Bi-directional **N**eighborhood-**A**ware Network) in Fig. 3.

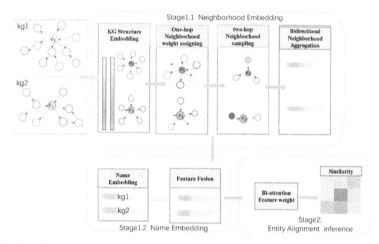

Fig. 3. The overall architecture of the Bi-directional Neighborhood-Aware Network

4.1 Neighborhood Embedding

We divide the neighbors of an entity into input and output subgraphs and design a bi-directional neighborhood aggregation method to combine the one-hop and two-hop useful neighborhoods' information.

KG Structure Embedding. Based on GCN, the representation vector of a node is updated and calculated by recursively aggregating and transforming the representation vectors of its neighboring nodes, and the output of the l-th layer feature matrix $H^{(l)}$ by the following convolutional computation:

$$H^{(l)} = \sigma\left(\left[\tilde{D}^{-\frac{1}{2}}\tilde{A}\tilde{D}^{-\frac{1}{2}}H^{(l-1)}W^{(l)}\right]\right) \qquad (1)$$

where σ is an activation function, $\tilde{A} = A + I$ is the adjacency matrix with self-connection, I is the identity matrix and $\tilde{D}$ is diagonal degree matrix of $\tilde{A}$. $W^{(l)}$ is a layer-specific trainable weight matrix in the l-th layer.

One-hop Neighborhood Weight Assignment. Not all neighbors are equally important to the central entity. Our main idea is that the more central entity with its neighbor appear in the same context, the more important neighbor is. We calculate the importance of the j-th input neighbor e_j to the central entity e_i through the function f_{ij_in}

$$f_{ij\_in} = \frac{|e_i|(e_j, r, e_i) \in T|}{|e|(e_j, r, e_i) \in T, j \in N_{i,1(in)}|} \tag{2}$$

$$h_{i,1(in)}^{(l)} = ReLU\left(\sum_{j \in N_{i,1(in)}} f_{ij\_in} W_l h_j^{(l-1)}\right) \tag{3}$$

$h_{i,1(in)}^{(l)}$ is the one-hop input neighbor vector layer of the entity e_i, $|.|$ represents the number of triples. $N_{i,1(in)}$ is the set of one-hop input neighbors of the entity e_i. T is the relational triples about e_i. W_l is a shared weight matrix. The calculation of the one-hop output neighbors $h_{i,1(out)}^{(l)}$ is the same as $h_{i,1(in)}^{(l)}$.

Two-hop neighborhood sampling. Only using local one-hop neighbors can't fully express entities. Especially for long-tail entities, introducing two-hop neighbors can further enrich their connections with useful neighbors and alleviate the heterogeneity of the KGs. However, the noise will inevitably be caused when introducing two-hop neighbors. So we selectively sample two-hop neighbors to solve this problem. Given an entity e_i, the probability to sample its two-hop neighbor e_j is determined by:

$$p(h_j|h_i) = \frac{\exp\left(h_i W_{sl} h_j^T\right)}{\sum_{j \in N_{i,2(in)}} \exp\left(h_i W_{sl} h_j^T\right)} \tag{4}$$

By aggregating the information of two-hop neighborhoods, the hidden representation of entity e_i, at the l-th layer is denoted as $h_{i,2(in)}^{(l)}$, which is computed as follows:

$$h_{i,2(in)}^{(l)} = Relu\left(\sum_{k \in N_{i,2(in)}} p_{ik} W_2^{(l)} h_k^{(l-1)}\right) \tag{5}$$

$$p_{ik} = \begin{cases} p(h_k|h_i), & \text{if } p(h_k|h_i) > \theta \\ 0 & otherwise \end{cases} \tag{6}$$

$N_{i,2(in)}$ is the set of two-hop input neighbors of the central entity e_i. W_2 is a trainable matrix. θ is a hyperparameter to control more useful two-hop neighborhoods.

Bi-Directional Neighborhood Aggregation. IMEA [15] proved that distinguishing the relationship direction between neighbors can improve the alignment effect, and make up for the information of long-tail entities. Based on this inspiration, we firstly concatenate the embeddings of the one-hop and the two-hop neighbors to generate the bi-directional neighbor features of the entity. They are maintained by :

$$h_{i,1}^{(l)} = \left[\frac{h_{i,1}^{(l_{in})}}{|N_{i,1(in)}|}; \frac{h_{i,1}^{(l_{out})}}{|N_{i,1(out)}|}\right] \tag{7}$$

$$h_{i,2}^{(l)} = \left[\frac{h_{i,2}^{(l_{in})}}{|N_{i,2(in)}|} ; \frac{h_{i,2}^{(l_{out})}}{|N_{i,2(out)}|} \right] \tag{8}$$

where $h_{i,1}^{(l)}, h_{i,2}^{(l)}$ are the output vectors of the l-layer of one-hop and two-hop neighbors of the entity, and $|N(i,2)^{in}|$ is the number of two-hop input neighbors of the entity e_i. Where [;] means vector concatenation.

$$g\left(h_{i,2}^{(l)}\right) = ReLU\left(M_1 h_{i,2}^{(l)} + b\right) \tag{9}$$

To reduce the propagation of neighbor noise, following previous work [3], we also introduce the gating mechanism $g\left(h_{i,2}^{(l)}\right)$ to control the output of one-hop and two-hop neighbors, and keep the useful information learned from the interaction. M_1 and b are a trainable weight matrix and bias vector respectively. Finally, the one-hop and two-hop neighbors are combined to generate the neighbor feature representation of the central entity e_i. Its representation $h_i^{(l)}$ is computed as follows:

$$h_i^{(l)} = g\left(h_{i,2}^{(l)}\right) h_{i,1}^{(l)} + \left(1 - g(h_{i,2}^{(l)})\right) h_{i,1}^{(l)} \tag{10}$$

4.2 Entity Name Embedding

The names of entities are the most intuitive way to preliminarily distinguish whether entity pairs are aligned. In real life, most entities have name attributes, which is often common in long-tail entities. GMNN [4] and RDGCN [5] directly use the name as the input vector of KG embedding instead of random initialization, which hides the rich semantics of the entity. To better improve the information of the long-tail entity, we propose names as an independent embedding module. [21] studies the differences between fastText and GloVe in applying word embedding models. But Bert uses massive text corpus data by self-supervised learning to obtain an excellent feature representation for words. Based on this, we just add an appropriate output layer pre-training on Bert to generate a high-quality name vector n_{e_i} for entity e_i.

4.3 Feature Fusion with Bi-attention

Different features reflect different information about entities. The importance of each feature is different for entities in different situations. Especially for long-tail entities, compared with information of neighbors, the name signal of the entity is more representative. Inspired by the work [12], we propose a novel feature fusion module in Fig. 4 to dynamically calculate different attention weights for each feature of entities.

Feature Matrix Construction. To get the final feature representation of the entity e_i. We concatenate structure embedding h_i and name vector n_{e_i} of the entity to generate feature matrix $F_{e_i} = \left[h_i; n_{e_i}\right] \in \mathbb{R}^{2 \times d}$. where d is the dimensionality of embedding.

Bi-Attention Similarity Matrix Calculation. To establish feature interactions between entity pairs and highlight different features. Based on the Attention Flow Layer in [12]. Between entity pairs $\{(e_1, e_2), e_1 \in E_1, e_2 \in E_2\}$, the attention weight features are derived from a sharing similarity matrix $S \in \mathbb{R}^{2 \times 2}$. S_{ij} indicates similarity between -th feature of the entity e_1 and j-th feature of the entity e_2, it is computed by

$$S_{ij} = \alpha\left(F_{e_1}^{i:}, F_{e_2}^{j:}\right) \tag{11}$$

where $F_{e_1}^{i:}$ is the i-th row vector of the entity e_1, and $F_{e_2}^{j:}$ is j-th row vector of the entity e_2. $i = 1,2; j = 1,2$; Then, $\alpha(u, v) = w_s^\top(u; v; (u \circ v))$ is a scalar function of trainable similarity, where $w_s^\top \in \mathbb{R}^{2d}$ is a trainable vector. (;) means vector concatenation. $\circ$ is elementwise multiplication.

Weight Assignment. To obtain the related feature attention vector of e_1 about e_2. The Bi-attention weight is calculated column-wise by the matrix S into the softmax layer, and then the result matrix is input row-wise into the average aggregation layer to generate the attention vector β. Finally, we multiply the similarity score of specific weight between entity pairs to yield the final similarity score by

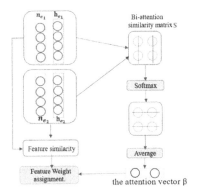

Fig. 4. Bi-attention feature fusion

$$Sim(e_1, e_2) = Sim_s(e_1, e_2) \cdot \beta^s + Sim_n(e_1, e_2) \cdot \beta^n \tag{12}$$

$sim_s(e_1, e_2)$, $sim_n(e_1, e_2)$ are the cosine similarity of the structural features and the name features for each given entity pair. Where β^s and β^n are the corresponding attention values of structure and name similarity obtained by vector β respectively.

4.4 Alignment and Training

Through pre-aligned seed pairs, we make the two knowledge graphs share parameters and unify them into the same vector space for alignment. We hope aligned entity pairs have a higher similarity and negative entity pairs have a lower similarity. A margin-based score function as the training objective as following:

$$L = \sum_{(e_i, e_j) \in A} \sum_{\left(e_i', e_j'\right) \in A'} \max\left\{0, \gamma - sim\left(e_i, e_j\right) + sim\left(e_i', e_j'\right)\right\} \tag{13}$$

We indicate that the positive alignment seed is A, the negative sample seed set A' is generated by replacing the nearest neighbors of the entities in the pre-aligned entity pair. The hyperparameter $\gamma > 0$ is used to control the loss boundary. The similarity score $sim(.)$ is obtained from formula (12) in Sect. 4.3.

5 Experiments

5.1 Experiment Setting

Datasets. We conduct expriments on three cross-lingual datasets from DBP15K, which are DBP$_{ZH-EN}$, DBP$_{JA-EN}$, DBP$_{FR-EN}$. Each dataset contains 15 thousand reference alignment links with popular entities from English to Chinese, Japanese and French respectively. And use the same split as previous works, 30% for training and 70% for testing.

Metrics. We use Hits@k and MRR as evaluation metrics. Hits@k is to measure the proportion of correctly aligned entities appearing in the top-k candidate rankings to source entities. MRR denotes the average reciprocal ranks of ground truth results.

Implementation Details. We use 2-layer GCN to embed KGs. The dimension of GCN hidden layer and name embedding are both 300. The learning rate is set to 0.001, and $\theta = 0.4$ is adopted for the threshold selection of two-hop sampling neighbors. The margin threshold γ is set to 0.9. The number of negative samples of each positive sample is set to 25. For entity names, we use Google Translate to convert all no-English entity names into English.

Comparison Methods. To be fair, we compare BNA with the existing GNN-based models. GCN-Align [2], AliNet [3], MuGNN [7], HGCN [6] and GMNN [4].

5.2 Main Result

From Table 1, the experimental results show that BNA outperforms the previous baselines on three cross-lingual data sets. All Hits@1 achieve more than 70.1%, Hits@10 are above 85.4%, and MRR are all above 72.5%. BNA shows a larger improvement over structure-based models. GMNN and HGCN also suggest that entity names provide useful clues for entity alignment. However, they are inferior to our model, because we introduce the entity's name embedding independently. In addition, we analyze the results in DBP$_{FR-EN}$, due to GMNN can encode the global matching state so the performance of BNA is slightly weaker than it.

Table 1. Results comparison on entity alignment

Method	DBP ZH-EN			DBP JA-EN			DBP FR-EN		
	Hits @1	Hits @10	MRR	Hits@1	Hits @10	MRR	Hits@1	Hits @10	MRR
GCN-Align	41.3	74.4	54.9	39.9	74.5	54.6	37.3	74.5	53.2
AliNet	53.9	82.6	62.8	54.9	83.1	64.5	55.2	85.2	65.7
MuGNN	49.4	84.4	61.1	50.1	85.7	62.1	49.5	87.0	62.1
HGCN	67.0	77.0	71.0	**76.3**	86.3	80.0	82.3	88.7	85.4
GMNN	67.9	78.5	69.4	74.0	87.2	78.9	89.4	**95.2**	**91.3**
BNA	**70.1**	**85.4**	**72.5**	75.1	**88.2**	**81.2**	**89.9**	91.8	89.7

5.3 Ablation Study

We construct three ablation studies to demonstrate the effectiveness of each component in BNA. Table 2 shows the relevant results. (1)BNA(w/o name) means that only use the neighbor information to embed entity. After the module of name embedding being removed, the result decreases in hits@10 by 10% on DBP ZH-EN, which verifies that name information can bring significant signals for entity alignment. (2) BNA(w/o sample) means that the all two-hop neighbors are introduced without considering sampling. An 11.6% drops in hits@1 on DBP ZH-EN confirms our BNA can reduce irrelevant noisy neighbors to the central entity. (3) BNA(w/o Bi-att) fuses equivalently each feature of the entity, regardless of the feature assignment based on similarity attention, and it drops by 5.9% in hit@10 on DBP15k ZH-EN. It shows feature fusion module can better highlight the importance of each feature.

Table 2. Results of ablation study

Methods	DBP15k ZH-EN			DBP15k JA-EN			DBP15k FR-EN		
	Hits @1	Hits @10	MRR	Hits @1	Hits @10	MRR	Hits @1	Hits @10	MRR
BNA (w/o name)	60.3	74.4	63.3	66.7	81.3	76.4	81.6	82.3	80.6
BNA (w/o sample)	58.5	78.3	68.2	70.9	80.7	70.2	76.2	81.1	77.2
BNA (w/o bi-att)	66.4	79.5	65.4	71.0	83.3	77.8	83.7	88.2	82.4
BNA	**70.1**	**85.4**	**72.5**	**75.1**	**88.2**	**81.2**	**89.9**	**91.8**	**89.7**

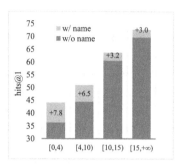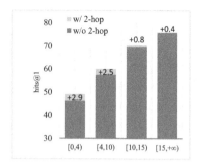

Fig. 5. Hits@1 results by degrees interval in different components

5.4 Evaluation by Degrees Interval

As a GCN-based model, BNA can learn sufficient embedding representation for entity by neighbor structure and name information. Thus, we compare the performance of components under different degree distributions. We divide the degree distribution of entities into four intervals on the data set DBP15K(V1) FR-EN from [20]. The result is shown in Fig. 5. In the case of the low degree distribution, long-tail entities benefit more from the name representations, reaching the highest improvement of 7.8% in the [0,4). With the degree distribution increasing, the effect of the name module gradually decreases. This demonstrates our feature fusion module dynamically adjusts the features. Especially for those entities which gain less information from neighbors. Because rich neighbors can enrich the representation of the central entity, then the signal of names will be less important. In addition, the introduction of two-hop neighbors also makes long-tail entities benefit more from structural information.

5.5 Robustness on Datasets

We compare our BNA with GNN-based models GMNN and HGCN on dense data set DBP15K(V1) and the sparse data set DBP15K(V2) FR-EN. As shown in Fig. 6, the alignment results of the three models on the sparse data set all suffer a drop, Because sparse structure information will provide few neighbor cues. Especially for GMNN and HGCN perform a higher difference due to they are sensitive to degree distribution. Even though GMNN and HGCN both use the embedding of entity names, our BNA surpasses them and performs more stably. On the one hand, these two models can't represent the name embedding of entity independently like BNA, and they depend more on knowledge graph structure. On the other hand, our model can adjust dynamically the structure and name feature so that they can complement each other.

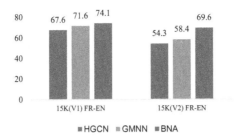

Fig. 6. Hits1 results on different datasets

6 Conclusion

In this paper, we propose a Bi-directional Neighborhood-Aware Network (BNA) for entity alignment. We not only solve the neighbor noise problem but also introduce two-hop bi-directional neighbors to enrich the contextual semantics of entities. We also learn the embedding of name to complement entities with few neighbor information. Besides, we propose a bi-attention feature fusion network that can dynamically adjust the similarity weights of different features. Experiments on three cross-lingual data sets show that our model obtains satisfactory performance, especially in aligning long-tail entities, BNA also achieves superior results.

Acknowledgment. This work was supported by the National Natural Science Foundation of China (62072086, 62172082, 62072084), the Fundamental Research Funds for the central Universities (N2116008).

References

1. Bordes, A., Usunier, N., Garcia-Duran, A., Weston, J., Yakhnenko, O.: Translating embeddings for modeling multirelational data. In: NeurIPS, pp. 2787–2795 (2013)
2. Wang, Z., Lv, Q., Lan, X., Zhang, Y.: Cross-lingual knowledge graph alignment via graph convolutional networks. In: EMNLP, pp. 349–357 (2018)
3. Sun, Z., Wang, C., Hu, W., Chen, M., Dai, J., Zhang, W., Qu, Y.: Knowledge graph alignment network with gated multi-hop neighborhood aggregation. In: AAAI **34**(01) pp. 222–229 (2020)
4. Xu, K., et al.: Cross-lingual knowledge graph alignment via graph matching neural network. In: ACL, pp. 3156–3161 (2019)
5. Wu, Y., Liu, X., Feng, Y., Wang, Z., Yan, R., Zhao, D.: Relation-aware entity alignment for heterogeneous knowledge graphs. In: IJCAI, pp. 5278–5284 (2019)
6. Wu, Y., Liu, X., Feng, Y., Wang, Z., Zhao, D.: Jointly Learning Entity and Relation Representations for Entity Alignment. In: EMNLP-IJCNLP, pp. 240–249 (2019)
7. Cao, Y., Liu, Z., Li, C., Li, J., Chua, T.-S.: Multi-channel graph neural network for entity alignment. In: ACL, pp. 1452–1461 (2019)
8. Wu, Y., Liu, X., Feng, Y., Wang, Z., Zhao, D.: Neighborhood matching network for entity alignment. In: ACL, pp. 6477–6487 (2020)
9. Zhang, Q., Sun, Z., Hu, W., Chen, M., Guo, L., Qu, Y.: Multi-view knowledge graph embedding for entity alignment. In: IJCAI, pp. 5429–5435 (2019)

10. Pei, S., Yu, L., Hoehndorf, R., Zhang, X.: Semi-supervised entity alignment via knowledge graph embedding with awareness of degree difference. In: WWW, pp. 3130–3136 (2019)
11. Chen, M., Tian, Y., Yang, M., Zaniolo, C.: Multilingual knowledge graph embeddings for cross-lingual knowledge alignment. In: IJCAI, pp. 1511–1517 (2017)
12. Seo, M., Kembhavi, A., Farhadi, A., Hajishirzi, H.: Bidirectional attention flow for machine comprehension. arXiv preprint 1611.01603 (2016)
13. Liu, F., Chen, M., Roth, D., Collier, N.: Visual Pivoting for (Unsupervised) Entity Alignment. In: AAAI 35(5), pp. 4257–4266 (2021)
14. Zeng, W., Zhao, X., Wang, W., Tang, J., Tan, Z.: Degree-aware alignment for entities in tail. In: SIGIR, pp. 811–820 (2020)
15. Xin, K., Sun, Z., Hua, W., Hu, W., Zhou, X.: Informed multi-context entity alignment. In: WSDM. pp. 1197–1205 (2022)
16. Sun, Z., Hu, W., Li, C.: Cross-Lingual Entity Alignment via Joint Attribute-Preserving Embedding. In: d'Amato, C., et al. (eds.) ISWC 2017. LNCS, vol. 10587, pp. 628–644. Springer, Cham (2017). https://doi.org/10.1007/978-3-319-68288-4_37
17. Jiang, S., Nie, T., Shen, D., Kou, Y., Yu, G.: Entity Alignment of Knowledge Graph by Joint Graph Attention and Translation Representation. In: Xing, C., Fu, X., Zhang, Y., Zhang, G., Borjigin, C. (eds.) WISA 2021. LNCS, vol. 12999, pp. 347–358. Springer, Cham (2021). https://doi.org/10.1007/978-3-030-87571-8_30
18. Chen, M., Tian, Y., Chang, K.-W., Skiena, S., Zaniolo, C.: Co-training embeddings of knowledge graphs and entity descriptions for cross-lingual entity alignment. In: IJCAI, pp. 3998–4004 (2018)
19. Liu, Z., Cao, Y., Pan, L., Li, J., Chua, T.: Exploring and Evaluating Attributes, Values, and Structures for Entity Alignment. In: EMNLP, pp. 6355–6364 (2020)
20. Sun, Z., et al.: A benchmarking study of embedding-based entity alignment for knowledge graphs. In: Proceedings of the VLDB Endowment 13(11), pp. 2326–2340 (2020)
21. Azzalini, F., Jin, S., Renzi, M., Tanca, L.: Blocking techniques for entity linkage: a semantics-based approach. Data Science and Engineering 6(1), 20–38 (2020). https://doi.org/10.1007/s41019-020-00146-w
22. Sun, Z., Hu, W., Zhang, Q., Qu, Y.: Bootstrapping entity alignment with knowledge graph embedding. In: IJCAI, pp. 4396–4402 (2018)

SAREM: Semi-supervised Active Heterogeneous Entity Matching Framework

Jinxiu Du[⊠], Tiezheng Nie, Wenzhou Dou, Derong Shen, and Yue Kou

School of Computer Science and Engineering, Northeastern University, Shenyang 110169, China
du971130@163.com

Abstract. Entity matching is a key technique in data quality research, which refers to the identification of records that refer to the same real-world entity in different data sources. This paper introduces SAREM, a semi-supervised entity matching framework for heterogeneous data. We first obtain effective feature vectors using an embedding approach that combines semantic and relational information, and this approach can be used for long sequences. Deep learning requires much-labeled data, which is very costly and time-consuming. In this paper, we address the problem by using a dropout layer for data augmentation and propose an active learning method that is more suitable for entity matching. We also address the classical challenges of deep active learning by reducing human intervention and improving model performance. We experiment with six public benchmark datasets, and the results clearly show that our method outperforms DeepER and DeepMatcher on all datasets. Our method can achieve comparable effectiveness to SOTA entity matching methods with a smaller amount of data, achieve the goal of cost reduction, and outperform SOTA entity matching methods on large datasets with long sequences.

Keywords: Entity matching · Semi-supervised learning · Active learning · Heterogeneous data

1 Introduction

In large data mining projects, it is often necessary to integrate data from multiple data sources to improve data quality, and entity matching (following is described by EM) is one of the very important steps [1]. It can improve the quality of integrated data by combining data describing the same entities into a clean, unified data set [2]. EM was born out of the fact that real-world data often has the problem of dirty data [14], making data quality an urgent issue. The task of EM [3] is to detect which descriptions refer to the same entity. Table 1 shows an example of EM, each record consists of 5 attributes, each field may be a different data type and the attribute values may differ significantly or even be missing.

Traditional EM methods compare two records at the attribute level based on whether the string similarity between aligned attributes reaches the threshold [15]. But it is difficult to have the same schema for real datasets, and the values of the same entity

© The Author(s), under exclusive license to Springer Nature Switzerland AG 2022
X. Zhao et al. (Eds.): WISA 2022, LNCS 13579, pp. 77–88, 2022.
https://doi.org/10.1007/978-3-031-20309-1_7

may be very different in form. As shown in Table 1, "18th International Conference on Web Information Systems and Applications" and "WISA 2021" are different forms. This traditional approach fails to solve this problem because it ignores contextual semantic information. Although deep learning-based EM methods [5] have solved the problems of schema and semantic information, methods like [6] still only regard records as text, and will lose the structural information in the records, leading to ignoring the rich relationships between different records. The graph neural.

Table 1. Show an example of two records referring to the same entity

Field	R1	R2
Title	Study of deep matching for aligning large knowledge bases	null
Authors	william grosky	william
Affiliations	University of Cambridge	null
Date	2022 08 16	2022 08
Pages	pp.572–580	572–580
Conference	18th International Conference on Web Information Systems and Applications	WISA 2021

network-based approach [17] will represent entities by constructing graphs and then learn the similarity between records. But it is error-prone, a wrong attribute-level node will often lead to a wrong graph structure, which leads to unreliable embedding.

Secondly, deep learning-based methods typically require large amounts of labeled data but the annotation process is labor-intensive. Unsupervised learning methods [27] can solve this problem, however, the performance of unsupervised learning methods is limited as they do not include any supervised signals and even slightly incorrect data can lead to incorrect results [18]. Using active learning [19] methods to proactively propose which data to annotate, reduces the cost of annotation and retains the powerful learning capability of deep learning.

In summary, our contributions are mainly as follows:

1. We propose SAREM, **S**emi-supervised **A**ctive hete**R**ogeneous **E**ntity **M**atching Framework, solving EM problems in dirty and missing data scenarios with pre-trained transformer-based language models.
2. We propose a feature extraction method that uses an embedding method that integrates semantic and relational information to obtain effective feature vectors, and this method can preserve important information of long series in the datasets and improve the performance of the model for long series on EM tasks.
3. Experiments on six public benchmark datasets clearly show that the method outperforms DeepER and DeepMatcher on all datasets. SAREM achieves comparable effectiveness to SOTA EM methods with smaller data and reduces cost and outperforms on datasets with long sequences.

2 Related Work

2.1 Entity Matching

Entity matching has been extensively developed as a classical research problem in data integration. A series of approaches based on probabilistic [4] and crowd-sourcing have been derived for structured data. For unstructured data, deep learning-based methods like DeepMatcher [5] and Ditto [6] have also achieved good results. As early as 2018, Ebraheem et al. [7] proposed to use deep learning to solve the EM problem, and then Thirumuruganathan et al. [8] proposed to define the EM problem within the new design space of deep learning. EM methods based on deep learning have become the key research direction in recent years.

[Niel Kooli et al. 2018] studied the EM method of the database based on the deep neural network, using N-gram embedding and a DNN model to decide matched or unmatched.

[Chen Zhao et al. 2019] proposed an end-to-end fuzzy EM method using pre- trained models and transfer learning, and this paper developed a new hierarchical neural network architecture to build end-to-end, high-quality EM systems.

[Cheng Fu et al. 2019] proposed an MPM model. It selects the best similarity measure for heterogeneous attributes and solves the problem of how to select appropriate similarity measures for different attributes.

[Nie Hao et al. 2019] proposed a deep sequence-to-sequence EM model, modeled EM as a token-level sequence-to-sequence matching task. However, because many words are OOV or numbers, and the context is usually sparse, it is still a big challenge.

[Zhang Dongxiang et al. 2020] applied the attention mechanism in the model, made full use of the semantic context of embedded vectors to describe the text, and proposed a comprehensive multi-context attention framework, with remarkable effect.

Then, [14] proposed a pre-trained transformer language model for EM, which further improved the performance. [6] develop three optimization techniques to improve the matching ability, but it relies on the existing NER model heavily and has a poor effect on long-sequence data because the input sequence needs to be within 512 when using BERT.

2.2 EM Based on Active Learning

The supervised learning method needs large training data. However, it is difficult to obtain labeled data, which requires experts to manually label. The active learning method can obtain a classifier with better performance with fewer training samples. It selects the most useful unlabeled samples by query algorithm, then adds them to the labeled data pool after being labeled, then continues to train the model.

[Kasai Jungo et al. 2019] proposed a method using transfer learning and active learning to solve low-resource problems. The purpose of joining active learning is to select some examples with high information to fine-tune the model. However, when the number of labeled samples is small, the performance still needs to be improved.

[Vamsi Meduri et al. 2020] provide a unified benchmark framework of active learning for EM, which allows users to combine different learning algorithms, to point out which active learning combinations have good effects on EM.

[Alex Bogatu et al. 2021] to solve the task-specificity and user-dependence in the process of EM, perform unsupervised representation learning, and reduce the cost of labeling through the active learning method.

[Arjit Jain et al. 2022] proposed an extensible active learning method to maximize the recall and accuracy of the block. In the method, an index framework divided by committees is used.

However, deep active learning faces many challenges, including insufficient labeled samples, model uncertainty, and inconsistent processing pipelines [20].

Our approach uses word embedding and graph embedding to capture semantic similarity and relationships between entities, solving the information loss problem [6, 13]. We use data augmentation, which solves the problem of insufficient labeled samples and improves the robustness of the model. The model is trained with semi-supervised learning in an active learning loop, and the strategy selects both the most uncertain samples according to the current model and those that represent the distribution of the datasets. Since the query strategy is mostly based on a fixed feature representation, the step of feature extraction and matching are separated in our method, solving the problem of inconsistent processing pipelines.

3 Problem Statement and Definition

We first give the definitions of EM and then elaborate on EM and its basic steps.

Definition 1. Entity matching. EM, also known as entity resolution, record linkage, or duplicate detection, refers to the process of linking records from different data sources [1]. Given two tables S and T, $s \in S, t \in T$, which is composed of one or more attributes, is a tuple pair in $S \times T$ and represents the Cartesian product of S and T. The purpose is to identify duplicate records in it. A record usually contains multiple attributes in the form of text, as shown in table 1. Each record has attributes such as *Title, Authors, Affiliations, Date, Pages, Conference*, etc. Many EM methods determine whether to match by comparing the similarity of aligned attributes.

Basic Step. The traditional method divides the problem into three main steps: block [16], comparison, and decision [21]. But EM methods based on deep learning usually replace comparison and decision with feature engineering and similarity learning [22]. The purpose of the block is to divide the data into smaller groups, provide candidate record pairs, and reduce the number of records participating in the comparison. Feature Engineering is characterized by embedding the text into vectors. Similarity learning refers to learning by building an appropriate deep learning neural network model with the vectors generated by feature engineering steps.

4 The Framework: Sarem

SAREM performs two tasks in the active learning cycle. One is the training of the matcher: (1) Firstly, start from the labeled data, obtains the flipped feature vectors by

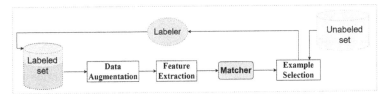

Fig. 1. The overall framework

data augmentation and feature extraction, and (2) input the vectors into the next layer network to train the matcher. The second is the expansion of labeled data. Unlabeled examples are selected to be labeled by selection strategies, and added to the labeled dataset. By expanding the labeled datasets, the matcher can be updated. The framework of SAREM includes data augmentation, feature extraction, matcher, example selection, and labeling, as shown in Fig. 1.

4.1 Data Augmentation

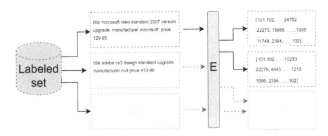

Fig. 2. Use the different hidden dropout masks for data augmentation.

To make the deep learning model achieve better performance and robustness, besides adding more labeled data for training, another important way is data augmentation. The research on image data augmentation has been very mature, but the data augmentation for text data is often task-specific. In the past, the text data augmentation methods used in EM used methods such as vocabulary replacement, reverse translation, random noise injection, etc. Here, we use the dropout layer to get the noise and replace the source data with the enhanced data though model-wise rather than data-wise. The specific method is to embed the source data by using different hidden dropout layers for N times, and then we can get N sentences with similar semantics but not identical ones, as shown in Fig. 2.

4.2 Feature Extraction

There are some problems when using pre-trained language models for EM. When the sequence exceeds 512, although some important information can be retained through TF-IDF, a lot of information still will be lost. To solve the problem, in feature extraction, we feed long sequences into the model in groups and do feature extraction through downsampling.

The idea of pooling originates from the downsampling process of CNN, a pooling layer is usually followed by a conv layer. The most important eigenvalues in a region are used to replace the whole region, thus realizing the function of data dimension reduction. Therefore, it is also possible to select important eigenvalues in long sequences through pooling. Each long sequence is divided into several segments. Because of the existence of special marks, each segment is divided into 510 tokens. They are put into a batch and fed into Bert. Then, the outputs are grouped according to the input, and the vectors of the same group are subjected to Max-Pooling so that they are downsampled into one-dimensional vectors, as shown in Fig. 3.

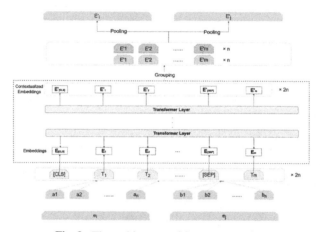

Fig. 3. The architecture of feature extraction.

However, we only have semantic information but lack the relationship between attributes, so we use graph embedding to get the relationship between attributes. We first construct a relational graph, as shown in Fig. 4. Graph G can be formally expressed as $G = (V, E)$, V and E represent a set of nodes and edges respectively. There are three types of nodes in G, namely tuple level nodes, attribute level nodes, and token level nodes. If there is an inclusive relationship between records and attributes, or between attributes and tokens, there is an edge between them. At the token level, tokens with similar values that appear simultaneously in the same sliding window are also connected. Next, the graph embedding vector of graph G is obtained by using GNN, the feature vector and graph vector are connected and input into the full connected layer and a softmax output layer makes the decision 0 or 1.

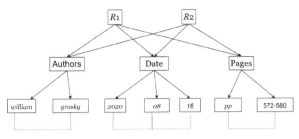

Fig. 4. Relation graph

4.3 Example Selection

4.3.1 Query-By-Committee

We use QBC based on vote entropy, which selects the samples with the largest entropy in the probability distribution. We train random forest models as the members of committee. It gives the prediction by averaging the class probabilities for each member and choosing the most likely one. After a few active learning queries, the hypotheses straighten out their disagreements and they reach consensus. As shown in Eq. (1).

$$x_{VE} = \arg\max_x -[\frac{V(y_1)}{C} \cdot \ln \frac{V(y_1)}{C} + \frac{V(y_2)}{C} \cdot \ln \frac{V(y_2)}{C}] \tag{1}$$

where $V(y_i)$ represents the number of members who voted for Class I, and C represents the total number of members.

4.3.2 Uncertainty Sampling

Besides QBC, other metrics are also available to measure the uncertainty of the prediction. SAREM supports the uncertainty sampling based on conditional entropy:

$$x_H = \arg\max_x -\sum_y p_\theta(y|x) \ \log p_\theta(y|x) \tag{2}$$

where p indicates the matching probability of two records. The greater conditional entropy, the greater uncertainty.

4.3.3 Information Density

When using uncertain sampling, only uncertain samples can be selected, and cannot take the data structure into account, the samples representing the distribution of data sets are ignored. To solve this problem, we use information density which uses the core set to represent the distribution of the feature space, and selects samples similar to each data element, as shown in Eq. (3).

$$x_{ID} = \arg\max_x (\frac{\text{sim}(x, x^{(1)}) + \text{sim}(x, x^{(2)})}{2})^\beta \tag{3}$$

where β is the exponential parameter, and $x^{(i)}$ is the representative element of class I.

4.3.4 Badge

We use BADGE which selects samples by using the computer illusion gradient embedding Eq. (4), so that it can consider both uncertainty and diversity in a batch, and keep a balance between them.

$$x_{BA} = \frac{\partial}{\partial \theta_{out}} \ell(f(x; \theta), \hat{y})|_{\theta=\theta_0} \tag{4}$$

where $f(x; \theta)$ is a neural network classifier, $\hat{y}$ is the most likely label of X predicted by this classifier. ℓ is a loss function, and the calculated result is labeled with the higher result.

4.4 Semi-supervised Task Learning

The feature extraction step can obtain the joint embedding $E(e, e')$ of sentence embedding and graph embedding of record e and e' It can be converted into the matching probability by the combination of the full connected layer and sigmoid function:

$$P(y = 1|(e, E(e, e'))) = (1 + \exp(-Fw(E(e, E(e, e')))))^{-1} \tag{5}$$

In the similarity learning stage, the forward step is performed twice for each training batch. Because data augmentation usually doesn't change the probability distribution of output, it will get two different forward results. The training goal is to reduce the distance between two predictions to improve the consistency of the model under disturbances. Therefore, the loss function contains two parts, the cross entropy loss of labeled data and the mean square error of these two forward results of unlabeled data, as shown in Eq. (6):

$$L = \lambda 1 \sum d_{CE} + \lambda 2 \sum L_{MSE} \tag{6}$$

$$LCE = \frac{1}{N} \sum_{i}^{N} \log(1 + \exp(-Fw(E\Theta(e1^i, e2^i)))) \tag{7}$$

$$dMSE = \frac{1}{C} \sum_{k=1}^{C} (f\theta(x)k - f\theta(\hat{x})k)^2 \tag{8}$$

where the first term is cross entropy, as shown in Eq. (7), which is only used to evaluate the loss of labeled data, and the second term is mean square error (MSE), as shown in Eq. (8), which is used to realize uniform regularization.

5 Experiments

5.1 Experiment Dataset

We used six public datasets, from ER benchmark datasets and Magellan datasets. We divide them into three categories: structured datasets, unstructured datasets and dirty datasets. Most of them were large datasets and we obtained candidate pair datasets after block [6] for experiments. The details of the datasets are shown in Table 2.

Table 2. These datasets are from the ER Benchmark datasets and the Magellan datasets, divided into 3 categories. The datasets marked with ∗ are generated from the clean version.

Type	Datasets	Domains	size
Structured	Amazon-Google	Software	11460
	DBLP-ACM	Citation	12,363
	Walmart-Amazon	Electronics	10242
Dirty	DBLP-ACM*	Citation	12,363
Textual	Abt-Buy	Product	9575
	company	Company	112632

5.2 Implementation Details and Evaluation Metrics

We implemented all the experiments in PyTorch1.9 and the Transformers library. We use FP16 optimizer to speed up the training and prediction speed. In the experiment, we use RoBERTa as the default transformer encoder, and then connect a full connected layer with *Tanh* activation function and a softmax binary classifier. We use Adam optimizer, and the learning rate is 3e-5, with linear decreasing, batchsize is 64 and epoch is 40. The active learning cycle defaults to 15, and the label budget B defaults to 512. Uncertainty sampling is used to compare with BADGE. Finally, we use F1 score to measure the effect.

5.3 Compared Methods

We compare SAREM with the SOTA EM solution as following.

DeepER: DeepER obtains vector representation by a simple averaging over the Glove embeddings per attribute and then uses a feedforward neural network to perform the binary classification.

DeepMatcher: DeepMatcher customizes the RNN architecture to aggregate the attribute values, then aligns the aggregated representation of the attributes, and used FastText to get word embeddings.

Ditto: Ditto is SOTA of EM method, which uses BERT to extract features, and uses three optimization techniques: injecting domain knowledge, text summarization and data augmentation to improve performance.

5.4 Experimental Results

Overall Performance: Table 2 shows the experimental results on six public datasets. The F1 scores of SAREM are higher than DeepER and DeepMatcher on the six datasets. It achieves the same effect as Ditto using less data and surpasses Ditto on Company. As expected, SAREM inherited the strong understanding ability of the pre-trained language model, so the effect on the DBLP-ACM* surpassed DeepER and DeepMatcher without BERT. When the performance of DeepER and DeepMatcher in DBLP-ACM* decreased, SAREM and Ditto were not affected.

Table 3. F1 scores on six public datasets, *size* is the size of datasets used by DeepER ﹨ DeepMatcher ﹨ Ditto, *size** is the size of datasets used by SAREM

Type	Datasets	DeepER	DM	Ditto	SAREM	size	size*
Structured	Amazon-Google	56.08	69.30	**75.58**	75.20	11460	8200
	DBLP-ACM	97.63	98.40	**98.99**	**98.99**	12363	8200
	Walmart-Amazon	50.62	66.90	**86.76**	86.10	10242	8200
Dirty	DBLP-ACM*	89.62	98.10	**99.03**	**99.03**	12363	8200
Textual	Abt-Buy	42.99	62.80	89.33	**89.85**	9575	8200
	Company	62.17	92.70	93.85	**94.10**	112632	8200

Fig. 5. F1 scores about different feature extraction solutions

Effectiveness of the Feature Extraction: WE compare our feature extraction with the methods in DeepER, DeepMatcher and Ditto. Specifically, we compare the FastText used by Deep Matcher, the Glove used by DeepMatcher and the feature extraction method based on three optimization techniques used by Ditto with our feature extraction. The results are shown in Fig. 5. It can be seen that there is little difference among several methods in structured data. Our method is the best for text datasets because we relax the sequence length limit, so that semantic information will not be lost, and graph embedding retains relational information.

Selection Strategies for Active Learning: WE compare different example selection strategies with SAREM and the baseline is random sampling, choosing samples at random from the candidate set. We obtain the F1 scores after 15 active learning rounds on the 6 datasets with different selection strategies, as shown in Table 3. We find that BADGE beat all other strategies and establishes the effectiveness for active learning (Table 4).

Table 4. Comparison of SAREM with different selection strategies on F1 scores after 15 rounds of active learning, SAREM support all of the following selection strategies.

Datasets	Random	Uncertainty	QBC	ID	BADGE
Amazon-Google	63.3	75.2	75.5	75.3	82.8
DBLP-ACM	96.6	98.9	98.1	99.0	99.1
Walmart-Amazon	58.3	86.1	76.1	86.1	90.2
DBLP-ACM*	97.6	99.0	99.0	99.1	99.2
Abt-Buy	75.2	89.8	80.8	90.0	92.5
Company	92.0	94.1	93.8	94.8	95.0

6 Conclusion

This paper introduces SAREM, a semi-supervised EM framework for heterogeneous data. SAREM uses the embedding method that integrates semantic information and relational information to get effective feature vectors and performs well on large datasets with long sequences. We also use the dropout layer for data augmentation. Our active learning method solves the classic challenge of deep active learning, reduces manual intervention, improves the performance of the model, and achieves the goal of cost reduction. In the future, we will continue to optimize SAREM in combination with the characteristics of semi-supervised learning.

Acknowledgment. This work was supported by the National Natural Science Foundation of China (62072086, 62172082, 62072084), the Fundamental Research Funds for the central Universities (N2116008).

References

1. Christen, P.: Data Matching - Concepts and Techniques for Record Linkage, Entity Resolution, and Duplicate Detection. Data-Centric Systems and Applications Description, pp. 1–270. Springer, Heidelberg (2012). https://doi.org/10.1007/978-3-642-31164-2
2. Chaudhuri, S., Chen, B.-C., Ganti, V., Kaushik, R.: Example-driven design of efficient record matching queries. In: PVLDB 2007, pp. 327–338 (2007)
3. Christen, P.: The data matching process. In: Christen, P. (ed.) Data Matching, pp. 23–35. Springer, Heidelberg (2012). https://doi.org/10.1007/978-3-642-31164-2_2
4. Fellegi, I., Sunter, A.: A theory for record linkage. J. Am. Stat. Assoc. **64**, 1183–1210 (1969)
5. Mudgal, S., et al.: Deep Learning for Entity Matching: A Design Space Exploration. (2018)
6. Li, Y., Li, J., Suhara, Y., et al.: Deep entity matching with pre-trained language models. PVLDB. **14**(1), 1–7 (2021)
7. Ebraheem, M., Thirumuruganathan, S., Joty, S., et al.: Distributed representations of tuples for entity resolution. Proc. VLDB Endow. **11**(11), 1454–1467 (2018)
8. Mudgal, S., Li, H., Rekatsinas, T., et al.: Deep learning for entity matching: a design space exploration. In: Proceedings of the 2018 International Conference on Management of Data, pp. 19–34. ACM (2018)

9. Kooli, N., Allesiardo, R., Pigneul, E.: Deep learning based approach for entity resolution in databases. In: Nguyen, N.T., Hoang, D.H., Hong, T.-P., Pham, H., Trawiński, B. (eds.) Intelligent Information and Database Systems. LNCS (LNAI), vol. 10752, pp. 3–12. Springer, Cham (2018). https://doi.org/10.1007/978-3-319-75420-8_1

10. Zhao, C., He, Y.: Auto-EM: end-to-end fuzzy entity-matching using pre-trained deep models and transfer learning. In: Proceedings of the World Wide Web Conference (2019)

11. Fu, C., Han, X., Sun, L., et al.: End-to-end multi-perspective matching for entity resolution. In: Twenty-Eighth International Joint Conference on Artificial Intelligence (2019)

12. Nie, H., Han, X., He, B., et al.: Deep sequence-to-sequence entity matching for heterogeneous entity resolution. In: Proceedings of the 28th ACM International Conference on Information and Knowledge Management, pp. 629–638 (2019)

13. Zhang, D., Nie, Y., Wu, S., et al.: Multi-context attention for entity matching. In: Association for Computing Machinery (2020)

14. Teong, K.-S., Soon, L.-K., et al.: Schema-agnostic entity matching using pre-trained language models. In: Proceedings of the 29th ACM International Conference on Information and Knowledge Management (2020)

15. Kooli, N.: Data Matching for Entity Recognition in OCRed Documents. Lorraine University, Thesis, Defense (2016)

16. Azzalini, F., Jin, S., Renzi, M., Tanca, L.: Blocking techniques for entity linkage: a semantics-based approach. Data Sci. Eng. 6(1), 20–38 (2020). https://doi.org/10.1007/s41019-020-00146-w

17. Li, B., Wang, W., Sun, Y., et al.: GraphER: token-centric entity resolution with graph convolutional neural networks. In: AAAI (2020)

18. Wu, R., Chaba, S., Sawlani, S., Chu, X., Thirumuruganathan, S.: ZeroER: entity resolution using zero labeled examples. In: Proceedings of the 2020 ACM SIGMOD International Conference onManagement of Data (SIGMOD 2020), 14–19 June 2020, Portland, OR, USA, 16 p. ACM, New York (2020)

19. Settles, B.: Active Learning Literature Survey. Technical report. University of Wisconsin-Madison Department of Computer Sciences (2009)

20. Ren, P., et al.: A survey of deep active learning. ACM Comput. Surv. 54(9), 40 (2021)

21. Elmagarmid, A.K., Ipeirotis, P.G., Verykios, V.S.: Duplicate record detection: a survey. IEEE TKDE. 19(1), 1–9 (2007)

22. Bogatu, A., Paton, N.W., Douthwaite, M., Davie, S., Freitas, A.: Cost–effective variational active entity resolution. In: 2021 IEEE 37th International Conference on Data Engineering (ICDE), pp. 1272–1283 (2021)

23. Kasai, J., Qian, K., Gurajada, S., et al.: Low-resource deep entity resolution with transfer and active learning. Meeting of the Association for Computational Linguistics (2019)

24. Vamsi Meduri, Lucian Popa, Prithviraj Sen, and Mohamed Sarwat.: A Comprehensive Benchmark Framework for Active Learning Methods in Entity Matching[C]. In Proceedings of the 2020 ACM SIGMOD International Conference on Management of Data.2020

25. Bogatu, A., Paton, N.W., Douthwaite, M., Davie, S., Freitas, A.: Cost–effective variational active entity resolution. In: 2021 IEEE 37th International Conference on Data Engineering (ICDE) (2021)

26. Jain, A., Sarawagi, S., Sen, P.: Deep indexed active learning for matching heterogeneous entity representations. PVLDB. 15(1), 31–45 (2022)

27. Xu, W., Sun, C., Xu, L., Chen, W., Hou, Z.: Unsupervised entity resolution method based on random forest. In: Xing, C., Fu, X., Zhang, Y., Zhang, G., Borjigin, C. (eds.) Web Information Systems and Applications. LNCS, vol. 12999, pp. 372–382. Springer, Cham (2021). https://doi.org/10.1007/978-3-030-87571-8_32

LocRDF: An Ontology-Aware Key-Value Store for Massive RDF Data

Yunxiao Zhang, Jinghan Li, Xueyang Liu, Rong Cheng, Yiran Hu, and Xin Wang$^{(\boxtimes)}$

College of Intelligence and Computing, Tianjin University, Tianjin, China
wangx@tju.edu.cn

Abstract. With the rapid development of the Semantic Web, the scale of RDF graphs surges. To describe ontology information, RDFs and OWL are endorsed by W3C, which further enhances the expressiveness of RDF graphs. A great challenge of managing RDF graphs is how to store massive data and efficiently reason ontology information at query time. There are two main issues with the existing RDF graph storage systems: 1) the relational data model is mainly used as the underlying storage architecture, which not only leads to exceeding the storage capacity, but also may incur high overhead while performing complex queries or multi-join queries; 2) the ontology reasoning module is either relatively independent of storage layer or used as an upper-layer application of storage and query system, causing redundancy and inefficiency in query. To address these issues, we present LocRDF, a novel storage system for RDF graphs via key-value store supporting ontology reasoning. LocRDF integrates ontology information into the underlying storage scheme with the application of a fixed-length interval encoding, promoting the efficiency of ontology reasoning at runtime. Experimental results on LUBM datasets show that extended ontology reasoning on large-scale RDF graphs scarcely affects query performance which is even significantly better than the existing state-of-the-art RDF query engines.

Keywords: RDF graph · Key-value store · Ontology-aware

1 Introduction

Resource Description Framework (RDF) [11] is a graph-based data model that describes the core data of Semantic Web applications. RDF consists of triples of the form (s, p, o) where s, p, and o are the abbreviations of subject, predicate, and object. Due to the user-friendly and expressive characteristics of the RDF data model, the application of RDF continues expanding in recent years. For example, in the fields of knowledge fusion [12], precision marketing [3], and semantic search [5], the use of RDF provides users with more intelligent services. As of 2021, the number of RDF triples in the DBpedia project is up to 850 million. Therefore, it is essential to explore the storage, query, and reasoning of RDF to efficiently handle such a large-scale RDF dataset.

X. Zhao et al. (Eds.): WISA 2022, LNCS 13579, pp. 89–101, 2022.
https://doi.org/10.1007/978-3-031-20309-1_8

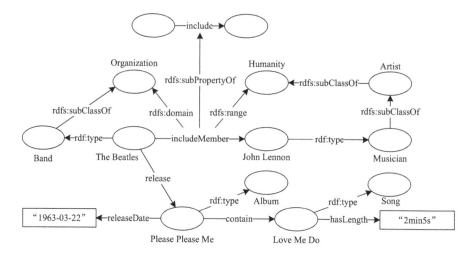

Fig. 1. A music knowledge RDF graph about The Beatles

Relational storage scheme and key-value storage scheme are two main methods of storing RDF graphs [13]. Different relational storage schemes share several common drawbacks: high costs while storing and querying, the restriction on the number of tables, etc. [14]. In comparison, key-value storage scheme satisfies the current requirements of large-scale data storage, high concurrent query, and high fault tolerance, remedying for the defects of relational storage scheme.

Implicit triples in RDF datasets, i.e., reasoning information, play a quite critical role. A given dataset of explicit triples can be extended with more valuable information by ontology reasoning rules [6]. However, the current researches on reasoning fail to integrate ontology information into the storage layer, resulting in inefficient query performance. Therefore, this paper proposes an RDF storage system that pushes the ontology information down to the storage layer, which can achieve efficient triple queries with ontology reasoning.

Contributions. Our contributions in this paper can be summarized as follows:

1. A novel storage system for RDF graphs via key-value store is proposed, which can support ontology reasoning and push reasoning information down to the storage layer, making ontology reasoning efficient at runtime.
2. Interval encoding, a fixed-length, space-saving encoding method is devised. Additionally, we integrate ontology information into the coding on the basis.
3. Extensive experiments on several datasets are conducted to evaluate the effectiveness and efficiency of our storage system. The results show that LocRDF outperforms Jena [2] and AllegroGraph [1] in processing ontology reasoning at query time.

Paper Organization. The paper is organized as follows. The related work is summarized in Sect. 2. Several formal definitions are given in Sect. 3. Section 4 gives detailed description of the storage scheme, query processing, and the reasoning method of our system. Section 5 presents experimental results and Sect. 6 concludes.

2 Related Work

WebPIE [10] uses multiple computing nodes to perform forward reasoning and materialization. However, the forward reasoning leads to high cost of storage. QueryPIE [9] adopts backward reasoning. It converts reasoning problems into query problems by decomposing the original queries into subqueries and constructing an AND-OR tree during the reasoning process. Jena and AllegroGraph both provide a lightweight RDF reasoning engine, but their operations depend on existing RDF storage. The aforementioned researches only involve the reasoning of data regardless of data storage, resulting in redundancy in the reasoning process: the data expansion caused by forward reasoning occupies a large amount of storage space; the queries constructed by backward reasoning make the data transmission redundant.

In addition, Stylus [4] proposes a compact key-value storage scheme. But the reasoning of Stylus is realized as an upper-level module of the system. It restructures the query into multiple subqueries as per the backward rules, making reasoning less efficient. Neo4j [8] is a mainstream graph storage system as well. It completes reasoning by rewriting the query in combination with reasoning rules. Both Stylus and Neo4j regard reasoning as an upper-layer application of storage system, which is unable to efficiently integrate ontology information with storage.

In summary, existing reasoning modules are either relatively independent, or used as upper-layer applications of storage and query systems. The basic process of reasoning in these systems is nothing more than forward reasoning to all implicit triples or constructing queries as per the backward reasoning. This will incur overhead of storage space and query time. Therefore, this paper focuses on integrating semantic information into the storage, combining the underlying storage and query system with the reasoning system to reduce redundancy and promote reasoning efficiency.

3 Preliminaries

In this section, we introduce several basic background definitions, including RDF triple classification for RDF graph [13] and ontology which are used in our algorithms and system. An example RDF graph depicted in Fig. 1 describes a music knowledge graph about The Beatles.

Definition 1 (Triple classification). *Let C be the set of classes of triples. $C = \{entity, prop, edge\}$ is used to denote the type of triples: namely entity* *triples, predicate description triples, and edge triples. Let L be an finite set representing literals. Define function $\varphi : T \rightarrow C$ that maps RDF triples to its corresponding type, where T represents a finite set of RDF triples.*

$$\varphi(t) = \begin{cases} entity, & when \ t \in \{t = (s,p,o) \mid (s,p,o) \in T \wedge p = \textbf{\textit{rdf:type}}\} \\ prop, & when \ t \in \{t = (s,p,o) \mid (s,p,o) \in T \wedge o \in L\} \\ edge, & when \ t \in \{t = (s,p,o) \mid (s,p,o) \in T \wedge o \notin L \wedge p \neq \textbf{\textit{rdf:type}}\} \end{cases}$$

An ontology reasoning system which only considers the five key properties in RDFs: `rdf:type(type)`, `rdfs:subClassOf(sc)`, `rdfs:subPropertyOf(sp)`, `rdfs:range(range)`, and `rdfs:domain(domain)`, has been proved to contain the main features of RDFs and capture the essence of RDF [6]. The ontology reasoning rules are shown in Table 1 and the definition of an ontology can be formally given as:

Definition 2 (Ontology). *An ontology is a 6-tuple $O = (C, P, SC, SP, \mathcal{D}, \mathcal{R})$, where*

1) C is a finite set of classes,
2) P is a finite set of predicates,
3) SC is a set of triples that declares a class to be a subclass of another, i.e., if SC contains $\langle c_i, \textbf{\textit{rdfs:subClassOf}}, c_j \rangle$, c_i is a subclass of c_j,
4) SP is a set of triples that declares a property to be a subproperty of another,
5) $\mathcal{D} \colon P \rightarrow C$ is a mapping that associates a property with a definite class of its subject, i.e., if $p \in P$ and $\mathcal{D}(p) = d_i$, the type of the subject of p is d_i,
6) $\mathcal{R} \colon P \rightarrow C$ is a mapping that associates a property with a definite class of its object.

4   LocRDF

In this section, we propose the encoding, storage, query and reasoning process of LocRDF.

Table 1. Ontology reasoning rules

#	Body	Head
R1	$(A,\mathsf{sp},B)\ (B,\mathsf{sp},C)$	(A,sp,C)
R2	$(A,\mathsf{sp},B)\ (X,A,Y)$	(X,B,Y)
R3	$(A,\mathsf{sc},B)\ (B,\mathsf{sc},C)$	(A,sc,C)
R4	$(A,\mathsf{sc},B)\ (X,\mathsf{type},A)$	(X,type,B)
R5	$(A,\mathsf{dom},B)\ (X,a,Y)$	(X,type,B)
R6	$(A,\mathsf{range},B)\ (X,a,Y)$	(Y,type,B)
R7	$(A,\mathsf{dom},B)\ (C,\mathsf{sp},A)\ (X,C,Y)$	(X,type,B)
R8	$(A,\mathsf{range},B)\ (C,\mathsf{sp},A)\ (X,C,Y)$	(Y,type,B)

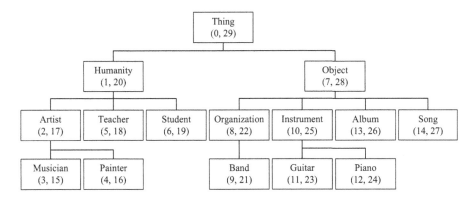

Fig. 2. The encoded class tree of the exemplary knowledge graph in Fig. 1

4.1   Encoding

As shown in Table 1, ancestry relationships between classes (properties) play an important role in the process of reasoning. We propose interval encoding for class (property) trees which has two distinct advantages: 1) the length of coding is independent of the tree size and the node position; 2) the decoding process of ontology information is quite efficient.

The process of interval encoding is listed in Algorithm 1. The input of the algorithm is an ontology O of an RDF graph. The codings of classes and properties are stored in two hash tables, μ and η, respectively. Algorithm 1 consists of three parts: 1) (lines 1–10) the algorithm first iterates over the triples in SC (SP) and builds a class (property) tree; 2) (lines 11–12) a pre-order traversal and a post-order traversal are performed from the root node, and the traversal orders of subtrees are both from left to right; 3) (lines 13–22) for each node in a class (property) tree, assuming that the total number of tree nodes is n, the pre-order code pre of the node is encoded as $0 \sim n-1$ as per the processed $prelist$, and the post-order code $post$ of the node is encoded as $n \sim 2n-1$ as per the processed $postlist$; the pair $(pre, post)$ is used as the coding of that node; the hash table μ (η) is constructed and output. The example of an encoded class tree is depicted in Fig. 2. For the given two classes (properties), their semantic relationship can be identified by the class (property) tree, and the relationship between the corresponding nodes can be decoded as per Theorem 1.

Definition 3 (Coding-containment). *Let the coding of class n_1 and n_2 be $(pre_1,\ post_1)$ and $(pre_2,\ post_2)$. We define the binary relationship $(pre_1, post_1)$ coding-contains $(pre_2, post_2)$ if and only if $post_1 > post_2$ and $pre_1 < pre_2$, denoted as $(pre_1,\ post_1) \sqsupset (pre_2,\ post_2)$.*

Theorem 1. *Let the coding of class n_1 and class n_2 be $(pre_1,\ post_1)$ and $(pre_2,\ post_2)$. If $(pre_1, post_1) \sqsupset (pre_2, post_2)$, then n_1 is the ancestor of n_2.*

Algorithm 1. Interval Encoding

Input: Ontology $O = (C, P, SC, SP, \mathcal{D}, \mathcal{R})$
Output: The hash table of class and classID μ (propertyID hash table η is calculated similarly)
1: *root* is the root node of the tree
2: **for** $\langle s_i, p_i, o_i \rangle \in SC$ **do**
3:    $o_i.children \leftarrow o_i.children \cup \{s_i\}$
4:    **if** s_i in $root.children$ **then**
5:       $root.children \leftarrow root.children \backslash \{s_i\}$
6:    **end if**
7:    **if** o_i not in $root.children$ **then**
8:       $root.children \leftarrow root.children \cup \{o_i\}$
9:    **end if**
10: **end for**
11: $preList \leftarrow preOrder(root)$
12: $postList \leftarrow postOrder(root)$
13: $code \leftarrow 0$
14: **for** c in $preList$ **do**
15:    $\mu[c].preCode \leftarrow code$ // encode the pre-order code of the interval encoding
16:    $code \leftarrow code + 1$
17: **end for**
18: **for** c in $postList$ **do**
19:    $\mu[c].preCode \leftarrow code$ // encode the post-order code of the interval encoding
20:    $code \leftarrow code + 1$
21: **end for**
22: **return** μ

4.2   System Storage

For an RDF graph with ontology information, both the RDF graph and ontology information need to be stored via specific storage schemes [15]. As depicted in Fig. 3, we design a key-value storage scheme in an effort to facilitate the effectiveness and efficiency of the ontology reasoning.

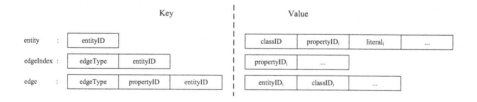

Fig. 3. The storage scheme for LocRDF

An RDF graph can be viewed as a directed graph consisting of entities and edges. Each entity in the RDF graph will have a unique entityID and class interval encoding classID. Particularly, a literal is stored as an attribute describing

---

Algorithm 2. Data Loading

Input: RDF graph G, Ontology $O = (C, P, SC, SP, \mathcal{D}, \mathcal{R})$, triple classification
     hash table φ, the class interval encoding hash table u, the property interval encod-
     ing hash table η;
Output: All key-value pair ANS
1: // domain range preprocessing;
2: **for** $\langle s_i, p_i, o_i \rangle \in G$ **do**
3:     **if** p_i has domain **then**
4:        $insert \langle s_i, \text{rdf:type}, \mathcal{D}(p_i) \rangle$
5:     **end if**
6:     **if** p_i has range **then**
7:        $insert \langle o_i, \text{rdf:type}, \mathcal{R}(p_i) \rangle$
8:     **end if**
9: **end for**
10: // data import;
11: **for** $\langle s_i, p_i, o_i \rangle \in G$ **do**
12:    **if** $\varphi(\langle s_i, p_i, o_i \rangle) = entity$ **then**
13:       $entity[s_i].key.entityID \leftarrow t.s$;
14:       $entity[s_i].value.classID \leftarrow u(o_i)$
15:    **end if**
16:    **if** $\varphi(\langle s_i, p_i, o_i \rangle) = prop$ **then**
17:       add $\langle p_i, o_i \rangle$ to $entity[s_i].value$
18:    **end if**
19:    **if** $\varphi(\langle s_i, p_i, o_i \rangle) = edge$ **then**
20:       add p_i to $edgeIndex[\langle EDGE_{out}, s_i \rangle].value$
21:       add s_i to $edgeIndex[\langle EDGE_{in}, o_i \rangle].value$
22:       add $\langle o_i, u(o_i) \rangle$ to $edge[\langle EDGE_{out}, p_i, s_i \rangle].value$
23:       add $\langle s_i, u(s_i) \rangle$ to $edge[\langle EDGE_{in}, p_i, o_i \rangle].value$
24:    **end if**
25: **end for**
26: **return** $entity + edgeIndex + edge$

---

the subject in the value of the *entity* key-value pair. The relationship between entityID and its classID is stored in *entity* key-value pair as well. The *edgeIndex* key-value pair stores all predicates associated with each entity, and edgeType identifies incoming edge or outgoing edge. The *edge* key-value pair stores edge information. The propertyID indicates the predicate represented by the edge. If the stored edge is an outgoing edge, entityID indicates the subject of the edge. When the stored edge is an incoming edge, entityID indicates the object of the edge. An array of $\langle$enityID, classID$\rangle$ will be stored in the value, which is the other end of the edge. Thus, it can be concluded that a record identifies the set of all edges with the same predicate starting from or ending in an entity. As for the classID in the value, it will be used for quick ontology reasoning during query.

The data-loading algorithm is shown in Algorithm 2. The input of the algorithm is an RDF graph G and its corresponding class and property interval encoding hash tables μ and η. The RDF graph with its ontology information will be stored into the set of key-value pairs ANS. Algorithm 2 consists of two parts:

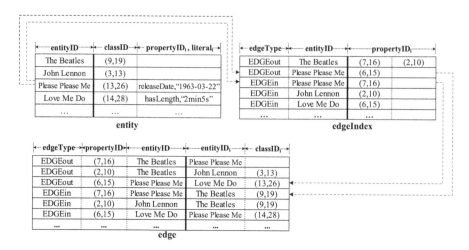

Fig. 4. The storage scheme of the exemplary knowledge graph in Fig. 1

1) (lines 1–9) the algorithm first preprocesses `rdfs:range` and `rdfs:domain`, i.e., converting it into `rdf:type`; 2) (lines 10–26) the algorithm iterates over each triple in G. The entity triples and prop triples are stored in *entity* key-value pair. For an edge triple, we add edge index in *edgeIndex* key-value pair and store the object in *edge* key-value pair. The storage scheme of the exemplary knowledge graph in Fig. 1 is depicted in Fig. 4.

4.3 Query and Reasoning

Based on the proposed storage structure, four basic data operators are designed as follows:

$getO(s, p)$: obtains the set of objects associated with s and p

$getS(o, p)$: obtains the set of subjects associated with o and p

$getP_{Out}(s)$: obtains the set of predicates associated with s

$getP_{In}(o)$: obtains the set of predicates associated with o

Given s and p, the operator $getO(s, p)$ ($getS(s, p)$) retrieves all the objects (subjects) consisting of two parts: 1) when the required triple is an edge triple, we obtain the object through querying *edge* and *edgeIndex* key-value pair; 2) when the required triple is a *prop* triple, we only need to access the value of the corresponding *entity* key-value pair. The operator $getP_{Out}(s)$ and $getP_{In}(s)$ are similar to $getO(s, p)$.

Through the aforementioned operators, the system can perform basic query operations. However, it still cannot perform effective ontology reasoning. Two more data operators are designed as follows:

$filter_e(e, ce)$: determines whether the entity e satisfies the constraint ce

$filter_p(p, cp)$: determines whether the property p satisfies the constraint cp

Table 2. Translating query triple patterns

Patterns	Translated operations
$\{\langle s, p, ?o^3\rangle, \varnothing, \varnothing, co^2\}$	$\{o \mid o \in getO(s,p) \wedge filter_e(o, co)\}$
$\{\langle s, ?p, ?o\rangle, \varnothing, cp, co\}$	$\{\langle p, o\rangle \mid p \in getP_{out}(s) \wedge filter_p(p, cp) \wedge o \in$ $getO(s,p) \wedge filter_e(o, co)\}$
$\{\langle ?s, p, o\rangle, cs, \varnothing, \varnothing\}$	$\{s \mid s \in getS(o,p) \wedge filter_e(s, cs)\}$
$\{\langle ?s, ?p, o\rangle, cs, cp, \varnothing\}$	$\{\langle s, p\rangle \mid p \in getP_{in}(o) \wedge filter_p(p, cp) \wedge s \in$ $getS(o,p) \wedge filter_e(s, cs)\}$
$\{\langle s, ?p, o\rangle, \varnothing, cp, \varnothing\}$	$\{p \mid p \in getP_{out}(s) \wedge filter_p(p, cp) \wedge o \in getO(s,p)\}$
$\{\langle ?s, p, ?o\rangle, cs, \varnothing, co\}$	$\{\langle s, o\rangle \mid s \in S^1 \wedge filter_e(s, cs) \wedge o \in getO(s,p) \wedge filter_e(o, co)\}$
$\{\langle ?s, ?p, ?o\rangle, cs, cp, co\}$	$\{\langle s, p, o\rangle \mid s \in S \wedge filter_e(s, cs) \wedge p \in$ $P \wedge filter_e(p, cp) \wedge o \in getO(s,p) \wedge filter_e(o, co)\}$

^1S and P are the sets of all the subjects and predicates in the dataset.
^2cs, cp, and co are the type constraints on subject, property, and object.
$^3?x$ represents the unknown item to be queried.

With the aforementioned six data operators, we can efficiently handle query patterns with ontology information constraints in Table 2, which can process one-hop queries and reduce the redundant communication between the query engine and underlying database. As for multi-hop queries, they can be combined by one-hop queries flexibly.

5   Experimental Evaluation

In this section, extensive experiments based on datasets of different sizes are conducted to evaluate the performance of LocRDF, using open source database Nebula Graph [7].

5.1   Experiment Setup and Datasets

The experiments are deployed on a single machine, which has 6-core Intel® Core™ i7-9750H CPU @ 2.60 GHz system, with 16 GB of memory, running Ubuntu 20.04.4 LTS.

In our experiments, we employ an RDF data benchmark LUBM, which is based on an ontology called Univ-Bench for the university domain. We compare the storage performance and query efficiency of LocRDF with the existing state-of-the-art RDF query engines: Jena 3.10.0 [2] and AllegroGraph 7.3.0 [1]. We choose the free version of AllegroGraph that supports LUBM1 to LUBM30. Table 3 shows the statistical information of the massive datasets in our experiments.

5.2   Experimental Results

Extensive experiments are conducted to verify the advantages of our system in terms of repository size and query efficiency.

Table 3. Statistical information of the datasets

Dataset	Number of triples	Number of vertexes	Size (MB)			
			Origin	LocRDF	Jena	AllegroGraph
LUBM10	1311976	207616	223	195	431	853
LUBM20	2772419	437958	473	187	622	1185
LUBM30	4094438	646550	700	264	802	1330
LUBM40	5476274	865022	937	324	991	–
LUBM50	6866224	1083818	1200	1099	1200	–

Repository Size. As shown in Table 3, LocRDF performs significantly better than Jena and AllegroGraph in storage. The dataset requires a larger storage space in Jena and AllegroGraph than its original size, while LocRDF has a compact storage scheme that improves the compression effect of the dataset. As the size of the dataset increases, LocRDF prevails greater in storage space. Experiments on LUBM30 show that the storage size of LocRDF is only 32.9% of Jena, 37.7% of the original dataset size and 19.8% of AllegroGraph.

Query Efficiency. We evaluate the time required to execute the same query sentence to compare the query efficiency of LocRDF, Jena, and AllegroGraph. The query time of all query sentences depicted in the figures is the average time of five executions of the same query sentences. The experiment consists of the following two parts: 1) time comparison of LocRDF executing the same query sentence w/o the reasoning function; 2) time comparison between LocRDF, Jena and AllegroGraph executing the same reasoning query sentences.

1) LocRDF w/o Ontology Reasoning. We select two test query sentences for experimentation. Q_1 and Q_2 find all triples when given s and p or o and p. The two query sentences are executed w/o reasoning enabled as depicted in Fig. 5. The experimental results demonstrate that the ontology reasoning based on interval encoding does not decelerate query performance, and the execution time is basically the same.

2) LocRDF VS. Jena and AllegroGraph. We select six test query sentences for experimentation. Q_3 and Q_4 query all triples given s or o while Q_6 and Q_7 additionally specify the type of s or o. Q_5 and Q_8 query all satisfied triples given p on the basis of Q_3 and Q_6, respectively. We compare the query time of the aforementioned six query sentences between LocRDF, Jena and AllegroGraph, as depicted in Fig. 6.

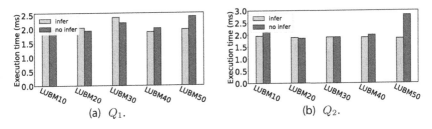

Fig. 5. Comparison of query time w/o ontology reasoning(Q_i is the ith query sentence in experiment)

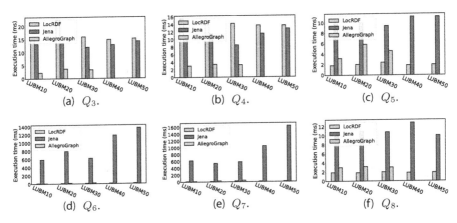

Fig. 6. Comparison of query efficiency between LocRDF, Jena and AllegroGraph when ontology reasoning is enabled (query$_i$ is the ith query sentence in experiment)

The experimental results indicate that when the size of dataset increases from LUBM10 to LUBM50, the average query time between LocRDF, Jena and AllegroGraph does not increase significantly in Q_3 and Q_4, and the execution time of LocRDF is nearly the same as Jena's and AllegroGraph's. Also, the execution of query sentences that require the ontology reasoning information, such as Q_6 and Q_7, are quite time-consuming in Jena and AllegroGraph because of massive join operations, while LocRDF uses interval encoding to reduce join operations via the use of $getO$, $getS$, $getP_{Out}$, and $getP_{In}$ operators, greatly improving the execution efficiency. The experimental results demonstrate that LocRDF performs significantly better in Q_6 and Q_7 on all datasets, and the average query time is reduced by 98% compared to Jena's and almost 50% compared to AllegroGraph's in LUBM30's experiment.

In terms of the two query sentences of Q_5 and Q_8, it can be found that LocRDF has better query efficiency when performing multi-constraint ontology reasoning, which is approximately an order of magnitude faster than Jena and AllegroGraph.

6   Conclusion

This paper proposes a novel storage system for RDF graphs via key-value store supporting ontology reasoning and presents interval encoding for class (property) trees. A prototype system of our storage scheme is designed and implemented based on Nebula Graph. Extensive experiments on the LUBM benchmark datasets verify the effectiveness and efficiency of the storage scheme and queries supporting ontology reasoning.

Acknowledgement. This work is supported by the National College Students' Innovation and Entrepreneurship Training Program of China (202110056120).

References

1. Allegrograph. https://franz.com/agraph/allegrograph/
2. Carroll, J.J., Dickinson, I., Dollin, C., Reynolds, D., Seaborne, A., Wilkinson, K.: Jena: implementing the semantic web recommendations. In: Proceedings of the 13th International World Wide Web Conference on Alternate Track Papers & Posters, pp. 74–83 (2004)
3. Fu, X., Ren, X., Mengshoel, O.J., Wu, X.: Stochastic optimization for market return prediction using financial knowledge graph. In: 2018 IEEE International Conference on Big Knowledge (ICBK), pp. 25–32. IEEE (2018)
4. He, L., et al.: Stylus: a strongly-typed store for serving massive RDF data. Proc. VLDB Endow. **11**(2), 203–216 (2017)
5. Li, Y.: Research and analysis of semantic search technology based on knowledge graph. In: 2017 IEEE International Conference on Computational Science and Engineering (CSE) and IEEE International Conference on Embedded and Ubiquitous Computing (EUC), vol. 1, pp. 887–890. IEEE (2017)
6. Muñoz, S., Pérez, J., Gutierrez, C.: Minimal deductive systems for RDF. In: Franconi, E., Kifer, M., May, W. (eds.) ESWC 2007. LNCS, vol. 4519, pp. 53–67. Springer, Heidelberg (2007). https://doi.org/10.1007/978-3-540-72667-8_6
7. The Nebula Graph Team: Nebula graph database manual 2.0. https://docs.nebula-graph.io/2.0/
8. The Neo4j Team: The neo4j manual v3.4 (2018). https://neo4j.com/docs/developpermanual/current/
9. Urbani, J., van Harmelen, F., Schlobach, S., Bal, H.: QueryPIE: backward reasoning for OWL horst over very large knowledge bases. In: Aroyo, L., et al. (eds.) ISWC 2011. LNCS, vol. 7031, pp. 730–745. Springer, Heidelberg (2011). https://doi.org/10.1007/978-3-642-25073-6_46
10. Urbani, J., Kotoulas, S., Maassen, J., Van Harmelen, F., Bal, H.: WebPIE: a web-scale parallel inference engine using mapreduce. J. Web Semant. **10**, 59–75 (2012)
11. W3C: Resource description framework. https://www.w3.org/RDF/
12. Wang, H., Fang, Z., Zhang, L., Pan, J.Z., Ruan, T.: Effective online knowledge graph fusion. In: Arenas, M., et al. (eds.) ISWC 2015. LNCS, vol. 9366, pp. 286–302. Springer, Cham (2015). https://doi.org/10.1007/978-3-319-25007-6_17
13. Wang, X., Zou, L., Wang, C., Peng, P., Feng, Z.: Research on knowledge graph data management: a survey. J. Softw. **30**(7), 2139–2174 (2019)

14. Wylot, M., Hauswirth, M., Cudré-Mauroux, P., Sakr, S.: RDF data storage and query processing schemes: a survey. ACM Comput. Surv. (CSUR) **51**(4), 1–36 (2018)

15. Zhang, R., Liu, P., Guo, X., Li, S., Wang, X.: A unified relational storage scheme for RDF and property graphs. In: Ni, W., Wang, X., Song, W., Li, Y. (eds.) WISA 2019. LNCS, vol. 11817, pp. 418–429. Springer, Cham (2019). https://doi.org/10. 1007/978-3-030-30952-7_41

Multi-view Based Entity Frequency-Aware Graph Neural Network for Temporal Knowledge Graph Link Prediction

Jinyu Zhang, Derong Shen[✉], Tiezheng Nie, and Yue Kou

Northeastern University, Shenyang 110004, China
{shenderong,nietiezheng,kouyue}@cse.neu.edu.cn

Abstract. Inferring missing facts in temporal knowledge graph (TKG) is a funda-mental and challenging task. Existing models typically use representation learn-ing to solve this problem. However, most of these models fall short of capturing multi-hop structural information and general preferences for future emerging facts when implementing the representation of target nodes. In addition, most of them use recurrent neural networks to achieve the aggregation of temporal information, which is not only less scalable as the time step increases, but also fails to explicitly address the problem of temporal sparsity of entity distribution in TKG. To address the above problems, we present a MEFGNN (Multi-view based Entity Frequency-aware Graph Neural Network) framework that learns node embedding to capture structural evolution of TKG by combining Multi-view Graph Neural Network (MGNN) and Entity Frequency-aware Attention Network (EFAN). Experiments on three real datasets show that MEFGNN outperforms state-of-the-art methods, our ablation study also validates the effectiveness of MGNN and EFAN.

Keywords: Temporal knowledge graph · Graph neural network · Representation learning · Link prediction

1 Introduction

Temporal knowledge graph (TKG) stores facts about what happened in real world and when the facts happened. Figure 1 shows a TKG consists of snapshots with differ-ent time stamps. Each snapshot is a heterogeneous graph containing different types of nodes such as countries (e.g., Iran, Pakistan), leaders (e.g., Trump, Macron) and political events (e.g., Protestor), and the nodes in the graph are connected by different edges (e.g., Prase, Critize). Over time, nodes and edges of graph appear and disappear, which leads to the evolution of graph structure. Compared with static knowledge graph (KG) that ignores temporal annotation of facts, TKG is more suitable for real scenarios and thus has received a lot of attention in recent years. However, like KG, TKG is far from com-plete. Therefore, predicting missing facts in TKG has become an increasingly important research task.

The inference task over a TKG aims at predicting the facts at moment t following two problem setups: given a TKG on a specific time period, if moment t is inside given

© The Author(s), under exclusive license to Springer Nature Switzerland AG 2022
X. Zhao et al. (Eds.): WISA 2022, LNCS 13579, pp. 102–114, 2022.
https://doi.org/10.1007/978-3-031-20309-1_9

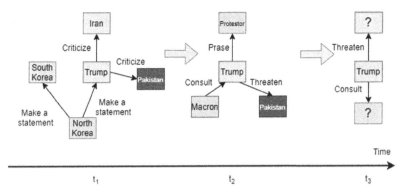

Fig. 1. An example of temporal knowledge graph

knowledge graph time period, the task is a link prediction of interpolation type; if t is beyond given time period, the inference task is a link prediction of extrapolation type. In this paper, we focus on the extrapolation setting which is particularly important in TKG inference because it helps to predict knowledge graph of future timestamps and upcoming events. There is a paucity of research on extrapolation inference, and in-depth research is needed.

Most of representation learning frameworks for TKG are implemented by jointly modeling both structural neighborhood and temporal structural evolution dimensions. [1, 2, 7, 19] modeling structural neighborhood feature of TKG by GNN-based methods, which easily integrates direct neighbor features of target nodes. However, GNN-based methods ignore the information of edges and multi-hop structure between nodes, which can cause semantic loss. [2, 13, 18] learn temporal evolution information of TKG by RNN [3]. Although RNN has strong parameter representation, it is less scalable with increasing time steps. Most importantly, the distribution of entities in TKG has time sparsity. For example, in a political event dataset, a subject relationship pair *(Obama, visit, ?)* may be more frequent in 2013 than in 2012. Therefore, a model could get more reference information in 2013 to answer where Obama would visit in 2014. However, current methods cannot cope with the serious challenge of temporal sparsity of entity distribution in TKG.

To address these issues, we introduce Multi-view based Entity Frequency-aware Graph Neural Network (MEFGNN) framework, which combines Multi-view Graph Neural Network (MGNN) and Entity Frequency-aware Attention Network (EFAN). Specifically, Multi-view Graph Neural Network first constructs structural features from both local and global perspectives in a network centered on each target node. Local features consider both the edges between nodes and multiple structural features. Global feature generalizes global information from the whole snapshots. MGNN is able to take into account the graph structure of TKG to efficiently learn the embedding of nodes and relations in each subgraph of TKG. Besides, in order to model time-evolving properties of TKG and to solve inhomogeneity problem caused by entity discrete distribution, we propose a new Entity Frequency-aware Attention Network, the core component of which is a temporal frequency based mask function that can focus on information about active snapshots determined by the frequency of entity occurrences. By combining attention mechanism with specific mask function, Entity Frequency-aware Attention Network is able to flexibly model temporal correlation of entity representations in different snapshots.

In this paper, node representations of TKG are generated by stacking on two dimensions: structural neighborhoods and temporal evolution. MGNN can extract features from a local network centered on each target node, while EFAN captures the evolution of TKG over multiple time steps by flexibly weighting historical representation. We summarize the main contributions of this paper as follows:

(1) We propose a Multi-view Graph Neural Network (MGNN) for heterogeneous graph embedding, which implements target node feature extraction in both local and global views. The method not only constructs rich structural features from a local network centered on each target node, but also reflects general preference for upcoming facts by global feature.

(2) We propose an Entity Frequency-aware Attention Network (EFAN) modeling temporal evolution information of TKG. To the best of our knowledge, this is the first model that explicitly addresses temporal sparsity of entity distribution in the TKG extrapolation link prediction task.

(3) Experiments show that our proposed framework shows a significant improvement over all baseline methods, which demonstrates the superiority of our method in obtaining a more accurate representation of TKG.

2   Related Work

Static Knowledge Graph Representation Learning. Much research exists on representation learning methods for static KG aim to represent each element of the KG as a low-dimensional vector and preserving its inherent semantics. Typically, these methods involve a decoding method that scores candidate facts based on entity and relationship embedding. Positive triples receive higher scores than negative samples. There are two main typical methods, namely translation methods and semantic matching methods. TransE [4] is a typical translation method that considers additive operations between head and tail entities. Based on TransE, many improved methods have been proposed, such as TransH [5] and TransR [6]. Besides, DistMul [7] uses a bilinear diagonal model and weighted element dot product to learn entity embedding. RESCAL [9] uses restricted Tucker decomposition for KG representation learning. Other semantic matching methods have been proposed, such as HoIE [11] and ComplEx [10]. In addition to the above two types of methods, many new representations of KG have been proposed in recent years. Some researchers have tried to learn KG representations based on graph neural networks, such as GCN [21], R-GCN [14], and GAT [15].

Temporal Knowledge Graph Representation Learning. Temporal knowledge graph representation learning methods aim to learn embedding of each element in TKG while considering the time of occurrence of facts. TAE [16] is the first method to integrate temporal information between relations into TKG embedding. Based on TransH, HyTE

[17] embeds entities and relations into a time-specific hyperplane. TA-DistMult [18] constructs temporal relational embeddings for each element by encoding temporal annotations using an RNN model. DyRep [1] implements neighbor node aggregation through simple splicing. EvolveGCN [19] uses RNN to evolve the parameters of GCN model at each moment. RE-NET is to achieve aggregation of snapshot history information through hidden state summary of RNN. Although these methods have achieved significant performance on TKG link prediction task, most of them ignore multi-hop structural information and summarize snapshot history information by RNN. As the time step increases, the scalability of RNN is poor. Most importantly, all of them are unable to explicitly address the problem of temporal sparsity of entity distribution. Therefore, in this work, we develop Multi-view Graph Neural Network fully exploits the structural neighborhood information of TKG in both global and local perspectives, and we propose Entity Frequency-aware Attention Network to explicitly solve the problem of temporal sparsity of entity distribution.

3 Definition

This section first introduces several related concepts, and then formalizes the definition of TKG link prediction.

Definition 1. Heterogeneous Graph. A heterogeneous graph is defined as a graph $G = (\nu, \mathcal{E})$, consisting of a set ν of nodes and a set $\mathcal{E}$ of edges. A heterogeneous graph can also be represented as a node type mapping function $\phi: \nu \rightarrow \mathcal{A}$ and an edge type mapping function $\psi: \mathcal{E} \rightarrow \mathcal{R}$. $\mathcal{A}$ and $\mathcal{R}$ are predefined node types and edge types, respectively, with $|\mathcal{A}| + |\mathcal{R}| > 2$.

Definition 2. Temporal Knowledge Graph. Temporal knowledge graph can be represented as a series of static snapshots $G = \{G^1, G^2, \ldots, G^{|T|}\}$, where each snapshot contains facts that occur simultaneously. $G^t = \{(s_i, r_i, o_i, t)\}$ in which $s_i \in \nu$ and $o_i \in \nu$ are subject entity and object entity, $r_i \in R$ is the relationship and $t \in T$ denotes the time when these facts occur.

Definition 3. Temporal Knowledge Graph Link Prediction. The goal of TKG link prediction is to predict the missing s or o of quadruple (s, r, o, t), which can be divided into two types of tasks. One task is to make prediction for head entity s given relation r and tail entity o, and another type task is to make prediction for tail entity o given relation r and head entity s at moment t. In order to unify the method evaluation criteria, we focus on the second task.

4 Proposed Approach

In this section, we introduce our proposed method MEFGNN. As shown in Fig. 2, our method takes a sequence of snapshots $\{G^1, G^2, \ldots, G^{|T|}\}$ as input. First, MGNN obtains the representation of each target node in each snapshot. Then, EFAN integrates the representation of the same node in different snapshots. Finally, as target nodes embedding are obtained, we can predict missing tail entity by a score function.

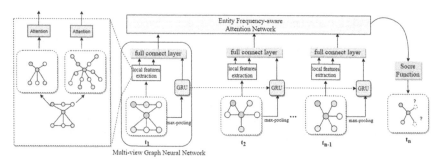

Fig. 2. The overall framework

4.1 Multi-view Graph Neural Network

The first key component is Multi-view Graph Neural Network (MGNN), which generates entity embedding based on snapshots G^t within each time step t. To fully characterize each target entity s (or o), we compute its embedding in both global and local perspectives, including two types of snapshot local features and one historical global feature. Then, to filter out important information, an attention mechanism is applied to each type of snapshot local features individually. The historical global feature can represent general preferences for emerging facts in the future. Snapshot local features can directly describe the knowledge associated with target entity. The local and global features complement each other. Finally a representation of each entity is obtained through a fully connected layer.

4.1.1 Features Extraction

Local Triple Feature. Triple structure is the basic structure of TKG, which is used to describe the binary relationships between entities. The triple structure involving the target node s illustrates the direct association between s and its neighbors, so it is meaningful to integrate such basic characteristic. For a quadruple (s, r_i, o_i, t) involving target node s, we can extract its local triple feature which can be expressed as follows:

$$h^t_{s_{T_i}} = h_{r_i} \star h_{o_i} \tag{1}$$

h_{r_i} is the initial embedding of relation r_i and h_{o_i} is the initial embedding of entity o_i. We obtain local triple feature $h^t_{s_{T_i}}$ of target node s at moment t via circular correlation operation $\star: R^d \times R^d \to R^d$, which is implemented in HolE. It has higher parameter expressiveness, whose specific implementation is as follows:

$$[\mathbf{a} \star \mathbf{b}]_i = \sum_{k=0}^{n-1} [\mathbf{a}]_k \cdot [\mathbf{b}]_{(k+i) \bmod n} \tag{2}$$

Finally, by extracting local triple features involving s, we can obtain a set $\left\{ h^t_{s_{T1}}, h^t_{s_{T2}}, \ldots, h^t_{s_{T_{|N^t_{sT}|}}} \right\}$, $\left| N^t_{sT} \right|$ is the number of local triple features of s in G^t.

Local Path Feature. Local Path Feature is one of the two types of snapshot local features, which are able to reflect multi-hop relationships between two entities. For each path feature-based neighbor entity u_i in G_s^t which is the neighborhood of s at timestamp t, we can find a relation path from s to u_i. The path can be denoted as $\left(s, r_{1_i}, \ldots, r_{n_i}, u_i\right)$, n represents the length of the path, which is set to 2 due to experimental results. We can implement local path feature encoding by RotatE [20] based on relational rotation in complex spaces. RotatE provides a way to capture the information implied by sequence structure of paths. Given $p(s, u_i) = (v_0, v_1 \ldots v_n)$ with $v_0 = u_i$ and $v_n = s$, let r_i be the relationship between v_i and v_{i+1}, $\odot$ is the elementwise product, and h_{r_i} be the representation of r_i, the relational rotation encoder is formulated as:

$$h_{s_{Pi}}^t = f(p(s, u_i)) = f\left\{h_{v_0}, h_{r_0}, h_{v_1} \ldots, h_{r_{n-1}}, h_{v_n}\right\} \tag{3}$$

$$o_0 = h_{v_0} \tag{4}$$

$$o_i = h_{v_1} + o_{i-1} \odot h_{i-1} \tag{5}$$

$$h_{s_{Pi}}^t = \frac{o_n}{n+1} \tag{6}$$

Finally, by extracting local path features involving s, we can obtain a set $\left\{h_{s_{P1}}^t, h_{s_{P2}}^t, \ldots, h_{s_{P|N_{sp}^t|}}^t\right\}$, $\left|N_{sp}^t\right|$ is the number of local path features of s in G^t.

Global Feature. Global feature H_t generalizes the global information from the whole snapshots until timestamp t, reflecting general preferences for emerging facts. The global feature can be expressed as follows:

$$H_t = \text{GRU}(g(G_t), H_{t-1}) \tag{7}$$

$$g(G_t) = \max\left(\left\{g\left(N_t^{(s)}\right)\right\}_s\right) \tag{8}$$

$g(G_t)$ is the aggregation of all facts in G_t, which is an max-pooling operation over all $g\left(N_t^{(s)}\right)$. $g\left(N_t^{(s)}\right)$ is the local structure information extracted from the neighborhood of target node s in snapshot G_t, with $N_t^{(s)}$ standing for facts to which s relates at timestamp t. We implement the revolution of global feature H_t through GRU [22], and we consider H_t as additional feature which acts as a bridge connecting different entities.

4.1.2 Features Fusion

Intra-feature Fusion. After obtaining the snapshot local features sets $\left\{\left\{h_{s_{Ti}}^t\right\}, \left\{h_{s_{Pi}}^t\right\}\right\}$, we adopt attention mechanism to weighted sum each feature independently. The key idea is that different features would contribute to target node's representation in different degrees. The weighted snapshot local feature can be expressed as follows:

$$\alpha_i^t = \frac{\exp(\boldsymbol{W}\boldsymbol{h}_{s*i}^t)}{\sum_{j=1}^{|N_*|} \boldsymbol{W}\boldsymbol{h}_{s*i}^t} \tag{9}$$

$$\boldsymbol{h}_{s*}^t = \sum_{i=1}^{|N_*|} \alpha_i^t \boldsymbol{h}_{s*i}^t \tag{10}$$

Here α_i^t is the parameterized attention vector for $\boldsymbol{h}_{s*i}^t$ ($\boldsymbol{h}_{s*i}^t$ is a local triple feature $\boldsymbol{h}_{s_{Ti}}^t$ or local path feature $\boldsymbol{h}_{s_{Pi}}^t$). $|N_*|$ is the length of each set, and $\boldsymbol{W} \in R^{d \times d}$ is weight matrix.

Inter-feature Fusion. Finally, we concatenate the obtained three kinds of features and use a fully connected layer to obtain the representation of entity s in G^t as follows:

$$\boldsymbol{h}_s^t = \sigma\left(\boldsymbol{W}\left[\boldsymbol{h}_{s_T}^t, \boldsymbol{h}_{s_P}^t, H_t\right] + \boldsymbol{b}\right) \tag{11}$$

By applying MGNN to target entity s in different snapshots, we can obtain a sequence of output representation for different snapshots, which can be expressed as $\left\{\boldsymbol{h}_s^1, \boldsymbol{h}_s^2, \ldots, \boldsymbol{h}_s^{|T|}\right\}$, where $|T|$ is total number of snapshots, $\sigma(.)$ is an activation function.

4.2 Entity Frequency-Aware Attention Network

The second key component proposed in this paper is Entity Frequency-aware Attention Network (EFAN), whose main task is to integrate the representation of entity information across time. Affected by the time span between snapshots, entity representations with long time span tend to have low relevance. The amount of data in different snapshots is also heterogeneous. For this reason, we propose EFAN to flexibly capture the relevance of entity representations in different snapshots. For each target entity s, MGNN has a sequence of output representations $\left\{\boldsymbol{h}_s^1, \boldsymbol{h}_s^2, \ldots, \boldsymbol{h}_s^{|T|}\right\}$ as input to Entity Frequency-aware Attention Network. Attention is executed at each time step of the entity embedding h_s^t to produce time-dependent entity representations $Z_{s,t}$.

$$q_{t,t_j} = \frac{(h_s^t W_q)(h_s^{t_j} W_k)^T}{\sqrt{d}} \tag{12}$$

$$\beta_{t,t_j} = \frac{m_s(t, t_j)\exp(h_s^j)}{\sum_{k<t} m_s(t, k)\exp(h_s^k)} \tag{13}$$

$W_q, W_k, W_v \in R^{d \times d}$ denote the linear projection matrix in transformer [23] layer, and β_{t,t_j} is attention weight matrix obtained from attention function. When $t < t_j$, attention weight $\beta_{t,t_j} \to 0$, which will ensure that only active temporal entity representations are assigned non-zero weights. Entity representations $Z_{s,t}$ is also adjusted according to the number of subject relationship pair involving s. If the number of entities associated with target node in a snapshot is higher, it is more helpful for link prediction of missing facts. Moreover, the reference value of information decays as time span increases. Combining the above two parts, the mask function m of Entity Frequency-aware Attention Network is specifically defined as:

$$m_s(t, t_i) = \begin{cases} 0, & t < t_j \\ exp\left(-\frac{\gamma(t-t_j)}{f_s^t + \lambda f_{s,r}^t}\right), & otherwise \end{cases} \tag{14}$$

f_s^t and $f_{s,r}^t$ represent subject frequency and subject relationship pair frequency at moment t, respectively. λ is a balance coefficient of f_s^t and subject $f_{s,r}^t$, which set to 0.1. γ is a normalization coefficient of time span and temporal frequency pattern. Attention mechanism can also be extended to multiple heads, which helps to stabilize learning process and reduce high variance introduced by the heterogeneity of graphs. Based on mask function m, Entity Frequency-aware Attention Network is able to model temporal correlation of entity representations while effectively solving the problem of temporal sparsity of entity distribution in TKG. The output of each entity s in time t is represented as follows:

$$z_{s,t} = \sum_{j=0}^{t} \beta_{t,t_j}(h_s^{t-j}W_v) \tag{15}$$

The final representation $\{z_{s,1}, z_{s,2}, \ldots, z_{s,|T|}\}$ of each entity s we can obtain in different snapshots considering structural neighborhoods and temporal evolution of TKG.

4.3 Decoder and Training

$\phi(.)$ Denotes score function of triple, and let DEC denote any appropriate decoding function of KG, e.g., the TransE decoder. The score function can be defined as follows:

$$\phi(s, r, o, t) = DEC(\mathbf{Z}_{s,t}, \mathbf{Z}_r) \tag{16}$$

$\mathbf{Z}_{s,t}$ represents embedding of subject entity s and $\mathbf{Z}_r$ is embedding obtained from learning of the relation r. To train model using this scoring function, parameters of learned model are learned using gradient-based small batch optimization. For each triple $\eta = (s, r, o)\epsilon D^t$, we sample a set of negative sample entities $D_\eta^- = \{o'|(s, r, o') \notin D^t\}$, and cross-entropy loss is defined as follows:

$$L = -\sum_{t=1}^{T} \sum_{\eta\epsilon D^{(t)}} \frac{\exp(\phi(s, r, o, t))}{\sum_{o'\epsilon D_\eta^-} \exp(\phi(s, r, o', t))} \tag{17}$$

5 Experiments

5.1 Experiment Dataset

We used an event-based dataset ICEWS18 and two time-associated datasets with meta-facts, WIKI and YAGO. Information about datasets is summarized in following table (Table 1).

Table 1. Statistics of datasets.

Datasets	Entity	Relation	Fact		
			Train	Valid	Test
ICEWS18	23033	256	8.27 MB	1.03 MB	1.11 MB
WIKI	12554	24	8.83 MB	1.20 MB	1.12 MB
YAGO	10623	10	2.71 MB	2.71 MB	384.6 KB

5.2 Evaluation Metrics

For each test fact (s, r, o, t), we replace object entity with all possible candidate entities. Then, candidate facts and original object facts are ranked in descending order of their scores. We use two metrics to evaluate performance of each model. One is mean reciprocal rank (MRR) defined as $MRR = \frac{1}{|Test|} \sum_{(s,r,o,t) \in Test} \frac{1}{rank(s,r,o,t)}$, and higher MRR indicates better model performance. Another metric defined as $Hits@N = \frac{1}{|Test|} \sum_{(s,r,o,t) \in Test} ind(rank(s, r, o, t) \leq N)$, where $ind()$ is 1 if the inequality holds and 0 otherwise.

5.3 Baseline Methods

- **R-GCN** [14]: This method introduces relation-specific conversions that depend on the type of edge and direction of edge, which is the first work to use GCN for modeling relational data.
- **DistMult** [7]: This method uses bilinear diagonal to learn facts embedding, whose scoring function captures only pairwise interactions between head entity s and tail entity o components along same dimension.
- **ConvE** [12]: This method uses two-dimensional convolution on embedding, which is the simplest multilayer convolutional architecture for link prediction.
- **TA-DistMult** [18]: This method constructs element temporal embedding by encoding temporal annotations using an RNN model, and uses DistMult as scoring function.
- **HyTE** [17]: This method represents timestamp as a hyperplane and projects entity and relation representations onto hyperplane to obtain new representation
- **EvolveGCN** [19]: This method captures the evolution of snapshot sequence through GCN and uses RNN to evolve the GCN parameters.
- **DyRep** [1]: This method proposes an algorithm for inductive learning, which no longer learns fixed embedding of nodes, but a way of node embedding, so that new node embedding can be easily obtained even if structure is changed.
- **RE-NET** [2]: This method employs recurrent event encoder to encode past facts, and uses R-GCN aggregator to model the connection of facts at the same timestamp.

5.4 Performance Comparison

Overall Performance: As shown in Fig. 2, we compared our proposed method with other baseline methods. In comparison with our method, static methods perform poorly because they don't take into account time factor. Moreover, MEFGNN outperforms other dynamic methods, including TA-DistMult, HyTE, and other dynamic methods, which also shows the effectiveness of the MEFGNN. DyRep+MLP does not fully utilize multi-hop structural information. RE-NET is structured with R-GCN aggregators and RNN encoders, whose scalability is poor as the time step increases, so it does not perform as well as our method. The improvement of Hits@3 on the YAGO dataset is the highest, which may be because YAGO dataset has a problem of temporal sparsity of entity distribution. According to experiments, our proposed framework shows a significant improvement over all baseline methods, which demonstrates the superiority of our method in obtaining more accurate representation of the TKG (Table 2).

Table 2. Comparison of different methods on three datasets for link prediction. The best results in each column is boldfaced.

Method	ICEWS18			WIKI			YAGO		
	MRR	H@3	H@10	MRR	H@3	H@10	MRR	H@3	H@10
R-GCN	23.19	25.34	36.48	37.57	39.66	41.90	41.30	44.44	52.68
DistMult	22.16	26.00	42.18	46.12	49.81	51.38	59.47	60.91	65.26
ConvE	36.85	39.92	50.54	47.55	49.78	49.92	62.66	63.36	65.57
TA-DistMult	25.83	31.57	44.96	48.09	49.51	51.70	61.72	63.32	35.19
HyTE	7.31	7.50	14.95	43.02	45.12	49.49	23.16	45.74	51.94
EvolveGCN	16.59	18.32	34.01	0.09	0.03	0.10	0.07	0.00	0.04
DyRep+MLP	9.87	10.66	18.66	11.61	12.73	21.65	5.87	6.54	11.97
RE-NET	42.92	**45.47**	55.80	51.97	52.07	53.91	65.16	65.63	68.08
EFGNN	**43.32**	44.88	**56.58**	**52.84**	**52.60**	**54.18**	**65.64**	**67.37**	**68.38**

Ablation Study: As shown in Table 3, our method is superior to baseline method with different graph neural network models. We replaced MGNN with GAT, GCN respectively. Since GAT and GCN only consider adjacent entities and ignore relationship information, they are much lower than our method, which shows the necessity of considering multi-hop structural information. We compared our method with three variants without local triple features, local path features, and global features, respectively. First, all these variants fail to outperform our original method, which indicates that all three features are valid and contribute to the final performance. In addition, the variant without triple features has the largest performance degradation, since triple features provide the most intuitive correlation between entities and their neighbors.

Table 3. Performance of different variants of our model for link prediction.

Variants	MRR	H@1	H@3	H@100
GCN – EFAN	36.32	19.66	22.36	54.36
GAT – EFAN	38.27	24.35	28.61	56.61
No local triple feature	36.31	34.46	35.61	49.92
No local path feature	42.45	36.19	44.09	58.33
No global information feature	43.68	36.24	44.74	58.20
MEFGNN	44.32	36.61	44.88	58.58

5.5   Parameter Analysis

Dimension of Embedding Representation. We analyze the performance of MEFGNN framework for three datasets varying dimensionality of embedding representation. As shown in Fig. 3(a), the performance of method increases and then decreases for each dataset as dimension d increases. This is because when d is too small, framework is unable to obtain rich information from TKG, and when d is too large, overfitting problem will occur. Therefore it is more appropriate to set dimension d to 200.

Multi-head Attention. To analyze the benefits of modeling via multi-headed attention, we independently vary number of structural heads in EFAN in the range {1, 2, 4, 8, 16}. From Fig. 3(b), we observe that EFAN benefits from multi-head attention. The performance stabilizes with 8 attention heads, which appears sufficient to capture graph evolution from multiple latent aspects.

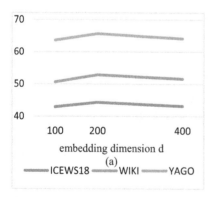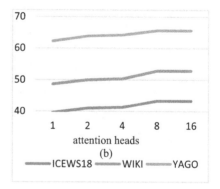

Fig. 3. Influence of embedding dimension and attention heads.

6   Conclusion and Outlook

In this paper, we propose MEFGNN framework for TKG link prediction. Our method emphasizes the importance of multi-hop structural information and historical global feature. We first learn the representation of node structure information using MGNN, and then integrate the representation of entity information across time using EFAN. This temporal information integration explicitly addresses the problem of temporal sparsity of entity distribution in TKG. We tested our proposed method on a link prediction task with three benchmark datasets. The experimental results demonstrate the superiority of our method and the effectiveness of various components. In future work, we will explore continuous time generalization to include more fine-grained temporal variations.

Acknowledgment. This work was supported by the National Natural Science Foundation of China (62172082, 62072084,62072086), the Fundamental Research Funds for the central Universities (N2116008).

References

1. Anonymous: Learning representation over dynamic graph. arXiv preprint arXiv:2106.01678 (2016)
2. Jin, W., Qu, M., Jin, X., et al.: Recurrent event network: autoregressive structure inferenceover temporal knowledge graphs. arXiv preprint arXiv:1904.05530 (2020)
3. Zaremba, W., Sutskever, I., Vinyals, O.: Recurrent neural network regularization. arXiv preprint arXiv:1409.2329 (2014)
4. Bordes, A., Usunier, N., Garcia-Duran, A., et al.: Translating embeddings for modeling multi-relational data. In: Curran Associates Inc., pp. 1–9 (2013)
5. Wang, Z., Zhang, J., Feng, J., et al.: Knowledge graph embedding by translating on hyperplanes. In: AAAI, pp. 1112–1119 (2014)
6. Lin, Y., Liu, Z., Sun, M., Liu, Y., Zhu, X.: Learning entity and relation embeddings for knowledge graph completion. In: AAAI, pp. 2181–2187 (2015)
7. Li, C., Zhai, R., Zuo, F., Yu, J., Zhang, L.: Mixed multi-channel graph convolution network on complex relation graph. In: Xing, C., Fu, X., Zhang, Y., Zhang, G., Borjigin, C. (eds.) WISA 2021. LNCS, vol. 12999, pp. 497–504. Springer, Cham (2021). https://doi.org/10.1007/978-3-030-87571-8_43
8. Yang, B., Yih, W, T., He, X., et al.: Embedding entities and relations for learning and inference in knowledge bases. arXiv preprint arXiv:1412.6575 (2014)
9. Nickel M., et al.: A three-way model for collective learning on multi-relational data. In: International Conference on Machine Learning, pp. 438–445(2011)
10. Trouillon, T., Welbl, J., Riedel, S., Gaussier, É., Bouchard, G.: Complex embeddings for simple link prediction. In: ICML, pp. 2071–2080 (2016)
11. Nickel, M., Rosasco, L., Poggio, T.A.: Holographic embeddings of knowledge graphs. In: AAAI, pp. 1955–1961 (2016)
12. Dettmers, T., Minervini, P., Stenetorp, P., Riedel, S.: Convolutional 2D knowledge graph embeddings. In: AAAI, pp. 1811–1818 (2018)
13. Seo, Y., Defferrard, M., Vandergheynst, P., Bresson, X.: Structured sequence modeling with graph convolutional recurrent networks. In: Cheng, L., Leung, A.C.S., Ozawa, S. (eds.) ICONIP 2018. LNCS, vol. 11301, pp. 362–373. Springer, Cham (2018). https://doi.org/10.1007/978-3-030-04167-0_33

14. Schlichtkrull, M., Kipf, T.N., Bloem, P., van den Berg, R., Titov, I., Welling, M.: Modeling relational data with graph convolutional networks. In: Gangemi, A., et al. (eds.) ESWC 2018. LNCS, vol. 10843, pp. 593–607. Springer, Cham (2018). https://doi.org/10.1007/978-3-319-93417-4_38

15. Velikovi, P., et al.: Graph attention networks. arXiv preprint arXiv:1710.10903 (2018)

16. Jiang, T., et al.: Encoding temporal information for time-aware link prediction. In: EMNLP, pp. 2350–2354 (2016)

17. Dasgupta, S.S., Ray, S.N., Talukdar, P.P.: HyTE: hyperplane-based temporally aware knowledge graph embedding. In: EMNLP, pp. 2001–2011 (2018)

18. García-Durán, A., Dumani, S., Niepert, M.: Learning sequence encoders for temporal knowledge graph completion. arXiv preprint arXiv:1809.03202 (2018)

19. Pareja A., Domeniconi G., Chen J., et al.: EvolveGCN: evolving graph convolutional networks for dynamic graphs. arXiv preprint arXiv:1902.10191 (2019)

20. Sun Z., Deng Z H., Nie J Y., et al.: RotatE: knowledge graph embedding by relational rotation in complex space. arXiv preprint arXiv:1902.10197 (2019)

21. Kipf, T, N., Welling, M.: Semi-Supervised Classification with Graph Convolutional Networks. arXiv preprint arXiv:1609.02907 (2016)

22. Cho K., Merrienboer, B, V., et al.: Learning phrase representations using RNN encoder-decoder for statistical machine translation. arXiv preprint arXiv:1406.1078 (2014)

23. Vaswani, A., Shazeer, N., Parmar, N., et al.: Attention is all you need. arXiv preprint arXiv: 1706.03762 (2017)

Semantic Reasoning Technology on Temporal Knowledge Graph

Jianuo Li[(✉)], Feng Zhao, and Hai Jin

National Engineering Research Center for Big Data Technology and System,
Services Computing Technology and System Lab, Cluster and Grid Computing Lab,
School of Computer Science and Technology, Huazhong University of Science and
Technology, Wuhan, China
jianuoli@hust.edu.cn

Abstract. Semantic reasoning techniques based on knowledge graphs have been widely studied since they were proposed. Previous studies are mostly based on closed-world assumptions, which cannot reason about unknown facts. To this end, we propose the *Two-Stage Temporal Reasoning Model* (TSTR) for reasoning about future facts. In the first stage, probability of future facts occurring is reasoned using repeated information in history. In the second stage, the semantics of the neighborhood nodes are aggregated using the structural encoder and the temporal information is captured using the temporal encoder. The predicted probabilities are obtained by the decoder. Finally, the candidate entity probabilities of the two-stage reasoning are weighted to achieve the prediction of the two-stage fusion. We tested the performance of the TSTR on public datasets and the results demonstrated the effectiveness of the TSTR.

Keywords: Temporal knowledge graph · Semantic reasoning · Recurrent neural network

1 Introduction

Currently, most research is mainly based on closed-world assumptions [1], focusing on the static knowledge graph. However, real-world events and facts are often time-dependent [2]. The introduction of temporal information is essential for the knowledge graph, and to represent the knowledge of temporal associations. Facts are usually coupled with timestamps to indicate that the facts are valid in the temporal knowledge graph.

Interpolation [3] and extrapolation [4] could be divided based on the TKG. Interpolation targets missing facts in the range of known facts, and extrapolation targets future time ranges for future missing facts. Most of the existing work focuses on interpolation, however, extrapolation helps to predict future occurring facts and is more challenging.

At present, there are already models based on extrapolation reasoning. However, these approaches have the following challenges: (1) Insufficient use of historical information. Most of the models have used recurrent neural networks,

© The Author(s), under exclusive license to Springer Nature Switzerland AG 2022
X. Zhao et al. (Eds.): WISA 2022, LNCS 13579, pp. 115–125, 2022.
https://doi.org/10.1007/978-3-031-20309-1_10

but almost these models ignored distant historical factual information and cannot use distant historical information to model entities and relationships. (2) Ignoring node importance. Most of the current models ignored the node importance issue with respect to the neighbor structure information. (3) Modeling of inactive entities. An entity is active if it has at least one neighboring entity. Entities and relationships are dynamically changing according to time, and only a small number of entities are active under each timestamp, and it is a challenge to model inactive entities. In order to solve the above problems, we focus on extrapolation and propose the TSTR model. TSTR model consists of two stages, historical reasoning and temporal reasoning respectively. Finally, we fuse the two stages to achieve reasoning about future facts

Contributions. Our contributions are summarized as follows:

- We propose the TSTR model, which includes two stages of historical reasoning and temporal reasoning, and integrate two approaches to accomplish the reasoning task together.
- We design and implement an attention-based multi-relational graph encoder, which takes into account the influence of relation on node feature extraction and aims to aggregate information from multi-relation and multi-hop neighbors.
- We design interpolation optimization algorithms for inactive entities to solve the temporal sparsity problem that exists in the temporal knowledge graph. In addition, a weighted attenuation mechanism is designed to explain the diminishing effect of historical facts.

2   Related Work

Static Knowledge Graph Reasoning. Bordes et al. proposed the famous translation model TransE in 2013 [5], although TransE model is a classical model, it ignores the diversity of relations and can only deal with one-to-one problems. TransH model [6] proposes to map multi-relational entities and relations into the hyperplane space with different representations of entities corresponding to different relations. The purpose is to fuse the mapping properties between relations. The RotatE [7] model is based on rotation, which inspired by the Euler decomposition function. Besides the RotatE model, the rotation models also include QuatE [8] and DihEdral [9], all of which utilize the rotation property. The RESCAL model [10] is the pioneer of bilinear models, and the model is mainly based on the idea of semantic matching, which measures the size of the probability of triples holding by computing the potential semantic information. DisMult [11] improves the RESCAL model by adding rule-based mining with good link prediction performance. The ComplEx [12] introduces the complex number space into the bilinear model for the first time. In addition, CrossE [13], TuckER [14] and other bilinear models have also achieved good results. ConvKB [15], ConvE [16], and ConvR [17] are based on convolutional neural networks. ParamE proposed to map entities to feature spaces, InteractE made improvements for feature interaction aspects.

However, all of the above methods ignore the temporal information and have weak reasoning ability for time-related knowledge graphs.

Temporal Knowledge Graph Reasoning. More and more researchers in knowledge graph reasoning techniques focus on the pattern of facts evolving over time, thus taking temporal factors into consideration. Leblay proposed the TTransE [18] in 2018, which simply extends the TransE-based knowledge representation learning approach. Dasgupta proposed the HyTe model [19], which treats each timestamp as a hyperplane, combining temporal information in the entity-relationship vector space. RE-NET proposes a recurrent event network and defines a recurrent event encoder. Zhu proposed CyGNet [20], which simulates historical recurrent facts using both replication and generation modes to learn from past recurrent facts. RE-GCN [21] proposes a GCN-based recurrent network with a 17 to 82 times faster prediction runtime compared to the RE-Net model. DACHA [22] is a pairwise graph convolutional network-based approach that considers the effect of different temporal relationships on future facts, modeling local and global relationships separately through historical relationship encoders. TITer [23] is a temporal knowledge graph reasoning model based on temporal paths, which integrates temporal information into a reinforcement learning framework.

3   Model

3.1   Problem Definition

The set of facts under the same timestamp is defined as the temporal subgraph G_t. A quadruple can be represented as (s, r, o, t), where s denotes the head entity, r denotes the relationship, o denotes the tail entity, and t denotes the timestamp. TSTR can calculate the probability of each fact occurring at future moments, then rank all possible facts according to their probabilities. We focus on predicting tail entities, it can also be simply extended to predict issues such as head entities and relationships.

3.2   Framework

TSTR model consists of two stages, which is structured as shown in Fig. 1. The quaternions (s, r, o, t) with temporal information are cut into consecutive temporal subgraphs according to the timestamps, and the reasoning of the two stages of TSTR is completed based on the temporal subgraphs.

Stage 1: Historical reasoning. Given specific head entities and relations, the corresponding tail entities are mined as candidate entities in the temporal subgraphs, and the table of historical information in each temporal subgraph is obtained. Through the incremental algorithm, the frequency of candidate entities recurring in history under time can be obtained, and the probability of candidate entities is inferred based on the frequency of factual occurrences.

Stage 2: Temporal reasoning. We encode the quadruplets in the temporal knowledge graph, where the encoder mainly consists of a structural encoder

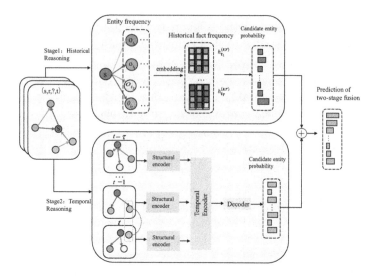

Fig. 1. TSTR model architecture

and a temporal encoder. Finally, we complete temporal reasoning through the decoder.

4   Method

4.1   Historical Reasoning

The historical information can be divided according to the time stamps. Specifically, for each training set, we need to obtain the historical information of all entities and relationships at t_T, $h_{t_T}^{(s,r)}$ is a multi-hot encoded vector containing the set of all tail entities corresponding to s and r in G_T, and each corresponding tail entity is represented by 1 in the history fact information and the other entities are represented by 0. In addition, we adopt the incremental learning, in which the historical information of each previous moment $\{h_{t_1}^{(s,r)}, h_{t_2}^{(s,r)} \ldots h_{t_T-1}^{(s,r)}\}$ is retained at t_T. Similar to recursion, to get $h_{t_T}^{(s,r)}$ only needs to accumulate the results of previous moments of training to get the historical information at t_T. By establishing the historical information under each timestamp, temporal subgraph can be downscaled to provide support for the subsequent calculation.

Under the timestamp t_k, assuming the existence of a specific head entity s and relationship r, the corresponding historical information $h_{t_k}^{(s,r)}$ can be obtained. We create a correction index vector v, which can be created to make corrections to the known facts in the temporal knowledge graph. The index vector v is generated using the fully connected layer MLP, and the vector size of v is the same as the size $h_{t_k}^{(s,r)}$, W_c and b_c are trainable parameters, $W_c \in R^{3d^*N}$, $b_c \in R^N$, N is the number of entities.

$$v = \tanh\left(W_c\left[s, r, t_k\right] + b_c\right) \tag{1}$$

To reduce the effect of the fact that the tail entity does not constitute the fact with s and r, we correct the known facts in $h_{t_k}^{(s,r)}$. The above entities are regarded as uninteresting entities, increasing the gap between uninteresting entities and entities that constitute facts with s and r. So we modify the index value of the uninteresting entity in v, the probability of the occurrence of the uninterested entity in the future is minimized, and the final modified historical information is denoted by $c_{t_k}^{(s,r)}$.

When given a query $(s, r, ?, t_k)$, the modified historical information $c_{t_k}^{(s,r)}$ can be obtained, and using the $softmax$ function, the probability $p(h)$ of the occurrence of each candidate entity in the historical information at t_k can be predicted as follow:

$$p(h) = softmax(c_{t_k}^{(s,r)}) \tag{2}$$

4.2 Temporal Reasoning

Structural Encoder. We design a multi-relational graph encoder based on the attention mechanism (RGAT encoder). RGAT encoder structure is shown in Fig. 2.

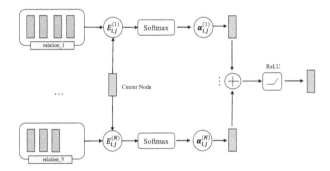

Fig. 2. Structural encoder architecture

The input layer of the RAGT encoder is a vector with N entity node feature values. The feature of the entity node e_i at timestamp t is represented as $h_{i,t}$. After the propagation layer, $x_{i,t}$ can be obtained in the output layer. To distinguish the different types of relationships, we introduce a relationship-based intermediate expression value for the nodes, using a weight matrix $W^{(r)}$ to assign intermediate values $g_{i,t}^{(r)}$, $G^{(r)} = \left[g_{1,t}^{(r)}, g_{2,t}^{(r)} \cdots g_{N,t}^{(r)}\right]$ to the nodes under each relationship.

The attention coefficient is calculated for the neighboring nodes of the central node. The attention coefficients between each pair of nodes are relatively independent, and the importance of node e_j to the central node e_i under the relation r denoted by E_{ij}. $LeakyReLu$ function is used on nonlinearization process. In addition, we normalize the coefficients of e_i. Finally, the output feature vector $x_{i,t}$ of the central node e_i is obtained by the activation function, as defined below:

$$x_{i,t} = \sigma \left(\sum_{j \in \mathcal{N}_i} Softmax(E_{ij}^{(r)}) h_{j,t} \right) \qquad (3)$$

Temporal Encoder. The design of the temporal encoder is shown in Fig. 3. The static vector representation of the entity e_i in the temporal subgraph is obtained. Subsequently, the static vector representation of entity nodes in consecutive timing subgraphs is integrated using GRU. $s_{i,t-1}$ represents the hidden layer output at the $t-1$ moment. The hidden state of the entity node at each moment and the output $s_{i,t}$ at moment t can be obtained at the hidden layer. To calculate the attention weight coefficients of entity nodes before moment t, the temporal encoder inputs S_i to the attention weight calculation layer and multiplies the corresponding parameter matrices W_q, W_k, and W_v to obtain the three vector matrices Q, K, V, respectively. Putting S_i respectively, the vector $s_{i,t}$ are calculated to obtain the attention weights $\alpha_{i,j}$ of entity node e_j and entity node e_i at the moment of t. Finally, the attention weights of entity node e_i can be obtained in dynamic vector $z_{i,t}$ as follow:

$$z_{i,t} = \sum_{j} \alpha_{i,j} v_{i,t} \qquad (4)$$

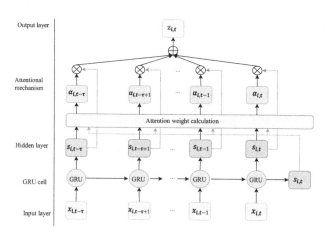

Fig. 3. Temproal encoder architecture.

Update for Inactive Entities. An inactive entity node is a random isolated node with no neighbors in the temporal subgraph, and the structure encoder only encodes based on the neighbors. After τ time steps, at the next active timestamp $t + \tau$, the entity nodes will share the vector representation of the previous moment. So we propose the vector update method for inactive entities.

Let t^- denote the last active timestamp of entity e_i before the current timestamp, and the static vector of the corresponding entity node under its timestamp is denoted as x_{i,t^-}. According to the memory mechanism of human brain, when reasoning with "slow thinking", it is better to reason based on the recent facts, and the facts of long time will be blurred in the memory. The decay rate of θ_{i,t^-} is a monotonically decreasing decay rate with respect to the time difference, ranging from 0 to 1. The updated vector $\hat{x}_{i,t}$ of the inactive entity combines the static vector representations under both timestamps x_{i,t^-} and $x_{i,t}$. $x_{i,t}$ of the inactive entity is replaced by $\hat{x}_{i,t}$, which is calculated as follows:

$$\theta_{i,t^-} = \exp\left\{ -\max\left(0, \lambda_x \left| t - t^- \right| + b_x \right) \right\} \tag{5}$$

$$\hat{x}_{i,t} = \theta_{i,t^-} x_{i,t^-} + \left(1 - \theta_{i,t^-}^x \right) x_{i,t} \tag{6}$$

4.3 Prediction for Two-Stage Integration

TSTR uses the TransE decoder, and the decoder is represented by DEC, then the score probability $p(t)$ of all candidate entities can be obtained by the TransE decoder. $z_{s,t}$, z_r, $z_{o,t}$ are vector of head entities, relations, and tail entities.

$$p(t) = \sigma\left(DEC\left(z_{s,t}, z_r, z_{o,t} \right) \right) \tag{7}$$

For each triple $\eta = (s, r, o) \in G(t)$, a set of negatively sampled triple $D_\eta^- = \left\{ o' \mid (s, r, o') \notin G^{(t)} \right\}$ is selected, where $\phi(.)$ denotes the score of a set of triple, and the cross-entropy loss function L is defined as:

$$L = -\sum_{t=1}^{T} \sum_{\eta \in G^{(t)}} \frac{\exp(\phi(s, r, o, t))}{\sum_{o' \in G_\eta^-} \exp\left(\phi\left(s, r, o', t \right) \right)} \tag{8}$$

Both historical reasoning and temporal reasoning calculate the candidate entity probabilities, and we combine the results of the two stages of TSTR to weight the candidate tail entity probabilities. In addition, we introduce a weighting factor ϖ to adjust the weights of the results of the two inference models, $\varpi \in [0, 1]$.

$$p(o \mid s, r, t) = \varpi * p(h) + (1 - \varpi) * p(t) \tag{9}$$

$$o_t = \operatorname{argmax}_{o \in E} p(o \mid s, r, t) \tag{10}$$

5 Experiments

5.1 Experiment Setup

Datasets and Metrics. We use three publicly available benchmark datasets, ICEWS14 [3], ICEWS18 [24], and GDELT [25]. The ICEWS dataset consists mainly of political facts. ICEWS18 and ICEWS14 are sub-datasets of ICEWS, extracted from two time periods respectively, and divided by a time interval of 24 hours. GDELT is derived from the news media and divided by a time interval of 15 minutes. Three datasets are divided into three sub-datasets by timestamps. In order to evaluate the TSTR model fairly and comprehensively, three measures: MRR (*Mean Reciprocal Ranks*), Hits@1, and Hits@10, are used to evaluate the model experimentally.

Baselines. We compare TSTR with traditional models and temporal models respectively.

- TransE. TransE is a translation-based model whose main idea is to reduce the dimensionality of entities and relations to vector space.
- RotatE. It is a rotation-based knowledge representation learning model, using the rotation feature.
- RGCN. The model introduced GNN to the domain of relational network-based knowledge graphs.
- ConvE. The model applied CNN to knowledge representation learning model based on knowledge graphs, and uses 2d convolution.
- HyTe. HyTe takes into account the important value of temporal information by projecting the triples onto the hyperplane.
- RE-NET. RE-NET obtains temporal evolution information, which is a classical model for extrapolation tasks based on temporal knowledge graphs.
- CyGNet. CyGNet is divided into two modes: copy and generation, and has achieved better results in learning knowledge from facts that have appeared frequently in the past.

5.2   Model Comparison

Performance Comparison. We evaluate the TSTR model using an extrapolated entity link prediction task. The performance of the TSTR model and other models on the temporal datasets are shown in Table 1.

Table 1. Results (in percentage) on three datasets

Methods	ICEWS14			ICEWS18			GDELT		
	MRR	H@1	H@10	MRR	H@1	H@10	MRR	H@1	H@10
TransE	18.6	1.1	47.1	17.5	2.5	43.9	16.3	0.0	42.2
RotatE	27.5	18.0	47.2	15.5	7.0	33.9	5.2	1.1	12.3
RGCN	27.1	18.4	44.2	17.0	8.7	34.0	10.5	4.5	22.5
ConvE	30.9	21.7	50.1	24.8	15.1	44.9	17.2	10.3	32.2
HyTe	11.5	5.6	22.5	7.3	7.3	7.5	6.5	0.0	18.5
RE-NET	38.9	29.3	57.5	28.4	18.4	47.9	19.1	11.3	34.3
CyGNet	36.5	27.4	54.4	26.8	17.1	45.7	18.2	10.6	31.5
TSTR	**40.4**	**30.5**	**60.2**	**29.7**	**19.2**	**49.5**	**18.2**	**10.2**	**32.8**

It can be found that although the HyTe model takes into account the temporal order information, its actual performance is still poor. Both RE-NET model and CyGNet model achieve better performance results. TSTR performs significantly better than other benchmark comparison models on three datasets, and TSTR performs better on the ICEWS14 dataset and ICEWS18 dataset. On average, with MRR improvement of more than 3%, Hits@1 improvement of more than 4%, and Hits@10 improvement of more than 4%, validating the effectiveness of the TSTR model.

TSTR did not achieve good results on the GDELT dataset. Most of the data in the GDELT dataset are relatively abstract concepts, and the represented entities are not targeted. these abstract concepts bring great difficulties to the extrapolation task of the TSTR model, which is the key reason for the poor performance of models such as TSTR on this dataset.

Table 2. Results (in percentage) by different stages of TSTR model on three datasets

Models	ICEWS14	ICEWS18	GDELT
TSTR-Stage1-only	28.7	19.8	12.3
TSTR-Stage2-only	36.4	25.9	15.2
TSTR	40.2	29.1	17.8

Ablation Study. We validate the performance of the two stages of TSTR using MRR on three datasets separately, and present the results in Table 2. Using only the stage 1 module cannot achieve the desired effect. In addition, stage 1 module outperforms on three datasets, which indicates modeling the historical repeated fact information is still useful for the extrapolation task of the TSTR model.Using only stage 2 also has an impact on the model effect, indicating that stage 2 is important for TSTR. Because stage 2 not only captures the dynamic evolution information of the facts using the temporal encoder, but also models the entity features in depth using the structural encoder.

6   Conclusion

We propose an approach to solve the task of inference of knowledge graphs oriented to temporal association from a novel cognitive perspective, and design a two-stage temporal inference model. Although good results were achieved, there are still some limitations and need to go further to improve it. We propose some ideas for future work, in which a path-based inference approach can be investigated to perform multi-hop inference in a large amount of historical information by means of reinforcement learning to find more relevant information to assist inference, and the interpretability of the model can be enhanced.

References

1. Cadoli, M., Lenzerini, M.: The complexity of propositional closed world reasoning and circumscription. J. Comput. Syst. Sci. **48**(2), 255–310 (1994)
2. Wang, Y., Gao, S., Li, W., Jiang, T., Yu, S.: Research and application of personalized recommendation based on knowledge graph. In: Proceedings of the Eighteenth WISA, pp. 647–658 (2021)
3. Trivedi, R., Dai, H., Wang, Y., Song, L.: Know-evolve: deep temporal reasoning for dynamic knowledge graphs. In: Proceedings of the 34th International Conference on Machine Learning, pp. 3462–3471 (2017)
4. Goel, R., Kazemi, S.M., Brubaker, M.A., Poupart, P.: Diachronic embedding for temporal knowledge graph completion. In: Proceedings of the Thirty-Fourth AAAI Conference on Artificial Intelligence, pp. 3988–3995 (2020)
5. Bordes, A., Usunier, N., García-Durán, A., Weston, J., Yakhnenko, O.: Translating embeddings for modeling multi-relational data. In: Advances in Neural Information Processing Systems, pp. 2787–2795 (2013)
6. Wang, Z., Zhang, J., Feng, J., Chen, Z.: Knowledge graph embedding by translating on hyperplanes. In: Proceedings of the Twenty-Eighth AAAI Conference on Artificial Intelligence, pp. 1112–1119 (2014)
7. Sun, Z., Deng, Z., Nie, J., Tang, J.: Rotate: knowledge graph embedding by relational rotation in complex space. In: Proceedings of the 7th International Conference on Learning Representations (2019)
8. Zhang, S., Tay, Y., Yao, L., Liu, Q.: Quaternion knowledge graph embeddings. In: Advances in Neural Information Processing Systems, pp. 2731–2741 (2019)
9. Xu, C., Li, R.: Relation embedding with dihedral group in knowledge graph. In: Proceedings of the 57th Conference of the Association for Computational Linguistics, pp. 263–272 (2019)
10. Nickel, M., Tresp, V., Kriegel, H.: A three-way model for collective learning on multi-relational data. In: Proceedings of the 28th International Conference on Machine Learning, pp. 809–816 (2011)
11. Yang, B., Yih, W., He, X., Gao, J., Deng, L.: Embedding entities and relations for learning and inference in knowledge bases. In: Proceedings of the 3rd International Conference on Learning Representations (2015)
12. Trouillon, T., Welbl, J., Riedel, S., Gaussier, É., Bouchard, G.: Complex embeddings for simple link prediction. In: Proceedings of the 33rd International Conference on Machine Learning, pp. 2071–2080 (2016)
13. Zhang, W., Paudel, B., Zhang, W., Bernstein, A., Chen, H.: Interaction embeddings for prediction and explanation in knowledge graphs. In: Proceedings of the Twelfth ACM International Conference on Web Search and Data Mining, pp. 96–104 (2019)
14. Balazevic, I., Allen, C., Hospedales, T.M.: Tucker: tensor factorization for knowledge graph completion. In: Proceedings of the 2019 Conference on Empirical Methods in Natural Language Processing and the 9th International Joint Conference on Natural Language Processing, pp. 5184–5193 (2019)
15. Nguyen, D.Q., Nguyen, T.D., Nguyen, D.Q., Phung, D.Q.: A novel embedding model for knowledge base completion based on convolutional neural network. In: Proceedings of the 2018 Conference of the North American Chapter of the Association for Computational Linguistics: Human Language Technologies, pp. 327–333 (2018)
16. Dettmers, T., Minervini, P., Stenetorp, P., Riedel, S.: Convolutional 2D knowledge graph embeddings. In: Proceedings of the Thirty-Second AAAI Conference on Artificial Intelligence, pp. 1811–1818 (2018)

17. Jiang, X., Wang, Q., Wang, B.: Adaptive convolution for multi-relational learning. In: Proceedings of the 2019 Conference of the North American Chapter of the Association for Computational Linguistics: Human Language Technologies, pp. 978–987 (2019)
18. Leblay, J., Chekol, M.W.: Deriving validity time in knowledge graph. In: Proceedings of the Companion of the The Web Conference 2018, pp. 1771–1776 (2018)
19. Dasgupta, S.S., Ray, S.N., Talukdar, P.P.: HyTE: hyperplane-based temporally aware knowledge graph embedding. In: Proceedings of the 2018 Conference on Empirical Methods in Natural Language Processing, pp. 2001–2011 (2018)
20. Zhu, C., Chen, M., Fan, C., Cheng, G., Zhang, Y.: Learning from history: modeling temporal knowledge graphs with sequential copy-generation networks. In: Proceedings of the Thirty-Fifth AAAI Conference on Artificial Intelligence, pp. 4732–4740 (2021)
21. Li, Z., et al.: Temporal knowledge graph reasoning based on evolutional representation learning. In: Proceedings of the 44th International ACM SIGIR Conference on Research and Development in Information Retrieval, pp. 408–417 (2021)
22. Chen, L., Tang, X., Chen, W., Qian, Y., Li, Y., Zhang, Y.: DACHA: a dual graph convolution based temporal knowledge graph representation learning method using historical relation. ACM Trans. Knowl. Discov. Data **16**(3), 46:1–46:18 (2022)
23. Sun, H., Zhong, J., Ma, Y., Han, Z., He, K.: TimeTraveler: reinforcement learning for temporal knowledge graph forecasting. In: Proceedings of the 2021 Conference on Empirical Methods in Natural Language Processing, pp. 8306–8319 (2021)
24. García-Durán, A., Dumancic, S., Niepert, M.: Learning sequence encoders for temporal knowledge graph completion. In: Proceedings of the 2018 Conference on Empirical Methods in Natural Language Processing, pp. 4816–4821 (2018)
25. Qiao, F., Chen, K.: Correlation and visualization analysis of large scale dataset GDELT. In: Proceedings of the 2016 International Conference on Identification, Information and Knowledge in the Internet of Things, pp. 68–72 (2016)

Multiple-Granularity Graph
for Document-Level Relation Extraction

Jiale Zhang, Mengqi Liu, and Lizhen Xu$^{(\boxtimes)}$

Department of Computer Science and Engineering, Southeast University,
Nanjing 21189, China
{220201870,lzxu}@seu.edu.cn, 1210199758@qq.com

Abstract. As an important branch of relation extraction, document-level relation extraction aims to extract relations among entities in the document. Compared with sentence-level relation extraction, document-level relation extraction relies on the implicit semantic connections between sentences to obtain relation between entities. In this paper, we propose Multiple-Granularity Graph (MGG) to extract relations within document, which uses encoding module, graph construction module, relation inference module and classification module. In the graph construction module, we define four node types, which are mention node, entity node, sentence node and document node. Based on these nodes, seven edge types are defined, which are Mention-Mention edge, Mention-Edge edge, Mention-Sentence edge, Mention-Document edge, Entity-Sentence edge, Sentence-Sentence edge and Entity-Entity edge. We then use path reasoning mechanism with attention mechanism to calculate the representation of Entity-Entity edge. Finally we judge the relation between entities by classification module. The experiment result shows that this method has achieved 62.77%, 63.11% and 62.94% in accuracy, recall and F1 value on the task of document-level relation extraction. Compared with other traditional methods, MMG can significantly improve the performance of document-level relation extraction.

Keywords: Document-level relation extraction · Document graph · Path reasoning mechanism

1 Introduction

As one of the research directions of natural language processing, the main goal of Relation Extraction (RE) is to identify semantic relations between entities. RE plays an important role in many artificial intelligence fields, such as knowledge graph construction [1], text generation, question answering system.

There are two mainstream research directions in relation extraction-sentence-level relation extraction and document-level relation extraction. Compared with sentence-level relation extraction, there are three difficulties in document-level relation extraction [2]. Firstly, there are more entities in document-level relation extraction text, so the relations between different entity pairs need to be considered. Secondly, entity might appear in many sentences in different form. Finally,

X. Zhao et al. (Eds.): WISA 2022, LNCS 13579, pp. 126–134, 2022.
https://doi.org/10.1007/978-3-031-20309-1_11

the transitivity of the relation needs to be taken into consideration. There are two main research methods for document-level relation extraction: document graph-based methods and sequence-based methods. We propose a Multi-Granularity Graph (MGG) model based on the document graph method. It obtains the relationship between entity pairs through the path reasoning mechanism.

2   Related Work

In recent years, many scholars and researchers pay large attention to document-level relation extraction. Yao [3] et al. published a large-scale document-level relation extraction dataset, where more than 40.7% of relation can only be extracted from multiple sentences.

For sequence-based method, Jia [4] et al. proposed a document-level n-ary relation extraction model that incorporates representations learned at different levels of text and sub-relationships in a document, which can significantly improve recall. Zhou [5] et al. proposed adaptive thresholding and local context pooling techniques to solve multi-label and multi-entity problems.

For document graph-based method, Song [6] et al. kept the original graph structure and directly modeled the entire document graph using a graph-state LSTM. Verga [7] et al. proposed a Transformer-based model, and then Sahu [8] et al. turned the Transformer into a graph convolutional neural network GCN. Both models only consider one target entity pair per document, and rely on external parsing tools to build document graphs.

The graph-based approach is easier to capture the interaction information between entities than the sequence-based approach. This paper uses heterogeneous document graph to better capture different dependencies between different nodes. In the document graph, nodes represent one of mentions, entities, sentences and document and the representation of edge depends on the types of nodes that the edge is connected to.

3   Proposed Model

As illustrated in Fig. 1, the proposed model consists of four modules, that is encoding module, graph construction module, relation inference module and classification module.

3.1   Encoding Module

In the encoding module, each word in the sentences of the input document is firstly transformed into a dense vector representation, i.e., a word embedding. Then these vectorized sentences are fed into the encoder, which is a BiLSTM [9, 10]. BiLSTM is a combination of forward LSTM and backward LSTM. BiLSTM calculates the input sequence in order and reverse order to obtain two different hidden layer representations, and then obtains the final hidden layer feature representation by vector splicing.

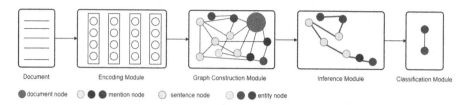

Fig. 1. The overall architecture of MGG. Firstly, an encoding module converts the input document into a contextualized representation of each word. Then, the graph construction module constructs nodes and edges. Next, the inference module searches the optimal path in order to judge the relation between nodes. Finally, the classification module gets the final relation. Edge types are not distinguished for simplicity.

3.2 Graph Construction Module

In the graph construction module, we define the vectorized representations for nodes and edges based on the contextualised representations from encoder. This section consists of two parts, node construction and edge construction. We compose the representations of the graph nodes in the first sub-layer and the representations of the edges in the second.

Node Construction. We define four different types of nodes in the document graph, which are mention nodes (M) n_m, entity nodes (E) n_e, sentence nodes (S) n_s and document node (D) n_d. The representation of each type of node has its own computation. Mention nodes represent different names of the same entity in the document. For each mention node, its representation is the average of representation of words that it contains. Entity nodes represent unique entities in the document. For each entity node, its representation is the average of representation of mentions that it corresponds. Sentence nodes represent sentences in the document. For each sentence node, its representation is the average of representation of words that it contains. Document node represent the input document. For document node, its representation is the average of represent of words it contains.

In summary, the representations of mention nodes, entity nodes, sentence nodes and document node are calculated as follows:

$$n_m = [avg_{w_i \in m}(w_i); t_m] \tag{1}$$

$$n_e = [avg_{m_i \in e}(m_i); t_e] \tag{2}$$

$$n_s = [avg_{w_i \in s}(w_i); t_s] \tag{3}$$

$$n_d = [avg_{w_i \in d}(w_i); t_d] \tag{4}$$

where t_m, t_e, t_s and t_d denote corresponding node type embedding respectively. The reason why they are used is to differentiate the nodes in document graph.

Edge Construction. After the nodes are built, we construct undirected edge between nodes. We cannot know the relation between entities, so we cannot connect directly one entity node to another entity node. Our goal is to find the optimal entity-entity path through these defined edges. The pre-defined edges are following:

Mention-Mention Edge (MM): mentions in a sentence might have implicit interaction connection. Base that, we connect the MM edge between mention nodes if they occur in the same sentence. And the representation of MM edge consists of the representation of mention nodes and the distance of between mentions, that is $e_{mm} = [n_{m_i}; n_{m_j}; dis_{m_i,m_j}]$.

Mention-Sentence Edge (MS): We connect the MS edge between mention node and sentence node if the mention appear in the sentence. And we concatenate the representation of mention node and the representation of sentence node as the representation of MS edge, that is $e_{ms} = [n_m; n_s]$.

Mention-Entity Edge (ME): We connect the ME edge between mention node and entity node if the mention corresponds to the entity. And the representation of ME edge consists of the representation of mention node and the representation of entity node, that is $e_{me} = [n_m; n_e]$.

Mention-Document Edge (MD): document node aims to encode the overall document information. In our opinion, the document node can interact with mention nodes in order to reduce the long distance between nodes that are distant from the other. We connect the MD edge between any mention node and document node. And we concatenate the representation of mention, the representation of sentence that contains this mention and the representation of document, that is $e_{md} = [n_m; n_s; n_d]$.

Sentence-Sentence Edge (SS): We define SS edge to obtain the non-local information. We connect the SS edge between any two sentence nodes. And the representation of SS edge consists of the representation of sentence nodes and the distance between sentences, that is $e_{ss} = [n_{s_i}; n_{s_j}; dis_{s_i,s_j}]$.

Entity-Sentence Edge (ES): Like MS edge, we connect the ES edge between entity node and sentence node if at least one mention that corresponds to this entity occur in the sentence. And we concatenate the representation of entity and the representation of sentence as the representation of ES edge, that is $e_{es} = [n_e; n_s]$.

In order to make the dimensions of different edges equal, we use different linear reduction function for different edge representation:

$$u_x = W_x \cdot e_x \qquad (5)$$

where $W_x \in R^{d_x \times d}$ is a learned matrix and $x \in [MM, MS, ME, MD, SS, ES]$.

3.3   Inference Module

In the inference module, we introduce path reasoning mechanism. After edge construction, there are six types of edges except entity-entity (EE) edge. The role of path reasoning mechanism is to update existing edges and generate EE edge representation. For this purpose, we use three-step algorithm to encode interaction between node and edge in the document graph and model EE edge presentation.

For the first step, we aim to generate a path between node i and node j basing on intermediate node k. we firstly use a transformer function to combine the representation of $e_{i,k}$ and the representation of $e_{k,j}$. Then we combine all existing path between node i and node j through node k. The node i, node j, node k can be any one of mention node, entity node, sentence node, document node. The transformer function is as follows:

$$f(e_{i,k}, e_{k,j}) = \sigma(e_{i,k} \odot (W \cdot e_{k,j})) \tag{6}$$

where σ is a non-linear sigmoid function, $W \in R^{d_z \times d_z}$ is a learned parameter matrix, $\odot$ represents element-wise multiplication and $e_{i,k}$ is the representation of edge between node i and node k.

For the second step, we construct a path representation represented by the representation of head node i, tail node j and k-th path $f(e_{i,k}, e_{k,j})$.

$$P_{i,j}^k = [n_{e_i}; n_{e_j}; f(e_{i,k}, e_{k,j})] \tag{7}$$

We introduce attention mechanism [12], using the entity pair(e_i, e_j) as query, to fuse the information of different paths between node i and node j.

$$s_k = \sigma([n_{e_i}; n_{e_j}] \cdot W_l \cdot P_{i,j}^k) \tag{8}$$

$$\alpha_k = \frac{e^{s_k}}{\sum_t e^{s_t}} \tag{9}$$

$$P_{i,j} = \sum_k \alpha_k P_{i,j}^k \tag{10}$$

where α_i denotes the normalized attention weight for k-th path. Therefore, this module will pay more attention to useful paths. And σ is an activation function.

For the last step, we aggregate the original edge representation and the new updated edge representation with a non-linear function.

$$e_{i,j} = \beta e_{i,j} + (1 - \beta) P_{i,j} \tag{11}$$

where $\beta \in [0,1]$ is a contribution indicator, if β is high, the contribution of original edge representation is high.

3.4   Classification Module

In the classification module, we use a softmax classifier to classify the relation between entities. This classifier uses the representation of EE edge between entity nodes in the graph to compute the classification result of corresponding entities.

$$y = softmax(W_e \cdot e_{ee} + b_e) \tag{12}$$

where $W_e \in R^{r \times d_z}$, $b_e \in R^{r \times 1}$, they are learned parameters of the classification module and r is the number of relation categories.

4   Experiment and Result Analysis

4.1   Dataset and Evaluation Indicators

This paper uses a dataset to test the document-level relation extraction effect of the proposed model. The dataset is DocRED dataset which was created by Yao et al. [2] DocRED is constructed from Wikipedia and Wikidata and contains 96 types of relations, 132, 275 entities, and 56, 354 relational facts in total. Documents in DocRED have about 8 sentences on average. More than 40.7% relation facts can only be extracted from multiple sentences. 61.1% relation instances require various reasoning skills such as logical reasoning. 93.4% intra-sentential relations can be inferred based solely on their cooccurred sentences. We follow the standard split of the dataset, 3, 053 documents for training, 1, 000 for development, and 1, 000 for testing.

In order to evaluate the performance of MGG model, we calculate the precision rate P, recall rate R and F1, and compare with other models. As is shown in Table 1 and following formulas, P denotes the number of correct predictions in the data predicted as keywords. R denotes the number of correct predictions in the data that are truly keywords.

F1 Score is an indicator used in statistics to measure the accuracy of a binary classification model. It takes into account both the precision and recall of the classification model. F1 score can be regarded as a harmonic average of the model's precision and recall, with a maximum value of 1 and a minimum value of 0. Only when precision and recall both are high, F1 score will be high.

Table 1. Necessary parameters for evaluation indicators.

True situation	Predict situation	
	Positive	Negative
Positive	TP (true positive)	FN (false negative)
Negative	FP (false positive)	TN (true negative)

$$P = \frac{TP}{TP + FP} \tag{13}$$

$$R = \frac{TP}{TP + FN} \tag{14}$$

$$F1 = \frac{2 \times P \times R}{P + R} \tag{15}$$

4.2 Experiment Result and Analysis

This paper proposes a document-level relation extraction model based on a document graph. We select 4 deep neural network models related to document-level relation extraction to compare with the model proposed in the paper. The introduction about four models is as follows:

1) GCNN (Graph Convolutional Neural Network) model, proposed by Sahu [8] et al., is employed to encode the graph and a bi-affine layer aggregates all mention pairs.
2) BERT model, proposed by Wang [12] et al., replaces the BiLSTM as encoder on the dataset. It could use the data to adjust the parameters.
3) BERT-Two-Step model, proposed by Wang [12] et al., is similar to BERT model, but it first predicts whether two entities have a relation and then predicts the specific target relation.
4) SIRE-BERT model, proposed by Zeng [13] et al., uses different methods to represent intra- and inter-sentential relations and the self-attention mechanism to model the logical reasoning process.

According to the evaluation indicators described in Sect. 4.1, Table 2 shows the experimental results of five document-level relation extraction models on the DocRED dataset.

Table 2. Comparison result of different document-level relation extraction models

Model	P/%	R/%	F1/%
GCNN	48.47	54.30	51.22
BERT	50.89	57.09	53.81
BERT-Two-Step	52.06	56.24	54.07
SIRE-BERT	60.12	59.25	59.68
MGG	62.77	63.11	62.94

From the comparison results in Table 2, it could be seen easily that the MGG model has significantly improved the precision rate and recall rate of document-level relation extraction.

In summary, the multiple-granularity graph model proposed in the paper can more effectively improve document-level relation extraction result than some traditional deep neural network models.

5   Conclusion

Extracting inter-sentence relations and conducting relational reasoning are challenging in document-level relation extraction. In this paper, we introduce multiple-granularity graph (MGG) model to better cope with document-level relation extraction. MGG utilizes a heterogeneous document graph to model mentions, entities, sentences and document. It also uses path reasoning mechanism to infer entity-entity relations. Experimental results on the large-scale human-annotated dataset, DocRED, show that MGG outperforms traditional deep neural network models.

References

1. Guo, Q., et al.: Constructing Chinese historical literature knowledge graph based on BERT. In: Xing, C., Fu, X., Zhang, Y., Zhang, G., Borjigin, C. (eds.) WISA 2021. LNCS, vol. 12999, pp. 323–334. Springer, Cham (2021). https://doi.org/10.1007/978-3-030-87571-8_28
2. Shuang, Z., Runxin, X., Baobao, C., Lei, L.: Double graph based reasoning for document-level relation extraction. In: Proceedings of the 2020 Conference on Empirical Methods in Natural Language Processing, pp. 1630–1640. Association for Computational Linguistics, Online (2020)
3. Yuan, Y., et al.: DocRED: a large-scale document-level relation extraction dataset. In: Proceedings of the 57th Annual Meeting of the Association for Computational Linguistics, pp. 764–777. Association for Computational Linguistics, Florence, Italy (2019)
4. Robin, J., Cliff, W., Hoifung, P.: Document-level N-ary relation extraction with multiscale representation learning. In: Proceedings of the 2019 Conference of the North American Chapter of the Association for Computational Linguistics: Human Language Technologies, pp. 3693–3704. Association for Computational Linguistics, Minneapolis, Minnesota (2019)
5. Wenxuan, Z., Kevin, H., Tengyu, M., Jing, H.: Document-level relation extraction with adaptive thresholding and localized context pooling. In: Proceedings of the AAAI Conference on Artificial Intelligence, vol. 35, no. 16, pp. 14612–14620 (2021)
6. Linfeng, S., Yue, Z., Zhiguo, W., Daniel, G.: N-ary relation extraction using graph-state LSTM. In: Proceedings of the 2018 Conference on Empirical Methods in Natural Language Processing (2018). https://doi.org/10.18653/v1/D18-1246
7. Patrick, V., Emma, S., Andrew, M.: Simultaneously self-attending to all mentions for full-abstract biological relation extraction. In: Proceedings of the 2018 Conference of the North American Chapter of the Association for Computational Linguistics: Human Language Technologies, pp. 2226–2235. Association for Computational Linguistics, Brussels, Belgium (2018)
8. Sunil, K.S., Fenia, C., Makoto, M., Sophia, A.: Inter-sentence relation extraction with document-level graph convolutional neural network. In: Proceedings of the 57th Annual Meeting of the Association for Computational Linguistics, pp. 4309–4316. Association for Computational Linguistics, Florence, Italy (2019)
9. Hochreiter, S., Schmidhuber, J.: Long short-term memory. Neural Comput. **9**(8), 1735–1780 (1997)
10. Schuster, M., Paliwal, K.K.: Bidirectional recurrent neural networks. IEEE Trans. Sig. Process. **45**(11), 2673–2681 (1997)
11. Bahdanau, D., Cho, K., Bengio, Y.: Neural machine translation by jointly learning to align and translate. arXiv preprint arXiv:1409.0473 (2014)
12. Hong, W., Christfried, F., Rob, S., Nilesh, M., William, W.: Fine-tune Bert for DocRED with two-step process. arXiv preprint arXiv:1909.11898 (2019)

13. Shuang, Z., Yuting, W., Baobao, C.: SIRE: separate intra- and inter-sentential reasoning for document-level relation extraction. In: Findings of the Association for Computational Linguistics, pp. 524–534. Association for Computational Linguistics, Online (2021)

A Fine-Grained Anomaly Detection Method Fusing Isolation Forest and Knowledge Graph Reasoning

Jie Xu and Jiantao Zhou[✉]

College of Computer Science, Engineering Research Center of Ecological Big Data, Ministry of Education, National and Local Joint Engineering Research Center of Mongolian Intelligent Information Processing Technology, Inner Mongolia Cloud Computing and Service Software Engineering Laboratory, Inner Mongolia Social Computing and Data Processing Key Laboratory, Inner Mongolia Discipline Inspection and Supervision Big Data Key Laboratory, Inner Mongolia Big Data Analysis Technology Engineering Laboratory, Inner Mongolia University, Hohhot, China
32009010@mail.imu.edu.cn

Abstract. Anomaly detection aims to find outliers data that do not conform to expected behaviors in a specific scenario, which is indispensable and critical in current safety environments related studies. However, when performing outlier detection on a large-scale multidimensional dataset, most of the traditional methods make no distinctions between classes of outliers and lack the reasoning ability and the explainability of the analysis results, which leads to the low accuracy of global outlier detection and the inability to effectively identify local outliers. In this paper, we propose a fine-grained anomaly detection method which well utilizes isolation forest and knowledge graph reasoning tactics. First, a better extracting data feature method and a more reasonable weighted strategy are used to find global outliers based on the traditional isolation forest algorithm, and then we construct a custom rule base according to the in-depth research and analysis of global abnormal data, and at last the ontology knowledge is reasoned based on such rule base to achieve the detection of local outliers. Through analysis of extensive experiment results, our method could effectively discover more abnormal data without big loss of time costs, has strong generalization, and correspondingly improves the performance for abnormal detection.

Keywords: Anomaly detection · Isolation forest · Knowledge graph reasoning · Rule base

1 Introduction

Anomaly detection, a significant data mining task, is proposed to find data that do not conform to expected behaviors from the perspective of features and

X. Zhao et al. (Eds.): WISA 2022, LNCS 13579, pp. 135–142, 2022.
https://doi.org/10.1007/978-3-031-20309-1_12

patterns in a specific scenario [1]. For different fields, the detection of anomalies often provides significant and actionable information [2–4]. These applications demand anomaly detection algorithms with high detection performance and fast execution.

Efficient approaches in the field of anomaly detection have performed in different perspectives and dimensions, achieved better results, but most anomaly detection methods make no distinctions between classes of outliers and do not consider the situation where two types of anomalies exist in the same time. Actually, the global outliers may be human error or unexpected activity. The local outliers are considered more interested and surprised that typically originate through some internal repeated process. Besides, most existing model-based approaches to anomaly detection lack the reasoning ability and the explainability to reveal the underlying reasons for the analysis results, which greatly reduces the credibility of the model and the accuracy of the analysis results. Therefore, if we can comprehensively consider global and local outliers, and then combine static knowledge in the knowledge graph with dynamic reasoning, we can discover outlier information at a more fine-grained level.

This paper studies a fine-grained anomaly detection method fusing isolation forest and knowledge graph reasoning. Firstly, the global anomaly detector uses an improved isolation forest method to find global anomalous data. To achieve this, two major innovations of this approach are: (i) kurtosis is used to extract data features, we select features that are more in line with Gaussian distribution, and recombine them as input data; (ii) for the isolation forest algorithm, considering the difference in anomaly detection ability between isolated binary trees, in this paper, the standard deviation of the path length of the isolated binary tree is used to improve the weighted calculation formula of the anomaly score. Secondly, the expert judgment detector refines the discovered anomalies into rules and builds a rule base according to the judgment of relevant domain experts. Finally, the knowledge graph reasoning detector provides an improved custom rule inference engine to find local anomalous information.

2   Related Work

This work is related to existing efforts on traditional anomaly detection methods and knowledge graph reasoning anomaly detection. Here we first introduce the related work in the two aspects.

Ruff et al. [5] discussed existing anomaly detection methods and evaluation indicators. As well as outlined critical open challenges and future research. Breunig et al. [6] proposed the LOF algorithm which is a density-based outlier detection method. Liu et al. [7] proposed the isolation forest algorithm which aims to explicitly isolate anomalies instead of profiles normal points. Liu et al. [8] provided the extension which is taken a step further in the split criterion isolation forest (SCiForest), proposing not only doing the splits by random hyperplanes but also determining the split threshold by a deterministic criterion. Hariri et al. [9] proposed the extended isolation forest algorithm (EIF), which highlights biases that are introduced by the way that data splits are created and resolves issues with the assignment of anomaly score to given data points. Actually, the accuracy of the anomaly detection of the isolated forest algorithm is also related to the number of isolated binary trees and the difference between the isolated binary trees.

Li et al. [10] discussed knowledge graph reasoning issues in anomaly detection and gave an overview of presented approaches and achievements in recent years. Actually, the knowledge graph can be considered a special data structure for knowledge storage. Although knowledge reasoning does not have formal semantics, it can reason by applying RDFS or OWL rules to a knowledge graph [11]. For example, Pujara et al. [12] have proven that the ontology represented by OWL EL is suitable for being transformed into a KG and performing reasoning on it efficiently. Proctor et al. [13] provided an introduction to Drools what it is and how it has been extended for reasoning for the complex event. Marx et al. [14] presented a simpler, rule-based fragment of multi-attributed predicate logic. Therefore, when the rules, statistical features and ontology constraints are effective, the method of ontology reasoning has a relatively high accuracy [15].

3   Model

In this section, we present a fine-grained anomaly detection model by fusing isolation forest and knowledge graph reasoning. The overview illustration of the proposed model is shown in Fig. 1.

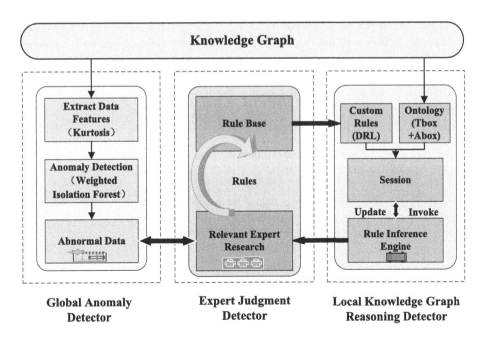

Fig. 1. The overview of proposed abnormal detection model.

3.1   Global Anomaly Detector

To efficiently mine global outliers in massive data, we design a global anomaly detector and propose a kurtosis weighted isolation forest algorithm (KWIF).

The Description of the Proposed KWIF Algorithm. First, feature selection plays a critical role when facing multidimensional data abnormal detection. We utilize kurtosis to analyze each attribute and judge whether the feature can be used as the input of the algorithm according to the pre-defined threshold. Second, Isolation Forest is an efficient method for global anomaly detection, but ignores the difference in the anomaly detection ability between the constructed isolated binary trees. Therefore, we adopt a path weighting strategy to improve the global detection capability of isolation forests. The proposed KWIF algorithm is a two-stage process that works as follows:

Training Stage. The training stage builds isolation trees (itrees) using the divided training set by the Kurtosis. First, build isolated binary trees according to the method. Then compute the standard deviation of the path lengths of each isolated binary tree (σ_{itree}). The σ_{itree} is defined as:

$$\sigma_{itree} = \sqrt{\frac{1}{n} \sum_{k=1}^{n} (h_k - \mu)^2} \tag{1}$$

where n is the total number of leaf nodes, and μ is the mean path length of all leaf nodes of the tree. Moreover, we use σ_{itree} to normalize the set of path length standard deviations to provide a weight value (W_{itree}) for each isolated binary tree in iforests. Finally, we calculate the weighted average path length of the sample points ($E(h(\mathbf{x}))$), which is defined as:

$$E(h(x)) = \sum_{i=1}^{n} w_{itree} \times h_i \tag{2}$$

where W_{itree} is the set of weight values, h_i is the path length of each sample point x in each isolated binary tree.

Testing Stage. The testing stage aims to compute anomaly scores from the weighted average path length of each instance for each instance. The anomaly score S is defined as($c(n)$ is the average of $h(x)$ given n.):

$$S(x, n) = 2^{-\frac{E(h(x))}{c(n)}} \tag{3}$$

3.2   Expert Judgment Detector

After finding global abnormal points through the weighted isolation forest algorithm, how to judge whether the discovered data is true outliers and get feedback.

So we design an expert judgment detector. We first analyze outliers through the experience of experts or teams, and then laws hidden behind the exceptions are summarized into production rules by formal methods. Table 1 gives some examples of our production rules. Finally, we store production rules in rule base (including Domain Rule Base, Rule Base Query Unit and Rule Base Submission Unit) as input to subsequent custom rule inference engines.

Table 1. Rules

Number	Rule description	Production rules
Rule1	Hospitalizations per-month over 5	when Person (h-frequency > 5) then Possible overtreatment

3.3 Local Knowledge Graph Reasoning Detector

Although anomalous data have been identified, they are still incomplete and sparse. So we design a local knowledge graph inference detector. The custom rule inference engine tool used in this paper is Drools, which mainly provides a rule inference engine and rule definition language. The inputs of the custom rule inference engine are the Ontology (TBox, ABox) and the DRL file of custom rules (containing all user-defined rules). We pass the two parts into the Session to invoke the rule inference engine for inference. Then, the Session is updated with the inference results to complete the detection of local outliers. Besides, we optimize the Rete-00 algorithm through two strategies, which improve the matching efficiency while reducing the memory space occupation. First, the condition part is sorted according to the frequency of occurrence, which increases the number of shared nodes in the Rete network. Second, a cost function is introduced to the rules, and the rules with higher costs are optimized and executed first.

4 Experiments

In this section, we first introduce the experimental setup details, then report the results and analysis.

4.1 Experiment Settings

Datasets. First, the Satellite (the low-dimensional case) and the Shuttle (the high-dimensional case) from the UCI repository are adopted to test the performance of the proposed KWIF algorithm. Second, we use the True Dataset to check the validity of the whole model. The True Dataset is the real medical data after desensitization. That is provided by the Discipline Inspection and Supervision Information Center of a prefecture-level city. These data record hospitalization information about all hospitals in a city (a month in 2019), which are unlabeled data and contain 5000 pieces related to more than 50 attributes.

4.2   Evaluation on the KWIF Algorithm

Evaluate on Detection Accuracy. Table 2 presents the results of the proposed algorithm compared with two algorithms LOF, IF and an improved algorithm EIF. First, KWIF shows better detection results than LOF and IF. Especially on the Shuttle, the accuracy rate reaches about 99%. Because the KWIF uses kurtosis to extract feature and assigns different weights to isolated binary trees with different anomaly detection capabilities. Second, KWIF achieves a comparable performance with EIF on the dataset Satellite. Because the EIF uses axis-parallel hyperplanes to resolve issues with the assignment of anomaly scores to given data points. However, the problem with EIF is that it takes more running time than KWIF (running time in Fig. 2 gives the result).

Table 2. Accuracy and F-Measure metrics of experimental methods

Test	Satellite		Shuttle	
Models	Accuracy	F-Measure	Accuracy	F-Measure
LOF	0.7464	0.6312	0.9036	0.3145
IF	0.7123	0.5421	0.9864	0.9128
EIF	0.7618	0.6416	0.9891	0.9278
KWIF	**0.7497**	**0.6328**	**0.9896**	**0.9289**

Evaluate on Generalization Ability and Running Time. The generalization ability is the ability of the outlier detection method to adapt to unknown data. It can be seen from (a) in Fig. 2 that KWIF and EIF find a closely equal number of anomalies in the real datasets. Both of them achieve good performance. A possible reason is that in the real dataset, KWIF and EIF modify the scoring function which makes the score evaluation of each sample point more accurate. From (b) in Fig. 2, the running time required for KWIF is slightly higher than IF but much lower than LOF and EIF.

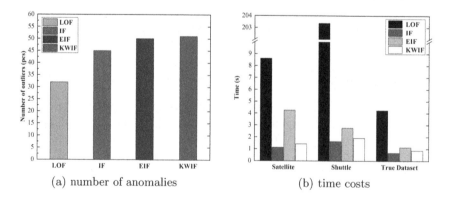

(a) number of anomalies   (b) time costs

Fig. 2. Generalization ability and running time

4.3   Evaluation on the Proposed Whole Model

Qualitative Analysis. Table 3 presents some results of the global anomaly. The global outliers may be human error or unexpected activity. For anomaly1, there are negative numbers in the hospitalization expenses. For anomaly2, the number of days and frequency of hospitalization is high, but at one's own expense and the reimbursement amount of medical insurance are 0, which may be the phenomenon of hanging bed hospitalization. Moreover, when observing anomaly3 or anomaly4 alone, we can not find the issue. In fact, the two patients have different hospitalization days but the settlement amount is the same. Then we also find out that the primary doctor for both patients is the same person, which may be a medical fraud problem. According to expert guidance, we summarize these anomalies into rules and store them in the rule base. Finally, we found nearly 200 local outliers (they are seemingly normal data inside the data), but there are 80 outliers that duplicate the global outliers. The local outliers are considered more interested and surprised that typically originate through some repeated process(e.g.: Doctor-patient collusion fraud, Group medical fraud).

Table 3. Global abnormal data instances

Attribute name	Anomaly1	Anomaly2	Anomaly3	Anomaly4
Gender	1	1	1	0
Days in hospital	5	30	15	10
Number of hospitalizations	5	77	31	31
Number of doctor's orders	8	15	8	5
Total settlement amount	−11263.43	9113.87	9984.43	9984.43
Amount actually received	−11263.43	9113.87	9984.43	9984.43
Total at one's own expense	−2791.31	0	6686.14	6686.14
Medical insurance co-payment	−8412.72	0	3298.29	3298.29
Other incurred amounts	0	0	0	0
Modifications	1	0	2	4
Abnormal score	−0.212854	−0.017882	−0.016741	−0.016473

5   Conclusion

In this paper, we have proposed a fine-grained anomaly detection method fusing isolation forest and knowledge graph reasoning. Firstly, we obtain global anomaly information through the proposed KWIF algorithm, and then analyze the deep-seated reasons behind the global outliers, we construct the custom rule base. Finally, effectively discover local outliers through a custom rule inference engine. Through analysis of extensive experiment results, our method could fine-grained find outliers without a big loss in running time, which greatly improves the detection accuracy of unbalanced data.

Acknowledgement. This work is supported by the National Natural Science Foundation of China under Grant No. 62162046, the Inner Mongolia Science and Technology Project under Grant No. 2021GG0155, the Natural Science Foundation of Major Research Plan of Inner Mongolia under Grant No. 2019ZD15, and the Inner Mongolia Natural Science Foundation under Grant No. 2019GG372.

References

1. Campos, G.O., et al.: On the evaluation of unsupervised outlier detection: measures, datasets, and an empirical study. Data Min. Knowl. Disc. **30**(4), 891–927 (2016). https://doi.org/10.1007/s10618-015-0444-8
2. Feng, Y., et al.: Anti-money laundering (AML) research: a system for identification and multi-classification. In: Ni, W., Wang, X., Song, W., Li, Y. (eds.) WISA 2019. LNCS, vol. 11817, pp. 169–175. Springer, Cham (2019). https://doi.org/10.1007/978-3-030-30952-7_19
3. Anandakrishnan, A., Kumar, S., Statnikov, A., et al.: Anomaly detection in finance: editors' introduction. In: KDD 2017 Workshop on Anomaly Detection in Finance, pp. 1–7. PMLR (2018)
4. Khraisat, A., Gondal, I., Vamplew, P., Kamruzzaman, J.: Survey of intrusion detection systems: techniques, datasets and challenges. Cybersecurity **2**(1), 1–22 (2019). https://doi.org/10.1186/s42400-019-0038-7
5. Ruff, L., Kauffmann, J.R., et al.: A unifying review of deep and shallow anomaly detection. Proc. IEEE **109**(5), 756–795 (2021)
6. Breunig, M.M., Kriegel, H.P., Ng, R.T., Sander, J.: Lof: identifying density-based local outliers. In: Proceedings of the 2000 ACM SIGMOD International Conference on Management of Data, pp. 93–104 (2000)
7. Liu, F.T., Ting, K.M., Zhou, Z.H.: Isolation forest. In: 2008 Eighth IEEE International Conference on Data Mining, pp. 413–422. IEEE (2008)
8. Liu, F.T., Ting, K.M., Zhou, Z.-H.: On detecting clustered anomalies using SCiForest. In: Balcázar, J.L., Bonchi, F., Gionis, A., Sebag, M. (eds.) ECML PKDD 2010. LNCS (LNAI), vol. 6322, pp. 274–290. Springer, Heidelberg (2010). https://doi.org/10.1007/978-3-642-15883-4_18
9. Hariri, S., Kind, M.C., Brunner, R.J.: Extended isolation forest. IEEE Trans. Knowl. Data Eng. **33**(4), 1479–1489 (2019)
10. Li, Z., Jin, X., Zhuang, C., Sun, Z.: Overview on graph based anomaly detection. J. Softw. **32**(1) (2021)
11. Wang, Y., Wang, X., Liu, B.: Incremental validation of RDF graphs. In: Xing, C., Fu, X., Zhang, Y., Zhang, G., Borjigin, C. (eds.) WISA 2021. LNCS, vol. 12999, pp. 359–371. Springer, Cham (2021). https://doi.org/10.1007/978-3-030-87571-8_31
12. Pujara, J., Miao, H., Getoor, L., Cohen, W.W.: Ontology-aware partitioning for knowledge graph identification. In: Proceedings of the 2013 Workshop on Automated Knowledge Base Construction, pp. 19–24 (2013)
13. Proctor, M.: Drools: a rule engine for complex event processing. In: Schürr, A., Varró, D., Varró, G. (eds.) AGTIVE 2011. LNCS, vol. 7233, pp. 2–2. Springer, Heidelberg (2012). https://doi.org/10.1007/978-3-642-34176-2_2
14. Marx, M., Krötzsch, M., Thost, V.: Logic on mars: ontologies for generalised property graphs. In: IJCAI, vol. 2017, pp. 1188–1194 (2017)
15. Ji, S., Pan, S., Cambria, E., Marttinen, P., Philip, S.Y.: A survey on knowledge graphs: representation, acquisition, and applications. IEEE Trans. Neural Netw. Learn. Syst. **33**(2), 494–514 (2021)

Design of Trademark Recommendation System Based on Knowledge Graph

Siling Feng, Xunyang Ji, and Mengxing Huang[✉]

School of Information and Communication Engineering, Hainan University, Haikou,
People's Republic of China
{fengsiling,huangmx09}@hainanu.edu.cn

Abstract. Today, knowledge graphs are used in more and more places. Especially in the field of recommendation systems, the knowledge graph shines brightly. In the past, the core of recommendation system is algorithm. Now, the emergence of the knowledge graph improves the working efficiency of the recommendation system as well as its accuracy. The system to be designed in this paper is about the design of trademark recommendation system based on knowledge graph. The steps to set up the system are as follows: The first step is to set up the working environment required by the system; The second step is to conduct a simple data preprocessing on the collected trademark data sets according to their classification to form a reliable CSV file importing knowledge map of trademark data. The third step is to connect the online Neo4j graph database and import the obtained CSV file into the graph database to form the knowledge graph about trademarks. The fourth step is to find interested trademarks and their classification according to the trademark knowledge map, and output their names, classification and other information. Step 4, use content-based recommendation algorithm to test the output trademark and get the recommendation result; The fifth step is to use PyQt5 plug-in to make a platform to display pictures and display the recommended trademarks. In general, this project is to make a trademark recommendation system with simple visualization function based on knowledge graph.

Keywords: Knowledge graph · Design of trademark recommendation system · Content-based recommendation algorithm

1 Introduction

Since Google released the knowledge graph in 2012, the technology of knowledge graph has developed rapidly, its theoretical system has become increasingly perfect, and its application effect has become increasingly obvious. After Google released the knowledge graph, it marked the emergence of the knowledge graph. At the same time, it also means that the old era of knowledge engineering is slowly passing, and the new historical chapter of knowledge engineering — big data knowledge engineering has been outlined [1]. Driven by the knowledge graph technology, the grand picture of intelligent upgrading and transformation of all walks of life is gradually unfolding. Looking back more than half a century, symbolism theory of artificial intelligence represented by physical

X. Zhao et al. (Eds.): WISA 2022, LNCS 13579, pp. 143–152, 2022.
https://doi.org/10.1007/978-3-031-20309-1_13

symbol system and knowledge engineering practice centered on expert system complement each other and shine brilliantly. Going back to more than 2000 years ago, the three sages of Ancient Greek philosophy created a great era of logical thinking, whose thoughts directly or indirectly affected the historical continuation and development from symbolism to knowledge engineering and then to knowledge mapping [2].

In recent years, with the development of Internet technology, the technology research and application of knowledge graph have been deepened and improved, and the technology of knowledge graph has attracted extensive attention from industry and academia. Recently, a large number of theoretical and research achievements, as well as a number of excellent engineering practice cases, have emerged in the field of knowledge mapping. Secondly, with the development of e-commerce, more and more excellent achievements for the benefit of human beings are constantly entering people's lives, including electronic products and daily necessities. With the continuous improvement of people's living standards, people's material needs also put forward higher standards, then often good companies to build a good reputation with excellent quality products can be more easily favored by consumers. Among them, as a legal representative of a company and a symbol of a product, a trademark can not only give people a sense of beauty, but also have a significant role in indicating the source of products or services, and can also play a role in distinguishing producers, operators or service providers. If it is a well-known trademark, it is not only considered to represent large companies and enterprises, but also means that it is protected by the legal right to exclusive use of trademarks across categories, which is more likely to be recognized by consumers.

At this time, a trademark recommendation system with excellent performance and considerable efficiency can provide users with more suitable recommendations. When the knowledge graph is added to display, it is easier for users to know the trademarks of this company, which is convenient for their daily life and shopping. The trademark recommendation system based on the knowledge graph starts from the user end. It first constructs a considerable knowledge graph according to the trademark data set to facilitate the user to know which trademarks exist under the company. Then, it searches for the trademarks of the company it likes and presents other trademarks of the company according to the recommendation of the company it belongs to. Therefore, the company industry to which the trademark belongs can be extracted to make better recommendations to users.

2 Construction of Knowledge Graph and the Design and Implementation of Recommendation System

The whole system is divided into four modules:

1. Data pre-processing module: trademark data sets are pre-processed to achieve the following effects: First, a unique LogoID is formed for each trademark. Second, add and classify each trademark in the CSV file according to the 19 classified in advance, which is called label in the data set. Third, record the local address of each trademark into the data set.

2. Knowledge graph construction module: Upload CSV files generated after pre-processing to Neo4j map database to build knowledge map. After it is built, it can return all the trademarks of a certain category that the user wants to know.

3. Recommendation module: determine the local name of the trademark to be searched, import the generated data set into the recommendation algorithm for training, make recommendations by referring to the label of each trademark as the key attribute, and finally return the name of the recommended trademark, the label to which it belongs, and the local home.

4. Display module: users first search in the search box made by Tkinter, import the name of the searched trademark into the recommendation module, and return the recommendation result and information after calculation according to the recommendation algorithm. Finally, the local address of the recommended trademark is automatically imported to PyQt5 to display the recommended trademark result (Fig. 1).

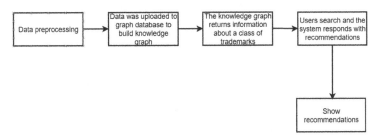

Fig. 1. Flowchart of trademark recommendation system

A. *Data pre-processing*

The data set used in this design is the open source data set downloaded from CSDN. The data set includes classified trademark pictures instead of processed CSV files. What is needed to import the graph database to build the knowledge map is a data CSV file, so I have the basic idea of data processing. Firstly, the file name, local address, and category of the original trademark image dataset are exported to a TXT file via Python. Secondly, the generated data TXT file is processed by Python code, and their information is divided into four attributes, namely logoID, name, label and home. Each attribute is separated by commas (because CSV files are essentially table files separated by commas).Finally, TXT files with four attributes of trademarks are converted into CSV files through WPS tables, and the encoding format is set as UTF-8. In this way, we get the CSV file of trademark data, and the data pretreatment is completed (Figs. 2, 3 and 4).

B. *Knowledge graph construction*

When we get the data CSV file out, it means we can import the data set into the Neo4j graph database. First of all, we need to enter the neo4j. Bat console in the command

apple	2022/4/18 21:19	文件夹
Armani	2022/4/18 21:19	文件夹
BMW	2022/4/18 21:19	文件夹
Chanel	2022/4/18 21:19	文件夹
Clover	2022/4/18 21:19	文件夹
crocodile	2022/4/18 21:19	文件夹
Huawei	2022/4/18 21:19	文件夹
Mercedes-Benz	2022/4/18 21:19	文件夹
Microsoft	2022/4/18 21:19	文件夹
Nike	2022/4/18 21:19	文件夹
Omega	2022/4/18 21:19	文件夹

Fig. 2. Data set partial raw data

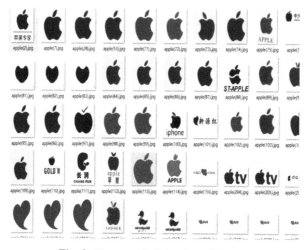

Fig. 3. Data set of all Apple trademarks

line of CMD to start the Neo4j graph database. Then we open the browser and enter the url http://localhost:7474 in the browser to open the online Neo4j graph database [3]. When the web page is displayed normally and the user name and password preset are entered successfully, it indicates that Neo4j is successfully started normally. Because Neo4j graph database has its own Cypher language, although the syntax is simple and intuitive, it is not easy to process. In order to facilitate the operation, I used the Special Py2neo library in Python to help me complete the construction of knowledge graph. The Py2neo library allows you to use Python directly on Pycharm to manipulate the Neo4j graph database, create nodes, and so on. Once connected to the Py2neo library, we can control the graph database in Python and return all the label information the user needs to know about (Figs. 5, 6, 7 and 8).

C. *Recommendation module*

This design recommendation module is based on content recommendation algorithm. Content-based recommendation algorithm is a common algorithm in large recommendation systems. After a long time of development, it has mature principles and powerful

	A	B	C	D	E	F	G
1	logoid	title	label	home			
2	1	apple(100)	apple	E:/LOGIO/logo/apple/apple(100).jpg			
3	2	apple(101)	apple	E:/LOGIO/logo/apple/apple(101).jpg			
4	3	apple(102)	apple	E:/LOGIO/logo/apple/apple(102).jpg			
5	4	apple(103)	apple	E:/LOGIO/logo/apple/apple(103).jpg			
6	5	apple(104)	apple	E:/LOGIO/logo/apple/apple(104).jpg			
7	6	apple(105)	apple	E:/LOGIO/logo/apple/apple(105).jpg			
8	7	apple(106)	apple	E:/LOGIO/logo/apple/apple(106).jpg			
9	8	apple(107)	apple	E:/LOGIO/logo/apple/apple(107).jpg			
10	9	apple(108)	apple	E:/LOGIO/logo/apple/apple(108).jpg			
11	10	apple(109)	apple	E:/LOGIO/logo/apple/apple(109).jpg			
12	11	apple(110)	apple	E:/LOGIO/logo/apple/apple(110).jpg			
13	12	apple(111)	apple	E:/LOGIO/logo/apple/apple(111).jpg			
14	13	apple(112)	apple	E:/LOGIO/logo/apple/apple(112).jpg			
15	14	apple(113)	apple	E:/LOGIO/logo/apple/apple(113).jpg			
16	15	apple(114)	apple	E:/LOGIO/logo/apple/apple(114).jpg			
17	16	apple(156)	apple	E:/LOGIO/logo/apple/apple(156).jpg			
18	17	apple(2).	apple	E:/LOGIO/logo/apple/apple(2).jpg			
19	18	apple(204)	apple	E:/LOGIO/logo/apple/apple(204).jpg			
20	19	apple(205)	apple	E:/LOGIO/logo/apple/apple(205).jpg			
21	20	apple(206)	apple	E:/LOGIO/logo/apple/apple(206).jpg			
22	21	apple(207)	apple	E:/LOGIO/logo/apple/apple(207).jpg			
23	22	apple(208)	apple	E:/LOGIO/logo/apple/apple(208).jpg			
24	23	apple(209)	apple	E:/LOGIO/logo/apple/apple(209).jpg			
25	24	apple(210)	apple	E:/LOGIO/logo/apple/apple(210).jpg			
26	25	apple(211)	apple	E:/LOGIO/logo/apple/apple(211).jpg			
27	26	apple(212)	apple	E:/LOGIO/logo/apple/apple(212).jpg			
28	27	apple(213)	apple	E:/LOGIO/logo/apple/apple(213).jpg			
29	28	apple(215)	apple	E:/LOGIO/logo/apple/apple(215).jpg			
30	29	apple(216)	apple	E:/LOGIO/logo/apple/apple(216).jpg			

Fig. 4. CSV file generated after data pretreatment

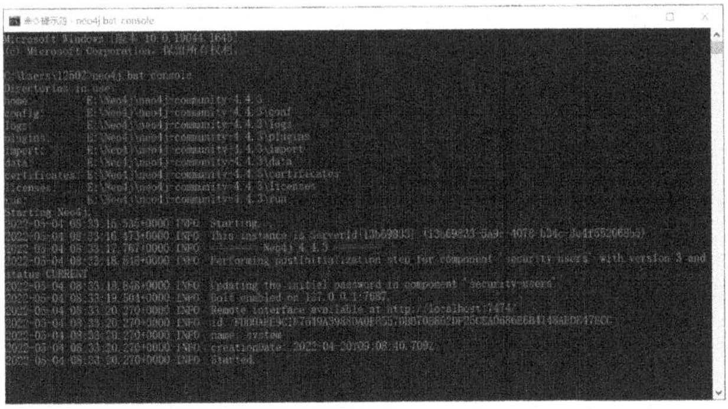

Fig. 5. CMD starts the Neo4j graph database

functions to support the recommendation system. For example, when Taobao recommends commodities to users, most of them also adopt such an algorithm. Content-based Recommendation (CB), which means to extract and recommend Content based on the target. For example, after a user searches for the trademark Apple of Apple, the recommendation system will recommend other trademarks of Apple to the user based on the content Apple [4]. This algorithm does not need to be based on the user's rating, evaluation and opinion on the project, but rather obtains the user's information from the content, features and categories of the project through machine learning, and finally generates the recommendation results for users.

The content-based recommendation algorithm mainly includes three steps: First, extract the attributes or keywords of the content. Second, use data about a user's preference for other products to generalize the user's preference (this step is actually optional).

Fig. 6. Neo4j successfully started the display page

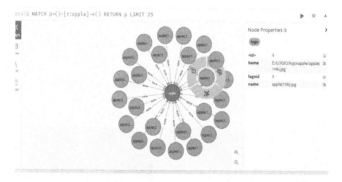

Fig. 7. Knowledge graph of Apple trademarks

```
E:\python39\Scripts\python.exe E:/python39/bishe/picture.py
[apple(Node('logo', home='E:/LOGIO/logo/apple/apple(100).jpg', logoid='1', name='apple(100).jpg'),
```

Fig. 8. Label is apple's trademark information

Third, compare the characteristics between other projects and the original project. If there is user preference data, it can be added. Finally, make a comprehensive recommendation to users.

In this paper, content-based recommendation algorithm is used to calculate the similarity between items using TF-IDF matrix. TF-IDF is a statistical method used to assess the importance of a word to one of the documents in a document set or a corpus. The importance of a word increases with the number of times it appears in the document, but decreases inversely with the frequency of its occurrence in the corpus. Various forms of TF-IDF weighting are often applied by search engines as a measure or rating of the degree of correlation between files and user queries [5]. Term frequency (TF) refers to the frequency with which a given word appears in a given document. This number is normalized to term count to prevent it from favoring long files. (The same word in a long document may have a higher number of words than in a short document, regardless of its importance.) The importance of the word ti in a particular document can be expressed as:

$$tf_{i,j} = \frac{n_{i,j}}{\sum_k n_{k,j}} \tag{1}$$

In the above formula, ni and j are the number of occurrences of target words in the whole document, while the denominator is the sum of the number of occurrences of all words in the whole document. And Inverse Document Frequency (IDF) is a measure of the universal importance of a word. The IDF of a particular word can be obtained by dividing the total number of files by the number of files containing the word and taking the logarithm base 10 of the resulting quotient:

$$idf_i = \log \frac{|D|}{|\{j : t_i \in d_j\}|} \tag{2}$$

where, the numerator represents the total number of files in the corpus, and the denominator represents the total number of files containing the target word TI. If the term is not in the data set, the denominator is zero. TF-IDF with high weight can be generated by high frequency words in a particular data set file and low file frequency of this word in the whole data set. Therefore, TF-IDF can filter out unimportant words and retain keywords, such as the characteristics and attributes of the item and its category [6]. This system uses TfIdfVectorizer in Python sklearn library to calculate TF-IDF matrix, uses TF-IDF matrix to filter out unimportant words, and takes attribute label in data set as the key word of this algorithm (Fig. 9).

```
logo = pd.io.parsers.read_csv('E:/Neo4j/neo4j-community-4.4.3/import/logo1.csv')#导入AlogoCSV文件
logo.head()
logo['label'].head()
tfidf = TfidfVectorizer(stop_words='english')
logo['label'] = logo['label'].fillna('')
tfidf_matrix = tfidf.fit_transform(logo['label'])
tfidf_matrix.shape
```

Fig. 9. The TF-IDF matrix is calculated and the label in the dataset is taken as the key word

Then we use the cosine distance to calculate the correlation between the two trademarks. Since we have obtained the TF-IDF vector, the linear_kernel() is directly used here to calculate the dot product to get the cosine distance. Finally, use panda.series() to create an index for the list of trademarks to facilitate the mutual index between the following trademarks and logoID (Fig. 10).

```
cosine_sim = linear_kernel(tfidf_matrix, tfidf_matrix)#计算余弦距离来判断两个商标直接的相关性
indices = pd.Series(logo.index,index=logo['title']).drop_duplicates()#drop_duplicates () 作用是删除重复行
```

Fig. 10. Computes cosine distance and indexes trademarks

Finally, the results of the above code processing are imported into the function in charge of the recommendation module to finally return the information related to the recommendation trademark. In this way, the system's recommendation module is complete (Fig. 11).

```
def get_recommendationhome(title,consine_sim=cosine_sim):#返回所推荐图片的名字和地址
    idx = indices[title]
    sim_scores = list(enumerate(cosine_sim[idx]))
    sim_scores = sorted(sim_scores, key=lambda x: x[1], reverse=True)
    sim_scores = sim_scores[1:2]
    logo_indices = [i[0] for i in sim_scores]#列表解析式
    return print(logo['title'].iloc[logo_indices],'\n',logo['home'].iloc[logo_indices])



def get_recommendation(title,consine_sim=cosine_sim):#输出推荐图片的地址
    idx = indices[title]
    sim_scores = list(enumerate(cosine_sim[idx]))
    sim_scores = sorted(sim_scores, key=lambda x: x[1], reverse=True)
    sim_scores = sim_scores[1:2]
    logo_indices = [i[0] for i in sim_scores]#列表解析式
    x=list(logo['home'].iloc[logo_indices])
    return x.pop()
```

Fig. 11. Recommended module code

D. *Display module*

In order to meet the requirement of using a visual window to display the logo, the PyQt5 module in Python is used here to help me build an image display window. PyQt5 is the latest version of PyQt. The main modules include QtCore,QtGui, QtWidgets and so on. It can be used cross-platform, can use mature IDES for interface design, automatically generate corresponding Python code, and has a series of window controls with different functions. After PyQt5 reads the home of the returned recommended trademark (i.e. the local address of the trademark), PyQt5's QDialog plug-in is used to complete the operation of creating the display trademark window. Qt in the use of QDialog to create a window, it creates a window is divided into two kinds, one is the module window, through exec() to start, the module window can only be manually closed, not automatically closed; The other is a non-module window, which is started by show() and the module window closes automatically. The manual closed module window is used to display the recommended trademark.

In addition to making a visual window module to display trademarks, it is necessary to make the same visual window interface as the search bar for recommended trademarks. Since the plug-in of PyQt5 module for making Windows can only be called once and cannot be reused to make another window, I used Tkinter module which can also make visual Windows for the function of search window. Tkinter is a Python module for making Windows. Tkinter, as a specific GUI interface that comes with Python, is an image window and a GUI interface that can be edited, which can conveniently realize many intuitive functions with GUI [7]. To make the search box, THIS time I used the Entry widget to make a single-line text Entry field for the search bar, and the Button widget to set two buttons, "Search" and "re-enter." The function of the "search" button is that when the user enters the original trademark in the search bar, the search button will be connected to the recommendation module in front of the displayed trademark module to return the trademark result and display the trademark picture. The "Re-enter" button is set up for users to make a second recommendation. It can clear the text in the search bar and re-enter it (Figs. 12, 13 and 14).

Trademark recommendation system
BMW(26).jpg
Search
re-enter

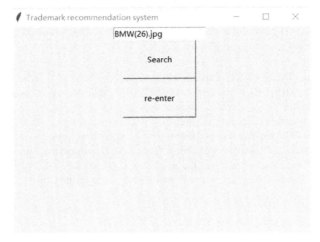

Fig. 12. Use Tkinter module to make the search window

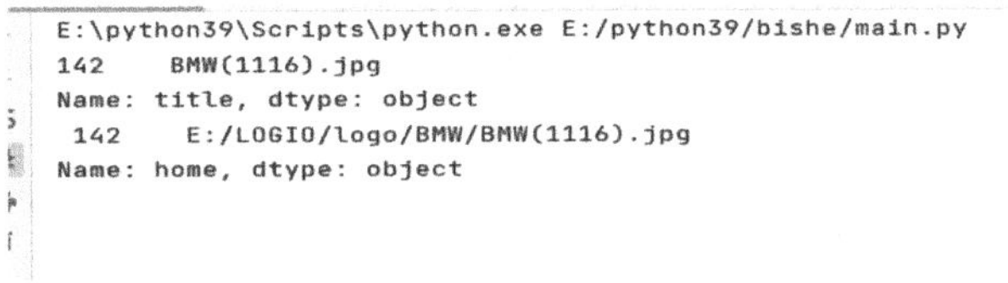

E:\python39\Scripts\python.exe E:/python39/bishe/main.py
142   BMW(1116).jpg
Name: title, dtype: object
142   E:/LOGIO/logo/BMW/BMW(1116).jpg
Name: home, dtype: object

Fig. 13. Returns information about recommended trademark

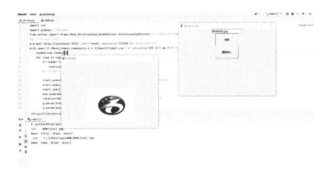

Fig. 14. Searching for BMW logo returns recommended trademark

3 Summary

The main content of this paper is to build a brand recommendation system with simple visualization function based on knowledge graph. The specific contents are as follows: First, the collected original data images are processed and converted into data CSV files with text information. Second, import the data CSV file into the Neo4j graph database to build a knowledge map of trademarks. Thirdly, content-based recommendation algorithm is adopted, TF-IDF matrix is used to filter out unimportant words in the data set, and important attribute label is reserved as keywords for recommendation. Fourthly, PyQt5 and Tkinter are used to make a visual window module, which is used as a trademark display module and a search module respectively.

Acknowledgment. This research is supported by national key research and development projects 2018YFB1404400, 2018yfb11404400 and 2018yfb1703403, natural science foundation of Hainan Province 2019cxtd400, key research and development project of Hainan Province zdyf2019115, National Natural Science Foundation of China 61865005, national optoelectronic Laboratory of Wuhan 2020wnlokf001, key research and development project of Hainan Province zdyf2019020, National Natural Science Foundation of China, 62062030, Department of education of Hainan Province, hnky2019-22. At the same time, the author thanks Hainan Key Laboratory of big data and intelligent services and other colleagues of Hainan Green smart Island Collaborative Innovation Center for their help.

References

1. Wang, Y., Gao, S., Li, W., Jiang, T., Yu, S.: Research and application of personalized recommendation based on knowledge graph. In: Xing, C., Fu, X., Zhang, Y., Zhang, G., Borjigin, C. (eds.) WISA 2021. LNCS, vol. 12999, pp. 383–390. Springer, Cham (2021). https://doi.org/10.1007/978-3-030-87571-8_33
2. Xiao, Y., Xu, B., Lin, X., et al.: Knowledge Graph Concepts and Techniques. Publishing House of Electronics Industry, Beijing (2020)
3. Lu, X., Zhang, Y., Qian, J.: Research on application of film knowledge graph based on graph database. Mod. Comput. (07), 76–83 (2016)
4. Chen, Y., Zhou, R., Zhu, W., Li, M., Yin, J.: Mining patent knowledge to achieve automatic extraction of keywords. J. Comput. Res. Dev. **53**(08), 1740–1752 (2016)
5. Tang, Y., Tang, J.: An improved TF-IDF text classification algorithm. Inf. Technol. Informatiz. (03), 13–16 (2022)
6. Pan, W.: Copyright protection of android applications based on text cosine similarity. China New Commun. **16**(02), 56–58 (2014)
7. Lessa, A.: Python Developer's Handbook. Publishing House of Electronics Industry, Beijing (2001)

Natural Language Processing

An Integrated Chinese Malicious Webpages Detection Method Based on Pre-trained Language Models and Feature Fusion

Yanting Jiang[1,2(✉)] and Di Wu[3,4]

[1] Chengdu Aeronautic Polytechnic, Chengdu 610100, China
jiangyanting@mail.bnu.edu.cn
[2] Sichuan University of Media and Communications, Chengdu 611745, China
[3] Shenzhen Institute of Advanced Technology, Chinese Academy of Sciences, Shenzhen 518000, China
[4] China University of Chinese Academy of Sciences, Shenzhen 518000, China

Abstract. This paper proposed an integrated Chinese malicious webpages detection method. Firstly, we collected and released a Chinese malicious webpages detection dataset called "ChiMalPages" containing URLs and HTML/JavaScript files, and specified the detailed types of malicious pages according to relevant laws. Secondly, we designed a feature template for Chinese webpages and ranked each feature's importance based on information gain of the Random Forest algorithm. Thirdly, we fine-tuned BERT on the external URLs classification task and text on webpages, respectively producing new models "BERT-URL" and "BERT-web-text". The performance of pre-trained models is obviously superior to the baseline models. Finally, we integrated features from manual templates, BERT-URL and BERT-web-text, and the classification F1 score reaches 79.84%, increasing by 7.37% compared with manually designed webpage features. Experiments proved that our method based on BERT is useful and not biased on detailed classes.

Keywords: Chinese malicious webpages detection dataset · URLs · Webpages text · Pre-trained language models · Feature importance ranking

1 Introduction

Internet has been an indispensable part of people's daily life. According to the statistical report from Chinese Internet Network Information Center, by the end of June 2021, the number of Chinese webpages has exceeded 335 billion [1].

However, as Chinese Internet booms, a number of malicious webpages have emerged. "Malicious webpages" is a general concept, which doesn't have precise definition and mainly involves: (1) webpages embedded with virus. (2) phishing webpages disguised as normal ones. (3) webpages including illegal content such as gambling, porn and illegal trades [2]. According to an incomplete survey, there are over 320000 discovered malicious Chinese websites and more pages in 2020 [3]. Malicious pages are suspected of cybercrime, being a big threat to people's privacy and property.

© The Author(s), under exclusive license to Springer Nature Switzerland AG 2022
X. Zhao et al. (Eds.): WISA 2022, LNCS 13579, pp. 155–167, 2022.
https://doi.org/10.1007/978-3-031-20309-1_14

Chinese malicious webpages are from websites whose registrars are in China [1]. And automatically detecting them is a challenging task. Firstly, most malicious pages are short-lived and related datasets are scarce, especially for Chinese webpages [4]. Secondly, webpages' features are complex: URL (Uniform Resource Locator), HTML pages, JavaScript (JS) codes and so on. And many previous works only focused on URLs [5–7]. Thirdly, disguise tricks of malicious pages would evolve and their features are not fixed. Traditional ways to detect malicious pages need improvements.

Recently pre-trained language models have shown their power on sequence modeling and transfer learning, in this paper, we proposed an integrated method to detect Chinese malicious webpages. The main contributions of this paper are as follows:

(1) We opened a malicious Chinese webpages detection dataset "ChiMalPages", which contains URLs, HTML/JS files of each webpage. And we specified detailed types of malicious pages, and cited relevant laws as evidence.
(2) For English webpage features proposed by previous works, we analyzed and ranked each feature's importance based on Information Gain of the Random Forest algorithm, finding some features for English pages may not be suitable for Chinese ones.
(3) We proved the powerfulness of pre-trained models (BERT [8], RoBERTa [9] et al.) on the URLs sequence modeling and classification task via transfer learning.
(4) We proposed an integrated method to detect malicious webpages, which combines pre-trained language models with the manual feature template. The experiment showed this integrated method is obviously superior to previous methods.

2 Related Work

2.1 Malicious Webpages Dataset Construction

In terms of malicious webpages data, it's not so hard to have access to malicious URLs. For example, "phishtank.org" is a widely-used website to download malicious URLs of English pages. The GitHub repository "CN-Malicious-website-list" also store enough malicious URLs of Chinese webpages [2]. However, it's difficult for researchers to get malicious pages' HTML and JS files [4]. This is mainly because the survival time of most malicious websites is quite short.

The data of related researches are shown in Table 1, from which we can tell that none of datasets are open to the whole research community.

Table 1. Previous Malicious webpages data information

Works	Num of malicious pages	Data open or not	Language	If specify pages' types
Wang [10]	500	Not	Mixed	Not
Gowtham [11]	1764	Not	English	Not
Xu [12]	850	Not	Mixed	Not
Ye [13]	2345	Not	Mixed	Not
Wei [14]	500	Not	Mixed	Not
Hu [15]	400	Not	English	Not
Wu [16]	1456	Not	Mixed	Not
Zhou [17]	249	Not	Chinese	Not
Chen [18]	856	Not	Mixed	Not
Ours	521	Open[a]	Chinese	Specified

[a]https://github.com/JiangYanting/Chinese_Malicious_Web_Pages_Dataset_And_Detection

2.2 Feature Extraction

Feature extraction is a process extracting useful information for malicious pages recognition. The webpage features can be divided into 2 classes: static and dynamic features. Dynamic features mainly involve browser actions, pages skipping relations, HTTP requests and so on[14]. Extracting dynamic features is very difficult and usually needs auxiliary techniques such as virtualization and Honeypot[19]. As a result, we mainly focus on static features. Static features are usually from pages' static information, including hosts information [20], URL features [21], web contents (such as HTML, JS tags) [22]. To extract these features is not so difficult.

Although URLs are not natural languages, they are similar with them in a way because they are both unstructured sequences [7]. URL Word embedding [16], character-level Convolutional Neural Networks (CNN) [23], Bidirectional Long-Short Term Memory (Bi-LSTM) [5] and attention mechanism [6] were applied to perform malicious URL feature selection and detection. However, these works have not combined features from webpage files and URLs.

2.3 Identification Methods

The main identification methods of malicious web pages include (1) blacklists, (2) heuristic rules, (3) machine learning algorithms.

The traditional way to identify malicious websites is blacklists [24]. Web browsers collect the human-reported URLs of malicious pages and store them into blacklists, then query whether the URLs of targeted pages are in blacklists. The main drawback of this way is delays in discovering new sites and incomplete coverage. According to Sheng et al. [25], for 47%–83% of all the phishing websites, after being found by people, it was over 12 h before they were stored into blacklists. This means harmful webpages would not be detected until they have caused damage.

The second way is heuristic rules. These heuristic rules often assume that the statistical or binary features of malicious websites are fixed. For large-scale web pages, this method would not only lead to high error rates, but also be hard to update because it relies on the domain knowledges [19].

The third way is machine learning algorithms, which regard malicious websites recognition as a supervised classification task. After extracting features from each web pages, Naïve Bayes, SVM, logistic regression [19] and Fully-Connected Networks (FCN) [23, 26] were constructed for classification.

In conclusion, there exist some aspects which need improvements in the previous works. Firstly, it's necessary to construct an open malicious websites recognition dataset for Chinese webpages. There exist huge differences between Chinese and English webpages. For example, Chinese PinYin is extremely rare in English pages' URLs. What's more, specifying the detailed types of malicious pages would make the dataset more fine-grained and practical. Secondly, datasets containing both URLs, HTML and JS files are scarce. It's obvious that judging a webpage benign or malicious only according to its URL is biased. Thirdly, it's noticeable that the natural language in webpages has been ignored in most of previous researches. Fourthly, pre-trained language models were not used to extract the webpages' features, which may be useful to capture potential features of the malicious pages.

3 Method Framework

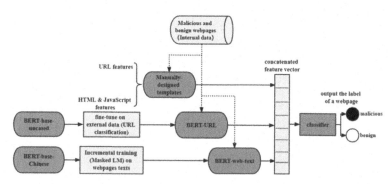

Fig. 1. General framework of malicious webpages detection

The framework of malicious webpages detection method is shown as Fig. 1. We planned to implement transfer learning mechanism of pre-trained language model BERT to detect malicious webpages. The datasets include two parts: external dataset and internal dataset. The purpose of the external dataset is to fine-tune BERT, and the internal dataset is aimed at testing the performance of our method. The feature vector of a webpage is from 3 channels: manually designed feature templates, BERT-URL and BERT-web-text. Each of them would be specified then.

3.1   Construction of Chinese Malicious Webpages Detection Dataset

The dataset "ChiMalPages" includes two parts: external dataset and internal dataset. The external dataset contains only URLs, whose goal is to fine-tune the pre-trained language models, and enable them to learn URLs representations. The internal dataset contains not only URLs but also HTML and JS files, whose goal is to evaluate the performance of our detection method finally.

Table 2. External and internal data

Data type	Contents	Scale
External data	Pure URLs	27000 benign URLs 27000 malicious URLs
Internal data	URLs, HTML and JavaScript files	954 benign pages, 521 malicious pages

According to similar works in English pages, benign samples are from Chinese websites with high traffic amount, such as "sohu.com", "mail.qq.com", "weibo.com", "pku.edu.cn" and "Tmall.com", which can be collected from websites "www.chinaz. com" and "www.alexa.cn". It's worth mentioning that websites traffic amount itself would not be included in samples, to enable our model to focus on pages' contents rather than predict webpages' traffic amount. Malicious URLs in external data are from "CN-Malicious-website-list" [2]. Malicious samples in internal data were collected from the exposure platform on the website "Security Union"[1]. Many Chinese netizens reported the malicious websites to this platform, which would further check them. Before the malicious pages went invalid, we had collected their URLs, HTML and JS files in time. There are no same URLs in internal and external data.

After analyzing data and consulting relevant lawyers, the 521 collected Chinese malicious pages can be divided into 5 types roughly. And their types, numbers, and laws evidence from the People's Republic of China (PRC) are shown in Table 3.

It is worth mentioning that the malicious pages classification is rough. In fact, some malicious pages lie on the border of classes and may have more than one label. For example, some gambling websites also have sexual suggestion, and some phishing web-pages also carry gambling and porn information on inconspicuous places. And many pages of the above 5 classes have excessive ads and popups, or force clients to download unknown software.

---

[1] https://jubao.anquan.org/exposure.

Table 3. Malicious pages types[2]

Malicious pages types	Num	Descriptions
1. Gambling	172	**[Suspicion]** Crime of gambling, opening a casino
2. PORN	84	**[Suspicion]** Crime of organizing or introducing prostitution, crime of spreading obscene articles
3. Phishing	221	**[Details]** Pretend to imitate webpages of Chinese government, education, academic and publication institutes. Imitate webpages of banks, E-mail, e-commerce, phone company, travel agencies and so on **[Suspicion]** Crime of illegally using network
4. Other pages breaking laws	43	**[Details]** Illegal trading of Taobao account, QQ account and phone number, setting up game servers without permission, selling game cheating tools, selling imitations **[Suspicion]** Crime of illegally using information network, destroying computer information system, infringing copyright and so on
5. Undesirable pages	33	**[Details]** These pages do not break laws obviously, but they concern excessive ads, popups, forcing downloading software and other problems

3.2 Manually Designed Feature Template

Many previous researches manually designed features for web pages. Chiew et al. [22] have summarized these feature items to a list. According to previous work on English webpages, we constructed a webpage feature template as Table 4 shows:

After constructing the template, we computed the importance values of each feature items (No. 0–No. 29) in the internal data based on the Random Forest (RF) classification algorithm. RF is a kind of ensemble learning, which generates various subsets of training data and trains various sub-classifiers Decision Tree (DT). Beginning with the root node, DT computes the Information Gain (IG) values of each feature item to determine the best partition feature, and partition samples to child nodes. With the partition process, the purity of samples in child nodes gradually improves, until the samples in child nodes belongs to the same class as far as possible. And after the voting of every classification tree, RF model outputs the predicted label [27].

Based on RF, the way to compute importance values is as follows: in a classification tree of RF, a node corresponds to a feature item f_item and the sample set D. And for the sample set D in a node, the num of classes is n, the num of samples is x, and the proportion of Class i ($i = 1, 2, ..., n$) is p_i, then the information entropy of D $Entropy_D$

---

[2] Some pages belong to more than one class, especially porn and gambling pages.

Table 4. Feature templates of webpages

No	Feature	Description
0	Num_Dots	Num of dots in URLs
1	Url_Length	Num of chars in URLs
2–7	Special_chars	Num of "-" "~" "&" "#" "_" "@" in URLs
8	Numeric	Num of numeric chars in URLs
9	IpAddress	If IP address is used in URLs
10	top-domain	If top-level domains (.com,.cn,.net) are used in URLs
11	2th domain	If second-level domains (.gov,.edu) are used in URLs
12	Sensitive	Num of sensitive words ("secure", "account", "login","signin", "comfirm", "banking") in URLs
13–24	HTML/JS tags	Num of tags "iframe" "eval" "setTimeout" "setInterval" "window.location" "window.open" "setAttribute""innerHTML""encodeURI""hidden" "display:none""download" in HTML / JS files
25	External URL	Num of URLs in HTML/JS files whose domains are different from that of this page's URL
26	JS_proportion	Proportion of JS files size in the webpage folder
27	Pic-Txt ratio	Ratio of num of images to length of text in webpages
28	Approval	If website approve information is provided in webpages
29	If_unicode	If Unicode chars appear in HTML/JS files

can be computed as follows:

$$Entropy_D = \sum_{i=1}^{n} p_i \log_2 p_i \tag{1}$$

The higher $Entropy_D$ is, the harder classification for D is. If the node containing D has m child nodes, and the information entropy of each child node is $Entropy_{sub\_j}$ ($j = 1, 2,..., m$), the num of samples in each child node is x_{sub_j}, then the importance value of the node and corresponding feature item f_item is IG_{f_item}:

$$IG_{f\_item} = x * Entropy_D \sum_{j=1}^{m} x_{sub\_j} Entropy_{sub\_j} \tag{2}$$

Similarly, the importance values of every feature in a tree can be gained. Then compute the average values of each feature item on all the trees of a RF model. After normalization, RF gained final importance values of 30 features (see Fig. 2).

From Fig. 2 we could tell that the 5 most important features in Table 4 are "External URL""Url_Length""Num_Dots""Pic-Txt ratio" and "JS_proportion". And "Pic-Txt ratio" played an important role in Chinese malicious pages detection in deed, which has

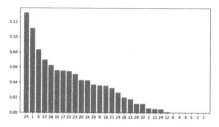

Fig. 2. Feature importance ranking[3]

not been noticed in the previous works. This indicated that many Chinese malicious web-pages may carry their sensitive contents via pictures instead of text, thus escaping being detected. The features in Fig. 2 whose importance ranks from 25th–30th are all relevant to URLs (feature No. 6, No. 4, No. 9, No. 5, No. 3, No. 7 in Table 4). These items were all related to URLs and borrowed from works about English malicious webpages detection. However, their importance is so little. This result reflected that the feature template aimed at English webpages cannot be copied directly to Chinese webpages. Besides, the disguise methods of malicious pages would change, so the feature engineering shouldn't never be updated.

3.3 BERT-URL Based on External Pure URLs Classification

In our task, to gain a vector representation good at differentiating benign and malicious URLs, we try to fine-tune "BERT-base-uncased" via external data in Table 2, which contains 27000 benign URLs and 27000 malicious URLs.

For the URL 2-class classification task, BERT extracted the feature vector v of the marker [CLS] on the top layer as the integrated representation of a URL, and then added a 768 * n Fully-connected layer W (n is the number of classes). Finally, through a softmax layer, the model output the probabilities that a URL belongs to each class c.

During training, the model would adjust W and parameters of 12 layers of BERT to maximize the probability corresponding to the true label.

In addition, we compared the external URLs classification performance of BERT with baseline models: the feature template in Table 4, URL char-level word2vec average vector, Fasttext and RoBERTa-base. The experimental results are shown in Sect. 4.1.

3.4 BERT-Web-text Based on Masked Language Model

To make BERT-base-Chinese encode the style of webpage language, we further trained it via Masked Language Model (Masked LM) [8], which learns semantic information through a cloze task. After Masked LM, BERT-base-Chinese produces new model "BERT-web-text".

---

[3] Numbers on the horizontal coordinates corresponds to numbers of feature items in Table 4.

4 Experimental Results and Analysis

4.1 External URLs Classification

We divided external data (27000 benign URLs and 27000 malicious URLs in Table 2) into the training, validation, and test data according to the proportion 8:1:1. All the training process followed the principle of Early Stopping to prevent overfitting. As for BERT, RoBERTa and DistilBERT, the initial learning rate is 2e−5, batch size is 16, dropout probability is 0.1. The URLs binary classification result is shown in Table 5.

Table 5. External URLs classification performance

Models	Accuracy on test data
URL Features in Table 4 except No. 3–7 and No. 9[a]	76.17%
word2vec (300-dimension)	78.39%
Fasttext (300-dimension)	84.97%
BERT-base-uncased	**87.01%**
RoBERTa-base	86.65%
DistilBERT-base-uncased	86.63%

[a]These feature items have little importance according to the analysis of Sect. 3.2

As Table 5 shows, BERT-base-uncased performed best among 6 models and it produced the fine-tuned model "BERT-URL". To verify the influence to URL representation after fine-tuning on the URL classification task, we extracted 768-dimension URL vectors from different BERT models, and visualized them by the PCA (Principal Component Analysis) dimension reduction algorithm. We selected 100 malicious URLs and 100 benign URLs from test data for an example. The visualization result was shown as Fig. 3 and Fig. 4[4].

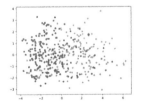
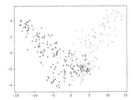

Fig. 3. URL vectors BERT-base-uncased   **Fig. 4.** URL vectors from BERT-URL (Color figure online)

We can find that from the initial model "BERT-base-uncased" to "BERT-URL", URLs belonging to the same class tend to cluster, and distance between different clusters are lengthened. This phenomenon indicated that when doing the external URL classification, the parameters of BERT itself were also optimized.

---

[4] The blue dots represent benign URLs. The red dots represent malicious URLs.

4.2   Chinese Malicious Webpages Detection on the Internal Data

We conducted classification experiments on internal data containing 954 benign webpages and 521 malicious webpages as Table 2 shows.

Random Forest (RF) and Fully-Connect Network (FCN) served as classifiers and the better one would be chosen. After 10-fold cross-validation, the detection results are the average on 10 test sets. And classifiers reaching the best performance were recorded. The results are in Table 6:

Table 6.   Chinese malicious webpage detection results on internal data

No	Feature extraction method	Acc (%)	F1 (%)
1	Table 4 except feature No. 3–7 and No. 9 which have little importance (hereafter abbreviated as **Template**)	80.61	72.47
2	URL features from **Template**	75.80	65.23
3	URL features from BERT-base-uncased	76.67	63.91
4	BERT-URL	81.09	72.25
5	**Template** + BERT-URL	82.58	75.04
6	BERT-URL + BERT-base-Chinese	84.67	78.41
7	BERT-URL + BERT-web-text	85.22	78.87
8	**Template** + BERT-URL + BERT-web-text	**85.76**	**79.84**

Some feature extraction methods are interpreted as follows:

"**Template**" means the webpages feature template based on Table 4, excluding feature items No. 3–7 and No. 9 which have little importance according to the analysis of Sect. 3.2. "BERT-URL" was from BERT-base-uncased fine-tuned on external URL classification task, as Sect. 4.1 described. "BERT-web-text" was from BERT-base-Chinese further pre-trained described in Sect. 3.4. Inspired by Sun et al. [28], if text length of a page is over 510, BERT extracted the representation of first 250 characters and the last 250 characters.

There are some findings according to Table 6. Firstly, when feature vectors have low dimensions, the classification performance of RF is superior to that of FCN. While the number of dimensions exceeds 1500 (No. 6 and No. 7 in Table 4), FCN becomes better.

Secondly, comparing experiment No. 2–No. 4, we found that the acc and F1 of BERT-URL increase by 5.29% and 7.02% compared with those of manually designed URL features. This indicated that after being fine-tuned on the external URLs, BERT-URL has the ability to distinguish malicious and benign URLs to a certain extent.

Thirdly, comparing experiment No. 1 with No. 7, although losing the HTML and JS features, the feature extraction method base on BERT-URL and BERT-web-text is still obviously superior to the feature **Template.** This result proved the usefulness of pre-trained language models on the malicious webpages detection task.

Fourthly, comparing experiment No. 1 with No. 8, after integrating the features from **Template**, BERT-URL and BERT-web-text, the performance reached the best level, indicating the feature fusion method we proposed plays a prominent part.

Finally, comparing experiment No. 4 with No. 7, No. 5 with No. 8, adding the webpages features from BERT-web-text can increase F1 by 6.62% and 4.8%. These experiments showing that natural language features in webpages cannot be neglected.

We selected one of the 10-fold cross-validation, and analyzed its detection results on detailed malicious webpages classes as Table 7 shows.

Table 7. Detailed malicious pages detection performance based on Feature fusion.

Webpages types	Num	Num of true positive
Benign webpages	96	86
Malicious-porn pages	9	8
Malicious-gambling pages	20	16
Malicious-phishing pages	20	16
Other pages breaking laws	10	8

From Table 7 we could tell that the feature fusion method performed well on various malicious pages, which proving this detection method is comprehensive and not biased.

4.3 Webpages High-Frequency Words Analysis

In addition, we counted the high-frequency words in malicious webpages and drew a wordcloud picture as Fig. 5. It's worth mentioning that regardless of context, some high-frequency words in malicious pages are hardly blamed in deed. For example, "成人" usually means "a mature, fully developed person legally responsible for his actions", while it may also serve as an euphemism for "porn" in some cases. As a result, taking context into consideration, BERT-web-text is more suitable for detecting sensitive contents than the simple character matching method based on sensitive vocabulary.

Fig. 5. Wordcloud of malicious webpages

5 Conclusion

Pre-trained language models have shown their power on sequence modeling [29]. This paper applied the pre-trained language models to malicious webpages detection, proposed the feature fusion method for Chinese malicious pages detection. Firstly, this paper released an open Chinese malicious webpages detection dataset, which contains external and internal data, specifying the detailed types of malicious pages and their legal risks. Secondly, we measured and ranked each feature's importance based on Information Gain as Fig. 2 shows, which can optimize the feature engineering. Thirdly, we fine-tuned BERT on external URLs classification and webpages text, producing new models "BERT-URL" and "BERT-web-text". The classification result showed the URL-encoding power of BERT obviously exceeds manual feature items, word2vec and Fasttext, even slightly exceeding RoBERTa and DistilBERT. Finally, combining features, the detection accuracy and F1 score reached 85.76% and 79.84%, respectively increasing by 5.15% and 7.37% compared with those of the machine learning method based on manual feature templates. These experiments proved that BERT works well on malicious webpages detection. Besides, BERT can tune webpage representation via transfer learning as webpages evolve. It extends the application fields of pre-trained models on cyber-security, and saves the cost of feature engineering, improving the efficiency of system update.

References

1. CNNIC. The 49th China Statistical Report on Internet Development. http://www.cnnic.cn/hlwfzyj/hlwxzbg/. Accessed 28 Mar 2022
2. Zzhihao. CN-Malicious-website-list. https://github.com/zzhihao2017/CN-Malicious-website-list. Accessed 28 Mar 2022
3. National Internet Emergency Center. 2020 China Internet Network Security Report. https://www.cert.org.cn/publish/main/17/index.html. Accessed 28 Mar 2022
4. Wan, M., Yao, H.: GAN model for malicious web training data generation. Comput. Eng. Appl. (6), 1–10 (2020)
5. Wang, H., Yu, L., Tian, S.W., et al.: Bidirectional LSTM Malicious webpages detection algorithm based on convolutional neural network and independent recurrent neural network. Appl. Intell. **49**(8), 3016–3026 (2019)
6. Peng, Y., Tian, S., Yu, L.: A joint approach to detect malicious URL based on attention mechanism. Int. J. Comput. Intell. Appl. **18**(3) (2019)
7. Sahoo, D., Liu, C., Hoi, S.C.H.: Malicious URL Detection using Machine Learning: A Survey. arXiv e-prints, 1701-7179 (2017)
8. Devlin, J., Chang, M., Lee, K., et al.: BERT: Pre-training of Deep Bidirectional Transformers for Language Understanding. arXiv e-prints, 1810-4805 (2018)
9. Liu, Y., Ott, M., Goyal, N., et al.: RoBERTa: A Robustly Optimized BERT Pretraining Approach. arXiv e-prints, 1907-11692 (2019)
10. Tao, W., Yu, S., Xie, B.: A novel framework for learning to detect malicious web pages. In: 2010 International Forum on Information Technology and Applications, vol. 2, pp. 353–357 (2010)
11. Gowtham, R., Krishnamurthi, et al.: A comprehensive and efficacious architecture for detecting phishing webpages. Comput. Secur. **40**, 23–37 (2014)

12. Xu, L.: A research of phishing detection technology based on deep learning. University of Electronic Science and Technology of China, ChengDu (2017)
13. Ye, Z.: Designing and application of a large-scale and fast malicious web page recognition method based on combination of Kafka and spark-streaming. Nanjing University of Posts and telecommunications, Nanjing (2019)
14. Wei, X., Cheng, W.: Malicious web page recognition based on feature fusion and machine learning. J. Nanjing Univ. Posts Telecommun. (Nat. Sci. Ed.) (5), 95–104 (2019)
15. Hu, Z., Wang, C., Wu, J., et al.: Malicious websites identification based on hyperlink analysis and classification rule. J. Inf. Resour. Manag. (1), 105–113 (2019)
16. Wu, H.: Research and implementation of activate defense technology for malicious crawlers. Beijing University of posts and telecommunications, Beijing (2019)
17. Zhou, W.: Machine learning based malicious webpage analysis. Shanghai Jiaotong University, Shanghai (2019)
18. Chen, B., Song, L.: Malicious webpage detection method for webpage content link hierarchy semantic tree. Comput. Eng. Appl. (11), 90–97 (2020)
19. Sha, H., Liu, Q., Liu, T., et al.: Survey on malicious webpage detection research. Chin. J. Comput. (3), 529–542 (2016)
20. Seifert, C., Komisarczuk, P., Welch, I., et al.: Identification of malicious web pages through analysis of underlying DNS and web server relationships. In: IEEE Conference on Local Computer Networks, pp. 935–941 (2008)
21. Spirin, N., Han, J.: Survey on web spam detection. ACM SIGKDD Explor. Newsl. **13**(2), 50 (2012)
22. Chiew, K.L., Tan, C.L., Wong, K.S., et al.: A new hybrid ensemble feature selection framework for machine learning-based phishing detection system. Inf. Sci. (484),153–166 (2019)
23. Le, H., Pham, Q., Sahoo, D., et al.: URLNet: Learning a URL Representation with Deep Learning for Malicious URL Detection. arXiv e-prints, 1802-3162 (2018)
24. Peng, P., Yang, L., Song, L.: Opening the blackbox of virustotal: analyzing online phishing scan engines. In: The Internet Measurement Conference, pp. 478–485 (2019)
25. Sheng, S., Wardman, B., Warner, G., et al.: An empirical analysis of phishing blacklists. In: 6th Conference on Email and Anti-Spam, CEAS 2009, Mountain View, CA, United states (2009)
26. Saxe, J., Berlin, K.: eXpose: A Character-Level Convolutional Neural Network with Embeddings For Detecting Malicious URLs, File Paths and Registry Keys. arXiv e-prints, 1702-8568 (2017)
27. Zhou, Z.: Machine Learning, pp. 178–181. Tsinghua University Press, Beijing (2016)
28. Sun, C., Qiu, X., Xu, Y., Huang, X.: How to fine-tune BERT for text classification? In: Sun, M., Huang, X., Ji, H., Liu, Z., Liu, Y. (eds.) CCL 2019. LNCS, vol. 11856, pp. 194–206. Springer, Cham (2019). https://doi.org/10.1007/978-3-030-32381-3_16
29. Yu, P., Wang, X.: BERT-based named entity recognition in Chinese twenty-four histories. In: Wang, G., Lin, X., Hendler, J., Song, W., Xu, Z., Liu, G. (eds.) WISA 2020. LNCS, vol. 12432, pp. 289–301. Springer, Cham (2020). https://doi.org/10.1007/978-3-030-60029-7_27

Code Comments Generation with Data Flow-Guided Transformer

Wen Zhou and Junhua Wu[✉]

School of Computer Science and Technology, Nanjing Tech University,
Nanjing 211816, China
wujh@njtech.edu.cn

Abstract. Code comments generation is a multi-disciplinary research, involving many different fields such as software engineering and natural language processing. In general, the automatic generation of code comments takes the source code as the input and its corresponding natural language description as outputs. In previous work, the code is regarded as a sequence and modeled by sequence models such as RNN, ignoring the data flow between variables and long-range dependencies in the code. Therefore, we propose a novel DFG-trans model to automatically generate code comments. Our model adds data flow guidance to the Transformer, which allows the model to focus on the flow of data between variables. This leads to better learning of semantic information in the code and alleviates the problem of long-distance dependencies. Experiments on a Java dataset show that our proposed model effectively improves the quality of code comments. Under various evaluation metrics, the scores of DFG-trans are better than some existing advanced models.

Keywords: Code comments · Natural language processing ·
Transformer · Data flow

1 Introduction

Code comments generation, also known as source code summarization, aims to generate natural language descriptions of code automatically. Effective code comments enable programmers to understand the function of a piece of the program quickly, thus greatly improving the reusability of code and the efficiency of programming. Writing code comments manually is time-consuming and inefficient. In addition, code comments need to be updated as the code changes. However, many existing code comments tend to be mismatched and outdated. Therefore, the research on code comments generation is essential and has attracted extensive attention.

Code comments generation is a multi-disciplinary research, involving many different fields such as software engineering and natural language processing (NLP). With the development of deep learning, deep neural networks are gradually introduced into program understanding, such as code generation [1]. Deep learning-based approaches to modeling code comment tasks are mainly seen as Neural Machine Translation (NMT). Most of them use an RNN or CNN-based encoder-decoder model, also known as the seq2seq model [2,3]. Specifically, the encoder encodes the source code into a vector representation of a certain length, which is then decoded by the decoder to generate the natural language description.

However, the traditional seq2seq model still has two limitations: (1) RNN or CNN-based sequence models cannot capture the long-range dependencies of code tokens; (2) the structural information of the code is not fully utilized in feature representation, and the use of AST to extract structural features brings some unnecessary deep structure, which leads to an overly complex model.

For the first point, we build the model based on Transformer, which captures the long-range dependencies of code tokens well [4]. Transformer was first proposed by Vaswani et al. [5] in 2017 and performs well in many NLP tasks, such as machine translation, text summarization, etc. Secondly, in order to make the model fully learn the structural information in the code, we choose the semantic-level graph structure, i.e. data flow, rather than the syntactic-level structure like AST, which allows for better learning of the code semantics and also makes the model more efficient. Data flow defines the flow of data between variables. Unlike AST, data flow pays more attention to the dependencies of data in code, which is the manifestation of code semantics.

In this paper, we propose a new model named DFG-Trans to generate code comments automatically, which is a Transformer variant. To incorporate the data flow into Transformer, we introduce a self-attention mechanism with data flow guided. The encoder layer guided by data flow is stacked with the basic encoder layer to form a new encoder, thus combining the sequential and structural information of the code. Our proposed approach not only alleviates the problem of long-range dependencies but also extracts structural features of the code well. When decoding, the decoder generates a probability distribution in the output vocabulary, and we use an improved beam search algorithm to sample the probability distribution to generate the most likely natural language description. To improve the accuracy of the comments generation, we normalize the length of the sampled sentences and add a coverage penalty to the scoring function.

2   Related Work

Basic Seq2seq Model. The basic seq2seq model regards code comments generation as a machine translation problem. Iyer et al. [6] first proposed code-NN, a seq2seq model with an attention mechanism, which uses long short-term memory (LSTM) neural network to constitute the basic unit of encoder and decoder. Liang and Zhu et al. [7] used Code-RNN for encoding, extracting features from the code and constructing vector representations. Hu et al. [8] trained the code comments model with transferred API knowledge and used a bi-encoder structure to generate comments for Java code. Unlike these works, Allamanis et al. [3] proposed a convolution-based neural network model that uses a CNN with the attention mechanism to learn translation invariable features in the code and the gated recurrent unit (GRU) as the decoder.

Structure-based Neural Networks. In recent years, more and more works have focused on structural features in code. Hu et al. [2] leveraged Structure-based Traversal (SBT) to traverse the AST of source code and convert it into a sequence that can be learned by RNN. LeClair et al. [9] proposed the ast-attendgru model, which comprehensively considers two types of code information: word-based representation and AST-based representation. Shido et al. [10] proposed an extended Tree-LSTM. Different from these approaches based on RNN, some works no longer parse AST as a flat input. LeClair et al. [11] used graph neural networks (GNN) based encoder to model the AST. Liu et al. [12] merged multiple different code representation methods (such as AST, control dependency graph, and program dependency graph) into a joint code property graph, and designed a new GNN model to encode the code snippet.

Other Noteworthy Works. Wan et al. [13] considered hybrid code representation and further used the deep reinforcement learning framework (i.e. actor-critic) to solve the exposure bias problem during decoding. Wei et al. [14] proposed a method based on dual learning, which trains the model by using the duality of comments generation task and code generation task. Ahmad et al. [4] used Transformer to generate code comments. Based on the work of them, we combine the Transformer with the data flow to improve the performance of the model on the task of automatic code comments generation.

3   Proposed Approach

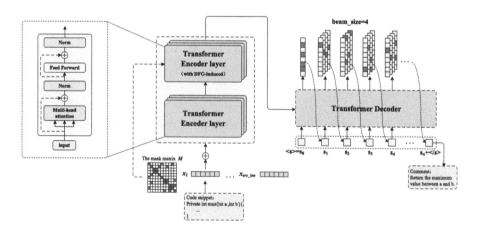

Fig. 1. The overall model architecture with data flow guided.

DFG-trans is a variant of Transformer with data flow guided. This section mainly introduces the following two aspects: the structured representation of data flow and the method of code comments generation with data flow guided. The Transformer encoder consists of two parts, where the improved self-attention mechanism is adopted in the data flow-guided encoder layer. Then the Transformer decoder employs the improved beam search algorithm to generate the final code comments for decoding. The overall model architecture is shown in Fig. 1.

3.1 Data Flow

Data flow is a graph that represents the dependencies between variables, in which nodes represent variables and edges represent the flow of data between variables. Unlike abstract syntax trees, the data flow of the code remains constant even under different abstract syntax. This graph structure can provide better semantics in the code for the machine translation model. Many programmers do not strictly follow the naming conventions when programming. Therefore, it is difficult for the model to understand the real semantics of the code by only modeling the code comments generation task as a machine translation problem. As shown in Fig. 2, after parsing the given source code into an abstract syntax tree, extract all variable nodes in the source code and construct the data flow [15]. The resulting data flow is represented by $G = (V, E)$, where V represents the set of nodes and E represents the set of edges. Each node in the data flow is denoted as $v_i^{pos(i)}$, and $pos(i)$ represents the position of the ith node in the sequence of source code. The matrix $M \in R^{src\_len \times src\_len}$ records variable data dependencies in the source code, where src_len represents the length of source code sequence.

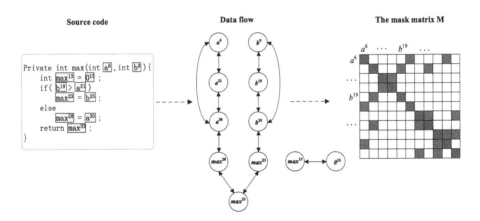

Fig. 2. The structured representation of code data flow.

$$M_{pos(i)pos(j)} = \begin{cases} 1 & \langle v_i, v_j \rangle \in E \text{ or } i = j \\ -\infty & \text{otherwise} \end{cases} \quad (1)$$

where $\langle v_i, v_j \rangle \in E$ indicates that there is an edge between v_i and v_j. $M_{pos(i)pos(j)}$ is equal to 1 when there is an edge between the ith node and the jth node or i equals j, otherwise M is denoted as negative infinity. The matrix M is introduced to the self-attention mechanism. If $M_{pos(i)pos(j)}$ equals 1, the attention score will remain unchanged, and if M equals negative infinity, the attention score will be calculated as 0. This allows the attention score between two tokens in the sequence that have no data dependency to be overwritten, thus making the model more attentive to the data flow relationships in the code and learning the semantics of the variables.

3.2   Transformer Variant Guided by Data Flow

As shown in Fig. 1, the overall architecture of DFG-Trans is mainly composed of an encoder and decoder. The encoder consists of a stack of N encoder layers, each of which consists of a multi-head self-attention mechanism layer and a fully connected feed-forward neural network layer [4]. These encoder layers are divided into two categories, one with data flow guided and the other without, so the calculation of attention score is different.

Self-Attention. We use sine and cosine functions to capture the relative position of each token and add it to the input embedding to obtain the final vector representation, which is used as the input of the self-attention layer.

The self-attention mechanism in Transformer is implemented by the calculation of three matrices: query Q, key K, and value V. The vector representation of the code sequence $X = [x_1, x_2...x_{src\_len}]$ is mapped to Q, K and V through three linear layers,

$$Q = QW_i^Q, K = KW_i^K, \quad V = VW_i^V \quad (2)$$

where W_i^Q, $W_i^K \in R^{d\_model \times d\_k}$, $W_i^V \in R^{d\_model \times d\_v}$ are the trainable weight matrices of the model, and the initial values of the Q, K, V matrices is X. The encoder layer without data flow guided adopts the self-attention mechanism proposed by Vaswani et al. [5], i.e. Scaled Dot-Product Attention. Matrices Q, K, and V are input into the multi-head self-attention layer, and the matrix of attention output C is calculated as:

$$C = \text{softmax}\left(\frac{QK^T}{\sqrt{d_k}}\right) V \quad (3)$$

Data Flow-Guided Masked Attention. In order to guide the model to focus on the data dependencies in the code and learn the code semantics, we define the data flow adjacency matrix M as the mask matrix (see 3.1). The improved self-attention mechanism is adopted in the encoder layers with data flow guided, and the self-attention output C' is computed as:

$$C' = \text{softmax} \left(\frac{M \cdot QK^T}{\sqrt{d_k}} \right) V \tag{4}$$

The matrix M records the data flow information in the code. When there are data dependencies between the $i-th$ token and the $j-th$ token, M_{ij} is equal to 1, otherwise M_{ij} is equal to negative infinity. After the calculation of the *softmax* function, the attention score between two tokens without data dependencies will be dropped out. After calculating the attention value of one head, the multi-head self-attention output is defined as follows:

$$MultiAttn(Q, K, V) = Concat\left(head_1, \ldots, head_h\right) W^o \tag{5}$$

The model concatenates the attention output of each head through $Concat$ function to obtain the final multi-head attention matrix, where $head_i$ denotes the output of the i-th head, h represents the number of heads, and W^o is a trainable parameter of the model.

Feed-forward. After residual connection and normalization, the output of the multi-head self-attention layer serves as the input H of the fully connected feed-forward network layer, which includes two linear transformations,

$$FEN(H) = RELU\left(HW_1 + b_1\right) W_2 + b_2 \tag{6}$$

where W_1, $W_2 \in R^{d\text{-}ff \times d\text{-}model}$, $b_1 \in R^{d\text{-}ff}$, $b_2 \in R^{d\text{-}model}$ are trainable parameters of the model and $ReLU$ function is used to activate the output between two linear transformations.

3.3   Improved Beam Search Decoding

The decoder generates the probability distribution of words in the natural language vocabulary based on the output of the encoder, and then uses a decoding algorithm to sample the probability distribution to generate the most likely comments. The greedy search and beam search [16] are usually used in decoding to obtain the target sentence. The strategy of greedy search is to select the token with the maximum probability at each time step until the terminator $\langle /s \rangle$ is output or the sentence reaches the maximum length. Beam search is a pruned breadth-first search algorithm, improved from greedy search algorithm. Within an affordable computational cost, beam search algorithm can obtain the relatively optimal sentence. In this paper, we use beam search to decode and generate natural language comments. The pseudocode is shown in Alg. 1, where ∘ denotes string concatenation and **V** is the target vocabulary.

Generally, the beam search algorithm takes log-likelihood as the score function. However, log-likelihood produces a negative probability, which accumulates with the increase of sentence length, resulting in lower (or more negative)

Algorithm 1. beam search

Input: source code $\mathbf{x}$, beam size k, maximum hypothesis length n

1: $Beam_0 \leftarrow \{\langle 0, BOS \rangle\}$
2: **for** $t \in \{1, ..., n-1\}$ **do**
3:      $Heap \leftarrow \emptyset$
4:      **for** $\langle s, \mathbf{y} \rangle \in Beam_{t-1}$ **do**
5:          **if** $\mathbf{y}.last() = EOS$ **then**
6:              $Heap.add(\langle s, \mathbf{y} \rangle)$
7:              continue
8:          **for** $y \in \mathbf{V}$ **do**
9:              $s \leftarrow score(\mathbf{x}, \mathbf{y} \circ y)$
10:             $Heap.add(\langle s, \mathbf{y} \circ y \rangle)$
11:     $Beam_t \leftarrow Heap.top(k)$
12: **return** $Beam_{n-1}.max()$

---

scores for long sentences [17]. Therefore, we normalize the length of sentences to improve the original heuristic algorithm. In addition, to make each token in the input sequence evenly noticed by the decoder, i.e. cover the entire input sequence, a coverage penalty is added to the score function. The improved scoring function $score(\mathbf{x}, \mathbf{y})$ is defined as:

$$\text{score}(\mathbf{x}, \mathbf{y}) = \frac{1}{L^\alpha} \log P\left(y_1, \ldots, y_L \mid \mathbf{x}\right) + CP(\mathbf{x}; \mathbf{y}) \tag{7}$$

We normalize the sentence by dividing the log-likelihood by L^α, where L represents the length of the sentence and $\alpha \in [0, 1]$ is a hyperparameter of the model. $CP(\mathbf{x}; \mathbf{y})$ is the coverage penalty function [17], which is specifically defined as:

$$CP(\mathbf{x}; \mathbf{y}) = \beta \cdot \sum_{i=1}^{L_{\mathbf{x}}} \log \left(\min \left(\sum_{j=1}^{L} P_{ij}, 1 \right) \right) \tag{8}$$

where P_{ij} represents the attention score between the jth target token and the ith input token, $L_{\mathbf{x}}$ refers to the length of the input sequence, and $\beta \in [0, 1]$ is the hyperparameter of the model.

4 Experiments

4.1 Datasets and Pre-processing

This paper conducts experiments on a Java dataset [2]. The dataset contains 69708 pairs of code-comments sequences, of which 89% are less than 200 in length and 95.45% are less than 50. The source code is parsed into an abstract syntax tree with a compiler tool, and then the variable information in leaves is extracted to construct data flow. The nodes in data flow contain the position of each variable in the code sequence in order to construct adjacency matrices. Code sequences and the adjacency matrices are then used for model training and testing.

4.2 Metrics and Training Details

Metrics. The three machine translation metrics used in the experiments are BLEU, METEOR, and ROUGE-L to evaluate the performance of the model. BLEU [18] is an evaluation metric for the accuracy-based similarity measure, analyzing the degree of simultaneous occurrence of the n-gram in candidate and reference. METEOR [19] calculates the harmonic average of accuracy and recall between candidate and reference. ROUGE-L [20] computes precision and recall based on the longest common subsequence, similar to BLEU.

Training Details. DFG-trans model is trained on GPU based on the PyTorch framework for a maximum of 250 epochs, which contains a 6-layer Transformer encoder and decoder. The embedding dimension of the word vector is set to 512 and the batch size is set to 16. The Adam optimizer is used to optimize the model during training, and the initial learning rate is 5e–4 with an attenuation rate of 0.95. We use the beam search algorithm when decoding, and set the beam size to 4.

4.3 Results and Analysis

Overall Results. We compare our proposed model (DFG-Trans) with six baselines reported in Ahmad et al. [4] and their proposed approach. Among them, Tree2Seq [21] and DeepCom [2] both take the structural information of the code into account, RL+Hybrid2Seq [13] uses deep reinforcement learning to solve the exposure bias during decoding, API+Code [8] leverages API sequences to assist code comments generation, Dual Model[14] trains the two tasks of code generation(CG) and code summarization(CS) simultaneously through a Dual training framework and utilizes the relationship between them to improve the performance of both sides. These models do a great deal of work in extracting the code semantics, but they do not consider the code semantics expressed by the data flow of the code.

The overall results of the model comparison are shown in Table 1, in which the results of the baseline models are quoted from Ahmad et al. [4]. According to the experimental results, the BLEU of DFG-Trans model is 45.25, the METEOR is 26.79, and the ROUGE-L is 55.26, all of which are better than other models. In particular, compared with the Transformer-based Approach proposed by Ahmad et al. [4], the three evaluation metrics improved by 0.67 BLEU, 0.36 METEOR, and 0.5 ROUGE-L points respectively.

Table 1. The results of our model compared with baseline.

Models	BLEU	METEOR	ROUGE-L
CODE-NN	27.60	12.61	41.10
Tree2Seq	37.88	22.55	51.50
RL+Hybrid2Seq	38.22	22.75	51.91
DeepCom	39.75	23.06	52.67
API+CODE	41.31	23.73	52.25
Dual model	42.39	25.77	53.61
Transformer	44.58	26.43	54.76
DFG-Trans	**45.25**	**26.79**	**55.26**

Ablation Study. Different from other models, DFG-trans takes into account both code sequence information and data flow information between variables. Therefore, in order to analyze the impact of code data flow on model performance, we carried out ablation experiments, respectively 1) only taking code sequence as input (w/o DFG), 2) with data flow guided but not splitting encoder (w/o Split). We compared these two groups of experiments with DFG-Trans, and added three metrics: precision, recall, and F1-Score, to further validate the performance of the model. The results are shown in Table 2.

Table 2. The results of ablation experiment.

Models	BLEU	METEOR	ROUGE-L	Precision	Recall	F1-score
w/o DFG	38.89	22.78	48.63	55.71	49.86	50.22
w/o Split	36.88	21.03	47.12	54.42	48.39	48.71
DFG-Trans	**45.25**	**26.79**	**55.26**	**60.35**	**53.33**	**54.49**

As shown in Table 2, the BLEU score of the model without data flow guided (w/o DFG) is only 38.89, which is 6.36 lower than that of DFG-trans, and other evaluation metrics are also far lower than that of DFG-trans. It indicates that the guidance of data flow enables the model to better learn the semantic information in the code, thus improving the accuracy of code comments generation. In addition, we observe that the result of the model with data flow guided but not splitting the encoder (w/o split) is even worse than the former (w/o DFG). The data flow adjacency matrix is used to cover the code sequence in the attention layer, which aims to make the model pay attention to the data flow relationship between variables. So that if this were done for all encoder layers, some of the other information in the code sequences would be lost, leading to the degradation in model performance. Table 3 presents an example of Java code from the test set, with different comments generated by the three models. It can be seen that the comments generated by DFG-trans are closer to the real comments, which indicates that DFG-trans can better understand the code semantics.

Table 3. The example of java code comments generation.

```
public int countTokens ( ) {
    int count = NUM ;
    boolean inToken = _BOOL ;
    for ( int i = pos , len = str.length( ) ; i < len ; i ++ ) {
        if ( delimiters.indexOf ( str.charAt (i) , NUM ) >= NUM ) {
            if ( returnDelimiters ) count ++ ;
            if ( inToken ) { count ++ ; inToken = _BOOL ; }
        } else {
            inToken = _BOOL ;
        }
    }
    if ( inToken ) count ++ ;
    return count ;
}
```

w/o DFG: returns the count of tokens remaining in the string .

w/o split: returns the count of tokens .

DFG-Trans: returns the count of unprocessed tokens remaining in the string .

True comm: returns the number of unprocessed tokens remaining in the string .

```
public final void goToNextPage ( ) {
    final boolean isLastPage ;
    isLastPage = viewPager.getCurrentItem ( ) == ( pages.size ( ) - 1 ) ;
    if ( !isLastPage ) {
        viewPager.setCurrentItem ( viewPager.getCurrentItem ( ) + 1 ) ;
    }
}
```

w/o DFG: navigate to the next page .

w/o split: navigate to the next, if not .

DFG-Trans: navigate to the next page, if not already at last page .

True comm: navigate to the next page, if not already on the last page .

Lengths Analysis. In order to analyze the influence of data flow on the long-range dependencies between code tokens, we conducted experiments with code sequences of different lengths as input, compared the performance of the model with data flow guided and its absence. The scores of BLEU and ROUGE-L are shown in Fig. 3.

Intuitively, the performance of DFG-Trans is relatively stable with the increase of code length, and there is no significant decline. In contrast, the BLEU and ROUGE-L scores of the model without data flow guided decrease significantly after the code length reaches 240. Especially when the code length is 330, the model performance reaches the lowest. It demonstrates that DFG-Trans can better alleviate the problem of long-range dependencies between code tokens than the basic Transformer model. And our proposed model generates better comments even for long code sequences.

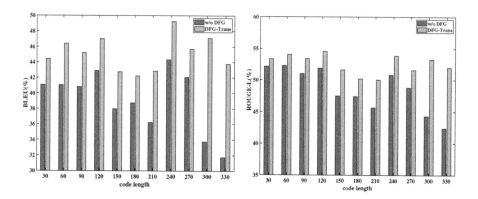

Fig. 3. BLEU and ROUGE-L score for different length of code.

5   Conclusion

In this paper, we propose a novel Transformer-based model for code comments generation, which improves Transformer by using the data flow-guided self-attention mechanism. Both code sequence information and data dependence are considered to further address the problem of long-range dependencies. In addition, the improved beam search algorithm further improves the accuracy of the generated natural language description. Experiments on a large Java dataset, compared with advanced baselines, demonstrate the superiority of our proposed model and illustrate that data flow can improve the code semantics understanding of the model. In future work, we will try to combine code control flow and data flow to extract code semantics, to further improve the performance of the model, and apply the existing results to other tasks in the field of program understanding, such as code generation, code search, etc.

Acknowledgments. This work was supported in part by the Natural Science Foundation of Jiangsu Province (21KJB520027).

References

1. Wang, Y., Wu, J.: The code generation method based on gated attention and interAction-LSTM. In: Xing, C., Fu, X., Zhang, Y., Zhang, G., Borjigin, C. (eds.) WISA 2021. LNCS, vol. 12999, pp. 544–555. Springer, Cham (2021). https://doi.org/10.1007/978-3-030-87571-8_47
2. Hu, X., Li, G., Xia, X., Lo, D., Jin, Z.: Deep code comment generation. In: 2018 IEEE/ACM 26th International Conference on Program Comprehension (ICPC), pp. 200–20010. IEEE (2018)
3. Allamanis, M., Peng, H., Sutton, C.: A convolutional attention network for extreme summarization of source code. In: International Conference on machine Learning, pp. 2091–2100. PMLR (2016)
4. Ahmad, W.U., Chakraborty, S., Ray, B., Chang, K.W.: A transformer-based approach for source code summarization. arXiv preprint arXiv:2005.00653 (2020)
5. Vaswani, A., et al.: Attention is all you need. Adv. Neural Inf. Process. Syst. **30** (2017)
6. Iyer, S., Konstas, I., Cheung, A., Zettlemoyer, L.: Summarizing source code using a neural attention model. In: Proceedings of the 54th Annual Meeting of the Association for Computational Linguistics (Volume 1: Long Papers), pp. 2073–2083 (2016)
7. Liang, Y., Zhu, K.: Automatic generation of text descriptive comments for code blocks. In: Proceedings of the AAAI Conference on Artificial Intelligence, vol. 32, no. 1 (2018)
8. Hu, X., Li, G., Xia, X., Lo, D., Lu, S., Jin, Z.: Summarizing source code with transferred API knowledge (2018)
9. LeClair, A., Jiang, S., McMillan, C.: A neural model for generating natural language summaries of program subroutines. In: 2019 IEEE/ACM 41st International Conference on Software Engineering (ICSE), pp. 795–806. IEEE (2019)
10. Shido, Y., Kobayashi, Y., Yamamoto, A., Miyamoto, A., Matsumura, T.: Automatic source code summarization with extended tree-lstm. In: 2019 International Joint Conference on Neural Networks (IJCNN), pp. 1–8. IEEE (2019)
11. LeClair, A., Haque, S., Wu, L., McMillan, C.: Improved code summarization via a graph neural network. In: Proceedings of the 28th International Conference on Program Comprehension, pp. 184–195 (2020)
12. Liu, S., Chen, Y., Xie, X., Siow, J.K., Liu, Y.: Automatic code summarization via multi-dimensional semantic fusing in gnn. arXiv preprint arXiv:2006.05405 (2020)
13. Wan, Y., et al.: Improving automatic source code summarization via deep reinforcement learning. In: Proceedings of the 33rd ACM/IEEE International Conference on Automated Software Engineering, pp. 397–407 (2018)
14. Wei, B., Li, G., Xia, X., Fu, Z., Jin, Z.: Code generation as a dual task of code summarization. Adv. Neural Inf. Process. Syst. **32** (2019)
15. Guo, D., et al.: Graphcodebert: Pre-training code representations with data flow. arXiv preprint arXiv:2009.08366 (2020)
16. Freitag, M., Al-Onaizan, Y.: Beam search strategies for neural machine translation. arXiv preprint arXiv:1702.01806 (2017)
17. Wu, Y., et al.: Google's neural machine translation system: Bridging the gap between human and machine translation. arXiv preprint arXiv:1609.08144 (2016)
18. Papineni, K., Roukos, S., Ward, T., Zhu, W.J.: Bleu: a method for automatic evaluation of machine translation. In: Proceedings of the 40th Annual Meeting of the Association for Computational Linguistics, pp. 311–318 (2002)

19. Banerjee, S., Lavie, A.: Meteor: an automatic metric for MT evaluation with improved correlation with human judgments. In: Proceedings of the ACL Workshop on Intrinsic and Extrinsic Evaluation Measures for Machine Translation and/or Summarization, pp. 65–72 (2005)
20. Lin, C.Y.: Rouge: A package for automatic evaluation of summaries. In: Text Summarization Branches Out, pp. 74–81 (2004)
21. Eriguchi, A., Hashimoto, K., Tsuruoka, Y.: Tree-to-sequence attentional neural machine translation. arXiv preprint arXiv:1603.06075 (2016)

Coreference Resolution with Syntax and Semantics

Zhiyuan Ma[1], Dan Meng[2], Chao Kong[1], Liang Zhou[1], Mengfei Li[1],
and Wan Tao[1(⊠)]

[1] School of Computer and Information, Anhui Polytechnic University, Wuhu, China
{mzy,lmf,lqj}@stu.ahpu.edu.cn, {kongchao,lzhou,taowan}@ahpu.edu.cn
[2] OPPO Research Institute, Shenzhen, China
mengdan@oppo.com

Abstract. Recent years have witnessed a widespread increase of interest in coreference resolution (CR) in the natural language processing (NLP) community. While most research efforts have focused on the span-based method, there has been relatively little research dedicated to CR with non-span method. Arguably, the non-span representation method like token representation can also be applied to reduce computing overhead by ignoring to many predicted spans. However, these methods are suboptimal in doing so, science it can hardly use the document syntax and semantics information well. This work addresses the research gap of CR with syntax and semantics. We present a new solution CRS_2, short for Coreference Resolution with Syntax and Semantics to capture syntax and semantics information simultaneously. Technically speaking, we made two contributions: (1) We employ a word-level method to capture coreference resolution between tokens, which can reduce the cost of computation. (2) We enhance token representation by incorporating syntax and semantics information from neighborhood tokens to improve span prediction performance. We perform extensive experiments on real dataset OntoNotes 5.0 covering the Chinese and English parts, empirically verifying the effectiveness and efficiency of CRS_2.

Keywords: Coreference resolution · Word-level · Non-span representation · Token enhance

1 Introduction

As one of the most important and basic tasks in the field of natural language processing, coreference resolution aims to aggregate all mentions pointing to the same entity in the real world. In the actual task, firstly, the text composed of words is processed, the mentions corresponding to entities are predicted from all possible spans of document, then whether each mention has a corresponding antecedent is found, and all mentions referring to the same entity will be aggregated. Finally several coreference clusters are extracted. Since it can extract the

Z. Ma, D. Meng and C. Kong—These authors contribute equally to this work.

implicit information from document, coreference resolution is often used in many downstream tasks of natural language processing, such as machine translation task [1], question and answer system [2], text classification task [3], information network [4]. As an important component of large-scale intelligent information system, coreference resolution model needs higher performance and certain operation efficiency to meet the requirements of online real-time processing.

Common coreference resolution systems use the mention ranking method: for each span, calculate its coreference score with the antecedents, and the antecedent with the highest score ranking will be predicted to have a coreference relationship with the span. About such methods, if the number of tokens in the text is $O(n)$, the corresponding number of spans will be $O(n^2)$, and the complexity of predicting the final coreference relationship between entities and mentions is $O(n^4)$, which causes the model a great challenge in efficiency. To improve this situation, pruning technology is proposed to prune the number of spans to reduce the computing overhead.Lee et al. retained m high score spans [5], and proposed C2F-COREF method, which was improved by calculating and retaining K high score antecedents of each span based on a simple and calculable score [6]. In addition, extracting syntax features from the syntaxc analysis tree is also very common in the early research of coreference resolution. Ge et al. proposed to rank the candidate antecedents by encoding the pronouns given by the parsing tree based on the pronoun resolution algorithm [7], and realized the construction of path related features based on the parsing tree by using the word sequence and dependency tags in the path between the given pronoun and its candidate antecedents. It obtains the path information between pronouns and can be used to measure the coreference possibility between pronouns and antecedents. At the same time, syntax information has also been applied to anaphora detection tasks using the method based on the tree kernel: Kong et al. designed a variety of path related features, such as the path between the root node and the current mention [8,9], which provides a new method for coreference resolution to capture potential text information. However, few attempts have been made to evaluate the utility of syntax and semantics in the neural coreference model. CorefQA [10], as the best coreference resolution method at present, regards the coreference resolution task as a machine reading comprehension problem, regards the mention extraction in the text as predicting the range of target, and uses a complete transformer to calculate the antecedent of each span, resulting in a large computing overhead of the model. Many of the above methods are based on spans for coreference resolution tasks. The introduction method of establishing the relationship between tokens such as component tree and relational graph makes the model obtain the potential information of the text better. However, the cost is that the model is more huge and complex, the requirements of data processing are higher, the computing cost of the system is greater.

In order to make the coreference resolution model more compact, concise and efficient, and allow it to be easily integrated into a larger natural language processing system, this paper proposes to separate mention extraction and coreference prediction from the spans based coreference resolution method. For the task of coreference prediction, we deal with it at the word level. First, we calculate all tokens and their previous tokens to obtain high-score tokens that are closely related to this token, and then prune them to reduce the calculation difficulty of later work. These tokens are calculated with the target token to obtain the coreference score. The final highest score token is regarded as the antecedent of the target token to predict the coreference relationship. Compared with the prediction method based on spans, the model can avoid calculating a large number of span representations generated by the combination of tokens, which can reduce the computing cost. For the mention extraction task, we strengthen the matrix representation of the target token by using the implicit information in the neighbor tokens of the target token. The subsequent coreference mention extraction operation will be performed only on those words found to have a coreference relationship with some other words, and assist in the mention extraction operation with the help of the enhanced token representations. In short, this paper proposes a lightweight model, which not only considers the efficiency of the model, but also strengthens the performance of the model with the help of syntax and semantics information between contexts.

2   Related Work

2.1   End-to-End Coreference Resolution

Lee et al. [6] have proposed a model aims to learn a possibility distribution P on all possible antecedents Y of each span i:

$$P\left(y_j\right) = \frac{e^{s(i,y_j)}}{\sum_{y' \in Y(i)} e^{s(i,y')}}, \tag{1}$$

where $s(i, y_j)$ is pairwise coreference score of span i and j. For all spans i, it includes the spans on the left and a specific fictional antecedent e, and a fixed score is $s(i, e) = 0$.

The pairwise coreference scores of span i and j are composed of the sum of coarse coreference scores and fine coreference scores. There is the following formula for this: $s_m(i)$ is used to indicate whether i is a mention; $s_c(i,j)$ denotes the coarse coreference score of span i and j; $s_a(i,j)$ represents the fine coreference score of span i and j, which are calculated with feed forward neural networks (FFNN) as follows:

$$\begin{aligned} s_m(i) &= FFNN_m\left(\boldsymbol{g}_i\right), \\ s_c(i,j) &= \boldsymbol{g}_i^T \boldsymbol{W}_c \boldsymbol{g}_j, \\ s_a(i,j) &= FFNN_a\left(\left[\boldsymbol{g}_i, \boldsymbol{g}_j, \boldsymbol{g}_i \odot \boldsymbol{g}_j, \phi\right]\right), \end{aligned} \tag{2}$$

where $\boldsymbol{g}_i$ is the vector representation of span i, ϕ is a vector of pairwise features, such as the distance between spans, whether they come from the same speaker, etc. The span representation is concatenated by a start and end token embedding in the context, the weighted sum of all tokens in the span and a feature vector:

$$\mathbf{g}_i = \left[\mathbf{x}_{START(i)}, \mathbf{x}_{END(i)}, \hat{\mathbf{x}}_i, \phi(i)\right]. \tag{3}$$

Attention mechanism is using to catch the weights to calculate $\hat{x}_i$. The model also used the weighted and updated span representation of its previous representation to allow the clustering information to be used for a second iteration of calculating coreference score. However, literature [11] has proved that its impact on performance is negative.

2.2 Coreference Resolution of Non-span Representations

Recently, Kirstein et al. have proposed a modification method of mention ranking model without span representations [12]. They calculate the start and end subtoken representations in each sequence: $X_s = \text{ReLU}(W_s \cdot X), X_e = \text{ReLU}(W_e \cdot X)$.

The resulting vector is used to calculate the coreference antecedent score of each span, and there is no need to construct a complex span representations. This method is competitive with other coreference ranking methods in performance and is more efficient. Nonetheless, the theoretical size of its antecedent score matrix is also $O(n^4)$, so the method still needs pruning approach.

2.3 Coreference Resolution of Associative Component Syntax

Fan et al. have combined with the method of component syntax, proved that the use of component tree containing more important information has a positive effect on the task of coreference resolution and is more natural [13]. They proposed a graph-based method to integrate the syntax structure of components and obtain semantic information to strengthen the task of coreference resolution. Extract the tokens information from text to build a component tree, use the edges with direction attributes between nodes to transfer implicit information of context, and update the component nodes in different directions by GAT:

$$
\begin{aligned}
h_{c_i}^{ff} &= GAT\left(h_{c_i}, h_{c_j} \mid c_j \in N_{c_i}^{tf}\right), \\
h_{c_i}^{tb} &= GAT\left(h_{c_i}, h_{c_j} \mid c_j \in N_{c_i}^{tb}\right),
\end{aligned}
\tag{4}
$$

where c_i indicates the target node, c_j represents neighbor nodes of target node which are combined by the edge of type t. $N_{c_i}^{tf}$ and $N_{c_i}^{tb}$ represent set of left and right neighbor nodes embedding representation. The updated target component node is represented as: $h_{c_i}^t = h_{c_i}^{tf} + h_{c_i}^{tb}$. The component nodes transmit syntax and other semantic information to the token nodes to optimize the token representation, and then dynamically fuse the enhanced token representation through the gate mechanism, which is used to calculate the span embedding and coreference score.

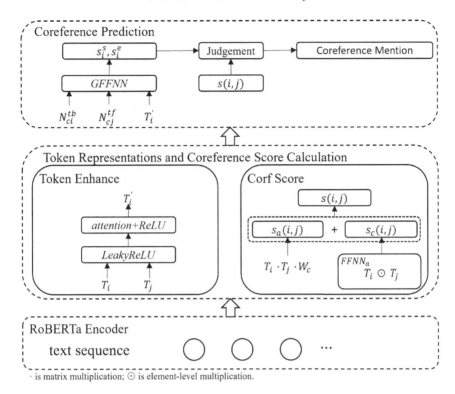

Fig. 1. The overall architecture of the CRS_2 model.

3   Methodology

In order to solve the problem of low efficiency of the system and insufficient combination of syntax and semantic information in the extraction of mention spans, CRS_2 method is proposed in this paper. The specific process is shown in Fig. 1. The embedding representation of each token in the text sequence is obtained through the encoder of the pretrained model and then will be used in two parts: (1) select the target token from the tokens representation and input it into the Coref Score module for calculation. After preprocessing, coarse coreference score calculation and fine coreference score calculation, the highest score antecedent of the target token is predicted; (2) The target token and its neighbor tokens representation are also used in the Token Enhance module to capture syntax, semantics and other information. Input the information obtained by the two modules into the Conference Prediction module to jointly calculate the most likely start and end position of the target span and finally obtain the coreference mention.

3.1   Token Representation

In order to calculate the matrix representation of each token in the document, the document is input into the pretrained encoder to obtain the context embedding matrix representation of each subtoken, and then the weighted sum of all subtokens is calculated through the softmax normalized exponential function to finally obtain the token representation for calculating the coreference score.

3.2   Coreference Score Calculation

Inspired by the work of Lee et al. [6], this paper uses bilinear scoring function to compute the k most similar antecedents of each token, so as to avoid calculating the coreference score of all previous tokens for each token. In this way, the computation of identifying antecedents for each token can be reduced, and the complexity of the subsequent calculation of the model will also be reduced. The coarse antecedent score formula is as follows: $s_c(i,j) = \boldsymbol{T}_i \cdot \boldsymbol{W}_c \cdot \boldsymbol{T}_j$, where $\boldsymbol{T}_i$ and $\boldsymbol{T}_j$ respectively represent the matrix representation of two tokens, $\boldsymbol{W}_c$ is the corresponding parameter matrix. Then construct the pairwise matrix, each matrix represents the embedding of two tokens and the embedding of features between two tokens. The fine antecedent score is obtained by inputting it into a feed-forward neural network: $s_a(i,j) = FFNN_a([\boldsymbol{T}_i, \boldsymbol{T}_j, \boldsymbol{T}_i \odot \boldsymbol{T}_j, \boldsymbol{\phi}])$, where $\odot$ represents element-level multiplication and $\boldsymbol{\phi}$ represents the vector representation of pairwise features between two tokens. The final coreference score is defined as the sum of the two scores: $s(i,j) = s_c(i,j) + s_a(i,j)$. The candidate antecedent with the highest positive score is assumed to be the predictive antecedent of each token. If a token does not get the antecedent with a positive score after calculating all possible antecedents, the token will be summarized as no antecedent.

3.3   Representation with Syntax and Semantics

When acquiring the embedding representation of each word in the document, only using the word-level processing method will make the model unable to capture more useful syntax and semantic information, resulting the information representation of each token limited, which is not conducive to the subsequent mention extraction task. In order to improve the information representation of tokens, the attention mechanism is used for each token to selectively integrate the information of its neighbor tokens:

$$a_{ij} = \text{softmax}\left(\sigma\left(\mathbf{a}^T[\boldsymbol{W}\boldsymbol{T}_i; \boldsymbol{W}\boldsymbol{T}_j]\right)\right),$$

$$\boldsymbol{T}'_i = \|_{k=1}^K \text{ReLU}\left(\sum_j \alpha_{ij}^k \boldsymbol{W}^k \boldsymbol{T}_j\right),$$

$$(5)$$

where K is the number of headers in the attention mechanism, $\boldsymbol{T}_i$ and $\boldsymbol{T}_j$ is the embedding representation of the target token i and its neighbor token j respectively, and $\mathbf{a}^T$, $\boldsymbol{W}$ and $\boldsymbol{W}^k$ are trainable parameters. σ is the LeakyRelu activation function proposed by Xu et al. [11]. $\|$ and $[\cdot;\cdot]$ represent the concatenation of information. T_i' indicates the updated target token representation. This method considers the situation that multiple words in the text form one mention, and improves the performance of subsequent mention prediction and extraction by strengthen token representation with syntax and semantics.

3.4   Mention Extraction

Based on the token enhance module and coreference score calculation module, the possible coreference expression prediction is carried out. Firstly, predict the span with a token as the head, connect all tokens in the sentence where the token is located with it, find the span by predicting the most likely start and end tokens in the sentence, and standardize the result of span prediction with help of the syntax and semantic information contained in the token itself. Then through a feed-forward neural network composed of two output channels (start and end scores) and a convolution layer with a kernel size of three to obtain the predicted coreference expression. For a token i:

$$
\begin{aligned}
s_i^s &= GFFNN\left(T_i', t_i \mid t_i \in N_{c_i}^{tb}\right), \\
s_i^e &= GFFNN\left(T_i', t_j \mid t_j \in N_{c_j}^{tf}\right),
\end{aligned}
\tag{6}
$$

where T_i' is the enhanced representation of the target token, $N_{c_i}^{tb}$ and $N_{c_j}^{tf}$ respectively represent the set of token representations before and after the target token, t_i and t_j represent one of the token representations respectively. GFFNN is the training layer which is used to compute score, s_i^s and s_i^e represent the probability of start and end token corresponding to target token.

Since the boundary of a reasonable span should be between the tokens in this span and the tokens of other spans, the coreference token relationship will be further used in the span prediction specification. For a token that is not a start or end token in the predicted span, the corresponding start and end token scores are also calculated. If its score is the same as the prediction result of other tokens in the predicted span, it will be regarded as a token in its span. If the start and end tokens of the predicted span are different, the coreference relationship obtained by the coreference score calculation module is used to help judge the span range: if the token has a different coreference relationship with the previous token in the span, the span is divided into two spans; If the token is regarded as having no antecedent, it is regarded as a token in the span. In the predicted spans, the end token will not be divided into a single span. In the prediction process, the tokens on the left side of the target token are not considered as the end token of the span, and the tokens on the right side of the target token are not considered as the start token of the span. Finally, the coreference expression of the prediction is obtained through calculation.

Table 1. Descriptive statistics of dataset.

Type	Mentions num	Mention pairs num
WL	163,104	62,803,841
SL	2,263,299	13,970,813,822

3.5 Model Training

In order to capture the coreference relationship between tokens in the data and the semantic relationship between neighbor tokens, the OntoNotes 5.0 training data set is transformed into the connection of a single word. To train the span extraction ability of the model, the data set is constructed from word to span. Inspired by the work of Lee et al. [5], using negative log marginal likelihood (NLML) as its base loss function to calculate the coreference loss of tokens, and only the optimal coreference cluster will be used to calculate: $L_{\text{NLML}} = -\log \prod_{i=0}^{N} \sum_{y' \in Y(i) \cap GOLD(i)} P(y')$, where $P(y')$ refers to the probability of correct prediction, $Y(i) \cap GOLD(i)$ indicating the tokens come from the intersection of the correct coreference set and the predicted coreference set.

Following the work of Vladimir et al. [14], binary cross-entropy (BCE) is used as an additional regularization factor: $L_{COREF} = L_{NLML} + \alpha L_{BCE}$, where α is an adjustable parameter. Since the pairs of pairwise tokens are classified independently in the coreference prediction stage, which can encouraged the model to output higher scores for all coreference representations. Set α to 0.5 to make NLML loss better than BCE loss. The span extraction module is trained by using cross-entropy loss on all start and end scores in the same sentence, and the loss of the two parts is summed to jointly train the coreference score calculation module and mention extraction module.

4 Experiments

4.1 Experimental Settings

Dataset. In order to evaluate the model, the data set selected in this paper is the Chinese and English parts of OntoNotes 5.0 data set [15]. The English corpus is divided into 2802, 343 and 348 documents for training, validation and testing respectively, while the Chinese corpus is divided into 1810, 252 and 218 documents for training, validation and testing. As shown in Table 1, the number of mentions and the number of mention pairs under the word-level and span-level methods of the OntoNotes 5.0 dataset are summarized. For each mention, only the mentions on the left side is considered as a possible pairwise anaphora relationship. It is clear that the number of mention pairs generated is comparatively large compared to the existing number of mentions.

Evaluation Metrics. The model was evaluated using three coreference resolution metrics: MUC, B^3, and $CEAF_{\phi_4}$. Calculate the average F1 score (AVG. F1) of the three, obtain it using the latest official script, and compare its performance with the results reported by the authors of other coreference resolution models. This experiment is implemented using the pytorch framework with Linux system. The proposed model is trained for 20 epochs on the Tesla P100 with 48G memory. The batch size of the input model is 512 and the running time is about 6 h. Most parameters of the experiment follow the settings of SpanBERT-base [11]. The hidden layer size of feed-forward neural network is 1024, the number of layers is 1, the maximum span length is 32, the number of coarse coreference score antecedents is 30, and the model learning rate is 3e-4.

Baselines. To verify the effectiveness of the proposed model, we used several representative models for comparison: E2E-COREF [5] is the first end-to-end neural coreference resolution model, first generates and encodes all spans in a text sequence, further obtains all antecedents and scores corresponding to each span, and judges whether they belong to the same cluster. C2F-COREF [6], in order to solve the problem that it is difficult for high-dimensional information to retain long-range information in dialogue, the model introduces mention pruning strategy from coarse-grained to fine-grained, fully considers the previous information, uses multiple rounds of iteration and attention mechanism to calculate the coreference probability. SpanBERT [16] uses the span encoder SpanBERT to better predict the coreference expression. CorefQA [10] uses the machine reading comprehension framework so that the task of coreference resolution is regarded as a question and answer task based on span prediction. Coref-HGAT [13] constructs heterogeneous graph attention network and introduces dependent syntax and semantic role tags. [17] have proposed a neural network-based model combining entity-level information, which trains deep neural network to construct pairwise distributed representation of cluster with same entity, so as to capture entity-level information and generate high-dimensional vector representation for coreference cluster pairs. [18] have strengthened the end-to-end coreference resolution model. It is proposed to use the component tree as a restriction method to filter legal candidate mentions, which can reduce the calculation consumption of the model and encode the node information in the syntax tree conversion sequence. SpanBERT + CM [11] verifies and explains the effectiveness and limitations of higher-order inference for coreference resolution, and improves the performance of the task by combining SpanBERT and cluster fusion methods. s2e [12] is a lightweight end-to-end collaborative reference model, which eliminates the dependence on span representation, manual features and heuristics so that the model is simpler and more efficient.

4.2   Performance Comparison

The model proposed in this paper is trained and tested on the Chinese and English parts of OntoNotes 5.0 dataset. According to the result published by other workers in the paper, the performance comparison can be obtained as shown in Table 2. The bottom three lines are part of the Chinese experiment, and the others are English experiments. It can be seen that in the English part,

Table 2. Performance comparison.

Method	MUC			B³			CEAF$_{\phi_4}$			
	P	R	F1	P	R	F1	P	R	F1	Avg.F1
E2E-COREF	78.4	73.4	75.8	68.6	61.8	65.0	62.7	59.0	60.8	67.2
C2F-COREF	81.4	79.5	80.4	72.2	69.5	70.8	68.2	67.1	67.6	73.0
SpanBERT-base	84.3	83.1	83.7	76.2	75.3	75.8	74.6	71.2	72.9	77.4
CorefQA + SpanBERT-base	85.2	**87.4**	**86.3**	**78.7**	76.5	**77.6**	**76.0**	75.6	**75.8**	**79.9**
coref-HGAT + SpanBERT-base	85.3	85.0	85.2	77.9	77.7	77.8	75.6	74.1	74.8	79.3
CRS₂ + RoBERTa-base	**85.8**	84.6	85.2	76.5	**78.1**	77.3	72.5	**75.9**	74.2	78.9
SpanBERT-large	85.8	84.8	85.3	78.3	77.9	78.1	76.4	74.2	75.3	79.6
coref-HGAT + SpanBERT-large	87.2	86.7	87.0	81.1	80.5	80.8	78.6	77.0	77.8	81.8
CorefQA + SpanBERT-large	**88.6**	**87.4**	**88.0**	**82.4**	**82.0**	**82.2**	**79.9**	**78.3**	**79.1**	**83.1**
CRS₂ + RoBERTa-large	87.2	87.0	87.1	80.3	81.4	80.8	78.3	77.7	78.0	81.9
Clark and Manning (2016)	73.9	65.4	69.4	67.5	56.4	61.5	62.8	57.6	60.1	63.7
Kong and Jian (2019)	77.0	64.6	70.2	**70.6**	54.7	61.6	64.9	55.4	59.8	63.9
CRS₂ + RoBERTa-large	**78.1**	**77.3**	**77.7**	70.4	**62.2**	**66.3**	**67.8**	**62.0**	**64.9**	**70.0**

we split the test into the separation and comparison of basic and large-scale pretrained models. In the basic model, the CorefQA + SpanBERT-base model based on question and answer way has advantages in some scores and has the highest Avg.F1. Compared with some suboptimal scores, CRS_2 model takes the lead of 0.5% in the accuracy score of MUC test, 0.4% in the recall rate of B^3 test, and 0.4% in $CEAF_{\phi_4}$. The recall rate of the test is 0.3%, which indicates that CRS_2 can also have significant performance on the basis of the basic pretrained model. When the large-scale pretrained model is used in the English part, the CorefQA + SpanBERT-large model has obvious leading advantages. The CRS_2 model is competitive compared to other methods and its performance in the recall rate part of different tests is second only to the optimal system. In the Chinese part of the data set, Clark et al. captured entity-level information by training a deep neural network and had good performance that year. Kong et al. used the component tree to obtain the implicit information of the text and took the lead in the accuracy score of the B^3 test. The CRS_2 model has achieved the overall leading score in the test of the Chinese part and has the highest Avg.F1, which is greatly improved compared to the suboptimal method. In general, CRS_2 model modifies the pruning technology used in the span based coreference resolution model, which makes this method have a high recall rate, and is also competitive with the comparative model in other tests.

4.3   Efficiency Comparison

As mentioned above, word-level associations on the OntoNotes 5.0 dataset contain more than one hundred and sixty thousand coreference expressions, and the number of different coreference pairings reaches more than sixty million. When using spans for coreference resolution, the number of corresponding two will be

Table 3. Efficiency comparison.

Method	Time (s)
SpanBERT-base	85
SpanBERT + CM	92
SpanBERT + CM -HOI	52
s2e	44
CRS$_2$	41

hard to compute. Therefore, while considering performance, it is also necessary to improve efficiency. In order to prove the efficiency comparison between CRS_2 model and other methods, the following tests are carried out: the inference time of different models on the same test set is counted, and the efficiency gap of each model is compared and analyzed. Table 3 shows the inference time of different models in seconds(s). It can be seen that compared with the method based on span representation, CRS_2 can reduce the inference time by about half, which is consistent with the content described above. Comparing the model proposed in this paper with other models, it can be seen that CRS_2 can reduce the time consumption. On the contrary, the span-based coreference resolution model will use pruning strategy to reduce the computational complexity before calculating the coreference score. In order to improve efficiency, it will also cause the loss of correct coreference expression. At the same time, we can found from Table 3 that high-order inference (HOI) has high requirements for time, and other methods use pruning strategies to varying degrees to try to find a balance between efficiency and performance. The inference here is only for reference. The method in this paper should be combined with others coreference resolution system for comparison, which is more convincing.

5   Conclusion

This paper proposes a coreference resolution method that integrates syntax and semantic information, which can reduce the computational consumption and maintain certain performance. Based on the word-level score calculation and pruning approach, this method predicts the coreference situation between different tokens, reduces the difficulty of model calculation, strengthens the token representation, better captures the hidden information in the data and improves the performance of mention prediction.

In future work, token representation and syntax fusion will be further studied in order to better obtain the original information and the association between words in the text.

Acknowledgment. This work was supported in part by the National Natural Science Foundation of China Youth Fund (No. 61902001), the Open Project of Shanghai Big Data Management System Engineering Research Center (No. 40500-21203-542500/021), the Industry Collaborative Innovation Fund of Anhui Polytechnic University-Jiujiang District (No. 2021cyxtb4), and the Science Research Project of Anhui Polytechnic University (No. Xjky072019C02, No. Xjky2020120, No. Xjky2022147). We would also thank the anonymous reviewers for their detailed comments, which have helped us to improve the quality of this work. All opinions, findings, conclusions and recommendations in this paper are those of the authors and do not necessarily reflect the views of the funding agencies.

References

1. Liu, X., et al.: On the complementarity between pre-training and back-translation for neural machine translation. In: EMNLP, pp. 2900–2907 (2021)
2. Abduljabbar, Z.A., et al.: Provably secure and fast color image encryption algorithm based on s-boxes and hyperchaotic map. IEEE Access **10**, 26257–26270 (2022)
3. Wang, C., Wang, J., Qiu, M., Huang, J., Gao, M.: Transprompt: towards an automatic transferable prompting framework for few-shot text classification. In: EMNLP, pp. 2792–2802 (2021)
4. Kong, C., Chen, B., Li, S., Chen, Y., Chen, J., Zhang, L.: GNE: generic heterogeneous information network embedding. In: WISA 2020, pp. 120–127 (2020)
5. Lee, K., He, L., Lewis, M., Zettlemoyer, L.: End-to-end neural coreference resolution. In: EMNLP, pp. 188–197 (2017)
6. Lee, K., He, L., Zettlemoyer, L.: Higher-order coreference resolution with coarse-to-fine inference. In: NAACL-HLT, pp. 687–692 (2018)
7. Ge, N., Hale, J., Charniak, E.: A statistical approach to anaphora resolution. In: ACL, pp. 161–170 (1998)
8. Kong, F., Zhou, G., Qian, L., Zhu, Q.: Dependency-driven anaphoricity determination for coreference resolution. In: COLING, pp. 599–607 (2010)
9. Kong, F., Zhou, G.: Combining dependency and constituent-based syntactic information for anaphoricity determination in coreference resolution. In: PACLIC, pp. 410–419 (2011)
10. Wu, W., Wang, F., Yuan, A., Wu, F., Li, J.: Corefqa: coreference resolution as query-based span prediction. In: ACL, pp. 6953–6963 (2020)
11. Xu, L., Choi, J.D.: Revealing the myth of higher-order inference in coreference resolution. In: EMNLP, pp. 8527–8533 (2020)
12. Kirstain, Y., Ram, O., Levy, O.: Coreference resolution without span representations. In: ACL/IJCNLP), Virtual Event, August 1–6, 2021, pp. 14–19 (2021)
13. Jiang, F., Cohn, T.: Incorporating syntax and semantics in coreference resolution with heterogeneous graph attention network. In: NAACL-HLT, pp. 1584–1591 (2021)
14. Dobrovolskii, V.: Word-level coreference resolution. In: EMNLP, pp. 7670–7675 (2021)
15. Pradhan, S., Luo, X., Recasens, M., Hovy, E.H., Ng, V., Strube, M.: Scoring coreference partitions of predicted mentions: a reference implementation. In: ACL, pp. 30–35 (2014)

16. Joshi, M., Chen, D., Liu, Y., Weld, D.S., Zettlemoyer, L., Levy, O.: Spanbert: improving pre-training by representing and predicting spans. Trans. Assoc. Comput. Linguistics **8**, 64–77 (2020)
17. Clark, K., Manning, C.D.: Improving coreference resolution by learning entity-level distributed representations. In: ACL, pp. 643–653 (2016)
18. Kong, F., Fu, J.: Incorporating structural information for better coreference resolution. In: IJCAI, pp. 5039–5045 (2019)

Self-adaptive Context Reasoning Mechanism for Text Sentiment Analysis

Shuning Hou[1], Xueqing Zhao[1(✉)], Ning Liu[1], Xin Shi[1], Yun Wang[2],
and Guigang Zhang[2]

[1] Shaanxi Key Laboratory of Clothing Intelligence, School of Computer Science,
Xi'an Polytechnic University, Xi'an 710048, China
zhaoxueqing@xpu.edu.cn
[2] Institute of Automation, Chinese Academy of Science, Beijing 100190, China

Abstract. With the rapid development of contemporary e-commerce
and social media, the need of better understanding and exploring users'
evaluations on e-commercial products is becoming urgent and crucial, which results in the emergence of a new research hot-spot aiming at analysing and mining latent features of customer reviews on e-
commercial products. In order to analyse the associations among semantic features with different lengths in e-commercial reviews, a text sentiment analysis method, named as ALBERT-SACR, is proposed based
on self-adaptive context reasoning mechanism in this paper. Firstly, the
global contextual features are extracted using the Transformer blocks of
ALBERT. Then, semantic features with different lengths are extracted
on the basis of multi-channel CNN combined with self-attention mechanism to perform context reasoning and adaptive adjustment of relational
weights. Finally, a fully connected neural network is used for sentiment
classification. public Chinese datasets Waimai and Shopping, where the
performance of our method is qualitatively compared with five other
methods. Simulation results verify both the effectiveness and efficiency
of our proposed ALBERT-SACR, and the adaptive nature of our proposed SACR is effective in contextual inference for semantic features of
different lengths.

Keywords: Sentiment analysis · ALBERT · Transformer ·
Self-adaption · Context reasoning · Multi-channel CNN · Self-attention

1 Introduction

With the development of the internet, online entertainment has become a normal part of people's daily life, and it is convenient for people to express their
opinions and share their knowledge from the internet. It is available to predict
political candidates [1], monitor online violence or bullying [2], and analyse public
sentiment in the COVID-19 [3] based on users' comments on social media. Nowadays, online shopping, cultural sharing or other models are aiming at precisely

X. Zhao et al. (Eds.): WISA 2022, LNCS 13579, pp. 194–205, 2022.
https://doi.org/10.1007/978-3-031-20309-1_17

acquiring users' needs by mining user behaviors, relevant applications include interest prediction, recommendation system [15,20] and some other personalized services. In platforms of e-commerce, people can express their sentiments by writing reviews about products. So users' reviews factor is one of the most important considerations for mining user behavior to provide better personalized services, which results in the emergence of a research hotspot in terms of sentiment analysis for e-commercial reviews.

As with the development of deep learning, the encoding and decoding of machine translation evolved from the complex recurrent and convolutional neural networks to simple Transformer [14] models which are based entirely on attention mechanisms and exhibited better performance. In the case of language processing, some Transformer based models are proposed with relatively good performance, such as OpenAIGPT [13], Bidirectional Encoder Representations from Transformers (BERT) [4], etc. There are also improvements based on BERT in sentiment analysis, such as CG-BERT [18], QACG-BERT [18], and MF-BERT [9], but these models always hold a large number of parameters. The self-supervised model named as ALBERT (short of A Little Bidirectional Encoder Representations from Transformers) [8] separates the parameter sizes of the hidden layer and lexical table embedding based on BERT, it also uses cross-layer parameter sharing to prevent the parameters from growing with the depth of the network, above improvements significantly reduce the number of parameters and improves the parameter efficiency of BERT. Since same words in different contexts may sometimes represent different meanings, Wu et al. [17] combined ALBERT with BiLSTM under the consideration of contextual semantic associations.

Above mentioned models don't consider the contextual semantic associations among semantic features of different lengths, and there are few sentiment classifications on Chinese datasets. To address these critical issues we propose a Chinese sentiment analysis model named as ALBERT-SACR (Self-adaptive Context Reasoning Mechanism based on ALBERT). Our main contributions are as follows: (1) We propose a sentiment analysis method based on self-adaptive context reasoning mechanism, named as ALBERT-SACR; (2) we propose a self-adaptive context reasoning mechanism to learn semantic features of different lengths; (3) we compare with five other methods to verify the effectiveness of our methods. Moreover, we further discuss how CNN channel numbers in our proposed methods influence classification results.

2   Related Works

ALBERT [8] is a modification of BERT [4] which mainly focuses on the parameter optimization. Ding et al. [5] performed sentiment analysis based on ALBER on a movie review dataset and compared with five classification algorithms. Wu et al. [17] used ALBERT-Att-BiLSTM for sentiment analysis which has been proved better than W2V-Att-CNN and W2V-Att-LSTM. Liu et al. [11] chose CNN to capture features based on ALBERT. Zhang et al. [21] combined

ALBERT with Capsule network to obtain the relationship between global features and local features. Ye et al. [19] proposed ALBERT-CNN to extract global features by pre-trained model ALBERT and local features by convolutional network, which makes full use of the information in sentences and improves the accuracy of sentiment analysis. In this paper, ALBERT is chosen as the base model to extract global contextual features of sentences.

Multi-channel CNN, represented by TextCNN (short of Text Convolutional Neural Networks) [7], can extract text features of different lengths. Guo et al. [6] selected multi-weight channel combined with 4-channel TextCNN for feature extraction. Li et al. [10] chose 4-channel TextCNN to extract local features and BiLSTM to extract global features, and they were combined to process data annotation in parallel way to improve the accuracy of sentiment analysis. Most of relevant efforts are put on English sentiment analysis, we attempt to concentrate on the extraction of Chinese semantic features with different lengths by multi-channel CNN.

Self-attention mechanism can learn syntactic and semantic information, which is primarily used in Transformer for semantic understanding and learning. Lv et al. [12] calculated context associations and relevance between context and aspect by self-attention mechanism, which can better capture the semantic features of short texts. Wei et al. [16] acquired sentences' weights using self-attention after BiLSTM. In this paper, self-attention mechanism is chosen to actively learn the relational weights among semantic features with different lengths and obtain context reasoning and self-adaptive convolutional fusion.

3   ALBERT-SACR

In this section, we present the overall framework of our ALBERT-SACR. Given an input sentence $S = \{word_1, word_2, \ldots, word_n\}$, n denotes the length of review and $word_i$ means the ith word in sentence S. The framework of our proposed ALBERT-SACR is shown in Fig. 1. The embedding layer firstly converts s into index lists Wid of word embedding matrix, and Seg of segment embedding matrix, respectively. Then, word embedding matrix, segment embedding matrix, together with the position embedding matrix representing position information of S generate the embedding output by adding the three types of matrices directly. Afterward, Transformer layer learns the global contextual features followed with the SACR layer which extracts the global contextual semantic features of different lengths and learns the contextual relations of them so as to get the fused features. The output layer is a fully connected layer which classifies the fused features.

3.1   Embedding

The embedding layer digitizes the input S and marks paragraph of $word_i$, and then embeds them to get Word embedding, Segment embedding and Position embedding. The process of digitization can be defined as follows:

$$Wid = \text{WordIndex}(S, list) \tag{1}$$

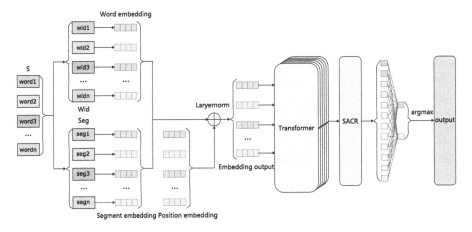

Fig. 1. The overall framework of ALBERT-SACR, the sentence S goes through the embedding layer, Transformer layer, SACR layer and output layer in sequence.

where $Wid = \{wid_1, wid_2, \ldots, wid_n\}$ is the index list of word embedding, *list* is the list of Chinese words provided by ALBERT, the WordIndex function transforms $word_i$ in a sentence S into corresponding index wid_i in the *list*. The paragraph position seg_i of $word_i$ is $word_i$'s paragraph marker, and paragraph marker of S is $Seg = \{seg_1, seg_2, \ldots, seg_n\}$. The embedding process can be defined as follows:

$$E = \text{layernorm}\left(E_{\text{word}} + E_{\text{segment}} + E_{\text{position}}\right) \tag{2}$$

$$E_{word} = \text{match}(Wid, \boldsymbol{wlist}) \tag{3}$$

$$E_{segment} = \text{match}(Seg, \boldsymbol{seglist}) \tag{4}$$

$$E_{position} = \text{match}(Pos, \boldsymbol{poslist}) \tag{5}$$

where E denotes the output of embedding layer, E_{word} denotes word embedding, E_{segment} denotes segment embedding of s, E_{position} denotes position embedding of s, the function layernorm denotes linear normalization to make the input values obey the standard normal distribution. *wlist*, *seglist*, *poslist* are the initialized tensor list of Wid, Seg, Pos, respectively. pos_i in $Pos = \{pos_0, pos_1, \ldots, pos_i\}$ denotes the position number where each word is located. The function match denotes that the value in Wid, Seg and Pos are matched with the index of *wlist*, *seglist*, *poslist*, respectively. The three one dimensional tensors matched in the *wlist*, *seglist* and *poslist* as Word embedding, Segment embedding and Position embedding of $word_i$ respectively.

3.2   Global Contextual Features

Global contextual features are extracted by Transformer that is used in A Little Bidirectional Encoder Representation from Transformers (ALBERT). Transformer contains multi-head attention layer and feedforward layer. Each layer of

Transformer can learn contextual features. After the weights are updated by multi-layers of Transformer, the associations among global contexts can be well extracted. Then, the following SACR layer learns the associations of semantic features, it contains both global contextual features and semantic features of different lengths, which can improve the learning effect of sentiment analysis. The structure of Transformer layer used in this paper is shown in Fig. 2(a). The output of the embedding layer $E = \{e_1, e_2, \ldots, e_m\}$ first goes into the multi-head attention layer, followed by the feed-forward neural network layer, and finally comes to the linear normalization. The multi-head attention is shown in Fig. 2(b), where each input e_i firstly gets the initialized matrices (query Q_i, key K_i and value V_i), and then learns the attention weights. The output of multi-head attention $O = \{O_1, O_2, \ldots, O_m\}$, where O_i is defined as follows:

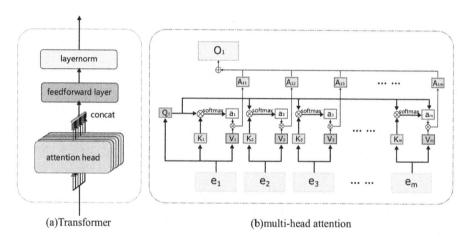

(a)Transformer   (b)multi-head attention

Fig. 2. The structure of transformer. (a) is overall structure of Transformer layer, (b) is multi-head attention layer of Transformer.

$$O_i = \sum_{j=1}^{m} A_{i,j} \tag{6}$$

$$A_{i,j} = \text{softmax}\left(\frac{Q_i \times K_j^{\text{T}}}{\sqrt{\lambda}}\right) \times V_j \tag{7}$$

where $A_{i,j}$ denotes the output weight that is learned by the ith input e_i after the multi-head attention, m is the number of attention heads, Q_i, K_i and V_i are the initialized matrices of query, key and value corresponding to the input e_i respectively, Q_i, K_i and V_i are constantly updated in the method training, T denotes the transpose of the matrix, λ is the size of a attention head, and the function softmax maps each value to the [0,1] interval. The feedforward neural network layer is defined as follows:

$$F = \text{gelu}\left(\boldsymbol{O} \times \boldsymbol{W}_1 + \boldsymbol{b}_1\right) \tag{8}$$

where $\boldsymbol{O}$ is the output of multi-head attention layer, $\mathbf{W}_1$ and $\mathbf{b}_1$ are the initialized matrices of weights and bias for $\boldsymbol{O}$ respectively, and the function of Gaussian linear activation gelu maps the value to zero if it is less than zero. The linear normalization layer ensures that the values in the network obey the standard normal distribution to avoid the vanishing gradient, and this layer is computationally defined as follows:

$$\gamma = \text{layernorm}\left(\boldsymbol{F} \times \boldsymbol{W}_2 + \boldsymbol{b}_2\right) \tag{9}$$

where γ is the output of the linear normalization layer.

3.3   Self-adaptive Context Reasoning Mechanism

Transformer can only get the global contextual features without considering the contextual associations between semantic features of different lengths. In this paper, we propose the Self-adaptive Context Reasoning mechanism (SACR) that can learn contextual associations of semantic features with different lengths and adaptively adjust the weights among those features. The structure of SACR is shown in Fig. 3, which contains multi-channel CNN and self-attention mechanism. The multi-channel CNN conducts operations of multi-channel convolution and max-pooling for the output of Transformer, and i-CNN indicates that the size of channel convolution is i, the number of channels in multi-channel CNN is 6, different size of convolution extracts text features with different lengths. Self-attention part firstly learns context reasoning among the semantic features with different lengths, and then adaptively adjusts the weights of fused convolution.

The convolution is defined as follows:

$$C_{x,y}^l = \text{relu}\left(\sum_{i=0}^{l}\sum_{j=0}^{k}\left(\gamma_{t+i,v+j} \times (\boldsymbol{W}_{3l})_{i,j} + (\boldsymbol{b}_{3l})_{i,j}\right)\right) \tag{10}$$

where $C_{x,y}^l$ denotes the value of tensor at the position (x, y) after the convolution of channel l, t and v denote the position where the convolution starts, $t = x \times s$, $v = y \times s$, s is the stride of convolution, l and k are the dimensional size of the convolution kernel, and the activation function of convolution is relu, respectively. After convolution, we get the tensor matrix $C^l = \left\{C_{0,0}^l, C_{0,1}^l, \ldots, C_{l,k}^l\right\}$. The max-pooling for C^l can be defined as follows:

$$M_{x,y}^l = \max\left(C_{t+0,v+0}^l, \ldots, C_{t+0,v+m}^l, \ldots, C_{t+n,v+0}^l, \ldots, C_{t+n,v+m}^l\right) \tag{11}$$

where $M_{x,y}^l$ is the tensor value of the position (x,y) after the max-pooling of channel l, n and m are the size of the pooling window. The function max gets the maximum form the given value. Afterwards, the results of the multi-channel

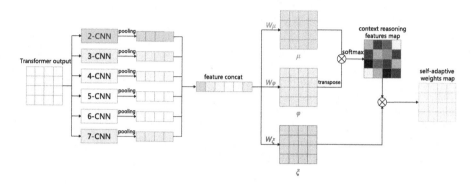

Fig. 3. The structure of SACR. Convolution and max-pooling are firstly performed to get semantic features with different lengths, and then the features are concatenated and passed through self-attention to get matrices $\boldsymbol{W}_\mu$, $\boldsymbol{W}_\varphi$ and $\boldsymbol{W}_\xi$, $\boldsymbol{W}_\mu$ and $\boldsymbol{W}_\varphi$ are learned to get context reasoning features map, context reasoning features map and $\boldsymbol{W}_\xi$ are learned to get the final self-adaptive features map.

CNN are concatenated to obtain $\boldsymbol{\eta}$, $\boldsymbol{\eta} = \left\{ \boldsymbol{M}^1, \boldsymbol{M}^2, \ldots, \boldsymbol{M}^\delta \right\}$, where $\boldsymbol{\delta}$ is the number of CNN channels. The context reasoning $\boldsymbol{CR}$ on $\boldsymbol{\eta}$ is defined as follows:

$$CR = \text{softmax}\left(\frac{\mu \times \varphi^{\text{T}}}{\sqrt{\sigma}}\right) \tag{12}$$

$$\mu = \eta \times \boldsymbol{W}_\mu + \boldsymbol{b}_\mu \tag{13}$$

$$\varphi = \eta \times \boldsymbol{W}_\varphi + \boldsymbol{b}_\varphi \tag{14}$$

where $\boldsymbol{\mu}$ and $\boldsymbol{\varphi}$ are the feature matrices of $\boldsymbol{\eta}$, the multiplication of $\boldsymbol{\mu}$ and $\boldsymbol{\varphi}^T$ can learn the contextual associations of semantic features with different lengths. The self-adaptive learning $\boldsymbol{SA}$ for contextual associations is defined as follows:

$$SA = CR \times \xi \tag{15}$$

$$\xi = \eta \times \boldsymbol{W}_\xi + \boldsymbol{b}_\xi \tag{16}$$

where $\boldsymbol{CR}$ is the output of context reasoning, $\boldsymbol{\xi}$ is the feature matrix of $\boldsymbol{\eta}$.

3.4 Output of Sentiment Analysis

The output layer classifies sentences based on their contextual associations among semantic features of different lengths. After SACR, this layer maps features to the same dimension as the number of classes using the fully connected layer. And the output of sentiment analysis is the index of maximum value of mapped features. Above processes can be defined as follows:

$$D = \text{layernorm}\left(\xi \times \boldsymbol{W}_4 + \boldsymbol{b}_4\right) \tag{17}$$

$$Out = \text{argmax}(\boldsymbol{D}) \tag{18}$$

where $\boldsymbol{\xi}$ is the output of SACR layer, $\boldsymbol{D}$ is a tensor of mapped features, Out is the final output of sentiment analysis, which is the index of the maximum in $\boldsymbol{D}$ extracted by the function argmax.

4 Simulation Experiments and Analysis

4.1 Experimental Settings

Datasets. In this paper, we use two public Chinese datasets Waimai[1] and Shopping[2], and the basic composition of the datasets is introduced in Table 1. The Waimai dataset contains only the after reviews of a takeaway platform. The dataset Shopping is the after-sale reviews of tablet.

Table 1. Basic composition of the datasets.

Datasets	Classes	Total	Train		Test	
			Positive	Negative	Positive	Negative
Waimai	2	11987	2800	5590	1200	2397
Shopping	2	10000	3500	3500	1500	1500

Hyperparameters. To make the performance comparison, we choose ALBERT [8], ALBERT-Dense, ALBERT-BiLSTM [17], ALBERT-CNN [19] and ALBERT-TextCNN as the comparative methods. Since both the comparative methods and our method ALBERT-SACR are based on ALBERT, the parameters of input layer, embedding layer and Transformer are set to be the same. Precisely, the input-size is 256, Embedding-size is 128, the number of Transformer layer is 6, the number of attention heads in multi-head attention is 12, the size of each attention head is 32, and the size of fully connected neural network is 768. The number of CNN channels is 6 and the size of convolutional kernel is [2–7] in ALBERT-TextCNN and our ALBERT-SACR.

Evaluation Metrics. We use four evaluation metrics in our experiments: accuracy (Acc), precision (P), recall (R) and F1 score (F1), which are defined as follows:

$$Acc = \frac{TP + TN}{TP + FP + TN + FN} \tag{19}$$

$$P = \frac{TP}{TP + FP} \tag{20}$$

$$R = \frac{TP}{TP + FN} \tag{21}$$

$$F1 = \frac{2 \times P \times R}{P + R} \tag{22}$$

---

[1] https://github.com/SophonPlus/ChineseNlpCorpus/tree/master/datasets/ waimai_10k.

[2] https://github.com/SophonPlus/ChineseNlpCorpus/tree/master/datasets/ online_shopping_10_cats.

where TP is the number of true positive samples, FP is the number of false positive samples, TN is the number of true negative samples, and FN is the number of false negative samples.

4.2 Comparative Experiments and Analysis

In order to verify the effectiveness of our ALBERT-SACR, five comparison methods, ALBERT, ALBERT-Dense, ALBERT-BiLSTM, ALBERT-CNN and ALBERT-TextCNN are used as comparative methods in this paper, and the results of our ALBERT-SACR were compared with optimal results of comparison models.

Table 2. Comparison results on Waimai and Shopping using six methods.

Methods	Waimai				Shopping			
	Acc	P	R	F1	Acc	P	R	F1
ALBERT	87.5%	80.4%	**82.2%**	81.6%	91.9%	91.3%	92.6%	91.9%
ALBERT-Dense	**88.0%**	82.2%	82.0%	**82.0%**	**92.6%**	**92.0%**	93.3%	**92.6%**
ALBERT-BiLSTM	87.6%	81.7%	81.1%	81.4%	91.6%	90.0%	**93.7%**	91.8%
ALBERT-CNN	87.8%	82.4%	80.7%	81.5%	91.5%	90.9%	92.3%	91.6%
ALBERT-TextCNN	87.5%	**83%**	78.7%	80.8%	91.6%	90.8%	92.7%	91.7%
ALBERT-SACR(ours)	**88.8%**	**88.1%**	76.9%	**82.1%**	**94.6%**	**94.4%**	**94.8%**	**94.6%**
Improvement	+0.8%	+5.1%	−5.3%	+0.1%	+2.0%	+2.40%	+1.1%	+2.0%

The comparison results are presented in Table 2. Our ALBERT-SACR is superior to the comparison methods on Waimai and Shopping. On dataset Waimai, Acc, P, R and F1 of our method are 88.8%, 88.1%, 76.9% and 82.1% respectively. In comparison methods, optimal results of Acc, P, R and F1 are 88.0%, 83.0%, 82.2% and 82.0% respectively. Compared to the comparison methods, Acc, P and F1 of our method improved 0.8%, 5.1% and 0.1% respectively.

On dataset Shopping, Acc, P, R and F1 of our method are 94.6%, 94.4%, 94.8% and 94.6% respectively, optimal results of Acc, P, R and F1 in comparision methods are 92.6%, 92.0%, 93.7% and 92.6% respectively. Compared to the comparison methods, Acc, P, R and F1 of our method are improved 2.0%, 2.40%, 1.1% and 2.0% respectively.

To summarize, our ALBERT-SACR achieves improvements on both Waimai and Shopping datasets, which validates the effectiveness of ALBERT-SACR. It indicates that self-adaptive context reasoning mechanism can improve the classification performance of text sentiment analysis by learning association among semantic features with different lengths. However, the performance on Waimai dataset is slightly less effective than Shopping dataset. This may be caused by unbalanced positive and negative samples of dataset Waimai.

4.3   Influence of CNN Channel Numbers

In order to explore influence of CNN channel numbers in SACR, we change the number of CNN channels to analyse the effect of SACR with different channels on two datasets. The results are shown in Fig. 4.

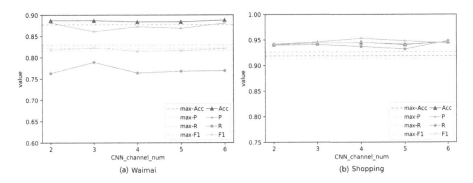

Fig. 4. Results of ALBERT-SACR on two datasets as CNN channel of SACR changes. In every chart, the horizontal coordinate is the number of CNN channels, and the vertical coordinate is value of the evaluation metric. The max-Acc, max-P, max-R and max-F1 are optimal values of Acc, P, R, and F1 in comparison methods, respectively. In horizontal coordinate, for multi-channel CNN, its convolution kernel size is [2,..., i+1] when its number of CNN channels is i.

Figure 4(a) illustrates the results of max-Acc, max-P, max-R and max-F1 as with CNN channel changing on Waimai dataset. In Fig. 4(a), max-Acc, max-P, max-R, and max-F1 are 88.0%, 83%, 82.2%, and 82.0%, respectively. During change of CNN channels, all Acc and P are higher than max-Acc and max-P respectively, all R values are lower than max-R, and F1 is lower than max-F1 only when the number of CNN channels is 4. With the change of CNN channel number, ALBERT-SACR is superior to the comparison methods no matter how CNN channel number changes; moreover, the optimal result is achieved when the number of CNN channels is 3x.

Figure 4(b) illustrates the results of max-Acc, max-P, max-R and max-F1 as with CNN channel changing on Shopping dataset. In Fig. 4(b), max-Acc, max-P, max-R, and max-F1 are 92.6%, 92.0%, 93.7%, and 92.6% respectively, and all Acc, P, R, and F1 exceed max-Acc, max-P, max-R, and max-F1 respectively. It shows that ALBERT-SACR is preferred over comparison methods on dataset Shopping, especially when CNN channel number is 6.

In conclusion, ALBERT-SACR outperforms all comparison methods when the number of CNN channels varies. The optimal CNN channel number varies on different datasets, it may be due to the various distribution of text lengths in different datasets. Besides, the variations of Acc and F1 are relatively smooth when the CNN channels change, it reflects that the adaptive nature of our proposed SACR is effective in contextual inference for semantic features of different lengths.

5   Conclusion and Future Works

In order to analyse the associations among semantic features with different lengths in e-commercial reviews, we proposed a text sentiment analysis method ALBERT-SACR on the basis of self-adaptive context reasoning mechanism in this paper. Our ALBERT-SACR firstly learns global contextual features using Transformer in ALBERT; and then, contextual associations between semantic features of different lengths are generated using the SACR layer proposed in this paper.In simulation experiments, we compare our method with five other methods which are ALBERT, ALBERT-Dense, ALBERT-BiLSTM, ALBERT-CNN and ALBERT-TextCNN on two datasets Waimai and Shopping. Moreover, we further discuss how CNN channel numbers in our proposed methods influence classification results. The simulation results show that ALBERT-SACR outperforms others; besides, the optimal CNN channel number varies on different datasets, it may be due to the various distribution of text lengths in different datasets.

In our future work, we will continue to expand the datasets. Moreover, we would like to explore the intrinsic impact factors of our method and further optimize the network structure in terms of improving both effectiveness and efficiency.

Acknowledgement. This work was supported by CCF Opening Project of Information System (CCFIS2021-03-01) and the Natural Science Project of Education Department of Shaanxi Province (No.21JK0646).

References

1. Awwalu, J., Bakar, A.A., Yaakub, M.R.: Hybrid N-gram model using Naïve Bayes for classification of political sentiments on Twitter. Neural Comput. Appl. **31**(12), 9207–9220 (2019). https://doi.org/10.1007/s00521-019-04248-z
2. Balakrishnan, V., Khan, S., Arabnia, H.: Improving cyberbullying detection using twitter users' psychological features and machine learning. Comput. Secur. **90**, 101710 (2020). https://doi.org/10.1016/j.cose.2019.101710
3. Chen, H., et al.: Country image in covid-19 pandemic: a case study of china. IEEE Trans. Big Data **1** (2020). https://doi.org/10.1109/TBDATA.2020.3023459
4. Devlin, J., Chang, M.W., Lee, K., Toutanova, K.: Bert: pre-training of deep bidirectional transformers for language understanding, pp. 4171–4186. Association for Computational Linguistics (2019). https://doi.org/10.18653/v1/N19-1423
5. Ding, Z., Qi, Y., Lin, D.: Alberta-based sentiment analysis of movie review. In: 2021 4th International Conference on Advanced Electronic Materials, Computers and Software Engineering (AEMCSE), pp. 1243–1246 (2021). https://doi.org/10.1109/AEMCSE51986.2021.00254

6. Guo, B., Zhang, C.X., Liu, J., Ma, X.: Improving text classification with weighted word embeddings via a multi-channel textcnn model. Neurocomputing **363** (2019). https://doi.org/10.1016/j.neucom.2019.07.052

7. Kim, Y.: Convolutional neural networks for sentence classification. In: Proceedings of the 2014 Conference on Empirical Methods in Natural Language Processing (2014). https://doi.org/10.3115/v1/D14-1181

8. Lan, Z., Chen, M., Goodman, S., Gimpel, K., Sharma, P., Soricut, R.: Albert: a lite bert for self-supervised learning of language representations. ArXiv:abs/1909.11942 (2020)

9. Li, M., Chen, L., Zhao, J., Li, Q.: Sentiment analysis of Chinese stock reviews based on BERT model. Appl. Intell. **51**(7), 5016–5024 (2021). https://doi.org/10.1007/s10489-020-02101-8

10. Li, Z., He, L., Guo, W., Jin, Z.: Research on sentiment analysis method based on weibo comments. East Asian Math. J. **37**(5), 599–612 (2021)

11. Liu, W., Pang, J., Li, N., Zhou, X., Yue, F.: Research on Multi-label Text Classification Method Based on tALBERT-CNN. Int. J. Comput. Intell. Syst. **14**(1), 1–12 (2021). https://doi.org/10.1007/s44196-021-00055-4

12. Lv, Y., et al.: Aspect-level sentiment analysis using context and aspect memory network. Neurocomputing **428** (2020). https://doi.org/10.1016/j.neucom.2020.11.049

13. Radford, A., Narasimhan, K.: Improving language understanding by generative pre-training (2018)

14. Vaswani, A., et al.: Attention is all you need, pp. 6000–6010 (2017)

15. Wang, Y., Yu, H., Wang, G., Xie, Y.: Cross-domain recommendation based on sentiment analysis and latent feature mapping. Entropy **22**, 473 (2020). https://doi.org/10.3390/e22040473

16. Wei, J., Liao, J., Yang, Z., Wang, S., Zhao, Q.: Bilstm with multi-polarity orthogonal attention for implicit sentiment analysis. Neurocomputing **383** (2019). https://doi.org/10.1016/j.neucom.2019.11.054

17. Wu, Y., He, J.: Sentiment analysis of barrage text based on albert-att-bilstm model, pp. 152–156 (2021). https://doi.org/10.1109/PRAI53619.2021.9551040

18. Wu, Z., Ong, D.: Context-guided bert for targeted aspect-based sentiment analysis (2020)

19. Ye, X., Xu, Y., Luo, M.: Albertc-CNN based aspect level sentiment analysis. IEEE Access **1** (2021). https://doi.org/10.1109/ACCESS.2021.3094026

20. Yin, Z., Kou, Y., Wang, G., Shen, D., Nie, T.: Explainable recommendation via neural rating regression and fine-grained sentiment perception. In: Xing, C., Fu, X., Zhang, Y., Zhang, G., Borjigin, C. (eds.) WISA 2021. LNCS, vol. 12999, pp. 580–591. Springer, Cham (2021). https://doi.org/10.1007/978-3-030-87571-8_50

21. Zhang, M., Wang, S., Yuan, K.: Sentiment Analysis of Barrage Text Based on ALBERT and Multi-channel Capsule Network, pp. 718–726 (2022). https://doi.org/10.1007/978-3-030-89698-0_74

AOED: Generating SQL with the Aggregation Operator Enhanced Decoding

Yilin Li[1], Xuan Pan[1], Dongming Zhao[2], Minhui Wang[3], and Yanlong Wen[1(✉)]

[1] College of Computer Science, Nankai University, Tianjin 300350, China
allinleeme@outlook.com, panxuan@dbis.nankai.edu.cn, wenyl@nankai.edu.cn
[2] Artificial Intelligence Laboratory, China Mobile Communication Group Tianjin
Co., Ltd, Tianjin, China
waitman_840602@163.com
[3] China Mobile Communication Group Tianjin Co., Ltd, Tianjin, China
wangmh_nk@126.com

Abstract. NL2SQL is a translation task that converts natural language queries to SQL. We revisit the popular NL2SQL models and find that the accuracy of aggregation operator prediction remains a bottleneck of current NL2SQL models. We present a novel statistics-based approach called AOED, which stands for Aggregation Operator Enhanced Decoding, to help predict aggregation operator. AOED is a carefully designed mechanism that takes full advantage of the statistical information of the aggregation keywords in the natural language query to help improve the prediction accuracy of the aggregation operator. Experiments on the WikiSQL dataset show that our model outperforms the state-of-the-art model SQLova and NL2SQL-RULE by 3.4% and 0.7% on overall SQL results in the logical form accuracy and by 0.2% and 0.7% on aggregation operator result.

Keywords: Semantic parsing · Natural language processing · Text-to-SQL · Deep learning

1 Introduction

Nowadays, numerous amounts of information are stored as structured data in relational databases. And a need for a human-friendly way to access this information is becoming more urgent. For instance, the government needs an effortless way to record, review, and modify financial revenue and expenditure records. However, mastering a query language makes it difficult for a non-specialist to access the precious information stored in the database.

The concept of natural language interface to database (NLIDB) was introduced in 1995 to address this issue. With the development of query language, SQL has become the most widely used one. Focusing on one aspect of NLIDB, Zhong *et al.* [15] proposed the problem of Natural Language to SQL (NL2SQL). NL2SQL is one of the hot topics of semantic parsing. Its purpose is to parse natural language descriptions into SQL queries.

© The Author(s), under exclusive license to Springer Nature Switzerland AG 2022
X. Zhao et al. (Eds.): WISA 2022, LNCS 13579, pp. 206–215, 2022.
https://doi.org/10.1007/978-3-031-20309-1_18

Recently, the models based on deep learning techniques have substantially improved the results of the NL2SQL task. Furthermore, existing NL2SQL solutions reach a bottleneck because of poor prediction accuracy of aggregation. Even the popular models suffer a problem that aggregation's prediction accuracy is about 5% or lower.

Intuitively, the aggregation operations in SQL always map to some keywords in natural language questions (NLQs). For instance, generating the MAX operator is probable when "maximum" appears in the corresponding NLQ.

Based on the observation, we propose a statistics-based model. The basic idea is to mark the particular keywords in NLQ and make the model to learn subtle relationships between the keywords and NLQ to improve the prediction accuracy of the aggregation operator. So the proposed model firstly collects the keyword information and encodes the keywords strongly associated with aggregation operations (KSAA) as a feature vector for NLQ encoding.

The rest of this paper is organized as follows. Section 2 gives an overview of related work in this field. Then we describe our methodology in Sect. 3. We report the quantitative results of our experiments and comparison to previous models in Sect. 4. Finally, we conclude the paper and introduce the future work in Sect. 5.

2   Related Work

2.1   Semantic Parsing

Parsing natural language into source code written in a programming language has long been an open topic in natural language processing after the introduction of NLIDB. Early models mostly take rule-based methods, such as Ginseng [1] and SODA [2]. Recently, deep learning has become a popular technique for NLP tasks. Especially, Transformer proposed by Vaswani et al. [11] further lays the foundation for applying deep learning in NLP. WikiSQL is a large English semantic parsing dataset proposed by Zhong et al. [15] with their model SEQ2SQL. Significantly, WikiSQL is an order of magnitude larger than previous semantic parsing datasets. TableQA [10] is a cross-domain Chinese NL2SQL dataset. It gives a more comprehensive assessment of NL2SQL models by requiring the generation of SQL to various forms of expression for condition values.

2.2   Sketch-Based Deep Learning Models

The SQLNet [13] presents a sketch-based approach. The sketch contains a dependency graph to determine the SQL parsing order, which means some sub-modules of the code prediction rely on previous sub-modules parsing results. For instance, the aggregation operator prediction relies on the result of the select column information. The HydraNet [7] breaks down the NL2SQL task into column-wise ranking and decoding and finally assembles the column-wise outputs into a SQL query. But the prediction result of aggregation operator accuracy is about 6% lower than other SQL operators.

BERT [3] is a large pre-trained language model which is the first finetuning-based representation model achieving state-of-the-art on many sentence-level and token-level NLP tasks. Then SQLova [5] takes advantage of BERT by bringing the best of previous NL2SQL approaches together with BERT. It uses six sub-modules in the decoding layer for prediction, and it is the first NL2SQL model which achieves human performance on WikiSQL dataset. Guo *et al.* [4] further improved SQLova [5] by applying an approach that uses match information of NLQ and table cells and match information of NLQ and table column headers. Furthermore, NL2SQL-RULE [4] remains almost the same prediction accuracy of aggregation as SQLova. Execution Guided Decoding (EG) is a technique in which the candidate output SQL list is fed into the executor, then discard the outputs that return empty or have grammar errors. Many sketch-based models use EG to eliminate the potential grammar errors during the decoding process. However, the aggregation operator prediction is overly dependent on the results of the table column's sub-module, so the performance of the aggregation operator's sub-module is always unsatisfactory.

3 Methodology

3.1 Model Overview

In this section, we give an overview of the proposed model. The overall architecture of the model is organized as Fig. 1. It consists of two primary modules: the encoding and decoding layers. We fuse KSAA in the decoding layer to strengthen the capability for aggregation operator parsing.

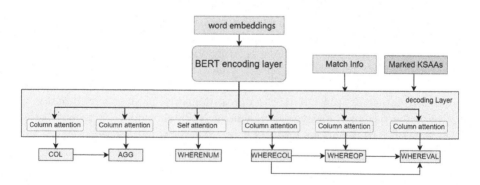

Fig. 1. Architecture of the proposed model.

In the encoding layer, we use BERT to obtain the context-aware and table-aware representations of words E in NLQs. The input of the BERT model is indicated as Eq. 1 as follows:

$$E = [\text{CLS}], N_1, N_2, \ldots, N_n, [\text{SEP}],$$
$$hd_1^1, hd_2^1, \ldots, hd_{n_1}^1, [\text{SEP}],$$
$$hd_1^2, hd_2^2, \ldots, hd_{n_2}^2, [\text{SEP}],$$
$$hd_1^m, hd_2^m, \ldots, hd_{n_m}^m, [\text{SEP}]$$

$$(1)$$

where N_i denotes the i-th token of natural language query, hd_q^p denotes the q-th token of the p-th table header, and n_p is the length of the p-th table header. The output of the encoding layer consists of two vectors, Q for question encoding and H for table header encoding.

We use the marked KSAAs as part of the input in the decoding layer. As shown in Fig. 1, the decoding layer consists of several sub-modules to construct SQL's different parts, as SQLova [5] did. The arrows denote the dependency relationship of these sub-module outputs. In each sub-module of the decoding layer, the KSAAs are input along with BERT output. The output of the sub-module can be described as Eq. 2, where P is the prediction result of the sub-module and QV and HV stores match the information of NLQ and table content and NLQ and table header as NL2SQL-RULE [4] do. QV is a vector whose length equals NLQ, and HV is a vector whose length is equal to table headers. We omit the detail here and use F to refer to the decoding layer operation.

$$P = \text{F}(Q, H, QV, HV, KSAA) \tag{2}$$

To mask the KSAAs in NLQs, we first analyze the keywords that have high correlations to the aggregation operators. Then we encode the KSAA in NLQ as a feature vector. Finally, we take the KSAA vector as part of the input into the decoding layer for aggregation operator generation.

3.2   Statistics of KSAA

To determine the KSAAs, we count the frequency of tokens appearing in NLQs. Figure 2 shows the statistic of the keywords strongly correlated to the aggregation operators. The count of "highest" covers 42.5% of all the NLQs whose corresponding SQLs have the MAX aggregation operator. And "average" covers even 67.3%.

We also analyze the relationship between these keywords and their corresponding aggregation operators. Figure 3 shows the probability of a particular aggregation operator when KSAAs appear. Except for the keyword "games", almost all keywords indicate strong relationships with specific aggregation operators. For example, the aggregation operator "MIN" probability is 99.4% when "earliest" appears in NLQ. Thus, paying attention to these keywords positively affects the parsing result of the aggregation operators.

As illustrated before, KSAAs are picked according to their frequency of appearance. Misspellings can lead to unexpected results because the model considers the incorrectly spelled word and the right one to be two different words.

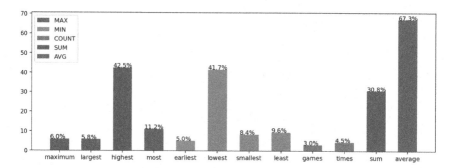

Fig. 2. The statistic of the keywords strongly correlated to SQL's aggregation operators.

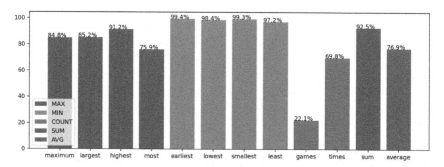

Fig. 3. The probability of the keywords and their correlated aggregation operators.

3.3   Aggregation Operator Enhanced Decoding

Based on SQLova [5] and NL2SQL-RULE [4], our model uses an encoding layer and a decoding layer and uses sub-modules to predict different parts of SQL with a sketch-based method.

Different from SQLova [5] and NL2SQL-RULE [4], we fuse the marked KSAAs as an extra input into the decoding layer to help the model better focus on the keywords correlated to aggregation operation. To obtain the strongly correlated keywords from NLQs, we use Algorithm 1 to extract the KSAAs and denote them as T_i.

Then we construct the KSAA vector as Algorithm 2, where Str^{aggop} refers to the string of aggregation operators such as "avg" and "count". Note that even though T^{hd} is similar to T. It helps exclude some special situations when the aggregation operation is done. If "avg" appears in table headers, marking the KSAAs is not suggested because maybe the aggregation operation has already appeared when filling the table. So we use T^{hd} to exclude these situations.

Algorithm 1. Extract KSAAs from NLQs

Input: $NLQs$
Output: a KSAAs set T
Set T a list of empty dictionary
Set $noise$ a empty list
for $sample$ in $dataset_{train}$ **do**
  $op = sample['sql']['agg']$
  for tok in $sample['nlq']$ **do**
    $tok =$ "the" $+ tok$ if tok is superlative_adj else tok
    $T[op][tok] += 1$
  end for
end for
for T_i in T **do**
  $T_i = \text{sort}(T_i, \text{reverse}=\textbf{true})$
  $T_i =$ get the first $p\%$ of T_i
end for
$noise =$ symmetric difference of $T_1, T_2, \ldots, T_i$
for T_i in T **do**
  $T_i = T_i$ - $noise$
end for
$T = T_1 \cup T_2 \cup T_3, \ldots, T_i$

Algorithm 2. Construct KSAA Vector

Input: KSAAs set T and tokenized NLQs $question\_toks$
Output: marked KSAA vector V
$T^{hd} = T + Str^{aggop}$
for tok in $question\_toks$ **do**
  if tok is superlative_adj **then**
    $V[\text{tok}] = 6$
  end if
  if tok in T **and** tok not in T^{hd} **then**
    $V[\text{tok}] = 5$
  end if
end for

Besides, we find that paying attention to the highest order adjectives helps the model better understand the semantics. Here we use a simple regular expression to match adjectives of the highest order. For example, if a word has a format of "*est", such as "greatest" and "highest", then it will be marked as adjectives of the highest order.

Finally, we take the feature vector as an extra input to the decoding layer. A scattered tensor G concatenates this feature vector as an index into the last dimension of encoding layer output of question Q, as shown in Eq. 3:

$$G = \text{scatter}(V)$$
$$M = G \oplus Q$$

$$(3)$$

where $\oplus$ denotes the concatenation operation, and M represents the concatenated output of Q and G.

The sub-modules in the decoding layer share the same input for prediction. In these sub-modules, we use M and the encoding layer output of table header H as the input, then output the probability of the respective target. Here we take the select column as an example. As shown in Eq. 4, we first separately encode the M and the table header H encodings by LSTM networks. We use W_i to denote different affine transformations. After several affine transformation and activation functions, we take the probability p as the prediction result.

$$
\begin{aligned}
Q, H &= \text{LSTM}(M_n, H) \\
s &= H^T W_1 Q \\
C &= \sum \text{softmax}(s)Q \\
K &= W_2 \tanh(W_3 H \oplus W_4 C) \\
p &= \text{softmax}(K)
\end{aligned}
\tag{4}
$$

4   Experiment

4.1   Experiment Settings

The dataset we use is WikiSQL consisting of 80654 hand-annotated instances, including natural language questions, SQL queries, and SQL tables extracted from 24241 HTML tables from Wikipedia. The dataset is split into three parts, the training set with 56355 samples, the development (dev) set with 8421 samples, and the test set with 15878 samples. Different from other dataset such as CSpider [9] and TableQA [10]. The query operation is limited to only one relatively small table.

Our experiment consists of three phases: collecting the aggregation keywords, marking the KSAAs, and model training. We trained our model on one NVIDIA 1080ti GPU. We use the adam algorithm as the optimizer and set the batch size and the learning rate to 8 and 0.001, respectively.

We compare our model with the following models: (1) SQLNet [13] is a sketch-based model; (2) SQLova [5] combines BERT [3] with traditional NL2SQL model; (3) IE-SQL [8] uses extraction-linking approach; (4) HydraNet [7] leverages a pretrained language model in Text-to-SQL; (5) NL2SQL-RULE [4] uses a matching method to enhance decoding; (6) SeaD [14] proposes a denoising approach and it is current state-of-the-art model on WikiSQL [15]. We use the logical form accuracy (LF) and the execution accuracy (X) as the metrics.

4.2   Overall Result

The results are shown in Table 1. Our model outperforms most compared models. Compared with NL2SQL-RULE, our model achieves about 0.55% improvement in two metrics on the dev set. And when compared with SeaD, our model achieves about 0.1% improvement in the logical form accuracy and 0.5% improvement in the execution accuracy on the dev set without Execution Guided Decoding (EG) [12]. The marked KSAAs explicitly emphasize the aggregation semantics in NLQ sentences, which allow effective training. Besides, the time AOED uses to collect KSAAs from the dataset is no more than three minutes on a mobile CPU, Ryzen 5800U.

Table 1. Accuracy on the development (dev) and test sets.

Model	dev LF	dev X	test LF	test X
SQLNet	–	69.8%	–	68.0%
SQLova	81.6%	87.2%	80.7%	86.2%
IE-SQL	84.6%	88.7%	84.6%	88.8%
HydraNet	83.6%	89.1%	83.8%	89.2%
NL2SQL-RULE	84.3%	90.3%	83.7%	89.2%
SeaD	84.9%	90.2%	84.7%	90.7%
AOED(ours)	**85.0%**	**90.7%**	**84.6%**	**90.3%**
AOED+EG(ours)	**85.8%**	**91.8%**	**85.3%**	**91.4%**

As shown in Table 2, we further compare our model's sub-module prediction accuracy with others. We improve about 0.55% on the prediction accuracy of the aggregation operator compared to the baseline NL2SQL-RULE. It's noticeable that AOED achieves the best performance on AGG prediction on the dev set, whether with EG or not. And AOED also outperforms NL2SQL-RULE in the WNUM and WOP predictions slightly. The results prove that the marked superlative adjectives effectively improve the semantic understanding of operators.

Table 2. Comparison of detailed accuracy on dev set. COL, AGG, WNUM, WCOL, WOP, and WVAL stand for select-column, select-aggregation, where-number, where-column, where-operator, and where-value, respectively.

Model	COL	AGG	WNUM	WCOL	WOP	WVAL
SQLova, dev	97.3%	90.5%	98.7%	94.7%	97.5%	95.9%
NL2SQL-RULE, dev	97.4%	90.0%	99.1%	97.9%	98.1%	97.6%
AOED(ours), dev	**97.2%**	**90.7%**	**99.3%**	**98.0%**	**98.3%**	**97.6%**
NL2SQL-RULE+EG, dev	97.4%	90.4%	98.9%	97.9%	97.7%	97.9%
AOED+EG(ours), dev	**97.2%**	**90.8%**	**98.9%**	**97.9%**	**97.7%**	**98.1%**

4.3 Ablation Study

We also studied the model's performance when separating the marked KSAAs from the model. Table 4 exhibits the results. The impact mainly focuses on aggregation prediction because the KSAAs, strongly associated with aggregations such as "smallest" and "average", improve the model sensitivity to aggregation operations. Without the KSAAs, the model is challenging to capture the aggregation semantics from the NLQs.

Table 3. Comparison of detailed accuracy on the test set.

Model	COL	AGG	WNUM	WCOL	WOP	WVAL
SQLova, test	97.3%	90.5%	98.7%	94.7%	97.5%	95.9%
AOED, test	**96.9%**	**90.8%**	**99.0%**	**97.7%**	**97.8%**	**97.1%**
AOED+EG, test	**96.9%**	**90.9%**	**98.4%**	**97.3%**	**97.1%**	**97.4%**

Table 4. Accuracy before and after ablation.

Ablation	dev LF	dev X	COL	AGG	WNUM	WCOL	WOP	WVAL
AOED	85.0%	90.7%	97.2%	90.7%	99.3%	98.0%	98.3%	97.6%
- marked KSAAs	84.5%	90.1%	97.3%	90.5%	99.2%	97.8%	98.1%	97.2%

5 Conclusion and Future Work

This paper proposes a novel statistics-based method to focus on specific operator predictions in the NL2SQL tasks. We design an automatic process to extract the keywords strongly associated with aggregation operations from NLQs and use these keywords to strengthen the capability for aggregation operator parsing. Our method improves the existing model's performance on WikiSQL, outperforming the previous state-of-the-art model NL2SQL-RULE by 0.4% in the aggregation prediction accuracy.

However, the proposed model needs a keyword extraction pre-processing, which relies heavily on many high-quality training samples. In the future, We will explore the unsupervised keyword extraction method [6] to try to decrease the number of required samples. Besides, the regular expression we use here is so simple that it cannot distinguish nouns with a pattern like adjectives of the highest order. For instance, our model cannot correctly recognize nouns like "vest", so the regular expression harms our method somewhat. We will also try tokenizers and use their lexicon library to help better match adjectives of the highest order and reduce the error rate of recognition and dependency on the quantity and quality of samples.

Acknowledgement. This research is supported by Chinese Scientific and Technical Innovation Project 2030 (No. 2018AAA0102100), National Natural Science Foundation of China (No. 62077031). We thank the reviewers for their constructive comments.

References

1. Bernstein, A., Kaufmann, E., Kaiser, C., Kiefer, C.: Ginseng: A Guided Input Natural Language Search Engine for Querying Ontologies. In: Jena User Conference (2006). (issue: May)
2. Blunschi, L., Jossen, C., Kossmann, D., Mori, M., Stockinger, K.: SODA: generating SQL for business users. Proc. VLDB Endowment **5**(10) (2012). https://doi.org/10.14778/2336664.2336667
3. Devlin, J., Chang, M.W., Lee, K., Toutanova, K.: BERT: pre-training of deep bidirectional transformers for language understanding. In: NAACL HLT 2019–2019 Conference of the North American Chapter of the Association for Computational Linguistics: Human Language Technologies - Proceedings of the Conference. vol. 1 (2019). https://doi.org/10.18653/v1/N19-1423
4. Guo, T., Gao, H.: Content enhanced bert-based text-to-sql generation. arXiv preprint arXiv:1910.07179 (2019)
5. Hwang, W., Yim, J., Park, S., Seo, M.: A comprehensive exploration on wikisql with table-aware word contextualization. arXiv preprint arXiv:1902.01069 (2019)
6. Jin, Y., Chen, R., Xu, L.: Text keyword extraction based on multi-dimensional features. In: Wang, G., Lin, X., Hendler, J., Song, W., Xu, Z., Liu, G. (eds.) WISA 2020. LNCS, vol. 12432, pp. 248–259. Springer, Cham (2020). https://doi.org/10.1007/978-3-030-60029-7_23
7. Lyu, Q., Chakrabarti, K., Hathi, S., Kundu, S., Zhang, J., Chen, Z.: Hybrid Ranking Network for Text-to-SQL. Tech. Rep. MSR-TR-2020-7, Microsoft Dynamics 365 AI (2020). https://www.microsoft.com/en-us/research/publication/hybrid-ranking-network-for-text-to-sql/
8. Ma, J., Yan, Z., Pang, S., Zhang, Y., Shen, J.: Mention extraction and linking for SQL query generation. In: EMNLP 2020–2020 Conference on Empirical Methods in Natural Language Processing, Proceedings of the Conference (2020). https://doi.org/10.18653/v1/2020.emnlp-main.563
9. Min, Q., Shi, Y., Zhang, Y.: A pilot study for Chinese SQL semantic parsing. In: EMNLP-IJCNLP 2019–2019 Conference on Empirical Methods in Natural Language Processing and 9th International Joint Conference on Natural Language Processing, Proceedings of the Conference (2019). https://doi.org/10.18653/v1/d19-1377
10. Sun, N., Yang, X., Liu, Y.: Tableqa: a large-scale chinese text-to-sql dataset for table-aware sql generation. arXiv preprint arXiv:2006.06434 (2020)
11. Vaswani, A., et al.: Attention is all you need. In: Advances in Neural Information Processing Systems, vol. 2017-December (2017). (iSSN: 10495258)
12. Wang, C., et al.: Robust text-to-sql generation with execution-guided decoding. arXiv preprint arXiv:1807.03100 (2018)
13. Xu, X., Liu, C., Song, D.: Sqlnet: generating structured queries from natural language without reinforcement learning. arXiv preprint arXiv:1711.04436 (2017)
14. Xuan, K., Wang, Y., Wang, Y., Wen, Z., Dong, Y.: SeaD: end-to-end Text-to-SQL Generation with Schema-aware Denoising. arXiv preprint arXiv:2105.07911 (2021)
15. Zhong, V., Xiong, C., Socher, R.: Seq2sql: generating structured queries from natural language using reinforcement learning. arXiv preprint arXiv:1709.00103 (2017)

Multi-granularity Chinese Text Matching Model Combined with Bidirectional Attention

Mengqi Liu[✉], Jiale Zhang, and Lizhen Xu

Department of Computer Science and Engineering, Southeast University, Nanjing 21189, China
220201893@seu.edu.cn

Abstract. Text matching is an important task in natural language processing field, and it is widely used in intelligent question answering, information retrieval and other fields. With the rise of deep learning, text matching has gradually shifted from the traditional word similarity method to the neural network research. Currently, text matching methods based on deep learning are mainly divided into representation-based text matching models and interaction-based text matching models. Among them, the representation-based text matching model encodes two sentences separately, which easily loses the semantic focus and makes it difficult to measure the importance of context between sentences. The interaction-based text matching model ignores global information, and thus affects the global matching effect. Based on this, this paper proposes a multi-granularity text matching model combined with bidirectional attention (MGBA). The model enables two text tensors to interact in advance through the bidirectional attention mechanism, and then extracts the multi-granularity features of text tensors through LSTM and CNN, so that the model can focus on different levels of text information, thus solving the problem that traditional deep models tend to lose semantic focus and ignore global information in the process of text matching.

Keywords: Text matching · Multi-granularity · Bidirectional attention

1 Introduction

Natural language processing is a hot research direction in the field of artificial intelligence [1], and text matching is one of the important tasks of natural language processing, which aims to judge the semantic similarity between two given text fragments. Meanwhile, many other tasks of natural language processing can be directly abstracted as text matching tasks. For example, information retrieval [2] can be regarded as the matching between query items and documents, and the essence of intelligent question answering [3] is also the matching between questions and candidate answers or standard questions. Therefore, research on text matching is crucial.

The wide application of deep learning in the field of natural language processing has created new opportunities for text matching technology. The deep learning model can encode the text to obtain the feature representation of the text, and then use RNNs or CNNs to further process the text feature sequence to obtain the semantic representation tensor of the text. Then, it can be determined whether the two texts are similar by calculating the similarity value of the two text tensors and according to the similarity value. According to different methods of further processing text feature representation, deep text matching models can be divided into representation-based text matching models [4] and interaction-based text matching models [5]. The text representation-based model encodes the two texts separately, which can better extract the information of each text, but it is easy to lose the semantic focus. Models based on text interaction can capture deep semantic information by interacting with text feature sequences in advance, but ignore the global sentence comparison information. Based on this, this paper proposes a multi-granularity text matching model combined with bidirectional attention.

2 Related Work

2.1 Representation-Based Models

The representation-based text matching model focuses on the construction of the text representation layer. The basic idea is to construct a twin-tower model, and use a multi-layer neural network to extract the semantic features of the two texts and then perform text matching. The pioneer of this kind of model is DSSM [4], proposed by Microsoft. Subsequently, in order to further improve the ability of the representation layer to extract local features, Microsoft proposed CNN-based CDSSM [6]. To solve the problem that CNN cannot capture long-distance dependencies of sentences, Palangi [7] et al. proposed an LSTM-based text matching model LSTM-DSSM. In addition, in order to obtain text feature representations at different granularity levels of text, Yin [8] et al. proposed a multi-granularity convolutional neural network model (MultiGranCNN), which uses convolution kernels of different sizes to capture features at different levels of text. Wan [9] et al. proposed a multi-view recurrent neural network (MV-LSTM), which takes the expression of each position of the LSTM as a sentence expression centered on the current position, and then constructs a similarity matrix based on multiple expressions, finally calculates the similarity of two texts based on the similarity matrix. Zhang [10] et al. proposed a Chinese-oriented multi-granularity mixture model. Since the Bert [11] pre-training model was proposed, Bert-based models have also emerged. Reimers [12] proposed the BERT version of the Twin Towers model.

2.2 Interaction-Based Models

The interaction-based model abandons the idea of encoding texts separately. It adds an interaction layer before the presentation layer to interact with the two text feature representations to obtain interaction information, and then performs subsequent modeling based on interaction information. The classical interaction-based matching model mainly includes ARC-II [13] proposed by Hang et al., which uses 1D convolution to focus on the adjacent word vector representations to obtain two matrices and then combines two text matrices into a 3D tensor as the interaction result. The MatchPyramid [5] model is similar to ARC-II, but uses a matching matrix as the interactive way. Gong [14] uses multi-head attention to encode two text sequences to interact word vectors.

3   Proposed Model

The structure of multi-granularity Chinese text matching model combined with bidirectional attention (MGBA) proposed in this paper is roughly as shown in the Fig. 1. The pre-trained glove word embedding layer projects two pieces of text into word vectors, and then the Bi-Attention layer is used to interact two tensors to obtain the interaction matrix. Next, feature extraction layer uses MV-LSTM and CNNs to extract the multi-granularity features of inputs. Finally, the obtained tensors are concatenated to calculate matching value through the matching layer. Each part of the model is described in detail below.

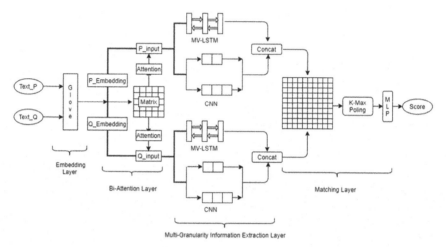

Fig. 1. Structure of MGBA model

3.1   Embedding Layer

The word embedding layer is a matrix. Its function is to project the words represented of text into high-dimensional dense vectors. The essence of embedding layer is a search function. According to the serial number of the dictionary where the word is located to find the row vector corresponding to the embedding matrix, and the vector is the project result of the word.

This paper selects the pre-trained Glove model as the embedding layer. For the embedding matrix $E \in R^{kd}$, k is the number of embedded words, and d is the dimension of the word vector. For $P_{Text} = \{p_1, p_2, \cdots p_m\}$ and $Q_{Text} = \{q_1, q_2, \cdots, q_n\}$, Glove can project them into matrix $P \in R^{md}$ and $Q \in R^{nd}$, where m, n are the lengths of the two pieces of text after word segmentation.

3.2   Bi-attention Layer

The attention mechanism was initially introduced into the neural network in order to focus on more critical information to the current task and reduce the attention to other information, thereby improving the efficiency and accuracy of task processing. The bidirectional attention mechanism [15] was originally proposed by Seq et al. in the reading comprehension task, and its excellent performance has proved its effectiveness. In the text matching task, the importance of the words in the sentence to the matching task is different. In order to better prominent the role of important words in subsequent processing, it is necessary to assign different weights to different words. The bidirectional attention mechanism is essentially Chinese word-level interactive attention, evaluating the importance of one word to another text. The specific calculation process of word-level attention is as follows:

$$\alpha_i^Q = \sum_{j=1}^{n} \frac{exp(e_{ij})}{\sum_{k=1}^{n} exp(e_{ik})} Q_j \forall i \in [1, 2, \cdots, m] \tag{1}$$

$$\beta_j^P = \sum_{i=1}^{m} \frac{exp(e_{ij})}{\sum_{k=1}^{m} exp(e_{kj})} P_i \forall j \in [1, 2, \cdots, n] \tag{2}$$

where: e_{ij} is the similarity between the i-th word in P and the j-th word in Q, which can be calculated by dot product or other methods. Q_j is the j-th row in matrix Q. α_i^Q is obtained by weighted summation of each word in the text Q, indicating the matching information between the i-th word in the text P and all words in the text Q, that is, the weight of the word relative to another piece of text. β_j^P is the same as α_i^Q.

Specifically, the outputs $P \in R^{m \cdot d}$ and $Q \in R^{n \cdot d}$ from the embedding layer can be calculated by the following formulas:

$$Matrix = P \cdot M \cdot Q + b \tag{3}$$

$$A_{P2Q} = sum_{col}(\sigma_{col}(Matrix) \cdot Q) \tag{4}$$

$$A_{Q2P} = sum_{row}(\sigma_{row}(Matrix) \cdot P) \tag{5}$$

Among them, M represents the weight matrix, and σ is the softmasx function. A_{P2Q} and A_{Q2P} are the corresponding attention. Finally, the weighted representation of the text can be calculated as follows:

$$P_A = P \cdot A_{P2Q} \tag{6}$$

$$Q_A = Q \cdot A_{Q2P} \tag{7}$$

3.3   Multi-granularity Information Extraction Layer

RNN is one of the most commonly used models when dealing with time series problems using deep learning. However, since the output of RNN at time t only depends on the hidden state at time $t-1$, it is difficult to deal with the long-term dependency problem of long texts. Therefore, in order to solve the long-term dependency problem, the long short-term memory neural network LSTM is proposed. LSTM introduces a special gate mechanism to control the circulation and loss of features.

Both traditional RNN and LSTM transfer information from front to back, which has limitations in many tasks. To solve this problem, bidirectional LSTM is designed. The idea is to input the same sequence to the forward and backward two LSTMs respectively, and then concatenate the hidden layers of the two networks together as the final output. The structure of MV-LSTM used in this paper is the same as BiLSTM, but the network model believes that the optimal matching position of a sentence is not just the last position, so multi-position matching is necessary, that is, Multi-View (MV). Unlike BiLSTM, which always takes the hidden layer result of the last step as the final output, this model treats the output of each step of the BiLSTM as the semantic representation of the text centering at that location. Therefore, the matching result of sentences depends on the combination degree of sentence representations in multiple locations, that is, multi-granularity matching. At the same time, considering that BiLSTM cannot fully capture the features of phrases and long phrase structure of sentences, this paper introduces convolutional neural network with convolution windows of 2 and 3 on the basis of MV-LSTM to obtain the phrases and long phrase features of sentences respectively. Finally, the feature tensors extracted by MV-LSTM and CNN is concatenated for the calculation of the matching layer.

3.4 Matching Layer

After the input texts pass the multi-granularity feature extraction layer, the result of splicing matrices represent different levels feature information of the sentence. Therefore, the matching layer needs to further evaluate the multi-angle information and highlight the features that have a greater impact on the matching results. The similarity matrix is used for calculation. The sketch map is shown as Fig. 2:

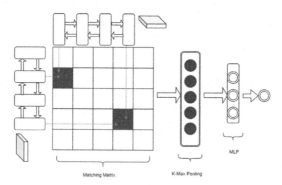

Fig. 2. Structure of matching layer

where M is the weight matrix used to re-weight two sentence interactions. The specific calculation method for each item refers to the method of the original paper of MV-LSTM. The equation is as follows:

$$s(u, v) = u^T \cdot M \cdot v + b \qquad (8)$$

Then, use the K-Max Pooling layer to perform the pooling operation on the matching matrix. Unlike the max pooling layer, the K-Max Pooling layer retains the largest K values, which can better retain the feature information. Then, the result of the pooling layer is input into the MLP to obtain the matching result.

4   Experiment and Result

4.1   Dataset and Contrast Models

This paper selects the dataset CCKS 2018 provided by WeBank for experiments to verify the effectiveness of the model proposed in this paper. The CCKS 2018 dataset is a real Chinese customer service corpus. The goal is to determine whether two given questions have the same meaning. The dataset provides 100,000 pairs of training text, 10,000 pairs of validation text, and 10,000 pairs of test text, in which the ratio of synonymous to non-synonymous pairs is approximately 1:1.

At the same time, this paper selects MV-LSTM, MultiGranCNN, MatchPyramid, Bert and MGF [16] for comparison. MV-LSTM and MultiGranCNN use LSTM and CNN, respectively, to extract multi-granularity features on text. MGF extracts the semantic characteristics of Chinese word granularity, and better optimizes the semantic matching of downstream sentences through the fusion of multi-granularity features. Match-Pyramid represents interaction model. These models cover most of the mainstream text matching models.

4.2   Metrics and Result

This paper uses the accuracy rate, recall rate and F1 value as evaluation indicators, and the calculation formula is as follows:

$$P = \frac{TP}{TP + FP} * 100\% \tag{9}$$

$$R = \frac{TP}{TP + FN} * 100\% \tag{10}$$

$$F1 = \frac{2 * P * R}{P + R} * 100\% \tag{11}$$

The main experimental parameters of the model in this paper are: 300-dimensional pre-trained Glove model, the dimension of the MV-LSTM hidden layer is 150, and the dropout value is 0.5. At the same time, in order to realize splicing the output of MV-LSTM and the output of the convolutional neural network, the stride size is 2 for both CNNs with different convolutional windows. The experiment selects Adam as the optimizer.

The final experimental results are as follows:

Table 1. Experimental comparison results on CCKS 2018

Model	P/%	R/%	F1%
MatchPyramid	73.84	72.82	73.32
MV-LSTM	75.22	75.13	75.17
MultiGranCNN	74.86	76.12	75.48
MGF	74.22	74.13	74.17
BERT	75.86	77.12	76.48
MGBA	76.76	77.34	77.05

5 Experiment and Result

This paper proposes a multi-granularity text matching model combined with bidirectional attention (MGBA). The model uses the bidirectional attention mechanism to fuse the representation model and the interaction model to make up for the original defects of the two models. Then, MV-LSTM and CNN are used to extract multi-granularity features from multiple perspectives on the text feature representation, and obtain richer semantic information of the text. Finally, it outperforms other mainstream representation-based and interaction-based text matching models on the CCKS 2018 dataset.

References

1. Xu, L., Li, S., Wang, Y., et al.: Named Entity Recognition of BERT-BiLSTM-CRF Combined with Self-attention. Springer, Cham (2021)
2. Li, H., Xu, J.: Semantic matching in search. Found. Trends Inf. Retr. 7(5), 343–469 (2014)
3. Xue, X., Jeon, J., Croft, W.B.: Retrieval models for question and answer archives. In: Proceedings of the 31st Annual International ACM SIGIR Conference on Research and Development in Information Retrieval, SIGIR 2008, Singapore, 20–24 July. ACM (2008)
4. Huang, P.S., He, X., Gao, J., et al.: Learning deep structured semantic models for web search using clickthrough data. In: Proceedings of the 22nd ACM international conference on Information & Knowledge Management, pp. 2333–2338 (2013)
5. Liang, P., Lan, Y., Guo, J., et al.: Text Matching as Image Recognition (2016)
6. Shen, Y., He, X., Gao, J., et al.: Learning semantic representations using convolutional neural networks for web search. In: Proceedings of the companion publication of the 23rd international conference on World wide web companion (2014)
7. Palangi, H., Deng, L., Shen, Y., et al.: Semantic modelling with long-short-term memory for information retrieval. Comput. Sci. (2015)
8. Yin, W., Schütze, H.: MultiGranCNN: an architecture for general matching of text chunks on multiple levels of granularity. In: Meeting of the Association for Computational Linguistics & the International Joint Conference on Natural Language Processing (2015)
9. Wan, S., Lan, Y., Guo. J., et al.: A deep architecture for semantic matching with multiple positional sentence representations. AAAI Press (2015)
10. Zhang, X, Lu, W., Zhang, G., et al.: Chinese sentence semantic matching based on multi-granularity fusion model (2020)

11. Devlin, J., Chang, M.W., Lee, K., et al.: BERT: pre-training of deep bidirectional transformers for language understanding (2018)
12. Reimers, N., Gurevych, I.: Sentence-BERT: sentence embeddings using Siamese BERT-networks. In: Proceedings of the 2019 Conference on Empirical Methods in Natural Language Processing and the 9th International Joint Conference on Natural Language Processing (EMNLP-IJCNLP) (2019)
13. Hu, B., Lu, Z., Hang, L., et al.: Convolutional neural network architectures for matching natural language sentences. Adv. Neural Inf. Process. Syst. 3 (2015)
14. Gong, Y., Luo, H., Zhang, J.: Natural language inference over interaction space (2017)
15. Seo, M., Kembhavi, A, Farhadi, A., et al.: Bidirectional attention flow for machine comprehension (2016)
16. Zhang, X.. et al.: Chinese sentence semantic matching based on multi-granularity fusion model. In: Pacific-Asia Conference on Knowledge Discovery and Data Mining. Springer, Cham (2020)

World Wide Web

Data Fusion Methods with Graded Relevance Judgment

Yidong Huang[1], Qiuyu Xu[1], Yao Liu[1], Chunlin Xu[2], and Shengli Wu[1(✉)]

[1] School of Computer Science, Jiangsu University, Zhenjiang, China
`swu@ujs.edu.cn`
[2] School of Computer Science, Guangdong Polytechnic Normal University,
Guangzhou, China

Abstract. Data fusion methods have been widely used in many information retrieval tasks. Its performance is affected by many factors including the data fusion algorithm used, the component retrieval systems involved, relevance judgment, the metrics used for evaluation, and others. Previously, data fusion research mainly focused on the data fusion methods and the component retrieval systems involved, but other factors such as relevance judgment and the metrics used for evaluation have not been addressed. As a matter of fact, relevance judgment is an important issue that affects many aspects of information retrieval and data fusion. The assumption of binary relevance judgment has been taken for all the previous research work in data fusion. However, this assumption is simplified and not satisfactory in many cases. Instead, graded relevance judgment is more general and able to deal with more complicated requirements. In this paper, we investigate data fusion methods, especially linear combination, to work with graded relevance judgment. Necessary updates are given for using those methods in the new situation. Experimented with two data sets in TREC, we find that data fusion is still an effective technology for performance improvement in general. Many of them are very competitive in a controlled environment, and linear combination with weights trained by multiple linear regression is the most stable in a more complicated environment.

Keywords: Data fusion · Information retrieval · Graded relevance judgment

1 Introduction

In the last two decades, data fusion has been investigated by many researchers and data fusion methods have been widely used in many information retrieval tasks. It is witnessed by the large number of papers published for the applications of them in various IR tasks, and also those fusion-based retrieval systems that participating in IR competitions such as TREC[1], CLEF[2], and NTCIR[3].

---

[1] Text REtrieval Conference (TREC) is held annually by the national institute of standards and technology, USA. Its web site is located at https://trec.nist.gov.

[2] http://www.clef-initiative.eu/.

[3] https://research.nii.ac.jp/ntcir/index-en.html.

X. Zhao et al. (Eds.): WISA 2022, LNCS 13579, pp. 227–239, 2022.
https://doi.org/10.1007/978-3-031-20309-1_20

Fusion performance is affected by many factors including the algorithm used, the component retrieval systems involved, relevance judgment, the metrics used for evaluation, and others. Previously, data fusion research mainly focused on the data fusion methods and the component retrieval systems involved, but other factors such as relevance judgment and the metrics used for evaluation have not been addressed.

Relevance judgment is an important issue that affects many aspects of information retrieval including the fusion-based approach. In previous data fusion study, only binary relevance judgment has been considered for the design and application of data fusion methods. However, such an assumption is simplified and not satisfactory in many cases. Instead, graded relevance judgment is more general and able to deal with more complicated requirements.

For example, in many tasks in TREC, documents were evaluated with multiple levels of relevance. In some of its Web (1999–2004, 2009–2014) and Microblog tracks (2011–2015), documents were classified as "highly relevant", "relevant", and "irrelevant". Then they were given scores of 2, 1, 0, respectively. Furthermore, in its Decision and Health Misinformation tracks (2019–2021), apart from relevance judgment, correctness and credibility of documents were also taken into account. Spam or misleading documents could be harmful. Especially relevant but harmful documents are even worse than irrelevant documents, thus a negative score (-1 or -2) was given to them. More effort is required to deal with these new problems for information retrieval systems. New metrics are also required to evaluate such results.

In this paper, we investigate how to adapt data fusion methods, especially linear combination, to work with multiple levels of relevance. A group of data fusion algorithms are reviewed and evaluated. Necessary changes are given for them to deal with the new situation. To our knowledge, this research issue has not been addressed before.

The rest of this paper is organized as follows: related work is discussed in Sect. 2. In Sect. 3, four measures are defined for graded relevance judgment. Then a group of data fusion methods are detailed in Sect. 4, with necessary updates for the new situation. Section 5 presents experimental settings and results of all the methods involved. Section 6 concludes the paper.

2   Related Work

Data fusion methods can be divided into two categories: equally-treated and biased methods. As their names indicate, equally-treated methods treat each component retrieval system equally while biased methods treat each component retrieval system in a different way. CombSum [4], CombMNZ [4], Reciprocal Rank Fusion [2], Borda Count [1], Condorcet Fusion [12] are typical equally-treated methods, while linear combination [19], Weighted Borda Count [1], Weighted Condorcet Fusion [20] are biased method. Equally-treated methods are easy to use, while biased methods require more effort for their application. Usually training is required to set appropriate weights for all the component retrieval systems. It is suitable for various situations in which equally-treated methods do not perform well.

Data fusion methods can also be divided into two categories based on the information required: score-based methods and ranking-based methods. Assuming that all the retrieved documents are associated with relevance scores, score-based methods combine scores for every document by using specific formula and rank them accordingly. CombSum, CombMNZ, and linear combination are score-based methods. Comparing and ranking all the documents, ranking-based methods include Borda Count, Weighted Borda Count, Condorcet Fusion, Weighted Condorcet Fusion, and Reciprocal Rank Fusion are ranking-based methods.

A large group of data fusion methods are enclosed under the umbrella of linear combination. Mainly different methods can be used for weight assignment, which is the key point for linear combination. Two factors, performance of component retrieval systems and dissimilarity among component retrieval systems (results), contribute significantly to the performance of fusion results. Either or both factors can be considered for the weight assignment. Therefore, methods can be divided into four categories: naive methods that consider neither of them (type I), considering component system performance only (type II), considering dissimilarity among component systems/results only (type III), and considering both factors (type IV). CombSum [4], CombMNZ [4], Borda Count [1], and Reciprocal Rank Fusion [2] are type I methods, while PosFuse [7], SlideFuse [8], MAPFuse [7], performance-related weights [21] are examples of type II methods, Correlation methods [22] are type III methods, and linear combination with weights trained by stratified sampling [17], regression [19] or various optimization methods [5,15,24,25] are type IV methods. Each type of method is appropriate under a certain circumstance.

To analyse fusion performance is another issue that has been investigated. In [23], a few factors including average performance of all component systems, performance variance of all component systems, dissimilarity among all component results, number of systems involved, were identified as factors that impact fusion performance significantly. On the other hand, [11,13,14] also investigate fusion performance prediction from another angle. All of them are extensions of typical query performance prediction methods for ordinary information retrieval systems [3]. They try to rank a group of queries based on their estimated performance.

In this piece of work, we adapt data fusion methods to make them suitable for graded relevance judgment. A number of data fusion methods, including CombSum, Reciprocal Rank Fusion, PosFuse, SlideFuse, a few different types of linear combination, are investigated.

3   Four Measures for Graded Relevance Judgment

Results evaluation is an important aspect in information retrieval. In this section we introduce four measures for graded relevance judgment. First of all, it is straightforward to expand the situation of binary relevance judgment into graded relevance judgment. In graded relevance judgment, documents are divided into $t + 1$ categories: grades t, $t - 1$, ..., 0 ($t \geq 2$). The documents in grade t are the most relevant, which are followed by the documents in grade $t - 1$, $t - 2$,...,1, and the documents in grade 0 are irrelevant. One primary assumption we take for these documents in various grades is: any document in grade t is regarded as 100% relevant and 100% useful to users, and any document in grade i ($i < t$) is regarded as $i/t\%$ relevant and $i/t\%$ useful to users.

Normalized Discounted Cumulative Gain (NDCG, or NDCG@k) is given in [6]. For a ranked list of documents, each ranking position is assigned a given weight. The top ranked documents are assigned the heaviest weight since they are the most convenient for users to read. A logarithmic function-based weighting schema was assigned with an exception for k top-ranked documents, which are assigned an equal weight of 1; then for any document ranked i that is greater than k, its weight is $w(i) = ln(k)/ln(i)$. Considering a document list up to m documents, its discount cumulated gain (DCG, or DCG@k) is defined as

$$DCG = \sum_{i=1}^{m} (w(i) * gr(d_i))$$

where $gr(d_i)$ is the grade that d_i is in. DCG can be normalized using a normalization coefficient DCG_best, which is the DCG value of the best result lists. Therefore, we have:

$$NDCG = \frac{1}{DCG\_best} \sum_{i=1}^{t} (w(i) * gr(d_i)) = \frac{DCG}{DCG\_best} \tag{1}$$

NDCG was defined for graded relevance judgment, but many other commonly used metrics, such as average precision (AP) and recall-level precision (RP), were defined in the condition of binary relevance judgment. Now we try to make them suitable for graded relevance judgment Let us assume that, in the whole collection, there are $total\_n$ documents whose grades are above 0, and $total\_n = z_1 + z_2 + ... + z_t$, where z_i denotes the number of documents in grade i.

It is straightforward to generalize P@10 as

$$P@10 = \frac{1}{10 * t} \{ \sum_{i=1}^{10} gr(d_i) \} \tag{2}$$

Before discussing the generalization of other metrics, let us introduce the concept of the best result list. For a given information need, a result list L is best if it satisfies the following two conditions:

- all the documents whose grades are above 0 appear in the list;
- for any document pair d_i and d_j, if d_i is ranked in front of d_j, then $gr(d_i) \geq gr(d_j)$.

Many result lists can be the best at the same time since more than one document can be in the same grade and the documents in the same grade can be ranked in different orders. In any best result list L, a document's grade is only decided by its rank; or for any given rank, a document's grade is fixed. Therefore, we can use $gr\_best(d_j)$ to refer to the grade of the document in ranking position j in one of the best result lists. We may also sum up the grades of the documents in top z_t, top $(z_t + z_{t-1})$,..., top $(z_t + z_{t-1} +...+ z_1)$ positions for any of the best result lists (these sums are the same for all the best result lists):

$$sb_t = \sum_{i=1}^{z_t} gr(d_i)$$

$$sb_{t-1} = \sum_{i=1}^{z_t+z_{t-1}} gr(d_i)$$

$$...$$

$$sb = sb_1 = \sum_{i=1}^{z_t+z_{t-1}+...+z_1} gr(d_i)$$

For RP, first we only consider the top z_t documents and use $\frac{1}{sb_t} \sum_{j=1}^{z_t} gr(d_j)$ to evaluate their precision, next we consider the top $z_t + z_{t-1}$ documents and use $\frac{1}{sb_{t-1}} \sum_{j=1}^{z_t+z_{t-1}} gr(d_j)$ to evaluate their precision, continue this process until finally we consider all top $total\_n$ documents by $\frac{1}{sb_1} \sum_{j=1}^{z_t+z_{t-1}+...+z_1} gr(d_j)$. Combining all these, we have

$$RP = \frac{1}{t}\{\frac{1}{sb_t} \sum_{j=1}^{z_t} gr(d_j) + \frac{1}{sb_{t-1}} \sum_{j=1}^{z_t+z_{t-1}} gr(d_j) + ... + \frac{1}{sb_1} \sum_{j=1}^{z_t+z_{t-1}+...+z_1} gr(d_j)\}$$

$$(3)$$

Note that in the above Eq. 3, each addend inside the braces can vary from 0 to 1. There are n addends. Therefore, the final value of RP calculated is between 0 and 1 inclusive.

Next let us discuss average AP. It can be defined as

$$AP = \frac{1}{total\_n} \sum_{i=1}^{total\_n} \frac{\sum_{j=1}^{i} gr(d_{t_j})}{\sum_{j=1}^{i} gr\_best(d_{t_j})} \tag{4}$$

where t_j is the ranking position of the j-th document whose grade is above 0, $\sum_{j=1}^{i} gr(d_{t_j})$ is the total sum of grades for documents up to rank t_i, and $\sum_{j=1}^{i} gr\_best(d_{t_j})$ is the total sum of grades for documents up to rank t_j in the best result. Considering all these $total\_n$ documents in the whole collection whose grades are above 0, AP needs to calculate the precision at all these document levels $(t_1, t_2,..., t_{total\_n})$. At any t_i, precision is calculated as $\frac{\sum_{j=1}^{i} gr(d_{t_j})}{\sum_{j=1}^{i} gr\_best(d_{t_j})}$, whose value is always in the range of 0 and 1. MAP can be defined over a group of queries as in the condition of binary relevance judgment.

4   Data Fusion Methods with Graded Relevance Judgment

In this section we detail a number of data fusion methods and also modification required for them to use properly under the condition of graded relevance judgment. Suppose there is a collection of documents C and for a given query q, all n retrieval systems ir_i are used to do the retrieval and their resultant lists are evaluated. For linear combination, the global score of any document d is calculated by

$$g(d) = \sum_{i=1}^{n} w_i * s_i(d) \tag{5}$$

where w_i is the weight set to ir_i and $s_i(d)$ is the score that d obtains from retrieval system ir_i. After the calculation, all the documents can be ranked by their global score $g(.)$ to obtain the fused results.

Both CombMNZ and RRF (Reciprocal Rank Fusion) use a special form of Eq. 5 to calculate scores by setting all w_i to be 1. For CombSum, all $s_i(d)$ are normalized raw scores from each retrieval system by the zero-one method. For RRF, it transforms rank into score by $s_i(d) = 1/(60 + rank_i(d))$, where $rank_i(d)$ is the ranking position of d in the result list of ir_i.

MAPFuse [7] can also be defined. For each weight w_i, we take its MAP value from a group of training queries. The only difference is that MAP is calculated in a different way. Similarly, we can define P@10Fuse. Instead of using MAP values, we use P@10 values as their weights.

In the weighting of MAPFuse and P@10Fuse, only performance is considered. As a matter of factor, both component system performance and dissimilarity affect fusion performance, MAPFuse and P@10Fuse can be enhanced accordingly. We obtain MAP*DissFuse and P@10*DissFuse. In P@10*DissFuse, each weight w_i is the conduct of $p10_i$ and $diss_i$, where $p10_i$ is the p@10 value of the system and $diss_i$ is the average dissimilarity of the retrieval system in question (ranked list as result) and all other retrieval systems (ranked lists as results). There are a lot of different ways of calculating dissimilarity between the ranked lists. In this piece of work, we use rank-based overlap to measure the similarity between the two results lists [16].

For a ranked list of u, let u_i be its element at rank i, and $u_{(i:j)}$ be its elements from rank i to j. At depth d, the intersection of lists u and v is:

$$I_d = u_{(1:d)} \cap v_{(1:d)} \tag{6}$$

The agreement between u and v is the ratio of overlaps to depth d:

$$A_d = (|I_d|)/d = (|u_{(1:d)} \cap v_{(1:d)}|)/d \tag{7}$$

The RBO (rank-biased overlap) distance metric is defined as:

$$RBO(u, v, b) = (1 - b) \sum_{d=1}^{\infty} b^{(d-1)} * A_d \tag{8}$$

Among them, b is an adjustable parameter, $0 < b < 1$. Finally, the dissimilarity of two ranked lists is defined as

$$Diss(u, v) = 1 - 1/RBO(u, v, b) \qquad (9)$$

In this work, we set b to 0.9, the same as in [16]. For a group of ranked lists l_i $(1 < i < n)$, the average dissimilarity between one and the rest is

$$Diss(i) = \frac{1}{n-1} \sum_{j=1 \wedge i \neq j}^{n} Diss(l_i, l_j) \qquad (10)$$

Now we can define P@10-Diss-Fuse. Let ir_i's weight be $w_i = P@10(i) * Diss(i)$. Here $P@10(i)$ is its P@10 values, and $Diss(i)$ is the average dissimilarity between ir_i and other $n - 1$ retrieval systems, over a group of queries in the training set.

Multiple linear regression can be used to train weights [18]. With binary relevance judgment, documents are divided into two types: relevant and irrelevant. Now with graded relevance judgment, documents can be divided into more categories accordingly.

Suppose that we have a collection of l documents, a group of m queries, and a group of n information retrieval systems. For each query q^i $(1 \leq i \leq m)$, all information retrieval systems ir_j $(1 \leq j \leq n)$ provide their estimated relevance scores to all the documents in the collection for each of the queries. Therefore, we have $(s_{1k}^i, s_{2k}^i, ..., s_{nk}^i, y_k^i)$ for $i = (1, 2, ..., m)$, $k = (1, 2, ..., l)$. Here s_{jk}^i stands for the score assigned by retrieval system ir_j to document d_k for query q^i; y_k^i is the judged relevance score of d_k for query q^i. If binary relevance judgment is used, then it is 1 for relevant documents and 0 otherwise. If $(n+1)$ graded relevance judgment is used, then we have $n+1$ categories. it is i/n for documents at grade level i and 0 otherwise.

$Y = \{y_k^i; i = (1, 2, ..., m), k = (1, 2, ..., l)\}$ can be estimated by a linear combination of scores from all component systems. Consider the following quantity

$$\mathcal{G} = \sum_{i=1}^{m} \sum_{k=1}^{l} [y_k^i - (\hat{\beta}_0 + \hat{\beta}_1 s_{1k}^i + \hat{\beta}_2 s_{2k}^i + ... + \hat{\beta}_n s_{nk}^i)]^2$$

when $\mathcal{G}$ reaches its minimum, the estimation is the most accurate. $\beta_0, \beta_1, \beta_2, ...,$ and β_n, the multiple linear regression coefficients, are numerical constants that can be determined from observed data.

In the least squares sense the coefficients obtained by multiple linear regression can bring us the optimum fusion results by the linear combination method, since they can be used to make the most accurate estimation of the relevance scores of all the documents to all the queries as a whole [19]. β_j can be used as weights for the fusion of retrieval systems ir_j $(1 \leq j \leq n)$.

PosFuse [7] and SlideFuse [8] are quite different from the above methods. For the training data set comprising a collection of documents and a group of queries, each retrieval system ir_i performs the retrieval. Then, combining the result lists for m queries, we calculate the posterior probability of being relevant for documents at each ranking position. For example, consider that 3-graded relevance judgment is used for a group of 50 queries. For rank position one, assume that 15 of the documents are in grade 2, another 15 in grade 1, and the rest 20 are in grade 0, then a score of $(15 * 1 + 15 * 0.5)/50 = 0.45$ will be given to any document at ranking position 1 later at the fusion stage, or $s_i(d) = 0.45$ if d is top-ranked. The same process applies to documents at any other ranking positions. Instead of calculating scores based on one ranking position, SlideFuse defines a sliding window that includes the position itself and also some neighbouring positions, thus the scores generated from overlapping windows are smoother. In this study, the window size is 3. It reduces to 2 for top and bottom rank positions. We can expect it to achieve better results when the training data set only includes a small group of queries. For both PosFuse and SlideFuse, we may also use Eq. 1 to perform the fusion, with all weights w_i set to 1 and $s_i(d)$ decided by the ranking position of d.

5 Experimental Settings and Results

In this section, we present the setting and results of the experiment carried out to validate the proposed methods. Two data sets are used. They are the TREC Microblog Track in 2013 and 2014 [9,10]. In 2013, 71 runs from 20 groups were submitted. There were 60 queries. In 2014, plus two baseline runs, 75 runs from 21 groups were submitted. 55 queries were used. The same collection, Tweets2013, was used in both cases. It included 243 million tweets crawled from the public Twitter sample stream between 1 February and 31 March 2013 (inclusive). In their relevance judgment file, documents are given a score of 2, 1, and 0 for relevant, modestly relevant, and non-relevant documents, respectively.

From each data set, we chose two groups of runs for our experiment. For the first group, eight top performers in P@10 were chosen. However, to guarantee certain diversity among those selected, we took up to two runs submitted from each participant. Because it is more likely that those runs submitted by the same participant are usually more similar than those runs submitted by different participants. For the second group, we tried to include all the runs submitted. Only a few runs were removed because there are duplicates. See Tables 1 and 2 for the statistics of those two groups. More details about the first group chosen in both TREC 2013 and 2014 are shown in Appendix. Note that all the metrics used in this section are for graded relevance judgment, as defined in Sect. 3. Through comparing its average and maximum, we can find that in both cases of 2013 and 2014, all eight runs in group 1 are very good and close in performance, while the runs in group 2 vary significantly in performance.

For the first group of eight runs, all the runs were used for training and testing. For the second group of 65 (or 74) runs, we divided all the queries (60 or 55) into two parts: odd-numbered and even-numbered. The odd-numbered queries were used as training data and the even-numbered queries were used as testing data, and vice versa.

Table 1. Statistics of two groups of runs selected in TREC 2013)

Group	P@10	RP	NDCG	MAP
Ave. Group 1 (8 runs)	0.7221	0.4295	0.5842	0.6200
Max. Group 1 (8 runs)	0.8017	0.5172	0.6101	0.6597
Ave. Group 2 (65 runs)	0.5200	0.2834	0.4253	0.4990
Max. Group 2 (65 runs)	0.8017	0.5172	0.6101	0.6597

Table 2. Statistics of two groups of runs selected in TREC 2014)

Group	P@10	RP	NDCG	MAP
Ave. Group 1 (8 runs)	0.8048	0.5330	0.7303	0.8467
Max. Group 1 (8 runs)	0.8255	0.5727	0.7593	0.8934
Ave. Group 2 (74 runs)	0.6436	0.3996	0.5830	0.7129
Max. Group 2 (74 runs)	0.8255	0.5727	0.7593	0.8982

Table 3. Fusion performance of the runs in the Microblog Track in TREC 2013)

Method	8 Runs				65 Runs			
	P@10	RP	NDCG	MAP	P@10	RP	NDCG	MAP
Best	0.8017	0.5172	0.6101	0.6597	0.8017	0.5172	0.6101	0.6597
CombSum	0.8100	0.5448	0.6634	0.7571	0.6983	0.4150	0.5957	0.7389
RRF	0.8017	0.5254	0.6631	0.8124	0.6883	0.4003	0.5877	0.7734
LN-MLR	**0.8300**	**0.5674**	0.6712	0.8152	**0.8317**	**0.5836**	**0.6914**	**0.8842**
PosFuse	**0.8300**	0.5609	**0.6875**	**0.8267**	0.6950	0.4232	0.5940	0.8163
SlideFuse	0.8250	0.5576	0.6824	0.8249	0.6917	0.4198	0.5917	0.8179
MAPFuse	0.8200	0.5440	0.6742	0.8105	0.7083	0.4197	0.6001	0.8064
P10Fuse	0.8117	0.5352	0.6678	0.8114	0.7033	0.4117	0.5988	0.7911
P10-Diss-Fuse	0.8217	0.5476	0.6713	0.8105	0.7050	0.4171	0.6011	0.8029

Table 4. Fusion performance of the runs in the Microblog Track in TREC 2014)

Method	8 Runs				74 Runs			
	P@10	RP	NDCG	MAP	P@10	RP	NDCG	MAP
Best	0.8255	0.5727	0.7593	0.8934	0.8255	0.5727	0.7593	0.8934
CombSum	0.8922	0.6255	0.8237	0.9608	0.8127	0.5343	0.7368	0.8123
RRF	0.8902	0.6207	0.8218	0.9639	0.8073	0.5418	0.7404	0.8517
LN-MLR	0.8843	**0.6373**	0.8174	0.9648	**0.8418**	**0.6089**	**0.7683**	**0.9104**
PosFuse	0.8922	0.6325	**0.8304**	**0.9748**	0.8018	0.5494	0.7364	0.8679
SlideFuse	**0.8941**	0.6285	0.8191	0.9723	0.8036	0.5479	0.7301	0.8679
MAPFuse	0.8882	0.6240	0.8164	0.9645	0.8127	0.5534	0.7406	0.8747
P10Fuse	0.8882	0.6207	0.8197	0.9640	0.8073	0.5485	0.7383	0.8679
P10-Diss-Fuse	0.8882	0.6234	0.8188	0.9629	0.8109	0.5537	0.7434	0.8805

Eight fusion methods, including CombSum, RRF, linear combination with weights trained by multiple linear regression (LN-MLR), PosFuse, SlideFuse, MAPFuse, P@10Fuse, and P@10-Diss-Fuse, were used to fuse them. Tables 3 and 4 present the fusion performance of the two data sets, respectively. The figures are the average of all 60 or 55 queries.

For the fusion of eight runs, we can see that all fusion methods perform better than the best run that participates in the fusion in all four metrics. This is a very good achievement because all the runs (component retrieval systems/results) are very good in performance. It is more challenging to achieve this by fusing very good retrieval systems than fusing not very good retrieval systems. One major reason for this achievement is the careful selection of all those runs.

On the other hand, the fusion results for the second group are very different. LN-MLR is the only one that performs well and better than all the other fusion methods. In this case, there are more component retrieval systems and their performances vary considerably, the situation is more complicated. It shows that LN-MLR is able to deal with it well, but that may not be manageable for the others.

6   Conclusions

In this paper, we have presented a group of data fusion methods that can be used in the condition of graded relevance judgment. Experimented with two datasets from TREC, it shows that all the data fusion methods are effective in performance improvement if component retrieval systems are selected carefully. On the other hand, linear combination with multiple linear regression for weights training is able to deal with complicated situations. In general, it conforms the validity of the adapted data fusion methods in the condition of graded relevance judgment.

In our future work, we plan to consider a further condition when negative grades are involved in relevance judgment. Both metrics and data fusion methods need to be updated to accommodate the new environment.

7   Appendix

See Tables 5 and 6

Table 5. Information of the eight selected runs in TREC 2013

Run	P@10	RP	NDCG	MAP
FSsvm	**0.8017**	0.5004	**0.6101**	0.6328
Avgrank	0.7867	0.5033	0.6002	0.5989
Direrank	0.7750	**0.5172**	0.6099	**0.6597**
PrisRun2	0.7217	0.3843	0.5980	0.6019
PrisRun4	0.7067	0.3853	0.5962	0.6086
PKUICST1	0.6650	0.3721	0.5612	0.5944
PKUICST3	0.6650	0.3827	0.5564	0.6130
QCRI4	0.6550	0.3905	0.5419	0.6507

Table 6. Information of the eight selected runs in TREC 2014

Run	P@10	RP	NDCG	MAP
ECNURankLib	0.8145	0.5427	0.7401	0.8730
ECNUSVM2013	0.8091	0.5357	0.7320	0.8016
HPRF1020RR	0.7891	0.5086	0.7016	0.8495
ICARUN1	0.7891	0.5434	0.7288	0.8542
PKUICST2	**0.8255**	0.5583	0.7569	0.8700
PKUICST3	0.8200	**0.5727**	**0.7593**	**0.8934**
PRF1030RR	0.7945	0.5036	0.7052	0.8386
Pris2014a	0.7964	0.4988	0.7181	0.7935

References

1. Aslam, J.A., Montague, M.: Models for metasearch. In: Proceedings of the 24th Annual International ACM SIGIR Conference, New Orleans, Louisiana, USA, pp. 276–284, September 2001
2. Cormack, G.V., Clarke, C.L.A., Büttcher, S.: Reciprocal rank fusion outperforms Condorcet and individual rank learning methods. In: Proceedings of the 32nd Annual International ACM SIGIR Conference, Boston, MA, USA, pp. 758–759, July 2009
3. Cronen-Townsend, S., Zhou, Y., Croft, W.B.: Predicting query performance. In: Järvelin, K., Beaulieu, M., Baeza-Yates, R.A., Myaeng, S. (eds.) SIGIR 2002: Proceedings of the 25th Annual International ACM SIGIR Conference on Research and Development in Information Retrieval, Tampere, Finland, 11–15 August 2002, pp. 299–306. ACM (2002)

4. Fox, E.A., Koushik, M.P., Shaw, J., Modlin, R., Rao, D.: Combining evidence from multiple searches. In: The First Text REtrieval Conference (TREC-1), Gaithersburg, MD, USA, pp. 319–328, March 1993
5. Ghosh, K., Parui, S.K., Majumder, P.: Learning combination weights in data fusion using genetic algorithms. Inf. Process. Manag. **51**(3), 306–328 (2015)
6. Järvelin, K., Kekäläinen, J.: Cumulated gain-based evaluation of IR techniques. ACM Trans. Inform. Syst. **20**(4), 442–446 (2002)
7. Lillis, D., Zhang, L., Toolan, F., Collier, R., Leonard, D., Dunnion, J.: Estimating probabilities for effective data fusion. In: Proceeding of the 33rd International ACM SIGIR Conference on Research and Development in Information Retrieval, Geneva, Switzerland, pp. 347–354, July 2010
8. Lillis, D., Toolan, F., Collier, R., Dunnion, J.: Extending probabilistic data fusion using sliding windows. In: Macdonald, C., Ounis, I., Plachouras, V., Ruthven, I., White, R.W. (eds.) ECIR 2008. LNCS, vol. 4956, pp. 358–369. Springer, Heidelberg (2008). https://doi.org/10.1007/978-3-540-78646-7_33
9. Lin, J., Efron, M.: Overview of the TREC-2013 microblog track. In: Voorhees, E.M. (ed.) Proceedings of The Twenty-Second Text REtrieval Conference, TREC 2013, Gaithersburg, Maryland, USA, 19–22 November 2013. NIST Special Publication, vol. 500–302. National Institute of Standards and Technology (NIST) (2013)
10. Lin, J., Wang, Y., Efron, M., Sherman, G.: Overview of the TREC-2014 microblog track. In: Proceedings of The Twenty-Third Text REtrieval Conference, TREC 2014, Gaithersburg, Maryland, USA, 19–21 November 2014. NIST Special Publication, vol. 500–308. National Institute of Standards and Technology (NIST) (2014)
11. Markovits, G., Shtok, A., Kurland, O., Carmel, D.: Predicting query performance for fusion-based retrieval. In: Chen, X., Lebanon, G., Wang, H., Zaki, M.J. (eds.) 21st ACM International Conference on Information and Knowledge Management, CIKM 2012, Maui, HI, USA, 29 October– 02 November 2012, pp. 813–822. ACM (2012)
12. Montague, M., Aslam, J.A.: Condorcet fusion for improved retrieval. In: Proceedings of ACM CIKM Conference, McLean, VA, USA, pp. 538–548, November 2002
13. Roitman, H.: Enhanced performance prediction of fusion-based retrieval. In: Proceedings of the 2018 ACM SIGIR International Conference on Theory of Information Retrieval, ICTIR 2018, Tianjin, China, 14–17 September 2018, pp. 195–198. ACM (2018)
14. Roitman, H., Kurland, O.: Query performance prediction for pseudo-feedback-based retrieval. In: Piwowarski, B., Chevalier, M., Gaussier, É., Maarek, Y., Nie, J., Scholer, F. (eds.) Proceedings of the 42nd International ACM SIGIR Conference on Research and Development in Information Retrieval, SIGIR 2019, Paris, France, 21–25 July 2019, pp. 1261–1264. ACM (2019)
15. Sivaram, M., Batri, K., Mohammed, A.S., Porkodi, V., Kousik, N.V.: Data fusion using Tabu crossover genetic algorithm in information retrieval. J. Intell. Fuzzy Syst. **39**(4), 5407–5416 (2020)
16. Webber, W., Moffat, A., Zobel, J.: A similarity measure for indefinite rankings. ACM Trans. Inf. Syst. **28**(4), 20:1–20:38 (2010)
17. Wu, S.: Applying statistical principles to data fusion in information retrieval. Expert Syst. Appl. **36**(2), 2997–3006 (2009)
18. Wu, S.: Data Fusion in Information Retrieval. Springer, Heidelberg (2012). https://doi.org/10.1007/978-3-642-28866-1
19. Wu, S.: Linear combination of component results in information retrieval. Data Knowl. Eng. **71**(1), 114–126 (2012)

20. Wu, S.: The weighted Condorcet fusion in information retrieval. Inf. Process. Manag. **49**(1), 114–126 (2013)
21. Wu, S., Bi, Y., Zeng, X., Han, L.: Assigning appropriate weights for the linear combination data fusion method in information retrieval. Inf. Process. Manag. **45**(4), 413–426 (2009)
22. Wu, S., McClean, S.: Data fusion with correlation weights. In: Losada, D.E., Fernández-Luna, J.M. (eds.) ECIR 2005. LNCS, vol. 3408, pp. 275–286. Springer, Heidelberg (2005). https://doi.org/10.1007/978-3-540-31865-1_20
23. Wu, S., McClean, S.: Performance prediction of data fusion for information retrieval. Inf. Process. Manag. **42**(4), 899–915 (2006)
24. Xu, C., Huang, C., Wu, S.: Differential evolution-based fusion for results diversification of web search. In: Web-Age Information Management - 17th International Conference, WAIM 2016, Nanchang, China, 3–5 June 2016, Proceedings, Part I, pp. 429–440 (2016)
25. Xu, Q., Wu, S.: Improving medical record search performance by particle swarm optimization based data fusion techniques. In: Xing, C., Fu, X., Zhang, Y., Zhang, G., Borjigin, C. (eds.) WISA 2021. LNCS, vol. 12999, pp. 87–98. Springer, Cham (2021). https://doi.org/10.1007/978-3-030-87571-8_8

Highway Accident Localization Based on Virtual Fence for Intelligent Transportation Systems

Jinbo Li[1,2], Guanghui Wang[1,2(✉)], Fang Zuo[1,2], and Xin He[1,2]

[1] Henan International Joint Laboratory of Intelligent Network Theory and Key Technology, Henan University, Kaifeng 475000, China
gwang@vip.henu.cn
[2] School of Software, Henan University, Kaifeng 475000, China

Abstract. It is important to determine the mile marker information of traffic accident (i.e., *accident localization*) for highway repairs and emergency rescue services in intelligent transportation systems. Since traditional localization methods usually target at reporting the coordinates of the accident, they obtain neither the driving direction of the accident vehicle nor the mile marker on actual highway. In order to address the above issue, in this paper, we propose to study the highway accident localization problem using virtual fence for intelligent transportation systems. First, the virtual fence of the highway is modeled for accident localization, including virtual area and virtual mile marker constructions. Then, an novel accident logic localization algorithm (ALLA) is proposed to determine the mile marker of the accident. The accident information can be automatically reported by using the WeChat applet. The driving direction and the mile marker can be determined by using the virtual fence. Finally, the performance of the proposed method is evaluated by using experimental simulations. The results show that the proposed method is able to efficiently localize highway accidents with low cost and high accuracy.

Keywords: Intelligent transportation systems · Highway accident localization · Virtual fence

1 Introduction

With the substantial increase in travel rates and motor vehicle ownership, traffic accidents are common, and the corresponding economic losses and casualties are serious in transportation systems [1]. By 2020, highway traffic accidents

This work was supported in part by the Henan Provincial Major Public Welfare Project (201300210400), the China Postdoctoral Science Foundation (2020M672211 and 2020M672217), the Key Scientific Research Projects of Henan Provincial Colleges and Universities (21A520003), and the Key Technology Research and Development Program of Henan (182102210106, 212102210090, 212102210094, 222102210133, and 222102210055).

X. Zhao et al. (Eds.): WISA 2022, LNCS 13579, pp. 240–253, 2022.
https://doi.org/10.1007/978-3-031-20309-1_21

has ranked third in the global list of major causes of death and disability [2–6]. In addition to statistics on deaths and injuries, it has also resulted in billions of dollars in medical costs and lost productivity. At the same time, the location information of traditional latitude and longitude has insufficient significance to highway repairs and emergency rescue services. On the contrary, how to determine the mile marker of traffic accident (i.e., *accident localization*) based on the user's coordinates with the electronic map is a key problem for intelligent transportation systems. For example, mentioning a user's current location as Tokyo Disney-land is more meaningful than the coordinates $35.6328° N$, $139.8806° E$ [7,8]. Therefore, it is necessary to study the highway accident localization problem for intelligent transportation systems.

Accurate and efficient highway accident localization is an practical and challenging problem. When a highway traffic accident occurs, it is necessary to obtain the mile marker of the accident vehicle for the highway repairs and emergency rescue services. In traditional localization methods, the mile marker information of the accident cannot be directly obtained from the information of the latitude and longitude. In order to determine this logical localization more accurately and efficiently, the existing work still faces the challenges of high labor, hardware costs and an inaccurate description of accident localization. Therefore, it is essential to propose an accurate and efficient highway accident localization method for intelligent transportation systems [9–12].

Some existing work on accident localization has been conducted, which can be categorized into mobile phone emergency rescue methods, accident vehicle survey, rescue methods based on satellite localization and accident reporting methods based on quick response (QR) code [13–16]. In the highway scenario, the mobile phone rescue localization time is long, and the alarm cannot carry other additional information [15]. The error of satellite localization is around 1 km for accident investigation and rescue, and the localization time is about 10 s [16]. Thus, the localization range is large and the localization time is long for highway repairs and emergency rescue services. In 2022, the method of accurate detection and alert system (current location) using global positioning system, GPS has high hardware cost and can only obtain longitude and latitude information [17]. Furthermore, under the accident reporting system based on QR codes, both of the cost of making QR code plates and the cost of long-term maintenance labor and materials are too high. Therefore, the above methods cannot accurately and efficiently carry out accident localization, which meet no requirement of highway repairs and emergency rescue services in the highway scenario.

In this paper, we propose an accurate and efficient highway accident localization method based on virtual fence for intelligent transportation systems. The method includes accident reporting, accident management and an accident logic localization algorithm (ALLA). The main idea is to use the virtual fence to determine the direction of the latitude and longitude coordinates of the accident. The virtual fence includes virtual area and virtual mile marker constructions on the highway. Virtual mile marker corresponds to the mile marker on the actual road, which is used to determine the mile marker location of the accident vehicle. Moreover, the distance between the accident point and the accident-prone point is prioritized to determine the mile marker. The accident information can be automatically reported by using the WeChat applet. The contribution of this paper is summarized as follows.

- We study the practical accident localization problem using virtual fence for highway repairs and emergency rescue services in intelligent transportation systems. A highway accident localization method is proposed to accurately and efficiently determine the mile marker of the accident.
- We construct the virtual fence including the virtual area and virtual mile marker for highway scenario to enable the proposed highway accident localization method to avoid high cost on labor and hardware. We also propose an accident logic localization algorithm ALLA to quickly determine the driving direction of an accident so as to enable the accuracy of the proposed highway accident localization method.
- We conduct experimental simulations to evaluate the proposed highway accident localization method. The results show that the proposed method is able to efficiently localize highway accidents with low cost and high accuracy.

The organization of the paper is as follows. Section 2 introduces the related work. Section 3 presents the system model and problem formulation. The proposed highway accident localization method is detailed in Sect. 4. The performance is evaluated in Sect. 5. Section 6 concludes the paper.

2   Related Work

Existing work is categorized into mobile phone emergency rescue methods, accident vehicle survey, rescue methods based on satellite localization and accident reporting methods based on QR code. First, the mobile emergency rescue notification system includes mobile phones and voice bhighwaycast servers. When pressing the emergency rescue key on the mobile phone, the mobile phone will receive the satellite localization coordinates through the satellite localization module and generate the corresponding rescue code through the rescue code module. After that, the satellite localization coordinates and rescue code are sent to the voice bhighwaycast server to generate a rescue notice [15]. Then, a rescue method based on satellite localization consists of a user mobile platform and a rescue service platform with satellite localization, which has wireless communication and mapping functions. The user mobile platform uses satellite localization and map functions to send accident localization information and interact with the rescue workers [16]. Furthermore, an accident reporting system based on QR code implements the API provided by the WeChat applet to generate the QR code and QR code card. When an accident occurs, scan the QR code installed on the highway to fill in other relevant information for reporting. In 2022, Butkar, U. D et al. proposed the accurate detection and alert system (current location) using global positioning system. In this method, the vehicle position is collected in real time through GPS, and the data are transmitted through SMS through GSM to locate the vehicle equipment. The GPS module provides the longitude and latitude coordinates of the target vehicle sent through the Internet of things. [17].

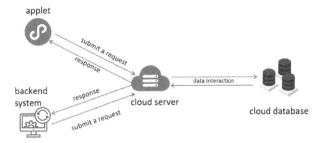

Fig. 1. System architecture diagram.

In the highway environment, the mobile phone rescue localization time is long, and the alarm can not carry other additional information [15]. The accuracy of satellite localization for accident investigation, rescue and method localization is 1 km, and the localization time is 10 s [16]. The localization range is too large and the localization time is too long. The two methods can not calculate the real mile marker corresponding to the accident point. For the accident reporting system based on QR codes, both of the cost of making the QR code plates and the cost of later maintenance labor and materials are too high. The method has high hardware cost and can only obtain longitude and latitude information [17]. Based on the above problems, our system designs the intelligent perception of highway accidents based on virtual fence, including drawing a virtual fence, initializing virtual fence information, using a wechat applet to report accidents, judging the reported localization information and the backend system responses.

3 System Model and Problem Formulation

3.1 System Model

The highway accident localization system consists of an applet, a backend system, cloud server and cloud databases. The applet side can view accident information and report it to the server side. The cloud server side deploys the backend API interface to accept requests and responses. The backend system can manage the virtual fence and handle the accident information. Cloud databases are settled to store users' data. The whole system architecture is shown in Fig. 1 as above.

This accident localization method includes initialization, accident localization determination and backend response. The flow chart is shown in Fig. 2. The virtual fence drawing module is set up in the backend management system, using the API provided by Baidu Map to draw the virtual fence in the polygon area and initialize the virtual mile marker information. Virtual mile marker information includes virtual mile marker, driving direction, the belonging level and determination distance. After the accident occurs, the applet in WeChat is used to report the accident. The applet will automatically obtain the coordinate

Fig. 2. Accident logic localization flow chart.

Table 1. Important notations

Symbol	Definition
l	The level of the mile marker
lat_i^l	The latitude coordinates of the i th virtual mile marker of l
lng_i^l	The longitude coordinates of the i th virtual mile marker of l
dir	The driving direction of the vehicle
P_i^l	The virtual marker of the i th virtual mile marker of l
MD	The current distance minimum value
MP	The virtual marker that corresponds to the current distance minimum value
dr_i^l	The determination rule for the i th virtual mile marker of l

information and some personal information. Click the submit button to submit the data to the server side. The accident information will be determined after the applet reports the accident, which includes the driving direction and the subordinate mile marker.

The corresponding mile marker of the backend system beats, and the system emits an alarm sound after determining the accident. The corresponding mile marker can be clicked to switch to the accident management page. The backend system will stop beating and the alarm will turn off after dealing with the accident. In the end, the WeChat message will be sent to the relevant workers. The important symbols used in the paper are shown in Table 1 above.

3.2 Virtual Fence Model

The virtual fence includes the virtual area and the virtual mile marker constructions. The virtual construction includes drawing the virtual fence and obtaining the virtual mile marker.

Construction of Virtual Area. The polygon area drawing function provided by the map navigation software is implemented to divide the highway into upstream and downstream areas. The drawn areas are stored in the database. As shown in Fig. 3, the upstream area virtual fence is divided into three parts: the highway area, the non-highway area and the isolation area. Similarly, the virtual fence in the downstream area is also divided into the above three parts. In practice, the distance outside the highway area can be set to exceed 150 m.

Virtual Mile Marker Construction. There are two cases for the construction of the virtual mile marker. If having the latitude and longitude coordinates

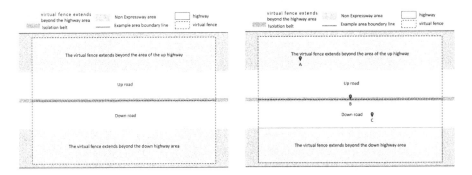

Fig. 3. Schematic diagram of virtual fence.

Fig. 4. Schematic diagram for judging the driving direction of the vehicles in accident.

Table 2. Decision rules

Level	Decision rules	Describe
1	$Dis \leq 50\,\mathrm{m}$	$Kx + i * 100(x \in (0, N), i = 1, ..., 9)$
2	$Dis < 50\,\mathrm{m}$ (Based on actual data)	$Sx + m(x \in (0, N)), m \in (0, 1000)$

corresponding to the mile marker and other relevant information, we can use the form of template batch import to import the normal mile marker. If there is no existing mile marker coordinate information, we can use the mile marker initialization function in the WeChat applet to make related staff initialize the mile marker information online. For virtual mile marker, level and determination rule are set, where level represents the level of virtual mile mark. 1 is normal, and the determination distance is set to 50 m, which appears as $K0 + 100$. The Algorithm 1 yields the determination distance, which appears as $K2$, and 2 represents the accident-prone locations. Dis represents the distance between the accident site and the mile marker, which is calculated according to the latitude and longitude coordinates [20] as shown in Table 2.

After the initialization of the normal mile marker, the existing accident information is imported into the backend system to aggregate the upstream and downstream accident coordinates accordingly to generate the accident-prone mile marker. After the generation, the accident-prone site mile marker and the subordinate direction will be stored in the database. It should be noted that, by setting the grade of mile marker, the time of subsequent determination of mile marker can be accelerated. For example, an accident is most likely to occur at the place where the mile marker is located. Therefore, level 1 is preferred when determining the accident reporting point.

3.3 Problem Formulation

Under the highway accident model and the virtual fence model, we need to design a highway accident localization method, using the accident coordinates to calculate the driving direction of the vehicle and the mile marker where it belongs. At the same time, the method should meet the following requirements.

- Accuracy: The mileage station determined by the system using the coordinates of the accident report should be the same as the actual mileage station.
- Efficiency: The speed for judging the accident localization should be fast and the cost time should be limited in 1 s.
- Economy: The implement of the method should save the labor cost, material cost, repairing cost and installing cost.

4 Highway Accident Localization

4.1 Accident Location Determination

The determination of the accident localization includes three parts: reporting the accident, determining the driving direction of the vehicle and determining the mile marker of the vehicle.

Report Accident. A reporting platform based on WeChat applet is developed in this system. When a traffic accident occurs, open the WeChat applet and click on the report page, the coordinates will be obtained automatically. After filling in the personal information with the permission of users, upload the accident picture and click the submit button, the accident will be submitted automatically.

Determine the Driving Direction of the Vehicle. The relevant data will be sent to the backend through the WeChat applet. There are many mature methods to judge the relationship between points and polygon, among which the ray method is widely used because of its simple implementation. Lead a ray from the target point A, if the marker of intersections with all edges of the polygon is odd, point A is in the polygon. If even, point A is outside the polygon.

As shown in Fig. 4, there are three points A, B, and C. Point A is in the area beyond the upside highway, and the judgment result should be the uplink. Point C is on the downhighway, and its judgment result should be the downlink. Point B is at the junction of the uplink virtual fence and the downlink virtual fence, where the vehicle is judged to have rushed through the barrier and into the isolation zone. In the case of point B, the user should either fill in the driving direction manually or communicate with the relevant departments by telephone.

Determine the Mile Marker of the Vehicle. The identification of the mile marker of the accident vehicle is divided into two steps. The first step is to determine whether the accident occurred in an accident prone area. If it is in the accident-prone range, the accident is determined to belong to the accident-prone localization mile mark. The second step implements jump search to quickly determine the normal mile marker if the accident is not within the range of an accident-prone area. As shown in Fig. 5 below, the $Pi(i = 1, 2, 3, 4, 5)$ indicates the localization of the accident.

- For accident point $P1$, assuming that the judgment distance of mile marker $S0 + 72$ is 25 m and the distance $d(P1, S72)$ between mile marker $P1$ and $S0 + 72$ is 20 m, then $P1$ belongs to mile marker $S0 + 72$.
- For $P2$, the car has driven out of the highway. If the judgment distance of the normal highway mile marker is 50 m, and the distance $d(P2, K0 + 200)$ between $P2$ and the mile marker $K0 + 200$ is not more than 50 m, it can be directly judged that $P2$ belongs to $K0+200$. If the distance $d(P2, K0+200)$ is greater than 500 m, the final calculation results indicate that the nearest mile marker is still $K0 + 200$.
- For $P3$, if the vehicle did not drive out of the highway, and the distance $d(P3, K0 + 200)$ between $P3$ and the mile marker $K0 + 200$ is less than or equal to 50 m, then $P3$ belongs to $K0 + 200$.
- For $P4$, the vehicle did not drive out of the highway who drove along the upward path. The distance from $P4$ to mile marker $K0 + 200$ is equal to the distance from $P4$ to mile marker $K0 + 300$. That is, $d(P4, K0 + 200) = d(P4, K0 + 300)$, and $P4$ belongs to $K0 + 300$.
- For $P5$, the vehicle who drove the downside highway did not leave the highway. The distance from $P5$ to mile marker $K0 + 200$ is equal to the distance from $P5$ to mile marker $K0 + 300$, that is, $d(P5, K0 + 200) = d(P5, K0 + 300)$. Then $P5$ belongs to $K0+200$. As can be seen from the determination results of $P4$ and $P5$: take the mile markers on the upward highway $Kx + i*100$ and $Kx + (i + 1) * 100$ as examples, accident reporting point p exists, $d(p, Kx + i * 100) = d(p, Kx + (i+1) * 100)$. The accident reporting point p belongs to the mile marker $Kx + (i + 1) * 100$. The down road is the opposite.

Backend System Response. Highlight the mile marker corresponding to the mile marker in the interface and issue an alarm, prompting the staff to send notification information to the rescue personnel for rescue. When the jumping mile marker is detected, the accident management page will be displayed. View the accident marker, mile marker and coordinate information, and then send the relevant information about the accident to the relevant rescue departments and rescue staff through the enterprise WeChat. After sending the enterprise WeChat message, the logo is no longer beats, and the alarm is closed.

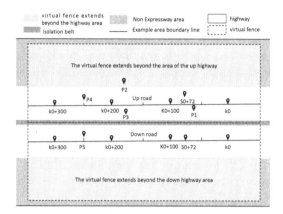

Fig. 5. Schematic diagram of the mile marker determination.

4.2   Algorithm Design

The whole process of accident determination consists of two parts: determining the direction of the accident and the mile marker of the accident. The accident determination method is shown in Algorithm 1. Because the number of ordinary mile markers may be too many, it may take a longer time to search all the normal mile markers. The mile marker designed in the system corresponds to the actual mile marker one by one. If the distance between the accident point and a certain point within 1200 m, we omit some points to accelerate the process of judgement. Therefore, the jump retrieval algorithm was designed. The jump retrieval algorithm is show by ①–⑤. The time complexity of the algorithm is $O(n)$, where n is the number of the mile markers on the highway.

5   Performance Evaluation

5.1   System Settings

The accident reporting platform is implemented by Wechat applet, and the backend system is deployed on a 2-core 2G cloud server. Both the MATLAB software and Python were run on computers configured as AMD R7 5800H @ 3. 20 HZ and 16. 00 GB RAM. The data source for this experiment is the data from parts of the Lianhuo highway in Kaifeng City, Henan Province. There are 100 normal virtual mile markers and 5 virtual mile marks in accident-prone areas, and the total number of accidents is 100, including 60 accidents in accident-prone areas. In the efficiency test and economic analysis, the proposed method is compared with the satellite position method (SPM) and scan the QR code method (SQRM) [16].

Algorithm 1. Accident Logical Localization Algorithm (ALLA)

Input: Accident coordinate m;

Output: dir and the mile marker p;

1: dir=InPolygon (m). Get alldir=InPolygon(m). Get all P_i^l's lng_i^l and lat_i^l based on the dir of travel where $l = 1$.

2: Calculate the distance between m and P_i^l, and record it as $d(m, P_i^l)$.

3: **if** $d(m, P_i^l)$ satisfies dr_i^l **then**

4:     Return dir and P_i^l.

5: **else**

6:     Compare each $d(m, P_i^l)$, record the smallest $d(m, P_i^l)$ in MD and record theP_i^l corresponding to the smallest $d(m, P_i^l)$ in MP.

7: **end if**

8: $dir = InPolygon(m, dir).i = 1$.

9: Get all lng_i^l and lat_i^l based on the dir of travel where $l = 2$. ①

10: Calculate $d(m, P_i^l)$ record it as d. ②

11: Calculate $d(m, P_i^l + (-)d/100)$ and record it as $d1$, Calculate $d(m, P_i^l + (-)d/100 + 1)$ and record it as $d2$. ③

12: **if** $d1 < d2$ **then**

13:     $i = i + d/100$ and the retrieval direction is less than i . ④

14: **else**

15:     $i = i + d/100 + 1$ and the retrieval direction is greater than i . ⑤

16: **end if**

17: Repeat Step ①-⑤ and record the smallest $d(m, P_i^l)$ in MD and record the P_i^l corresponding to the smallest $d(m, P_i^l)$ in MP, if $d(m, P_i^l)$ satisfies dr_i^l, return dir and P_i^l else return MP and MD.

5.2   Accuracy Evaluation

During the simulations, accuracy is defined as whether the positioning results are as expected. We simulated the error between the uploaded accident coordinates and the real coordinates. The error is denoted by α. α is in meters. Figure 6 is the change of accuracy and error rate under the cases $\alpha = 1, 2, 4, 8, 10$.

5.3   Efficiency Evaluation

The efficiency evaluation includes user operation's time, network communication's time and algorithm calculation's time. Table 3 shows the comparisons of our method, the satellite position method (SPM) and the QR code method (SQRM) [16]. The user operation's time used in the method of this paper includes the time that the user opens the applet, and fills in some accident information such as mobile phone marker, name and coordinate system. The localization speed of the satellite localization method is within 10 s, which is the calculation time of the algorithm in the table. The user operation time is longer because the satellite localization does not involve the automatic filling function. The user operation's time of scanning the QR code is mainly reflected in the scanning and identification time. The determination time of SQRM is less than that of ALLA,

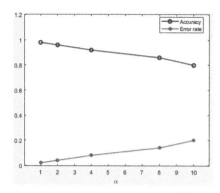

Fig. 6. The accuracy of accident position determination

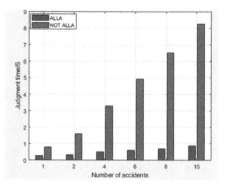

Fig. 7. The time consumption of ALLA and NALLA to determine the accident position.

Table 3. Communication efficiency

Method	User operation's time/s	Network communication's time/s	The algorithm calculation's time/s
ALLA	10	1.4	0.188
SPM	16	1.4	10
SQRM	13	1.4	0

because the QR code method binds the information which does not need to be determined.

It can be seen from Table 3 that the method proposed in this paper has advantages over the overall efficiency of satellite localization and scanning the QR code. The method proposed in this paper takes less time than the traditional telephone location method. Traditional telephone localization takes much longer for the first localization than for the method presented here. As shown in Fig. 7, When using the ALLA algorithm and not using the ALLA algorithm (NALLA), the judgment time is compared when the accident volume is 1, 2, 4, 6, 8 and 10 respectively. NALLA refers to the cycle traversal to calculate the distance between each station and the accident coordinate. The determination speed of ALLA is much faster than that of NALLA. In summary, the method presented in this paper has high efficiency.

5.4 Economic Evaluation

Using the QR code to report accidents, the cost of a QR code card is more than 500 Yuan [21]. If 200 QR code cards are needed, the cost will be more than 100, 000 Yuan. The installation of these QR code cards also requires labor costs and later maintenance costs should also be considered. Uwb and other localization devices also need a high cost. A uwb device usually needs a few hundred yuan

[22]. Using satellite localization, the method has three platforms that may require three or more servers.

The cost required in the proposed method includes one server and one registered domain name. Taking Tencent Cloud 2-core 2G server in April 2022 as an example, the lower cost is about 130 Yuan/year [23]. The method in this paper minimizes material and maintenance costs. The results of economic evaluation comparison are shown in 200 mile markers as shown in Table 4 above, which we can see the reduction of cost beating in this system.

Table 4. Economic assessment

Method	Total cost of materials	Computer hardware cost
ALLA	0	130 Yuan/Year
SPM	13500 Yuan/receiver	390 Yuan/Year
SQRM	200 Yuan/QR code	130 Yuan/Year

6   Conclusion

Determining the mile marker information of a traffic accident (i.e., *accident localization*) is significant for highway repairs and emergency rescue services in intelligent transportation systems. Traditional localization methods usually target at reporting the coordinates of the accident, thus obtaining neither the driving direction of the accident vehicle nor the mile marker on actual highway. In this paper, we propose a highway accident localization method using virtual fence. First, the virtual fence of the highway is constructed for accident localization, including virtual area and virtual mile marker. Then, an novel accident logic localization algorithm (ALLA) is proposed to determine driving direction and the mile marker of an accident. The accident information can be automatically reported by using the WeChat applet. Finally, the proposed method is evaluated by using experimental simulations, which demonstrates that the proposed method is able to efficiently localize highway accidents with low cost and high accuracy.

Funding. The Elite Postgraduate Students Program of Henan University, China (grant numbers SYLYC2022149 and SYLYC2022147).

References

1. Soehodho, S.: Public transportation development and traffic accident prevention in Indonesia. IATSS Res., S0386111216300085(2017)
2. Mannering, F.: Temporal instability and the analysis of highway accident data. Anal. Methods Accid. Res. **17**, 1–13 (2018)

3. Tsala, S.A.Z., et al.: An in-depth analysis of the causes of road accidents in developing countries: case study of Douala-Dschang Highway in Cameroon. J. Transp. Technol. **11**(3), 455–470 (2021)
4. Adanu, E. K., Agyemang, W., Islam, R., Jones, S.: A comprehensive analysis of factors that influence interstate highway crash severity in Alabama. J. Transp. Saf. Secur., 1–25 (2021)
5. Hu, X., Wang, G., Jiang, L., Ding, S., He, X.: Towards efficient learning using double-layered federation based on traffic density for internet of vehicles. In: Xing, C., Fu, X., Zhang, Y., Zhang, G., Borjigin, C. (eds.) WISA 2021. LNCS, vol. 12999, pp. 287–298. Springer, Cham (2021). https://doi.org/10.1007/978-3-030-87571-8_25
6. Gichaga, F.J.: The impact of road improvements on road safety and related characteristics. IATSS Res. **40**(2), 72–75 (2017)
7. Rosayyan, P., Subramaniam, S., Ganesan, S.I.: Decentralized emergency service vehicle pre-emption system using RF communication and GNSS-based geo-fencing. IEEE Trans. Intell. Transp. Syst. **22**(12), 7726–7735 (2020)
8. Tang, S., Yu, Y., Zimmermann, R., Obana, S.: Efficient geo-fencing via hybrid hashing: a combination of bucket selection and in-bucket binary search. ACM Trans. Spat. Algorithms Syst. (TSAS) **1**(2), 1–22 (2015)
9. Verdon, M., Horton, B., Rawnsley, R.: A case study on the use of virtual fencing to intensively graze Angus heifers using moving front and back-fences. Front. Anim. Sci. **2**, 663963 (2021)
10. Rey-Merchán, M.D.C., Gómez-de-Gabriel, J.M., López-Arquillos, A., Fernández-Madrigal, J.A.: Virtual fence system based on IoT paradigm to prevent occupational accidents in the construction sector. Int. J. Environ. Res. Public Health **18**(13), 683 (2021)
11. Chen, S., Gao, S.: Implementation method of the virtual electronic fence in the map service engine. Geospat. Inf. **18**(1), 81–84 (2020)
12. Li, Y., Wang, G., Zuo, F.: Efficient privacy preserving single anchor localization using noise-adding mechanism for Internet of Things. In: Xing, C., Fu, X., Zhang, Y., Zhang, G., Borjigin, C. (eds.) WISA 2021. LNCS, vol. 12999, pp. 261–273. Springer, Cham (2021). https://doi.org/10.1007/978-3-030-87571-8_23
13. Alfarraj, O., Baihan, A., Baihan, M.: Design and development of a smart sensing kit for the detection of accident location using smartphone. Sens. Lett. **13**(5), 365–370 (2015)
14. Rastogi, S., kumar Gautam, A., Vaish, A., Umar, M.: Web based accident reporting and tracking system (2019)
15. Jian, X.: Mobile phone emergency rescue notification system and methods. CN 101340700 A (2007)
16. Xia Shu branch.: Investigation, rescue system and method based on satellite localization. CN 102930715 B (2012)
17. Butkar, U.D., Gandhewar, N.D.: Accident detection and alert system (current location) using global positioning system. J. Algebraic Stat. **13**(3), 241–245 (2022)
18. Park, S., Son, S.O., Park, J., Oh, C., Hong, S.: Using vehicle data as a surrogate for highway accident data. In: Proceedings of the Institution of Civil Engineers-Municipal Engineer, vol. 174, no. 2, pp. 67–74. Thomas Telford Ltd. (2021)
19. Liang, G., Sun, X., Zhang, Y., Chen, M., Zhang, W.: Identifying expressway accident black spots based on the secondary division of road units. Promet Traffic Transp. **33**(5), 731–743 (2021)
20. Han, M.: Knowledge of latitude and longitude to calculate the two-point exact distance. Technol. Commun. (11), 2 (2011)

21. JD localization devices Page. https://i-item.jd.com/10040601321386.html. Accessed 10 May 2022
22. TaoBao. https://uland.taobao.com/. Accessed 7 May 2022
23. Tencent Cloud Server Page. https://cloud.tencent.com/act/pro/seckill_season? fro-m=16848. Accessed 20 May 2022

Design and Implementation of Analyzer Management System Based on Elasticsearch

Jingqi Sun[1,3], Peng Nie[1(✉)], Licheng Xu[2], and Haiwei Zhang[1]

[1] College of Cyber Science, Nankai University, Tianjin 300350, China
sunjingqi@dbis.nankai.edu.cn, {niepeng,zhhaiwei}@nankai.edu.cn
[2] College of Computer Science, Nankai University, Tianjin 300350, China
xulicheng@dbis.nankai.edu.cn
[3] Tianjin Key Laboratory of Network and Data Security Technology, Tianjin, China

Abstract. Elasticsearch is one of the most popular full-text search and analytics engine. It can store, search, and analyze big volumes of data in near real time. Before searching, Elasticsearch will build multiple inverted indexes for the data. The analyzer plays a crucial role in this process. An appropriate analyzer can segment text into semantically meaningful words, which can significantly improve the query accuracy. However, the default analyzer has limited performance in the Chinese context. Existing methods generally replace the default analyzer with manual configuration to optimize the query effect. The cost of service downtime caused by manually updating analyzers is often unacceptable in production environments. Based on Elasticsearch's Restful-API, we have implemented a framework for dynamic configuration of analyzers in a cluster environment. The framework supports common Chinese analyzers and provides a visual interface. Experiments show that the framework proposed in this paper reduces the update and maintenance time cost of the analyzer in the online environment by 94% compared to manual update. At the same time, compared with the default analyzer configuration of Elasticsearch, the accuracy of the system based on this framework is improved by 30%.

Keywords: Analyzer configuration · Cluster management · Information system

1 Introduction

With the development of Internet technology, information retrieval technology is also developing. As a lightweight full-text search engine with a complete community ecosystem, Elasticsearch has been widely used in various enterprises [1]. However, when dealing with non-Latin languages, Elasticsearch uses the heuristic minimum granularity segmentation scheme by default, ignoring the semantics in the original text, resulting in a decrease in query accuracy [2]. To solve this problem, a Chinese analyzer is usually used to segment Chinese documents instead of using the default configuration [3]. After word segmentation, the semantics of the text can be considered to improve the precision of the query [4]. Traditional word segmentation configuration methods usually rely on sending HTTP requests to the cluster to update Chinese analyzers. The association between indexes and analyzers leads to a large update cost and reduces development efficiency. Further, Elasticsearch lacks authentication and authority management mechanisms like the MySQL database [5], so the system has low security and is prone to data leakage [6]. To solve the above problems, we designed and implemented an Elasticsearch-based analyzer management platform. The contributions of this paper are:

- We designed and implemented an Elasticsearch-based analyzer automatic update and index management framework, which can help users quickly configure and test word segmentation, achieve hot update of analyzers and self-refactoring of indexes, and adapt to online environment better than traditional methods.
- We introduced the RBAC permission model into the framework, designed a perfect permission system, and solved the data security problem brought by Elasticsearch.
- We provide a visual Elasticsearch cluster management function, which can quickly obtain the real-time status of the cluster and dynamically change the configuration of the cluster, which greatly improves the development efficiency.
- Finally, we conduct an empirical study on the system. The test results show that the framework proposed in this paper can shorten the configuration time and improve the precision of queries compared with the traditional framework. The introduction of this framework can improve the precision rate by about 30%.

2   Related Work

An apparent difference between Elasticsearch and traditional relational databases is that the former is an unstructured NoSQL database, so many concepts such as indexes, documents, etc. are not intuitive enough for us, while Elasticsearch's visualization tools solved this problem. It can intuitively display the structure and content of various data, which is convenient for our understanding. At present, the mainstream cluster management tools include Elasticsearch-Head, Dejavu, Kibana, etc. Elasticsearch-Head is a plugin used to monitor the running state of the Elasticsearch [7]. It obtains data from the cluster and displays it through the HTTP RestfulAPI provided by Elasticsearch, but it has simple functions and a simple interface. Kibana is an officially provided data visualization tool. It can visualize vast amounts of data in Elasticsearch using charts and graphs. It often appears in big data analysis scenarios [8]. However, Kibana does not have many functions in cluster management, index management, etc., which is challenging to meet the daily development needs. Dejavu is a web-UI for Elasticsearch. It has made many optimizations for the visualization of cluster data and supports online preview and update of data, as well as data import and export [9]. But the feature of Dejavu is quite simple for it only provides the feature of data operation.

In summary, the above tools are still basic cluster management tools. They doesn't support some specific features, such as the configuration of the analyzer described in this article. At the same time, these tools rely on Elasticsearch's HTTP service, so the relevant permissions must be opened before they can be used, which may lead to some security risks.

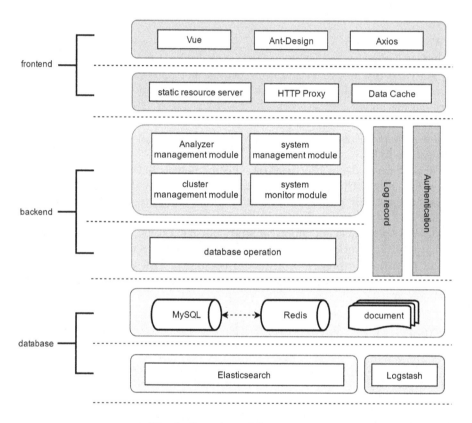

Fig. 1. Overview of the system.

3   Concept of Design

3.1   Overview of System

The system provides a function to configure the analyzer automatically. The target users of the system are developers. By using this system, it can significantly improve the development efficiency. After logging into the system, users can select and submit an appropriate analyzer. The system will automatically generate the analyzer configuration and upload it to the Elasticsearch cluster. Then, the system will rebuild the index of the relevant document and ensure that the application running based on Elasticsearch is always available during the rebuilding process.

We adopts the traditional Browser/Server architecture(B/S architecture). The most significant advantage of B/S architecture is convenience and cross-platform. Users can use the system as long as they have installed a browser [10]. From the development perspective, once the development is completed, it can run on anywhere, which significantly reduces the development cost. The architecture design of the system adopts hierarchical model (see Fig. 1). In hierarchical model, each layer performs its duties. The upper layer depends on the lower layer, and the upper layer does not need to care about the implementation of the lower layer [11]. It only needs to call the interface provided by it to decouple the system functions and improve the maintainability of the system.

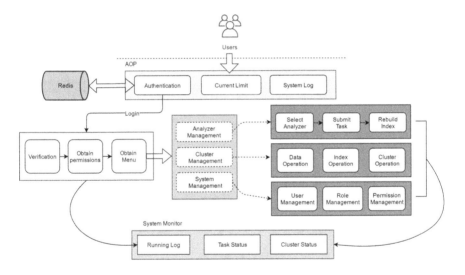

Fig. 2. Module relationship and execution process.

3.2 System Design

This system can be divided into four modules: analyzer management module, cluster management module, system monitor module, system management module. Each module is independent of each other, and the execution process of the system is shown in the figure (see Fig. 2). The analyzer management module is the core function module of the system, which completes the selection, installation and configuration process of the analyzer in Elasticsearch. The system management module plays a significant role in the system, mainly accomplishing the functions of authentication, user management to guarantee the security of the system. The remaining modules are mainly auxiliary to the operation of the system. The cluster management module and the system monitoring module are responsible for monitoring the Elasticsearch cluster environment, monitoring the system's running status and system operation environment.

Analyzer Management Module. The analyzer management module supports automatically install an analyzer or generate an analyzer based on a custom configuration. The system can automatically install analyzers include analyzer plugins currently certified by Elasticsearch officially. Users can apply analyzers to clusters by selecting, downloading, and installing them directly online. User-defined analyzers are also supported, and configurations are generated to join the cluster. To achieve the automatic configuration process, we used an asynchronous task processing method (see Fig. 3).

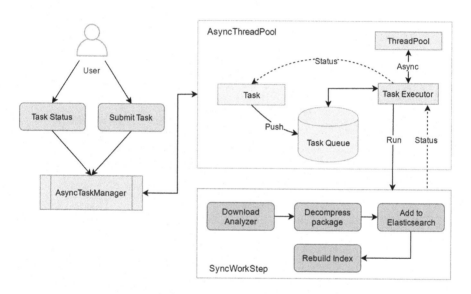

Fig. 3. Asynchronous task management model

After the user selects the analyzer and confirms, an asynchronous task is automatically generated, and the task information is submitted to the asynchronous task manager and saved in a queue. At the same time, the parameter information of the task is communicated to the task executor, which initiates the specific task process. We abstracted the entire process into four steps: Analyzer downloading, Package unpacking, Analyzer installing, and Index rebuilding. After these four steps, the analyzer will take effect in the new index.

The word segmentation effect of some analyzers depends on their dictionary. We provide online dictionary configuration function for IK-Analyzer. Users can directly operation the dictionary to improve the precision of the query.

Cluster Management Module. In order to further improve the availability of the system, we implemented the cluster management module. This module can be divided into three sub-modules: index management, data operation, cluster monitoring. The index management module supports index operations of the Elasticsearch, such as index deletion, index creation and other functions. The data operation module provides a visible interface for user to catch the data in Elasticsearch.

In production environment, when configuration updates, services should running as well to avoid problems while users using the system. The easiest way to solve this problem is to set a timed task, such as updating the configuration of the cluster in the middle of the night. However, this can cause data lag. Because configuration updates often do not take effect immediately. To avoid these problems, we designed a non-downtime index rebuilding method (see Fig. 4). After the user initiates an index rebuild request, we first alias the source index and generate a new index that uses the updated analyzer. When the new index is built, migrate the data from the old index through the reindex API. Then alias the new index, and delete the old. The entire process is not visible to the user like a black box. When the system receives a query request, it selects an available index to execute the user's query. Services are always available to users.

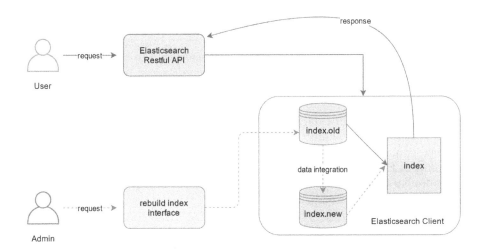

Fig. 4. The process of rebuilding the index

System Management Module. Based on based on the role-based access control (RBAC) model, we designed this system management module. We define various resources as different permissions, and then assign corresponding permissions to roles. [14]. Each different user will have its own set of roles and can only operate authorized resources.

The module can be divided into two sub-modules, the authentication module and the permissions management module. The authentication module allows users to sign in to the system through their accounts and authorizes users of the system. At the same time, to prevent the same account from being reused, the system restricts users from using the system online at the same time. After the user login to the system, the system will query the database for the permissions and display pages that match the user's permissions. The logon status is cached in the in-memory database Redis so that when the user closes the browser, the session will not be destroyed and will be saved in Redis for a period of time. Users can directly enter the system without having to sign in again to improve their experience.

The permission management section uses the RBAC(Role-Based Access Control) model [12], where the user-role-permission relationship is many-to-many, each user has multiple roles, each role has multiple permissions, and the permissions are the sum of all the permissions of all its roles [13]. Moreover, the permissions are mainly reflected in data access, page access and operation. All operations in the system are for authorized users.

Monitor Module. In order to make the system run stably for a long time, we realize a monitor module for the system. The monitor module is invisible for users but very important for the system. We provide tow main function in this module. They are log record and environment monitor. Logging is very important for any information system. When there is an exception to the system, we can find problems in the log in time. At the same time, we can record the action of users using the system, from what we can find much valuable information. In our system, we mainly record the sign info and the operation log of users. These are benefit for the maintenance and management of the system.

The environment monitor is for the developers. They can quickly obtain the parameters of the runnable environment. It provides a way to observe the lower environment from above services. We realize it based on Spring-Boot-Actuator, which provides a series API to get the running status and environmental parameters of the current system [15].

4   System Implementation

4.1   Development Environment

The system adopts the development method of separation of front and back ends. The front end is developed using Vue+Less+axios+nginx. Among them, the most important role is Node.js. Node.js's package manager NPM can help us quickly download and install the expansion packages and configuration items needed for front-end development [16].

The backend uses the technology stack of Java+Spring Boot+MySQL+ Redis+ Elasticsearch. The backend development environment needs to install the necessary environment. For example, Java needs the support of JDK (Java Development Toolkit), and Spring Boot needs Maven and Spring CLI to complete package management and project creation [17]. The specific development environment and development tools are shown in Table 1.

Table 1. Development environment and tools.

Item	Content
Operation System	Windwos 10
Language	Java, Vue, Less, JavaScript, Painless
Development tools	IntelliJ IDEA, WebStorm
DataBase	MySQL5.7, Redis
Middleware	Elasticsearch 6.6.2
Web Server	Apache Tomcat8.0, Nginx
Package Management	Apache Maven 3.5.2

4.2   Implementation

According to the system design, the analyzer management module is the core of the system (see Fig. 5). It mainly consists of four parts: plug-install, custom analyzer, analyzer testing, and dictionary configuration. This module does not need to use a database for data storage but interacts directly with the cluster through the Java API provided by Elasticsearch [18].

Plug-install. Plug-install feature provides the ability to install the online ananlyzer plug-ins to users. At present, the system supports analysis-ik, analysis-synonym, analysis-angj, analysis-kuromoji and other tokenizers. Users can directly search for word segmentation plugins online and download them. The system will automatically install the plug-in to the Elasticsearch plug-in directory and perform automatic configuration.

Custom Analyzer. The custom analyzer function supports users to configure personalized tokenizers. In Elasticsearch, the analysis process will actually go through three steps: character filtering (char_filter), tokenize (tokenizer), and token filtering (token_filter). In the character filtering stage, specific characters in the text will be removed, such as tags in HTML, XML and other texts; in the tokenize stage, the text after the filtered characters will be segmented, and different tokenizers can be selected during tokenizing. The standard tokenizer will be used by default. At the token filtering stage, the token generated is processed

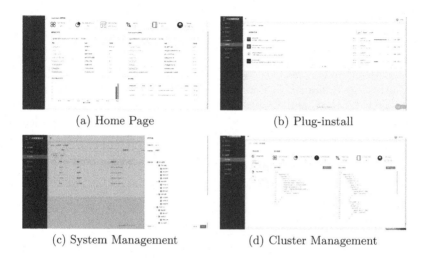

(a) Home Page   (b) Plug-install

(c) System Management   (d) Cluster Management

Fig. 5. The implementation of the system

again, such as stop words, synonyms, etc. This page provides a form, users can complete the content of the form, configure the corresponding parameters, and finally generate a complete Json sentence of the custom tokenizer.

Analyzer Testing and Dictionary Configuration. In order to test the effect of the analyzer, the system implements the analyzer testing module. The function of this module is relatively simple. The page is divided into upper and lower parts, with input boxes and test buttons at the top. Users can select the index and the analyzer in the index, enter the text, and then click the "test" button to obtain the text information after segmentation.

5   Experimental Results

We implement the system according to the design proposed in Sect. 3. In order to analyze the efficiency improvement brought by our designed framework when the analyzer is hot updated, we use a real-world dataset for testing. The main content of the experimental data set is cases from hospitals, including about 300K records.

We use IK-analyzer as the optional analyzers in the framework, and take the time required by manually updating the analyzers by four professional Elasticsearch engineers and use the average time required for them to update the analyzers as baseline. In the analyzer hot update experiment, 10 experiments were carried out on our framework and manual update respectively.

Figure 6 shows the experimental results of the time required for updating the analyzers. The experiments show that the introduction and implementation of this system have significant effects on improving development efficiency, optimizing the management of Elasticsearch clusters, and improving the security of

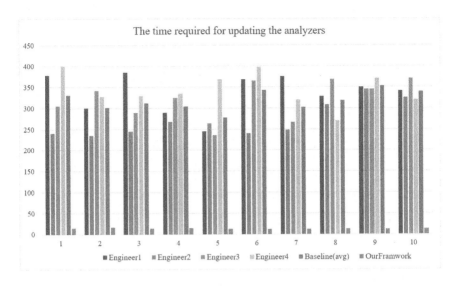

Fig. 6. The time required for updating the analyzers

Elasticsearch clusters. Compared with the time required for manual update, by using the framework proposed in this paper, the time can be shortened to 15s, which reduces the time consumption by about 94%.

Further, we compare the query precision rate P, recall rate R and F1 value of the search system based on our framework and Elasticsearch's default configuration as baseline in this dataset. P is the ratio of the number of related documents in the result to the number of returned results. R is the ratio of the number of related documents to the number of all related documents in the result [19] (see Table 2).

Table 2. Parameters required for evaluation indicators.

	Related documents	Non-related documents
Related documents	TP	FP
Non-related documents	FN	TN

The experimental results (see Table 3) show that by introducing our framework, the precision of the query can be effectively improved. It can be seen that in the default mode of Elasticsearch, the recall rate can reach 100% because the semantics of the vocabulary are not considered, but the precision rate is relatively low. By introducing the framework of this paper, the query accuracy can be improved. Compared to Elasticsearch's default mode, the F1-score is improved about 30%.

Table 3. Comparison of different top-K.

	P	R	F1
baseline(k=20)	0.64	1.00	0.78
our framework(k=20)	0.92	0.86	0.88
baseline(k=50)	0.58	1.00	0.73
our framework(k=50)	0.84	0.78	0.80
baseline(k=200)	0.32	1.00	0.48
our framework(k=200)	0.72	0.69	0.71
baseline(k=500)	0.18	1.00	0.31
our framework(k=500)	0.59	0.61	0.60

Finally, we perform a functional comparison with the existing Elasticsearch management tools (see Table 1). It can be seen that compared with the existing system, our framework can cover more functions and better meet the development needs.

Table 4. Comparison with other tools

Features	Our System	dejavu	ES-head	Kibana
Port	9300	9200	9200	9200
Modern UI	Ant-Design V2.1	React 16.6	JQuery 1.6.1, slightly stodgy	Node.JS, Hapi, Jade
Browser features	CRUD, data filters, full-text search	CRUD, data filters	Read Data, full-text search	Read View, visualizations, charting
Data Import/export	Support for JSON	support for JSON,CSV	No	Only Export
Analyzer Installation	Support auto installation	No	No	No
Custom Configuration for Analyzer	Visually build and test	No	No	No
Authority system	Yes	No	No	Need X-Path

6   Conclusions and Future Work

In this paper, We designed and implemented an Elasticsearch-based analyzer automatic update and index management framework, which can help users quickly configure and test word segmentation, achieve hot update of analyzers and self-refactoring of indexes, and adapt to online environment better than traditional methods. Experimental results show that this framework can significantly improve the development efficiency and make the query results more accurate. And we introduced the RBAC permission model into the framework, designed a permission system, and solved the data security problem brought by Elasticsearch.

However, there are still some deficiencies in the framework that need to be improved. For example, the current framework only supports some Chinese analyzers, such as IK-Analyzer, which is not compatible with other languages and cannot complete the multi-language version of the analyzer configuration. In addition, the permissions can be further refined to improve the security and availability. In the future work, we will further improve the system aiming at the above shortcomings.

References

1. Elasticsearch Guide. https://www.elastic.co/guide/index.html.. Accessed 19 Mar 2022
2. Divya, M.S., Goyal, S.K.: ElasticSearch: an advanced and quick search technique to handle voluminous data [J]. Compusoft **2**(6), 171 (2013)
3. Qiu, Q., Xie, Z., Liang W., Li, W.: DGeoSegmenter: a dictionary-based Chinese word segmenter for the geoscience domain[J]. Comput. Geosci. **121**, 1–11 2018
4. Huihui, S., Xin. N.: Research and Implementation of Chinese Automatic Word Segmentation System Based on Complex Network Features [J]. Wireless Commun. Mobile Comput. **2022**, 1–10 (2022)
5. Lu, H., Hong, Y., Yang, Y., Duan, L., Badar, N.: Towards user-oriented RBAC model. In: Wang, L., Shafiq, B. (eds.) DBSec 2013. LNCS, vol. 7964, pp. 81–96. Springer, Heidelberg (2013). https://doi.org/10.1007/978-3-642-39256-6_6
6. Takase, W., Nakamura, T., Watase, Y., et al.: A solution for secure use of Kibana and Elasticsearch in multi-user environment[J]. arXiv preprint arXiv:1706.10040 (2017)
7. Wei, B., Dai, J., Deng, L., et al.: An Optimization Method for Elasticsearch Index Shard Number[C]. In: 2020 16th International Conference on Computational Intelligence and Security (CIS). IEEE, pp. 191–195 (2020)
8. Reelsen A. Using elasticsearch, logstash and kibana to create realtime dashboards J]
9. Varis, J.: Enhancing a product's development and debugging with supporting product development [J] (2020)
10. Jibiao, J., Xinghua, Q.I.: Design and realization of online query system of Chinese-English terminology of Chinese medicine [J]. China Terminol. **24**(2), 92 (2022)
11. Concepcion, A.I., Zeigler, B.P.: DEVS formalism: a framework for hierarchical model development[J]. IEEE Trans. Softw. Eng. **14**(2), 228–241 (1988)
12. Cruz, J.P., Kaji, Y., Yanai, N.: RBAC-SC: role-based access control using smart contract [J]. Ieee Access **6**, 12240–12251 (2018)
13. William Stallings. Role-Based Access Control in Computer Security. https://www.informit.com/articles/article.aspx?p=782116.. Accessed 19 Mar 2022
14. Pi, C., Nie, P., Feng, Y., Xu, L.: Design and implementation of annotation system for character behavior & event based on spring boot. In: Xing, C., Fu, X., Zhang, Y., Zhang, G., Borjigin, C. (eds.) WISA 2021. LNCS, vol. 12999, pp. 756–763. Springer, Cham (2021). https://doi.org/10.1007/978-3-030-87571-8_66
15. Chinotan. SpringBoot Actuator. https://my.oschina.net/u/3266761/blog/2960. Accessed 19 Mar 2022

16. Dhulavvagol, P.M., Bhajantri, V.H., Totad, S.G.: Performance Analysis of Distributed Processing System using Shard Selection Techniques on Elasticsearch [J]. Procedia Comput. Sci **167**, 126–136 (2020)
17. Fan Zhang., F.: Design and Implementation of Physical Education Video Teaching System Based on Spring MVC Architecture [C]. Durham University, Suffolk University. In: Proceedings of the 4th International Conference on Information and Education Innovations (ICIEI 2019. Durham University, Suffolk University: SCIence and Engineering Institute (SCIEI), pp. 117–120 (2019)
18. Coronel, J.B., Mock, S.: Designsafe: using elasticsearch to share and search data on a science web portal[M]. In: Proceedings of the Practice and Experience in Advanced Research Computing 2017 on Sustainability, Success and Impact, pp. 1–3 (2017)
19. Arora, M., Kanjilal, U., Varshney, D.: Evaluation of information retrieval: precision and recall [J]. Int. J. Indian Cult. Bus. Manage. **12**(2), 224–236 (2016)

A Hybrid Model for Spatio-Temporal Information Recognition in COVID-19 Trajectory Text

Haoyu Yu[1], Xuan Pan[1], Dongming Zhao[2], Yanlong Wen[1(✉)], and Xiaojie Yuan[1]

[1] College of Computer Science, Nankai University, Tianjin 300350, China
1811456@mail.nankai.edu.cn, panxuan@dbis.nankai.edu.cn,
{wenyl,yuanxj}@nankai.edu.cn
[2] Artificial Intelliqence Laboratory, China Mobile Communication Group Tianjin Co., Ltd, Tianjin, China

Abstract. Since the outbreak of the COVID-19 epidemic at the end of 2019, the normalization of epidemic prevention and control has become one of the core tasks of the entire country. Health self-examination by checking the trajectory of diagnosed patients has gradually become everyone's basic necessity and essential to epidemic prevention. The COVID-19 patient's spatio-temporal information helps to facilitate the self-inspection of the masses of whether their trajectory overlaps with the confirmed cases, which promotes the epidemic prevention work. This paper, proposes a named entity recognition model to automatically identify the time and place information in the COVID-19 patient trajectory text. The model consists of an ALBERT layer, a Bi-GRU layer, and a GlobalPointer layer. The previous two layers jointly focus on extracting the context's characteristics and the semantic dependencies. And the GlobalPointer layer extracts the corresponding named entities from a global perspective, which improves the recognition ability for the long-nested place and time entities. Compared to the conventional name entity recognition models, our proposed model has high effectiveness because it has a smaller parameter scale and faster training speed. We evaluate the proposed model using a dataset crawled from the official COVID-19 trajectory text. The F1-score of the model has reached 92.86%, which outperforms four traditional named entity recognition models.

Keywords: COVID-19 · Spatio-temporal information · Named entity recognition · A lite BERT

1 Introduction

Since entering the normalization stage of epidemic prevention and control, active and regular self-health check has become a part of people's daily lives. Means of self-health assessment include checking the color of the health code, receiving

X. Zhao et al. (Eds.): WISA 2022, LNCS 13579, pp. 267–279, 2022.
https://doi.org/10.1007/978-3-031-20309-1_23

notifications from the prevention and control department, and comparing the text information of the official notification Corona Virus Disease 2019 (COVID-19) patient's activity trajectory. Among them, the only way to conduct active self-screening is to check whether one coincides with the trajectory of confirmed cases.

However, the COVID-19 trajectory text contains duplicate or ambiguous location descriptions, resulting in some people's being reluctant to read and omitting the critical information. As the trajectory text of coronavirus disease has a fixed format with a detailed description, the location and time information are the most frequently changed entities. It is suitable for extracting the time and place information by the named entity recognition models. So the automatic spatio-temporal information recognition would substantially reduce the time cost to obtain the patient's spatio-temporal data in the trajectory, which improves the trajectory query efficiency and promotes epidemic prevention works.

Until now, many researchers have done a lot of work on identifying Chinese location and time entities by the named entity recognition (NER) model. The mainstream framework is building deep learning neural networks and training them through the supervised learning process on the related corpus. Most models are basically built around the Bidirectional Encoder Representations from Transformers (BERT) [9] pre-training language model and fine-tuned for downstream tasks using transfer learning methods. Through the analysis of 20,000 actual diagnosis track texts, We find that the sentence patterns are relatively fixed. But many duplicate names in the detailed description of time and place need to be distinguished. And most of the time and place entities involved are nested entities, which brings a challenge to understand the contextual semantic dependencies for entity recognition models. However, most of the current mainstream models have problems such as large parameter scales, slow training speed, and poor recognition of nested words. Around these problems, we have changed the model structure to better complete the NER task.

In this paper, we propose a deep learning model based on a hybrid framework consisting of the A Lite Bidirectional Encoder Representations from Transformers (ALBERT) pre-training language model [15], the Bidirectional Gated Recurrent Unit (Bi-GRU) network [5], and the GlobalPointer [22] framework, which is used for the temporal and location named entities recognition in the COVID-19 trajectory text. The ALBERT pre-training language model is a variant of the BERT model with an optimized parameter scale that efficiently performs bidirectional semantic training on the original text. The Bi-GRU layer complements the location information provided by the ALBERT location encoding for contextual dependencies. This module makes a more complete and accurate semantic-dependent feature and relative location information with a selectively forgetting mechanism for irrelevant contexts. The GlobalPointer layer processes the identification of nested words better than the traditional conditional random field (CRF) method. The richer relative position information makes the GlobalPointer more sensitive to the length and span of nested entities in time and location that appear in large numbers in the COVID-19 trajectory text and promotes the spatio-temporal information recognition performance.

2   Related Work

2.1   Rule-Based Methods

Rule-based NER systems rely on manually formulated rules which can be designed based on domain-specific gazetteers and syntactic lexical patterns. Well-known systems include LaSIE-II [12], NetOwl [14], Facile [3], and SAR [1]. It is also the first introduced method in Chinese named entity recognition. In 1997, Zhang et al. [28] formulated the corresponding matching rules by extracting and summarizing the regular features of institutional nouns such as university names. It achieved a good recognition effect with a correct rate and a recall rate of 97.3% and 96.9%, respectively, by using the maximum matching method to analyze and extract entities. However, this kind of rule-based method excessively relies on manual rule-making and dictionary-making to realize named entity recognition leading to problems such as poor comprehensiveness, ambiguous words, and difficulties in updating and migrating. The utilization efficiency achieved by the labor cost of a large amount of input is low, so it has gradually become a complementary method to improve other means.

2.2   Statistical-Based Methods

The method based on statistics assigns the attribute of statistical probability to the entities appearing in the text and then converts the entity extraction into sequence annotation by using the maximum probability to replace the best result. The basic models of sequence annotation include hidden Markov model (HMM) [19], maximum entropy model [2], decision tree model [17] and conditional random field model [24]. Yu et al. [27] stacked multi-layer hidden Markov models to predict label results and subdivide them into each kind of entities. Chen et al. [4] introduced conditional random fields to the task of Chinese named entity recognition. In 2021, Su [22] proposed a Global Pointer model that is faster and has better effects on nested entities than the CRF model by integrating location information to handle the entity classification task. Although this statistical-based method gives the model certain portability and robustness and can be applied to processing the large-scale corpus, its effect is overly dependent on the richness of artificially formulated features. It is at present mostly used as a specific layer in deep learning methods to assist in large-scale named entity recognition tasks.

2.3   Deep Learning-Based Methods

The method based on deep learning makes up for the problems of strong dependence on manual annotation and difficulty in generalization and migration, etc. for the above two methods. Collobert et al. [7] proposed named entity recognition based on the CNN-CRF model, but it is difficult to use a convolutional neural network to solve the work of variable length input and interdependence; In 2015, Huang et al. [11] combined the Bidirectional Long Short-Term Memory

(Bi-LSTM) neural network able to acquire input information features bidirectionally with the CRF layer. It solved the problems of long-term information acquisition and context feature extraction. In 2018, Devlin *et al.* [9] proposed the BERT pre-training language model. It consists of multiple decoder structures based on the core self-attention mechanism in Transformers, which is proposed by Vaswani *et al.* [23]. Each word vector in the input vector would obtain the feature information of all word vectors in the sequence through the pre-training language model processing. Xu *et al.* [26] proposed a Chinese named entity recognition research method based on the Bert-BiLSTM-CRF model combined with self-attention and achieved a good result. Subsequently, it was found that BERT has the problem of the explosive growth of parameter scale with the increase of BERT model size. Liu Y *et al.* proved in RoBERTa [16] method that the Next Sentence Prediction (NSP) training task in BERT would have a negative effect on neural network learning moreover. Lan Z *et al.* [15] proposed the ALBERT pre-training language model, which further reduced the number of parameters, improved the model performance, and shortened the training time. Based on it, Deng *et al.* [8] constructed ALBERT-Bi-LSTM-CRF model and applied it on the Chinese named entity recognition task. What is worth mentioning is that most traditional NER research only concentrates on flat entities and ignores nested entities. Thus, researchers like Shen *et al.* [20] proposed relevant methods to deal with the recognition of nested entities.

The foregoing content is the sorting and introduction of related work in the field of named entity recognition in this paper. It has been found that in recent years, the related named entity recognition research tends to use the hybrid neural network structure to improve the model effect. Especially since the BERT language pre-training model was proposed, the main focus of work has been on the optimization of the BERT model itself and its auxiliary design. Besides, there are many model methods, in addition to those mentioned above, which perform well.

3   Methodology

3.1   Overview

This section gives an overview of our hybrid neural network model established to identify, extract and automatically label the time and location names in the COVID-19 trajectory text. As shown in Fig. 1, the proposed model consists of an ALBERT layer, a Bi-GRU layer, and a GlobalPointer layer. The ALBERT layer trains the embedding vector through the decoder layer so that each word fuses the feature information of all other words. It preprocesses text data into character encoding and sequence encoding. Then the Bi-GRU layer extracts contextual features by learning the preprocessed word vectors and fuses the dependent features between the information to form a sequence of feature vectors with relatively complete information. Based on the output feature vectors, the GlobalPointer layer predicts the labels from a global perspective by incorporating relative position information and calculating the serial correlation as a score for the entity label.

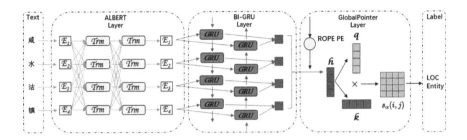

Fig. 1. The frame work of the proposed model

3.2   ALBERT Layer

The ALBERT pre-training language model [15] has been improved in two main directions based on BERT. It uses embedding matrix decomposition and cross-layer parameter sharing to reduce the number of parameters and compress the model size. Besides, it replaces the NSP task with the Sentence-Order Prediction (SOP) task to judge whether the context is coherent. Through decomposing the embedding matrix, the equal binding between word embedding E and hidden layer H is released, which can further reduce the parameter scale without affecting the performance of the model. As shown in Eq. 1:

$$O(V \times H) \to O(V \times E + E \times H) \tag{1}$$

Cross-layer parameter sharing reduces the number of parameters in the hidden layer by sharing all network parameters. It maximizes the parameter size and makes the network parameters more stable. At the same time, in order to avoid excessive regularization caused by shared parameters, ALBERT chose to remove Dropout. The SOP task has a smaller granularity of information units by ordering and reversing two consecutive context sentences respectively, and then predicting whether the overall word order is correct. It also solves the problem of confusing topic prediction and context order judgment in the NSP task.

Before passing the training set trajectory text into the ALBERT pre-training language model, each character needs to be initialized as an embedding vector, which contains position information by using the sine and cosine function for position encoding:

$$PE_{(pos,2i)} = \sin\left(pos/10000^{2i/d_{\mathrm{model}}}\right)$$
$$PE_{(pos,2i+1)} = \cos\left(pos/10000^{2i/d_{\mathrm{model}}}\right) \tag{2}$$

The embedding vector is iteratively processed by the encoder structure layer composed of **Trm**, and the number of layers of the encoder structure [23] is determined by the size of the selected ALBERT model version. The core method of the encoder is the self-attention mechanism, which linearly transforms the embedding vector with different weights to obtain the query matrix Q, the key matrix K, and the value matrix V:

$$\text{Attention}(Q, K, V) = \text{softmax}\left(\frac{QK^T}{\sqrt{d_k}}\right) V \tag{3}$$

After residual connection, normalization, and activation, the information features related to the context of each vector can be extracted. It makes each character contain the information features of all other characters in the vector sequence. Through the Masked Language Model (MLM) and SOP training tasks, ALBERT completes the text preprocessing by calculating loss corresponding to actual sentence labels and back-propagating the gradients for training. Finally, it forms output vector sequences containing a large amount of feature information.

3.3   Bi-GRU Layer

GRU [5] is a variant of Bidirectional Long Short-Term Memory (LSTM) [10], and both of them appeared to solve the problems of gradient disappearance and gradient explosion in the face of long-distance information capture in traditional recurrent neural networks. Compared with LSTM, GRU changes the input gate, forget gate, and output gate into update gate z_t and reset gate r_t. It also merges the unit state and output into one state h.

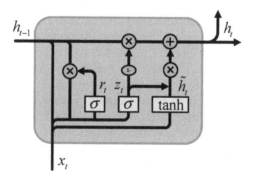

Fig. 2. Gated recurrent unit [18]

As shown in Fig. 2, the forward calculation formula of its state and output is:

$$\mathbf{z}_t = \sigma \left(W_z \cdot [\mathbf{h}_{t-1}, \mathbf{x}_t] \right)$$
$$\mathbf{r}_t = \sigma \left(W_r \cdot [\mathbf{h}_{t-1}, \mathbf{x}_t] \right)$$
$$\tilde{\mathbf{h}}_t = \tanh \left(W \cdot [\mathbf{r}_t \circ \mathbf{h}_{t-1}, \mathbf{x}_t] \right) \quad (4)$$
$$\mathbf{h}_t = (1 - \mathbf{z}_t) \circ \mathbf{h}_{t-1} + \mathbf{z}_t \circ \tilde{\mathbf{h}}_t$$

The Bi-GRU network is a special recurrent neural network model composed of multiple GRU units, which can be divided into forward layers and reverse layers. It can read the information features of the above and the following at the same time from the vector sequence preprocessed by ALBERT. To avoid the omission of semantic dependenciesand, it splices together bidirectional sequences of vectors as a output vector. At the same time, it uses Dropout to remove certain dimensions from the data to avoid possible overfitting. The input sequence $x = (x_1, x_2, \cdots, x_n)$ is processed by the forward GRU and the backward GRU respectively to obtain the corresponding sequence $\left(\overrightarrow{h}_1, \overrightarrow{h}_2, \cdots, \overrightarrow{h}_n \right)$ and $\left(\overleftarrow{h}_1, \overleftarrow{h}_2, \cdots, \overleftarrow{h}_n \right)$. Correspondingly splice the above two sequences to get the output vector $h_k = \left[\overrightarrow{h}_k; \overleftarrow{h}_k \right]$ which contains the bidirectional feature information at the corresponding position of x_k. As shown in Eq. 5, it calculate the probability P of the corresponding label of each word by the activation function σ, where W denotes the weight corresponding to the sequence vector, h denotes the output vector after splicing, and c denotes the corresponding offset:

$$P = \sigma(W h + c) \quad (5)$$

3.4   GlobalPointer Layer

GlobalPointer [22] predicts entity tags from a global perspective by incorporating location information. It transforms the named entity recognition problem into a multi-label classification problem of selecting k real entities from possible $n(n + 1)/2$ candidate entities in a sequence of length n. It uses the basic principle which is similar to the attention mechanism to convert the correlation of the sequence into the score of the entity. The GlobalPointer layer performs linear transformation on the encoded vector sequence $[h_1, h_2, \cdots, h_n]$ from Bi-GRU layer to obtain the vector sequence $[q_{1,\alpha}, q_{2,\alpha}, \cdots, q_{n,\alpha}]$ and $[k_{1,\alpha}, k_{2,\alpha}, \cdots, k_{n,\alpha}]$. Based on Rotational Position Encoding (RoPE) method, the relative position information is integrated by multiplying the matrix R_i that satisfies the relation $R_i^\top R_j = R_{j-i}$. Finally, multiply the sequence matrices to get the score of entity $s_\alpha(i, j)$:

$$s_\alpha(i, j) = \left(R_i q_{i,\alpha} \right)^\top \left(R_j k_{j,\alpha} \right) = q_{i,\alpha}^\top R_i^\top R_j k_{j,\alpha} = q_{i,\alpha}^\top R_{j-i} k_{j,\alpha} \quad (6)$$

It improves the generalization ability of the model by using the Softmax+Cross Entropy method [21] which considers multi-label classification problem as a pairwise comparison of the target category score and the non-target category score instead of multiple binary classification problems. The loss function of GlobalPointer applied to multi-label classification is:

$$L_{loss} = \log \left(e^{s_0} + \sum_{(i,j) \in P_\alpha} e^{-s_\alpha(i,j)} \right) + \log \left(e^{s_0} + \sum_{(i,j) \in Q_\alpha} e^{s_\alpha(i,j)} \right) \quad (7)$$

It uses the property of $logsumexp$ to make the maximum value of the difference between the score of all non-target entities $\{s_1, \cdots, s_{t-1}, s_{t+1}, \cdots, s_n\}$ and the score of the target entity $\{s_t\}$ less than zero as far as possible. Hence, the score of the target entity is greater than that of each non-target class. At the same time, set the threshold s_0 to 0 so that the scores of the target entities are greater than 0 and those of the non-target entities are less than 0 to distinguish the target entities from all entities. In the formula, P_α denotes the head and tail set of all entities of type α in the sample; Q_α denotes the head and tail set of all non-entities or entities of type non-α in the sample:

$$\Omega = \{(i,j) \mid 1 \le i \le j \le n\}$$
$$P_\alpha = \{(i,j) \mid t_{[i:j]} \in \alpha\} \quad (8)$$
$$Q_\alpha = \Omega - P_\alpha$$

All sequence fragments $t_{[i:j]}$ that satisfy $s_\alpha(i,j) > 0$ are entities of type α. Compared with the real label, the training of the model is completed by returning the loss function to update and iterate the parameters of the model. The loss function and evaluation index of the GlobalPointer model have entity granularity, which can score labels more closely to the application scenario. It solves the problem that CRF is too strict for entity prediction, and also enables the model to have better recognition ability for nested entities.

4   Experiments

4.1   Experiment Settings

Dataset. The dataset includes 7000 records from the official text of the activity trajectories of patients diagnosed with COVID-19 since 2020. After the filtering process, we retain the unduplicated records and label them manually. Finally, we generate a dataset containing 3000 records with 423,174 words, in which 2,400 records are used as training sets, 300 records are used as validation sets, and the last 300 records are used as test sets. The labeled named entities include the time entity and the location entity. The symbols in the labeling and their descriptions are shown in Table 1:

Table 1. Symbols and descriptions

Symbols	Descriptions
B_LOC	Initial word for the location entity
I_LOC	Non-initial word for the location entity
B_T	Initial word for the time entity
I_T	Non-initial word for the time entity
O	Other word

Implementation Details. We use Apple M1 Pro with a 10-core CPU and 16-core GPU chips for the model training. The programming language is Python3.9, and the development platform adopts Pytorch 1.10.2. In addition, some of the compared models are trained with NVIDIA GeForce RTX 2070 Super GPU with the Python 3.6.8 programming language and the TensorFlow 1.14.0 development platform. The ALBERT pre-trained language model with the albert_zh_base version. We use the test set and choose the traditional metrics precision, recall and F1 values for the model evaluation.

4.2   Baselines

We use four typical NER models for comparison.

Bi-LSTM+CRF. [25] This is a basic named entity recognition model. It has a Bi-LSTM network to capture the features and a CRF layer to output the prediction results.

Bi-LSTM+Attention+CRF. [13] This model has the attention network to solve the memory compression problem on LSTM.

ALBERT+Bi-LSTM+CRF. This model uses the ALBERT pre-trained language model to enhance the ability to capture context-related features in downstream tasks with fine-turning.

ALBERT+Bi-GRU+CRF. This model replaces the Bi-LSTM network with the Bi-GRU network to reduce the parameter scale.

The training word embeddings without ALBERT in the compared models are pre-trained using the Word2Vec [6] framework from the Google open-source tool. We use the dropout mechanism to reduce the overfitting.

4.3   Results

Table 2. Results of different models

Models	Precision (%)	Recall (%)	F1-score (%)
Bi-LSTM+CRF	87.32	85.47	86.39
Bi-LSTM+Attention+CRF	90.13	88.30	89.21
ALBERT+Bi-LSTM+CRF	91.96	89.57	90.74
ALBERT+Bi-GRU+CRF	91.76	90.11	90.93
ALBERT+Bi-GRU+GlobalPointer	**93.37**	**92.65**	**93.01**

Table 2 shows the prediction results. Our proposed model achieves the best performance among the compared models. Specifically, by comparing the ALBERT+Bi-LSTM+CRF model to the ALBERT+Bi-GRU+CRF model, although the Bi-GRU and Bi-LSTM networks in the model play the same role in extracting the context's characteristics and semantic dependence, the Bi-GRU network has a more lightweight parameter scale than that of Bi-LSTM network. And by comparing the GlobalPointer to the CRF framework, the F1 scores of our models improved 2.08% on the test set compared to the ALBERT+Bi-GRU+CRF model. Because the GlobalPointer framework introduces the positional encoding information, this is similar to the self-attention mechanism, which has a much finer granularity in recognizing a large number of entities with nested structures.

In addition, we found that whether to introduce the attention mechanism is a crucial factor affecting the performance of the above models. The effect of the Bi-LSTM+CRF model without the attention mechanism is even worse than the Bi-LSTM+Attention+CRF model which just has the worst effect in the kind of models with the attention mechanism. The neural network model further introduced the ALBERT layer for pre-trained effects better than the pre-training model with the Word2Vec method because the ALBERT pre-training language model can better obtain contextual semantic dependency information, and then more feature information predicts entities more accurate.

4.4   Ablation Study

Table 3. Ablation study of the Bi-GRU network

Models	Precision(%)	Recall(%)	F1-score(%)
ALBERT+Bi-GRU+GlobalPointer	93.37	92.65	93.01
ALBERT+GlobalPointer	91.04	90.76	90.90

Here we evaluate the Bi-GRU layer through an ablation study. Table 3 shows the experimental results. The results of the analysis table show that the Bi-GRU layer improves the effect of the overall model by 2.11%. In fact, the location information of the entity obtained only through the ALBERT layer is incomplete because the self-attention mechanism in ALBERT only calculates the position information obtained from the position embedding vector but ignores understanding the dependencies among other words. So we need to use the Bi-GRU layer further to extract the location features and semantic dependencies in the sequence to obtain a comprehensive location representation for the subsequent scoring layers.

5   Conclusion and Future Work

This paper proposes a hybrid neural network for simultaneously identifying time and location entities in the text of the COVID-19 trajectory. The text contains many long-nested spatio-temporal entities. In response to this challenge, we use the ALBERT pre-training model and the Bi-GRU network to extract the entity's context-dependent features and apply the GlobalPointer framework to score the long-nested entities from a global perspective. Experiments are conducted on the actual confirmed cases' trajectory text. The results demonstrate the model outperforms the traditional NER models, showing its effectiveness in spatio-temporal information recognition. In the future, we will further optimize the encoding layers to exploit deep semantic relations and enlarge the scale of the training data set for improvement. Also, we will develop a visualization system to illustrate the patients' spatio-temporal trajectories to make an effort to the epidemic prevention works.

Acknowledgements. This research is supported by National Natural Science Foundation of China (No. U1936206). We thank the reviewers for their constructive comments.

References

1. Aone, C., Halverson, L., Hampton, T., Ramos-Santacruz, M.: Sra: Description of the ie2 system used for muc-7. In: Seventh Message Understanding Conference (MUC-7): Proceedings of a Conference Held in Fairfax, Virginia, April 29-May 1, 1998 (1998)
2. Berger, A., Della Pietra, S.A., Della Pietra, V.J.: A maximum entropy approach to natural language processing. Comput. linguist. **22**(1), 39–71 (1996)
3. Black, W.J., Rinaldi, F., Mowatt, D.: Facile: Description of the ne system used for muc-7. In: Seventh Message Understanding Conference (MUC-7): Proceedings of a Conference Held in Fairfax, Virginia, April 29-May 1, 1998 (1998)
4. Chen, W., Zhang, Y., Isahara, H.: Chinese named entity recognition with conditional random fields. In: Proceedings of the Fifth SIGHAN Workshop on Chinese Language Processing, pp. 118–121 (2006)
5. Cho, K., Van Merriënboer, B., Gulcehre, C., Bahdanau, D., Bougares, F., Schwenk, H., Bengio, Y.: Learning phrase representations using rnn encoder-decoder for statistical machine translation. arXiv preprint arXiv:1406.1078 (2014)

6. Church, K.W.: Word2vec. Natural Lang. Eng. **23**(1), 155–162 (2017)

7. Collobert, R., Weston, J., Bottou, L., Karlen, M., Kavukcuoglu, K., Kuksa, P.: Natural language processing (almost) from scratch. J. Mach. Learn. Res. **12**(ARTICLE), 2493–2537 (2011)

8. Deng, B., Cheng, L.: Chinese named entity recognition method based on albert (2020)

9. Devlin, J., Chang, M.W., Lee, K., Toutanova, K.: Bert: Pre-training of deep bidirectional transformers for language understanding. arXiv preprint arXiv:1810.04805 (2018)

10. Hochreiter, S., Schmidhuber, J.: Long short-term memory. Neural comput. **9**(8), 1735–1780 (1997)

11. Huang, Z., Xu, W., Yu, K.: Bidirectional lstm-crf models for sequence tagging. arXiv preprint arXiv:1508.01991 (2015)

12. Humphreys, K., et al.: University of sheffield: Description of the lasie-ii system as used for muc-7. In: Seventh Message Understanding Conference (MUC-7): Proceedings of a Conference Held in Fairfax, Virginia, April 29-May 1, 1998 (1998)

13. Jin, C., Shi, Z., Li, W., Guo, Y.: Bidirectional lstm-crf attention-based model for chinese word segmentation. arXiv preprint arXiv:2105.09681 (2021)

14. Krupka, G., Hausman, K.: Isoquest inc.: description of the netowl^{TM} extractor system as used for muc-7. In: Seventh Message Understanding Conference (MUC-7): Proceedings of a Conference Held in Fairfax, Virginia, April 29-May 1, 1998 (1998)

15. Lan, Z., Chen, M., Goodman, S., Gimpel, K., Sharma, P., Soricut, R.: Albert: A lite bert for self-supervised learning of language representations. arXiv preprint arXiv:1909.11942 (2019)

16. Liu, Y., et al.: Roberta: A robustly optimized bert pretraining approach. arXiv preprint arXiv:1907.11692 (2019)

17. Magerman, D.M.: Statistical decision-tree models for parsing. arXiv preprint cmp-lg/9504030 (1995)

18. Olah, C.: Understanding lstm networks (2018). https://colah.github.io/posts/2015-08-Understanding-LSTMs

19. Rabiner, L., Juang, B.: An introduction to hidden markov models. IEEE ASSP Mag. **3**(1), 4–16 (1986)

20. Shen, Y., Ma, X., Tan, Z., Zhang, S., Wang, W., Lu, W.: Locate and label: A two-stage identifier for nested named entity recognition. arXiv preprint arXiv:2105.06804 (2021)

21. Su, J.: Generalize softmax + cross-entropy to multi-label classification problems (2020). https://spaces.ac.cn/archives/7359

22. Su, J.: Globalpointer:deal with nested and non-nested ner in a unified way (2021). https://spaces.ac.cn/archives/8373

23. Vaswani, A., et al.: Attention is all you need. In: Advances in Neural Information Processing Systems, vol. 30 (2017)

24. Wallach, H.M.: Conditional random fields: An introduction. Technical Reports (CIS) p. 22 (2004)

25. Xu, L., et al.: Cluener 2020: Fine-grained name entity recognition for chinese. arxiv 2020. arXiv preprint arXiv:2001.04351

26. Xu, L., Li, S., Wang, Y., Xu, L.: Named entity recognition of BERT-BiLSTM-CRF combined with self-attention. In: Xing, C., Fu, X., Zhang, Y., Zhang, G., Borjigin, C. (eds.) WISA 2021. LNCS, vol. 12999, pp. 556–564. Springer, Cham (2021). https://doi.org/10.1007/978-3-030-87571-8_48

27. Yu, H.k., Zhang, H.p., Liu, Q., Lv, X., Shi, S.: Chinese named entity identification using cascaded hidden markov model. J. Commun. **27**(2), 87–94 (2006)
28. Zhang, X., Wang, L.: Identification and analysis of chinese organization names. J. Chinese Inform. Process. **11**(4), 22–33 (1997)

A Research on the Theory and Technology of Trusted Transaction in Modern Service Industry

Xin Wei[1], Yi Zhu[1], Jian Zhang[1], Guigang Zhang[2], Chao Li[1(✉)], Yong Zhang[1] ⓘ, and Chunxiao Xing[1] ⓘ

[1] Beijing National Research Center for Information Science and Technology, Tsinghua, Beijing 100084, China
li-chao@tsinghua.edu.cn
[2] Institute of Automation, Chinese Academy of Sciences, Beijing 100084, China

Abstract. This paper mainly discusses the theory and related technology of trusted transactions in modern service industry. Three prohibition principles of trusted transactions in modern service industry are summarized by lucubrate. Through detailed research on the blockchain technology, especially the Consortium Blockchain, we find that its innovation and characteristics meet all the requirements of trusted transactions. After that, we discuss the cryptography and computer principles and technologies behind the three prohibition principles of blockchain supporting trusted transactions in details, implemented on which the smart contract technology is analyzed. On this basis, we put forward the overall framework of trusted transaction based on blockchain technology, and take the e-commerce platform as an example, and the basic design on multilink and microservice framework. Ultimately, we conduct study on big data credit investigation and credit evaluation for trusted transactions under the blockchain system, and make a preliminary discussion on the dynamic penetrating supervision model based on big data of blockchain transactions.

Keywords: Trusted transactions · Blockchain · Three prohibition principles · Penetrating supervision · Smart contract

1 Introduction

As the booming of the global internet, the vast ocean of our universe has become a "global village", and the modern service industry transactions grow increasingly convenient. Although modern information technology is finding wider and wider application in modern communication and transaction, the essential issues of trusted transaction in modern service industry has not been improved substantially. Modern information technology has raised higher requirements for trusted transactions.

In the past, international and domestic transactions with banks as bookkeeping and transfer intermediaries were based on state and bank credit, requiring both parties or multiple parties to trade on the basis of banks, which have three main requirements for each party:

1. Immutability: the transaction accounts can't be changed or altered by either party upon the consensus reached by both parties (signature, seal, etc.).
2. Unforgeable: the transaction accounts cannot be forged by anyone or any institution.
3. Undeniable: the transaction record or contract execution cannot be denied or prevented by either party without the alteration in the case of mutual agreement on the transaction or contract.

The "Three Prohibition Principles" are the basis of trusted transactions. And intermediaries such as banks must ensure that multiple disbursements of funds are prevented.

Blockchain is a new technology applied in Satoshi Nakamoto's paper "Bitcoin: A Peer-to-peer Electronic Cash System" [1], which is becoming the strategic frontier technology that each country competes to develop. As a boom value Internet technology, it is inevitable to combine with modern service industry and internet finance. It technically realizes the "three prohibition principles" (Immutability, Unforgeable, Undeniable) and fundamentally eliminates the "double payment problem" in transactions. In 2015, Ethereum realized smart contracts [2, 3], which pushing blockchain technology on a new level, taking a big step forward for trusted transactions in modern services, and laying a solid foundation for the realization of trusted transactions. Blockchain technology mainly has the following several innovations:

- Model innovation: Blockchain pioneered a transaction model that reduced intermediation, and open up a new model for tamper-modifiable data, undeniable, and traceable applications.
- Technological innovation: Blockchain is a new application model combined by distributed data storage, point-to-point transmission, consensus mechanism, encryption algorithm and other computer technologies with a new structure.
- Application (convergence) innovation: Blockchain technology can be applied in financial transactions, certificate storage, traceability, "Internet of Things", medical and health care, smart city, anti-security identification, supply chain management, data sharing and transaction, digital identity and other fields. The technical characteristics can be used to play an important role in blockchain + (mainly in blockchain technology) or in + blockchain (using one or several characteristics of blockchain technology).

2   Theoretical and Technological Basis of Trust Transactions

Trusted transactions in modern service industry involve internet financial payment, electricity rapid transactions, service smart contract and other modern means. Because of the fund flow and the exchange of goods and services, the process is still necessary to guarantee the transaction and service records unforgeable, immutability, undeniable and traceable in addition to basic safety and efficiency. The emergence of the blockchain, especially the alliance chain, not only realizes the efficient transaction and security guarantee of the modern service industry, but also realizes the "three prohibition principles" of the transaction ledger under the Internet conditions. The main theories and technologies of the blockchain and alliance chain include: modern cryptography theory and technology, consensus algorithm, and distributed complete ledger storage technology. The application of smart contracts on the blockchain provides a reliable basis for the automatic execution of contracts.

2.1 Cryptographic Principle and Technology

In 1976, Whitefield Diffie and Martin Hellman put forward the idea of public key cryptography, which marked the birth of modern cryptography and was a milestone event in the history of international cryptography. Modern cryptography theory and its derived related technologies are one of the three major supporting technologies that constitute the blockchain system, and they are the practical theoretical and technical foundation of the blockchain technology. Blockchain technology relies on encryption technology to achieve data security, digital certificates are mainly used to identity authentication, ensuring that the signed data is undeniable or traceable. Under the alliance chain, these two technologies are internally connected and play a crucial role in the "three prohibition principles" of the blockchain ledger [4, 5].

At present, ECDSA and EdDSA are the most widely used signature algorithms based on ECC in the world. The national standard of digital signature algorithm based on Chinese elliptic curvilinear is SM2, and that of digital certificate is SM9.

2.2 Consensus Mechanism

The consensus mechanism is to formulate a series of rules to act. This rule includes transaction rules and data consistency rules. Trusted transactions in modern services require blockchain systems to solve the compliance and consistency of transaction data. Compliance means that all records entering the ledger must be regular, and irregular transactions cannot be entered or should be marked as invalid. Consistency means that all accounting nodes must keep the same transaction data, only by which transaction data can be ultimately prevented from being tampered with the help of password technology. And the correct transaction data can be effectively and timely restored when part of the node data is illegally tampered with [6]. The main consensus algorithms are:

lottery-based consensus: proof of workload mechanism (POW), proof of equity (POS); voting-based consensus: practical Byzantine general algorithm PBFX, Paxos algorithm; fault-tolerance-based consensus: RAFT and KAFKA.

2.3 Distributed Storage Technology

Distributed storage is one of the other pillar technologies of blockchain systems which decentralize the storage of data in multiple separate devices [7]. Distributed storage technology in the blockchain has the following effects:

Ensure consistency of data;

Prevent illegal tampering;

Guarantee data availability: Prevent transactions system and data from unavailability after failure of some nodes in the system;

Distinguish fault tolerance: Distributed storage transactions system still works when some nodes stop working due to failure.

2.4 Smart Contract

The concept of smart contract was introduced by Nick Szabo in 1995. The development of smart contract has gone through three stages of concept, realization concept and rapid development. Although Nick Szabo proposed and described the application scenarios of smart contracts, he failed to find the right way to implement them. Until 2015, the emergence of Ethereum made smart contracts to be convenient and quickly realized [8, 9].

Smart contracts often require input data to trigger the execution conditions, which and then output the execution results. The traditional Internet uses different network protocols, which leads that the smart contracts cannot directly read the data on the Internet. To solve the problem of data acquisition, the "prediction machine" is needed, which is a middleware connecting blockchain to the traditional Internet. It can access data from the Internet and actively feeds it back to smart contracts on the blockchain.

3 Trusted Transactions Overall Framework

3.1 Overall Framework

The trusted transaction overall framework is designed for enhance the transactions reliability, convenience, accuracy and security in modern service industry. It can greatly enhance the data protection degree and transaction credibility, and its scalable application scenarios can be extended to other industries, such as item traceability, public management, supply chain and other fields. Figure 1 below is the overall framework of trusted transactions.

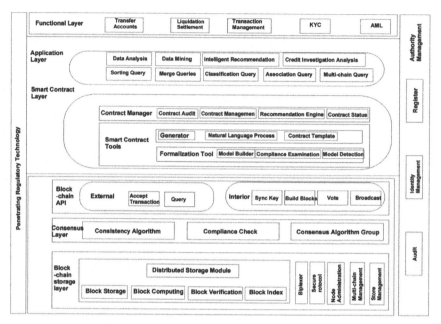

Fig. 1. Overall framework of trusted transactions

3.2 Trusted Transaction Foundation Design Based on Blockchain

At the storage layer, we design multiple chains to achieve a combination of flexibility and efficiency: Membership Blockchain, Transactions Blockchain, Commodity Blockchain, and Evaluation Blockchains:

Member Blockchain (MBC) mainly authorize and manage users by the authority management department. These users participate in transactions through authorized digital certificates and are the subject of trusted transactions.

Transaction Blockchain (Transaction Blockchain, TBC) is used to record the transactions of all parties, which involves in the commodity ID or service ID, the transfer amount of transaction funds, the signature of the transaction, and the signature of the endorsement node in some cases.

Commodity chain (Commodity Blockchain, CBC) is the description of goods, which mainly includes the main parameters of the commodity. For transaction efficiency and transaction block restrictions, transaction chain only recorded the ID in order to be able to trace and to prevent the transaction of commodity description and parameters from different disputes.

Evaluation Blockchain (EBC) mainly records the score and use evaluation by member users. Through the analysis of these data and machine learning methods, services can be provided for the credit and commodity recommendation of merchant members.

On the consensus algorithm selection, our framework is alliance chain, which can choose efficient and safe algorithm as practical Byzantine general (PBFT) consensus algorithm and consistency RAFT algorithm, and do not need to use waste energy workload (POW) algorithm. We can design the algorithm pool, and choose different efficient algorithm according to different needs.

Many services and products need the support of smart contract. Such as Go, java, or everything with a Turing-complete linguistic environment was recommended. Microservice framework provides flexibility, scalability, scalability, and high availability, so smart contracts will be structured based on microservice organizations. The permission management involving the choice of digital certificates and signature algorithms, and has a crucial impact on the security and performance of the whole system. National standard SM2 and SM9 was used as the standards for digital signatures and digital certificates, and the CdDSA algorithm was transformed to improve security and overall system efficiency.

4 Trusted Transactions Services: Big Data Credit Investigation and Credit Evaluation

The credit economy is the advanced form of social economy. Credit assessment by enterprises or individuals is crucial in modern transactions. How to build on modern service transactions based on blockchain? How to establish a real-time modern service credit investigation and credit evaluation structure that can achieve high service credit investigation accuracy and regulatory performance? We discuss the following issues on the basis of the e-commerce platform:

4.1 Research on Dynamic Credit Information Collection, Management and Maintenance Methods of Trusted Transactions in Modern Service Industry

There are many sources of credit investigation and credit data in modern service industry, including transaction data (TBC) basing on blockchain, user basic information (MBC), and others data (including government, bank and other chain data).

By adopting the service credit investigation data sharing mechanism based on the idea of open data, constructing the license alliance chain, collecting static and dynamic data of credit investigation, and integrating strong, medium and weak data sources, redundant data of credit investigation data collection can be reduced, the operation cost of service credit investigation and the cost of resource contention can be reduced, and the problem of service credit investigation information island can be solved. Isolation verification, cross verification, data desensitization and other technologies can improve the application scope of data, and at the same time ensure the security of credit investigation data and the data privacy [10].

In the data preprocessing stage, the unstructured data needs to be defined effectively, transformed into structured data, and normalized. For structured data, ETL technology is used to extract and put characteristic values with high correlation between credit investigation and credit into the blockchain of credit investigation after numerical transformation to prevent artificial tampering. For the new credit investigation data obtained in the future, it can be added to the subsequent blockchain of credit investigation and indexed according to the account address to prevent the disclosure of sensitive data.

4.2 Research on Establishing Dynamic Service Credit Evaluation Model by Using On-Chain Transaction Data and Online Available Credit Data Through Regression, Bayesian, Decision Tree and Other Machine Learning Methods

The processing framework based on feature selection is adopted to mainly analyze multi-trusted data sources. The choice of model training adopts regression, Bayesian classification algorithm, decision tree and deep learning algorithm based on artificial neural network or the combination of more than two algorithms. Through a large amount of data training and comparison, high accuracy models are deployed in the credit investigation framework and regulatory framework.

Applying the above credit prediction method and the data analyze processing platform, and using techniques such as statistics, machine learning algorithm to evaluate risk, we can increase the analysis of customer needs. Through multi-angle and multi-dimensional segmentation of user groups, we make a comprehensive analysis of users' integrity as far as possible, and help credit investigation agencies to control risk, cross check, prevent fraud and evaluate credit.

4.3 Research on Applying Enterprise or Individual Credit Information Obtained from Evaluation Model to Modern Service Industry Supervision, and Re-establishing Dynamic Penetrating Supervision Model Based on Blockchain Transaction

High-credit entities can be given a higher credit limit in supervision and increase the amount and cycle of payments in smart contract validation. In the case of credit-crossing entities, transactions will be reduced or banned, and the application of smart contracts will be subject to stricter rules. By adding the supervision mechanism into the consensus algorithm, the credit investigation model and credit evaluation results can be supervised in real time.

4.4 Research on Dynamic Updating the Credit Model Mechanism and the Real-Time Credit Supervision Model

In order to ensure the real-time credit evaluation of the entity (enterprise and individual users), the credit evaluation system will update the users credit rating timely when new entity data and transactions are generated. The framework is designed with high performance to ensure that high-frequency transactions runs smoothly. Meanwhile, higher security and privacy are needed to protect data. We improved the performance by studying the storage methods on and off the chain to reduce the data size on the chain, higher performance can be realized by isolation validation, sharding, layering or a combination of multiple methods.

5 Summary

This paper discussed on trusted transaction framework of blockchain, theoretical and technical support foundation, smart contract technology application, trusted big data credit investigation, credit evaluation model and other models, we conduct research on the improvement of the credibility, convenience, scalability, supervision, accuracy and security of transactions in modern service industry, and lays a solid foundation for the further establishment of modern service industry trusted transactions platform to provide theoretical and technical support.

Acknowledgement. This work was supported by National Key R&D Program of China (2018YFB1402701).

References

1. Nakamoto, S.: Bitcoin: A Peer-to-Peer Electronic Cash System (2008)
2. Buterin, V.: Ethereum: a next-generation smart contract and decentralized applica-tion platform (2015). https://github.com/ethereum/wiki/wiki/White-Paper. Last accessed 14 Apr 2022
3. Tern, S.: Survey of smart contract technology and application based on blockchain. Open J. Appl. Sci. **11**, 1135–1148 (2021)
4. Diffie, W., Hellman, M.: New directions in cryptography. IEEE Trans. Inf. Theor. **22**(6), 644–654 (1976)
5. Hook, D., Eaves, J.: Java Cryptography Tools and Techniques. Leanpub (2019)
6. Hyperledger Architecture, Vol. 1: Introduction to Hyperledger Business Blockchain Design Philosophy and Consensus. https://www.hyperledger.org/learn/white-papers. Last accessed 14 Apr 2022
7. Zhao, X., Lei, Z., Zhang, G., Zhang, Y., Xing, C.: Blockchain and distributed system. In: Wang, G., Lin, X., Hendler, J., Song, W., Xu, Z., Liu, G. (eds.) WISA 2020. LNCS, vol. 12432, pp. 629–641. Springer, Cham (2020). https://doi.org/10.1007/978-3-030-60029-7_56
8. Szabo, N.: The idea of smart contracts (1997). https://nakamotoinstitute.org/the-idea-of-smart-contracts/. Last accessed 14 Apr 2022
9. Hyperledger Architecture, Vol. II: Smart Contracts. https://www.hyperledger.org/learn/white-papers. Last accessed 14 Apr 2022
10. Mao, X., Li, X., Guo, S.: A blockchain architecture design that takes into account privacy protection and regulation. In: Xing, C., Fu, X., Zhang, Y., Zhang, G., Borjigin, C. (eds.) WISA 2021. LNCS, vol. 12999, pp. 311–319. Springer, Cham (2021). https://doi.org/10.1007/978-3-030-87571-8_27

Causal Effect Estimation Using Variational Information Bottleneck

Zhenyu Lu[1], Yurong Cheng[1(✉)], Mingjun Zhong[2], George Stoian[2], Ye Yuan[1], and Guoren Wang[1]

[1] School of Computer Science, Beijing Institute of Technology, Beijing 100081, China
{yrcheng,yuan-ye}@bit.edu.cn
[2] Department of Computing Science, University of Aberdeen, Aberdeen, UK
mingjun.zhong@abdn.ac.uk, george_stoian@protonmail.ch

Abstract. Causal inference is to estimate the causal effect in a causal-relationship when intervention is applied. Precisely, in a causal model with binary interventions, i.e., control and treatment, the causal effect is simply the difference between the factual and counterfactual. The difficulty is that the counterfactual may never been obtained which has to be estimated and so the causal effect could only be an estimate. The key challenge for estimating the counterfactual is to identify confounders which effect both outcomes and treatments. A typical approach is to formulate causal inference as a supervised learning problem and so counterfactual could be predicted. Including linear regression and deep learning models, recent machine learning methods have been adapted to causal inference. In this paper, we propose a method to estimate Causal Effect by using Variational Information Bottleneck (CEVIB). The promising point is that VIB could be able to naturally distill confounding variables from the data, which enables estimating causal effect by only using observational data. We have compared CEVIB to other methods by applying them to three data sets showing that our approach achieved the best performance.

Keywords: Causal inference · Causal effect · Variational information bottleneck · Confounding variables · Intervention

1 Introduction

Causal inference [7,11,12] is a task to estimate the effect in a causal relationship when an intervention action is made, which could be used to guide decision-making. In general, causal inference is to estimate the level of outcome changes when their causes are intervened. For example, in medical science we require to estimate the causal effect of a treatment to understand the effectiveness of the treatment applied for a particular disease; in social media, we could use causal inference to infer key users from chats and web data [2,5]. Causal inference is hard because we could only observe the factual, but would never be able to obtain the counterfactual. The key to estimate the causal effect is to estimate the

X. Zhao et al. (Eds.): WISA 2022, LNCS 13579, pp. 288–296, 2022.
https://doi.org/10.1007/978-3-031-20309-1_25

counterfactual, and so the causal effect is simply the difference between factual and counterfactual. A possible approach for causal inference is the randomized controlled trails (RCT), which is a standard approach in clinical trials. But it could cost expensive and would be impossible in some situations. In RCT, large numbers of participants may be required to control the effect of confounding factors. It is crucial to identify the confounding variables and then eliminate their effects on the outcomes so that the effect of the treatment could be correctly estimated.

Simply, causal inference could be formulated as a linear regression model, where the outcome is represented as the linear regression of the treatment and any other feature variables including confounding variables [7]. The major disadvantage of linear regression approach is its limited ability to handle big data, which is the favorable circumstance in the big data era. Recent advances in deep learning techniques provide an excellent opportunity to utilise large observational data to estimate causal effects. In literature, a couple of deep neural network (DNN) approaches have been proposed for causal inference and all these methods attempt to represent the causal inference as a supervised learning problem. The important aspect for such approach is that once the causal inference is posed as a supervised learning problem, the established supervised learning algorithms could then be easily applied boosting the research in this area.

In this paper, we consider to use neural network inspired by variational information bottleneck [1] to not only fit a model to the observed data, but also address hidden confounders that affects treatments or outcomes. Based on this approach, we develop a novel regularization framework that forces the model to "forget" some hidden parameters which are not confounders and learning to extract the confounding variables from data. We present experiments in Sect. 5 that demonstrate the advantages of this model and show that it outperforms state-of-the-art models in a variety of datasets.

2   Causal Effect Models

Suppose that a dataset $D = \{X_i, T_i, Y_i\}_{i=1}^N$ is given, where X represents the covariates of a subject, e.g., the health status; T represents the treatment applied to the patient, e.g., a medication; and Y represents the outcome after the treatment is applied. Note that we consider a binary variable for T and so $T \in \{0, 1\}$. For a control subject i, since $T_i = 0$ is implemented, the outcome y_{i0} is called the factual outcome which is observed; however, $T_i = 1$ was never implemented, the potential outcome y_{i1} is called the counterfactual outcome which is thus never able to be observed. For computing the causal treatment effect for a subject i, i.e., $y_{i1} - y_{i0}$, it is required to know both factual and counterfactual outcomes of a subject, but the counterfactual outcome is not obtained. Our objective is to estimate the average treatment effect (ATE) denoted by ψ and individual treatment effect (ITE) when a treatment T is applied to a patient.

For a subject i with a d-dimensional covariate $X_i \in R^d$ and its outcome being $Y_i \in R^1$ after a treatment T_i is applied, we assume that the observed covariates

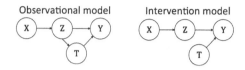

Fig. 1. The observation model (left) and the intervention model (right). X are covariates, Z are the latent confounders, T is a treatment, and Y is an outcome.

X_i include all possible causes for outcome and treatment. Those causal variables for both treatment and outcome are hard to identify. We therefore assume the confounders to be a collection of latent variables which could be distilled from covariates. Figure 1 (left) represents an observational data model using this assumption, in which X, T and Y are observed, and Z are the confounders. To estimate causal effect, it needs to use the intervention model (the right graph in Fig. 1), in which the intervention treatment is applied which is not effected by any confounder.

So the ITE could be represented as $ITE(x_i) = E_{Y|X,do(T_i=1)}[Y|X = x_i, do(T_i = 1)] - E_{Y|X,do(T_i=0)}[Y|X = x_i, do(T_i = 0)]$. The ATE is then the expectation with respect to all the individuals and so $\psi = E_X[ITE(X)] = E_X[E_{Y|X,do(T_i=1)}[Y|X = x_i, do(T_i = 1)] - E_{Y|X,do(T_i=0)}[Y|X = x_i, do(T_i = 0)]]$.

3   Related Works

In this section, we briefly discuss causal effect estimation methods. Statistical methods have been proposed for causal effect estimation. For example, regression methods fit the treatment assignment and as well as the covariates to represent the outcomes [3,11]. Sample re-weighting methods aim to correct the treatment assignment using observational data in order to overcome the subject selection bias [13]. Other methods include doubly-robust approaches which combine covariate adjustment with propensity score weighting [3,4].

In recent years, many machine learning methods have been applied to the problem of estimating the potential outcomes and treatment effects. Most of them could be viewed as discriminative (or supervised) learning methods. For example, Bayesian Additive Regression Trees (BART) [6] and Causal Forests [17] have been used to estimate causal effect. The work in [8] proposed Balanced Linear Regression (BLR) and Balanced Neural Networks (BNN) to learn a balanced covariate representation for causal outcomes. Other approaches include GANITE which uses generative adversarial networks [19] to estimate causal effect.

Methods that have similarities with our CEVIB are [10], CEVAE [9] and Dragonnet [15]. CEVAE and Dragonnet have a similar neural networks structure to that of TARNet [14]. Note that the method proposed in [10] is also using VIB, but it is using a different architecture to ours and also the main aim of that approach is to deal with missing data.

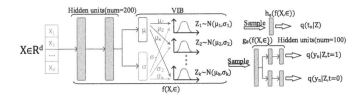

Fig. 2. The architecture framework for estimating causal effect using variational information bottleneck.

4 Variational Information Bottleneck for Causal Inference

Under the sufficient assumptions provided in Sect. 2, the observed data $D = \{X, T, Y\}$ were generated from the observational data model illustrated in Fig. 1 with confounding variables Z. According to our observational data model, we hope that all the confounding variables could be distilled from the data X because X contain all the possible confounders. We employ the idea of variational information bottleneck (VIB) which attempts to distill sufficient information from X which maximumly inform both T and Y and our VIB network architecture (CEVIB) is shown in the Fig. 2. Our framework is most relevant to the Dragonnet [15]. CEVIB provides an end-to-end procedure for predicting treatment and outcome. We firstly use neural network to learn the representation $Z(x) \sim N(\mu^k, \sigma^k)$ following a Normal distribution, where k denotes the dimension of hidden confounders. The hidden confounders are regularized by information bottleneck and predict the treatment and outcome.

4.1 Training the Observational Model

In this section, we apply the idea of VIB to train the observation model using the data. Denote $\widetilde{Y} = \{Y, T\}$. Our aim is to optimize the following problem which attempts to maximizing the mutual information between $\widetilde{Y}$ and the confounders Z whist threshold the mutual information between the covariates X and Z:

$$\max_{\theta} I(Z, \widetilde{Y}; \theta), \text{subject to}, I(X, Z; \theta) \leq I_c, \tag{1}$$

where $I(Z, \widetilde{Y}; \theta) = \int p(Z, \widetilde{Y}) \log \frac{p(Z, \widetilde{Y})}{p(Z)p(\widetilde{Y})} d\widetilde{Y} dZ$ represents the mutual information, and I_c is a constant. This is equivalent to maximizing the following objective function $R_{IB}(\theta) = I(Z, \widetilde{Y}; \theta) - \beta I(X, Z; \theta)$. Instead we could maximize the following lower bound

$$R_{IB}(\theta) = I(Z, \widetilde{Y}; \theta) - \beta I(X, Z; \theta)$$
$$\geq \int p(x)p(\widetilde{y}|x)p(z|x) \log q(\widetilde{y}|z) dx d\widetilde{y} dz - \beta \int p(x)p(z|x) \log \frac{p(z|x)}{r(z)} dx dz$$
$$\approx \frac{1}{N} \sum_{n=1}^{N} \left\{ \int p(z|x_n) \left[\log q(\widetilde{y_n}|z) - \beta \log \frac{p(z|x_n)}{r(z)} \right] dz \right\} = L. \tag{2}$$

where $q(y|z)$ is a variational approximations to $p(y|z)$ and $r(z)$ is a variational approximation to $p(z)$. In the causal effect model (see the observation model in Fig. 1), our outputs given the latent variable has the form $q(\widehat{y}|z) = q(y|z, t = 1)^t q(y|z, t = 0)^{1-t} q(t|z)$. So the lower bound can be represented as

$$
L = \frac{1}{N} \sum_{n=1}^{N} \left\{ \int p(z|x_n) \left[t_n \log q(y_n|z, t_n = 1) + (1 - t_n) \log q(y_n|z, t_n = 0) + \log q(t_n|z) - \beta \log \frac{p(z|x_n)}{r(z)} \right] dz \right\}
$$

$$
= \frac{1}{N} \sum_{n=1}^{N} \left\{ \int p(z|x_n) \left[t_n \log q(y_n|z, t_n = 1) + (1 - t_n) \log q(y_n|z, t_n = 0) + \log q(t_n|z) \right] dz + \beta K L[p(z|x_n)||r(z)] \right\}
$$

$$
= L_1 + L_2 \tag{3}
$$

where $L_1 = \frac{1}{NM} \sum_{nm} [t_n \log q(y_n|z_{nm}, t_n = 1) + (1 - t_n) \log q(y_n|z_{nm}, t_n = 0) + \log q(t_n|z_{nm})]$ where $z_{nm} = f(x_n, \epsilon_m)$ and $L_2 = \frac{\beta}{N} \sum_{n=1}^{N} KL[p(z|x_n)||r(z)]$. Suppose the latent variable z has dimension K, we assume a diagonal multivariate Normal distribution, $p(z|x_n) = \prod_{k=1}^{K} p(z_k|x_n) = \prod_{k=1}^{K} N(z_k|\mu_k(x_n), \sigma_k^2(x_n))$ where μ_k and σ_k are neural networks. Therefore, we can use change of variables to draw a sample $z_k = f_k(x_n, \epsilon) = \mu_k(x_n) + \sigma_k(x_n)\epsilon$ where $\epsilon \sim N(0, 1)$. Denote $f(x_n, \epsilon) = [f_1(x_n, \epsilon), \cdots, f_K(x_n, \epsilon)]^T$. This gives the $L1$. We define the following distributions for outcome and treatment variables: $\log q(y_n|f(x_n, \epsilon_m), t_n) \approx -(y_n - g_\theta^{t_n}(f(x_n, \epsilon)))^2$ and $\log q(t_n|f(x_n, \epsilon_m)) \approx -(t_n - h_\theta(f(x_n, \epsilon)))^2$.

Therefore, our task is then to learn the neural networks $g_\theta^{t_n}(.)$, $h_\theta(.)$, $\mu_k(.)$ and $\sigma_k(.)$. For the L_2, we use the identity $KL(N(\mu, \Sigma)||N(0, 1)) = \frac{1}{2} \sum_{k=1}^{K} (\sigma_k^2 + \mu_k^2 - 1 - ln(\sigma_k^2))$ where $\mu = (\mu_1, \cdots, \mu_K)$ and $\Sigma = diag(\sigma_1^2, \cdots, \sigma_K^2)$. So $L_2 = \frac{\beta}{N} \sum_{n=1}^{N} \left[\frac{1}{2} \sum_{k=1}^{K} (\sigma_k^2(x_n) + \mu_k(x_n) - 1 - ln(\sigma_k^2(x_n))) \right]$.

Our CEVIB is to train to maximize L_1 and L_2 using observational data, and so all the conditional distributions are trained, which can then be used in the intervention model to estimate causal effects.

4.2 Estimating Causal Effects

After training, the average treatment effect τ can be calculated using $\tau = \int Y p(Y|do(T = 1)) dY - \int Y p(Y|do(T = 0)) dY$. We can then apply the $do-calculus$ according to the intervention model structure. $p(Y|do(T = t)) = \int p_{do(T=t)}(Y, Z, X, T = t) dZ dX = \int p_{do(T=t)}(Y|Z, X, T = t) p_{do(T=t)}(Z, X, T = t) dZ dX = \int p_{do(T=t)}(Y|Z, T = t) p(Z|X) p(X) p_{do(T=t)}(T = t) dZ dX = \int p(Y|Z, T = t) p(Z|X) p(X) dZ dX \approx \int q(Y|Z, T = t) p(Z|X) p(X) dZ dX$.

Where we have applied the approximation $p(Y|Z, T = t) \approx q(Y|Z, T = t)$. Note that all these conditionals involving computing $p(Y|do(T = t))$ have been trained using observational model. We then draw samples from $p(Y|do(T = t))$ which will be used to calculate the average treatment effect. To draw samples from $p(Y|do(T = t))$, we do the following process

- $x^{(i)} \sim p(X)$, i.e., draw a sample $x^{(i)}$ from $p(X)$
- $z^{(i)} \sim p(Z|x^{(i)})$
- $y_t^{(i)} \sim q(Y|z^{(i)}, T = t)$

This provides a Monte Carlo estimate for the average treatment effect $\hat{\tau} = \frac{1}{N_1} \sum_{i=1}^{N_1} y_1^{(i)} - \frac{1}{N_0} \sum_{i=1}^{N_0} y_0^{(i)}$. We will use this method to estimate causal effects in the following experiments.

Table 1. The results of applying various models to IHDP and Twins data. Best results are denoted in bold. Note that $within - s$ means that training data were used for both training and prediction and $out - of - s$ means that the models were trained on training data and tested on the test data.

	Datasets(Mean +- std)							
	IHDP				Twins			
Method	$\sqrt{\epsilon_{PEHE}^{within-s}}$	$\epsilon_{ATE}^{within-s}$	$\sqrt{\epsilon_{PEHE}^{out-of-s}}$	$\epsilon_{ATE}^{out-of-s}$	$\sqrt{\epsilon_{PEHE}^{within-s}}$	$\epsilon_{ATE}^{within-s}$	$\sqrt{\epsilon_{PEHE}^{out-of-s}}$	$\epsilon_{ATE}^{out-of-s}$
BLR	$5.8\pm.3$	$.72\pm.04$	$5.8\pm.3$	$.93\pm.05$	$.312\pm.003$	$.0057\pm.0036$	$.323\pm.018$	$.0334\pm.0092$
BART	$2.1\pm.1$	$.23\pm.01$	$2.3\pm.1$	$.34\pm.02$	$.347\pm.009$	$.1206\pm.0236$	$.338\pm.016$	$.1265\pm.0234$
CF	$3.8\pm.2$	$.18\pm.01$	$3.8\pm.2$	$.40\pm.03$	$.366\pm.003$	$.0286\pm.0035$	$.316\pm.011$	$.0335\pm.0083$
BNN	$2.2\pm.1$	$.37\pm.03$	$2.1\pm.1$	$.42\pm.03$	$.325\pm.003$	$.0056\pm.0032$	$.321\pm.018$	$.0203\pm.0071$
CFRW	$\mathbf{.71\pm.0}$	$.25\pm.01$	$\mathbf{.76\pm.0}$	$.27\pm.01$	$.315\pm.007$	$.0112\pm.0016$	$.313\pm.008$	$.0284\pm.0032$
CEVAE	$2.7\pm.1$	$.34\pm.01$	$2.6\pm.1$	$.46\pm.02$	$.341\pm.006$	$.0065\pm.0040$	$.373\pm.012$	$.0679\pm.0212$
TARNet	$.88\pm.0$	$.26\pm.01$	$.95\pm.0$	$.28\pm.01$	$.317\pm.005$	$.0108\pm.0017$	$.315\pm.003$	$.0151\pm.0018$
GANITE	$1.9\pm.4$	$.43\pm.05$	$2.4\pm.4$	$.49\pm.05$	$\mathbf{.289\pm.005}$	$.0058\pm.0017$	$\mathbf{.297\pm.016}$	$.0089\pm.0075$
Dragonnet	$1.31\pm.4$	$.14\pm.02$	$1.32\pm.5$	$.21\pm.04$	$.322\pm.001$	$.0092\pm.0078$	$.317\pm.001$	$.0074\pm.0092$
Dragon-tarreg	$1.22\pm.3$	$.14\pm.01$	$1.30\pm.3$	$.20\pm.05$	$.322\pm.0017$	$.0060\pm.0088$	$.318\pm.002$	$.0060\pm.0101$
CEVIB	$.85\pm.1$	$\mathbf{.12\pm.01}$	$.92\pm.2$	$\mathbf{.15\pm.02}$	$.320\pm.0003$	$\mathbf{.0016\pm.0009}$	$.315\pm.001$	$\mathbf{.0055\pm.0018}$

5   Experiments

5.1   The Data Sets and Training Setup

Due to the difficulty of gathering treatment effects factuals for both control and treatment, evaluating the methods for estimating causal effects may have to use synthetic or semi-synthetic data sets. Our experiments will apply CEVIB to three semi-synthetic benchmark data sets: **IHDP**[1] **Twins**[2] and **ACIC**[3]

To train CEVIB, we use the architecture in Fig. 2. For both IHDP and ACIC experiments, we randomly split each file data into test/validation/train with proportion 63/27/10 and repeat the procedure for 25 times. For Twins experiments, we randomly split the data into test/validation/train with proportion 56/24/20 and repeat the procedure for 50 times.

To evaluate the performance, we report the results of the following metrics for each data set, which are the absolute error in average treatment effect

---

[1] IHDP data is available at https://github.com/Osier-Yi/SITE/tree/master/data [18].

[2] Twins data is at https://github.com/jsyoon0823/GANITE/tree/master/data. [19].

[3] We use the scaling folder in ACIC to evaluate our methods. The ACIC data set is available at https://github.com/IBM-HRL-MLHLS/IBM-Causal-Inference-Benchmarking-Framework/tree/master/data/LBIDD. [16].

$\epsilon_{\text{ATE}} = |\frac{1}{N} \sum_{i=1}^{N} (y_1^{(i)} - y_0^{(i)}) - \frac{1}{N} \sum_{i=1}^{N} (\hat{y}_1^{(i)} - \hat{y}_0^{(i)})|$, the Precision in Estimation of Heterogeneous Effect (PEHE) $\epsilon_{\text{PEHE}} = \frac{1}{N} \sum_{i=1}^{N} ((y_1^{(i)} - y_0^{(i)}) - (\hat{y}_1^{(i)} - \hat{y}_0^{(i)}))^2$.

We compared CEVIB with Dragonnet [15] and causal effect variational auto-encoder(CEVAE) [9]. We also compared to Dragon-tarreg which is a Dragonnet using a targeted regularization method based on the augmented inverse probability weighted(AIPW) estimator. In addition, we also compared CEVIB with BLR [8], BART [6], CForest [17], BNN [8], CFR_{wass} and $TARNet$ [14], and GANITE [19].

Table 2. The results for the methods applied to ACIC data.

Method	ACIC(Mean +- std)			
	$\sqrt{\epsilon_{PEHE}^{within-s}}$	$\epsilon_{ATE}^{within-s}$	$\sqrt{\epsilon_{PEHE}^{out-of-s}}$	$\epsilon_{ATE}^{out-of-s}$
CEVAE	131 ± 1.8	50.2 ± 1.3	118 ± 2.2	51.9 ± 1.4
Dragonnet	67.8 ± 2.0	25.8 ± 2.1	68.2 ± 2.6	26.3 ± 2.1
Dragon-tarreg	62.5 ± 2.0	24.8 ± 3.1	62.2 ± 2.4	24.5 ± 3.0
CEVIB	$\mathbf{43.9 \pm 0.8}$	$\mathbf{14.2 \pm 1.7}$	$\mathbf{44.3 \pm 1.8}$	$\mathbf{14.7 \pm 1.9}$

5.2   Results

For comparison purpose, various methods described in the previous subsection are applied to the data sets IHDP and Twins. For the error metrics ATE and PEHE, both the within samples and out of samples were computed. The results are shown in Table 1 which shows that our CEVIB outperforms all of other methods on the data IHDP in terms of ATE and very competitive in terms of PEHE. On the Twins data set, it indicates that CEVIB has one of the best results. Due to computational limits, on the ACIC data, we compared CEVIB to CEVAE, Dragonnet, and Dragon-tarreg. The results are shown in the Table 2. The results clearly indicate that our CEVIB outperforms all the other methods across all the error metrics.

6   Conclusions

In this paper, we have proposed a variational information bottleneck (VIB) approach to estimate causal effects. The interesting point of using VIB is that VIB is able to distill compact information representing the latent confounders from data. Representing confounding variables is important because the effect of confounding variables on outcome can then be integrated out so that the effect of treatment could be purified. We showed that causal inference could be represented as a prediction problem. We have used VIB to train a model to predict both factual and counterfactual outcomes and so the causal effect can be calculated. The proposed algorithm CEVIB was applied to three data sets and compared to other methods showing that our method outperformed other approaches in most of the experiments.

References

1. Alemi, A.A., Fischer, I., Dillon, J.V., Murphy, K.: Deep variational information bottleneck. In: Proceedings of the International Conference on Learning Representations (ICLR) (2017)
2. Alvari, H., Shaabani, E., Sarkar, S., Beigi, G., Shakarian, P.: Less is more: Semi-supervised causal inference for detecting pathogenic users in social media. In: Companion Proceedings of The 2019 World Wide Web Conference, pp. 154–161 (2019)
3. Athey, S., Imbens, G., Pham, T., Wager, S.: Estimating average treatment effects: supplementary analyses and remaining challenges. Am. Econ. Rev. **107**(5), 278–81 (2017)
4. Benkeser, D., Carone, M., Laan, M.V.D., Gilbert, P.: Doubly robust nonparametric inference on the average treatment effect. Biometrika **104**(4), 863–880 (2017)
5. Chen, H., Dong, Y., Gu, Q., Liu, Y.: An end-to-end deep neural network for truth discovery. In: Wang, G., Lin, X., Hendler, J., Song, W., Xu, Z., Liu, G. (eds.) WISA 2020. LNCS, vol. 12432, pp. 377–387. Springer, Cham (2020). https://doi.org/10.1007/978-3-030-60029-7_35
6. Chipman, H.A., George, E.I., McCulloch, R.E.: Bart: Bayesian additive regression trees. Annals Appl. Stat. **4**(1), 266–298 (2010)
7. Gelman, A., Hill, J.: Data analysis using regression and multilevel/hierarchical models. Cambridge University Press (2006)
8. Johansson, F., Shalit, U., Sontag, D.: Learning representations for counterfactual inference. In: Proceedings of Machine Learning Research, vol. 48, pp. 3020–3029. PMLR, New York, New York, USA (2016)
9. Louizos, C., Shalit, U., Mooij, J., Sontag, D., Zemel, R., Welling, M.: Causal effect inference with deep latent-variable models. In: Proceedings of the 31st International Conference on Neural Information Processing Systems, pp. 6449–6459 (2017)
10. Parbhoo, S., Wieser, M., Wieczorek, A., Roth, V.: Information bottleneck for estimating treatment effects with systematically missing covariates. Entropy **22**(4) (2020)
11. Pearl, J.: Causality. Cambridge University Press, Cambridge (2009)
12. Peters, J., Janzing, D., Schölkopf, B.: Elements of causal inference: foundations and learning algorithms. The MIT Press (2017)
13. Rosenbaum, P.R.: Model-based direct adjustment. J. American Stat. Assoc. **82**(398), 387–394 (1987)
14. Shalit, U., Johansson, F.D., Sontag, D.: Estimating individual treatment effect: generalization bounds and algorithms. In: International Conference on Machine Learning, pp. 3076–3085. PMLR (2017)
15. Shi, C., Blei, D.M., Veitch, V.: Adapting neural networks for the estimation of treatment effects. In: Proceedings of the 33rd International Conference on Neural Information Processing Systems, pp. 2507–2517 (2019)
16. Shimoni, Y., Yanover, C., Karavani, E., Goldschmnidt, Y.: Benchmarking framework for performance-evaluation of causal inference analysis. arXiv e-prints, pp. arXiv-1802 (2018)

17. Wager, S., Athey, S.: Estimation and inference of heterogeneous treatment effects using random forests. J. American Stat. Assoc. **113**(523), 1228–1242 (2018)
18. Yao, L., Li, S., Li, Y., Huai, M., Gao, J., Zhang, A.: Representation learning for treatment effect estimation from observational data. Adv. Neural Inform. Process. Syst. **31**, 2633–2643 (2018)
19. Yoon, J., Jordon, J., Van Der Schaar, M.: Ganite: Estimation of individualized treatment effects using generative adversarial nets. In: International Conference on Learning Representations (2018)

Fault Diagnosis of Web Services Based on Feature Selection

Yue-Mei Xi, Zhi-Chun Jia$^{(\boxtimes)}$, Fei-Xiang Diao, Yun-Shuo Liu, and Xing Xing

School of Information Science and Technology, Bohai University, Jinzhou 121013, China
jiazhichun@qymail.bhu.edu.cn

Abstract. In order to better perform fault diagnosis on web services and help users to accurately detect service faults, this paper proposes a web service fault diagnosis model based on anomaly detection (ADWSFD). At present, the outlier detection integration framework of embedded feature selection, which combines outlier scoring and feature selection, plays an important role in detecting outlier performance. We use this framework to score failures of the services participating in the experiment. Specifically, by unifying the attribute selection and fault scoring into a loss function of pairwise comparison and sorting, a reliable service attribute subset is established, and failure scoring of services based on it. In order to improve the reliability of the model, we propose to use a self-paced learning algorithm to achieve service attribute weighting. On this basis, we use the distance-based service fault scoring method to judge its impact on service fault diagnosis, and the validity of the web service fault diagnosis model based on anomaly detection is verified through experimental analysis.

Keywords: Web service · Fault diagnosis · Abnormal detection · Self-paced learning · Nearest neighbour

1 Introduction

In recent years, service providers have realized that providing value-added services on the Internet can bring potential high profits. In the process of service work, many service failures will be encountered. The failure is an abnormal condition that may lead to failure at the component, equipment or shortcoming [1]. Most fault detection methods focus on supervised learning [2, 3], however, in practical applications, collecting all abnormal or faulty data is very difficult, so this paper proposes an unsupervised learning service fault diagnosis model.

This paper proposes a web service fault detection model based on anomaly detection [4–6], by unifying attribute selection and fault scoring into a loss function of pairwise comparison, a reliable subset of service attributes is established [7], service fault scoring is carried out based on it. And we use self-learning [8] method to realize service attribute weighting, and use distance-based fault scoring method to judge its impact on service fault diagnosis. The main contributions of this paper are as follows:

© The Author(s), under exclusive license to Springer Nature Switzerland AG 2022
X. Zhao et al. (Eds.): WISA 2022, LNCS 13579, pp. 297–304, 2022.
https://doi.org/10.1007/978-3-031-20309-1_26

- This paper uses an outlier detection integrated framework with embedded feature selection to make service fault diagnosis for abnormal services.
- This paper uses LeSiNN to calculate service failure scores.
- We derive a self-paced learning attribute selection algorithm to eliminate the negative effects of unreliable attribute.

2 Related Work

At present, according to the characteristics of service faults, researchers have proposed many diagnostic frameworks and detection and diagnosis for service faults.

Bui et al. [2] proposed a fault detection method for multi-layer web applications in cloud computing based on fuzzy one-class support vector machine. Using fuzzy logic membership functions, they designed an OCSVM [9] to reduce the effect of outliers.

Cai, Li et al. [10] propose a graph matching-based empirical method for RCA to automatically identify which applications bring performance anomalies in a service in near real-time without the need to insert application code, and introduce them in the graph. Relative importance weights to locate key nodes and key lines.

Zahid [11] formalized the collaborative distributed business process (BP) fault resolution problem, they leverage information from existing fault-free BPs that use similar services to resolve faults in user-developed BPs.

3 Preliminary Study

3.1 Self-paced Learning

A self-paced learning algorithm [8] is to select samples with high confidence that the predicted value is close to the real value [12]. There is training set $D = (x_1, y_1), (x_2, y_2), \ldots, (x_n, y_n)$ and learning model z.

$$\min Z(w, v) = \sum_{i=1}^{n} v_i z(x_i, y_i, w) - \lambda \sum_{i=1}^{n} v_i \tag{1}$$

where w are the parameters to be optimized, and λv_i is the self-step regularization term. If w is fixed, the optimal v_i is determined by the fixed form:

$$v_i = \begin{cases} 1, & z(x_i, y_i, w) < \lambda \\ 0, & otherwise \end{cases} \tag{2}$$

4 Web Service Fault Diagnosis Model

4.1 Definitions and Notations

$S = \{s_1, s_2, \ldots, s_n\}$ is the set of services in the model, where n is the number of services, $s_i = \{s_{i1}, s_{i2}, \ldots, s_{il}\}$ for each service, and l is the number of service attribute.

We initially divide services into two groups, faulty service $S^+ = \{s_1^+, s_2^+, \ldots, s_{n^+}^+\}$ and normal service $S^- = \{s_1^-, s_2^-, \ldots, s_{n^-}^-\}$, n^+ and n^- are the number of faulty services and normal services, respectively. There are relatively important attributes in the service, and there are unimportant attributes, so set the service attribute weight vector: $W = \{w_1, w_2, \ldots, w_t\}$, t is the number of groups of service attribute weights, where one of the set of service attribute weights is: $w_1 = \{w_{11}, w_{12}, \ldots, w_{tl}\}$. $E = \{e_1, e_2, \ldots, e_t\}$ is the loss function for service attribute weights. Service Failure Score: $FS = \{fs_1, fs_2, \ldots, fs_n\}$. Rely on service failure score FS.

4.2 The Framework of ADWSFD

The ADWSFD model first obtains the initial service failure score, and selects the faulty service candidate set according to the score, and then compares the two services and attribute selection algorithm based on self-paced learning. After screening, the group reliable service attribute weight is obtained, and then the service failure scoring and service attribute selection are then combined, a reliable service failure score is finally obtained and ranked.

Web Service Failure Scoring Algorithm. In this paper we use the distance-based fault point detector LeSiNN [13] to score fault values for services.

$$f\_s(p, s) = \frac{1}{n} \sum_{i=1}^{n} nn\_dist_i(p, s_i) \tag{3}$$

where $nn\_dist_i(p, s_i)$ is the nearest neighbor distance. The service attribute weight is embedded into the fault scoring function $f\_s(p, s) \rightarrow f\_s(p, s, w)$. We use the Euclidean distance [14] formula to calculate distance.

$$dist(s_i, s_j, w) = \sqrt{\sum_{k=1}^{l} w_k * (x_{ik} - x_{jk})^2} \tag{4}$$

Service Attribute Selection. Next, we introduce the specific algorithm.

Filter the Initial Set of Failed Service Candidates. Using the threshold set by Chebyshev's inequality [15] to filter out a set of failure service candidates [16].

$$S^+ = (s, p|f\_s(p, s) > \mu + \varepsilon\sigma, s \in S) \tag{5}$$

where μ and σ are the mean and standard deviation of the initial fault value score $f\_s(p, s)$, respectively, and $\varepsilon \geq 0$ is a user-defined threshold rate.

Comparing the Two Services to Obtain the Loss Function. We design a method to compare the ranking of the two services [17] to construct the loss function that maximizes the score of the faulty service higher than the score of the normal service:

$$H(f\_s) = \frac{1}{n^+ n^-} \sum_{i=1}^{n^+} \sum_{j=1}^{n^-} \theta(f\_s(p, s_i^+) > f\_s(p, s_j^-)) \tag{6}$$

$$\theta = \begin{cases} 1, f\_s(p, s_i^+) > f\_s(p, s_j^-) \\ 0, \qquad otherwise \end{cases} \tag{7}$$

In this paper we use the unlabeled service S instead of S^-, We define the selected data as $\overset{\Delta}{S^+} = \{\overset{\Delta}{s_1^+}, \overset{\Delta}{s_2^+}, \ldots, \overset{\Delta}{s_{m^+}^+}\}$ and $\overset{\Delta}{S} = \{\overset{\Delta}{s_1}, \overset{\Delta}{s_2}, \ldots, \overset{\Delta}{s_m}\}$. We add a sparsity constraint (l_1-norm) to the objective function. In addition, because θ is not continuous in the objective function, the logistic loss function is used instead. $f(a) = 1/1 + \exp(-a)$. Therefore, the new objective function is obtained as:

$$\min_w \frac{1}{m^+} \frac{1}{m} \sum_{i=1}^{m^+} \sum_{j=1}^{m} 1/(1 + \exp(f\_s(p, \overset{\Delta}{s_i^+}, w) - f\_s(p, \overset{\Delta}{s_j}, w))) + \phi l_1(w) \tag{8}$$

where $\phi = 10^{-4}$, $F_w(p, \overset{\wedge}{s_i^+})$ is the loss function of s_i^+.

Attribute Selection Algorithm Based on Self-paced Learning. Because the initial service failure score is calculated based on all service attributes, the selected service failure may be unreliable. Combining Eq. (1) and Eq. (8) to obtain an attribute selection algorithm:

$$\min E(v, w) = \frac{1}{m^+} \sum_{i=1}^{m^+} (v_i F_w(p, \overset{\wedge}{s_i^+}) - \lambda v_i) + \phi l_1(w) \tag{9}$$

when w is fixed, v_i takes the following values:

$$v_i = \begin{cases} 1, F_w(p, \overset{\wedge}{s_i^+}) < \lambda \\ 0, \quad otherwise \end{cases} \tag{10}$$

We constrain λ to prevent self-learning from selecting services with high loss values.

$$\lambda^c = \begin{cases} \mu(F_{w^{c-1}}) + \sigma(F_{w^{c-1}}), \qquad c = 1 \\ \max\{\lambda^{c-1}, \mu(F_{w^{c-1}}) + \sigma(F_{w^{c-1}})\}, c > 1 \end{cases} \tag{11}$$

where c is the number of iterations in the self-paced learning process, $F_{w^{c-1}}$ is the loss of all data in $c - 1$ iterations, μ and σ^2 are the mean and variance of these data.

Finally, the service attribute weights of t groups are obtained, $W = \{w_1, w_2, \ldots, w_t\}$.

Reliable Service Failure Scoring. We define a threshold for service attribute weights to filter out more important attribute weights [18]:

$$Q_j = \{q_i | q_i \in Q, (w_{ji}/\max(w_j)) > \gamma\} \tag{12}$$

where $\gamma = 0.05$. Finally we get a reliable service failure scoring function [19]:

$$FS_k(s) = \sum_{i=1}^{t} r^i \psi(f\_s(p, s, Q_i)) \tag{13}$$

where $\psi(f\_s(p, s, Q_i) = \frac{f\_s(p,s,Q_i)}{\sum_{j=1}^{n} f\_s(p,s_j,Q_i)}$, $r^i = \exp(-e_i)/\sum_{j=1}^{t} \exp(-e_j)$.

Algorithm 1. ADWSFD algorithm
Input: Service set S
Output: Service failure score $RFS_k(s)$ for each service s
1:   Randomly select a normal service sample p
2:   Calculate the initial service failure score of service S by Eq. (3)
3:   Obtain the faulty service candidate set S^+ by Eq. (5)
4:   **for** $i = 1 \rightarrow t$ **do**
5:       Randomly filter m^+ services from the faulty service candidate set S^+
6:       Randomly filter m services from the faulty service candidate set S
7:       When v=1, get w_0 by Eq. (9)
8:       **repeat**
9:           Update λ by Eq. (11);
10:           Update v by Eq. (10);
11:           Update w by Eq. (9);
12:       **until** convergence;
13:       Select appropriate subsets of service attributes Q,by Eq. (12);
14:   **end for**
15:   Get a reliable service failure score $FS_k(s)$,by Eq. (13);
16:   **return** $FS_k(s)$

5   Implementation and Evaluation

5.1   Experiment Setup

Datasets. In our experiments, we use the QWS dataset, which has 11 service attributes and 364 real web services.

Parameters Setting. All experiments are performed on a CPU of 3.00 GHz and a physical memory of 8.00 GB, and all parameters are implemented in python3.9. Other parameter settings: $m^+ = 10$, $m = 2m^+$, $t = 2(n^+/m^+)$, $\varepsilon = 2$, $\phi = 0.00001$.

Evaluation Methods. Average metrics used were accuracy, AUC value, and ROC curve. The closer the ROC curve [19] is to the upper left corner, the higher the accuracy of the model. AUC is the area under the ROC curve. AUC = 1, it is a perfect model. AUC = [0.85, 0.95], works well. AUC = [0.7, 0.85], the effect is average (Table 1).

Table 1. Confusion matrix.

Actual/Predict	Positive	Negative
Positive	Ture Positive(TP)	False Positive(FN)
Negative	False Negative(FP)	Ture Negative(TN)

$$Accuracy = (TP + TN)/(TP + FN + FP + TN) \tag{14}$$

$$AUC = (1 + TPR - FPR)/2 \tag{15}$$

5.2 Model Analysis

The Setting of Parameter m. This section conducts empirical experiments to investigate how the number of unlabeled samples affects AUC. We fixed $m = 2m^+$, $m = 10, 14, 20, 26, 30$. Use the QWS data set to do ten experiments and average the result to obtain a convincing evaluation. It can be clearly observed from the Fig. 1. That the service fault diagnosis effect of our proposed model is the best at $m^+ = 10$, and the following experiments are performed at $m^+ = 10$.

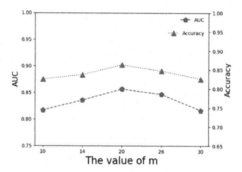

Fig. 1. Impact of m in QWS dataset

AUC and Accuracy of the Model. Since we randomly selected a normal service, the service failure score results obtained in each experiment are not uniquely, we conduct multiple experiments to get the mean value. Accuracy and AUC value of this service fault diagnosis model are 0.8557 and 0.8603 respectively. According to the standard of AUC to judge the pros and cons of the prediction model, the fault detection effect of the fault detection model based on anomaly detection is great.

ROC Curve. We take $m = 10, 14, 20, 26, 30$ and their ROC curves were drawn. Obviously, the closer the ROC curve is to the upper left corner, the better the effect of our fault detection model. From the comprehensive analysis of AUC, ROC curve and fault diagnosis accuracy, ADWSFD fault diagnosis model has better effect and can effectively judge fault service (Fig. 2).

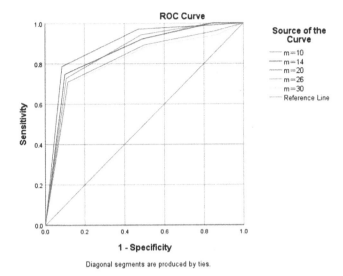

Fig. 2. ROC curve

6 Conclusion and Future Research Directions

In this paper, we apply an embedded anomaly detection framework combining outlier scoring and feature selection to web service fault diagnosis, and propose the ADWSFD fault diagnosis model. Obtain a reliable subset of service attributes through the initial failure score, and then use it to get a reliable service failure score. Experiments are carried out on the dataset QWS, and the effectiveness of the ADWSFD fault diagnosis model is verified by analysis.

Acknowledgements. This paper is partially supported by the National Natural Science Foundation of China under Grant No. 62172057 and No. 61972053, The Project is sponsored by "Liaoning BaiQianWan Talents Program" under Grant No.2021921024.

References

1. Wang, T., et al.: Workflow-aware automatic fault diagnosis for microservice-based applications with statistics. IEEE Trans. Netw. Service Manag. (99), 1–1 (2020)
2. Bui, K.T., et al.: A fault detection and diagnosis approach for multi-tier application in cloud computing. J. Commun. **22**(5), 399–414 (2020)
3. Li, P., et al.: Fault localization with weighted test model in model transformations. IEEE Access (99), 1–1 (2020)
4. Wang, G., Lin, X., Hendler, J., Song, W., Xu, Z., Liu, G. (eds.): WISA 2020. LNCS, vol. 12432. Springer, Cham (2020). https://doi.org/10.1007/978-3-030-60029-7
5. Kang, Z., et al.: Robust graph learning from noisy data. IEEE Trans. Cybern. (2018)
6. Kang, Z., et al.: Similarity learning via kernel preserving embedding. Proc. AAAI Conf. Artif. Intell. **33**, 4057–4064 (2019)

7. Cheng, L., et al.: Outlier detection ensemble with embedded feature selection. AAAI-20 3503–3512 (2020)
8. Pawan, M., et al.: Self-paced learning for latent variable models. In NIPS 1189–1197 (2011)
9. Hjx, A., Wtl, A.J.I.F.: Robust AdaBoost based ensemble of one-class support vector machines. Inf. Fusion **55**, 45–58 (2020)
10. Cai, Z., et al.: A real-time trace-level root-cause diagnosis system in alibaba datacenters. IEEE Access (99), 1–1 (2019)
11. Zahid, M.A., et al.: Collaborative business process fault resolution in the services cloud. IEEE Trans. Serv. Comput. (99), 1–1 (2021)
12. Ghasedi, K., et al.: Balanced self-paced learning for generative adversarial clustering network. IEEE/CVF CVPR (2020)
13. Pang, G., et al.: LeSiNN: detecting anomalies by identifying least similar nearest neighbours. In: IEEE International Conference on Data Mining Workshop (2016)
14. Song, H., Chen, Z.-c.: Multi-attribute decision-making method based distance and COPRAS method with probabilistic hesitant fuzzy environment. Int. J. Comput. Intell. Syst. **14**(1) (2021)
15. Pang, G., et al.: sparse modeling-based sequential ensemble learning for effective outlier detection in high-dimensional numeric data. In: Proceedings of Thirty-Second AAAI Conference on Artificial Intelligence (2017)
16. Zhang, C., Zhuang, C., Zheng, X., Cai, R., Li, M.: Stochastic model predictive control approach to autonomous vehicle lane keeping. Journal of Shanghai Jiaotong University (Science) **26**(5), 626–633 (2021). https://doi.org/10.1007/s12204-021-2352-y
17. Zhang, Q., Ren, F.: Prior-based bayesian pairwise ranking for one-class collaborative filtering. Neurocomputing **440**, 365–374 (2021). https://doi.org/10.1016/j.neucom.2021.01.117
18. Wang, F., et al.: Unsupervised soft-label feature selection. Knowl.-Based Syst. **219**(2), 106847 (2021)
19. Duc, K.T., et al.: Nonparametric estimation of ROC surfaces under verification bias. Revstat Stat. J. **18**(5), 697–720 (2020)

The Change of Code Metrics for Predicting the Label Change on Evolutionary Projects: An Empirical Study

Qiao Yu[1]($\boxtimes$), Shujuan Jiang[2], Yi Zhu[1], Hui Han[1], Yu Zhao[1], and Yuanpeng Jiang[3]

[1] School of Computer Science and Technology, Jiangsu Normal University, Xuzhou 221116, China
yuqiao@jsnu.edu.cn
[2] School of Computer Science and Technology, China University of Mining and Technology, Xuzhou 221116, China
[3] Library, China University of Mining and Technology, Xuzhou 221116, China

Abstract. Traditional software defect prediction approaches often focus on static code metrics. Software evolution could cause changes to the source code of the software, as well as changes to the code metrics and label. In recent years, researchers have proposed many process metrics for evolutionary projects, but they are mainly used to predict the defect proneness of software modules. Whether the change of code metrics (CCMs) could be used to predict the label change on evolutionary projects, and which CCMs are more correlated to the label change? To answer these questions, this paper proposes a framework for building the new datasets with CCMs and label change, and explores the correlations between CCMs and label change with feature ranking approaches. An empirical study is conducted on 40 versions of 11 open-source projects. The experimental results indicate that CCMs can predict the label change.

Keywords: Software defect prediction · Software evolution · Change of code metrics · Label change

1 Introduction

Software evolution is a dynamic and continuous process in the life cycle of software. For large-scale software systems, it is difficult for humans to develop software without any defects or bugs [1]. Therefore, it is important to predict and fix defects with limited resources. Software defect prediction aims to use historical data to predict the potential defects in software products, and then reasonably allocate the test resources and improve the efficiency of software testing. At present, software defect prediction has been one of the research focuses in software engineering [2, 3], which has great significance in improving software quality and software reliability.

Generally, traditional software defect prediction approaches often predict the software modules as defective or non-defective based on static code metrics. Software evolution could cause changes to the source code of the software, as well as changes to

X. Zhao et al. (Eds.): WISA 2022, LNCS 13579, pp. 305–313, 2022.
https://doi.org/10.1007/978-3-031-20309-1_27

the code metrics and label. The label change refers to the category change of software modules during the evolution process. For example, the label of one module may change from defective to non-defective, from non-defective to defective, or unchanged. However, only a few researchers focused on the label change of software modules. Jiang et al. [4] conducted an empirical study on the influence of process metrics on the change of defects for evolutionary projects. Motivated by this work, we consider that since code metrics can predict the defect proneness of a module, whether the change of code metrics (CCMs) could be used to predict the label change on evolutionary projects? Which CCMs are more correlated to the label change?

To answer these questions, this paper proposes a framework for building new datasets with CCMs and label change, and explores the correlations between CCMs and label change with feature ranking approaches.

The main contributions of this paper are listed as follows.

(1) A framework is proposed for building new datasets with CCMs and label change based on neighboring versions of evolutionary projects.
(2) An empirical study is conducted on 40 versions of 11 open-source projects. The experimental results indicate that CCMs can predict the label change.

The remainder of this paper is organized as follows. Section 2 summarizes the related work on software defect prediction with process metrics. Section 3 describes the details of the proposed framework. Section 4 shows the empirical study. Section 5 lists the threats to validity. Section 6 draws the conclusion.

2   Related Work

Traditional software defect prediction approaches often focus on static code metrics (also known as product metrics), such as McCabe metrics, Halstead metrics, and object-oriented CK metrics, which only reflect the static characteristics of software [5]. In practice, software defects often change with software evolution, and the prediction model based on code metrics may be ineffective for evolutionary projects [6].

In recent years, researchers have proposed many process metrics for evolutionary projects. For example, Kpodjedo et al. [7] proposed the design evolution metrics (DEM), including the numbers of added, deleted, and modified attributes, methods, and relations. The results showed that combining DEM with traditional metrics could improve the identification of defective classes, and DEM could predict more defects within a given size of code. Bhattacharya et al. [8] proposed the graph-based approach to capturing the product and process metrics. The results indicated that the graph metrics could be used to predict bug severity, maintenance effort, and defect-prone releases. Rahman and Devanbu [9] compared process metrics (such as code changes, developer/committer information, etc.) with code metrics, and the experimental results indicated that process metrics performed better than code metrics. Moreover, Madeyski and Jureczko [10] investigated the performance of four process metrics, including number of revisions (NR), number of distinct committers (NDC), number of modified lines (NML), and number of defects in the previous version (NDPV). They indicated that adding NDC could achieve better performance. Wang D. and Wang Q. [11] presented two types of evolution metrics, including version and code levels, and the experimental results showed the validity of these evolution metrics. Liu et al. [12] investigated the effectiveness of code churn based unsupervised model in effort-aware just-in-time defect prediction. Stanić and Afzal [13] investigated the performance of different combinations of code metrics and process metrics, and the results indicated that the combination of process metrics and static code metrics tended to improve the prediction performance. Yu et al. [14] presented two new process metrics from the defect rates of historical packages and the change degree of classes. The experiments were conducted on 33 versions of nine open-source projects. The results showed that adding the proposed process metrics could improve the performance of CVDP.

However, the above works focused on the defect proneness of a software module as defective or non-defective, but only a few researchers focused on the label change of software modules. Jiang et al. [4] studied the influence of process metrics on the change of defects for evolutionary projects. They divided the change of defect state into three categories: elimination of defects, introduction of defects, and others. The experimental results indicated that process metrics played different importance to the change of defect state.

In this paper, we aim to explore the feasibility of CCMs for predicting the label change and explore the correlations between CCMs and label change.

3   The Proposed Framework

For continuous versions of an object-oriented project, the classes may change during the evolution process. Thus, we need to process the original datasets to extract new metrics and categories for building new datasets.

This paper proposes a framework for building new datasets with CCMs and label change based on the common classes of neighboring versions. The framework is shown in Fig. 1.

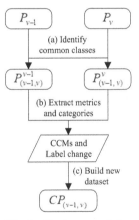

Fig. 1. The framework for building new datasets.

The framework includes three phases:

(a) Identify common classes

For object-oriented projects, the common classes of two neighboring versions have evolution data, but new classes do not have evolution data. To extract the CCMs, we need to process the original projects to get the common classes of neighboring versions.

We use the text matching approach to identify their common classes. As we all know, two classes with the same class name and full path can be regarded as common classes. But if one class is refactored (e.g. moving from one package to another) during the evolution process, which leads to the classpath changed but the class name remains unchanged. It should also be regarded as a common class. Therefore, considering the impact of refactoring classes, the class with the same class name will be regarded as the common class, even if the classpath is not the same. For example, 'bsh.BlockNameSpace' in jedit-4.2 is regarded as the common class with 'org.gjt.sp.jedit.bsh. BlockNameSpace' in jedit-4.3.

For two neighboring versions of one project (P_{v-1} and P_v), we can get their common classes, marked as $P^{v-1}_{(v-1,v)}$ and $P^{v}_{(v-1,v)}$. Where $P^{v-1}_{(v-1,v)}$ represents the common classes in P_{v-1}, and $P^{v}_{(v-1,v)}$ represents the common classes in P_v. Although $P^{v-1}_{(v-1,v)}$ and $P^{v}_{(v-1,v)}$ contain the same classes, their feature values are different.

(b) Extract metrics and categories

We extract new metrics and categories based on the common classes of two neighboring versions. The absolute changes of code metrics are designed as CCMs in formula (1).

$$CCMs_{ij}(v - 1, v) = \left|CM_{ij}(v - 1) - CM_{ij}(v)\right| \qquad (1)$$

where $CM_{ij}(v\text{-}1)$ and $CM_{ij}(v)$ indicate the values of j-th code metric for the i-th class in version v-1 and version v respectively.

This paper focuses on label change, so the categories of new datasets should indicate the change of label. Therefore, we define two categories for new datasets.

lc (label changed): the label of a class changes from non-defective to defective or from defective to non-defective.

lu (label unchanged): the label of a class unchanged, including defective to defective and non-defective to non-defective.

(c) Build new dataset

Based on CCMs and label change, we can build a new dataset, marked as $CP_{(v\text{-}1, v)}$. Especially, $CP_{(v\text{-}1, v)}$ has the same number of metrics as P_{v-1} and P_v.

4 Empirical Study

To show the validity of CCMs for predicting the label change, we conducted an empirical study on 11 open-source projects from the PROMISE Repository. All experiments were conducted on Open JDK 1.8 and Weka 3.8.

4.1 Experimental Datasets

We used 40 versions of 11 open-source projects, including ant, camel, ivy, jedit, log4j, lucene, poi, synapse, velocity, xalan, and xerces. The details of these projects are available in our recent work [15]. Especially, each project contains 20 code metrics at the class level as listed in [14]. The CCMs can be extracted based on these code metrics. Therefore, there are also 20 CCMs in new datasets, which are marked as c-wmc, c-dit, c-noc, and so on.

The new datasets with CCMs and label change (called CCMLC) are available on GitHub[1]. There are 29 datasets in CCMLC. The details are shown in Table 1. It shows the name of new dataset (columns 1 and 5). Then, it shows the numbers of all samples, lc samples, and lu samples respectively (columns 2–4 and 6–8). We find that the dataset size is reasonable, but the numbers of lc and lu are imbalanced, where lc is often less than lu.

---

[1] https://github.com/yuqiaoqkl/CCMLC.git.

Table 1. The details of CCMLC datasets.

Dataset	# All	# *lc*	# *lu*	Dataset	# All	# *lc*	# *lu*
ant-1.3–1.4	125	30	95	lucene-2.0–2.2	192	72	120
ant-1.4–1.5	168	48	120	lucene-2.2–2.4	235	93	142
ant-1.5–1.6	292	70	222	poi-1.5–2.0	229	128	101
ant-1.6–1.7	350	83	267	poi-2.0–2.5	314	202	112
camel-1.0–1.2	274	83	191	poi-2.5–3.0	382	122	260
camel-1.2–1.4	577	135	442	synapse-1.0–1.1	152	43	109
camel-1.4–1.6	861	152	709	synapse-1.1–1.2	219	71	148
ivy-1.1–1.4	110	57	53	velocity-1.4–1.5	156	72	84
ivy-1.4–2.0	227	34	193	velocity-1.5–1.6	210	90	120
jedit-3.2–4.0	265	58	207	xalan-2.4–2.5	700	301	399
jedit-4.0–4.1	293	53	240	xalan-2.5–2.6	768	293	475
jedit-4.1–4.2	295	61	234	xalan-2.6–2.7	862	454	408
jedit-4.2–4.3	327	47	280	xerces-1.2–1.3	436	96	340
log4j-1.0–1.1	98	20	78	xerces-1.3–1.4	401	218	183
log4j-1.1–1.2	104	65	39	/	/	/	/

4.2 Experimental Design

To show the validity of CCMs for predicting the label change, we designed the following research questions (RQs).

RQ1: How about the performance of CCMs for predicting the label change?
RQ2: Which CCMs are more correlated to the label change?

In our experiments, we used Random Forest [16] and AdaBoost [17] as the prediction models. We use their default parameters in Weka 3.8. Besides, Precision, Recall, F-measure, and AUC were selected as the performance indicators, which were commonly used in software defect prediction [18].

4.3 Experimental Results and Analysis

(1) The performance of CCMs for predicting the label change on evolutionary projects
We conducted within-project defect prediction with 10 times 10-fold cross-validation on CCMLC datasets. The results are displayed with the violin plots in Fig. 2. The statistical results with different ranges are listed in Table 2. We find that nearly half of the values are larger than 0.700 on Precision, Recall, and F-measure, and most values are larger than 0.600 on four performance indicators. We can conclude that the CCMs can predict the label change to a certain extent.

Besides, we also find that Random Forest performs better than AdaBoost. The reason may be that Random Forest is stable with class imbalance [19] and insensitive to the hyperparameters [20]. However, AdaBoost is sensitive to hyperparameters [20], and its performance can be improved with appropriate hyperparameters.

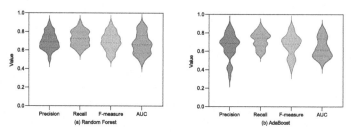

Fig. 2. The performance on CCMLC datasets.

Table 2. The statistical results with different ranges.

Range	Random Forest				AdaBoost			
	Precision	Recall	F-measure	AUC	Precision	Recall	F-measure	AUC
[0, 0.5)	0	0	0	1	4	0	3	1
[0.5–0.6)	3	2	4	8	4	3	4	12
[0.6–0.7)	14	10	11	11	9	9	8	8
[0.7–0.8)	7	11	9	4	9	11	11	6
[0.8, 1.0]	5	6	5	5	3	6	3	2

(2) The correlations between CCMs and label change

To measure the correlations between CCMs and label change, we selected Correlation [21] and Gain Ratio [22] as the feature ranking approaches. We use the heat map to show the correlations in Fig. 3. The darker color means that the correlation is higher. It shows that the correlations are various with different CCMs. Some CCMs are more related to the label change (eg. c-wmc, c-rfc, c-ce, c-npm, c-loc, c-amc, and so on). Besides, the ability to reveal correlations varies widely for different datasets.

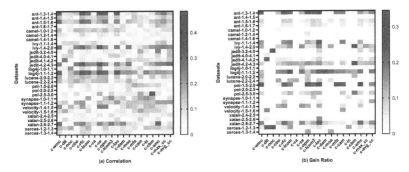

Fig. 3. The correlations between CCMs and label change.

5 Threats to Validity

There are several threats to the validity of experiments, which can be summarized into three aspects as follows.

(1) Construct validity

The dataset size is a threat to construct validity. The dataset size depends on the number of common classes. If the number of common classes is small, the new datasets may not work well. Especially, refactoring classes are considered when identifying the common classes, which also ensures the number of common classes. In our experiments, the dataset size is reasonable, which ensures the validity of experiments.

(2) Internal validity

The prediction model and its hyperparameters are threats to internal validity. We used Random Forest and AdaBoost as prediction models with default parameters in Weka 3.8. However, appropriate hyperparameters can improve the performance of models, especially for AdaBoost. Besides, we only select Correlation and Gain Ratio to show correlations. More feature selection approaches should be explored next.

(3) External validity

The quality of CCMLC datasets is a threat to external validity. We selected 40 versions of 11 object-oriented projects for building new datasets. They are commonly used in software defect prediction. However, we find that there are some noises in CCMLC datasets. For example, there are some samples with all values of CCMs zero, but the labels are different (*lc* or *lu*), which may affect the prediction performance. We will identify more noises and improve the quality of CCMLC datasets.

6 Conclusion

This paper proposed CCMs for predicting the label change based on neighboring versions of object-oriented projects, and explored the correlations between CCMs and label change. An empirical study was conducted on 40 versions of 11 object-oriented projects. The experimental results indicate that the proposed CCMs can predict the label change. Moreover, we can get the more correlated CCMs based on correlation analysis. Next, we will focus on these more correlated CCMs to improve the performance of software defect prediction.

Acknowledgments. This work was supported in part by the National Natural Science Foundation of China (61902161 and 62077029), the CCF-Huawei Populus Grove Fund (CCF-HuaweiFM202209), the Guangxi Key Laboratory of Trusted Software (kx201704), and the Research Support Program for Doctorate Teachers of Jiangsu Normal University (17XLR001).

References

1. Yuan, W., Wang, P., Guo, Y., He, L., He, T.: Mining the software engineering forums: what's new and what's left. In: Proceedings of Web Information Systems and Applications, pp. 513–524 (2020)
2. Hosseini, S., Turhan, B., Gunarathna, D.: A systematic literature review and meta-analysis on cross project defect prediction. IEEE Trans. Softw. Eng. **45**(2), 111–147 (2019)
3. Li, N., Shepperd, M., Guo, Y.: A systematic review of unsupervised learning techniques for software defect prediction. Inf. Softw. Technol. **122**, 106287 (2020)
4. Jiang, L., Jiang, S., Gong, L., Dong, Y., Yu, Q.: Which process metrics are significantly important to change of defects in evolving projects: an empirical study. IEEE Access **8**, 93705–93722 (2020)
5. Menzies, T., Milton, Z., Turhan, B., Cukic, B., Jiang, Y., Bener, A.: Defect prediction from static code features: current results, limitations, new approaches. Autom. Softw. Eng. **17**(4), 375–407 (2010)
6. Shatnawi, R., Li, W.: The effectiveness of software metrics in identifying error-prone classes in post-release software evolution process. J. Syst. Softw. **81**(11), 1868–1882 (2008)
7. Kpodjedo, S., Ricca, F., Galinier, P., Guéhéneuc, Y., Antoniol, G.: Design evolution metrics for defect prediction in object oriented systems. Empir. Softw. Eng. **16**(1), 141–175 (2011)
8. Bhattacharya, P., Iliofotou, M., Neamtiu, I., Faloutsos, M.: Graph-based analysis and prediction for software evolution. In: Proceedings of International Conference on Software Engineering, pp. 419–429 (2012)
9. Rahman, F., Devanbu, P.: How, and why, process metrics are better. In: Proceedings of International Conference on Software Engineering, pp. 432–441 (2013)
10. Madeyski, L., Jureczko, M.: Which process metrics can significantly improve defect prediction models? An empirical study. Softw. Qual. J. **23**(3), 393–422 (2015)
11. Wang, D., Wang, Q.: Improving the performance of defect prediction based on evolution data. J. Softw. **27**, 3014–3029 (2016)
12. Liu, J., Zhou, Y., Yang, Y., Lu, H., Xu, B.: Code churn: a neglected metric in effort-aware just-in-time defect prediction. In: Proceedings of International Symposium on Empirical Software Engineering and Measurement, pp. 11–19 (2017)
13. Stanić, B., Afzal, W.: Process metrics are not bad predictors of fault proneness. In: Proceedings of International Conference on Software Quality, Reliability and Security Companion, pp. 493–499 (2017)
14. Yu, Q., Jiang, S., Qian, J., Bo, L., Jiang, L., Zhang, G.: Process metrics for software defect prediction in object-oriented programs. IET Softw. **14**(3), 283–292 (2020)

15. Yu, Q., Zhu, Y., Han, H., Zhao, Y., Jiang, S., Qian, J.: Evolutionary measures for object-oriented projects and impact on the performance of cross-version defect prediction. In: Proceedings of Asia-Pacific Symposium on Internetware, pp. 192–201 (2022)
16. Breiman, L.: Random forests. Mach. Learn. **45**, 5–32 (2001)
17. Freund, Y., Schapire, R.E.: Experiments with a new boosting algorithm. In: Proceedings of the International Conference on Machine Learning, pp. 148–156 (1996)
18. Gong, L., Jiang, S., Jiang, L.: Research progress of software defect prediction. J. Softw. **30**, 3090–3114 (2019)
19. Yu, Q., Jiang, S., Zhang, Y., Wang, X., Gao, P., Qian, J.: The impact study of class imbalance on the performance of software defect prediction models. Chin. J. Comput. **41**(4), 809–824 (2018)
20. Tantithamthavorn, C., McIntosh, S., Hassan, A.E., Matsumoto, K.: The impact of automated parameter optimization on defect prediction models. IEEE Trans. Softw. Eng. **45**(7), 683–711 (2018)
21. Guyon, I., Elisseeff, A.: An introduction to variable and feature selection. J. Mach. Learn. Res. **3**, 1157–1182 (2003)
22. Karegowda, A., Manjunath, A.S., Jayaram, M.A.: Comparative study of attribute selection using gain ratio and correlation based feature selection. Int. J. Inf. Technol. Knowl. Manag. **2**(2), 271–277 (2010)

Identifying Influential Spreaders in Complex Networks Based on Degree Centrality

Qian Wang[✉], Jiadong Ren, Honghao Zhang, Yu Wang, and Bing Zhang

College of Information Science and Engineering, Yanshan University, Qinhuangdao 066000, Hebei, China
wangqianysu@163.com

Abstract. Identifying the most influential spreaders in complex networks is vital for optimally using the network structure and accelerating information diffusion. In most previous methods, the edges are treated equally and their potential importance is ignored. In this paper, a novel algorithm based on Two-Degree Centrality called TDC is proposed to identify influential spreaders. Firstly, the weight of edge is defined based on the power-law function of degree. Then, the node weight is calculated by the weight of its connected edges. Finally, the spreading influence of node is defined by considering the influence degree of the neighborhoods within 2 steps. In order to evaluate the performance of TDC, the Susceptible-Infected-Recovered (SIR) model is used to simulate the spreading process. Experiment results show that TDC can identify influential spreaders more effectively than the other comparative centrality algorithms.

Keywords: Complex network · Influential spreaders · Two-degree centrality

1 Introduction

In recent years, the importance of nodes is a hot subject for characterizing the structure and dynamics of complex networks. And it is an important problem for evaluating the spreading ability of nodes in network analysis. For example, in the process of epidemic spreading, the critical sources of infection can effectively control the spread of disease [1]. In the marketing industry, new products are sold through the customer networks, and the most influential customers are found to promote product sales for the maximum profits by recommending to their friends [2]. Therefore, it is of great significance for identifying the influential spreaders in complex networks [3].

In complex networks, there were some traditional centrality methods for controlling and analyzing the spreading dynamics over the past years, such as degree centrality (DC) [4], betweenness centrality (BC) [5], closeness centrality (CC) [6] and so on. And a variety of improved centrality algorithms were proposed to identify influential spreaders more accurately. Jackson et al. [7] proposed decay centrality (DYC) and it measures the closeness of a node to the rest of the nodes in the network, while the computational cost is relatively expensive by calculating the shortest distances from a target node to the rest of the nodes. K-shell decomposition (KS) [8] is a well-established method for analyzing

© The Author(s), under exclusive license to Springer Nature Switzerland AG 2022
X. Zhao et al. (Eds.): WISA 2022, LNCS 13579, pp. 314–326, 2022.
https://doi.org/10.1007/978-3-031-20309-1_28

the structure of large-scale graphs, and it views that the most influential spreaders are located in the core of the network. However, the k-shell decomposition assigns too many nodes in the same level, and then the importance of each node in the network cannot be identified. Liu et al. [9] improved k-shell decomposition method based on the distance from a target node to the network core, while the computational cost is relatively expensive by calculating the shortest distances from a target node to core nodes. Bae et al. [10] proposed a neighborhood coreness centrality, and the spreading influence of a spreader is quantified based on the k-shell values of its neighbors, the more neighbors of a spreader, the more powerful the spreader is. Ma et al. [11] proposed a gravity centrality and the contribution of neighbors to the centrality of a node is calculated by the gravity formula, the k-shell value of a node is viewed as its mass, and the shortest path distance between two nodes is used as their distance. Yang et al. [12] proposed an improved gravity model based on the k-shell algorithm, the location difference between nodes represented by the k-shell values is used as the attraction coefficient which adjusts the attractiveness of the central nodes in the network. Shang et al. [13] considered that only paying attention to the local static geographical distance between nodes would ignore the dynamic interaction between nodes in the real network, and proposed a gravity model based on effective distance. There are also other node rank algorithms for improving the rank performance [14, 15]. However, these algorithms do not consider the potential importance of edges, the edges are treated equally.

Actually, the potential importance of edges should be considered [16]. Intuitively speaking, the more steps of neighbors are taken into consideration, the more accurately we can identify an influential spreader. However, this can lead to considerable computational complexity, due to the large scale of the data with temporal and spatial variations, it is very difficult to collect the complete network information in some real-world networks [17]. Inspired by the theory of three degree of separation in social network [18], the connection of three degrees is "strong connection", which means, the influence of a node is strongly influenced by the neighbors no more than 3 steps, and the effect of the neighbors after 3 steps is weak. In this paper, by considering both the performance and time cost, the potential importance of edges within 2 steps is considered, and a novel influence algorithm called two-degree centrality (TDC) is proposed to identify influential spreaders. The SIR model [19] is used to simulate the spreading process, and the monotonicity index [8] is used to quantify the resolution of different centrality methods. The comparison for the spreading influence of top-k nodes between TDC and other methods are performed.

The rest of this paper is organized as follows. In Sect. 2, the definitions are given. The algorithm is proposed in Sect. 3. In Sect. 4, experiments are performed to verify the proposed algorithm. Conclusions is summarized in Sect. 5.

2 Definition

An unweighted and undirected network $G = (V, E)$ is considered, where V is the node set and E is the edge set of the network, $|V|$ and $|E|$ are the total number of nodes and the total number of edges in a network, respectively. Network structure is represented as an adjacency matrix graph $= (a_{ij})_{N*N}$, where $a_{ij} = 1$, if node i is connected with node j. Otherwise, $a_{ij} = 0$. Γ_i denotes the set of direct neighbors of node i.

Definition 2.1. (The Weight of Edge) The weight of edge E(i,j) is defined based on the power-law function of degree. The Weight of Edge (WE) is defined as follows.

$$WE(i, j) = \left(d_i \times d_j\right)^{\alpha} \tag{1}$$

where d_i is the degree of node i and α is a tuning weight parameter, adjusting the strength of the edge weight, this definition is supported by the empirical evidence of the real weighted networks [20]. In addition, in Refs [21], it is found that different networks have the strongest robustness against attack and random failure when $\alpha = 1$, in this paper, α is set to 1 based on the findings of researchers.

Definition 2.2. (The Weight of Node) The weight of node i is defined based on the sum of the weights of all the connected edges of node i. The Weight of Node (WN) is defined as follows.

$$WN(i) = \sum_{j \in \Gamma_i} WE(i, j) \tag{2}$$

where Γ_i is the set of direct neighbors(1- step neighbor) of node i.

Definition 2.3. (The Influence of Node) The influence of node i is defined by considering the influence degree of the neighbors within 2 steps and introducing the tuning parameter θ. The Influence of Node (INV) is defined as follows.

$$INV(i) = WN(i) + \theta \sum_{j \in \Gamma_i} WN(j) + \theta^2 \sum_{l \in \Gamma_{j \setminus i}} WN(l) \tag{3}$$

where θ is from 0 to 1 and Γ_i is the set of direct neighbors(1-step neighbor) of node i.

3 A Two-Degree Centrality Method

When the influence of a spreader is measured in unweighted complex networks. The importance of edge is different in many real-world networks, and the connection of each edge between nodes has a certain meaning. Therefore, the weight of each edge needs to be considered. In addition, on this basis, the weight of node is obtained. Finally, the influence of a spreader is not only related to itself, but also influenced by its neighbors. And the influence degree will gradually decreases as the number of steps increases. In this paper, the tuning parameter θ is introducing to adjust the influence degree of different neighbors and a two-degree centrality TDC is proposed for identifying influential spreaders in unweighted complex network.

TDC algorithm is to identify influential spreaders in complex networks. In unweighted network, the potential importance of each edge is different and the influential of a spreader is affected by its neighbors. Firstly, the power-law function of degree is used to measure the weight of edge (WE). Then, the weight of node (WN) is calculated by the sum of the weight of its connected edges. Finally, considering that the influence of node is strongly influenced by 1-step neighbors and 2-step neighbors, the effect of different neighbors is adjusted by introducing the tuning parameter θ, where the initialization process of tuning parameter θ is given in Sect. 4.3. Here, TDC is proposed to identify influential spreaders by taking the potential importance of edge and the influence degree of the neighbors within 2 steps into consideration.

A simple network is shown in Fig. 1. The calculation process of TDC is shown in Fig. 2. In the TDC algorithm, α is set to 1, and θ is set to 0.1.

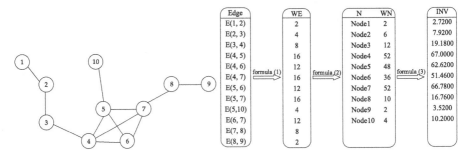

Fig. 1. A simple network.   Fig. 2. The calculation process of TDC.

Step 1: For the edge E(4,5), the weight is calculated according to Formula (1).

$$WE(4,5) = (d_4 \times d_5)^1 = 16$$

Step 2: For the node 4, the weight is calculated according to Formula (2).

$$WN(4) = \sum_{j \in \Gamma_i} (i,j) = 52$$

Step 3: For the node 4, the influence is calculated according to Formula (3).

$$INV(4) = WN(i) + \theta \sum_{j \in \Gamma_i} WN(j) + \theta^2 \sum_{l \in \Gamma_{j \setminus i}} WN(l) = 67$$

By using TDC algorithm, the influence of all nodes in the simple network is shown in Fig. 2. As we can see that $WN(4) = WN(7) = 52$, if only consider the weight of the node itself, and the influence of node 4 and node 7 can be distinguished when the weight of each edge is considered.

Node influence is sorted based on different methods of the simple network as shown in Table 1. It can be seen that the influence of each node cannot be well distinguished by DC and KS. And the node influence rank is almost the same by CC, DYC and TDC, except node 2 and node 10, which shows that TDC can identify influential nodes in networks feasibly and objectively. The superiority of TDC over CC and DYC will be further verified in Sect. 4.4.

Table 1. Rank of node influence with different methods

Rank	DC	CC	DYC($\delta = 0.1$)	KS	TDC
Rank1	4,5,7	4	4	4,5,6,7	4
Rank2	6	7	7	1,2,3,8,9,10	7
Rank3	2,3,8	5	5		5
Rank4	1,9,10	6	6		6
Rank5		3	3		3
Rank6		8	8		8
Rank7		2	2		10
Rank8		10	10		2
Rank9		9	9		9
Rank10		1	1		1

4 Experimental Analysis

4.1 Experimental Data

Three real networks are used to evaluate the performance of TDC. (1) Zachary: a Karate Club Network [22], the data contains a friendship network between members of the karate club. (2) Jazz: a collaboration network between Jazz musicians [23]; (3) Email: a network of email interchanges between members of a University [24]. Here are the basic topological properties of the three networks, n and m are the total number of nodes and edges respectively, $< k >$ and $< k^2 >$ are the average degree and the second-order average degree respectively, $\beta_{rand}^c = < k > / < k^2 >$ and β denote the spreading threshold and the spreading probability as shown in Table 2.

Table 2. THE basic topological properties of the three real networks

Network	n	m	$< k >$	$< k^2 >$	β_{rand}^c	β
Zachary	34	78	4.5882	35.6471	0.1287	0.15
Jazz	198	2742	37.6970	1070.2424	0.0352	0.05
Email	1133	5451	9.6222	179.1640	0.0537	0.07

4.2 The Monotonicity Index

The monotonicity index M(R) of ranking list X is defined to quantify the resolution of different centrality methods [9].

$$M(R) = \left[1 - \frac{\sum_{r \in R} n_r(n_r - 1)}{n(n - 1)} \right]^2 \quad (4)$$

where n is the size of network(ranking list R) and n_r is the number of nodes with the same rank r. The value of M(R) ranges from 0 to 1. If M(R) = 1, which means that the ranking list R is perfectly monotonic, each node is assigned a different influence value. Otherwise, all nodes are in the same rank as M(R) = 0. So a good method in ranking the influences of nodes should has a high M value.

4.3 The Sir Model

In complex networks, the SIR model has been widely applied in the research of disease, information and rumors spreading. And the SIR model is a commonly used tool to examine the spreading ability of nodes [19]. In this model, there are three states: S (Susceptible), I (Infected) and R (Recovered). (i) S is used to represent susceptible individuals who have not yet be infected; (ii) I is used to represent infected individuals who are capable of spreading the disease to other susceptible individuals; (iii) R is used to represent recovered individuals who already have been recovered and will not be infected again. At the beginning, only one node is infected and the other nodes are susceptible. During the propagation process, infected nodes randomly select their direct susceptible neighbors with probability β, and then enter the recovered state. When there are no infected nodes in the network, the spreading process will stop. At each step, the number of infected nodes is represented by F(t), it is obvious that F(t) increases with t, and will remain stable at last. And the spreading capacity of the initial infected nodes is quantified according to the number of infected nodes in the network when the spreading stops. The larger F(t) value of a node is, the stronger the spreading influence of the node is. By averaging over 500 independent experiments, the reliability of the results is ensured.

4.4 Iniitialization of the Tuning Parameter Θ

In order to find out the corresponding tuning parameter θ in the real network, the Kendall's tau coefficient is used to analyze the rank correlation of nodes between TDC and the SIR model. Here, the value of β is in the seventh column of Table 2. It considers a set of joint observations from two rank lists of X and Y. X is the node rank based on TDC with certain θ and Y is the node rank based on the SIR model. If $(x_i - x_j)(y_i - y_j) > 0$, the observations have concordant rank in X and Y. If $(x_i - x_j)(y_i - y_j) < 0$, they are considered to be inconsistent rank. If $(x_i - x_j)(y_i - y_j) = 0$, they have a same rank in X and Y. Then Kendall's tau coefficient τ is defined as follows [10].

$$\tau = \frac{n_c - n_d}{0.5n(n - 1)} \tag{5}$$

where n_c and n_d are the numbers of concordant pairs and inconsistent pairs, respectively. The higher the value of τ is, the more accurate the ranked list generated by a certain θ corresponds to TDC is.

Since different topological properties of networks may lead to different tuning parameter θ, which will affect the TDC performance a lot. So, the value of θ should not be too large or too small to make the influence degree of neighbors or node itself enhanced

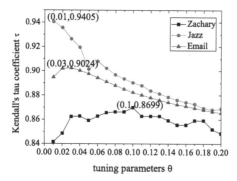

Fig. 3. Kendall's tau coefficient τ on different tuning parameters θ.

or weakened more. Therefore, the value of θ ranges from 0.01 to 0.2, and the degree of freedom is 0.01 gained by experimental analysis. The tuning parameter θ of different network are shown in Fig. 3, and TDC gets the best performance in the Zachary network when θ = 0.1; the best performance in the Jazz network when θ = 0.01; and the best performance in the Email network when θ = 0.03.

4.5 Effectiveness

In order to verify the effectiveness of TDC. The monotonicity M of different ranking methods is summarized in Table 3. As shown in Table 3, TDC achieves the maximum M value in three different networks. In the Zachary network, M(TDC) = 0.9542. In the Jazz network, M(TDC) = 0.9993. In the Email network, M(TDC) = 0.9999. Here, the resolution of TDC is the best, CC and DYC are also good, and the KS is relatively bad. The influence of each node can be well distinguished by TDC. The monotonicity index M shows the resolution, but cannot evaluate the spreading ability of nodes.

Table 3. The corresponding monotonicity in different network

Network	M(DC)	M(CC)	M(DYC)	M(KS)	M(TDC)
Zachary	0.7079	0.8993	0.9438	0.4958	0.9542
Jazz	0.9659	0.9878	0.9988	0.7944	0.9993
Email	0.8874	0.9988	0.9999	0.8088	0.9999

Next, in order to further verify the effectiveness of TDC, the rank of the top-k nodes is obtained. The SIR model is applied to evaluate the number of nodes infected by the top-k nodes, F(t) represents the number as mentioned above and t is the step value which is set to 4. The number of infected nodes are shown in Fig. 4. It is clearly that the curve of TDC is downward sloping more gently than DC, CC, DYC and KS.

Namely the spreading ability of top-k nodes based on TDC decreases more steadily with the increasing of k. It also illustrates that the top-k node rank of TDC is consistent with the ability of node infection. So the top-k nodes of TDC are more accurate than DC, CC, DYC and KS.

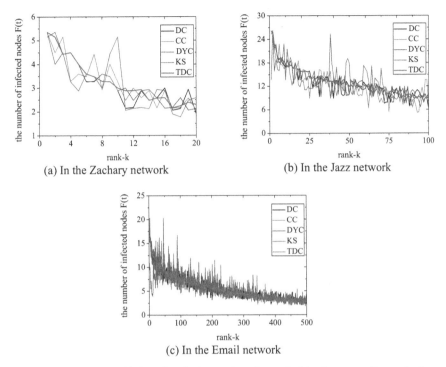

(a) In the Zachary network   (b) In the Jazz network

(c) In the Email network

Fig. 4. The number of infected nodes by top-k nodes ranked by four centrality methods.

Meanwhile, the top-10 nodes ranked by DC, CC, DYC, KS and TDC are listed in Table 4. The spreading ability of the different top-10 nodes between TDC and other centrality method are compared. The top-10 nodes are used as initial nodes to infect other nodes in the network, the infection situation of each step in the propagation process is analyzed, and the number of infected nodes in the network are compared when the spreading stops. The simulation results are shown in Figs. 5–7.

In the top-10 nodes, the same node in different algorithms does not need to be compared. For example, in the Zachary network, based on DC and TDC, different nodes in the top-10 nodes are 24 and 31, the experimental results show that the spreading ability of node 31 is stronger than that of node 24 as shown in Fig. 5 (a). So the top-10 node rank of TDC is more accurate than DC.

From Figs. 5–7, it is clear that the curve of TDC is above the other methods at the node propagation process of different networks. In other words, TDC is the best for the number of infected nodes at each step during the propagation process. What's more, when the spreading of the infected nodes stops, the final number of infected nodes based on TDC is larger than other four methods, especially in the Zachary networks and the Email networks. It further illustrates that TDC is the best

Table 4. The top-10 nodes ranked in the three networks

Rank	Zachary					Jazz					Email				
	DC	CC	KS	DYC	TDC	DC	CC	DYC	KS	TDC	DC	CC	KS	DYC	TDC
Rank1	34	1	1	34	1	67	67	67	4	7	105	333	299	105	105
Rank2	1	3	2	1	34	7	7	7	7	67	333	23	389	333	42
Rank3	33	34	3	33	33	20	23	23	12	20	16	105	434	23	16
Rank4	3	32	4	3	3	23	90	20	13	23	23	42	552	42	333
Rank5	2	9	8	2	2	90	93	90	14	18	42	41	571	41	23
Rank6	4	14	9	32	9	13	20	93	15	90	41	76	726	233	196
Rank7	32	33	14	4	14	18	74	13	18	13	196	233	756	76	41
Rank8	9	20	31	9	32	93	101	18	19	93	233	52	788	16	3
Rank9	14	2	33	14	4	109	125	109	20	19	21	135	885	196	76
Rank10	24	4	34	24	31	80	109	74	21	74	76	378	886	135	21

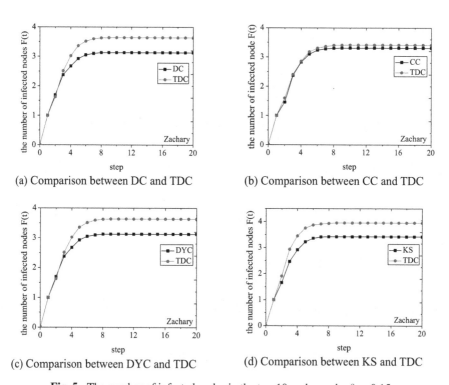

(a) Comparison between DC and TDC

(b) Comparison between CC and TDC

(c) Comparison between DYC and TDC

(d) Comparison between KS and TDC

Fig. 5. The number of infected nodes in the top-10 nodes under $\beta = 0.15$.

In DC, CC and KS, the edges are treated equally and their potential importance is ignored. In addition, DC is a straightforward local metric and less relevant. Since a node with a little highly influential neighbors may be more influential than many less influential neighbors. CC and DYC are global metrics and their computational complexity is considerable. KS assigns too many nodes in the same level, and then the importance of each node in the network cannot be identified. TDC considers the potential importance of edge and the influence degree of the neighbors within 2 steps. In a word, TDC can more accurately identify the spreading ability of top-10 nodes than DC, CC, DYC and KS. It also illustrates that TDC can be used in networks with different topological properties.

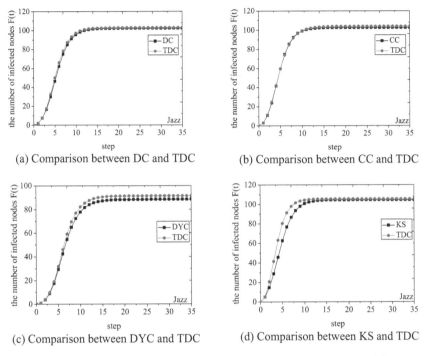

(a) Comparison between DC and TDC

(b) Comparison between CC and TDC

(c) Comparison between DYC and TDC

(d) Comparison between KS and TDC

Fig. 6. The number of infected nodes in the top-10 nodes under $\beta = 0.05$.

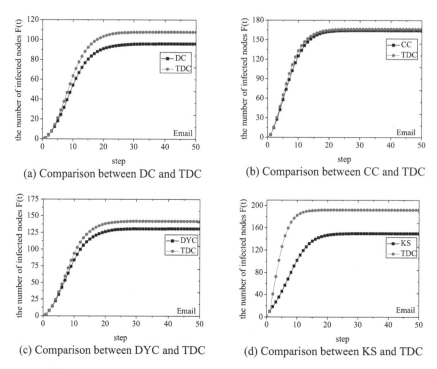

(a) Comparison between DC and TDC

(b) Comparison between CC and TDC

(c) Comparison between DYC and TDC

(d) Comparison between KS and TDC

Fig. 7. The number of infected nodes in the top-10 nodes under $\beta = 0.07$.

5 Conclusions

In this paper, TDC is proposed to identify influential spreaders in complex networks. The weight of each edge is defined, and the weight of node is further defined by the weight of its connected edges. The spreading influence of node is defined by considering the influence degree of the neighborhood within 2 steps. TDC is more effective than the algorithm that ignores the potential importance of edge. In order to evaluate the performance of the proposed algorithm, the monotonicity index is used to quantify the resolution of different centrality methods, the experiment illustrates that the influence of each node can be well distinguished by TDC than other four centrality methods DC, CC, DYC and KS. In addition, the SIR model is applied to simulate the spreading process in the real-world networks. Compared with the four centrality methods, the spreading ability of top-k nodes based on TDC decreases more steadily with the increasing of k. TDC can more accurately identify the spreading ability of top-10 nodes than the four centrality methods. In a word, TDC can identify influential spreaders more effectively than the four centrality methods.

Acknowledgment. This work is funded by the Natural Science Foundation of Hebei Province of China under Grant No. F2022203089 and F2022203026, the Science and Technology Project of Hebei Education Department under Grant Nos. QN2021145 and BJK2022029, the National Natural Science Foundation of China under Grant No.61807028. The authors are grateful to valuable comments and suggestions of the reviewers.

References

1. Qin, Y., Zhong, X., Jiang, H., Ye, Y.: An environment aware epidemic spreading model and immune strategy in complex networks. Appl. Math. Comput. **261**, 206–215 (2015)
2. Liu, Y., Deng, Y., Wei, B.: Local immunization strategy based on the scores of nodes. Chaos **26**(1), 013106 (2016)
3. Kang, J., Zhang, J., Song, W., Yang, X.: Friend Relationships Recommendation Algorithm in Online Education Platform. In: Xing, C., Fu, X., Zhang, Y., Zhang, G., Borjigin, C. (eds.) WISA 2021. LNCS, vol. 12999, pp. 592–604. Springer, Cham (2021). https://doi.org/10.1007/978-3-030-87571-8_51
4. Freeman, L.C.: Centrality in social networks conceptual clarification. Social Networks **1**(3), 215–239 (1978)
5. Freeman, L.C.: A set of measures of centrality based on betweenness. Sociometry **40**(1), 35–41 (1977)
6. Sabidussi, G.: The centrality index of a graph. Psychometrika. Psychometrika **31**(4), 581–603 (1966)
7. Jackson, M.O.: Social and Economic Networks. Princeton University Press (2010)
8. Zeng, A., Zhang, C.J.: Ranking spreaders by decomposing complex networks. Phys. Lett. A **377**(14), 1031–1035 (2012)
9. Liu, J.G., Ren, Z.M., Guo, Q.: Ranking the spreading influence in complex networks. Physica A Statistical Mechanics & Its Applications **392**(18), 4154–4159 (2014)
10. Bae, J., Kim, S.: Identifying and ranking influential spreaders in complex networks by neighborhood coreness. Physica A Statistical Mechanics & Its Applications **395**(4), 549–559 (2014)
11. Ma, L.L., Ma, C., Zhang, H.F., Wang, B.H.: Identifying influential spreaders in complex networks based on gravity formula. Physica A Statistical Mechanics & Its Applications **451**, 205–212 (2016)
12. Yang, X., Xiao, F.: An improved gravity model to identify influential nodes in complex networks based on k-shell method. Knowl.-Based Syst. **227**, 107198 (2021)
13. Shang, Q., Deng, Y., Cheong, K.H.: Identifying influential nodes in complex networks: Effective distance gravity model. Inf. Sci. **577**, 162–179 (2021)
14. Wang, J.: A novel weight neighborhood centrality algorithm for identifying influential spreaders in complex networks. Physica A Statistical Mechanics & Its Applications **475**, 88–105 (2017)
15. Dong, S., Zhou, W.: Improved influential nodes identification in complex networks. Journal of Intelligent & Fuzzy Systems **41**, 6263–6271 (2021)
16. Qiu, L., Zhang, J., Tian, X.: Ranking influential nodes in complex networks based on local and global structures. Appl. Intell. **51**(7), 4394–4407 (2021). https://doi.org/10.1007/s10489-020-02132-1
17. Pei, S., Muchnik, L., Andrade, Jr.: Searching for Superspreaders of Information in Real-world Social Media. Scientific Reports **4**, 5547 (2014)
18. Christakis, N.A., Fowler, J.H.: Social contagion theory: examining dynamic social networks and human behavior. Stat. Med. **32**(4), 556–577 (2013)
19. Castellano, C., Pastor-Satorras, R.: Thresholds for epidemic spreading in networks. Phys. Rev. Lett. **105**(21), 218701 (2010)
20. Barrat, A., Barthélemy, M., Vespignani, A.: Traffic-driven model of the world wide web graph. Lect. Notes Comput. Sci. **3243**, 56–67 (2004)
21. Wang, W.X., Chen, G.: Universal robustness characteristic of weighted networks against cascading failure. Physical Re-view E Statistical Nonlinear & Soft Matter Physics **77**(2), 026101 (2008)

22. Wei, D., Zhang, X., Mahadevan, S.: Measuring the vulnerability of community structure in complex networks. Reliab. Eng. Syst. Saf. **174**, 41–52 (2018)
23. Gleiser, P.M., Danon, L.: Community Structure in JAZZ. Adv. Complex Syst. **6**(4), 565–573 (2003)
24. Guimera, R., Danon, L., Diaz-Guilera, A., Giralt, F., Arenas, A.: Self-similar community structure in a network of human interactions. Phys. Rev. E: Stat., Nonlin, Soft Matter Phys. **68**(6), 065103 (2004)

Machine Learning

A Knowledge-Guided Method for Disease Prediction Based on Attention Mechanism

Yuanzhi Liang[1], Haofen Wang[2], and Wenqiang Zhang[1]([✉])

[1] Academy for Engineering and Technology, Fudan University, Shanghai, China
wqzhang@fudan.edu.cn
[2] College of Design and Innovation, Tongji University, Shanghai, China

Abstract. Disease prediction, aimed at predicting possible future diseases of patients, is a fundamental research problem in medical informatics. Many studies have proposed the introduction of external knowledge to enhance existing models with some effect, but since most of these studies only consider entities directly related to the patient, they fail to take full advantage of the correlation between entities in the knowledge graph. To this end, we propose a new approach, which uses medical knowledge graphs for multi-hop reasoning to guide the self-attention based transformer model for disease prediction. Specifically, our approach design a reinforcement learning algorithm to perform path reasoning in the knowledge graph to obtain explicit disease progression paths. Since there is a semantic gap between the Electronic Health Records (EHR) data and the knowledge path data, we feed them into two separate transformer encoders to obtain the embedding representation. In order to measure the importance of the different knowledge information in relation to the patient information, an attention module is introduced to obtain a global attention representation. Experimental results on the real-world medical dataset MIMIC-III show the superiority of the proposed approach compared to a series of state-of-the-art baselines. At the same time, multi-hop knowledge paths bring stronger interpretability for disease prediction.

Keywords: Disease prediction · Knowledge graph · Attention mechanism · Transformer

1 Introduction

Disease prediction is an important research in medical informatics, and precision medicine can enable patients to detect possible diseases as early as possible for better treatment, while alleviating the general situation of insufficient medical resources [2,4,7,9,13].

Although medical data mining has a very promising future, current deep learning in the field of disease prediction has encountered many challenges. Firstly, high-quality rare disease data are difficult to collect, and unclean data

X. Zhao et al. (Eds.): WISA 2022, LNCS 13579, pp. 329–340, 2022.
https://doi.org/10.1007/978-3-031-20309-1_29

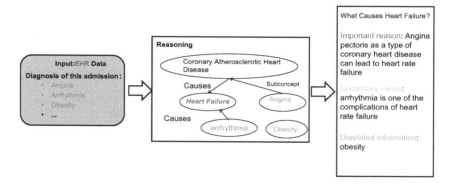

Fig. 1. An example of knowledge-based reasoning leading to disease prediction. Reasoning about a patient's clinical events from existing medical knowledge guides the final disease prediction.

cleaning makes it difficult to meet the high-quality data needed for deep learning, resulting in biased or even non-converging algorithms. Second, the black-box properties of existing deep learning models make it difficult to meet the requirements of disease prediction in terms of interpretability.

To this end, many researchers have proposed the introduction of external medical knowledge to provide some guidance to machine learning models [11,16]. DG-RNN [15] proposes to introduce knowledge as a complement to information sources, which introduces entities from the knowledge graph associated with patient features into the RNN network. MendMKG [14] proposes to use knowledge graph complementation techniques to improve the reliability of knowledge graphs.

However, existing methods fail to deeply understand the importance of different knowledge for disease prediction and fail to achieve better results and interpretability. To address this challenge, we propose a new learning approach designed to use the reinforcement learning algorithm to perform knowledge reasoning from existing medical knowledge to uncover hidden information between diseases and guide the self-attentive transformer-based model to make disease prediction, Fig. 1 is an example of using knowledge graph reasoning for disease prediction. When a doctor with knowledge of a patient's EHR information is given to him, he usually makes single- or multi-hop reasoning about that patient's EHR information based on some priori knowledge and obtains some new characteristics which, together with the original characteristics, make predictions about the patient's possible future diseases.

Specifically, we use the knowledge graph from existing medical knowledge, which records the relationships that exist between medical entities in the form of a triad, and perform prediction of possible disease progression through knowledge reasoning based on reinforcement learning. In order to find the correct disease path as much as possible, we trained a strategy-based reinforcement learning model to control the wandering of the agent. After inference, we consider knowledge graph paths as a carrier form of knowledge and as a complement to knowledge level information. In order to be able to consider both the information in the a priori knowledge and the information on the patient's characteristics, we embedding the EHR data and path data separately, and they are fed into two identical transformer encoder modules for independent encoding. In order to measure the importance of the different path information in relation to the patient information, an attention module E-P(EHR and Path) Attention is brought in to obtain a global attention representation. At the same time, explicit knowledge graph paths can also inform the interpretability analysis of predictions and enhance the reliability of the model. Extensive experiments on the MIMIC-III [5] dataset demonstrate the effectiveness of the proposed use of deep knowledge, and the results are even more pronounced on small datasets, which can be of great help in predicting rare diseases or diseases for which data are difficult to collect, while the model can also return appropriate medical knowledge as an explanation for the results made by the model.

The following points are our main contributions:

- We propose a novel knowledge reasoning guided disease prediction framework which deeply mines the information implicit in the knowledge graph through the reinforcement Learning algorithm.
- We propose E-P Attention module to measure the degree of importance between different paths and diseases in the knowledge graph as a way to obtain a more targeted coding representation.
- We evaluate the effectiveness of the proposed model on MIMIC-III. Experiments show that our results outperform all state-of-the-art baseline models, and a noteworthy point is that our proposed approach still maintains good performance in the absence of sufficient data volume.

2   Method

2.1   Preliminaries

We represent the characteristics that a patient has as $p_c \in \{0, 1\}$, where 1 means that the patient has a characteristic such as a disease or risk corresponding to the element. The characteristics of a patient on a single admission are then denoted as $v_t = \{p_1, p_2, \ldots, p_n\}$. Each of these features that the patient has can be linked to a pre-constructed knowledge graph $\mathcal{G}$. The knowledge graph $\mathcal{G}$ consists of an entity set ε and a relation set $\mathcal{R}$, We define the knowledge graph as $\mathcal{G} = \{(e_h, r, e_t) | e_h, e_t \in \varepsilon, r \in \mathcal{R}\}$, where each entity in ε represents a medical concept in the knowledge graph.

In the following sections, we use $\mathcal{G}$ to denote the pre-constructed knowledge graph, $p \in \varepsilon$ to denote the patient entity and $r \in \mathcal{R}$ to denote the relationship representing the pathway of disease progression.

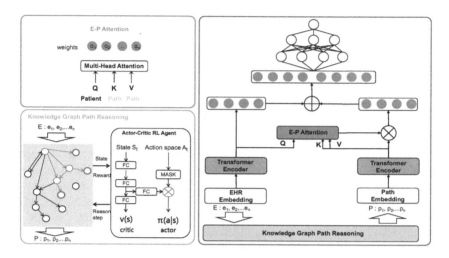

Fig. 2. The framework of our model, in which the Knowledge Graph Path Reasoning module uses reinforcement learning algorithms to extract pathway information from the knowledge graph based on the disease the patient is currently suffering from, is separately encoded independently using the transformer encoder, and the E-P Attention module introduces the attention mechanism to measure the importance between knowledge and EHR information to obtain a targeted feature representation.

2.2   Model Overview

The structure of the proposed method is shown in Fig. 2. The main components are the Knowledge Graph Path Reasoning module and the E-P Attention module. The Knowledge Graph Path Reasoning module is responsible for extracting medical knowledge from the knowledge graph based on the patient's current disease, embedding the patient's EHR information and the extracted knowledge path information into two identical transformer encoders for local encoding. The E-P Attention module is responsible for measuring the importance of the knowledge information in relation to the EHR information in order to obtain a more targeted encoding representation.

2.3   Knowledge Graph Path Reasoning

In order to find as reasonable a path as possible, we use the reinforcement learning algorithm to control the agent to navigate through the knowledge graph several times to better generate disease paths.

Drawing on the actor-critic structure, two-layer fully connected neural network are used as a representation x_t of the learning state s_t. The actor network is a layer of fully connected network used as a learning policy π_t, and value network for learning value v_t according to x_t. Policy π_t and value v_t is generating by following equation:

$$x_t = \sigma(\sigma(s_t W_1 + b_q)W_2 + b_2) \tag{1}$$

$$\pi_t = softmax((x_t W_p + b_p) \odot \mathcal{A}_t) \tag{2}$$

$$v_t = x_t W_v + b_v \tag{3}$$

where σ represents the nonlinear activation function, we use the rectified linear unit(ReLU) in this study, $\odot$ represent the Hadamard product. We define the state s_t of agent at time-step t as $s_t = (e_h, e_t, h_t)$, where e_h represents the source entity and e_t represents the entity currently located. The action $a \in \mathcal{A}_t$ of an agent we define as a tuple of edges and the node connected to it. We define the action space $\mathcal{A}$ as:

$$\mathcal{A}_t = \{(r, e) | (e_t, r, e) \in \mathcal{G}_c, e \notin \{e_1, e_2, \ldots, e_t - 1\}\} \tag{4}$$

We set the reward R_t as follows:

$$R_t = \begin{cases} 1 & \text{if } e_t \in \text{disease occurs at next admission} \\ 0 & \text{if } e_t \text{ is a disease not occurs at next admission} \\ -1 & \text{otherwise} \end{cases} \tag{5}$$

We have made it our goal to maximise the expected cumulative rewards.

$$J(\theta) = \mathbb{E}_\pi \left[\sum_{t=0}^{T-1} \gamma^t R_{t+1} \right] \tag{6}$$

where θ represents parameters in the actor-critic network, and γ is the decay coefficient.

2.4 Transformer Encoder

After obtaining patient EHR and knowledge pathway features Sequence $Z_e \in \mathbb{R}^{R \times C}$ and $Z_p \in \mathbb{R}^{R \times C}$, where R is the maximum number of features and C is the embedding dimension, use it as input learning feature representation from the transformer encoder respectively. For a knowledge path, we add up all the features on the path as its embedding. Unlike existing RNN model that have a limited receive field, the multi-layer Transformer [12] has a global receive field. Our transformer is pre-trained on the basis of BERT [3], the corpus is EHR data from MIMIC-III, and the pre-training task extends the masked language model (MLM) from BERT.

Knowledge-Aware Attention. Knowledge pathways from the knowledge graph are medically rich to help predict possible future illnesses of patients. To measure the relative importance of knowledge paths, we designed an attention network based on multi-headed attention, which allows the model to consider information from different subspaces at different locations. The formula for calculating attention is as follows:

$$Attn(Q, K, V) = softmax(\frac{QK^T}{\sqrt{d_k}})V \tag{7}$$

$$MultiHead(Q, K, V) = Concat(Attn_1, \ldots, Attn_H) \tag{8}$$

where queries, keys and values are transformed into Q, K, V by a mapping matrix, respectively, d_k is the dimension of queries and keys and H is the number of heads. In order to measure the importance of different knowledge pathways to patients, we designed EHR towards Path (E-P) attention to accomplish this goal. As shown in Fig. 2, queries are derived from the patient EHR representation e, keys and values are derived from the knowledge path representation p, and a weight α is assigned to each entity to indicate its importance by calculating the similarity of the patient's own characteristics to the knowledge path:

$$\alpha = softmax(\frac{QK^T}{\sqrt{d_k}})V, q = \alpha V \tag{9}$$

where q denotes the knowledge paths representation, α refers to the attention distribution.

After obtaining a globally encoded representation with attention, is fed into a full connected layer followed by the softmax function to predict the distribution P over disease label.

3 Experiments

To demonstrate the validity of our proposed method, we used the patient's EHR information to predict the patient's disease at the next admission.

3.1 Dataset and Evaluation Metrics

In this paper, we conducted experiments on the real-world dataset MIMIC-III[1], an authoritative public database collected at Beth Israel Deaconess Medical Center, which includes tens of thousands of patient hospital records and is already the benchmark dataset in the medical field. The statistics of the processed data are shown in Table 1. To test the performance of the proposed method with different data volumes, we set up multiple species of data volumes. To verify the robustness of our method, we chose three different diseases as the target for the disease prediction task, heart failure, hypertension and diabetes respectively. Finally, we divided the filtered data into a training set, a validation set, and two test set in a ratio of 7:1:1:1. We used open resources to manually construct the medical knowledge graph used for the experiments.

The Evaluation Metrics used for the experiments were AUC (macro-AUROC), which is widely used in the field of medical research.

3.2 Implementation Details

We used PyTorch[2] implementation of all the models and performed training on machines with NVIDIA TESLA V100 GPUs. For the training setup, we used

---

[1] https://mimic.mit.edu/.
[2] https://pytorch.org/.

Table 1. Statistics of dataset

Statistics	MIMIC-III
Number of patients	6386
Number of hospitalizations	10690
Max EHR feature length	129
Average EHR feature length	31.63
Max knowledge features	21
Average knowledge features	5.89

AdamW [6] for learning and set the learning rate and weight decay to 0.00001 and 0.000005 respectively. We set the dropout probability to 0.1 and the batch size to 8. If the AUC on the validation set did not improve in 5 consecutive epochs we would end the training early.

For the reinforcement learning agent for knowledge graph inference, we used GTX3090 to train for 9 min for 3 epochs to get the available agent. For baseline models, we keep the number of parameters of the same order of magnitude as our method to make a fair comparison. In particular, for Retain, Dipole and HiTANet the dimensionality of the hidden layer is set to 240. For the other baselines we use the original parameter settings from the public source code. The best result of baselines within 10 epochs are reported.

3.3   Baselines

We compared the proposed model with the following models, including machine learning models and deep learning models. Logistic regression(LR), Support Vector Machine(SVM) are classical machine learning models and we use them as machine learning model baseline. Recurrent Neural Networks(RNN) is an excellent deep learning model for processing sequential data and RNN+ is a two-layer Long short-term memory(LSTM) model and we use it as deep learning baseline. Retain [1], Dipole [10], DG-RNN [15] and HiTANet [8] is the state-of-the-art model for disease prediction. To verify the effectiveness of introducing multi-hop knowledge, we designed and implemented a version of Our-m that introduces only single-pick knowledge, and all concepts in the knowledge graph that are directly related to the features that the patients themselves have are added to the features. To prevent the effect of pre-training, we verified the effect of the transformer version Ours-a without loading pre-training parameters.

3.4   Experimental Results

In order to demonstrate the effectiveness of the proposed method, in this section we compare the proposed method with existing baseline methods for disease prediction. Table 2 shows the overall performance of our method for different sizes of data. Our proposed model outperforms the baseline model at all settings, which demonstrates the effectiveness of the proposed method. The overall performance of traditional machine learning methods is worse than that of deep learning methods. We speculate that there are two possible reasons for this. The first is the difference in the representation of medical events. Traditional methods use the high-dimensional one-hot representation. In contrast, deep learning methods use the medical concept embedding, which map each concept to a relatively low-dimensional vector, which can represent the clinical meaning of medical concepts. A second possible reason is that deep learning methods can better model risk prediction tasks for high and sparse data.

Table 2. AUC od three disease predictions on mimic

Task	Models	All	0.5	0.9	0.99
Heart failure	LR	74.91	73.49	68.64	52.50
	SVM	71.43	72.66	67.31	50.33
	RNN	76.84	73.76	68.98	45.19
	RNN+	77.13	74.46	69.16	54.37
	RETAIN	78.50	73.18	69.76	55.16
	Dipole	78.57	74.94	70.08	58.57
	DG-RNN	79.61	75.22	72.17	59.33
	HiTANet	80.36	75.81	72.24	59.76
	Ours-m	81.31	76.33	73.46	60.53
	Ours-a	83.46	80.12	75.03	67.33
	Ours	**84.07**	**81.13**	**77.16**	**68.00**
Diabetes	LR	75.03	76.65	72.93	48.50
	SVM	74.31	75.94	71.34	56.24
	RNN	79.16	77.13	75.61	53.33
	RNN+	81.07	78.64	76.13	51.19
	RETAIN	82.34	79.18	75.19	58.33
	Dipole	83.47	80.76	76.08	54.16
	DG-RNN	82.47	80.33	76.89	55.76
	HiTANet	83.76	80.47	76.18	55.91
	Ours-m	84.48	81.57	77.02	56.33
	Ours-a	87.04	83.13	79.42	57.50
	Ours	**88.72**	**85.42**	**82.90**	**75.00**
Hypertension	LR	65.34	64.71	62.76	58.71
	SVM	66.37	64.56	61.67	48.41
	RNN	67.17	65.92	61.89	55.45
	RNN+	68.02	66.84	62.05	58.50
	RETAIN	70.38	67.30	63.45	51.46
	Dipole	69.23	66.86	62.21	53.16
	DG-RNN	69.46	67.02	63.48	58.67
	HiTANet	70.81	67.83	63.97	51.33
	Ours-m	71.13	69.24	65.56	62.33
	Ours-a	71.43	70.87	**68.46**	63.45
	Ours	**72.94**	**71.34**	68.10	**65.64**

The effect of DG-RNN is better than the model without knowledge introduction but inferior to our model, which can show that knowledge introduction is effective and multi-hop knowledge can play a bigger role than single-hop knowledge.

With the knowledge reasoning module removed, Ours-m performed worse than Ours, suggesting that the multi-hop medical knowledge introduced was helpful. The performance of Ours-a is lower than Ours but higher than other baseline methods, indicating that our method does not rely entirely on the effect brought by pre-training. Note that our method is slightly less time efficient than the baseline method, but for the discreet task of disease prediction is worthwhile. At the same time, machine learning and deep learning methods have AUC between 0.5 and 0.6 with very small amounts of data, almost losing their predictive capabilities, but our method can maintain AUC values above 0.7, which is very important in the medical field where data resources are very precious. It is also shown that the proposed method can effectively use information from deep reasoning of knowledge graphs in disease prediction.

3.5 Multi-hop Analysis

Experiments demonstrate the effectiveness of introducing multi-hop relational entities for disease prediction tasks. However the introduction of entities with different number of hops in the knowledge graph was significantly different in terms of how helpful they were for prediction. In this section we have analysed different disease prediction tasks with different hop counts and the results are shown in Fig. 3.

Specifically, we set the maximum number of hops to 2, 3, 4 and 5 respectively to test their AUC performance. In all three disease prediction tasks, a maximum hop count of 2 showed the best results, indicating a preference for shorter knowledge chains in disease prediction, as too long knowledge chains would result in the model not being able to effectively identify its relationship with the original EHR features. We have analysed several cases and found that excessively long knowledge chains are often not medically meaningful, suggesting that we should introduce as many short knowledge chains as possible in disease prediction tasks.

3.6 Model Interpretability

The disease prediction task is characterized by a high requirement for interpretability, and simply giving prediction results does not meet the practical requirements for disease prediction. In this section, we examine the analysis of a specific case to illustrate the model's interpretability of prediction results. As shown in Table 3, this case study is about heart failure.

The first row explicitly shows the patient's EHR data for one visit. The trained reinforcement learning model made knowledge reasoning based on the data from this visit and three reasoning paths were obtained. The three knowledge mapping paths are the result of reasoning, as shown in rows 2,5,7 of the table. In the second row of the table, since angina is a type of coronary heart

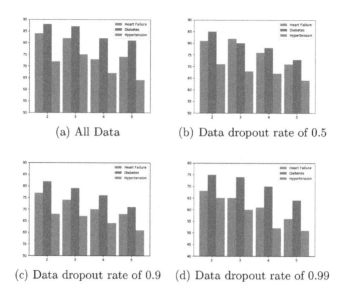

(a) All Data   (b) Data dropout rate of 0.5

(c) Data dropout rate of 0.9   (d) Data dropout rate of 0.99

Fig. 3. The impact of different hop counts on the model prediction results for different data volumes, using AUC as an indicator.

Table 3. Case study results of heart failure for showing the explicit interpretability that Our method has

Raw Data	Angina, Arrhythmia, Paroxysmal Supraventricular Tachycardia Obesity, Diabetes
Knowledge graph reasoning path	Angina $\xrightarrow{Subconcept}$ Coronary Atherosclerotic Heart Disease Coronary Atherosclerotic Heart Disease $\xrightarrow{Causes}$ Heart Failure
Explanation1	Angina is one of Coronary Atherosclerotic Heart Disease.
Explanation2	Prolonged inadequate coronary blood supply can lead to the development of ischaemic cardiomyopathy,which in turn can lead to the development of heart failure in patients
Knowledge graph reasoning path	Arrhythmia $\xrightarrow{Causes}$ Heart Failure
Explanation1	Patients with cardiac arrhythmias suffer from reduced systolic or diastolic function of the heart and reduced cardiac output resulting in inadequate blood supply to all tissues and organs of the body, which in turn causes sympathetic excitation resulting in heart failure.
Knowledge graph reasoning path	Diabetes $\xrightarrow{Causes}$ Peripheral Arterial Disease
Explanation1	Increased blood sugar in diabetes can lead to hardening of the arteries or to plaque, narrowing and, in severe cases, occlusion of the arteries

disease, which is the cause of heart failure, angina is among the important factors contributing to heart failure in the patient's data for this visit. Due to the angina is one of coronary atherosclerotic heart, and the prolonged inadequate coronary blood supply can lead to the development of ischaemic cardiomyopathy, which in turn can lead to the development of heart failure in patients. This explains why we are particularly concerned about the medical event of angina pectoris in the patient's current admission visit. The four knowledge paths in the table represent the potential state of the patient's current disease from the level of medical knowledge, which is very helpful in relation to prediction, in line with the physician's thinking habits, and can improve the interpretability of the model for both the patient and the physician.

4   Conclusion

In this paper, we introduce a new disease prediction method designed to obtain disease development paths using the reinforcement learning algorithm on knowledge graphs, local encoding by the transformer and introducing the attention mechanism to measure the importance of knowledge paths with respect to patient EHR features to obtain a targeted global representation. The experimental results on the real-world dataset MIMIC-III demonstrate the effectiveness of our proposed method, while also providing assistance in cases where the amount of data is insufficient. At the same time, multi-hop paths can provide a more reasonable interpretation of prediction results, which is very important for the application of disease prediction.

Acknowledgements. This work was supported by National Natural Science Foundation of China (No.62072112, 62176185), Scientific and Technological Innovation Action Plan of Shanghai Science and Technology Committee (No.20511103102), Fudan Double First-class Construction Fund (No. XM03211178).

References

1. Choi, E., Bahadori, M.T., Kulas, J.A., Schuetz, A., Stewart, W.F., Sun, J.: Retain: An interpretable predictive model for healthcare using reverse time attention mechanism. arXiv preprint arXiv:1608.05745 (2016)
2. Davenport, T., Kalakota, R.: The potential for artificial intelligence in healthcare. Future healthcare J. **6**(2), 94 (2019)
3. Devlin, J., Chang, M.W., Lee, K., Toutanova, K.: Bert: Pre-training of deep bidirectional transformers for language understanding. arXiv preprint arXiv:1810.04805 (2018)
4. Jiang, F., et al: Artificial intelligence in healthcare: past, present and future. Stroke Vasc. Neurol. **2**(4), 230–243 (2017)
5. Johnson, A.E., et al.: Mimic-iii, a freely accessible critical care database. Sci. data **3**(1), 1–9 (2016)
6. Kingma, D.P., Ba, J.: Adam: A method for stochastic optimization. arXiv preprint arXiv:1412.6980 (2014)

7. Liu, X., Zhao, R., Zhang, Y., Zhang, F.: Prognosis prediction of breast cancer based on CGAN. In: Xing, C., Fu, X., Zhang, Y., Zhang, G., Borjigin, C. (eds.) WISA 2021. LNCS, vol. 12999, pp. 190–197. Springer, Cham (2021). https://doi.org/10.1007/978-3-030-87571-8_16

8. Luo, J., Ye, M., Xiao, C., Ma, F.: Hitanet: Hierarchical time-aware attention networks for risk prediction on electronic health records. In: Proceedings of the 26th ACM SIGKDD International Conference on Knowledge Discovery & Data Mining, pp. 647–656 (2020)

9. Lysaght, T., Lim, H.Y., Xafis, V., Ngiam, K.Y.: Ai-assisted decision-making in healthcare. Asian Bioethics Rev. **11**(3), 299–314 (2019)

10. Ma, F., Chitta, R., Zhou, J., You, Q., Sun, T., Gao, J.: Dipole: Diagnosis prediction in healthcare via attention-based bidirectional recurrent neural networks. In: Proceedings of the 23rd ACM SIGKDD international conference on knowledge discovery and data mining, pp. 1903–1911 (2017)

11. Qiu, L., Gorantla, S., Rajan, V., Tan, B.C.: Multi-disease predictive analytics: A clinical knowledge-aware approach. ACM Trans. Manage. Inform. Syst. (TMIS) **12**(3), 1–34 (2021)

12. Vaswani, A., et al: Attention is all you need. In: Advances in Neural Information Processing Systems, vol. 30 (2017)

13. Wang, H., Cui, Z., Chen, Y., Avidan, M., Abdallah, A.B., Kronzer, A.: Predicting hospital readmission via cost-sensitive deep learning. IEEE/ACM Trans. Comput. Biol. Bioinform. **15**(6), 1968–1978 (2018)

14. Xu, X., et al.: Predictive modeling of clinical events with mutual enhancement between longitudinal patient records and medical knowledge graph. In: 2021 IEEE International Conference on Data Mining (ICDM), pp. 777–786. IEEE (2021)

15. Yin, C., Zhao, R., Qian, B., Lv, X., Zhang, P.: Domain knowledge guided deep learning with electronic health records. In: 2019 IEEE International Conference on Data Mining (ICDM), pp. 738–747. IEEE (2019)

16. Zhang, X., Qian, B., Li, Y., Yin, C., Wang, X., Zheng, Q.: Knowrisk: an interpretable knowledge-guided model for disease risk prediction. In: 2019 IEEE International Conference on Data Mining (ICDM), pp. 1492–1497. IEEE (2019)

Machine Reading Comprehension Based on Hybrid Attention and Controlled Generation

Feng Gao[1,2,3], Zihang Yang[1,2,3], Jinguang Gu[1,2,3(✉)], and Junjun Cheng[4]

[1] School of Computer Science and Technology, Wuhan University of Science and Technology, Wuhan 430065, China
616660865@qq.com
[2] Big Data Science and Engineering Research Institute, Wuhan University of Science and Technology, Wuhan 430065, China
[3] Laboratory of Content Organization and Knowledge Service for Rich Media Digital Publishing, Wuhan University of Science and Technology, Wuhan 430065, China
[4] China Information Security Evaluation Center, Beijing 100083, China

Abstract. With the development of natural language processing technology, machine reading comprehension has been widely used in various fields such as QA systems and Intelligence Engineering. However, despite numerous models are proposed in the general domain, there is still a lack of appropriate dataset and models for specific domains like anti-terrorism or homeland security to address this problem. Therefore, a Chinese reading comprehension data set in the field of anti-terrorism (ATCMRC) is constructed, and a generative machine reading comprehension model (AT-MT5) is proposed. ATCMRC was constructed in a semi-automated manner, and a domain-specific vocabulary is created based on the dataset to assist the AT-MT5. The model uses a hybrid attention layer is and a controlled answer generation layer to enhance text perception. Finally, the ATCMRC dataset and the AT-MT5 model is evaluated against existing approaches. The experimental results show that ATCMRC covers key issues in the domain and presents challenging MRC tasks for existent models, while AT-MT5 achieves better results in the domain specific dataset than the existing methods.

Keywords: Machine reading comprehension · Attention · Anti-terrorism · Dataset

1 Introduction

Machine reading comprehension (MRC) is an important task in natural language processing, which requires machines to read a text passage or document as context, and answer questions according to the context. Various MRC datasets have also been established, such as SQUAD [1], MS MARCO [2], and Dureader [3]. However, these large-scale data sets are not specific to the characteristics of the field of anti-terrorism.

In the field of anti-terrorism, the terrorist attacks are highly sudden and uncertain in time, space and region. There are also huge differences in the form of terrorism and the motives of terrorist attacks in different countries and regions. Its semantic features

© The Author(s), under exclusive license to Springer Nature Switzerland AG 2022
X. Zhao et al. (Eds.): WISA 2022, LNCS 13579, pp. 341–353, 2022.
https://doi.org/10.1007/978-3-031-20309-1_30

include both the category hierarchy relationship through the vertical structure of event classes, and the horizontal non-hierarchical relationship between What, Who, Why, How and other elements [4]. As a result, existing MRC models trained using these general datasets cannot perform well in the specific domain.

Apart from lacking domain-specific dataset and trained models, most existing MRC models focus on tasks like cloze Test, multiple choice and span extraction [5, 6]. However, in intelligence engineering related to anti-terrorism, intelligence officers need to summarize pieces of news, reports and other forms of articles on a daily basis, thus need to perform the task of generative answering, i.e., providing answers that may contain words that do not appear in the context. We refer to models solving this problem as Generative MRC, which do not only need to create correct answers, but also provide them in a user-friendly and human-understandable way, making it a challenging task. Moreover, Generative MRC in the domain need to identify questions that do not have answers in the context.

The Attention mechanism can improve the model's ability to capture text features [7]. While the generative MRC task emphasizes the association between context and problem, the bidirectional attention mechanism is exactly based on the interaction between context and problem for modeling, making the problem information and context information complementary. Since the anti-terrorism domain contains many proper nouns, the text attention based on professional knowledge can help the model to identify the domain features. Therefore, combining domain knowledge attention with question context bidirectional attention can improve the performance of the model in generative MRC tasks.

The answer of the extractive MRC model comes from the article fragment, the form does not conform to the human speech habits, and it cannot adapt to the complex problems in the anti-Terrorism domain. The generative MRC model often uses the beam search to select words that are not in the context, which improves the fluency of the answer. Controllable text generation methods can also be used to make the answer form more in line with the question, thereby improving the model performance.

Our contributions can be summarized as follows:

1. We collected anti-terrorism news from mainstream news media websites within the past 15 years. Based on the real needs of the field, a semi-automatic method is used to construct a Chinese machine reading comprehension dataset in the field of anti-terrorism (ATCMRC).
2. The generative reading comprehension model for the anti-terrorism domain (AT-MT5) is proposed. The model introduces a domain keyword list to enhance domain knowledge. The text perception capability of the model is improved by hybrid attention layer, and answer prediction is optimized using controlled text generation methods.

2   Related Work

Existing Datasets For Reading Comprehension. In the research of MRC techniques, common datasets include SQUAD, MS MARCO, and Dureader. Among them, SQUAD is a dataset of article paragraph types, the question and answer (Q&A) pairs are generated manually, and the answer is a fragment in the original text.MS MARCO uses the search engine to collect the questions and context, and the answers are generated manually. Dureader is the largest Chinese MRC dataset released by the Baidu team in 2017. Its data comes from Baidu Search and Baidu Know. The data is structured as a single question that corresponds to multiple contexts and contains both extractive and generative Q&A pairs. During the construction of ATCMRC, the data structure refers to the SQUAD. The extractive and generative Q&A pairs draw on the Dureader and MS MARCO.

Advanced MRC Model. Over the past few years, a large number of advanced NLP models have emerged. BERT [8] proposed by Devlin et al. uses MLM (masked language model) to overcome one-way limitation in traditional frameworks. And BERT is suitable for subtasks such as text classification, sequence labeling, or question answering, but it is not competent for machine translation and text generation tasks. The "Text-to-Text Transfer Transformer" (T5) model [9, 10] released by Raffel et al. (2020) converts NLP tasks such as translation, question answering, and classification into a unified "text-to-text" format [11]. T5 uses the Transformer's Encoder-Decoder structure and selects ReLU as the activation function, the attention layer adopts a fully-visible attention strategy and uses the denoising objective as the unsupervised objective function. As a multilingual version of the T5, mT5 [12] uses Gated-GELU instead of ReLU as the activation function, and pre-trains on unlabeled data without using methods such as dropout to improve model performance. Since mT5 is a multilingual model, its vocabulary contains 250,000 words, so the model is not accurate enough for Chinese word segmentation, and the processing of professional corpus also needs to be optimized. At the same time, since the mT5 model is a multi-task general framework, targeted improvements in MRC tasks can effectively improve the performance of the model in specific tasks.

Attention In MRC. The performance of MRC models depends heavily on the application of attention mechanisms. The Bi-DAF proposed by Carrie et al. [13] in 2017 is based on the interaction between context and question, using bidirectional attention from context to question and question to context. The problem and context of Bi DAF are summarized into a single feature vector, and the attention vector at each moment is related to the embedding of its previous layer, and can flow to the subsequent network. Inspired by the Bi-DAF model, the ATMT5 encoder incorporates a bi-directional attention layer, and this attention mechanism is designed to effectively enhance the model's ability to perceive the problem and context features.

Answer Generation. Commonly used text generation methods include copy generation mechanism, beam search [14] and so on. To make the text generated by the model more consistent with the task requirements, the text theme or sentiment can be controlled by target attributes or control codes. Conditional language generation methods improve the compatibility between text and tasks by focusing on topics, predicates, objects, etc. The controlled text generation method [15] guides the output of the model by combining the target input, which not only ensures the smoothness of the answer, but also makes the output more in line with the task requirements. This paper uses the question as the target content, the language model decoder output as the control content, then combines the beam search method to optimize the answer generation.

Unanswerable Task. The unanswerable question is a key subtask in the field of MRC. Hu et al. [16] normalize the no-answer probability using a multi-head network, and introduced an independent loss function to solve the reader no-answer problem. Back et al. [17] use the attention mechanism to compute the satisfaction scores of candidate answers to the semantic features of the question. However, the previous research mainly focused on the prediction of unanswerable questions by extractive models [18]. In this paper, the unanswerable probability is calculated based on the feature representation of the question. A new loss function is introduced into the generative model for joint training to help the model determine whether the question is answerable.

3 ATCMRC Dataset Construction

This section will introduce the construction method of Chinese MRC datasets in the field of anti-terrorism. By combining automatic pre-labeling and manual labeling, this method improves the labeling efficiency while ensuring the quality of Q&A pairs. Considering that the corpus characteristics of other professional fields are similar to the anti-terrorism field, with strong timeliness and many key entities, it makes the method can be applied to other professional domains.

3.1 Data Sources

First, use "anti-terrorism", "terrorist attack", "terrorist organization", and "violent conflict" as a collection of keywords to search on official media such as CCTV, Sohu News, and Baidu Information, and limit the time range to 15 years. 2000 pieces of anti-terrorism security news text data that meet the requirements were obtained from the search results. The data is cleaned by the following steps:

Preliminary data screening. News entries with news sources other than official media, missing headlines, published 15 years ago and news text less than 200 characters or greater than 1000 characters in length were removed.

Content filtering. News entries with completely duplicate entries and high overlap in content, as well as those with obvious logical errors and descriptions related to anti-terrorism security, are deleted. Then supplemented with a corresponding number of news that meet the criteria to ensure data volume.

Data formatting. To facilitate manual annotation, the news data are organized and the field names of each part are set with reference to CMRC2018. For example: ID of each news item, news field (Context), extracted question fields (Q1, Q2......Q5) and the corresponding answer fields (A1, A2......A5), the starting position of each answer in the text (Answer_start), the generative question field (GTquestion), the corresponding answer field (GTanswer).

3.2 Q and A Pair Labeling

Because the question and answer in the ATCMRC adopt the method of model pre-labeling and manual labeling, the data quality requirements are high. It is necessary to formulate clear and unambiguous labeling rules before labeling, so as to ensure that the standards for generating Q&A pairs are consistent and meet the requirements.

Labeling Rules. Randomly extract 50 news from the cleaned data, and conduct text feature analysis. combined with the opinions of experts in the anti-terrorism field, the following question-and-answer pairing rules are formulated:

- For question labeling, try not to use the original sentences in the news, and give priority to asking questions about key entities such as people's names, place names, time, and events in the news.
- The question-answer pairs should contain entity type, description type, right or wrong type, unanswerable type and multi-hop type.
- For generative answer, it should conform to the question-asking style and the answers should be as concise as possible to avoid excessive bias in model testing.

Question Pre-Labeling. The pre-labeling method uses the ERNIE-GEN [19, 20] Chinese pre-training model to generate some of the questions. The model is fine-tuned on the CMRC2018 dataset using the anti-terrorism domain keyword vocabulary (Subsect. 4.1), and the specific parameter settings are shown in Table 1. The news in the Context field (Subsect. 3.1) is used as the model input, use the fine-tuned model to generate corresponding questions, and fill in the Q1 field as the pre-labeling result of the question.

Table 1. Parameter settings of the ERNIE-GEN.

Batch_size	Learning_rate	Length_penalty	Log_interval	Noise_prob
5.0	5e-5	1.0	20.0	0.2

Manual Labeling. Manual labeling is performed according to the labeling rules. There were six annotators, working in groups of two, divided into 10 labeling rounds. After each round, members of the group checked each other to see if the Q&A pairs met the labeling rules. The unqualified pairs were marked out and the reasons were indicated, and the next marking round was handed over to the next group for re-marking (the unqualified pairs in the third group were re-marked by the first group). The last round of unqualified pairs will be corrected as the end of the marking process.

Examples of Datasets. A sample ATCMRC dataset is shown in Table 2, in which the Answer_start field of the starting position of the extracted answer is automatically calculated.

Table 2. Example of ATCMRC.

Context: According to US media reports on May 30 ...... Russian security forces said the six terrorist suspects were arrested in the Russian Republic of Ingushetia, "on suspicion of planning and carrying out the Stavropol bombing that took place on the 26th." It is reported that the six suspects are all members of an illegal armed organization in the North Caucasus. On the evening of the 26th......	
Q1: Russia's southern security forces arrest 6 suspects over alleged crime?	
A1: Stavropol bombing	Answer_start: 36
Q2: Where were these 6 terror suspects arrested in Russia?	
A2: Republic of Ingushetia	Answer_start: 68
Q3: Where are the six suspects members of illegal armed groups?	
A3: the North Caucasus	Answer_start: 157
GTquestion: What happened in the Russian city of Stavropol on May 26?	
GTanwser: Bombing near Ropol Palace of Sports and Culture in Stavropol	

4   AT-MT5 Model

This chapter improves the mT5 for generative tasks. Firstly, the vocabulary of anti-terrorism domains is constructed using temporal feature words and technical keyword matching algorithms to replace the vocabulary in the original model. Then, a hybrid attention layer is added before the model encoder to enhance the model's perception of textual features of questions and news, as well as the model's ability to capture key domain information. Answer prediction is optimized using controlled text generation method, so that the answer is generated in the form of the question. The output layer constructs a binary classifier based on the feature representation of the question to judge whether the question is answerable (see Fig. 1).

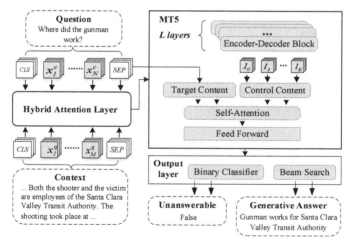

Fig. 1. AT-MT5 model

4.1 Construction of Domain Vocabulary

The Chinese pre-training model vocabulary of mT5 contains 50,000 words, covering a large number of commonly used Chinese words. However, there are many high-frequency proper terms in the field of anti-terrorism, such as "Taliban", "Islamic State", "suicide attack" and so on. And the news vocabulary in the field has strong timeliness. Therefore, we improved the Time feature word and technical keyword matching algorithm [21] based on the above features. The anti-terrorism domain corpus was divided using jieba [22] to obtain the original set of words V. The anti-terrorism domain keyword set S, the core entity vocabulary C, and the temporal feature word set T was then selected by combining expert opinions. Where the equivalence index E mainly considers the weight of association strength between t_i ($t_i \in T$) and w_i ($w_i \in W$), and the specific algorithm is described as follows.

Algorithm 1: Temporal feature word and domain keyword matching

Input: keywords in the field of anti-terrorism S,
temporal feature word set T [0..n-1],core entity vocabulary W, original vocabulary V;
Output: Vocabulary with strong semantic relevance W={S, T, C}, $w_i \in W$;

```
1    for i ← 0 to n-1 do
2        Vᵢ ← SamePeriod(V, eᵢ)
3        Cᵢ ← Vᵢ ∩ C
4        Fᵢ ← Vᵢ ∩ S
5        Kᵢ ← Fᵢ ∩ Cᵢ
6        if Kᵢ = ∅ then
7             eᵢ ← eᵢ₋₁
8        end
9        wᵢ ← MaxEquivalence (Kᵢ)
10   i ← i + 1
11   end
```

Finally, the 500 words in the output set W are reserved as special vocabulary in the field of anti-terrorism and security, and join the vocabulary of the mT5 model.

4.2 Hybrid Attention Layer

In order to highlight the domain features of text information, the input layer of AT-MT5 model introduces anti-terrorism domain vocabulary to assist in word separation. The domain keywords appearing in the news and questions are set as special_token. In the embedding layer, the learnable parameter w is added to the special_token as a weight, so as to enhance the model's attention to the characteristics of anti-terrorism fields. Finally, obtaining the news embedding representation x^c and the question embedding representation x^q.

Inspired by the Bi-DAF, the attention layer is divided into two modules: News-to-Question attention and Question-to-News attention to enhance the highly correlated semantic information in x^c and x^q.

The News-to-Question Attention module uses the inner product function to calculate the score S^c of x^c for each word in x^q. The normalized score of the inner product result is obtained by the softmax layer. Then use the normalized score to weight the question vector x^q to obtain the news perception vector A^q of the question.

$$S_i^c = \alpha(x_c, x_i^q) = x_c^T x_i^q \tag{1}$$

$$\beta_i^c = \frac{e^{S_i^c}}{\sum_j e^{S_j^c}} \tag{2}$$

$$A^q = (CLS, \beta_1^c x_1^q, \beta_2^c x_2^q, ..., \beta_{N_q}^c x_{N_q}^q, SEP) \tag{3}$$

The Question-to-News module takes news as the object of attention, and the other processes are the same as above. As a result, the problem-aware vector A^c of the news text is obtained. Finally, splicing A^q and A^c to obtain the embedded representation e of the model input.

$$e = \{CLS, \beta_1^c x_1^q, \beta_2^c x_2^q, ..., \beta_{N_q}^c x_{N_q}^q, SEP, \beta_1^q x_1^c, \beta_2^q x_2^c, ..., \beta_{N_c}^q x_{N_c}^c, SEP\} \tag{4}$$

4.3 Controlled Text Generation

Since the generation task is to obtain text answers whose semantics and form meet the requirements of the question. Therefore, in the model decoder part, the target text and control text are used to guide the model to generate answer.

The main process is to use the hidden state h_c output by AT-MT5 as the feature representation of the control text, and the embedded representation of the question as to the feature representation of the target text, see formula (6). Then fuse h_c and $x_{1:t-1}^q$ through self-attention, and splicing the Key and Value of h_c with the original.

$$K_q' = [K^c; K_q] \; V_q' = [V^c; V_q] \tag{6}$$

$$A = Soft \max(QK_q'^T)V_q' = Softmax(W)V_q' \tag{7}$$

Then input the self-attention result A into the feedforward layer to obtain the hidden state containing the control text features, and obtain the logtis(i) score of each word in the vocabulary at the i-th position in the answer through the fully connected layer. Then use softmax to get the probability representation of the j-th word in the vocabulary at the answer i position.

4.4 Unanswerable Question Prediction

AT-MT5 model obtains H_q by aggregating the News-To-Question representation with participation weight and the question in Sect. 4.2.

$$(w_1, w_2, \cdots, w_n) = softmax(A^q) \tag{8}$$

$$H^q = \sum_{i}^{n} w_i A_i^q \tag{9}$$

And build a binary classifier based on this new vector to predict the probability p^{UN} that the question cannot be answered, where w^{UN} is a trainable weight and b^{UN} is a trainable bias.

$$p^{UN} = \text{softmax}(w_{UN} Linear(H^q) + b_{UN}) \tag{10}$$

$$Loss = loss_{ans} + loss_{UN} \tag{11}$$

5 Experiments and Analysis

This section provides a comprehensive analysis of the ATCMRC dataset in terms of news text types, the number of question-answer pairs, and question types. Then the advanced pre-trained model is used to fine-tune and test on ATCMRC. Then, the performance of AT-MT5 is verified through comparative experiments and ablation experiments.

5.1 Data Analysis of ATCMRC

The ATCMRC dataset contains 2000 news, 7300 extractive Q&A pairs, and 2000 generative Q&A pairs, with a total number of characters of 4,766,264. The dataset is divided into the training set, development set and test set according to the ratio of 6:2:2.

The type of news content, which can be divided into three categories: terrorist attacks, military conflicts, and non-terrorist attacks. Various types of news data provide sufficient corpus for model training, which can effectively avoid the problem of insufficient model generalization ability due to single form and content. The distribution of specific news types is shown in Fig. 2.

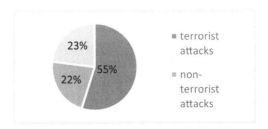

Fig. 2. News type distribution of ATCMRC

The Q&A pairs can be divided into entity type, descriptive type, yes and no type, and unanswerable type according to the type of questions, and the ratio of the number of each type of Q&A pairs is shown in Table 3.

Table 3. Question type distribution for Q&A pairs

1. Question Type	2. Extractive	3. Generative
4. Entity	5. 35%	6. 37%
7. Descriptive	8. 40%	9. 38%
10. Yes And No	11. 20%	12. 20%
13. Unanswerable	14. 5%	15. 5%

5.2 ATCMRC Experiment

This section uses BART, MT5, GPT-2 pre-trained models to fine-tune and test on this dataset, using ROUGE-L [23] as the evaluation metric for models on ATCMRC. ROUGE is mainly calculated based on the recall rate.

The parameter settings and test results of each pre-training model are shown in Table 4, where the human level is based on the answer given by experts in the anti-terrorism field, and ROUGE-L is calculated. The results show that the mT5 model performs the best, but there is still a large gap with the human level.

Table 4. Model parameter settings and test results.

Module	Vocabulary size	Hidden size	ROUGE-L
BART	50264	1024	28.83
MT5	50000	768	41.92
GPT-2	50257	768	34.26
human level	-	-	79.17

5.3 AT-MT5 Experiment

To verify the performance of the AT-MT5 model in various tasks in the field of anti-terrorism, the BART, MT5, and GPT-2 [24] models were used to conduct comparative experiments with AT-MT5 on ATCMRC. On the DuReader, AT-MT5 is compared with other models to verify its generalization ability. Finally, an ablation experiment is performed to analyze the effect of each module on the performance of AT-MT5.

Experiment Setup. AT-MT5 uses a vocabulary in the field of anti-terrorism, with a total number of 50,000 words. The tokenizer of the model sets special word segmentation rules, giving priority to the segmentation of special words in the field of anti-terrorism. The pre-trained model part uses MT5 for conditional generation, the maximum length of the input sequence is 512, the hidden layer dimension is 768, the learning rate is initially set to 5e-5. AT-MT5 uses the Adam optimizer, the num_beams of answer prediction module is 8, and the model is trained for 20 rounds with the specific parameter settings shown in Table 5.

Table 5. AT-MT5 parameter settings

Parameter	Value	Parameter	Value
vocab_size	50000	warm_up_ratio	0.1
learning_rate	0.00005	Answer_max_length	50
hidden_size	768	Max sequence length	512
dropout_rate	0.1	num_beams	8

Comparative Experiment. The AT-MT5, BART, and GPT-2 models are used for fine-tuning the ATCMRC and DuReader datasets, the answer generation adopts the beam search method, num_beams are set to 8, and the maximum length of the answer sequence is 50, with ROUGE-L as the evaluation metric. The experimental results are shown in Table 6, and the performance of the AT-MT5 model on AT-CMRC is significantly better than other models. The performance on DuReader dataset is also 2.49% higher than mt5. Then using the vector cosine similarity (VS) to evaluate the semantic similarity between the predicted answer and the real answer. The results show that the AT-MT5 significantly outperforms other models on the ATCMRC, but underperforms the other models on DuReader.

Finally, using the AT-MT5, BART, and GPT-2 models to test the Unanswerable question in the ATCMRC. The experimental results show that the prediction accuracy of the AT-MT5 is 66.11%, and the prediction accuracy of the BART and GPT-2 models are 60.56% and 57.14%, respectively.

Table 6. Comparative experimental results

DataSet \ Model		AT-MT5	mT5	BART	GPT-2
DuReader	ROUGLE-L	42.85	40.36	33.29	38.17
	VS	45.36	45.98	43.76	43.54
ATCMRC	ROUGLE-L	57.10	41.92	28.83	34.26
	VS	50.82	43.76	39.64	40.98

Ablation Experiments. The ablation experiment performed on the ATCMRC use ROUGE-L as the evaluation metric. The anti-terrorism domain word list and attention layer and controlled text generation module (CT) are removed respectively. Then analyze the impact of each module on the Model's acquisition of text and question information and the quality of the generated answer.

The results of ablation experiments are shown in Table 7, where H_ATT represents the hybrid attention layer for questions and news. First, remove the vocabulary in the fields of anti-terrorism, which leads to the inaccuracy of the model's segmentation of special words, resulting in a slight loss of model performance. After removing the bidirectional attention layer of questions and news, the performance of the model drops by 4.33%, which proves that the H_ATT can effectively help the model to enhance the semantic awareness of questions and news. Finally, the controlled text generation method is removed, and the model performance drops by 3.21%, which proves that this method can make the model generate answers more in line with the requirements of the question and closer to human language.

Table 7. Ablation experiment results

AT-MT5	MT5	-vocabulary	-H_ATT	-CC
57.10	41.92	55.61	53.37	53.89

6 Conclusion

This paper construct ATCMRC dataset based on text features and real needs in the field of anti-terroris. And testing the generative MRC task on ATCMRC using advanced pre-trained model. The results show that ATCMRC is more specialized and challenging than the general dataset in the anti-terrorism domain.

The AT-MT5 model in this paper uses the vocabulary in the field of anti-terrorism, and adds hybrid attention layer to the model. Determine whether the question is answerable based on the attention embedding representation of the question. And generate answers through controlled text generation methods. Comparative experiments show that the AT-MT5 outperforms the best performing model mT5 by 15.18% on ATCMRC. It also outperforms the mT5 by 2.49% on Dureader. The results of ablation experiments show that each optimization module can help the model to better obtain semantic information and make the answer closer to human language.

As future work, we will continue to expand the data volume of ATCMRC, improve the difficulty of Q&A pairs, and use more data set evaluation methods to analyze and improve ATCMRC. Knowledge Graphs can be introduced to enhance the understanding and reasoning ability of AT-MT5.

References

1. Rajpurkar, P., et al.: SQuAD: 100,000+ questions for machine comprehension of text. In: Proceedings of the 2016 Conference on Empirical Methods in Natural Language Processing (2016)
2. Nguyen, T., et al.: MS MARCO: A Human Generated MAchine Reading. Comprehension Dataset (2016)
3. He, W., et al.: Du Reader: a Chinese Machine Reading Comprehension Dataset from Real-world Applications (2017)
4. Zheng, H., et al:. Analysis and prospect of China's contemporary anti-terrorism intelligence perception. MATEC Web of Conferences. vol. 336. EDP Sciences (2021)

5. Guo, S., et al.: Frame-based neural network for machine reading comprehension. Knowledge-Based Systems **219**, 106889 (2021)
6. Cui, Y., et al.: Understanding attention in machine reading comprehension. arXiv preprint arXiv:2108.11574 (2021)
7. Xu, L., Li, S., Wang, Y., Xu, L.: Named Entity Recognition of BERT-BiLSTM-CRF Combined with Self-attention. In: Xing, C., Fu, X., Zhang, Y., Zhang, G., Borjigin, C. (eds.) WISA 2021. LNCS, vol. 12999, pp. 556–564. Springer, Cham (2021). https://doi.org/10.1007/978-3-030-87571-8_48
8. Devlin, J., et al.: Bert: Pre-training of deep bidirectional transformers for language understanding. arXiv preprint arXiv:1810.04805 (2018)
9. Raffel, C., et al.: Exploring the limits of transfer learning with a unified text-to-text transformer. arXiv preprint arXiv:1910.10683 (2019)
10. Vaswani, A., et al.: Attention is all you need. Advances in neural information processing systems pp. 5998–6008 (2017)
11. Li, F., et al.: Multi-task joint training model for machine reading comprehension. Neurocomputing **488**, 66–77 (2022)
12. Xue, L., et al.: mt5: A massively multilingual pre-trained text-to-text transformer. arXiv preprint arXiv:2010.11934 (2020)
13. Seo, M., et al.: Bidirectional attention flow for machine comprehension. arXiv preprint arXiv:1611.01603 (2016)
14. Scholak, T., Schucher, N., Bahdanau, D.: PICARD: Parsing incrementally for constrained auto-regressive decoding from language models. arXiv preprint arXiv:2109.05093 (2021)
15. Chan, A., et al.: CoCon: A Self-Supervised Approach for Controlled Text Generation (2020)
16. Hu, M., et al.: Read + Verify: Machine Reading Comprehension with Unanswerable Questions (2018)
17. Back, S., et al. NeurQuRI: Neural question requirement inspector for answerability prediction in machine reading comprehension. International Conference on Learning Representations (2019)
18. Fu, S., et al.: U-Net: Machine Reading Comprehension with Unanswerable Questions (2018)
19. Lin, D., Wang, J., Li, W.: Target-guided knowledge-aware recommendation dialogue system: an empirical investigation. Proceedings of the Joint KaRS & ComplexRec Workshop. CEUR-WS (2021)
20. Xiao, D., et al.: ERNIE-GEN: an enhanced multi-flow pre-training and fine-tuning framework for natural language generation. arXiv preprint arXiv:2001.11314 (2020)
21. Ye, C., Fuhai, L.: Research on the construction method of future technology vocabulary in technology roadmap. Modern Library and Information Technology **2013**(05), 59–63
22. Sun, J.: Jieba (Chinese for "to stutter") Chinese text segmentation: built to be the best Python Chinese word segmentation module 2013. https://github.com/fxsjy/jieba (2021)
23. Sellam, T., Das, D., Parikh, A.P.: BLEURT: Learning robust metrics for text generation. arXiv preprint arXiv:2004.04696 (2020)
24. Radford, A., et al.: Language models are unsupervised multitask learners. OpenAI blog **1**(8), 9 (2019)

Contextual Policy Transfer in Meta-Reinforcement Learning via Active Learning

Jingchi Jiang[1]([✉]), Lian Yan[2], Xuehui Yu[2], and Yi Guan[2]

[1] AIoT Research Center, Harbin Institute of Technology, Harbin 150001, China
jiangjingchi@hit.edu.cn
[2] Language Technology Research Center, Harbin Institute of Technology,
Harbin 150001, China

Abstract. In meta-reinforcement learning (meta-RL), agents that consider the context when transferring source policies have been shown to outperform context-free approaches. However, existing approaches require large amounts of on-policy experience to adapt to novel tasks, limiting their practicality and sample efficiency. In this paper, we jointly perform off-policy meta-RL and active learning to generate the latent context of the novel task by reusing valuable experiences from source tasks. To calculate the importance weight of source experience for adaptation, we employ maximum mean discrepancy (MMD) as the criterion to minimize the experience distribution distance between the target task and the adapted source tasks in a reproducing kernel Hilbert space (RKHS). Integrating source experiences based on active queries with a small amount of on-policy target experience, we demonstrate that the experience sampling benefits the fine-tuning of the contextual policy. Then, we incorporate it into a standard meta-RL framework and verify its effectiveness on four continuous control environments, simulated via the MuJoCo simulator.

Keywords: Meta reinforcement learning · Active learning · Uncertain sampling · MuJoCo

1 Introduction

Deep reinforcement learning that combines traditional reinforcement learning with nonlinear functional approximation [1,2], has made extensive advances in sequential decision-making problems. Although existing reinforcement learning can learn effective control strategies in complex environments, an individual policy per task often requires millions of interactions with the environment. Learning with large repertoires of behaviours is often not sample-efficient, which restricts the application of reinforcement learning algorithms in real scenarios.

To address this concern, transfer learning [3] reduces the number of samples required to learn a novel (target) task by reusing previously acquired knowledge

© The Author(s), under exclusive license to Springer Nature Switzerland AG 2022
X. Zhao et al. (Eds.): WISA 2022, LNCS 13579, pp. 354–365, 2022.
https://doi.org/10.1007/978-3-031-20309-1_31

from other related (source) tasks. Similarly, many studies in reinforcement learning focus on reuse policies [4,5]. To make such a transfer successful in practice, a learning agent should be able to identify which source policies are relevant in each state of the target environment, referred to as contextual transfer [6]. In recent years, various frameworks [7,8] have been proposed from different perspectives to tackle the policy transfer. Meta-reinforcement learning (meta-RL) is naturally well-suited to mine shared common contexts by making use of experience collected across a set of source tasks. Once trained, the meta-learned policy can be transferred to a related target task directly. However, most current meta-RL methods require online data during both meta-training and meta-testing, especially when there is a bias of behaviour distribution between the source task and target task. In the case of continuous glucose control, considering the difference in insulin sensitivity between young and elderly diabetic patients, the distribution of dosing actions is significantly different between them. To achieve model adaptation, the meta-learned dosing policy for elderly patients needs to be retrained by large amounts of interactive experience with young patients.

In this paper, we tackle the adaptative problem of meta-learned policy when the number of interactions with target task are limited. For example, due to ethical limitations, the dosing agent makes it difficult to conduct a large number of interactive experiments on human bodies for adaptation. Meanwhile, the lack of prior off-line experience also hinders to fine-tuning the meta-model. In other words, we cannot obtain enough online and offline experiences from the target task, but can only select valuable experiences from related source tasks. On the one hand, to compute the important weight of source experience for task adaptation, we adapt the covariate shift hypothesis to minimize the distance between the experience distributions of the target task and adapted source tasks. On the other hand, to further dynamically explore some informative experiences, we adapt the active learning approach to interact with source tasks. By minimizing the distance between the distributions of online and offline experiences, we not only expand the experience pool but also improve the exploration of meta-RL. The problem setting is summarized in Fig. 1. Integrating these two objectives into one context-based meta-RL framework [9], we propose a novel **Adaptive Meta Reinforcement Learning (AMRL)**. AMRL can achieve excellent sample efficiency and fast adaptation, and perform dynamic exploration strategies to perceive the behavioural uncertainty of target tasks. Finally, experiments on four continuous control environments demonstrate that the proposed approach surpasses state-of-the-art results in sample efficiency and time consumption. Furthermore, we visually analyse how our model conducts dynamic exploration to adapt rapidly target tasks and what source experiences are selected by active learning.

2   Problem Statement

Our approach is motivated by situations in which the agent based on source tasks can leverage prior (offline) experiences and explore environments to generate interactive (online) experiences for adapting a new (target) task. Based on

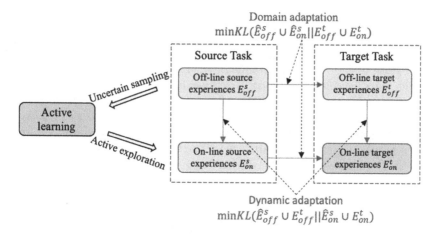

Fig. 1. Problem setting: target task experiences is insufficient, while a large amount of off-line experiences is available in the source task. The objective is to reuse source task experiences in order to train an effective model for the target task.

these experiences (including source prior experience E_{off}^s, target prior experience E_{off}^t, source interactive experience E_{on}^s, and target interactive experience E_{on}^t), we adopt a meta-RL framework to learn a latent representation as the context of the target task. It is assumed that $E_{off}^t \ll E_{off}^s$ and $E_{on}^t \ll E_{on}^s$, meaning that target task data is insufficient in both online and offline experiences. We also assume that offline experience is fixed in the whole learning procedure and online experience can be generated according to agent requirements. Sample efficiency of target task in the adaptation phase is central to our problem statement.

Markov Decision Process. We follow the previous formulation of reinforcement learning, where each task $\mathcal{T} = \{\mathcal{S}_0, \mathcal{P}, \mathcal{R}\}$ is a Markov decision process (MDP), consisting of an initial state distribution $\mathcal{S}_0 \to p(s_0)$, transition distribution $\mathcal{P} \to p(s_{t+1}|s_t, a_t)$, and reward function $\mathcal{R} \to r(s_t, a_t)$, and experience is defined as a four-tuple $\langle s_t, a_t, r_{t+1}, s_{t+1}\rangle$. Neither the transition $\mathcal{P}$ nor the reward function $\mathcal{R}$ are assumed to be known by the agent. According to the statistics of $E_{off}^s \cup E_{on}^s$ and $E_{off}^t \cup E_{on}^t$, the margin state distribution of source and target tasks is represented as $P_s(s)$ and $P_t(s)$, respectively. Similarly, the offline experience $E_{off}^s \cup E_{off}^t$ and online experience $E_{on}^s \cup E_{on}^t$ follow the margin state distribution $P_{off}(s)$ and $P_{on}(s)$. The objective of an agent is to find an optimal deterministic *policy* $\pi^* \to p(a_t|s_t)$ that maximizes the cumulative reward, defined as $Q^\pi(s, a) = \mathbb{E}_{s_i \sim \mathcal{P}, a_i \sim \pi}\left[\sum_{i=0}^{N} \gamma^i r_i \mid s_0 = s, a_0 = a\right]$, starting from an initial state-action pair (s_0, a_0).

Context-Based Meta-RL. In general, meta-RL consists of two processes: meta-training and meta-testing. For meta-training, known a set of similar training tasks $\mathcal{T}_{1\sim n}$, the agent collects experiences about n tasks by interacting with environments and learns a common policy that is applied to all tasks on condition of a *context* $\mathbf{c}$ of per task. Let $\mathcal{E}_{1\sim N}^{\mathcal{T}_i}$ be N experiences collected so far in the task $\mathcal{T}_i$ so that an inference network $q_\phi(\mathbf{c}_{\mathcal{T}_i} \mid \mathcal{E}_{1\sim N}^{\mathcal{T}_i})$ estimates the latent context $\mathbf{c}_{\mathcal{T}_i}$ of $\mathcal{T}_i$. For meta-testing, experiences including online and offline data on a novel task is fed into the inference network to obtain a contextual representation. The policy adapts to the novel task by conditioning on the task-specific context.

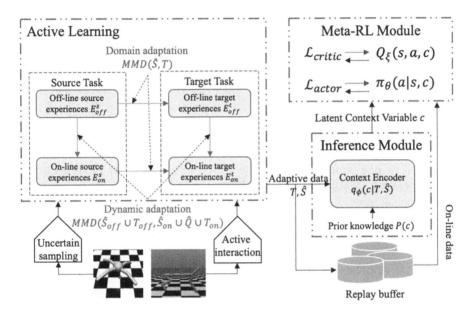

Fig. 2. The summary of adaptive meta reinforcement learning (AMRL). AMRL obtains online experiences from environments in ways of uncertainty sampling and active interaction. The importance weight of source experience is learned by minimizing the distance of maximum mean discrepancy (MMD) between distributions. Then, the context encoder q_ϕ uses adaptive experiences to infer the posterior over the latent context variable $\mathbf{c}$, which conditions the actor and critic, and is optimized with gradients from the critic, following the settings of soft actor-critic (SAC).

3 The Method

3.1 Latent Contexts with Active Learning

In this paper, we consider that the margin state distribution is different in the source and target tasks, $P_s(s) \neq P_t(s)$, while $P_{off}(s)$ is the same as $P_{on}(s)$ under sufficient experience. In the case of insufficient experiences, the key issue for adaptation is to accurately estimate the importance weight for each source experience and close the gap among distributions, $KL(P_s(s)\|P_t(s))$

and $KL(P_{off}(s)||P_{on}(s))$. Here, we rewrite the objective function of the target task.

$$\max_\theta \mathbb{E}_{s_i \sim \mathcal{P}, a_i \sim \pi} \left[\sum_{i=0}^{N} \gamma^i r_i \mid s_0 = s, a_0 = a \right]$$

$$\approx \max_\theta \sum_{l=1}^{M} \sum_{i=0}^{N} \gamma^i r\,(s_i, a_i)\, P_t(s_i)\pi_\theta(a_i|s_i)$$

$$= \max_\theta \sum_{l=1}^{M} \sum_{i=0}^{N} \gamma^i r\,(s_i, a_i)\, \frac{P_t(s_i)}{P_s(s_i)} P_s(s_i)\pi_\theta(a_i|s_i) \qquad (1)$$

$$\approx \max_\theta \sum_{l=1}^{M} \sum_{i=0}^{N} \gamma^i r\,(s_i, a_i)\, \frac{P_t(s_i)}{P_s(s_i)} \widetilde{P}_s(s_i)\pi_\theta(a_i|s_i)$$

$$= \max_\theta \sum_{l=1}^{M} \sum_{i=0}^{N} \gamma^i r\,(s_i, a_i)\, \frac{P_t(s_i)}{P_s(s_i)} \pi_\theta(a_i|s_i).$$

where M is the number of episodes and N is the number of steps in each episode. Note that the margin state distribution of the source task is replaced by its estimate $\widetilde{P}_s(s_i)$, which is a fixed value given sufficient source experiences. If source experiences are reused for training the policy of the target task, we expect to assign each source experience with an importance weight $\beta(s_i) = P_t(s_i)/P_s(s_i)$, in which the variable s_i is taken from the first element of each four-tuple experience. Here $P_t(s_i)$ and $P_s(s_i)$ denote the density functions of the margin state distribution of the target and source tasks, respectively. To avoid the density estimation, we can directly optimize the importance weights by minimizing the distance between the distributions of the target task and adapted source task. We employ maximum mean discrepancy (MMD) as the criterion to estimate the distance between different distributions. Specifically, the empirical estimate of MMD can be formalized as:

$$MMD(\widehat{S}, T) = \left\| \frac{1}{n_S} \sum_{\mathcal{E} \in S} \beta(\mathcal{E})\varphi(\mathcal{E}) - \frac{1}{n_T} \sum_{\mathcal{E} \in T} \varphi(\mathcal{E}) \right\|_{\mathcal{H}} \qquad (2)$$

where $S = E^s_{off} \cup E^s_{on}$ and $T = E^t_{off} \cup E^t_{on}$ represents the sets of source experience and target experience, respectively. n_S and n_T is the number of experiences. $\widehat{S} = \{\beta(\mathcal{E})\mathcal{E} \mid \mathcal{E} \in E^s_{off} \cup E^s_{on}\}$ is the set of adapted source experience, and $\varphi : \mathcal{E} \to \mathcal{H}$ is a mapping from the feature space to a reproducing kernel Hilbert space (RKHS). It is easy to observe that the MMD is actually measured with the distance between the means of the two domain experiences mapped into an RKHS. The target of task adaptation is to optimize the importance weights β by minimizing Eq. 2.

As discussed previously, we not only use target experiences to generate a comprehensive context of the target task but also supplement the weighted source experiences. The latent context of target task $\mathcal{T}_i$ is estimated by an inference network $q_\phi(c_{\mathcal{T}_i} \mid T, \widehat{S})$. The basic idea is that the distributions of source data and target data should be closed, such that the latent context of the target task trained will have good generalization ability. This implies that the importance weights for task adaptation are considered when performing distribution matching.

Moreover, based on the assumption that the offline state distribution $P_{off}(s)$ is the same as the online state distribution $P_{on}(s)$, we measure another distance as follows:

$$MMD(\widehat{S}_{off} \cup T_{off}, \widehat{S}_{on} \cup \widehat{Q} \cup T_{on}) \tag{3}$$

where experience sets consist of five parts: offline source experience and target experience, $\widehat{S}_{off}$ and T_{off}; online source experience and target experience, $\widehat{S}_{on}$ and T_{on}; and active experience $\widehat{Q}$. Here, the symbol $\widehat{\;}$ represents the experience set adapted with importance weights, e.g., $\widehat{S}_{off} = \{\beta(\mathcal{E})\mathcal{E} \mid \mathcal{E} \in E_{off}^s\}$. In addition to the normal interactions with source tasks, the agent will actively perform some exploratory actions on the condition of "unknown" states, which are embodied in some unusual offline experiences. The resulting active experiences $\widehat{Q}$ are considered part of online experiences to deepen the understanding of environments.

3.2   Exploration of Uncertain Sampling

At the end of each episode, we put the active experiences into a replay buffer. These experiences will participate in the learning of the target-task context. We thus incorporate an uncertainty term to explore "unknown" states. Specifically, we first obtain the predictive action a_t of each online experience $\langle s_t, a_t, r_{t+1}, s_{t+1} \rangle$. Then, the certainty of the state s_t for the current agent g is simply estimated with $|g(s_t)|$, indicating that a state with a confidence of predictive action closer to zero is more uncertain. Our target is to collect a small batch of off-line experiences with larger uncertainty and perform exploratory behaviours to reduce their uncertainty. By combining the objectives of distribution matching and uncertain sampling, we have the following framework for adaptive meta-RL with active learning:

$$\min MMD(\widehat{S}, T) + MMD(\widehat{S}_{off} \cup T_{off}, \widehat{S}_{on} \cup \widehat{Q} \cup T_{on}) + \lambda \alpha \beta |g_{S_{off}}|. \tag{4}$$

where λ is a tradeoff parameter for balancing the contributions of distribution matching and uncertainty, and α should be optimized to achieve a minimal value on $\alpha \beta |g_{S_{off}}|$. This framework can be rewritten in more detail as the following optimization problem:

$$
\begin{aligned}
\max_{\alpha, \beta} \| &\frac{1}{n_S} \sum_{\mathcal{E} \in S} \beta(\mathcal{E})\varphi(\mathcal{E}) - \frac{1}{n_T} \sum_{\mathcal{E} \in T} \varphi(\mathcal{E}) \|^2 + \| \frac{1}{n_{off}} (\sum_{\mathcal{E} \in S_{off}} \beta(\mathcal{E})\phi(\mathcal{E}) \\
&+ \sum_{\mathcal{E} \in T_{off}} \phi(\mathcal{E})) - \frac{1}{n_{on}} (\sum_{\mathcal{E} \in S_{on}} \beta(\mathcal{E})\phi(\mathcal{E}) + \sum_{\mathcal{E} \in S_{off}} (1 - \alpha(\mathcal{E}))\beta(\mathcal{E})\phi(\mathcal{E}) \\
&+ \sum_{\mathcal{E} \in T_{on}} \phi(\mathcal{E})) \|^2 + \lambda \sum_{\mathcal{E} \in S_{off}} \alpha(\mathcal{E})\beta(\mathcal{E})|g(\mathcal{E})| \\
s.t. \quad &\alpha(\mathcal{E}) \in \{0, 1\}, \forall \mathcal{E} \in S_{off}; \quad \beta(\mathcal{E}) \in [0, 1], \forall \mathcal{E} \in S
\end{aligned} \tag{5}
$$

Note that the binary constraints on α make the above problem NP-hard. By relaxing the constraints to let $\alpha(\mathcal{E}) \in [0, 1]$, the problem is biconvex and can be solved alternately with a guarantee of convergence. To optimize α with β fixed, the quadratic programming problem is exhaustively derived in [10].

3.3 Meta-Reinforcement Learning

Our work instead focuses on off-policy meta-learning and builds on a standard context-based meta-RL that incorporates the soft actor-critic algorithm (SAC) and a context encoder (inference network). The summary of our framework is presented in Fig. 2. By computing gradients for the inference network $q_\phi(c \mid T, \widehat{S})$ with reweighted source experiences, we optimise the parameters of the inference network $q_\phi(c \mid T, \widehat{S})$ jointly with the parameters of the critic $Q_\xi(s, a, c)$. The inference network encodes task context is trained by gradients from the Bellman update for the critic, and the critic loss can be written as:

$$\mathcal{L}_{critic} = \mathbb{E}_{\substack{(s,a,r,s') \sim \mathcal{P} \\ c \sim q_\phi(c|T,\widehat{S})}} \left[Q_\xi(s,a,c) - (r + \overline{V}(s', \bar{c})) \right]^2. \tag{6}$$

where $\overline{V}$ is a target network and $\bar{c}$ indicates that gradients are not being computed through it. The actor loss is nearly identical to SAC, with the additional dependence on c as a policy input.

$$\mathcal{L}_{actor} = \mathbb{E}_{\substack{s \sim \mathcal{P}, a \sim \pi_\theta \\ c \sim q_\phi(c|T,\widehat{S})}} \left[D_{\mathrm{KL}}\left(\pi_\theta(a \mid s, \bar{c}) \middle\| \frac{\exp\left(Q_\xi(s,a,\bar{c})\right)}{\mathcal{Z}_\theta(s)}\right) \right]. \tag{7}$$

where $\mathcal{Z}_\theta(s)$ is the normalized distribution of Q. Here, we use the reparameterisation trick to sample action: $a_t = f_\theta(\epsilon_t; s_t)$, where ϵ_t is a Gaussian distribution. Then, we can rewrite Eq.7 as

$$\mathcal{L}_{actor} = \mathbb{E}_{\substack{s \sim \mathcal{P}, \epsilon_t \sim \mathcal{N} \\ c \sim q_\phi(c|T,\widehat{S})}} \left[log\pi_\theta(f_\theta(\epsilon_t; s_t)|s_t) - Q_\xi(s_t, f_\theta(\epsilon_t; s_t)) \right]. \tag{8}$$

Note that the context used to infer $q_\phi(c \mid T, \widehat{S})$ is distinct from the experience used to construct the critic loss. During meta-training, we sample context batches separately from RL batches and weight the context batches from source tasks by active learning. Concretely, the actor and critic are trained with batches of transitions drawn uniformly from the entire replay buffer. The inference network is trained by weighted context batches.

4   Experiments

In our experiments, we introduce the experimental setup, including the environment, evaluation, and implementation details. Then, we conduct extensive ablation studies and efficiency experiments to analyse the effectiveness of our model.

4.1 Setup

Environment and Evaluation. We evaluate our approach on four common continuous control benchmarks of the MuJoCo[1] simulator, which requires adaptation across reward functions (Half-Cheetah-Vel, Ant-Goal (Normal), Ant-Goal (Differ), and Point-Robot). Each environment samples 20 different tasks from the task-goal space, including 15 training tasks (denoted by $\mathcal{T}_0 \sim \mathcal{T}_{14}$) and 5 testing tasks (denoted by $\mathcal{T}_{15} \sim \mathcal{T}_{19}$). All tasks have a horizon length of 200. For the four continuous control benchmarks of MuJoCo, we use the episode reward of sampling tasks to reflect the overall performance.

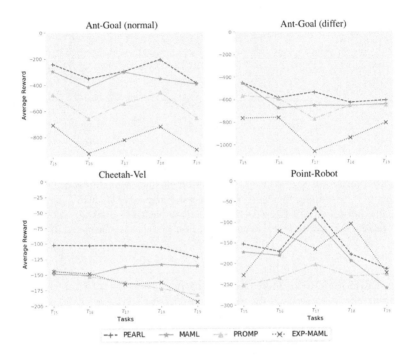

Fig. 3. Average returns achieved by different methods on the four common continuous control benchmarks of MuJoCo. The different lines in the figure correspond to the average returns obtained by different models over 100 tests. Our model significantly outperforms previous meta-RL methods on four continuous control benchmarks and consistently achieves better results than the best-performing prior methods in each setting.

4.2 Comparison with State-of-the-Art

More Excellent. To evaluate meta-testing tasks, we perform adaptation at the trajectory level, where the first trajectory is collected with context variable c sampled from the prior $p(c)$. Subsequent trajectories are collected with $c \sim$

---

[1] https://github.com/deepmind/mujoco.

$q_\phi(c|T, \widehat{S})$, where the context is aggregated over all trajectories. To compute final performance, we report the average returns of trajectories collected after all trajectories have been aggregated into the context. We compare our model with previous meta-RL methods, including the existing meta-RL methods PEARL [9], ProMP [11] , MAML [12] and EXP-MAML [13], using publicly available code. Figure 3 shows the comparison of state-of-the-art methods. Our method achieves better results on most MuJoCo environments.

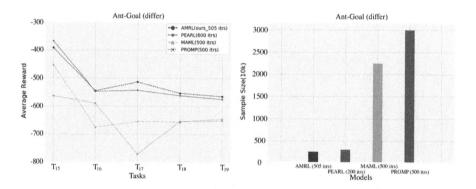

Fig. 4. Sample efficiency comparison. Our model outperforms state-of-the-art methods in terms of average return and sample efficiency across four continuous control environments. The sample efficiency can be improved by more than 16% compared with PEARL. Furthermore, PEARL with 600 iterations basically matches the effect of our model with 505 iterations, which demonstrates an increase in the speed of adaptation.

Sample Efficiency and Fast Adaptation. The reuse of source experience with weight can compensate for the lack of target experience. By fine-tuning the meta-model jointly with $\widehat{S}$, the sample efficiency of our model is greatly improved. For each testing task, our model uses only 21k target experiences to achieve meta-model adaptation. A total of 509k experiences from both the source and target tasks are involved in training, while the best-performing PEARL model among previous meta-RL methods needs 606k experiences to match the performance of ours. This proves that the testing task can be better controlled even with a small amount of target experience. By analysing the experimental results, we further found that the effect of PEARL after 600 iterations is basically the same as that of our model after 505 iterations (consisting of 500 training iterations and 5 active learning iterations). This illustrates that context-based RL with the active exploration strategy not only improves sample efficiency but also reduces time consumption. Figure 4 shows the number of samples required for the four comparison methods to achieve the same performance in the Ant-goal (Differ) environment.

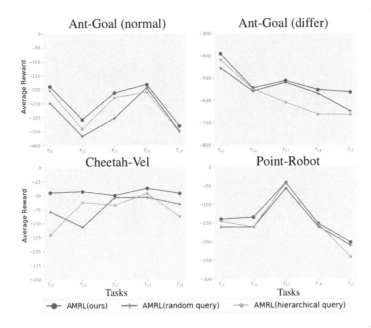

Fig. 5. Influence of active learning. The active exploration strategy of the meta-model is replaced by a random selection strategy and a hierarchical selection strategy. By testing in four benchmark tasks, the average returns of the meta-RL based on active learning are significantly better than the others. This proves that active learning is beneficial for improving the performance of the meta-model.

4.3 Influence of Active Learning

Ablation Studies. By calculating the importance weights of source-experiences, the self-adaptive E^s_{off} and E^s_{on} combined with target experiences are used to fine-tune the meta-RL. In this section, we mainly discuss the impact of active learning on the overall effect of the meta-model. As a comparison, we use a random selection strategy and hierarchical selection strategy to filter samples from source experiences S to replace the part of the active learning strategy. Figure 5 shows the average returns of the three strategies in the four continuous control benchmarks. We can see that the effect of the meta-model based on active learning is much better than that of the other two methods. Assigning self-adaptive weights to source experiences can filter out the valuable information that is better suited for the target task.

Why Does Active Learning Work? From the previous analysis, the active learning improves the performance of the meta-RL and sample efficiency. Then, we analyse the effectiveness of active learning from the perspective of experimental data. Considering the intuitiveness of the visualization, we take the case of transferring from training tasks $\mathcal{T}_4$, $\mathcal{T}_8$, $\mathcal{T}_{13}$ to testing task $\mathcal{T}_{18}$ in the Ant-Goal (Differ) environment as an example. Figure 6.a expresses the distribution

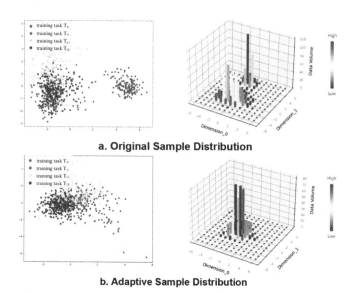

a. Original Sample Distribution

b. Adaptive Sample Distribution

Fig. 6. Experience distribution before and after active learning. The left half of the figure is the scatter representation of the experiences by dimensionality reduction. The right half intuitively shows the experience distribution. The sample density increases from dark blue to dark red. Adaptation based on active learning closes the gap between two experience distributions.

of experiences before active learning, while Fig. 4.b shows the distribution of adaptive experiences after active learning. We can observe that the source experiences weighted by active learning are closer to the target experiences, which alleviates the lack of samples in the target task. Combined with the adaptive source experiences, the target task can be given a more expressive context for adaptation.

5   Conclusion

In this paper, we propose a novel meta-RL algorithm, AMRL, which reuses adaptive source experiences to compensate for the lack of experience in target tasks, improving the sample efficiency. Our approach is particularly amenable to context-based meta-RL algorithms, as active learning solves the problems of distribution differentiation by minimizing mismatch between source and target experiences, which are joined to infer a latent context variable on which the policy is conditioned. In addition, an uncertain sampling strategy performs richer exploration behaviours that enhance adaptation efficiency of meta-RL algorithm. Experimental results for task adaptation show that our approach has surpassed state-of-the-art performance while requiring far less experience and time on a diverse set of continuous control benchmarks. Visualizing the experience distributions before and after adding the active learning module can further verify that the actively selected experience is more beneficial to improve the performance on the target task.

References

1. Mousavi, S.S., Schukat, M., Howley, E.: Deep reinforcement learning: an overview. In: Bi, Y., Kapoor, S., Bhatia, R. (eds.) IntelliSys 2016. LNNS, vol. 16, pp. 426–440. Springer, Cham (2018). https://doi.org/10.1007/978-3-319-56991-8_32
2. Li, P., Yin, Z., Li, F.: Quality control method for peer assessment system based on multi-dimensional information. In: Wang, G., Lin, X., Hendler, J., Song, W., Xu, Z., Liu, G. (eds.) WISA 2020. LNCS, vol. 12432, pp. 184–193. Springer, Cham (2020). https://doi.org/10.1007/978-3-030-60029-7_17
3. Zhuang, F., et al.: A comprehensive survey on transfer learning. Proc. IEEE **109**(1), 43–76 (2020)
4. Zang, X., Yao, H., Zheng, G., Nan, X., Kai, X., Li, Z.: Metalight: value-based meta-reinforcement learning for traffic signal control. Proc. AAAI Conf. Artif. Intell. **34**, 1153–1160 (2020)
5. Lin, L., Zhenguo, L., Xiaohong, G., Pinghui, W.: Meta reinforcement learning with task embedding and shared policy. arXiv preprint arXiv:1905.06527 (2019)
6. Taylor, M.E., Stone, P.: Transfer learning for reinforcement learning domains: a survey. J. Mach. Learn. Res. **10**(7), (2009)
7. Chen, X., Duan, Y., Chen, Z., Xu, H., Chen, Z., Liang, X., Zhang, T., Li, Z.: CATCH: context-based meta reinforcement learning for transferrable architecture search. In: Vedaldi, A., Bischof, H., Brox, T., Frahm, J.-M. (eds.) ECCV 2020. LNCS, vol. 12364, pp. 185–202. Springer, Cham (2020). https://doi.org/10.1007/978-3-030-58529-7_12
8. Yunhao, T., Tadashi, K., Mark, R., Rémi, M., Michalm, V.: Unifying gradient estimators for meta-reinforcement learning via off-policy evaluation. In: Advances in Neural Information Processing Systems, vol. 34 (2021)
9. Kate, R., Aurick, Z., Chelsea, F., Sergey, L., Deirdre, Q.: Efficient off-policy meta-reinforcement learning via probabilistic context variables. In: International conference on machine learning, pp. 5331–5340. PMLR, (2019)
10. Huang, S.-T., Chen, S.: Transfer learning with active queries from source domain. In IJCAI, pp. 1592–1598 (2016)
11. Rothfuss, J., Lee, Clavera, I., Asfour, T., Abbeel, P.: Promp: Proximal meta-policy search. arXiv preprint arXiv:1810.06784 (2018)
12. Finn, C., Abbeel, P., Levine, S.: Model-agnostic meta-learning for fast adaptation of deep networks. In: International Conference on Machine Learning, pp. 1126–1135. PMLR (2017)
13. Gurumurthy, S., Kumar, S., Sycara, K.: Mame: Model-agnostic meta-exploration. In Conference on Robot Learning, pp. 910–922. PMLR (2020)

Emotion Cause Pair Extraction Based on Multitask

Yilin Li[1], Dechen Gao[1], Yuezhong Liu[2], and Shenggen Ju[1(✉)]

[1] College of Computer Science, Sichuan University, Chengdu 610005, China
jsg@scu.edu.cn
[2] Enterprise Service, Commonwealth Bank of Australia, Sydney, NSW 2000, Australia

Abstract. Emotion cause pair extraction is a sub-task of sentiment analysis tasks, which aims to extract all emotion clauses in a given document and the cause clauses corresponding to the emotion. At present, only the method of sharing parameters at the bottom layer is used to build the connection between tasks, and the shared word encoder is used in word encoding, the phenomenon that the model pays different attention to emotional words and cause words is ignored, and the rich interactive relationship information between the three tasks cannot be fully utilized. This paper proposes a multi-task emotion cause pair extraction model based on feature fusion. The model learns the multi-information at the word level through the feature fusion module, and uses the two tasks of emotion clause extraction and cause clause extraction as auxiliary tasks. The extracted result is transformed into label information, and the method of label embedding is used to integrate into the generation of emotion cause pair representation, thereby improving the effectiveness of the emotion cause pair extraction.The experimental results of the model in the paper on the ECPE data set prove that the Mul-ECPE model has improved in the evaluation indicator of the three tasks compared to the previous series of models.

Keywords: Emotion cause pair extraction · Multitask · Feature fusion · Label embedding

1 Introduction

The emotion analysis task is a hot research area in the field of natural language processing. As comments and opinions on the Internet play an increasingly important role in influencing people's choices or events, the reasons for people's comments become more important. The objective of the Emotion Cause Extraction [1] (ECE) task is to discover the causes behind emotions. With the development of deep learning technology, deep learning is used to solve ECE task. Inspired by the Memory networks in the Question Answering System, Gui et al. [2] designed an ECE model based on the QA system. This method ignores the context of emotional words in the text. Therefore, Li et al. [3] designed a co-attention network to make full use of the information between clauses. Furthermore, Yu et al. [4] proposed to use hierarchical network to model the information at word, phrase and clause level. Xia et al. [5] also designed a hierarchical network

© The Author(s), under exclusive license to Springer Nature Switzerland AG 2022
X. Zhao et al. (Eds.): WISA 2022, LNCS 13579, pp. 366–379, 2022.
https://doi.org/10.1007/978-3-031-20309-1_32

based on Transformer, and combined the global prediction information with the relative position. Fan et al. [6] further introduced external emotional corpus as regular term in hierarchical network, and at the same time introduced regular term for relative position.

However, the ECE task requires marking the polarity of emotions in advance. The Emotion Cause Pair Extraction (ECPE) task is an improvement on ECE task. It does not need to mark the Emotion in advance. The task objective is to select the sentence where the emotion is and the sentence which causes the emotion. At present, the ECPE task is implemented by the deep learning model, and there are mainly two methods to solve the ECPE task: the two-step method and the end-to-end method. The two-step method is the traditional method for this task. First, all the emotion and cause clauses in the text are extracted to obtain two clause sets. A filter was then trained to pick the right emotion cause pair from the candidates.

However, it has the problem of error propagation. In addition, because the first step of the two-step method faces the emotion and cause clause extraction task, rather than the ECPE task, the effect of the model will be reduced.

In order to solve the problem, end-to-end architecture is widely used. Wu et al. [7] proposed a multi-task neural network to perform ECPE and two auxiliary tasks in a unified model. Fan et al. [8] compared the ECPE task to the directed graph generation task, and designed the transition-based ECPE model TransECPE. Wei et al. [9] proposed a more efficient model RANKCP, which used graph attention network to model document content and structure, solving the ECPE task from the perspective of ranking. In addition, in order to utilize the interactive information between different tasks, the parameter sharing method in multi-task learning is adopted to establish the relationship between the two sub-tasks and the emotion cause pair. Tang et al. [10], calculated the relationship score between emotion and cause clauses by using self-attention [11], extracted the emotion cause pair whose scores exceeded the threshold.

But the multi-task model cannot make full use of the rich interaction information between tasks, and the information transfer between tasks is also implicit. Label embedding has been proved to be an effective method in various fields and tasks. In the field of natural language processing, label embedding for text classification has been studied in the heterogeneous network of Tang et al. [12] and the multi-task learning environment of Zhang et al. [13]. Subsequently, Wang et al. [14] regarded the task of text classification as the problem of label and word joint embedding, and introduced an attention framework to measure the compatibility between text sequence and label embedding. Zhang et al. [15] proposed multi-task label embedding, which maps labels of each task to semantic vectors and uses a method similar to word embedding to process word sequences, so as to transform classification tasks into vector matching tasks.

The main contributions of this paper can be summarized as follows: (1) Construct the relationship between tasks by sharing parameters at the bottom level, share the results of emotion and cause extraction with the ECE task explicitly through label embedding. (2) Feature fusion is adopted to obtain feature representation with multiple relational information. (3) The model obtained optimal results on data set ECPE, which verified the validity of the model.

2   Related Work

Xia et al. [16] proposed an end-to-end solution to solve the error propagation problem of the two-step method, integrating the generation, pairing and final prediction of emotion cause pair representation into a joint framework. They designed a two-dimensional Transformer [17] and two variants to obtain the representation of emotion cause pair and captured the interaction between different emotions and causes in the process of generating vector representation.

Some researchers transfer the ideas of other tasks to ECPE tasks. Song et al.[18] regarded ECPE as a link prediction problem, to predict whether there is an edge from the emotion clause to the cause clause, and if there is, the two clauses constitute an emotion cause pair. Fan et al. [19] treated the ECPE task as a sequence labeling problem and proposed a multi-task sequence labeling model via label distribution refinement.

Most current graph convolution network rely on the structural information of graphs for classification. Li et al. [20] retained the structural information and feature similarity of graphs, and introduced the attention mechanism for classification. Chen et al. [21] applied a graph convolution network to the model and constructed an emotion cause pair graph to simulate three kinds of dependency relationships between candidate pairs in the local neighborhood, where each node represents a candidate emotion cause pair and the edge connecting two nodes represents the dependency relationship between two candidate pairs. Finally, the graph convolution network is used to extract these three types of edges so as to disseminate the context information in the graph.

3   Methodology

3.1   Task Description

The task of ECPE in this paper is to extract emotion clauses and corresponding cause clauses from a given text. A text case in the ECPE data set is given in Table 1. The text is divided into eight clauses according to punctuation marks, the sixth clause is taken as an emotion clause, the eighth clause is the corresponding cause clause. The final goal of ECPE task is to extract emotion cause pairs composed of (C5, C6).

Table 1. Data example.

Text	Class
C1:When I saw the reply written to me by the leaders of ministries and commissions that the proposal was adopted	\
C2:I knew I was doing my part for the development of the country	\
C3:Bai Jinyue, an ordinary worker of Xingtai Iron and Steel Corp., Ltd. in Hebei Province	\
C4:holding the letters of thanks that the ministries and commissions of the state have given him over the years	\
C5:told Chinanew reporters excitedly	Emotion Clause
C6:that my suggestions have been adopted by the state for 27 years	Cause Clause

3.2 Model Description

The main architecture of GAT-ECPE, an ECPE task model proposed in this paper, is shown in Fig. 1.

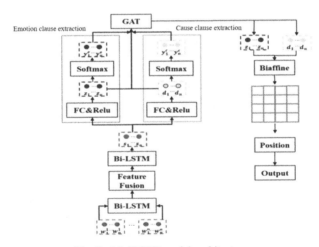

Fig. 1. MulECPE model architecture.

3.3 Input Layer

The input layer contains the word embedding and word encoding layer. Firstly, word embedding is carried out on the input text sequence, and the words which have been segmented are transformed into vectors to obtain the text sequence matrix. The i-th clause in document D can be represented by an embedding matrix $c_i = \left(w_1^i, \ldots, w_{|D|}^i \right)$.

Bi-LSTM is used to encode words. For each clause, Bi-LSTM is used to encode the context information at the word level, as shown in the following formula.

$$\left[h_i^1, \ldots, h_i^j, \ldots, h_i^{|c_i|}\right] = \left[Bi - LSTM\left(w_i^1, \ldots, w_i^j, \ldots, w_i^{|c_i|}\right)\right] \quad (1)$$

where w_i^j represents the word vector of the j-th word in the i-th clause in the document, and h_i^j represents the feature representation of the clause word level.

3.4 Feature Fusion Layer

After passing through the input layer, the feature representation at the word level is obtained. The feature representation of each clause of the text is shown in the following formula.

$$c_i = \left[h_{i,1}, \ldots, h_{i,n}\right] \quad (2)$$

Convolution kernels of different sizes are used to conduct one-dimensional convolution operations at the word level, so that the model can learn multiple features at the word level, as shown in the formula.

$$x_t^i = conv_t\left(c_1, \ldots, c_{|c|}\right) \quad (3)$$

where $conv_t$ represents the convolution operation, and t represents the size of the convolution kernel. Considering that the traditional connection cannot correctly process the language combination, the convolutional layer is densely connected:

$$x = [\oplus_t x_t] \quad (4)$$

After densely connection, the downstream layer of the convolution part can access the features generated by the upstream layer. This paper draws on the multi-scale feature attention mechanism proposed by Wang et al. [14], so that each position of the text adaptively selects features of different scales. The multi-scale feature attention mechanism consists of two steps: convolution aggregation and scale weighting (Fig. 2).

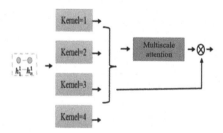

Fig. 2. Feature fusion.

First, the convolution aggregation aims to use a descriptor s_l^i to represent the features of the different scales obtained at position i. K convolution kernels are used in each convolution block. Each eigenvector is then represented by using a scalar.

$$s_l^i = F_{ensem}\left(x_l^i\right) \quad (5)$$

where F_{ensem} represents a function that sums all K elements of the input vector. And the output scalar can be used as the description of the eigenvector.

Secondly, after scale weighting, features of different scales can be weighted adaptively. The final feature representation calculation is as follows:

$$x_{atten}^i = \sum_{l=1}^{L} \alpha_l^i x_l^i \qquad (6)$$

3.5 Intra-clause Information Encoding Layer

In order to further learn the semantic order information in the document, Bi-LSTM is applied. The calculation method is shown in the formula:

$$s_i = [Bi - LSTM(a_i)] \qquad (7)$$

3.6 Label Embedding Layer

According to the Bi-affine attention mechanism, this paper takes the emotion clause as the core item and the cause clause as the dependency. The matrix composed of emotion and cause clauses are asymmetric and can sense direction. Two MLP are applied in this paper to generate the representation of emotion clauses and cause clauses respectively, as shown in the following formula:

$$z_i^e = \sigma \left(W^e h_i^e + b^e \right) \qquad (8)$$

$$z_i^c = \sigma \left(W^c h_i^e + b^c \right) \qquad (9)$$

where z_i^e represents the i-th emotion clause, and z_i^c represents the i-th cause clause.

The extraction of emotion and cause clause were embedded into the input of graph attention network, when processing the results extracted from the two auxiliary tasks, the results extracted from the emotion and the cause clause are first mapped to the vector representation by embedding matrix W:

$$Y_i^e = W^e y_i^e \qquad (10)$$

$$Y_i^c = W^c y_i^c \qquad (11)$$

where y represents the extraction result of clauses, Y is the transformed vector, and s is the context-encoded clause representation of the output of inter-clause encoding layer. After vector concatenating, they are used as the input of graph attention network to encode the representation of emotion and cause clause.

$$s_i^e = \left[s_i; Y_i^e \right] \qquad (12)$$

$$s_i^c = \left[s_i; Y_i^c \right] \qquad (13)$$

The concatenated vectors are transformed into emotion clause representation and cause clause representation respectively through the fully connected layer.

$$z_i = \sigma \left(W^e s_i^e + b^e \right) \tag{14}$$

$$d_i = \sigma \left(W^c s_i^c + b^c \right) \tag{15}$$

where z represents emotion clause and d represents cause clause.

3.7 Graph Attention Layer

Although the information between clauses has been learned through Bi-LSTM, it can only learn the single-hop relationship, while there may be a multi-hop relationship between emotion and cause. The graph attention mechanism can naturally capture the multi-hop relationship and integrate the higher-order information. In this chapter, the clause representation is updated by stacking two graph attention network layers.

The attention weight reflects correlation between clauses, which is learned by MLP. The attention between clauses is calculated by the following methods:

$$e_{ij}^{(t)} = w^{(t)^T} \tanh \left(\left[W^{(t)} z_i^{(t-1)}; W^{(t)} d_i^{(t-1)} \right] \right) \tag{16}$$

$$\alpha_{ij}^{(t)} = \frac{\exp \left(LeakyReLU \left(e_{ij}^{(t)} \right) \right)}{\sum_{k \in N(i)} \left(\exp \left(LeakyReLU \left(e_{ij}^{(t)} \right) \right) \right)} \tag{17}$$

Considering that sentiment clauses correspond to multiple cause clauses, the original neighbor aggregation method in the graph attention network is updated so that the multi-hop relationship information between two clauses can be better learned in the update. The formula is shown below:

$$z_i^t = g \left[W_1 \sum_{j \in N(i)} \left(\alpha_{ji}^t z_j^{t-1} + \alpha_{ij}^t d_j^{t-1} \right) + b_1 z_i^{t-1} \right] \tag{18}$$

$$d_i^t = g \left[W_2 \sum_{j \in N(i)} \left(\alpha_{ij}^t z_j^{t-1} + \alpha_{ij}^t d_j^{t-1} \right) + b_2 d_i^{t-1} \right] \tag{19}$$

In order to integrate these two representations into the matrix of emotion cause pairs for the final mission objective, we fold the emotion cause pairs into two parts. First, a Bi-linear like operation needs to be performed for each possible emotional cause pair. Second, Bi-affine transformations are used to deal with complex interactions, as shown in the formula:

$$M_{p,q} = \left(W^m z_p^e + b^m \right)^T z_q^c \tag{20}$$

All possible clause pairs in the original document d are regarded as candidates, assuming that the document length is $|d|$, then all of the possible emotion cause will form $|d| * |d|$ matrix, $M_{p,q}$ means the possible correct emotion cause pairs composed of the p-th emotion clause and the q-th cause clause.

The sigmoid function is further used to activate the emotion cause pair matrix:

$$\tilde{M}_{p,q} = g\left(M_{i,j}\right) \tag{21}$$

The location information between emotion and cause clause is also an important factor influencing the accuracy, emotion clause is close to cause clause, so the weight of the setting is bigger in the closer position, as shown the following formula.

$$A_{p,q} = \frac{|D| - |p - q| + \epsilon}{|D| + \epsilon} \tag{22}$$

where ϵ is the smooth term, integrates position matrix with the emotion cause pair:

$$\hat{M}_{p,q} = \tilde{M}_{p,q} \odot A_{p,q} \tag{23}$$

3.8 Extraction Results

In the E2EECPE task proposed by Song et al. [18], the emotion and cause clause extraction is regarded as two independent tasks. In this paper, the previous task is improved by adding the result of extraction of emotion and cause clauses into the representation of subsequent emotion and cause clauses.

The formulas of emotion and cause extraction of clauses are as follows:

First, the feature representation of the clause used for extraction is obtained:

$$z_i^{ae} = \sigma\left(W^{ae} h_i^{ae} + b^{ae}\right) \tag{24}$$

$$z_i^{ac} = \sigma\left(W^{ac} h_i^{ac} + b^{ac}\right) \tag{25}$$

Input to softmax layer after passing through two fully connected layers:

$$y^{ae} = softmax\left(\hat{W}^{ae} z_i^{ae} + \hat{b}^{ae}\right) \tag{26}$$

$$y^{ac} = softmax\left(\hat{W}^{ac} z_i^{ac} + \hat{b}^{ac}\right) \tag{27}$$

The method of obtaining the final result of emotion extraction and cause extraction is similar to the formula of auxiliary task, but the parameters are different:

$$\tilde{z}_i^e = \sigma\left(\tilde{W}^e h_i^e + \tilde{b}^e\right) \tag{28}$$

$$\tilde{z}_i^c = \sigma\left(\tilde{W}^c h_i^c + \tilde{b}^c\right) \tag{29}$$

where $\tilde{W}^e \in R^{d_z \times 2d_h}$, $\tilde{b}^e \in R^{d_z}$ and $\tilde{W}^c \in R^{d_z \times 2d_h}$, $\tilde{b}^c \in R^{d_z}$.

3.9 Prediction Layer

Compare $\hat{M}$ with the set threshold η to judge whether it is for emotion cause pair:

$$
\begin{cases}
\hat{Y}_{p,q} = 1\left(\hat{M}_{p,q} > \eta\right) \\
\hat{Y}_{p,q} = 0\left(\hat{M}_{p,q} \le \eta\right)
\end{cases}
\tag{30}
$$

The whole model structure can be trained by standard gradient descent. The loss is the combination of cross entropy and L2 regularization, and consists of two parts:

The first part is the loss function of task for emotion causes:

$$
L_{pair} = -\sum_{p,q} Y_{p,q} \log\left(\hat{M}_{p,q}\right) - \sum_{p,q} (1 - Y_{p,q}) \log\left(1 - \hat{M}_{p,q}\right)
\tag{31}
$$

The second part is the loss of emotion extraction and cause extraction for clauses:

$$
L_{class} = -\sum_{i} \left[\sum_{k} y_k^e \log\left(\hat{y}_k^e\right) + \sum_{k} y_k^c \log\left(\hat{y}_k^c\right) \right]
\tag{32}
$$

$y_k^e, y_k^c, Y_{p,q}$ represent the extraction of emotion clauses, the extraction of cause clauses, and the correct judgment result of ECPE.

If it is an emotion clause then $y^e = 1$, otherwise $y^e = 0$. Same rule for y^c, $Y_{p,q}$, y^e.

Two auxiliary tasks are also set, which also need to calculate the loss function:

$$
L_{aux} = -\sum_{i} \left[\sum_{k} y_k^e \log\left(y_k^{ae}\right) + \sum_{k} y_k^c \log\left(y_k^{ae}\right) \right]
\tag{33}
$$

The final training objective is composed of the above objective function, and coefficients are added to control its influence, as shown in the formula:

$$
L = L_{pair} + L_{class} + \beta L_{aux} + \lambda ||\theta||_2
\tag{34}
$$

β is used to adjust the influence of auxiliary tasks on the overall model, θ represents all parameters that is optimizable, and λ represents the parameters of L2.

The detailed approaches are stated in Algorithm 1.

Algorithm 1: MulGAT-ECPE

Input: text embedding vector w
Output: Emotion Cause Pair $\{E_p, C_q\}$

for i = 0->epoch **do**

    #Step 1:feature fusion and encode for input embedding
    h=Bi-LSTM(w)
    a=F_{fusion}(h) //multi-scale convolution operation is used for feature fusion
    s=Bi-LSTM(a) //Further study the semantic order information in the text
    #Step 2:emotion clause and cause clause extraction
    y_e, y_c=MLP(s)
    z,d=GAT(y_e, y_c)   //Capture multi-hop relationships and integrate higher-order information
    #Step 3:calculate emotion cause pairs matrix
    M=Biaffine(z,d)
    #Step 4:predict
    $\hat{Y} \leftarrow \begin{cases} 1 \ if \ M_{p,q} > \eta \\ 0 \ if \ M_{p,q} \leq \eta \end{cases}$
    return E_p, C_q
end for

---

4 Experiment

4.1 Experiment Setup

Dataset
The dataset was constructed by Xia et al. [16], each document contains only one emotion and its corresponding one or more causes. The average document length is 14.77, and the maximum length is 73. Table 2 shows the proportion of documents with different emotion cause pairs.

Table 2. Dataset information

The Document Contains The Number of Emotion Cause Pairs	Quantity	Proportion
One Pair	1746	89.77%
Two Pairs	177	9.10%
Three or More Pairs	22	1.13%

In order to better meet the task for ECPE, the documents with the same text content are integrated into one document. According to statistics, there are 1,945 paragraphs in total, including 2,167 emotion cause pairs, and the number of documents which only have one pair of emotion cause pair accounts for 89.77% of the total documents.

Experimental Parameters
In this paper, the experimental conditions are GTX2080Ti, PyTorch framework is used,

Word2Vec trained on Weibo is used as the word vector, the number of filters is set to 50. The specific experimental parameters are shown in Table 3.

Table 3. Dataset information.

Parameters	Value
Batch Size	16
Dimension of Word Vector	200
Dimension of Bi-LSTM Hidden State	300
Dimension of MLP	100
Layer of Graph Attention Network	2
Multi Head Attention	3
Smooth Item ϵ	1
L2 Regularized Coefficient λ	10^{-5}
Threshold η	0.3
Layer of Convolution	4
Kernel Size	(1,2,3,4)
K	128

4.2 Experimental Results and Analysis

Table 4 shows a specific document, and the final analysis result is shown in Fig. 3. A lighter color indicates a higher score for the candidate emotion cause pair.

Table 4. Case analysis.

Text	Class
C1:Relying on parents' support and the money saved by working in the school	\
C2:they invested more than 500,000 yuan in total	\
C3:and started everything from scratch	\
C4: to set up a chicken farm	\
C5:In 3 months, the sales have reached more than 500,000 yuan	\
C6:With such a good beginning	Cause Clause
C7:they are both proud and confident about the prospect	Emotion Clause

Figure 3 shows that all the output result values are small, this is because after sigmoid, the output result of the double affine attention layer is multiplied by the weight matrix, resulting in a lower score.

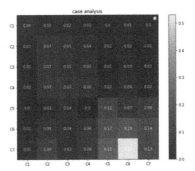

Fig. 3. Result analysis.

As shown in Fig. 4, the ablation experiment was set up to better analyze the effect of each module proposed in this paper.

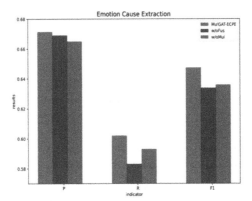

Fig. 4. Comparison of emotion cause pair extraction results.

According to the ablation experiment, when label embedding and feature fusion were added to the model in this paper, the three evaluation indexes of ECPE task were greatly improved. This is because label embedding makes up for the shortcoming of the shared feature parameters. In addition, with feature fusion, the indicators also increased, indicating that the multi-scale features obtained by this method can help the computer better understand emotion words and cause words.

In order to better verify the validity, the Mul-ECPE model is compared with other models in Table 4. Table shows that the proposed model is superior to the latest model in the F1 values of three tasks. The F1 value of emotion ECPE task was 0.8% higher than that of the latest model (Table 5).

Table 5. Experiment Results.

Model	Class	Emotion clause extraction			Cause clause extraction			Emotion cause clause extraction		
		P	R	F1	P	R	F1	P	R	F1
Inter-CE(2019)	Two Step	0.8458	0.8035	0.8263	0.6838	0.5754	0.6231	0.6780	0.5254	0.5896
Inter-EC(2019)	Two Step	0.8406	0.8097	0.8242	0.6989	0.5991	0.6426	0.6691	0.5503	0.6013
MAM(2021)	Two Step	0.8486	0.8123	0.8319	0.7035	0.6093	0.6516	0.6893	0.5604	0.6165
DQAN(2021)	Two Step	-	-	-	-	-	-	0.6733	0.6040	0.6362
RHNSC(2020)	End to End	-	-	-	-	-	-	0.6956	0.5871	0.6357
E2EECPE(2020)	End to End	0.8595	0.7915	0.8238	0.7062	0.6030	0.6503	0.6478	0.6105	0.6280
Hier-BiLSTM(2020)	End to End	0.6816	0.6629	0.7480	0.7227	0.5532	0.6248	0.6925	0.5371	0.6030
PairGCN(2020)	End to End	0.8587	0.7208	0.7829	0.7283	0.5953	0.6541	0.6999	0.5779	0.6321
MTNECP(2020)	Multi-Task	0.8662	0.8393	0.8520	0.7400	0.6378	0.6744	0.6828	0.5894	0.6321
LAE-Joint(2020)	Multi-Task	0.8810	0.7810	0.8260	-	-	-	0.6880	0.5960	0.6400
TDGC(2020)	Multi-Task	0.8080	0.8406	0.8256	0.6742	0.6534	0.6636	0.6515	0.6354	0.6394
MulGAT-ECPE	Text	0.8643	0.8421	0.8617	0.7310	0.6015	0.6814	0.6714	0.6070	0.6484

5 Conclusion

The Mul-ECPE model proposed in this paper is used for ECPE. This model makes full use of the effect of feature fusion, multi-scale information is allocated to the word vector representation so that the model can better understand the emotion word and cause word. The addition of label embedding can make full use of the rich interactive relationship information among the three tasks. Experimental results also show that the proposed model achieves good results on ECPE data sets. Existing methods remain at the clause level, so future work can be improved from the perspective of granularity level, so as to extract emotion cause pairs at a fine-grained level.

Acknowledgments. This work was supported by the Key projects of the National Natural Science Foundation of China (No.62137001).

References

1. Lee, S.Y.M., et al.: Emotion cause events: corpus construction and analysis. In: International Conference on Language Resources & Evaluation. DBLP (2012)
2. Gui, L., et al.: A Question Answering Approach for Emotion Cause Extraction (2017)
3. Li, X., et al.: A co-attention neural network model for emotion cause analysis with emotional context awareness. In: Proceedings of the 2018 Conference on Empirical Methods in Natural Language Processing (2018)
4. Yu, X., et al.: Multiple level hierarchical network-based clause selection for emotion cause extraction. IEEE Access (2019)
5. Xia, R., Zhang, M., Ding, Z.: RTHN: A RNN-Transformer Hierarchical Network for Emotion Cause Extraction (2019)

6. Fan, C., et al.: A knowledge regularized hierarchical approach for emotion cause analysis. In: Proceedings of the 2019 Conference on Empirical Methods in Natural Language Processing and the 9th International Joint Conference on Natural Language Processing (EMNLP-IJCNLP) (2019)
7. Wu, S., et al.: A multi-task learning neural network for emotion-cause pair extraction. ECAI IOS Press **2020**, 2212–2219 (2020)
8. Fan, C., et al.: Transition-based directed graph construction for emotion-cause pair extraction. In: Proceedings of the 58th Annual Meeting of the Association for Computational Linguistics (2020)
9. Wei, P., Zhao, J., Mao, W.: Effective inter-clause modeling for end-to-end emotion-cause pair extraction. In: Proceedings of the 58th Annual Meeting of the Association for Computational Linguistics (2020)
10. Tang, H., Ji, D., Zhou, Q.: Joint multi-level attentional model for emotion detection and emotion-cause pair extraction. Neurocomputing **409**, 329–340 (2020)
11. Vaswani, A., et al.: Attention is all you need. In: Proceedings of the 31st International Conference on Neural Information Processing Systems, pp. 6000–6010 (2017)
12. Tang, J., Qu, M., Mei, Q.: Pte: predictive text embedding through large-scale heterogeneous text networks. In: Proceedings of the 21th ACM SIGKDD international conference on knowledge discovery and data mining, pp. 1165–1174 (2015)
13. Zhang, Y., et al. Deconvolutional paragraph representation learning. In: Proceedings of the 31st International Conference on Neural Information Processing Systems, pp. 4172–4182 (2017)
14. Wang, G., et al.: Joint embedding of words and labels for text classification. In: Proceedings of the 56th Annual Meeting of the Association for Computational Linguistics (Volume 1: Long Papers), pp. 2321–2331 (2018)
15. Zhang, H., et al.: Multi-task label embedding for text classification. Proceedings of the Conference on Empirical Methods in Natural Language Processing. **2018**, 4545–4553 (2018)
16. Xia, R., Ding, Z.: Emotion-cause pair extraction: a new task to emotion analysis in texts. In: Proceedings of the 57th Annual Meeting of the Association for Computational Linguistics, 1003–1012 (2019)
17. Ding, Z., Xia, R., Yu, J.: ECPE-2D: emotion-cause pair extraction based on joint two-dimensional representation, interaction and prediction. In: Proceedings of the 58th Annual Meeting of the Association for Computational Linguistics, pp. 3161–3170 (2020)
18. Song, H., et al.: End-to-end emotion-cause pair extraction via learning to link. arXiv preprint arXiv:2002.10710 (2020)
19. Fan, C., et al.: Multi-task sequence tagging for emotion-cause pair extraction via tag distribution refinement. IEEE/ACM Transactions on Audio, Speech, and Language Processing **29**, 2339–2350 (2021)
20. Li, C., Zhai, R., Zuo, F., Yu, J., Zhang, L.: Mixed Multi-channel Graph Convolution Network on Complex Relation Graph. In: Xing, C., Fu, X., Zhang, Y., Zhang, G., Borjigin, C. (eds.) WISA 2021. LNCS, vol. 12999, pp. 497–504. Springer, Cham (2021). https://doi.org/10.1007/978-3-030-87571-8_43
21. Chen, X., Li, Q., Wang, J.: Conditional causal relationships between emotions and causes in texts. Proceedings of the Conference on Empirical Methods in Natural Language Processing (EMNLP) **2020**, 3111–3121 (2020)

Dynamic Alternative Attention for Visual Question Answering

Xumeng Liu[1], Wenya Guo[1($\boxtimes$)], Yuhao Zhang[2], and Ying Zhang[1]

[1] College of Computer Science, Nankai University, Tianjin 300350, China
{liuxumeng,guowenya}@dbis.nankai.edu.cn, yingzhang@nankai.edu.cn
[2] College of Cyber Science, Nankai University, Tianjin 300350, China
zhangyuhao@dbis.nankai.edu.cn

Abstract. In recent years, researchers have focused on Visual Question Answering (VQA) due to its numerous real-world applications. And visual attention mechanisms are widely used to assist answer prediction by selecting important regions. Nevertheless, few works consider the process of how the model progressively selects informative regions. To simulate the dynamic reasoning process of human beings, the existing method, AiR-M, decomposes the answer prediction process into a sequence of reasoning steps, in which each step contains a reasoning operation and a corresponding attention map. However, AiR-M neglects the variable number of reasoning steps for different questions and pads the reasoning step sequence with invalid steps, which introduces inaccurate information into answer prediction and thus limits the model performance. In this paper, we propose a Dynamic Alternative Attention model (DA^2) to address this problem. Specifically, DA^2 consists of a feature extraction module denoted as DA^2-f and a training module denoted as DA^2-t. DA^2-f is used to provide the answer prediction progress with more accurate visual information by adaptively filtering out the visual regions of invalid steps. And DA^2-t improves model training by masking out the attention maps corresponding to invalid steps in the objective function. Experimental results on the GQA dataset verify the effectiveness of our proposed method.

Keywords: Visual Question Answering · Reasoning operation · Visual attention

1 Introduction

The task of Visual Question Answering (VQA), proposed by Antol *et al.* in 2015 [3], has become one of the research focuses because of its wide downstream applications, such as assisting visually impaired people [5], developing image retrieval systems [24,31], scientific chart analysis [21,22].

VQA requires predicting an appropriate answer for the given image and question. To improve the performance of VQA models, various attention-based methods have been proposed to select question-relevant regions from images. However,

X. Zhao et al. (Eds.): WISA 2022, LNCS 13579, pp. 380–392, 2022.
https://doi.org/10.1007/978-3-031-20309-1_33

few approaches explore how the model progressively extracts the visual contents and achieves the corresponding answer. To this end, AiR-M [8] introduces the concept of **dynamic attention** and decomposes the answer prediction process into a sequence of reasoning steps. As shown in Fig. 1, each reasoning step contains an operation and a corresponding attention map. The operation denotes the atomic operation of the reasoning process, such as *select* and *query*, and the attention map highlights the operated objects. All attention maps belonging to the reasoning step sequence are added to generate an aggregated attention map. Finally, the aggregated attention map is used to calculate the final visual features and then predict the question's answer.

AiR-M pads the reasoning step sequences into a fixed length with invalid steps, in which invalid steps include a new operation type named *invalid* and the corresponding attention maps. In the attention maps for invalid steps, all the attention weights are initialized as the same value. However, AiR-M does not distinguish the introduced invalid steps and the original valid steps, introducing noise for feature extracting and training, thus limiting the model performance.

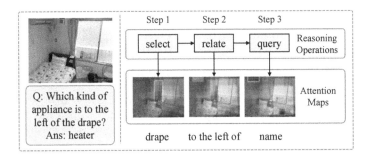

Fig. 1. An example for dynamic attention-based VQA. The model first decomposes the answering process into several reasoning steps. Then for each step, the operation and the attention map are predicted. For example, step 1 is to select the *drape* in the picture, so the first operation should be *select*, and the corresponding attention map should concentrate on the *drape*. Then step 2 and 3, each gets the relationship of *to the left of* and queries the *name* of the appliance.

In this paper, we propose a Dynamic Alternative Attention model (DA^2) to address the above issues. DA^2 can distinguish invalid operations and filter out the corresponding attention maps. Specifically, DA^2 consists of DA^2-f and DA^2-t for the process of feature extraction and model training, respectively. DA^2-f first judges the type of current operation for each step and filters out the attention maps corresponding to invalid operation from the final visual features for answer prediction. DA^2-t directly removes the attention maps for invalid operations from the corresponding objective function. Combining DA^2-f and DA^2-t allows the model to concentrate on learning visual attention corresponding to valid operations, making the learned visual attention more accurate, and thus improving the VQA performance. The main contributions are list as below.

1. We propose DA^2-f to adaptively filter out the attention maps corresponding to invalid operations during feature extracting, which can help achieve more accurate final visual features for answer prediction.
2. We propose DA^2-t to improve the loss function and help the model concentrate on learning attention maps corresponding to valid steps.
3. Experimental results on GQA [19] show that our proposed method can outperform existing state-of-the-art methods.

2 Related Work

Due to the wide real-world applications of VQA, various approaches have been applied to improve the model performance. For example, numerous methods introduce scene graphs to enhance the ability to understand the overall image structure [17,28,30]. Works [4,12,13,23] focus on inter-modal interaction. Wu *et al.* [35] transforms VQA into traditional QA by converting images to their caption. Many approaches introduce external knowledge to VQA [6,7,32–34], with natural language knowledge pieces [15] or knowledge graphs [16]. And [25, 38] propose new datasets with realistic application scenarios, etc.

Dynamic-reasoning-operation-based VQA. Researches show that a lot of models answer questions via the bias of datasets instead of reasoning properly [1,14]; thus, [20,39] make efforts to deal with the problem. But those works do not involve real-world data, thus limiting the performance in realistic downstream applications [19]. Therefore, the GQA dataset [19] is proposed to fix it. GQA contains real-world images, questions, and scene graphs for each image. Moreover, it provides a sequence of reasoning operations for each question.

Attention-based VQA. The attention mechanism in VQA refers to identifying RoIs in images that are relevant to the question, so those regions should be assigned with higher weights for answer prediction. Some models generate a single attention map to show the weights of regions [18,29,36], while some other methods produce multiple attention maps and assign different uses for each map [11,37]. Also, models HAN [27], ASM [40] and PAAN [26] supervised learn for attention maps. However, the above methods do not simulate the human dynamic reasoning process when performing the VQA task, leading to poor performance. And AiR-M [8], a dynamic attention-based model, decomposes the answer prediction process into a sequence of reasoning steps, then generates the corresponding operations and attention maps (see Fig. 1) with supervision. Yet, AiR-M introduces noise in the feature extraction and the training process, which limits the performance of the model.

3 Approach

In this section, we give an overview of our proposed Dynamic Alternative Attention (DA^2). First, using the extracted image and question features, DA^2-f predicts the operations and the single-step attention map for each reasoning step.

Next, DA^2-f adaptively aggregates single-step attention maps, uses the aggregated map to compute informative visual features and then predict the answer. Finally, DA^2-t is used to simultaneously train the prediction of answers, operation types, and the corresponding attention maps.

Specifically, our feature extraction module, DA^2-f, first extracts image and question features, v_i and q. DA^2-f then decomposes the answer prediction process into a sequence of reasoning steps with a specific length then predicts an operation and an attention map for each step. For the t-th step, DA^2-f first predicts the operation γ_t^P from the extracted question features q, and predicts the single-step attention map α_t^P from the predicted γ_t^P, q, and v_i. To calculate the aggregated attention map α^P, α_t^P for each step are selected adaptively with function $mask_F$. Finally, q, and v_i combined with α^P are used as the input of some fully connected layers and a softmax function to predict the answer ans^P. Then we train the model with our training module named DA^2-t. DA^2-t supervised learns α_t^P, γ_t^P, and ans^P, where α_t^P are masked by function $mask_T$. Ultimately, we get a VQA model which predicts reasoning steps with corresponding operations and attention maps to facilitate question answering.

3.1 Feature Extraction Module

We propose the feature extraction module named DA^2-f based on AiR-M [8]. In this module, reasoning operations are first predicted, and single-step attention maps corresponding to the operations are generated, then maps are selected by $mask_F$. With $mask_F$, more accurate aggregated attention maps and final visual features can be utilized to predict better VQA answers.

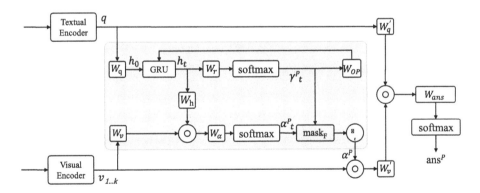

Fig. 2. The framework of DA^2-f. P denotes the predicted results. The predicted single-step attention maps α_t^P are filtered by $mask_F$, and the left ones are added to generate the aggregated attention map α^P.

As shown in Fig. 2, the DA^2-f first extracts image features v_i and text features q via UpDown [2] and GRU [9]. Then DA^2-f predicts a sequence of reasoning steps to imitate the reasoning process of human beings. For step t, an operation γ_t^P and a single-step attention map α_t^P are predicted. Specifically, first, the semantic representation of operation h_t is generated from q by another GRU. Then DA^2-f predicts the operation γ_t^P from h_t. And the single-step attention map α_t^P is predicted with h_t and v_i, in which α_t^P denotes the weights of detected objects from the image for step t. Next, we implement function $mask_F$ (see Fig. 3(a)) to adaptively mask the α_t^P then generate the aggregated attention map α^P:

$$\alpha^P = \sum_t mask_F(\gamma_t^P) \cdot \alpha_t^P, \tag{1}$$

$$mask_F(\gamma_t^P) = \begin{cases} 0 \ \ if \ \gamma_t^P \ is \ invalid, \\ 1 \ \ else, \end{cases} \tag{2}$$

where P stands for the predicted results generated by the model, and G stands for the ground-truth labels. With $mask_F$ shown in Eq. (2), the low-quality single-step attention maps corresponding to invalid predicted operations are filtered out, which helps improve the quality of α^P. With the product of v_i and α^P, final visual features are calculated. Finally, some fully connected layers and a softmax function are used to generate the final VQA answer ans^P.

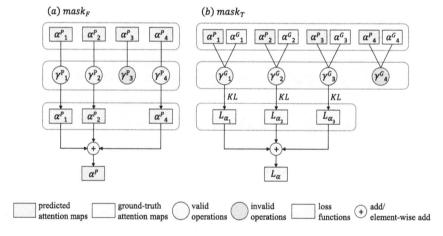

Fig. 3. Examples for $mask_F$ used in DA^2-f and $mask_T$ used in DA^2-t. For $mask_F$ in (a), the predicted single-step attention maps α_t^P are filtered out if the corresponding predicted operations γ_t^P are invalid. Those α_t^P which correspond to the valid γ_t^P are added together to generate the aggregated attention map α^P. For $mask_T$ in (b), just the attention maps α_t^P and α_t^G both corresponding to valid ground-truth reasoning operations γ_t^G, are calculated in the loss function.

Generally, $mask_F$ can filter out the noneffective predicted single-step attention maps α_t^P which cannot combine the valid operations. Therefore, α^P and final visual features can be more accurate for answer prediction.

3.2 Training Module

In this subsection, we propose a training module DA^2-t based on the loss function of AiR-M [8]. The loss function of DA^2-t has three penalty terms, which are:

$$L = L_{ans} + \theta L_\alpha + \phi L_\gamma, \tag{3}$$

where L_{ans}, L_α, and L_γ denote the loss corresponding to the answer, the attention map, and the reasoning operation, respectively. L_{ans}, L_α, and L_γ are added together with weights θ and ϕ. L_α and L_γ are the sum of loss for each reasoning step t:

$$L_\alpha = \sum_t L_{\alpha_t}, \tag{4}$$

$$L_\gamma = \sum_t L_{\gamma_t}. \tag{5}$$

As shown in Eq. (4) and Eq. (5), the loss terms of attention and operations are the sum of differences for each step.

Specifically, the loss items of answers and operations are calculated as:

$$L_{ans} = CE(ans^P, ans^G) = ans^G \log(ans^P), \tag{6}$$

$$L_{\gamma_t} = CE(\gamma_t^P, \gamma_t^G) = \gamma_t^G \log(\gamma_t^P). \tag{7}$$

As in Eq. (6) and Eq. (7), the loss of the answer and the operation parts are calculated with the cross-entropy function CE. For the attention part, as the invalid operation is used to pad the ground-truth operation sequences, function $mask_T$ (see Fig. 3(b)) is implemented to remove the introduced invalid information. The loss term of the single-step attention map is calculated as:

$$L_{\alpha_t} = mask_T(\gamma_t^G) \cdot \alpha_t^G \log(\frac{\alpha_t^G}{\alpha_t^P}), \tag{8}$$

$$mask_T(\gamma_t^G) = \begin{cases} 0 \ \ if \ \gamma_t^G \ is \ invalid, \\ 1 \ \ else. \end{cases} \tag{9}$$

As shown in Eq. (9), $mask_T$ removes those predicted attention maps α_t^P which correspond to the invalid ground-truth operations γ_t^G. Only the differences between γ_t^P and valid γ_t^G are considered.

As the invalid operations for initial padding do not possess corresponding attention maps in the real world, when the ground-truth γ_t^G is invalid, it is more important to make the predicted γ_t^P invalid. And, when γ_t^P is invalid, equaling to γ_t^G, α_t^P would not be added to the aggregated map α^P based on DA^2-f.

Conclusively, α_t^P is useless to be included in L_α when the corresponding ground-truth γ_t^G is invalid. And removing those useless α_t^P with $mask_T$, which target at invalid α_t^G, can make the model training focus on other efficient parts of the model. With $mask_T$, image regions corresponding to valid operations could be focused on more, which means the quality improvement of predicted attention maps. And for the whole VQA scene, narrowing the influence of L_α can increase the ratio of L_{ans} in the whole loss function L, which makes the model training concentrate on answer prediction, thus leading to better answers.

4   Experiments

4.1   Experiment Settings

Datasets. All experiments are performed on GQA[1] dataset. GQA contains images, questions, and scene graphs for each image. GQA has the following advantages:

1. All the images are obtained from real-world scenes, which facilitate the trained model to be better applied to realistic scenarios.
2. Each image comes with a scene graph of objects and relations, which helps models get the structural information of the whole picture.
3. Each question contains a sequence of reasoning operations for the answer prediction.

Evaluation Metrics. We evaluate the performance of our DA^2 model with the standard evaluation metric defined in [3]:

$$Acc(ans) = min(1, \frac{\#\{humans\ provided\ ans\}}{3}). \tag{10}$$

Implementation Details. Our method is improved based on AiR-M, and we train the model following the original setting proposed in [8].

4.2   Baselines

We compare our method with three categories of approaches: (1) attention mechanisms without attention supervision and dynamic attention such as Top-Down [2]; (2) supervised learning methods for attention but not combined with dynamic attention, such as HAN [27], ASM [40] and PAAN [26]; (3) a model that shares both attention supervision and dynamic attention named AiR-M [8]. Details for each baseline model are:

---

[1] https://cs.stanford.edu/people/dorarad/gqa/download.html.

1. **Top-Down** [2] uses two LSTMs to process information from visual and textual modalities respectively.
2. **PAAN** [26] introduces adversarial learning to learn attention maps supervised.
3. **HAN** [27] first implements low-rank bilinear pooling and several convolutional layers to generate coarse-grained attention maps, then selects important regions with GRU. HAN is trained on VQA-HAT [10] with supervised learning.
4. **ASM** [40] is an attention supervised learning method that first fuses the image and question embedding, then predicts attention maps with CNN.
5. **AiR-M** [8] decomposes answer prediction into sequences of reasoning steps and produces an operation and an attention map for each step. AiR-M performs supervised learning for answers, attention maps, and reasoning operations simultaneously.

4.3   Results

We compare DA^2 with the baseline models on GQA-testdev. As shown in Table 1, our DA^2 achieves the best performance. Specifically, both DA^2 and AiR-M perform better than the supervised models without dynamic attention, PAAN, HAN, and ASM. It shows that dynamic attention can improve the accuracy of answer prediction. Also, DA^2 and AiR-M outperform Top-Down. It proves that VQA models can work better with dynamic attention and attention supervision together. Moreover, DA^2 can work better than our based model, AiR-M, which means our proposed model can fix the invalid information introduced problem of AiR-M, thus improving the model performance.

Table 1. Comparison with baseline models on GQA-testdev

Models	Acc (%)
Top-Down	51.31
PAAN	48.03
HAN	49.96
ASM	52.96
AiR-M	53.46
DA^2	**53.80**

4.4   Ablation Study

To analyze the improvement of DA^2 performance, we ablate the model and compare the VQA accuracy. Table 2 shows the ablation study results, and AiR-M* denotes the reimplemented result with the supplied code from the original paper [8]. DA^2-f is 0.46% better than AiR-M*, which proves that DA^2-f works, and it is useful to adaptively delete the attention maps corresponding to invalid predicted operations and better final visual features can be generated. And compared to AiR-M*, DA^2-t has an improvement of 0.32%, which shows that DA^2-t can outperform AiR-M* by removing the attention maps corresponding to invalid ground-truth operations in the training process. Both the DA^2-f and the DA^2-t we proposed can improve VQA performance. Moreover, our DA^2, with DA^2-f and DA^2-t working together, achieve the best performance.

Table 2. Ablation study for DA^2-f and DA^2-t on GQA-testdev. AiR-M* denotes the reimplemented result with the supplied code from the original paper [8].

Models	Acc (%)
AiR-M*	52.83
DA^2-f	53.29
DA^2-t	53.15
DA^2 (DA^2-f + DA^2-t)	**53.80**

4.5   Visualization

Filtering Out Invalid Predicted Operations Corresponded Attention Maps. As Fig. 4(a), both the fourth reasoning steps predicted by AiR-M and DA^2 are invalid. It is obvious that the corresponding attention maps start to disperse and cannot concentrate on specific objects well. AiR-M cannot remove this kind of map in low quality. And our proposed DA^2-f module can filter out the invalid information, thus improving the quality of the aggregated attention maps.

Improvements of Single-step Attention Maps Quality. DA^2-t is proposed to improve the model training quality. And the loss function is streamlined by removing those predicted single-step attention maps which correspond to the invalid ground-truth operations. DA^2-t can make the model training focus on valid and efficient attention information more, thus providing more accurate aggregated attention maps for the answer prediction process. As shown in Fig. 4(b), AiR-M only notices *her* but not the real subject of the question, *towel.* Overall, via increasing the rate for attention training targets at those corresponding to valid operations, DA^2-t improves the accuracy of the attention map predicted for each step. In turn, it provides predicted information with higher quality for the downstream answer prediction process.

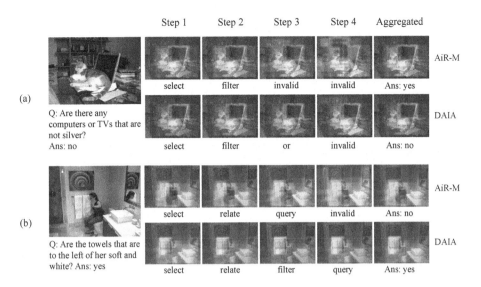

Fig. 4. Visualization of attention maps.

5   Conclusions

In this paper, we deal with the dynamic attention mechanism in VQA. The existing dynamic attention-based model, AiR-M, cannot perform well due to its introduction of invalid steps for padding without considering the length difference of reasoning step sequences. To address this problem, we propose DA^2, which can alternatively select the predicted attention maps during feature extracting and training. DA^2 includes DA^2-f for feature extracting and DA^2-t for model training. DA^2-f can offer more accurate final visual features to answer prediction. And DA^2-t can streamline the loss function and make the training process focus on more important information. Finally, experiments show that our DA^2, both DA^2-f and DA^2-t, can improve the performance of VQA.

Acknowledgements. This research is supported by the NSFC-Xinjiang Joint Fund (No. U1903128), and the Fundamental Research Funds for the Central Universities (No. 63223046).

References

1. Agrawal, A., Batra, D., Parikh, D., Kembhavi, A.: Don't just assume; look and answer: overcoming priors for visual question answering. In: CVPR, pp. 4971–4980. Computer Vision Foundation/IEEE Computer Society (2018)
2. Anderson, P., et al.: Bottom-up and top-down attention for image captioning and visual question answering. In: CVPR, pp. 6077–6086. Computer Vision Foundation / IEEE Computer Society (2018)

3. Antol, S., et al.: VQA: visual question answering. In: ICCV, pp. 2425–2433. IEEE Computer Society (2015)

4. Ben-younes, H., Cadène, R., Cord, M., Thome, N.: MUTAN: multimodal tucker fusion for visual question answering. In: ICCV, pp. 2631–2639. IEEE Computer Society (2017)

5. Bigham, J.P., et al.: Vizwiz: nearly real-time answers to visual questions. In: ACM, pp. 333–342 (2010)

6. Bordes, A., Usunier, N., García-Durán, A., Weston, J., Yakhnenko, O.: Translating embeddings for modeling multi-relational data. In: Advances in Neural Information Processing Systems, pp. 2787–2795 (2013)

7. Bruna, J., Zaremba, W., Szlam, A., LeCun, Y.: Spectral networks and locally connected networks on graphs. In: ICLR (2014)

8. Chen, S., Jiang, M., Yang, J., Zhao, Q.: AiR: attention with reasoning capability. In: Vedaldi, A., Bischof, H., Brox, T., Frahm, J.-M. (eds.) ECCV 2020. LNCS, vol. 12346, pp. 91–107. Springer, Cham (2020). https://doi.org/10.1007/978-3-030-58452-8_6

9. Cho, K., van Merrienboer, B., Bahdanau, D., Bengio, Y.: On the properties of neural machine translation: encoder-decoder approaches. In: EMNLP, pp. 103–111. Association for Computational Linguistics (2014)

10. Das, A., Agrawal, H., Zitnick, L., Parikh, D., Batra, D.: Human attention in visual question answering: do humans and deep networks look at the same regions? Comput. Vis. Image Underst. **163**, 90–100 (2017)

11. Fukui, A., Park, D.H., Yang, D., Rohrbach, A., Darrell, T., Rohrbach, M.: Multimodal compact bilinear pooling for visual question answering and visual grounding. In: EMNLP, pp. 457–468. The Association for Computational Linguistics (2016)

12. Gao, P., et al.: Dynamic fusion with intra- and inter-modality attention flow for visual question answering. In: CVPR, pp. 6639–6648. Computer Vision Foundation / IEEE (2019)

13. Gao, P., You, H., Zhang, Z., Wang, X., Li, H.: Multi-modality latent interaction network for visual question answering. In: ICCV, pp. 5824–5834. IEEE (2019)

14. Goyal, Y., Khot, T., Summers-Stay, D., Batra, D., Parikh, D.: Making the V in VQA matter: elevating the role of image understanding in visual question answering. In: CVPR, pp. 6325–6334. IEEE Computer Society (2017)

15. Gui, L., Wang, B., Huang, Q., Hauptmann, A., Bisk, Y., Gao, J.: Kat: a knowledge augmented transformer for vision-and-language. NAACL (2022)

16. Guo, Q., et al.: Constructing Chinese historical literature knowledge graph based on BERT. In: Xing, C., Fu, X., Zhang, Y., Zhang, G., Borjigin, C. (eds.) WISA 2021. LNCS, vol. 12999, pp. 323–334. Springer, Cham (2021). https://doi.org/10.1007/978-3-030-87571-8_28

17. Haurilet, M., Roitberg, A., Stiefelhagen, R.: It's not about the journey; it's about the destination: following soft paths under question-guidance for visual reasoning. In: CVPR, pp. 1930–1939. Computer Vision Foundation/IEEE (2019)

18. Huang, P., Huang, J., Guo, Y., Qiao, M., Zhu, Y.: Multi-grained attention with object-level grounding for visual question answering. In: ACL, pp. 3595–3600. Association for Computational Linguistics (2019)

19. Hudson, D.A., Manning, C.D.: GQA: a new dataset for real-world visual reasoning and compositional question answering. In: CVPR, pp. 6700–6709. Computer Vision Foundation / IEEE (2019)

20. Johnson, J., Hariharan, B., van der Maaten, L., Fei-Fei, L., Zitnick, C.L., Girshick, R.B.: CLEVR: a diagnostic dataset for compositional language and elementary visual reasoning. In: CVPR, pp. 1988–1997. IEEE Computer Society (2017)

21. Kafle, K., Price, B.L., Cohen, S., Kanan, C.: DVQA: understanding data visualizations via question answering. In: CVPR, pp. 5648–5656. Computer Vision Foundation / IEEE Computer Society (2018)

22. Kahou, S.E., Michalski, V., Atkinson, A., Kádár, Á., Trischler, A., Bengio, Y.: Figureqa: an annotated figure dataset for visual reasoning. In: ICLR. OpenReview.net (2018)

23. Li, L., Gan, Z., Cheng, Y., Liu, J.: Relation-aware graph attention network for visual question answering. In: ICCV, pp. 10312–10321. IEEE (2019)

24. Lin, X., Parikh, D.: Leveraging visual question answering for image-caption ranking. In: Leibe, B., Matas, J., Sebe, N., Welling, M. (eds.) ECCV 2016. LNCS, vol. 9906, pp. 261–277. Springer, Cham (2016). https://doi.org/10.1007/978-3-319-46475-6_17

25. Lu, J., Yang, J., Batra, D., Parikh, D.: Hierarchical question-image co-attention for visual question answering. In: Advances in Neural Information Processing Systems, pp. 289–297 (2016)

26. Patro, B.N., Anupriy, S., Namboodiri, V.: Explanation vs attention: a two-player game to obtain attention for VQA. In: AAAI, pp. 11848–11855. AAAI Press (2020)

27. Qiao, T., Dong, J., Xu, D.: Exploring human-like attention supervision in visual question answering. In: AAAI, pp. 7300–7307. AAAI Press (2018)

28. Shi, J., Zhang, H., Li, J.: Explainable and explicit visual reasoning over scene graphs. In: CVPR, pp. 8376–8384. Computer Vision Foundation / IEEE (2019)

29. Shih, K.J., Singh, S., Hoiem, D.: Where to look: focus regions for visual question answering. In: CVPR, pp. 4613–4621. IEEE Computer Society (2016)

30. Tang, K., Zhang, H., Wu, B., Luo, W., Liu, W.: Learning to compose dynamic tree structures for visual contexts. In: CVPR, pp. 6619–6628. Computer Vision Foundation / IEEE (2019)

31. Vo, N., et al.: Composing text and image for image retrieval - an empirical odyssey. In: CVPR (2019)

32. Wang, P., Wu, Q., Shen, C., Dick, A.R., van den Hengel, A.: Explicit knowledge-based reasoning for visual question answering. In: IJCAI, pp. 1290–1296. ijcai.org (2017)

33. Wang, P., Wu, Q., Shen, C., Dick, A.R., van den Hengel, A.: FVQA: fact-based visual question answering. IEEE Trans. Pattern Anal. Mach. Intell. **40**(10), 2413–2427 (2018)

34. Wu, F., Jing, X., Wei, P., Lan, C., Ji, Y., Jiang, G., Huang, Q.: Semi-supervised multi-view graph convolutional networks with application to webpage classification. Inf. Sci. **591**, 142–154 (2022)

35. Wu, J., Hu, Z., Mooney, R.J.: Generating question relevant captions to aid visual question answering. In: ACL, pp. 3585–3594. Association for Computational Linguistics (2019)

36. Xu, H., Saenko, K.: Ask, attend and answer: exploring question-guided spatial attention for visual question answering. In: Leibe, B., Matas, J., Sebe, N., Welling, M. (eds.) ECCV 2016. LNCS, vol. 9911, pp. 451–466. Springer, Cham (2016). https://doi.org/10.1007/978-3-319-46478-7_28

37. Yang, Z., He, X., Gao, J., Deng, L., Smola, A.J.: Stacked attention networks for image question answering. In: CVPR, pp. 21–29. IEEE Computer Society (2016)

38. Yu, Z., Yu, J., Fan, J., Tao, D.: Multi-modal factorized bilinear pooling with co-attention learning for visual question answering. In: ICCV, pp. 1821–1830 (2017)

39. Zhang, P., Goyal, Y., Summers-Stay, D., Batra, D., Parikh, D.: Yin and yang: balancing and answering binary visual questions. In: CVPR, pp. 5014–5022. IEEE Computer Society (2016)
40. Zhang, Y., Niebles, J.C., Soto, A.: Interpretable visual question answering by visual grounding from attention supervision mining. In: WACV, pp. 349–357. IEEE (2019)

A Biomedical Trigger Word Identification Method Based on BERT and CRF

Xinyu He[1,2,3]([⊠]), Jiayi Feng[1], Feiyan Sun[1], Mengfan Yan[1], Junjie Qian[1], Wenqian Dai[1], and Hongyu Wang[2]

[1] School of Computer and Information Technology, Liaoning Normal University, Dalian 116081, China
`hexinyu@lnnu.edu.cn`
[2] Information and Communication Engineering Postdoctoral Research Station, Dalian University of Technology, Dalian, Liaoning, China
[3] Postdoctoral Workstation of Dalian Yongjia Electronic Technology Co., Ltd., Dalian, Liaoning, China

Abstract. Biomedical trigger word identification is a challenging task in biomedical text mining, which plays a key role in improving biomedical research and disease prevention. The traditional trigger word identification methods rely too much on the establishment of dictionaries or rules, which lead to poor performance. Aiming at the above problems, this paper puts forward a BERT-CRF model. In this model, BERT is used to train the emission matrix in CRF model. By using the transformer architecture of BERT, the language model, is used to train. By using CRF layer, some constraints can be added to ensure that the final prediction results are valid. The experimental results on MLEE data set show that the proposed model achieves state-of-the-art performance with an F value of 82.71%, which indicates that this method is beneficial to improve the trigger word recognition performance.

Keywords: Trigger word identification · Word vector · BERT-CRF

1 Introduction

With the increasing number of biomedical literature collections, it is complex and inconvenient to select the knowledge that meets the research needs of system biologists from many biomedical literature collections. Therefore, the rapid and automatic extraction of biological events from biological literature is increasingly urgent in the biomedical field. At present, there are two main approaches to biomedical event extraction: The first kind is a pattern matching-based approach. The other kind is based on machine learning. In this case, we deeply study the subtask-trigger word recognition in machine learning.

The task of trigger word recognition is usually divided into two steps: firstly, it detects whether a word is a trigger word, and then determines the type of trigger word. Since errors in trigger word recognition can be accumulated in the process of subsequent element detection, the performance of trigger word recognition directly affects the final performance of event extraction. According to relevant literature, trigger word recognition is mainly divided into: statistical/dictionary-based approach, rule-based approach

© The Author(s), under exclusive license to Springer Nature Switzerland AG 2022
X. Zhao et al. (Eds.): WISA 2022, LNCS 13579, pp. 393–402, 2022.
https://doi.org/10.1007/978-3-031-20309-1_34

and machine learning-based approach. Buyko [1] used a dictionary-based approach. They manually count all the specific trigger words in sentences or texts in the corpus to construct a dictionary, which then matched the trigger words with the given corpus and finally filtered to obtain a relatively complete trigger word dictionary. Cohen et al. [2] adopted ontology-driven semantic analysis with manually added rules in the trigger word recognition stage. Word Sense Disambiguation (WSD) system [3] adopted a series of word sense disambiguation schemes to solve the ambiguity problem in the process of trigger word recognition. Machine learning based trigger word recognition can be automatically learned based on the training set, which trains trigger word recognition classifier by feature set, and transforms the recognition problem of trigger word into classification problem.

With the development of deep learning, Bi-LSTM + CRF model is widely used for trigger word recognition, and the effect is ideal. However, it lacks of rich word-level features. To solve this problem, we employed BERT instead of Bi-LSTM, thus the accuracy and training efficiency of sequence labeling have reached a new height.

In this paper, we propose a new trigger word recognition method based on BERT and CRF. In this method, word level features are considered during data processing; BERT is used for sequence tagging and word segmentation, and then the word vector is fused with all the word vectors to which the word belongs. In addition, because BERT feature extractor is the unique constraint condition of Transformer and CRF, the performance is greatly improved.

2   Relevant Work

Trigger word recognition is the primary task in biomedical event extraction. Studies have shown that more than 60% of the errors in event extraction tasks can be attributed to the trigger word recognition stage [4]. Therefore, a variety of research on trigger word recognition directions based on different methods have emerged in recent years. According to the related literature, there are three approaches to trigger word recognition: statistical/dictionary-based approach, rule-based approach and machine learning-based approach. In recent years, the application of machine learning is more and more extensive, and gradually applied to the field of biomedicine. Wei et al. [5] proposed a two-stage approach based on conditional random fields and SVM on the corpus of BioNLP'13, in which the CRF was used to label the trigger words and SVM was used to identify the type of trigger words. He et al. [6] proposed a two-stage method based on SVM and PA online algorithm. First, SVM was used to determine whether a trigger word candidate was a trigger word, and then PA algorithm was introduced to determine the specific trigger word type before. The superficial-based machine learning method has achieved good performance, however, these methods depend on NLP tool to extract artificial features. Therefore, the methods based on deep learning have gradually become the mainstream in recent years. Li et al. [7] adopted the neural network model training word vector as the basic feature for biomedical event extraction. Nie et al. [8] first applied in-depth learning to biomedical trigger word recognition by proposing a feedforward neural network model based on Skip-gram word vector. Ruahul et al. [9] viewed trigger word recognition as a sequence labeling problem, and classified each context representation obtained by coding the model of cyclic neural network they proposed. Diao et al. [10] proposed a hybrid neural network composed of SVM and BiLSTM for biomedical event trigger word recognition. Li et al. [11] constructed a trigger word recognition model based on dependency information and gated attention mechanism. Wang et al. [12] also took trigger word recognition as a sequence annotation task, and proposed a model based on LSTM-CRF and tested the effect of word vectors trained by different models. However, long and complex expressions often exist in biomedical literature, and LSTM-CRF takes a long time to process long texts. Therefore, we build a model based on BERT and CRF, which is much faster than LSTM-CRF training speed under the premise of sufficient computing resources.

3 Method

3.1 BERT Pre-training Model

The early static word vector models represented by Word2vec and GloVe, and dynamic word vector models, such as CoVe and ELMo based on context modeling are widely known, and the pre-training language model only refers to the language model that has undergone large-scale data training in advance. In 2018, the expression model based on deep Transformer represented by GPT and BERT appeared, and the word of pre-training language model really became widely known. The appearance of pre-training language model raises natural language processing to another new level. The appearance of pre-training model is also considered as a milestone in the field of natural language processing in recent years.

The BERT pre-training model has been successful on many natural language processing tasks. The input of BERT is expressed as a sentence or a sentence pair. In this paper, the input natural language sentence is converted into token sequence by WordPiece embeddings S. The [CLS] needs to be added at the beginning of the token sequence, and the embedding of the token of the final output [CLS] is used for classification task. As sentence pairs are integrated, we need to distinguish them in the token sequence in two ways [13]:

(1) A special token [SEP] is added between the tokens of two sentences in the token sequence;
(2) A segment embedding for learning is added for each token to distinguish whether the token belongs to sentence A or sentence B.

Specifically, the input to BERT consists of the sum of three parts: token embeddings (word vector), segment embeddings (sentence vector), and position embeddings (position vector). As shown in Fig. 1.

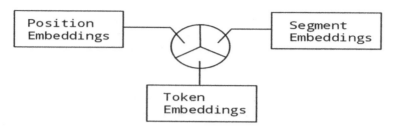

Fig. 1. Input composition of Bert

CRF Layer. CRF is a classical probability graph model. Let x and y be random variables, and P(Y|X) is the conditional probability distribution of y given x. If the random variable y forms a Markov random field, then the conditional probability distribution P(Y|X) is a conditional random field [14].

The linear chain conditional random field is defined as follows: Let $X = (X_1, X_2, ..., X_n)$, $Y = (Y_1, Y_2, ..., Y_n)$ Random variable sequences which are all linear expressions. if under the condition of given random variable sequence X, the conditional probability distribution P(Y|X) of random variable sequence Y constitutes a conditional random field, That is,

$$P(Y_i, |X, Y_1, Y_{i-1}, Y_{i+1}, \ldots, Yn) = P(Y_i|X, Y_{i-1}, Y_{i+1}) \tag{1}$$

where P(Y|X) is called linear chain conditional random field. $\rangle = 1, 2, \ldots n$, only one side is considered when $\rangle = 1$ and n.

Linear chain conditional random field formula: Let P(Y|X) be a linear chain conditional random field, then under the condition that the value of random variable X is x, the conditional probability of the value of random variable Y is y has the following form:

$$P(y|\S) = \frac{1}{Z(\S)} \exp(\sum_{i,k} \lambda_k t_k(y_{i-1}, y_i, \S, i) + \sum_{i,l} u_l s_l(y_i, \S, i)) \tag{2}$$

$$\text{Among} = Z(\S) \sum_y \exp(\sum_{i,k} \lambda_k t_k(y_{i-1}, y_i, \S, i) + \sum_{i,l} u_l s_l(y_i, \S, i)) \tag{3}$$

t_k, s_l: eigenfunction, λ_k, u_l: corresponding weight, $Z(\S)$: normalizing factor.

3.2 Data Pretreatment

The "BIO" notation method is used in this paper. For English text, the minimum input is a word. According to the requirements of BERT model, the maximum sequence length (max_seq_length) needs to be set in advance, and the sequence is padding according to this parameter. Figure 2 shows an example of annotation.

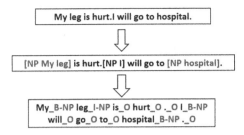

Fig. 2. An example of annotation

In addition, the CRF layer can learn some constraints, which may include [15]:

(1) The beginning of the sentence should be "B-" instead of "I-".
(2) In the mode like "B-label1i-label2i-label3 ...", categories 1, 2 and 3 should be the same entity category. For example, "B-n I-n" is correct, while "B-n I-v" is wrong, and "I-n I-v" is also wrong.

4   Experiment and Analysis

4.1   Data Sets and Evaluation Criteria

Data Set. In this paper, the MLEE dataset of biomedical event extraction is used for experiment. The MLEE dataset [16] contains event information at the cell, tissue, organ, and even whole organism level. In this paper, MLEE datasets are selected to annotate richer event information at other levels besides the molecular level to study biological mechanisms and pathological processes at different levels. The MLEE dataset defines 16 entities and 19 event structure types. The definitions of its 19 categories of biomedical events are shown in Table 1, including event types and participating event elements. The 19 biological event types, categorized into four broad categories, namely "Anatomical", "Molecular", "General", and "Planned", which describe biological processes such as anatomy, pathology, molecular and general, and treatment-related processes. The "Blood vessel development" event type may have no participating elements, and the elements of the "Regulation", "Positive Regulation", "Negative Regulation", and "Planned Process" event types can be another event; these four types of events are nested events.

Table 1. Event type definition

Event type groups	Event types	Event elements	
		Main event elements	Secondary event elements
Anatomical	Cell_proliferation	Theme	

(continued)

Table 1. (*continued*)

Event type groups	Event types	Event elements	
		Main event elements	Secondary event elements
	Development	Theme	
	Blood vessel development	Theme	AtLoc
	Growth	Theme	
	Death	Theme	
	Breakdown	Theme	
	Remodeling	Theme	
Event type groups	Event types	Event elements	
		Main event elements	Secondary event elements
Molecular	Synthession	Theme	
	Gene expression	Theme	
	Transeription	Theme	
	Catabolism	Theme	
	Phosphorylation	Theme	Site
	Dephosphorylation	Theme	SiteAloc/ToLoc/FromLoc
General	Localization	Theme	Site
	Binding	Theme	Site
	Regulation	Theme	Site
	Positive regulation	Theme	Site
	Negative regulation	Theme	Site
Planned	Planned process	Theme	Instrument

The MLEE dataset is divided into a training set, a verification set, and a test set. The detailed statistical information of each type of event in the data set is shown in Table 2. The main events include (Blood vessel development, Regulation, Positive regulation, etc.), and it can be seen that the number distribution of different event types is imbalanced, which makes extraction of biomedical events more difficult.

Evaluation Criteria. In this paper, the experimental model first uses the training set for training verification and adjustment of model parameters, and then uses the training set and the verification set and training to get the final model. In this paper, Precision (P), Recall(R), and F-score (F1) values are used as evaluation indicators. The specific calculation formulas are shown below. Where TP represents the number of instances in the prediction result for which all positive predictions are correct; FP denotes the number of instances of all negative case prediction errors in the prediction result; FN refers to

Table 2. Detailed statistics of the MLEE data set

Event types	Training set	Verification set	Test set	Total number
Cell_proifieration	72	16	45	133
Development	173	36	107	316
Blood vessel development	396	149	310	855
Growth	81	28	60	169
Death	41	18	38	97
Breakdown	26	19	24	69
Remodeling	21	2	10	33
Event types	Training set	Verification set	Test set	Total number
Synthesis	12	1	4	17
Gene expression	190	80	164	434
Transcription	15	6	16	37
Catabolism	18	3	5	26
Phosphorylation	18	11	4	33
Dephosphorylation	1	4	1	6
Localization	247	67	136	450
Binding	88	33	63	184
Regulation	424	93	256	773
Positive regulation	626	286	415	1327
Negative regulation	467	147	306	920
Planned process	295	154	196	645
Main events	3211	1153	2160	6524
Other events	85	22	46	153
All events	3296	1175	2206	6677

the number of instances of all positive prediction errors in the prediction result.

$$P = \frac{TP}{TP + FP} \tag{4}$$

$$R = \frac{TP}{TP + FN} \tag{5}$$

$$F1 = \frac{2 * P * R}{P + R} \tag{6}$$

4.2 Parameter Settings

Our framework is implemented in the environment of Python and Tensorflow, and the final training-related parameters (such as learning rate) are determined by nearly several hundred times. The dimension of the word embeddings generated by BioBERT is 768. The learning rate is selected 0.001 from the set of {0.1, 0.01, 0.001, 0.0001}. The optimizer chooses the AdamW (Adam Weight Decay Optimizer) and dropout is set to 0.5.

4.3 Experimental Results and Analysis

Table 3. Performance comparison of different models

Metrics	Training_epochs	Time (train + eval)	Accuracy (%)
BERT-CRF	5	49	80.28
CRF	50	47	60.73
CRF (embedding)	50	47	60.86
Bi-LSTM-CRF	50	255	72.56
Bi-LSTM (embedding)	50	218	72.12

To validate the superiority of the proposed model based on BERT and CRF, five models were compared in this paper, as shown in Table 3. Where embedding means input token uses pre-trained word embedding. It can be seen that the BERT-CRF model has the highest accuracy rate. Compared with the Bi-LSTM-CRF mainstream model, Bi-LSTM is replaced with BERT. Because the word level is considered, word segmentation is performed, and the word vector is fused with all the word vectors to which the word belongs, when training_epochs is 5, the time efficiency is improved, and the accuracy is improved by 7.72%; Compared with CRF model, BERT model is added for sequence annotation. In addition, since the BERT feature extractor is a Transformer, the accuracy of the experiment is even improved by 19.55%; Compared with that of CRF (embedding), the accuracy rate is improved by 19.42%. It can be seen that the training efficiency and performance of BERT-CRF model are very excellent.

Comparison of Trigger Word Recognition Performance with Different Methods.
In order to better verify the effectiveness of the proposed method, we compare the overall performance of the existing references for biological trigger word recognition with that of the proposed method, and the results are shown in Table 4. It can be seen that compared with Pyysalo et al. [16], which summarized the rich features such as context and dependency relationship, and adopted the classification method of SVM, the proposed model improves the precision by 9.49%, and the F value by 6.87%; Compared with the semi-supervised learning model proposed by Zhou et al. [17], this paper utilizes the unique constraint conditions of CRF conditional random fields to improve the precision by 8.11% and the F value by 5.82%; Compared with the method based on Nie et al. [8] neural network and word vector, the precision is improved by 9.24%, and the F value is improved by 5.48%; Compared with Wang [18] trigger word recognition method based on dependency word vector, because its expression efficiency is not high,

the improved precision by this method is increased by 7.01%, and the F value is increased by 5.61%. Compared with the hybrid neural network composed of SVM and BiLSTM proposed by Diao et al. [10], the F value is improved by 2.05%. Compared with BRNN w/GPA [11], F value using advanced dependency characterization learning module was improved by 1.34%. In general, performance our method improved overall compared with other methods.

Table 4. Performance comparison of different methods

Methods	Precision (%)	Recall (%)	F-value (%)
Pyysalo	70.79	81.69	75.84
Zhou	72.17	82.26	76.89
Nie	71.04	84.60	77.23
Zhou	75.35	81.60	78.32
Wang	73.27	81.35	77.10
Diao	80.03	81.54	80.66
BRNN w/GPA	81.33	81.42	81.37
Ours	80.28	85.31	**82.71**

5   Conclusion

In this paper, we propose a BERT-CRF model for the task of trigger word recognition to solve the problems of the mainstream model Bi-LSTM-CRF. Firstly, we construct the token embeddings (word vector), segment embeddings (sentence vector) and position embeddings (position vector) as the multi-information representation to form the BERT input. Furthermore, the properties of the word vector itself and the relationship between the word vectors are considered. Secondly, for the coding part of BERT's Transformer architecture, this paper defines and constructs two models respectively: a pre-training language model based on Mask mechanism and a bi-directional language model to train the language model by adding an extra text to predict whether the two parts of BERT's input text are continuous. Finally, some constraints are added through CRF layer to ensure that the final prediction results are valid. With these useful constraints, erroneous prediction sequences are greatly reduced and the accuracy of the results is greatly improved. The overall experimental process uses MLEE data sets. From the results, it can see that the proposed BERT-CRF method has achieved good performance on MLEE data sets.

Acknowledgments. This work is supported by the National Science Foundation of China (No. 62006108), General program of China Postdoctoral Science Foundation (No. 2022M710593), Liaoning Provincial Science and Technology Fund project (No. 2021-BS-201), and Natural Science research projects of Liaoning Education Department (No. LQ2020027).

References

1. Ren, W., Beard, R.W.: Consensus seeking in multiagent systems under dyanmically changing interaction topology. IEEE Trans. Autom. Control **50**(5), 655–661 (2005)
2. Du, J., Zhang, F., Yang, J., et al.: Mean consistency of multi-agent systems under random topology. Appl. Res. Comput. **29**(3), 1011–1013 (2012)
3. Donker, M., Frazzoli, E., Johansson, K.: Distributed event-triggered control with guaranteed ∞ gain and improved and decentralized event-triggering. IEEE Trans. Autom. Control **57**(6), 1362–1376 (2012)
4. Björne, J.: Biomedical event extraction with machine learning. TUCS Dissertations **178**, 1–121 (2014)
5. Wei, X., Zhu, Q., Lyu, C., et al.: A hybrid method to extract triggers in biomedical events. J. Digital Inf. Manage. **13**(4), 300–305 (2015)
6. He, X., Li, L., Liu, Y., et al.: A two-stage biomedical event trigger detention method integrating feature selection and word embeddings. IEEE/ACM Trans. Comput. Biol. Bioinf. **15**(4), 1325–1332 (2018)
7. Li, C., Song, R., Lialata, M., et al.: Using word embedding for bio-event extraction. In: Process of Workshop on Biomedical Natural Language Processing. Stroudsburg: Association for Computational Lin-guistics, pp. 121–126 (2015)
8. Nie, Y., Rong, W., Zhang, Y., et al.: Embedding assisted prediction architecture for event trigger identification. J. Bioinform. Comput. Biol. **13**(3) (2015)
9. Rahul, P.V.S.S., Sahu, S.K., Anand, A., et al.: Biomedical event trigger identification using bidirectional recurrent neural network based models. arXiv: Computation and Language, pp. 316–321 (2017)
10. Diao, Y., Lin, H., Yang, L., et al.: "FBSN": a hybird fine-grained neural network for biomedical event trigger identification. Neurocomputing **381**, 105–112 (2020)
11. Li, L., Zhang, B.: Exploiting dependency information to improve biomedical event detection via gated oplar attention mechanism. Neurocomputing **421**, 210–221 (2021)
12. Wang, Y., Wang, J., Lin, H., et al.: Biomedical event trigger detection based on bidirectional LSTM and CRF. Bioinformatics and Biomedicine, pp. 445–450 (2017)
13. Devlin, J., Chang, M.W., Lee, K., et al.: BERT:Pre-training of deep bidirectional trans-formers for language understanding. In: Proceedings of the 2019 Conference of the North American Chapter of the Association for Computational Linguistics: Human Language Technologies, Volume 1 (Long and Short Papers), pp. 4171–4186 (2019)
14. Xu, L., Li, S., Wang, Y., Xu, L.: Named entity recognition of BERT-BiLSTM-CRF combined with self-attention. In: Xing, C., Fu, X., Zhang, Y., Zhang, G., Borjigin, C. (eds.) WISA 2021. LNCS, vol. 12999, pp. 556–564. Springer, Cham (2021). https://doi.org/10.1007/978-3-030-87571-8_48
15. Tian, Z., Li, X.: Research on Chinese event detection method based on BERT-CRF model. Comput. Eng. Appl. **57**(11), 135–139 (2020)
16. Pyysalo, S., Ohta, T., Miwa, M., et al.: Event extraction across multiple levels of biological organization. Bioinformatics **28**(18), i575–i581 (2012)
17. Zhou, D., Zhong, D., He, Y.: Event trigger identiffication for biomedical events extraction using domain knowledge. Bioinformatics **30**(11), 1587–1594 (2014)
18. Wang, J., Zhang, J., Yuan, A., et al.: Biomedical event trigger detection by dependency-based word embedding. In: IEEE International Conference on Bioinformatics and Biomedicine. IEEE, pp. 429–432 (2015)

Tackling Non-stationarity in Decentralized Multi-Agent Reinforcement Learning with Prudent Q-Learning

Jianan Wei[1], Liang Wang[1(✉)], Xianping Tao[1,2], Hao Hu[1], and Haijun Wu[2]

[1] State Key Laboratory for Novel Software Technology, Nanjing University,
Nanjing 210023, People's Republic of China
`jawei@smail.nju.edu.cn`, {`wl,txp,myou`}`@nju.edu.cn`
[2] National Experimental Teaching Demonstration Center of Computer Science
Technology and Software Engineering, Nanjing University, Nanjing 210023,
People's Republic of China
`hjwu@nju.edu.cn`

Abstract. Multi-Agent Reinforcement Learning (MARL) is challenging due to the non-stationary issue of an agent's learning environment caused by multiple co-evolving agents, i.e., the uncertainty rises with multiple agents learning and evolving simultaneously. Though centralized learning can solve the problem to some degree, it is hard to expand into numerous agents or real-world applications. We focus on the native property of the non-stationary environment and use a independent method to alleviate its harmful influence. Here are the two main contributions of this article: 1) We formalize the non-stationary environment by supposing it follows a pessimistic transition, and we explain the existence of a lower bound of accumulated reward expectation that the agent can learn. 2) We devise a scheme called Prudent Q-Learning (PQL) based on independent learning for controlling multi-agent systems in non-stationary environments. We present both tabular and Deep Q-Learning based implementations of the proposed Prudent Q-Learning scheme.

Keywords: Non-stationary environment · Reinforcement learning · Independent learning

1 Introduction

A multi-agent system (MAS) consists of a set of autonomous and interactive agents that share the same environment and use sensors to perceive the environment and perform actions. However, learning is difficult for the agents in MAS due to partial observation and local viewpoints, and they may regard the environment as non-stationary [18]. In fact, it is the state uncertainty and uncertainty with other agents because the agents can not determine what the real state of the environment and other agents' actions are [17].

© The Author(s), under exclusive license to Springer Nature Switzerland AG 2022
X. Zhao et al. (Eds.): WISA 2022, LNCS 13579, pp. 403–415, 2022.
https://doi.org/10.1007/978-3-031-20309-1_35

In multi-agent reinforcement learning (MARL), the environment is often modeled as a Markov Game (MG) [10] which is defined by a tuple $(I, S, A, R, \mathcal{T})$ where I is a set of N agents, S is a set of states, $A = A_1 \times A_2 \ldots \times A_N$ is the set of joint action of all agents, $R = (R_1, R_2, \ldots, R_N)$ is the reward function and $\mathcal{T} : S \times A \to D(S)$ shows the transition probability between states.

The environment is defined as everything that interact with the agent. Thus, in an MAS, other agents are also a part of the environment from one agent's perspective [23]. In an MG, the $\mathcal{T}$ and R are correlated with the actions of all the agents, which means the environment can be modified by all the agents [1]. During the training process, the policy of every agent is changing over time [19], resulting in changing the environment together. In that case, the variation of $\mathcal{T}$ and R lead the environment not to hold the Markov property defined by MDP anymore, and in this time, we call the environment is **non-stationary**. From the perspective of one agent x of a MAS, other agents' actions are uncontrollable meanwhile hard to fully observe, which makes the $\mathcal{T}$ and R_x estimated by agent x outdated. This will also make the policy learned by agent x quickly become obsolete. That is why many conventional RL algorithms perform poorly and inefficiently in MAS.

Some centralized training methods that can be applied in solving non-stationarity [7,8]. Though these algorithms succeed in complex team competition problems and video games, they are highly dependent on high-performance hardware, central training node, and excellent parallel programming. Besides, some decentralized training methods implemented by computing and sharing additional information can also solve the non-stationary environment [20,21]. However, these methods need special training patterns or strong premises like full observation or communication, which makes time difficult to expand to a large number of agents.

In this paper, inspired by the idea of Minimax from García, J. et al. [4], we propose Prudent Q-Learning, and try to deal with the non-stationary environment in MARL. The core of our approach is to pessimistically estimate the environmental transition that always leads to the worst state. By learning such the worst case that happened before, the agents are able to choose prudent actions and avoid bad things caused by the non-stationary environment.

In summary, this paper makes the following contributions:

1) We formalize the non-stationary environment by supposing it follows a pessimistic transition and explain the existence of a lower bound of accumulated reward expectation which the agent can learn.
2) We designed a method Prudent Q-Learning by improving Q-Learning [25], and we test it in some MARL scenarios.
3) We also combine Prudent Q-Learning with DQN [16] and prioritized experience replay [22]. A relevant experimental scenario is set up to display its effectiveness.

2   Related Work

Varying Learning Rate: In MAS, an agent can be punished because of a bad choice of the other agent even if it has chosen an optimal action. Then this agent had better attach less importance to punishment after choosing an action that has been satisfying in the past [9]. So the agents should learn the positive step reward with a higher learning rate and the negative one with a lower one (Hysteretic Q-Learning [14]). We think oppositely as the agent should be more prudent about the punishment because the action is regarded as a bad thing even if the other agent causes it. The agent should adopt more prudent action to avoid being punished rather than compete with other agents.

Worst Case Criterion: Worst case criterion has been discussed in safe RL [4]. The learning targets are usually written as min-max form formulas in the paradigm. One of them considers the set of all interaction trajectories as $(s_0, a_0, s_1, a_1, ...)$ that may occur under a policy π. It regards the environment as an opponent that is trying to minimize the cumulative reward by leading the next state to a "worst case". The other one considers all the possible transition matrices related to the state and action. It pessimistically estimates the unknown transition possibility, which will always lead to a state with low rewards. In either case, the goal is to find a policy that maximizes the reward expectation under the minimum cumulative rewards. Some methods have been proposed to solve the worst case criterion, such as $\hat{Q}$-Learning [6] based on minimizing step reward and Cautious RL [5] using a pessimistic and optimistic safety padding to guarantee safety. We suppose that the uncertain environment is non-stationary and pessimistically assess the accumulated reward expectation. In other words, we believe that the transition matrix is time-varying regardless of what causes the worst case criterion.

Independent Learning: Independent RL was proposed by Ming T. et al. [24] where agents learn and make decisions individually in MAS. This kind of learning method is different from the centralized-training and decentralized-execution (MADDPG [12]) or the decentralized-training and decentralized-execution (PR2 [26]) paradigms. However, in many practical applications where a large number of agents are required, the joint observation and action spaces will grow exponentially when considering centralized training. Furthermore, the preconditions about the observing of the other agents or communication do not always exist, so it is impossible to model or consider all the opponent agents in decentralized training. Independent learning agents will always optimize their policy according to the varying environment and interact with their surroundings by relying on sensor information [15]. AS such, independent learning methods are generally more versatile and can be applied in uncertain environments [3]. We think independent learning is suitable for our work when considering an extension to the real-world application or operating efficiency.

3 Proposed Approach

3.1 Theoretical Derivation of PQL

We assume the real state transition followed by the non-stationary environment is τ which changes over time. On each agent's view, it is assumed that the environment follows a **pessimistic state transition** τ'. In this background setting, the agent will always encounter the **worst state** when given an initial state and action.

$Q'_\pi(s, a)$ denotes the estimated value of accumulated reward expectation of an agent with policy π when the environment follows transition τ'. And S_{sa} denotes the reachable states set when choosing action a in state s. Ideally, $Q'_\pi(s, a)$ can be written as

$$
\begin{aligned}
Q'_\pi(s_t, a_t) &= \mathbb{E}[r(s_t, a_t, s_{t+1}) + \gamma r(s_{t+1}, a_{t+1}, s_{t+2}) + \cdots + \gamma^{T-1-t} r(s_{T-1}, a_{T-1}, s_T)] \\
&= \mathbb{E}[r(s_t, a_t, s_{t+1})] + \gamma Q'_\pi(s_{t+1}, a_{t+1})
\end{aligned} \tag{1}
$$

where $t, t+1, ..., T$ denote the time step and $s_{t+1}, s_{t+2}, \cdots s_T$ follow the transition τ'. Under this circumstance we define the **worst state** as

$$
s' = \arg \min_{s' \in S_{sa}} \max_{a'} Q'_\pi(s', a') \tag{2}
$$

where $\max_{a'} Q'_\pi(s', a')$ is called the value of state s'. And the **pessimistic state transition** τ' can be written as

$$
\tau'(s'|s, a) = \mathbb{P}\{s' = \arg \min_{s' \in S_{sa}} Q'_\pi(s', a')\} = 1 \tag{3}
$$

where $a' = \arg \max_{a'} Q'_\pi(s', a')$. τ' is always a bad transition, and we believe that Q'_π in Eq. (1) is a lower bound of expectation of accumulated reward that follows the real state transition τ.

Then we use $W(s, a)$ to display the worst case the agent met under state s, action a and transition τ' which is defined as

$$
W(s, a) = \min_{s' \in S_{sa}} \max_{a'} Q'_\pi(s', a'). \tag{4}
$$

When the agent interacts with the real environment following the transition τ, we can obtain an interaction sequence $<s, a, r, s_{next}>$ and define its weight d as

$$
d = \max_{a''} Q'_\pi(s_{next}, a'') - W(s, a) \tag{5}
$$

which is used to update $W(s, a)$ if it is less than 0. But notice that $W(s, a)$ in Eq. (5) is found to converge into its definition in Eq. (4).

So in the ideal pessimistic transition τ', the update is not fully determined by the real-time transition $<s, a, r, s_{next}>$ because

$$
Q'_\pi(s, a) = Q'_\pi(s, a) + \alpha[r + \gamma \min_{s'} \max_{a'} Q'_\pi(s', a') - Q'_\pi(s, a)] \tag{6}
$$

and here s' in the the equation is not always s_{next} during the training. Notice that W is updated during the learning process, $Q'_\pi(s, a)$ will finally converge into its definition in Eq. (1).

Notice that the min-max formula should not be confused with other methods using the term Minimax, such as Minimax Q-Learning proposed by Zhu, Y. et al. [28] talking about zero-sum Markov games for only two agents and Minimax Regret used by Xu, L. et al. [27] talking about judgment criterion. Our approach is independent and can be easily applied to multiple agents. Our method will lead to a prudent policy, so we call it **Prudent Q-Learning (PQL)**.

3.2 Tabular-Based PQL

We can directly use a table to record the worst case an agent has ever met, and we denote it as W. It records the worst state reachable from (s, a). We will get a transition $<s, a, r, s_{next}>$ and calculate its weight following Eq. (5) when the agent interacts with the environment.

However, only the value of $\max_{a'} Q'_\pi(s_{next}, a')$ is not enough to learn from a transition. So W is retrofit to a 3-element tuple $<r, s', V>$ to record the transition and $W.V$ is used to record the value of the worst state instead. Each $W(s, a)$ is initially empty and assigned tuple $<r, s_{next}, \max_{a'} Q(s_{next}, a')>$ when meeting the transition $<s, a, r, s_{next}>$ for the first time (never seen state-action (s, a) before). Besides, $W(s, a)$ is also assigned this tuple if the weight d of the transition is less than 0 who is calculated as below:

$$d = \max_{a''} Q'_\pi(s_{next}, a'') - W(s, a).V. \tag{7}$$

Now consider the update of Q'_π. The agent needs to know the worst transition under (s, a) as we have discussed in Sec.3.1. Then the update is not performed with every transition the agent met according to the ideal transition τ' we supposed. That's to say, Q'_π is updated only when the worst next state is met as shown in Eq. (6). However, W is also updated during the learning process. The update of $Q'_\pi(s, a)$ is written as

$$Q'_\pi(s, a) = Q'_\pi(s, a) + \alpha[W(s, a).r + \gamma W(s, a).V - Q'_\pi(s, a)]. \tag{8}$$

However, Eq. (8) is too focused on the worst case, which makes the agents seldom use the transitions they met during training. We proposed two strategies to control the degree of pessimism. The first one is **repeat and probability update**. The agent will update n times when dealing with a transition. For each update, the agent will update as Eq. (8) with probability p (**pessimistic parameter**) and the same as Q-Learning did with $1 - p$. The another one is **time threshold** (denoted as t_h) which is attached to clear up outdated records of each episode. $W(s, a)$ will be updated by the current transition if it is updated t_h steps before. So the structure of W table is modified as $<r, s_{next}, \max_{a'} Q(s_{next}, a'), timestamp>$. The ultimate process is shown as Algorithm 1.

Algorithm 1. Tabular-Based Decentralized PQL for Each Agent

Input: Discount factor γ, learning rate α, timeliness threshold t_h, pessimistic parameter $p \in [0, 1]$, update n times for each step, max step limit T for each episode.

Initialization: set Q'_π to 0, W to empty and $Record$ to empty.

```
1:  for t=1,2,...T do
2:      Choose action a = arg max_a Q̂'_π(s, a) or randomly with probability ε
3:      Execute joint action in E and get r, s'
4:      V(s') = max_a' Q̂'_π(s', a')
5:      if W(s, a) is empty or V(s') < W_i(s, a).V or W(s, a).t < t − t_h then
6:          update W(s, a) with < r, s', V(s'), t >
7:          update Record(s, a)
8:      end if
9:      while repeatedly update for n times do
10:         \\ As long as Record(s, a) is not empty, and its value is worst case
11:         if Record(s, a) is worst case then
12:             update Q̂'_π(s, a) = Q̂'_π(s, a) + α(r + γV(s') − Q̂'_π(s, a))) with prob 1 − p
13:             < r_w, s_w, V_w, t >= W(s, a)
14:             update Q̂'_π(s, a) = Q̂'_π(s, a) + α(r_w + γV_w − Q̂'_π(s, a))) with prob p
15:         else
16:             update Q̂'_π(s, a) = Q̂'_π(s, a) + α(r + γV(s') − Q̂'_π(s, a)))
17:         end if
18:     end while
19: end for
```

Here the $Record$ table records W's update type (new encounter, worst case, and time out). When $p = 0$, the algorithm degenerates into Q-Learning; when $p = 1$, the algorithm is strictly pessimistic PQL; when $p = 0.5$, the algorithm updates Q according to current transition and worst case both for $\frac{n}{2}$ times.

4 Experiments

4.1 Criterion

Total reward of all agents is a baseline metric to evaluate the algorithms. Besides, **Collision** originally proposed in safe RL is often used to measure the degree of bad things [13]. Intuitively, the degree of bad things that happen when agents are trying to complete the task naturally fit into our definition of "worst case" in Sect. 3.1 caused by non-stationary issues during MARL training.

The fewer bad things happen means the better adaptation to the non-stationary environment. A prudent policy should avoid collisions, so the metrics about collisions are suitable for us to evaluate a policy. So we adopt the evaluation metrics as below. The smaller CT and CAC are, the more prudent the algorithm is.

- **Reward**: The direct learning target of RL algorithms.
- **Collision Agent Count (CAC)**: The number of agents that have encountered at least a collision in each episode.
- **Collision Times (CT)**: Each time two agents collide, collision times increase 2.
- **Finish Agent Count (FAC)**: The number of agents that have reached their destination in the given steps of each episode. This metric is to test whether the agent has achieved the expected goal.

4.2   Environment Settings

Swap Environment: Scenario *Swap* shown as Fig. 1(a) is designed to display the correctness and mechanism of PQL. Two agents' goal is to exchange locations. Each agent can move left, right, up, down, or stay in each step. Both agents will get a +1000 reward for achievement and -1000 for collision. The observation is each agent's coordinates. Besides, *Extended Swap* shown as Fig. 1(b) is constructed to further test our algorithm. But this time each agent can get their coordinates and 8-neighbourhood information instead of all coordinates.

Crossroad Environment: The scenario *Traffic* shown as Fig. 1(c) is constructed to test how the agent would handle potential collisions. Agents in the corner of the map will cross the intersection to the opposite side of the road. Each agent will get a +1000 reward for success and -1000 for collision. The observation setting is the same as *Extended Swap*. The action space size is two, as agent can move forward or stay. Besides, we construct a larger environment *Extended Traffic* shown as Fig. 1(d) for further test. Each agent will get a +1000 reward for crossing the road and a -330 punishment for collision.

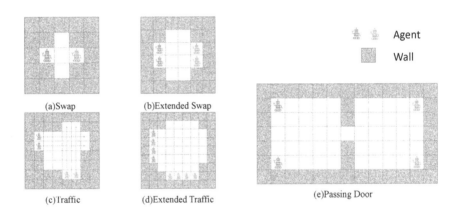

(a)Swap   (b)Extended Swap

(c)Traffic   (d)Extended Traffic

(e)Passing Door

Fig. 1. MARL environments

Passing Door Environment: The more complex scenario *Passing Door* inspired by Liu Y. et al. [11] is shown as Fig. 1(e). Four agents are located in the corner of the map that is divided into two rooms with a 1 × 1 grid connection. The agents on the left will pass through the connection and reach the right room. So do the ones on the right. The agent will get a -500 punishment for the first collision and -100 for the other times. A +1000 reward is given for reaching the opposite room only once. The action and observation setting are the same as *Extended Swap*.

4.3   Pessimistic Parameter Selection

Different pessimistic parameters ($p = 0.25, 0.5, 0.75, 1.0$) of PQL referred by Sect. 3.2 are tested. Each agent in these environments learns independently and does not share parameters. Each parameter experiment is repeated 3 times and 50,000 episodes for each time. *Traffic* and *Passing Door* are served as testing environments for the comparison. After that, we set learning rate to 0.001, discount factor to 0.95, random exploration probability to 0.15, t_h to the max step (300 in this scenario) of each episode, and the update times to 2 for each learning step.

The results of parameter comparison are listed in Table 1. The agents with a higher pessimistic parameter can perform better over reward and collision most of the time, as it corresponds to a more prudent policy, which leads the agents to avoid collision as much as possible. Though the goal completion rate fluctuates during the training in *Passing Door*, it is generally accepted as they both maintain a relatively high level. The relatively larger pessimistic parameter can enhance cautiousness to some degree at the cost of goal completion rate, so it is necessary to balance these two factors when considering application in other scenarios.

Table 1. Result of parameter comparison. The average value of parameters are listed in the table under different criteria.

Env	p	Reward	CAC	CT	FAC
Traffic	0.25	3696.62	0.300	0.363	4.0
	0.5	3715.12	0.281	0.284	4.0
	0.75	3646.11	0.347	0.254	4.0
	1.0	**3802.09**	**0.191**	**0.198**	4.0
Passing Door	0.25	2370.04	2.617	2.767	3.970
	0.5	2322.53	2.667	2.915	**3.972**
	0.75	2273.13	2.717	3.042	3.969
	1.0	**2571.40**	**2.355**	**2.181**	3.950

4.4   Overall Performance and Comparisons

In this part, we compared our algorithm with Q-Learning and Hysteretic Q-Learning. The experiment setting is the same as Sect. 4.3. For fairness, the Q table for comparison algorithms is updated twice for each step as PQL does.

The brief results are listed in Table 2. Though almost all algorithms have a similar goal completion rate, PQL has advantages on reward and collision most of the time. For the same reason, the PQL agents adopt more prudent policies, resulting in fewer collisions. Even in the most unstable scenario *Swap*, PQL is less volatile than Q-Learning and Hysteretic Q-Learning, which often fail abruptly many times during the training.

Table 2. Result of algorithm comparison. The average value of algorithms are listed in the table under different criteria. PQL stands for Prudent Q-Learning, QL for Q-Learning, and HY for Hysteretic Q-Learning.

Env	Alg	Reward	CAC	CT	FAC
Swap	QL	1300.68	0.339	0.339	1.640
	HY	1334.77	0.327	0.327	1.662
	PQL	**1490.49**	**0.254**	**0.254**	**1.745**
Ex Swap	QL	2181.06	1.064	1.814	**3.995**
	HY	2058.86	1.150	1.927	3.986
	PQL	**2431.42**	**0.882**	**1.560**	3.991
Traffic	QL	3696.33	0.300	0.304	4.0
	HY	3516.86	0.476	0.483	4.0
	PQL	**3806.49**	**0.186**	**0.194**	4.0
Ex Traffic	QL	7569.65	1.272	1.304	8.0
	HY	7486.56	1.517	1.556	8.0
	PQL	**7654.14**	**0.980**	**1.048**	8.0
Passing Door	QL	2296.22	2.664	3.065	3.975
	HY	2255.29	2.748	3.156	**3.986**
	PQL	**2594.54**	**2.407**	**2.133**	3.984

The experiments above can conclude that agents can obtain more prudent policies by properly learning some pessimistic experiences, thereby avoiding bad things caused by environmental instability. This is better than treating all experiences the same (Q-Learning) or simply learning optimistically (Hysteretic Q-Learning).

5 Extend to Deep RL

In this chapter, we explored expanding PQL to deep RL to improve the limitation of tabular-based method. The core difficulty lies in the expansion of W (a 4-element tuple table) proposed in Sect. 3.2.

Each transition has a weight d computed by Eq. (7). The smaller d is, the worse the transition will be. The core idea of PQL is to properly learn bad transitions and the calculation of d does not require the whole transition $<s, a, r, s_{next}>$ but only the value of the state s_{next} computed by $\max_{a''} Q'_\pi(s_{next}, a'')$. Q'_π table can be fit by a network as DQN does. Naturally, prioritized experience replay buffer (Sum Tree) [22] is used in Deep PQL which is similar as memory replay in NLP [2]. The transition with smaller weight will be chosen more frequently during experience replay. Besides, another network is needed to fit $W(s, a).V$ (hereafter denoted as $W_v(s, a)$) which is updated when $d < 0$.

However, d needs to be scaled as it might be too large or small. Denote $d^{sa}_{min-loc}$ and $d^{sa}_{max-loc}$ to represent the minimum and maximum weights that have ever appeared, for all $s' \in S_{sa}$, we have:

$$
\begin{aligned}
d^{sa}_{min-loc} &= \min(\max_{a''} Q'_\pi(s', a'') - W_v(s, a)) \\
d^{sa}_{max-loc} &= \max(\max_{a''} Q'_\pi(s', a'') - W_v(s, a))
\end{aligned}
\tag{9}
$$

where "loc" means local as W_v changes over time. And weight d is calculated as

$$
d^{sa}(s_{next}) = \max_{a''} Q'_\pi(s_{next}, a'') - W_v(s, a)
\tag{10}
$$

according to the transition $<s, a, r, s_{next}>$. At last, a simple linear scaling is used to compute the final weight as follow

$$
d'^{sa}_{loc}(s_{next}) = \frac{d^{sa}_{loc-max} - d^{sa}}{d^{sa}_{loc-max} - d^{sa}_{loc-min}}(\alpha - \beta) + \beta, \quad \beta < \alpha
\tag{11}
$$

where α and β are the upper and lower bounds of weight.

Referring to DQN, we use neural network to fit $Q'_\pi(s, a)$, $W_v(s, a)$, $d^{sa}_{min-loc}$ and $d^{sa}_{max-loc}$ and a target network $\hat{Q}^t_\pi(s, a; \theta^-)$. The parameters of the networks are θ, w, b_{min} and b_{max} respectively. The update of these networks can be implemented by Stochastic Gradient Descent. The ultimate implementation is shown as Algorithm 2.

We compare Deep PQL and Independent DQN on the scenario *Passing Door*. The experiment settings are the same as Sect. 4. We record the average data for every 50 tests. The learning rate of W, d_{min} and d_{max} network used in PQL is $1e - 4$. Beside, we set $\alpha = 3, \beta = 1$, replay buffer size to 3200 and batch size to 32. The final result is shown as Fig. 2. PQL has an advantage in the average

---

Algorithm 2. Deep PQL with Prioritized Experience Replay for Each Agent

Input: Discount factor γ, learning rate for all the networks, upper bound α and lower bound β of weight, batch size and Sum Tree capacity N_D for experience replay.
Initialization: Initialize $\theta, w, b_{min}, b_{max}$ with random weights, $\theta^- = \theta$. Sum Tree D is empty with capacity N_D.

1: **for** step t=1,2,...,T **do**
2:     Choose action $a = \arg\max_a \hat{Q}'_\pi(s, a; \theta)$ or randomly with probability ϵ.
3:     Execute joint action and observe s_{next} and r.
4:     $d^{sa} = \max_{a'} \hat{Q}'_\pi(s_{next}, a'; \theta) - \hat{W}(s, a; w)$.
5:     Update $\hat{W}(s, a; w)$ if $d^{sa} < 0$.
6:     Update $\hat{d}_{min-loc}(s, a; b_{min})$ and $\hat{d}_{max-loc}(s, a; b_{max})$ with d^{sa}.
7:     $d'^{sa}_{loc} = \frac{\hat{d}_{max-loc}(s,a;b_{max}) - d^{sa}}{\hat{d}_{max-loc}(s,a;b_{max}) - \hat{d}_{min-loc}(s,a;b_{min})}(\alpha - \beta) + \beta$.
8:     Store $< s, a, r, s' >$ in D with weight d'^{sa}_{loc}.
9:     Sample a batch-size transitions $< s_j, a_j, r_j, s_{j+1} >$ from D.
10:     $y_j = \begin{cases} r_j & \text{for terminal } s_{j+1} \\ r_j + \gamma \max_{a'} \hat{Q}^t_\pi(s_{j+1}, a'; \theta^-) & \text{for non-terminal } s_{j+1} \end{cases}$
11:     Perform a gradient descent step on $\left(y_j - \hat{Q}'_\pi(s_j, a_j; \theta)\right)^2$.
12:     Every C steps set $\hat{Q}^t_\pi = \hat{Q}'_\pi$.
13: **end for**

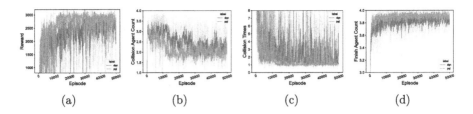

| (a) | (b) | (c) | (d) |

Fig. 2. Deep RL test result on Passing Door

Reward criterion from the 10,000th training episode though both algorithms are unstable. Although PQL has no significant advantage over DQN in Collision Times, fewer PQL agents have encountered collisions. Besides, PQL's target completion rate is higher than DQN. The result can illustrate that PQL based on deep reinforcement learning is feasible.

6   Conclusion

It is hard for agents to learn in non-stationary environments as almost everyone changes over time. To avoid bad things caused by the environment, the agents adopt the relatively prudent PQL we proposed. According to the experiment results, tabular-based PQL is more prudent than Q-Learning and Hysteretic Q-Learning to some degree in the scenarios we constructed while getting more average rewards and having a higher goal completion rate. Though PQL based on deep networks fluctuates during training, it performs better than DQN most of the time, which is generally satisfactory. More work still needs to be done on deep RL when considering the broader application of PQL.

Acknowledgement. This work is supported by the National Key R&D Program of China under Grant No. 2018AAA0102302, and the Collaborative Innovation Center of Novel Software Technology and Industrialization.

References

1. Canese, L., et al.: Multi-agent reinforcement learning: a review of challenges and applications. Appl. Sci. **11**(11) (2021). https://doi.org/10.3390/app11114948
2. Chen, Y., Wen, Y., Zhang, H.: Cost-effective memory replay for continual relation extraction. In: Xing, C., Fu, X., Zhang, Y., Zhang, G., Borjigin, C. (eds.) WISA 2021. LNCS, vol. 12999, pp. 335–346. Springer, Cham (2021). https://doi.org/10.1007/978-3-030-87571-8_29
3. Daskalakis, C., Foster, D.J., Golowich, N.: Independent policy gradient methods for competitive reinforcement learning. Adv. Neural Inf. Process. Syst. **33**, 5527–5540 (2020)
4. García, J., Fernández, F.: A comprehensive survey on safe reinforcement learning **16**(1), 1437–1480 (2015)

5. Hasanbeig, M., Abate, A., Kroening, D.: Cautious reinforcement learning with logical constraints. arXiv preprint arXiv:2002.12156 (2020)
6. Heger, M.: Consideration of risk in reinforcement learning. In: Cohen, W.W., Hirsh, H. (eds.) Machine Learning Proceedings 1994, pp. 105–111. Morgan Kaufmann, San Francisco (CA) (1994). https://doi.org/10.1016/B978-1-55860-335-6.50021-0
7. Iqbal, S., Sha, F.: Actor-attention-critic for multi-agent reinforcement learning. PMLR (2019)
8. Jaderberg, M., et al.: Human-level performance in 3d multiplayer games with population-based reinforcement learning. Science **364**(6443), 859–865 (2019)
9. Lauer, M., Riedmiller, M.: An algorithm for distributed reinforcement learning in cooperative multi-agent systems. In: Proceedings of the Seventeenth International Conference on Machine Learning, pp. 535–542 (2000)
10. Littman, M.L.: Markov games as a framework for multi-agent reinforcement learning. In: Proceedings of the Eleventh International Conference on International Conference on Machine Learning, pp. 157–163. ICML 1994, Morgan Kaufmann Publishers Inc., San Francisco, CA, USA (1994)
11. Liu, Y., Hu, Y., Gao, Y., Chen, Y., Fan, C.: Value function transfer for deep multi-agent reinforcement learning based on n-step returns. In: IJCAI 2019, AAAI Press (2019)
12. Lowe, R., Wu, Y., Tamar, A., Harb, J., Abbeel, P., Mordatch, I.: Multi-agent actor-critic for mixed cooperative-competitive environments. NIPS'17, Curran Associates Inc., Red Hook, NY, USA (2017)
13. Lütjens, B., Everett, M., How, J.P.: Safe reinforcement learning with model uncertainty estimates. In: 2019 International Conference on Robotics and Automation (ICRA), pp. 8662–8668 (2019). https://doi.org/10.1109/ICRA.2019.8793611
14. Matignon, L., Laurent, G.J., Le Fort-Piat, N.: Hysteretic q-learning: an algorithm for decentralized reinforcement learning in cooperative multi-agent teams. In: 2007 IEEE/RSJ International Conference on Intelligent Robots and Systems, pp. 64–69. IEEE (2007)
15. Matignon, L., Laurent, G.J., Le Fort-Piat, N.: Independent reinforcement learners in cooperative markov games: a survey regarding coordination problems. Knowl. Eng. Rev. **27**(1), 1–31 (2012)
16. Mnih, V., et al.: Human-level control through deep reinforcement learning. Nature **518**(7540), 529–533 (2015)
17. Oliehoek, F.A., Amato, C.: A Concise Introduction to Decentralized POMDPs. Springer, Heidelberg (2016). https://doi.org/10.1007/978-3-319-28929-8
18. Omidshafiei, S., Pazis, J., Amato, C., How, J.P., Vian, J.: Deep decentralized multi-task multi-agent reinforcement learning under partial observability. ICML'17 (2017)
19. Papoudakis, G., Christianos, F., Rahman, A., Albrecht, S.V.: Dealing with non-stationarity in multi-agent deep reinforcement learning. CoRR **abs/1906.04737** (2019). http://arxiv.org/abs/1906.04737
20. Rabinowitz, N., Perbet, F., Song, F., Zhang, C., Eslami, S.A., Botvinick, M.: Machine theory of mind. PMLR (2018)
21. Raileanu, R., Denton, E., Szlam, A., Fergus, R.: Modeling others using oneself in multi-agent reinforcement learning. PMLR (2018)
22. Schaul, T., Quan, J., Antonoglou, I., Silver, D.: Prioritized experience replay. In: Bengio, Y., LeCun, Y. (eds.) 4th International Conference on Learning Representations, ICLR 2016, San Juan, Puerto Rico, 2–4 May 2016, Conference Track Proceedings (2016). http://arxiv.org/abs/1511.05952

23. Sutton, R., Barto, A.: Reinforcement Learning: An Introduction, Second edition. MIT Press, Cambridge (2018)
24. Tan, M.: Multi-agent reinforcement learning: independent vs. cooperative agents. In: Proceedings of the Tenth International Conference on Machine Learning, pp. 330–337. Morgan Kaufmann (1993)
25. Watkins, C.J., Dayan, P.: Q-learning. Mach. Learn. **8**(3), 279–292 (1992)
26. Wen, Y., Yang, Y., Luo, R., Wang, J., Pan, W.: Probabilistic recursive reasoning for multi-agent reinforcement learning. CoRR abs/1901.09207 (2019). http://arxiv.org/abs/1901.09207
27. Xu, L., Perrault, A., Fang, F., Chen, H., Tambe, M.: Robust reinforcement learning under minimax regret for green security. PMLR (2021)
28. Zhu, Y., Zhao, D.: Online minimax q network learning for two-player zero-sum markov games. IEEE Trans. Neural Netw. Learn. Syst. (2020)

Deep Multi-mode Learning for Book Spine Recognition

Wanru Yang[ID] and Xiaohua Shi[(✉)][ID]

Library, Shanghai Jiao Tong University, Shanghai 200240, China
{wryang,xhshi}@sjtu.edu.cn

Abstract. Traditional library book inventory or shelf-reading costs a tremendous human resource to ensure the reading accuracy of books. This paper provides a deep multi-mode learning solution for book spine recognition in the library with higher reliability and lower cost. Our book recognition method includes improved Hough Transformation in image segmentation for book spine, optical character recognition with deep learning to extract text information on book spine, and combined text matching to the final inventory. The experiment shows that we achieve good performance in the overall accuracy of book spine recognition and the average time per spine.

Keywords: Deep learning · Book spine recognition · Computer vision · Image processing

1 Introduction

With the development of text recognition technology based on image processing and the popularization of smart devices, visual analysis's application field is becoming more extensive. A paper book contains much potential visual information which has yet to be explored and applied in library systems. The text recognition of the backs of books stored on the shelves is one of them.

This paper will present a deep multi-mode learning method for book spine recognition. It can be used to solve the problem of low efficiency and low accuracy of book inventory in the prior work. The multi-mode learning method includes improved Hough transformation in image segmentation for book spine, optical character recognition with deep learning to extract text information on book spine, and text matching to check the final inventory results.

2 Related Work

Computer vision technologies have been widely applied in smart library applications. Fowers *et al.* [2] present the development of a simple color enhancement feature descriptor called Color Difference-of-Gaussians SIFT for library inventory process automation. Shi *et al.* [8] propose a new image recognition method to identify books in libraries based on barcode decoding together with deep

X. Zhao et al. (Eds.): WISA 2022, LNCS 13579, pp. 416–423, 2022.
https://doi.org/10.1007/978-3-031-20309-1_36

learning and describe its application in library book identification processing. Nevetha *et al.* [6] propose a technique to improve book spine border detection by devising set of constraints based on structural properties that can be used to filter the detected line segments so as to obtain book spine boundaries.

Cutting the bookshelf image into multiple spine images includes detecting the bookshelf image edges using an edge detection algorithm, identifying straight lines in the bookshelf image, and cutting the bookshelf image into a plurality of spine images containing the spine according to the recognized straight line. We may use Hough Transform [5] for main feature extraction to find imperfect instances of objects in a specific type of shape through a voting procedure. This voting procedure is carried out in parameter space. In this parameter space, the candidate is obtained as a local maximum in the so-called accumulator space, obtained by the algorithm used to calculate the Hough transform Construct. The main advantage of the Hough transform is that it can tolerate the gaps in the feature boundary description and is relatively unaffected by image noise:

$$r = xcos\theta + ysin\theta, r \geq 0, 0 \leq \theta \leq \pi \tag{1}$$

where r is the distance from the origin to the nearest point on the line, and θ is the angle between the x-axis and the line connecting the origin and the nearest point. Therefore, each straight line of the image can be associated with a pair of parameters (r, θ). This parameter (r,θ) is called the Hough space and is used for a collection of two-dimensional straight lines.

3   Our Approach

To improve the accuracy and success rate of book spine detection, we propose combining spine localization and detection with deep learning OCR to achieve high accuracy real-time recognition of books. The whole processed model for Deep multi-mode learning is designed in Fig. 1.

3.1   Book Spine Extraction

Edge detection is a common problem in computer vision since edges can extract shapes of many things. It can keep the vital figure while reducing irrelevant information. Usually, noise suppression and accurate edge location cannot be met simultaneously. For classic algorithms, some can detect the soft edge, and some can survive strong noise. We test four operators to evaluate the results, Canny [1], Log, Prewitt, and Sobel.

For a given RGB image of one grid of bookshelf, we firstly convert it into gray style. After the conversion, we equalize the histogram of gray scale image to improve the contrast detail. In order to find the best operator that suits our library, we took 100 shelfgrid images(nearly 3000 books) as testcases to generate their edge detection results.

When we get the edge picture, we can do the hough transformation to vote for spine lines. Since book spines are vertical, we can set the angle is limited to

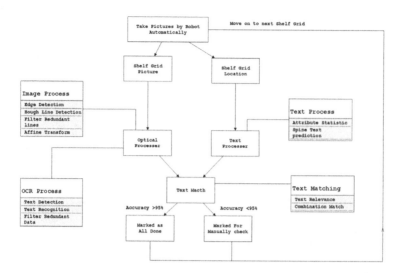

Fig. 1. The overall system workflow for Deep multi-mode learning.

detect only vertical images. As the books are placed left to right, they naturally shift to the right side, so we choose the θ in Eq. 1 to be [-20:0.2:20].

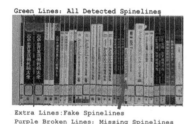

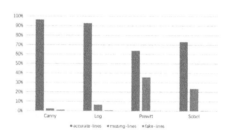

Fig. 2. Line calculate.   **Fig. 3.** Different operators comparison.

We do experiments on the same 100 cases to collect the accuracy of lines by a different operator. The lines are marked, and the line statistics are collected in Fig. 2. As shown in Fig. 3, we choose Canny operator for its best performance.

We need to adjust the threshold of the canny operator to get better performance. Since a shelf grid(one image) contains at most 45 books, we can set the Hough local maximum as 50. Then we need to adjust the voting threshold to decide whether a line can be added as a spine line. We do experiments on the same 100 cases to collect the error lines. Many may prefer to choose 0.38 as the lowest total error percentage to achieve the best experiment. However, in our condition, we consider that we can filter the fake lines. In our system, we choose 0.35 after Hough Maximun threshold comparison, the lowest missing lines percentage, rather than the total error percentage.

When we got the line set, we first sorted them from left to right by the x value of intersecting fonts that line intersect with the upper boundary. Since the book spines tend to be parallel lines, we check the intersection points of each pair of adjacent lines and remove the right line of those pair lines whose intersections are in the view of the image. We can also extend the area of intersection points.

We define a "near value" as half of the average width of the book spine to remove the parallel lines. We suppose the 75% of the image width is filled with book spines. Moreover, we got the original hough line sets S_h. We know that the actual line sets must be more miniature. Still, we can estimate the average width W_a a book spine in this shelf grid, and the "near value" V is:

$$V = 0.2 * W_a = \frac{0.375 * width(Image)}{length(S_h)} \tag{2}$$

Finally, we got the set of clean hough lines. We segment the image using these pairs of adjacent lines. Since the lines are not exactly vertical, we got a set of quadrilaterals instead of standard rectangles after the segmentation. So we use affine transformation [9] to make every book spine strictly vertical. Even though every book is designed for placed vertically, some horizontal texts are still added to the book spine design in order to convey more information to patrons. Our OCR recognition will solve that part.

3.2 Optical Character Recognition

We adopt PaddlePaddleOCR[1] to train our data and use it to do more accurate identification work. The function of the PP-OCR engine can be divided into text detection, detection frame correction, and word recognition. PP-OCR takes the construction of a large-scale Chinese and English recognition data set as an example to make itself practical. In that case, we can mark our dataset and train the PP-OCR to make it more accurate.

Before we enter the identifying part, we find out that some books, the spine of which are too thin, can give distortion to the final results. Therefore, before text recognition, we can improve it by extending the boundaries a little to expand the spine area horizontally. If we extend too much, we will get irrelevant data recognized, increasing the average time.

To choose whether a book is too thin, we can use the estimated average length W_a in Eq. 2 to check. If the spine width is smaller than the average, we will do the extension. We take tests on 20 shelf grids with nearly 700 books. We find out that even though the OCR recognition percentage on the spine increases with the extension, the redundancy text is misleading. Besides, the OCR times do matter in this case. We finally choose 10% as the thin book extension percentage.

---

[1] https://github.com/PaddlePaddle/PaddleOCR.

After we trained our dataset, we got higher accuracy on OCR (Fig. 4). The accuracy of OCR is evaluated as below:

$$\begin{cases} A_{CharRecogniztion} = \frac{length(result)-length(error)}{length(real)} \\ A_{TextLocation} = 1 - \frac{length(error)}{length(result)} \\ A_{TextArea} = \frac{|Area(realarea)-Area(detectedbox)|}{Area(realarea)} \end{cases} \tag{3}$$

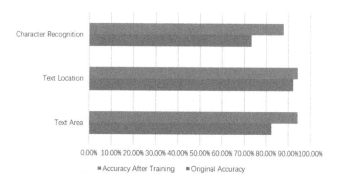

Fig. 4. Accuracy of OCR.

3.3 Text Processing

We analyze the book spine patterns and finally choose Publisher, Author, and irrelevant data to predict the pattern of spine information. We can use one classification to break them down. We know that, for one book, each attribute can only have two facts: whether it appears on the spine or not. For that condition, we can develop one classifier for each attribute, then combine them.

We randomly choose 100 shelfgrid (nearly 2000 books after removing duplicators) metadata, after removing the special characters, we extract its "Title","CLC","Publish Year","Publisher" as a total text to represent the book [3]. These metadata as the dataset to count the word frequency. We design three classifiers; C1 means where the author appears on the spine text; if it appears, we mark it as 1; otherwise, we mark it as 0. C2(Publisher) and C3(Irrelevant Data) are the same. We use the Jieba² word segmentation to obtain the text content of high-frequency word statistics, a total of 1628 high-frequency words, by eliminating some auxiliary words, modal particles, and high-frequency words unrelated to this paper.

We choose 20 words after TF-IDF calculation [7], the out vector is 20 dimensional. We define "Nearest" with a mathematical definition of distance d(V1, V2) in Table 1.

$$d(V1, V2) = \sqrt{\sum_{i=1}^{20}(V1_i - V2_i)^2} \tag{4}$$

Table 1. Sample text and it's nearest points

Sample Text	爱上制作|TN49 2013|刘蒙阒|人民邮电出版社	
C1	Text	Distance
0	Arduino一试就上手|TN49 2013|孙骏荣|科学出版社	0.4257
0	一起造物吧|TN49 2017|柴火创客教育项目组|人民邮电出版社	0.4301
1	不可不知的37种电子元器件|TN49 2017|张晓东|人民邮电出版社	0.4303
0	创客电子制作入门|TN49 2017|无线电编辑部|人民邮电出版社	0.4305
0	如何在数字时代如鱼得水|TN49 2014|查特菲尔德|山东文艺出版社	0.4478

As shown in the sample text, most of the C1 values in the top5 closest points vote for 0. Therefore we predict that the author's name will not appear on the sample book spine [10]. C2 and C3 go in the same way. Finally, we get the predicted spine text.

3.4 Text Matching

After we get the prediction of a book spine information, we can use the prediction with the OCR result to check the inventory. We consider that OCR results can specify each character area. However, because of the font difference, some of them cannot be related to the proper character correctly. At the same time, typesetting dramatically influences the character order of the final text. In summary, an appropriate OCR result has nearly the same length as the prediction; most characters overlap, and character order matters less.

Since Location Info plays a less important part in reproject picture, we can consider the OCR results in S_o and the prediction result in S_p as two-word sets; then, we use Dice similarity coefficient λ_D to get their relevance. Since the stain on the book spine might affect the OCR result, adding a lot of extra miscellaneous characters. We should reduce those noise. so we adjust the equation

$$\lambda'_D = \frac{2(S_o \cap S_p)}{S_p + S_p} = \frac{(S_o \cap S_p)}{S_p} \tag{5}$$

However, this situation can't apply to those whose spine text is very rich with irrelevant data and can contain the total text of another flat spine information in matching. So we introduced Levenshtein distance λ_L at the same time [4]. Therefore, we choose the five highest λ_D in the matching set and the one with the least λ_L.

In order to demonstrate an approach for expressing the accuracy and precision of book spine text mapping to the right book, our approach begins with an understanding of the difference between the OCR sequence and the prediction text from the KNN classifier. We come up with two different combinations.

1) Combine with the text area location info. Since our location info is 4 points of a rectangle, We tend to consider the Call Number is always on the vertical lowest position, and the title is the one with the largest text area.

---

[2] https://pypi.org/project/jieba/.

2) Combine with the books that lie on both sides. In order to minimize the error on short text matching, we advocate five books as a set for matching.

4   System Test

In our system test, a dataset was created from existing bookshelves in our library. Robots took images using a standard cellular phone camera (2048 × 1536 resolution). Pictures of shelf grids are shot one by one, and each image has the shelf grid location info to connect to the dataset. We concentrate on the book spine extraction, OCR, and the final Text Matching part. Book spines are boundaries of books that can be detected by edge detection. Since book spines are pairs of parallel lines, line features play an essential part in edge patterns.

As we know, the bigger the image is, the higher the OCR will be, which consumes a longer time at the same time. So we have to resize the image to a reasonable resolution so that each book will not cost longer than 500 milliseconds. To accelerate the process, we also try parallel computation, which needs higher performance computer.

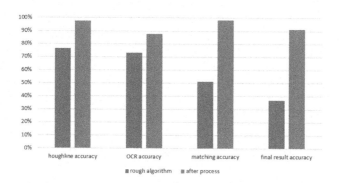

Fig. 5. Accuracy comparison.

In our experiments, we tried to increase accuracy in three major parts: spine lines, spine text OCR, and text matching. As shown in Fig. 5, all of them get significant improvement resulting in the higher overall accuracy of book spine recognition which is more than 90%, and the average time per spine is less than 300 milliseconds. The experimental results show that our system generally achieves a good recognition effect. Under proper conditions of images captured by robots, we can accurately recognize most of the books' spines in real-time.

5   Conclusion and Future Work

In this paper, we design and implement a book spine recognition system. The experiment shows that the overall accuracy of book spine recognition is more than 90%, and the average time per spine is less than 300 milliseconds. For further improvement in speed and accuracy, GPU and AI-aided image recognition can be integrated into the system. Our system achieves an accurate, real-time, low-cost book inventory system prototype. To make it more practical for use in different libraries, many aspects, such as mechanical structure, fast book spine image matching, accuracy and robustness under different illumination, spine image layout, and book placement, should be further improved in the coming days. We are now trying to build upon the prototype a system covering such functions as auto movement, book spine image capture and recognition, counting, pose evaluation, etc., i.e., a full system of smart book inventory.

Acknowledgements. This work was supported by the National Social Science Foundation President Project (Grant No. 20FTQB012).

References

1. Canny, J.: A computational approach to edge detection. IEEE Trans. Pattern Anal. Mach. Intell. **6**, 679–698 (1986)
2. Fowers, S.G., Lee, D.J.: An effective color addition to feature detection and description for book spine image matching. International Scholarly Research Notices 2012 (2012)
3. Guo, C., Xie, L., Liu, G., Wang, X.: A text representation model based on convolutional neural network and variational auto encoder. In: Wang, G., Lin, X., Hendler, J., Song, W., Xu, Z., Liu, G. (eds.) WISA 2020. LNCS, vol. 12432, pp. 225–235. Springer, Cham (2020). https://doi.org/10.1007/978-3-030-60029-7_21
4. Levenshtein, V.I., et al.: Binary codes capable of correcting deletions, insertions, and reversals. In: Soviet Physics Doklady, vol. 10, pp. 707–710. Soviet Union (1966)
5. Mukhopadhyay, P., Chaudhuri, B.B.: A survey of hough transform. Pattern Recognit. **48**(3), 993–1010 (2015)
6. Nevetha, M., Baskar, A.: Automatic book spine extraction and recognition for library inventory management. In: Proceedings of the Third International Symposium on Women in Computing and Informatics, pp. 44–48 (2015)
7. Ramos, J., et al.: Using tf-idf to determine word relevance in document queries. In: Proceedings of the First Instructional Conference on Machine Learning, vol. 242, pp. 29–48. Citeseer (2003)
8. Shi, X., Tang, K., Lu, H.: Smart library book sorting application with intelligence computer vision technology. Librar. Hi Tech **39**(1), 220–232 (2020)
9. Weisstein, E.W.: Affine transformation (2004). https://mathworldwolfram.com/
10. Zhang, H., Xie, X., Wen, Y., Zhang, Y.: A twig-based algorithm for Top-k subgraph matching in large-scale graph data. In: Wang, G., Lin, X., Hendler, J., Song, W., Xu, Z., Liu, G. (eds.) WISA 2020. LNCS, vol. 12432, pp. 475–487. Springer, Cham (2020). https://doi.org/10.1007/978-3-030-60029-7_43

DEAR: Dual-Level Self-attention GRU for Online Early Prediction of Sepsis

Yu Zhao[1,3], Yike Wu[4], Mo Liu[5], Xiangrui Cai[2,3(✉)], Ying Zhang[2,3], and Xiaojie Yuan[1,3]

[1] College of Cyber Science, Nankai University, Tianjin 300350, China
[2] College of Computer Science, Nankai University, Tianjin 300350, China
caixr@nankai.edu.cn
[3] Tianjin Key Laboratory of Network and Data Security Technology, Tianjin 300350, China
[4] College of Journalism and Communication, Nankai University, Tianjin 300350, China
[5] Centre for Computational Biology, Duke-NUS Medical School, 8 College Road, Singapore 169857, Singapore

Abstract. Sepsis is one of the leading causes of death in intensive care units (ICUs). Online early prediction of sepsis has the potential for application to support clinicians. Despite the great success of deep neural networks in modeling electronic health records (EHRs), many architectures are incapable to be applied to online early prediction scenarios due to two major limitations. First, they overlook the earlier signs of the disease which are vital for the early prediction of sepsis. Second, they are unable to provide interpretation of prediction results. To tackle the above limitations, we propose a Dual-level sElf-Attention Gated Recurrent Unit Networks model, DEAR. On the one hand, DEAR is able to directly identify and strengthen the important time steps from the precedent memory. Specifically, the cell of DEAR straightforwardly fusion the history of both feature level and temporal level into the current step. On the other hand, DEAR provides interpretability with multi-head attention. Experimental results in the real-world sepsis dataset demonstrate that our model outperforms state-of-the-art methods in terms of both utility and balanced accuracy. In addition, the visualization of multi-head attention weights also indicates that DEAR reveals the importance of different time steps in the early prediction of sepsis onset.

Keywords: Online early prediction of sepsis · Self-Attention · Gated recurrent unit

1 Introduction

Sepsis is a life-threatening acute systemic infection caused by the toxins produced by pathogenic bacteria dumped into blood circulation. It could cause tissue damage, organ failure, or even death [13,14,16]. Studies have shown that sepsis

X. Zhao et al. (Eds.): WISA 2022, LNCS 13579, pp. 424–435, 2022.
https://doi.org/10.1007/978-3-031-20309-1_37

deaths could be prevented with early diagnosis and treatment: each hour of delayed treatment would raise the mortality rate by 4–8% [7,12]. Therefore, the *online early prediction of sepsis* problem is instrumental to solve and has the potential to support clinicians. The online early prediction of sepsis with electronic health records (EHRs) is illustrated in Fig. 1. In online early prediction scenario, the model is expected to make predictions hourly with access to *only the observed records*, aiming at predicting septic 6–12 hours before the true onset.

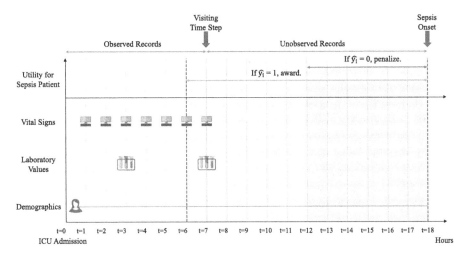

Fig. 1. Online early prediction of sepsis with EHRs. The utility measure demands the model to alert septic 6–12 h before the true onset time step. At each visit, the model makes prediction based on the observed records.

EHRs [19] include multivariate time series of almost regularly recorded vital signs and irregularly recorded laboratory values, as well as static demographic features. The *irregularly-sampled time series* is a sequence of records with not-equal intervals, such as the "Laboratory Values" in Fig. 1. With the progress of deep learning techniques, many deep neural networks were proposed to handle the irregularly-sampled multivariate time series of EHRs. There are two main-streams of methods. One mainstream of methods is *recurrent-based methods* [2,9], usually based on Recurrent Neural Networks (RNN). They process the sequential data conditioned on previous observed records recurrently. Another mainstream of methods is *global-based methods* [4,15,18], which process all time steps together. They utilize global information to capture the characteristics of time series to make predictions.

However, these architectures are not well compatible with the task of online early prediction of sepsis in two folds. First, they ignore the importance of early signs of disease. Earlier records may play an important role in the prediction in practice [6,10,17]. For instance, subtle changes in white blood cells and body temperature at an early state may indicate the onset of sepsis [6,7]. Neglecting the early records makes the model insensitive to early signs of disease and fails to alert in time. For recurrent-based models [2,9], earlier memory is more likely to be lost due to the well-known gradient vanishing problem, even in LSTM and GRU models which are designed to avoid this problem [1]. As a result, a record at a closer time step tends to have a greater impact on the result of prediction, which is incompatible with the task. For global-based methods [4,15,18], they all emphasize global information. However, under online early prediction scenario, the subsequent time steps are not available and they fail to make the most of their strengths. As a result, they lead to a decline in performance.

Second, the lacking of interpretation of some black-box models makes the model hard to apply in practical health care. To be on the safe side, the clinicians need to confirm whether the prediction is reasonable based on the judgment evidence provided by the model, instead of just following the result of a computational model.

In this work, we propose an approach for online early prediction of sepsis, **DEAR**, a <u>D</u>ual-level s<u>E</u>lf-<u>A</u>ttention Gated <u>R</u>ecurrent Unit model. It addresses the two problems respectively. On one hand, to enhance the importance of earlier records, we apply the multi-head self-attention to precedent memory in each time step, which identifies and strengthens key time steps in a step-by-step way. Consequently, the earlier subtle signs are accumulated to the following time steps and lead the model to predict in time. On the other hand, the multi-head self-attention also provides interpretation from multiple perspectives for each prediction. Furthermore, due to the irregularly-sampling problem of EHRs, we utilize *feature-level self-attention* to directly capture the irregular-pattern from inputs. Considering that the input features may be not expressive enough, we further apply *temporal-level self-attention* to the hidden states for capturing the temporal-pattern from a higher level. In this way, we form a *dual-level self-attention* to precedent memory at both *feature level* and *temporal level*.

Experimental results demonstrate that DEAR outperforms state-of-the-art methods on a real-world early sepsis prediction dataset. We evaluated on clinical utility, prediction balanced-accuracy, and AUROC metrics of each approach. The results demonstrate that DEAR achieves better performance in terms of utility and balanced-accuracy. Moreover, the model interpretation, cohort study, and parameter analysis demonstrate the interpretability and effectiveness of DEAR in online early prediction of sepsis.

2   Related Work

EHRs [3,5,20] are a large-scale and structured collection of multivariate time series of observations of patients, including multiple measurements of vital signs, laboratory and demographic features. EHRs are critical to assisting disease prediction, especially for computational methods to understand the medical trends and predict risks, including sepsis.

EHRs also raise challenges about irregular sampling, which means the records are not sampled with consistent intervals. Recent works mainly focus on tackling these limitations. There are two mainstream methods. One mainstream of methods is *recurrent-based methods* [2,9]. They process the sequences step by step, conditioned on previous observed records. In order to handle the irregularly-sampled records, some methods converted the irregular sampling to missing data problem. The intervals with no data sampled are considered missing. Che et al. provided several variants of Gated Recurrent Unit (GRU) networks [2]. The GRU model was extended to perform imputation. GRU-Simple augmented the input of GRU model with missing data mask and time intervals. GRU-Decay or GRU-D utilizes a missingness pattern with two trainable decay acting on both input values and hidden states. Some methods managed to support irregular sampling. Neil et al. proposed Phased-LSTM [9] to extend the Long Short Term Memory unit with a new time gate, which is inspired by biological neurons. It allows the model to process irregular events-based sequences.

Another mainstream of methods is *global-based methods* [4,15,18]. They process time steps all at once. In order to eliminate the constraint of recurrent operation of RNNs, Transformer [18] is proposed to extract global dependencies between input and output. Shukla et al. present an interpolation prediction networks (IP-Nets) [15], which captures the global structure of time series in an end-to-end manner. In order to scalability and efficiency, Horn et al. proposed SEFT [4] to form global encoding and aggregation of observed sets.

Online early prediction of sepsis is a study of utilizing multivariate time series of EHRs, aiming at automatically early predicting the disease onset at each time step, without access to future information. Utilizing EHRs to perform online monitoring of acute medical conditions such as septic shock is significant to reduce the mortality rate, and has the potential to apply in realistic clinical scenarios. In this paper, we focus on modeling irregularly sampled EHRs under online early prediction scenario.

3 Methodology

3.1 Problem Formulation

In this paper, we focus on the *online early prediction of sepsis* task, aiming at making positive predictions 6 h to 12 h before the septic time step.

Dataset. For each patient x, we represent a multivariate time series with D features of T length as $x = (x_1, x_2, ..., x_T)^\mathrm{T} \in \mathbb{R}^{T \times D}$, where $x_t \in \mathbb{R}^D$ represents the patient's records at time step t of all variables and x_t^d denotes the measurement of d-th variable of x_t. We denote the time-stamp of the t-th observation as $s_t \in \mathbb{R}$. As for the irregularly-sampling problem, we follow GRU-Simple [2] to convert to the missing data problem, where the intervals are considered missing. To denote which data is missing, we introduce a missing data mask $m = (m_1, m_2, ..., m_T)^\mathrm{T}$ corresponding to x where $m_t \in \{0,1\}^D$ is the notation of missing variables at time step t as Eq. 1. We also maintain time interval $\delta_t^d \in \mathbb{R}$ for d-th variable in time step t since its last observation as Eq. 2.

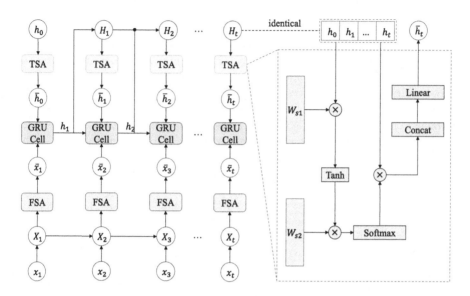

Fig. 2. The architecture of DEAR. The *TSA* denotes temporal-level self-attention and the *FSA* denotes feature-level self-attention. A detailed illustration of the self-attention mechanism is on the right side.

$$m_t^d = \begin{cases} 1, & \text{if } x_t^d \text{ is observed} \\ 0, & \text{otherwise} \end{cases} \tag{1}$$

$$\delta_t^d = \begin{cases} s_t - s_{t-1} + \delta_{t-1}^d, & t > 1, m_{t-1}^d = 0 \\ s_t - s_{t-1}, & t > 1, m_{t-1}^d = 1 \\ 0, & t = 1 \end{cases} \tag{2}$$

Input. For missing values, we absorb the setting of GRU-Simple [2]. We concatenate the mask indicators m_t and time interval δ_t with the input x_t as Eq. 3.

$$x_t \longleftarrow [x_t; m_t; \delta_t] \tag{3}$$

Output. For each patient x, we are given the labels $y = (y_1, y_2, ..., y_T) \in \{0, 1\}^T$, where $y_t = 1$ indicates the patient is in septic shock state at time step t, otherwise $y_t = 0$. The output objective is $\hat{y} = (\hat{y}_1, \hat{y}_2, ..., \hat{y}_T)$ where $\hat{y}_t$ is the predicted septic probability of the patient at the t-th time step.

3.2   Proposed Method

Overview. In this paper, we propose DEAR, a dual-level self-attention GRU model for online early prediction of sepsis. Specifically, we employ a GRU as the base architecture, following the recurrent-based methods. To overcome the limitations of overlooking early records and lacking interpretability, we further design a dual-level attention mechanism. The *feature-level self-attention* is used for capturing the pattern of irregularly sampling directly, while the *temporal-level self-attention* is used for capturing the pattern of patient's health conditions at a higher level. As a result, we form a *dual-level self-attention* to identify and strengthen the key time steps. More importantly, we apply the attention mechanism to each step, so that earlier signs of sepsis are recurrently enhanced and the model is more likely to notice the change in patient's condition.

The left side of Fig. 2 shows the architecture of DEAR. DEAR is composed of GRU cells with two self-attention blocks of temporal level and feature level. At time step t, the input x_t and its precedent inputs $\{x_i\}_{i<t}$ together form feature-level memory X_t. The memory is attentively weighted to a new input $\bar{x}_t$ through *feature-level self-attention (FSA)*. The temporal-level memory H_{t-1} formed by hidden states h_{t-1} and $\{h_i\}_{i<t-1}$ from GRU also transferred to a new state $\bar{h}_{t-1}$ through *temporal-level self-attention (TSA)*. Then GRU cell then recurrently accumulates the former information to the latter steps.

DEAR Unit. For time step t, we construct the feature-level memory is $X_t = (x_1, ..., x_t)$, where x_t is the health record of patient at time step t. The temporal-level memory is $H_{t-1} = (h_0, h_1, ..., h_{t-1})$, where $h_t \in \mathbb{R}^u$ is the output of GRU cell and u is the hidden layer size.

These two memories are applied to corresponding self-attention block $\bar{x}_t = FSA(X_t)$ and $\bar{h}_{t-1} = TSA(H_{t-1})$ to obtain current weighted input and state. The vectors $\bar{h}_{t-1} \in \mathbb{R}^u$ and $\bar{x}_t \in \mathbb{R}^D$ are the same sizes as standard GRU cell inputs h_{t-1} and x_t and can be directly handled by GRU. As a result, the GRU cell is formulated to Eq. 4. Compared with standard GRU cell $h_t = GRUCell(h_{t-1}, x_t)$, we replace the input by the weighted state and input.

$$h_t = GRUCell(\bar{h}_{t-1}, \bar{x}_t) \tag{4}$$

Self-attention Mechanism. The right side of Fig. 2 is a detailed illustration of the self-attention mechanism. In this section, we use temporal-level self-attention *TSA* as an example, as the operation of *TSA* and *FSA* is the same.

We absorbed the structured self-attention mechanism [8] to generate attention weights. Using precedent memories, we perform an attentive operation to enhance the impact of key time steps. We obtain the attention weights A through the Eq. 5, where $W_{s1} \in \mathbb{R}^{d_a \times u}$ is a weight matrix and d_a is the hyperparameter of the size of the attention layer.

Instead of performing single-head attention, it is found that multi-head attention is more effective in learning different components in one sequence. Thus we denote the matrix $W_{s2} \in \mathbb{R}^{r \times d_a}$ as a multi-head attention weight matrix where r means the hops of attention. Then we have the attention weight matrix $A \in \mathbb{R}^{r \times n}$

where each component means the importance of the state in the different hop of attention. Then we perform dot-product between A and H to get the context matrix $C = AH = (C_1, C_2, ..., C_r)$.

$$A = softmax(W_{s2} \tanh(W_{s1} H_n^{\mathrm{T}})) \tag{5}$$

Attentive Combination. To combine multi-head attention, we use *concat + linear* mechanism following Transformer [18]. First we concatenate the r context vectors C_i to one vector $c = [C_1, C_2, ..., C_r]$. Then through a linear layer, the vector is projected to a u-size vector h_n. This way, the computational cost of the following GRU would not change.

Similarly, we get the self-attention weighted input $\bar{x}_n$. At each step, we transfer the H_{t-1} and X_t to $\bar{h}_{t-1}$ and $\bar{x}_t$, then pass them to the GRU Cell for modeling.

Learning Objective. After GRU model the time series, we apply a dense layer on the sequence of GRU output hidden states and activate with the sigmoid function, to get the probability of future occurrence of sepsis $\hat{y}_t$ at each time step. We utilize the binary cross-entropy loss defined in Eq. 6 for binary classification, where y_t and $\hat{y}_t$ are the sepsis condition label of patient and predicted probability at time step t, respectively.

$$\mathcal{L} = \sum_{t=1}^{T} -(y_t \log(\hat{y}_t) + (1 - y_t) \log(1 - \hat{y}_t)) \tag{6}$$

4   Experiments

4.1   Experimental Settings

Dataset. We used EHR data from PhysioNet 2019 Sepsis Early Prediction Challenge [11]. The data was collected from three distinct U.S. hospital systems. We used the provided samples from 40,336 patients in our work. For each patient, up to 40 variables were recorded, as well as the binary label of sepsis onset. We split the dataset into training set (80%) and test set (20%) while preserving the same prevalence of sepsis 0.018 in each set. We further split 20% of the training set as validation set while the prevalence remains fixed.

Baselines. We compared the performance of DEAR t recurrent-based model: GRU-Simple [2], GRU-Decay [2], Phased-LSTM [9], and global-based model: Transformer [18], IP-Nets [15], SEFT [4].

Evaluation Measures. We use the utility as the main measure to assess the algorithm. Due to the imbalance of the dataset, we also report the class imbalance sensitive metrics: balanced-accuracy and AUROC.

- **Utility** Utility rewards the algorithm which succeeds in early predicting sepsis 6 h to 12 h before the onset and penalizes the algorithm which fails at detecting or makes a late prediction.

- **Balanced-Accuracy** Balanced-accuracy is used for the scenario where the classes are imbalanced. The septic label is severely abnormal in our case.
- **AUROC** AUROC (Area Under the Receiver Operating Curve) is to measure the model's ability to discriminate between septic and non-septic classes.

4.2 Results

Online Early Prediction Performance Results. We present the results of the early prediction of sepsis in Table 1. In terms of utility and balanced-accuracy, DEAR outperforms other models. It suggests that the dual-level self-attention mechanism can achieve better early prediction performance, even when the classes are highly imbalanced. Moreover, the standard deviation of DEAR is less than other methods in utility and balanced-accuracy, which demonstrates the stability of DEAR.

In terms of AUROC, DEAR reaches a comparable performance with other methods. This is reasonable considering the contradiction between AUROC and utility. On one hand, the utility of early prediction demands the model to predict positive before the sepsis truly onset when the label is negative. On the other hand, AUROC considers the early prediction phase as False Positive. Under early prediction setting, we relatively value the utility metric more.

It is worth noting that with access to all records, global-based methods IP-Nets and Transformer exceed all recurrent-based methods. We deduce that the future records may leak through the imputation step of IP-Nets or the layer normalization of Transformer. However, access to future records is highly incompatible with the online early prediction task.

Model Interpretation. Due to the step-by-step architecture, DEAR is able to identify the critical records at every time step. Figure 3 presents weights of all

Table 1. Performance of online early prediction of sepsis. We highlight the **best** result of each column. The gray[†] rows denotes that result are obtained with access to the future records. For fairness, we compare methods with no access to future records. The average results and standard deviations are obtained with 5 different random seeds.

Model	Utility	Balanced-Accuracy	AUROC
GRU-Simple [2]	24.5 ± 1.8	63.4 ± 1.0	**79.6 ± 0.9**
GRU-D [2]	7.1 ± 3.6	53.7 ± 2.0	72.9 ± 7.7
Phased-LSTM [9]	27.1 ± 1.9	64.9 ± 0.9	77.7 ± 0.1
Transformer[†] [18]	71.3 ± 1.4	91.2 ± 0.2	97.3 ± 0.2
Transformer [18]	−43.9 ± 10.0	53.6 ± 1.7	65.8 ± 3.7
IP-Nets[†] [15]	62.2 ± 1.3	87.1 ± 0.9	94.1 ± 0.4
IP-Nets [15]	−11.9 ± 4.0	63.8 ± 0.9	74.2 ± 1.2
SEFT [4]	20.1 ± 3.4	60.2 ± 1.8	67.9 ± 2.6
DEAR	**28.6 ± 0.9**	**66.0 ± 0.5**	77.5 ± 1.4

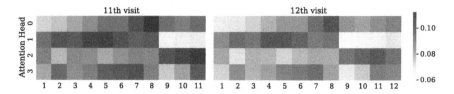

Fig. 3. Multi-head attention weights for the 11th and 12th time step were generated by DEAR. Each column represents the attention weights extracted from the multiple attention heads. Each row indicates the attention weights of the time step to the current time step. The darker color denotes the time step is more important to the current decision.

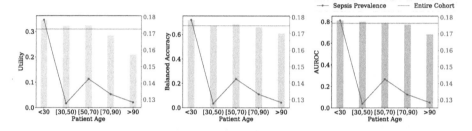

Fig. 4. Performance of cohorts at different ages. The patients in the test set were divided into 5 cohorts according to their ages. The red lines represent the performance of the entire cohort, and the green lines represent the sepsis prevalence. (Color figure online)

precedent memory extracted at the 11th and 12th time step from temporal-level self-attention. The patient was first labeled septic at the 18th hour and DEAR succeeds in early predicting at the 8th, 9th, 10th, 11th, and 12th hours. Our model DEAR shows interpretation advantages in the following aspects:

- At each visit, not only closer but also earlier time steps contributed to the current prediction. More specifically, in the 11th time step, some earlier time steps have darker colors, as well as closer time steps. It demonstrates that DEAR can identify the key time steps, including the earlier ones.
- The predicted septic time steps remain important in the next time steps. Using the 11th and 12th time steps as an example, after the model predicts sepsis at the 11th hour, the 11th time step remains relatively important at the 12th time step. It demonstrates that DEAR is able to accumulate the previous critical information to following time steps.
- The multi-head attention tends to focus on different components of the series. For example, the first attention head identifies the 8th time step important, and the third head identifies the 9th to 11th time steps important. It demonstrates the perspective diversity produced by multi-head attention.

Cohort Study. We roughly divided the patients in test set into 5 cohorts. Figure 4 visualizes the test results in different cohorts. The red lines represent the result of the entire cohort and the green lines represent the sepsis onset rate among all patients, i.e. sepsis prevalence.

We observe an overall decreasing trend in the performance with age. We deduce there are two folds of reasons. On the one hand, elderly patients tend to have worse and more complex physical conditions, even multiple diseases. The symptoms of other diseases could be similar to symptoms of sepsis, which is confusing for the model to identify. On the other hand, DEAR reaches higher performance with higher prevalence due to the tendency to predict positive, which is reasonable considering the severity of False Negative in clinical prediction.

4.3   Parameter Analysis

In this section, we explore the parameter of the self-attention mechanism. The results are demonstrated in Fig. 5.

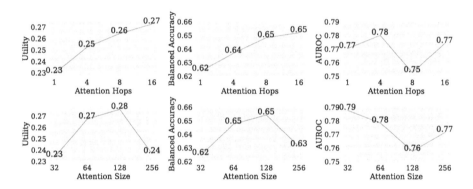

Fig. 5. Effect of hyperparameters in the self-attention mechanism. The first row depicts three metrics with different numbers of attention hop. The second row presents three metrics with different attention sizes.

Effect of Attention Hops r. We vary r from 1 to 16 with exponential growth. From the first row of Fig. 5 we observe that, utility and balanced-accuracy increase with more attention hops. Specifically, compared with $r = 1$, i.e. single-head attention, the model with multi-head attention showed better performance. The AUROC is not stable during the increase of r due to the contradiction between AUROC and utility.

Effect of Attention Size d_a. We vary the attention size d_a from 32 to 256 with exponential growth. From the second row of Fig. 5 we observe that utility and balanced-accuracy first increase until $d_a \geq 128$, then decrease. We can conclude that with larger hidden dimensions, the model is able to learn more complex information. However, with much larger dimensions, it is more easily to over-fit.

The effect to AUROC is the opposite of utility and balanced-accuracy due to the contradiction too.

5   Conclusion

Online early prediction of sepsis is instrumental to solve. To solve the problems of ignoring earlier signs and lack of interpretation of existing methods, we propose DEAR, a dual-level self-attention GRU model. On one hand, DEAR explicitly combines precedent memories from both feature-level and temporal-level, and accumulates the earlier signs in a step-by-step way. In specific, at each time step, DEAR Unit generates both feature-level and temporal-level self-attention weights to identify and strengthen the important previous steps. On the other hand, DEAR is able to provide interpretability with multi-head attention. The experimental results on a real dataset show that our approach outperforms state-of-the-art methods in utility and balanced-accuracy. The visualization of attention weights also indicates the interpretability of DEAR.

Acknowledgements. This research is supported by the National Natural Science Foundation of China (No. 62002178) and NSFC-Xinjiang Joint Fund (No. U1903128).

References

1. Baytas, I.M., Xiao, C., Zhang, X., Wang, F., Jain, A.K., Zhou, J.: Patient subtyping via time-aware LSTM networks. In: Proceedings of the 23rd ACM SIGKDD International Conference on Knowledge Discovery and Data Mining, pp. 65–74 (2017)
2. Che, Z., Purushotham, S., Cho, K., Sontag, D., Liu, Y.: Recurrent neural networks for multivariate time series with missing values. Sci. Rep. **8**(1), 1–12 (2018)
3. Häyrinen, K., Saranto, K., Nykänen, P.: Definition, structure, content, use and impacts of electronic health records: a review of the research literature. Int. J. Med. Inform. **77**(5), 291–304 (2008)
4. Horn, M., Moor, M., Bock, C., Rieck, B., Borgwardt, K.: Set functions for time series. In: International Conference on Machine Learning, pp. 4353–4363. PMLR (2020)
5. Jensen, P.B., Jensen, L.J., Brunak, S.: Mining electronic health records: towards better research applications and clinical care. Nat. Rev. Genet. **13**(6), 395–405 (2012)
6. Kenzaka, T., et al.: Importance of vital signs to the early diagnosis and severity of sepsis: association between vital signs and sequential organ failure assessment score in patients with sepsis. Intern. Med. **51**(8), 871–876 (2012)
7. Kumar, A., et al.: Duration of hypotension before initiation of effective antimicrobial therapy is the critical determinant of survival in human septic shock. Crit. Care Med. **34**(6), 1589–1596 (2006)
8. Lin, Z., et al.: A structured self-attentive sentence embedding. arXiv preprint arXiv:1703.03130 (2017)
9. Neil, D., Pfeiffer, M., Liu, S.C.: Phased lstm: accelerating recurrent network training for long or event-based sequences. In: Proceedings of the 30th International Conference on Neural Information Processing Systems, pp. 3889–3897 (2016)

10. Quinten, V.M., van Meurs, M., Ter Maaten, J.C., Ligtenberg, J.J.: Trends in vital signs and routine biomarkers in patients with sepsis during resuscitation in the emergency department: a prospective observational pilot study. BMJ Open **6**(5), e009718 (2016)
11. Reyna, M.A., et al.: Early prediction of sepsis from clinical data: the physionet/computing in cardiology challenge 2019. In: 2019 Computing in Cardiology (CinC), p. 1. IEEE (2019)
12. Seymour, C.W., et al.: Time to treatment and mortality during mandated emergency care for sepsis. N. Engl. J. Med. **376**(23), 2235–2244 (2017)
13. Seymour, C.W., et al.: Assessment of clinical criteria for sepsis: for the third international consensus definitions for sepsis and septic shock (sepsis-3). JAMA **315**(8), 762–774 (2016)
14. Shankar-Hari, M., et al.: Developing a new definition and assessing new clinical criteria for septic shock: for the third international consensus definitions for sepsis and septic shock (sepsis-3). JAMA **315**(8), 775–787 (2016)
15. Shukla, S.N., Marlin, B.: Interpolation-prediction networks for irregularly sampled time series. In: International Conference on Learning Representations (2018)
16. Singer, M., et al.: The third international consensus definitions for sepsis and septic shock (sepsis-3). JAMA **315**(8), 801–810 (2016)
17. Tennilä, A., Salmi, T., Pettilä, V., Roine, R.O., Varpula, T., Takkunen, O.: Early signs of critical illness polyneuropathy in icu patients with systemic inflammatory response syndrome or sepsis. Intensive Care Med. **26**(9), 1360–1363 (2000)
18. Vaswani, A., et al.: Attention is all you need. Adv. Neural Inf. Process. Syst. **30**, 5998–6008 (2017)
19. Wu, Y., Zhang, Y., Wu, J.: Configurable in-database similarity search of electronic medical records. In: Xing, C., Fu, X., Zhang, Y., Zhang, G., Borjigin, C. (eds.) WISA 2021. LNCS, vol. 12999, pp. 62–73. Springer, Cham (2021). https://doi.org/10.1007/978-3-030-87571-8_6
20. Yadav, P., Steinbach, M., Kumar, V., Simon, G.: Mining electronic health records (ehrs) a survey. ACM Comput. Surv. (CSUR) **50**(6), 1–40 (2018)

A Route Planning Method of UAV Based on Multi-aircraft Cooperation

Xiaofang Liu[1], Ke Wang[2], Xing Fan[1], Zhiqiang Xiong[1(✉)], Xiaoye Tong[1],
and Bo Dong[1]

[1] Wuhan Digital Engineering Institute, Wuhan 430205, China
fexechina@qq.com
[2] School of Computer and Artificial Intelligence, Zhengzhou University, Zhengzhou 450001,
Henan, China

Abstract. In modern warfare, cooperative operations of multiple UAVs has become a trend and been valued by various countries. However, as the numbers of UAVs increases, the cost of computing resources for the existing cooperative planning algorithm of multiple UAVs will increase exponentially, which is a heavy burden for hardware. In order to utilize the existing military electronic information system resources to complete the cooperative planning of multiple UAVs, this paper firstly proposes a dynamic configurable cooperative planning architecture and parallel processing method to meet the requirement of cooperative route planning. Then a dynamic load balancing route planning processing algorithm is designed, which can solve the problem of unbalanced load of each processing nodes in multi-computer system when the routing points are unevenly distributed in battlefield space. In the end, the experimental results show that the proposed method can significantly reduce the processing delay of multi-UAV route planning and improve the acceleration ratio of parallel processing.

Keywords: UAV · Task collaboration · Route planning · Multi-UAV system · Parallel processing method

1 Introduction

In recent years, UAV architecture, target search, routing communication and cooperative task allocation have been widely studied around the world. With the in-depth exploration of UAV operation, the cooperative completion of designated tasks by multiple UAVs has become the focus of research. Multi-UAV cooperative operation refers to the deployment of groups of small and medium-sized UAVs in an uncertain battlefield environment to carry out task allocation and route planning for UAV formation in a wide airspace [1], so as to complete a series of tasks such as search and strike [2]. At the same time, due to the limited capability of a single UAV, it can only control and make decisions based on its own sensors and local information, so the UAV formation needs to realize efficient cooperative task planning through inter aircraft communication [3].

Cooperative planning of multiple UAVs is an important topic for formation to complete cooperative search and strike [4], including constraint processing, heterogeneous elimination and task allocation of UAV cluster [5]. In the early stage, most of the research focused on central task planning and management, that is, the task assignment center station collects local information of each UAV and plans the UAV formation based on all the information. However, at present, the distributed task planning method has gradually become the mainstream. Document [6] considers the execution sequence between cooperative tasks, takes the task sequence as an important constraint of UAV task planning, and uses the improved particle swarm optimization algorithm to improve the planning efficiency. Document [7] proposed an online task allocation algorithm based on UAV task alliance, which requests neighbor UAVs to form a task alliance to deal with sudden tasks and re plan tasks. Based on the existing UAV command and control system, this paper attempts to connect multiple control stations through the network to form a multi-UAV processing structure, and solve the collaborative planning problem of multiple UAVs through multi-UAV collaborative processing.

2   The Architecture of Multi-UAV Task Planning and Processing

This paper designs a dynamic configurable high availability multi-UAV task planning system to meet the requirements of information infrastructure of military electronic information system. The architecture is shown in Fig. 1.

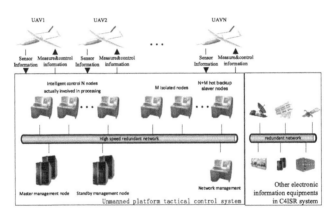

Fig. 1. Architecture of multi-UAV processing system.

The system uses COTS components to construct the high availability infrastructure of the cluster system, and constructs the hardware architecture of the system according to the principle of Isolated Redundancy to eliminate a single fault point of the system. The system is scalable and efficient in fault detection and processing, and provides a variety of fault detection methods. At the same time, it provides multi-level high availability support. According to the different characteristics of distributed applications of military electronic information system, it adopts the strategy of application integration to realize high availability from bottom to top. It can also provide real-time monitoring of system operation status and online reconstruction.

3   Distributed Collaborative Planning Process

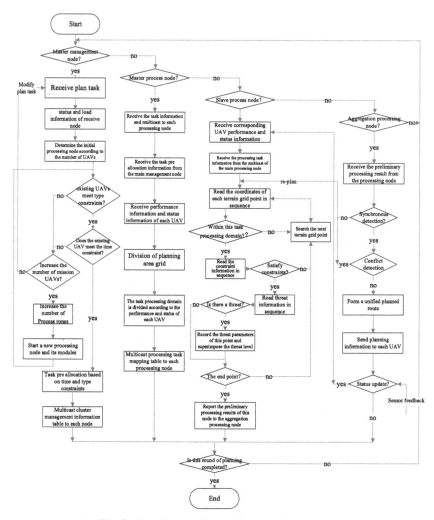

Fig. 2. Distributed collaborative planning process.

Taking the longitude and latitude of the starting point as the coordinate origin, the distribution space of waypoints is divided into a large sector with a radius of L according to the azimuth distribution of waypoints and threat sources relative to the coordinate origin. As shown in Fig. 2, the whole sector is meshed by taking Δl as length the unit and β as the radian. Among them, $\beta = 2 \arcsin \frac{\Delta l}{L}$, the value of Δl is determined by task requirements, area size and other factors.

The normal working flow of UAV distributed collaborative planning processing is shown in Fig. 2. The specific steps are as follows:

1) Each node of the system receives data from the network and updates the time constraint table, altitude constraint table, type constraint table, terrain constraint table, self constraint table, threat constraint table, meteorological constraint table, political constraint table according to the received data type. Processing task mapping table and cluster management information table;
2) The main management node pre allocates the tasks according to the time, type and system load status, decides whether to add a new processing node, and sends the cluster management information table to each node of the system;
3) The main processing node receives the task information and the task pre allocation information of the main management node, and multicast to each processing node. The main processing node divides the planning area grid according to the combat area, divides the processing area according to the UAV type, performance, time and system load, and sends the information processing task mapping table to each processing node;
4) Query the processing task mapping table from the processing node, conduct route planning for UAVs belonging to the local processing domain, and report the preliminary processing results of the node to the convergence processing node;
5) The aggregation processing node collects the preliminary processing results of each processing node, performs synchronization and conflict detection, synchronizes the processing results of each processing node, and detects whether there is a conflict. If there is a conflict, the relevant planning is returned to the relevant slave processing node for re planning. If there is no conflict, a collaborative planning table is formed;
6) The aggregation processing node distributes the collaborative planning table to each UAV management and control node.

4 Collaborative Planning Algorithm

4.1 Planning Method

The main operation steps of applying the improved ant colony algorithm to solve the route planning problem of a single UAV are as follows:

1) Initialize the ant colony AC.
2) Calculate the distance between any two waypoints in the waypoint set.
3) For $A_v \in AC$, do the following:

    a. If A_v has constructed a complete solution, reselect an artificial ant as the currently specified artificial ant.

    b. If the currently specified ant A_v has not constructed a complete solution, then Calculate the probability $P_{ij}^{(v)}(t)$ of A_v and adds the waypoint R_j to the taboo table of A_v.

4) $A_v \in AC$, update the pheromone on the corresponding path segment according to its constructed solution.
5) Record the iterative shortest path $Optimal_{local}$: If $Optimal_{local}$ is better than $Optimal_{global}$, update the global optimal solution, $Optimal_{global} \leftarrow Optimal_{local}$.

6) Judge whether the algorithm termination conditions are met. If not, turn to 3) execute a new iteration; If satisfied, the global optimal solution $Optimal_{global}$ is output. The specific calculation formula is as follows:

$$P_{ij}^{(v)}(t) = \begin{cases} \dfrac{\tau_{ij}^{\alpha}(t)\eta_{ij}^{\beta}(t)}{\sum\limits_{s=allowed_v} \tau_{is}^{\alpha}(t)\eta_{is}^{\beta}(t)} & , \ j \in allowed_v \\ 0 & , \ otherwise \end{cases} \tag{1}$$

$$\eta_{ij} = \frac{1}{d_{ij}} \tag{2}$$

where $\tau_{ij}(t)$ is the pheromone concentration on the path between waypoint R_i and waypoint R_j; η_{ij} is the heuristic information between the two waypoints; $allowed_v$ is the set of currently selectable subsequent waypoints A_v.

After completing the construction of a complete solution, The specific calculation method is shown in formula (4):

$$\tau_{ij}(t+1) = \lambda \times \tau_{ij}(t) + \sum_{v=1}^{m} Q \frac{d_{ij}^{(v)}}{\Delta l} \tag{3}$$

where λ is the pheromone attenuation coefficient, which meets $\lambda \in [0, 1]$, Q is the basic unit amount of pheromone released by artificial ants per unit time.

4.2 Cross Mutation

We use crossover and mutation to improve the processing efficiency of traditional ant colony algorithm. We adopt the method that the crossover probability p_c and mutation probability p_m automatically change with the fitness function value. And establish the following expressions of p_c and p_m:

$$p_c = \begin{cases} k_1(f - f_{min})/(f_{avg} - f_{min}) & f < f_{avg} \\ k_1 & f \geq f_{avg} \end{cases} \tag{4}$$

$$p_m = \begin{cases} k_2(f - f_{min})/(f_{avg} - f_{min}) & f < f_{avg} \\ k_2 & f \geq f_{avg} \end{cases} \tag{5}$$

where k_1 and k_2 are constants, $k_i \in [0, 1][0, 1]$. f_{avg} is the average objective function value of the current generation of evolutionary population. f is the smaller value of the objective function among the two cross individuals.

5 Load Balancing Strategy

As shown in Figs. 3 and 4. Taking the true north as 0 rad, divide the large sector of the distribution space into n sectors, so that the number of waypoints in each sector is roughly equal. $R_i(i = 0, 1, \cdots, n-1)$ represents the dividing point of the sector division of the planning task, and $R_0 = 0$, $C_j(j = 0, 1, \cdots, m)$ represents the corresponding processing node (primary planning node). Since the number of cross-border flights of UAV is generally small, generally only one processing node (convergence processing node) is required to perform the processing of synchronous detection and conflict detection.

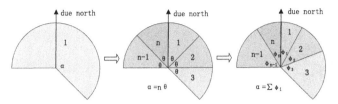

Fig. 3. Schematic diagram of collaborative planning fragmentation strategy.

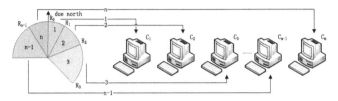

Fig. 4. Schematic diagram of collaborative planning and processing load scheduling.

The system load dynamic balancing algorithm is described as follows:

1) Start the planning processing task and complete the system initialization:

   (1) Initialize the node CPU load difference threshold Th_{cpu}, network card load difference threshold Th_{nic} and information processing delay threshold Th_{tp};
   (2) According to the scenario of this combat mission, the number of initialization processing nodes is n, and the reconnaissance airspace is divided into n equal sectors:

$$\begin{cases} R_0 = 0; \\ R_i = \frac{\alpha}{n}, i = 1, 2, \cdots, n-1 \end{cases} \tag{6}$$

2) Start the load balancing module and send an active query request for load information to each node;
3) Waiting for receiving node load information;
4) Whether to initiate load scheduling is determined according to whether the maximum value of the load difference on any two processing nodes (non sink nodes) in the system is greater than the given threshold, i.e.
   if
$$\left(\max\left(\left|L_{cpui} - L_{cpuj}\right|\right)\right) > Th_{cpu} \text{ or } \left(\max\left(\left|L_{nici} - L_{nicj}\right|\right)\right) > Th_{nic} \text{ or } \\ \left(\max\left(\left|T_{pi} - L_{pj}\right|\right)\right) > Th_{tp}$$
   {
   Re divide the sectors with roughly equal number of waypoints according to the current target distribution;
   Mapping the planning task to each processing node;
   Migrate the history related information of the cross region node to the current node;
   }.

Adjust the load difference threshold according to the system operation and turn 3).

6 Simulation

In the simulation test, the grid width Δl is 1000 m, 800 m, 600 m, 400 m, 200 m and 100 m respectively to increase the processing complexity. The reconnaissance route of 18 UAVs is planned in five cases: 1 management node and 2 processing nodes, 1 management node and 3 processing nodes, 1 management node and 4 processing nodes, 1 management node and 5 processing nodes, 1 management node and 6 processing nodes. In the test, the main management node, the main processing node and the aggregation processing node are reused by the management node. The experiment mainly tests the processing delay T, system acceleration ratio S_N and system efficiency E_n of the system.

Based on the above test results, the following conclusions are drawn through analysis:

1) The parallel algorithm of collaborative planning has high speedup ratio and system efficiency when the workload reaches a certain scale.
2) For the same number of processing units, with the increase of workload, the acceleration ratio will increase accordingly.
3) For the same workload (problem scale), with the increase of the number of processing units, the overall system efficiency will decrease accordingly (Figs. 5 and 6).

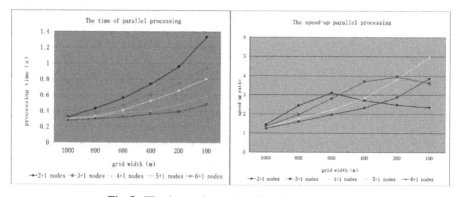

Fig. 5. The time and speedup of parallel processing.

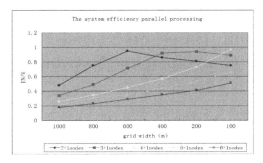

Fig. 6. The system efficiency of parallel processing.

References

1. Suresh, M., Ghose, D.: UAV grouping and coordination tactics for ground attack missions. IEEE Trans. Aerosp. Electron. Syst. **48**(1), 673–692 (2012)
2. Elloumi, S.: An efficient linearization for the constrained task allocation problem. Appl. Spectrosc. **56**(9), 1170–1175 (2015)
3. Li, J., Han, Y.: Optimal resource allocation for packet delay minimization in multi-layer UAV networks. IEEE Commun. Lett. **21**(3), 580–583 (2017)
4. Hu, X., Wang, G., Jiang, L., Ding, S., He, X.: Towards efficient learning using double-layered federation based on traffic density for internet of vehicles. In: WISA 2021, pp. 287–298 (2021)
5. Liu, L., Michael, N., Shell, D.A.: Communication constrained task allocation with optimized local task swaps. Auton. Robot. **39**(3), 429–444 (2015)
6. Li, J., Cao, X., Guo, D., Xie, J., Chen, H.: Task scheduling with UAV-assisted vehicular cloud for road detection in highway scenario. IEEE Internet Things J. **7**(8), 7702–7713 (2020)
7. George, J.M., Sujit, P.B., Sousa, J.B., et al.: Search strategies for multiple UAV search and destroy missions. J. Intell. Rob. Syst. **61**(1), 355–367 (2011)

An Improved Monte Carlo Denoising Algorithm Based on Kernel-Predicting Convolutional Network

Jiameng Liu[1,2], Fang Zuo[1,2(✉)], and Guanghui Wang[1,3(✉)]

[1] Henan International Joint Laboratory of Intelligent Network Theory and Key Technology, Henan University, Kaifeng 475000, China
`zuofang@henu.edu.cn, gwang@vip.henu.edu.cn`
[2] School of Software, Henan University, Kaifeng 475000, China
[3] Subject Innovation and Intelligence Introduction Base of Henan Higher Educational Institution-Software Engineering Intelligent Information Processing Innovation and Intelligence Introduction Base of Henan University, Kaifeng 475000, China

Abstract. Monte Carlo(MC) path tracing renders images with severe noise under one-sample-per-pixel conditions, and one of the challenges of denoising is to achieve high rendering quality with low time overhead. The Kernel prediction method is a high-quality denoising algorithm. In this paper, we propose an improved kernel prediction network algorithm to remove Monte Carlo noise with interactive rate. During training, we construct a variant of the multi-scale ResBlock structure to improve the denoising ability of our network. During deployment, we remove the residual connection by using a reserving and merging operation on ResBlock to equivalently transform multi-branch topology to a single-branch model. Experiments show that compared with previous methods, our method achieves a better trade-off between time expenditure and denoising quality. When removing Monte Carlo noise, we can improve the denoising quality without introducing high time overhead.

Keywords: Monte Carlo Noise · Deep learning · Image denoising · High realistic rendering

1 Introduction

Monte Carlo path tracing algorithm can solve the rendering equation and guarantee the unbiased solution of the integral to render the noise-free picture. However, in order to obtain a noise-free image, each pixel must be sampled multiple

This work was supported in part by the Henan Provincial Major Public Welfare Project (201300210400), the China Postdoctoral Science Foundation (2020M672211 and 2020M672217), the Key Scientific Research Projects of Henan Provincial Colleges and Universities (21A520003), and the Key Technology Research and Development Program of Henan (182102210106, 212102210090, 212102210094, 222102210133, and 222102210055), and the Graduate Education Innovation and Quality Improvement Action Plan project of Henan University(No. SYLJD2022008 and No. SYLKC2022028), 2022 Discipline Innovation Introduction Base cultivation project of Henan University.

X. Zhao et al. (Eds.): WISA 2022, LNCS 13579, pp. 444–452, 2022.
https://doi.org/10.1007/978-3-031-20309-1_39

times by accumulating random samples. The current GPU performance cannot directly render pictures at real-time rates. Therefore, how to obtain a higher rendering image quality ininteractive speed has become one of the challenges in academia. This paper adopts post-processing methods to solve the problem that MC algorithm is difficult to converge quickly [1–3]. First, we usually reduce the number of rays sampled per pixel for fast rendering, and then we reconstruct low-quality noisy images to restore high rendering quality.

Monte Carlo noise is fundamentally different from natural noise. It is impossible to simply use the denoising algorithm for ordinary images. Dabov exploited non-local image modeling and principal component analysis to denoise the pictures [4]. Kalantari's idea is to use a noise estimation metric to locally identify the amount of noise in different parts of the image [5]. Schied proposed to SVGF to distinguish noise and detail [6]. In recent years, deep learning technology has surpassed many traditional algorithms in the field of graphics [7,8]. Kalantari proposed a MLP method that training with a set of examples of noisy inputs and the corresponding reference outputs [9]. Bako and Vogels described the kernel-prediction(KP) network to evaluate the local weighting kernels to calculate each denoised pixel from its neighbors [10,11]. Chakravarty proposed an interactive approach based on recurrent denoising autoencoder architecture to denoising the picture [12]. Xu adopt GAN for denoising MC noises [13]. Zhang proposed an automatically decomposing noisy method to make kernel-predicting denoisers more effectively [14]. Tan utilized squeeze-and-excitation modules for the burst denoising to reduce blur. But most of the above methods only focus on offline rendering [15]. The rest can achieve high-speed noise reduction for drawn images, but the quality of the drawn images is too low.

Encouraged by Meng's multi-scale kernel prediction network [16], we propose an improved structure to better denoising the MC images. We add the Resnet block to the encoder stage because Veit demonstrated that residual connection can be seen as a collection of many paths of differing lengths, and use only short paths to achieve a deep network in training [17]. Then, to reduce the computational overhead caused by network reasoning, we use a reserving and merging (RM) operation to equivalently remove the residual connection [18]. In conclusion, our approach has the following contributions:

- During training, we construct a variant of the ResBlock structure to improve the denoising ability of our network.
- During deployment, we remove the residual connection by using a reserving and merging operation on ResBlock to equivalently transform multi-branch topology to a single-branch model.

2   Problem Statement

The goal of interactive MC denoising is to construct a 1spp noise image to an estimate high quality image that is as close as possible to a noise-free image that would be gained as a lot of samples goes to infinity. This estimate image is usually calculated by weighted averaging on a block X_p of per-pixel vectors around the neighborhood N_p to produce the filtered noise-free output at pixel p. And the parameters at each pixel can be written as:

$$C'_p = \sum_{q \in N_p} C_q w(p, q), \tag{1}$$

where C_q is the noise value of pixel p, C'_p is the denoised pixel value after filtering, N_p is the neighborhood centered on pixel p, and $w(p, q)$ is the weight of neighborhood pixel q to pixel p. The key of the denoising algorithm is to accurately predict the weight value $w(p, q)$, so as to reduce the estimation error between the predicted pixel C'_p and the truth pixel C_p as much as possible.

3 An Improved Kernel-Prediction Network Algorithm

The prominent feature of KP algorithm is to reconstruct the denoised pixels from neighbors by using the CNN to estimate the local weighting kernels. However, its rendering speed is too slow. So we use a noval operation to reduce the computational overhead and obtain high quality images.

3.1 RM Multi-scale Architecture

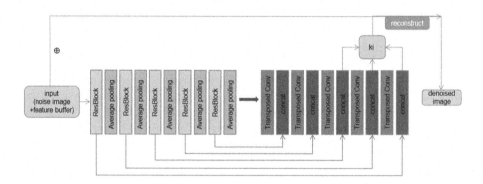

Fig. 1. An overview of RM multi-scale kernel-prediction network.

Figure 1 shows our RM multi-scale kernel prediction network. The encoder part of the network acts as a scale decomposition, with a structure of 5 layers of ResBLock and average pooling. The decoder part receives input information from the encoder part through a skip connection. After processing input information, it outputs prediction kernels of different resolutions. The decoder structure is 5 layers of transpose convolution and concatenate function. Specifically, in the scale decomposition phase, we input 1spp noise pictures from the dataset into the network and use a uniform 2×2 Downsampling to obtain the upper levels, and the auxiliary buffer including temporal accumulation maps does the same downsampling operation as the noise pictures. In the Scale Composition phase, we constructed a three-hierarchical scheme by mixing the results between the two operators to compute three different levels of kernels. Then we use the kernel to filter the noise input pictures at three different resolutions, and combine the filtered results from the lowest scale to the highest scale step by step to get the final noise reduction results. We mix the two adjacent levels of images with:

$$M_k = U_s(d_{k-1}) \times \alpha_k - U_s(D_s(d_k)) \times \alpha_k + d_k, \tag{2}$$

Here, M_k is a mixed picture, D_s is a downsampling function (2×2 average pooling), U_s is a nearest-neighbor upsampling function, and d_k is the denoised image from level $ki(i = 1, 2...)$. α_k is a scalar weight map for level k, which is generated with kernels.

3.2 Reserving and Merging Operation

The residual network can be viewed as a collection of many paths, not as a single super-deep network. Studies shown that the deep path contributing to the gradient during training is shorter than expected. The encoder process in U-net is actually a feature extraction process. If we pick up the encoder process alone, it will be very similar to VGG plain network, so we can make improvements on a lightweight U-net network to change the VGG-like structure to a resnet structure. The residual network can obtain better performance than the VGG-like network in training, but due to the multi-branch topology taking up too much computational overhead, the inference speed is slower than the VGG-like network [19]. So we can first train a multi-branch model with a resnet structure, record its trained parameters, and introduce a reserving and merging operation in the inference stage to remove the residual connection, and equivalently convert the previously trained parameters to the parameters of the single path model without residual connection. Then we use the transformed network for inference to reduce the running cost and improve the inference speed of the network.

Reserving: First, we insert a few Dirac initialized filters (output channels) defined as 4-dimensional matrices in Conv 1. The number of output channels is the same as the number of input channels of the convolution layer. In addition to using Conv 1, the input Feature map needs to be preserved using BN and ReLU. The Dirac initialized filters are shown below:

$$I_{c,n,i,j} = \begin{cases} 1 & if \quad c = n \quad and \quad i = j = 0, \\ 0 & otherwise \end{cases} \tag{3}$$

Merging: By extending the input channels in Conv2 and then dirac initialize these channels. The value z of the $i - th$ channel is obtained by summing the $i - th$ filter output and the $i - th$ input feature mapping in the original ResBlock. The entire Merging process can be expressed as follows:

$$Z_{c,h,w} = \sum_n^C \sum_i^K \sum_j^K \left(W_{c,n,i,j}^{R2} \times x_{n,h+i,w+j}^{R1+} + B_c^{R2} \right) + x_c^+, h, w. \tag{4}$$

Thus, with the reserving and merging operations, we can delete the residual connection while converting Resblock equivalently to a single-way branch model.

4   Experimental Setup

4.1   Dataset

We use the dataset exposed by Koskela, which is called BMFR dataset. The dataset consists of a set of 360 noisy 1-spp noise pictures and 4096-spp reference images, with a total of 6 scenes: classroom, living room, San Miguel, sponza, sponza glossy and sponza moving light. These scenes are configured with different materials, lighting and camera motion. Each scene has 60 consecutive frames. We use the same method as most previous works to accomplish the temporal accumulation pre-processing operation of a single noisy image by reprojection technology [20,21].

4.2   Training

For this dataset, we use the frame of one scene as the test data, and use the remaining images of other scenes as the training data. In addition, we randomly select a frame in a scene from the training data as the verification data. This means that we use a separate network to obtain the test results of each scenario, where the network uses only the data from other scenarios for training. Therefore, the performance of the test data can be used as an index of the generalization ability of the trained denoiser.

In the training phase, we use 128 * 128 resolution image patches. We randomly sample 60 such clips from each frame of the training data to form a training dataset, and extract 20 clips to establish a verification data set. In the deployment phase, the network input consists of frames with full resolution (1280 * 720) from the test data. Our denoising algorithm is implemented with tensorflow. According to the memory size of GPU, the size of mini batches we use in training is 20 and in testing is 10. We use Adam optimization algorithm to update the neural networr weights. We set the learning rate to 0.0001 and keep the default values of other parameters. On rtx2080ti gpu, the typical training time of training 100 generations in a test scenario is about 12.5 h.

5   Performance Evaluation

In order to better evaluate the experimental results, the denoising algorithms in this paper are compared with some general denoising algorithms. For public denoising algorithms, each algorithm has its own unique characteristics and processing methods due to different datasets, so the universality and portability of the algorithm is not strong. The final comparison algorithm chosen in this paper is as follows: Bako's KP algorithm [9], Chakravarty's RAE algorithm [11] and Meng's MR-KP algorithm [15]. KP algorithm is only an offline denoising algorithm used by Disney. In this paper, we use the optimized lightweight KP algorithm for comparison. RAE algorithm is integrated into OptiX 5.1 as a black box module. MR-KP is a multi-scale kernel prediction neural network that fuses three-layer resolution kernels to reconstruct the final image.

This paper uses PSNR and SSIM as the quantification index of the denoising result. As shown in the following image, we used five scenarios as training sets and another scenario as test sets. In the first two simple secens, there is no obvious difference between several algorithms because the relatively orderly texture account for a large proportion of the whole image. However, in the latter complex lighting scenarios, our algorithm is much better than other algorithms.

Analyzing several sets of noise reduction images, we can see that in the classroom, the RAE algorithm removes the jagged edges of chairs, but the wooden texture on chairs is blurred at the same time. KP methods can not preserve the wooden texture on chairs completely, and its PSNR is lower than MR-KP and RM-KP. The same is true for san-miguel scenes, where the RAE algorithm obscures the edges of the image even though it can achieve smooth noise reduction. For example, for opaque glass bottles, which have complex light reflection, noise reduction will also erase the high light reflection on the bottle body, resulting in the loss of high frequency details and spatial stereo.

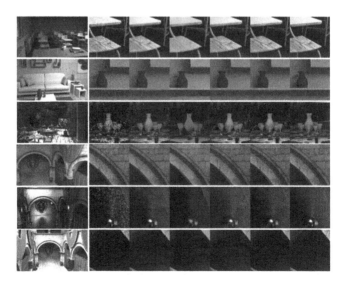

Fig. 2. From left to right are 1-ssp noisy pictures, KP, RAE, MRKP, OURS and reference pictures for 1-spp BMFR test data denoising image. Each row represents an independent experiment in which the scenarios shown (such as classrooms) are used for testing and the other five scenarios are used for training.

From the last three complex ilumination scenarios, we can see that the KP network has less noise reduction capability than MR-KP and RM-KP. For example, in the shaded part above the brass tube, the dark details are confused when there is a glossy reflection. Compared with RM-KP, MR-KP has lower PSNR and SSIM. And MR-KP restores the dark-light transition is darker and closer to reference pictures, while the dark details of MR-KP are brighter and the change of color transition is slightly unnatural. In the Sponza movinglight scene, since all the selected scenes are dark, it can be found that the RM-KP noise canceller can retain the information of the dark scene relatively well without losing too much detail subjectively.

Table 1. A comparision of average PSNR and SSIM values(higher is better) for evaluationg our trained denoisers on 1-spp test data.

Scene	PSNR				SSIM			
	KP	RAE	MR-KP	Ours	KP	RAE	MR-KP	Ours
Classroom	32.172	27.327	32.324	32.318	0.977	0.931	0.978	0.977
Living room	32.672	25.594	32.856	**32.859**	0.961	0.943	0.969	**0.969**
San Miguel	24.519	20.170	24.816	**24.855**	0.829	0.772	0.837	**0.852**
Sponza	30.672	24.713	30.955	30.936	0.968	0.865	0.974	0.973
Sponza-glossy	26.349	23.445	27.441	**28.177**	0.960	0.873	0.923	**0.935**
Sponza-moving light	25.042	22.303	25.398	**25.517**	0.941	0.807	0.948	**0.957**

In summary, our proposed algorithm achieves better results in most complex lighting environment. It can greatly improve the quality of image reconstruction and maintain the speed of image reconstruction. Due to the limited number of scene types in the actual training, when denoising the scenes with different lighting environments from the training samples, our method may cause some areas to be blurred and lose details.

6   Conclusion and Future Work

In this paper, we propose a MC denoising algorithm to solve the problems existing in high photorealistic rendering. Then we introduce a novel structure of RM-KP denoising algorithm. Finally, experiments show that the proposed algorithm achieves good results in removing MC noise and improves the quality of image reconstruction without introducing additional computational overhead. In the future, we will further explore the kernel prediction denoising method, and design a new network to better deal with the balance of speed and quality of MC denoising.

References

1. Hill, S., McAuley, S., Belcour, L., Earl, W., Harrysson, N., Georgiev, I.: Physically based shading in theory and practice. In: ACM SIGGRAPH 2020 Courses (SIGGRAPH 2020), Article 11, pp. 1–12 (2020)
2. Zwicker, M., et al.: Recent advances in adaptive sampling and reconstruction for monte Carlo rendering. Comput. Graph. Forum **34**(2), 667–681 (2015)
3. Işık, M., Mullia, K., Fisher, M., Eisenmann, J., Gharbi, M.: Interactive Monte Carlo denoising using affinity of neural features. ACM Trans. Graph. (TOG) **40**(2021), 1–13 (2021)
4. Dabov, K., Foi, A., Katkovnik, V., Egiazarian, K.O.: Image denoising with block-matching and 3D filtering. In: Electronic Imaging (2006)
5. Kalantari, N.K., Sen, P.: Removing the noise in monte Carlo rendering with general image denoising algorithms. Comput. Graph. Forum **32**, 93–102 (2013)
6. Schied, C., et al.: Spatiotemporal variance-guided filtering: real-time reconstruction for path-traced global illumination. In: Proceedings of High Performance Graphics (HPG 2017). Association for Computing Machinery, New York, NY, USA, Article 2, pp. 1–12 (2017)
7. Huo, Y., Yoon, S.: A survey on deep learning-based Monte Carlo denoising. Comput. Vis. Media **7**, 169–185 (2021)
8. Yao, X., Zhang, Z., Cui, R., Zhao, Y.: Traffic prediction based on multi-graph Spatio-Temporal convolutional network. In: Xing, C., Fu, X., Zhang, Y., Zhang, G., Borjigin, C. (eds.) Web Information Systems and Applications. WISA (2021)
9. Kalantari, N.K., Bako, S., Sen, P.: A machine learning approach for filtering Monte Carlo noise. ACM Trans. Graph. **34**(4) Article 122, 12 (2015)
10. Bako, S., et al.: Kernel-predicting convolutional networks for denoising Monte Carlo renderings. ACM Trans. Graph. **36**(4), Article 97, 14 (2017)
11. Vogels, T., et al.: Denoising with kernel prediction and asymmetric loss functions. ACM Trans. Graph. (TOG) **37**, 1–15 (2018)
12. Chaitanya, C.R.A., et al.: Interactive reconstruction of Monte Carlo image sequences using a recurrent denoising autoencoder. ACM Trans. Graph. **36**(4), Article 98, 12 (2017)
13. Xu, B., et al.: Adversarial Monte Carlo denoising with conditioned auxiliary feature modulation. ACM Trans. Graph. (TOG) **38**, 1–12 (2019)
14. Zhang, X., Manzi, M., Vogels, T., Dahlberg, H., Gross, M., Papas, M.: Deep compositional denoising for high-quality Monte Carlo rendering. Comput. Graph. Forum **40**, 1–13 (2021)
15. Tan, H., Xiao, H., Lai, S., Liu, Y., Zhang, M.: Denoising real bursts with squeeze-and-excitation residual network. IET Image Process. **14**, 3095–3104 (2020)
16. Meng, X., Zheng, Q., Varshney, A., Singh, G., Zwicker, M.: Real-time Monte Carlo Denoising with the Neural Bilateral Grid. EGSR. (2020)
17. Veit, A., Wilber, M.J., Belongie, S.: Residual networks behave like ensembles of relatively shallow networks. NIPS (2016)
18. Meng, F., Cheng, H., Zhuang, J., Li, K., Sun, X.: RMNet: equivalently removing residual connection from networks. ArXiv abs/2111.00687. (2021)
19. Ding, X., Zhang, X., Ma, N., Han, J., Ding, G., Sun, J.: RepVGG: making VGG-style ConvNets great again. In: 2021 IEEE/CVF Conference on Computer Vision and Pattern Recognition (CVPR), pp. 13728–13737 (2021)

20. Koskela, M., et al.: Blockwise multi-order feature regression for real-time path-tracing reconstruction. ACM Trans. Graph. (TOG) **38**, 1–14 (2019)
21. Fan, H., Wang, R., Huo, Y., et al.: Real-time Monte Carlo denoising with weight sharing kernel prediction network. Comput. Graph. Forum **40**(4), 15–27 (2021)

Research on Target Detection Method Based on Improved YOLOv5

Rongzhuang Wang[1,2(✉)], Jun Liu[1,2], and Hualiang Zhang[3,4,5]

[1] School of Computer Science and Technology, Shenyang University of Chemical Technology,
Shenyang 110142, China
1725444636@qq.com, liujun@syuct.edu.cn
[2] Liaoning Key Laboratory of Intelligent Technology for Chemical Process Industry,
Shenyang 110142, China
[3] Key Laboratory of Networked Control Systems, Chinese Academy of Sciences,
Shenyang 110016, China
[4] Shenyang Institute of Automation, Chinese Academy of Sciences, Shenyang 110016, China
zhanghualiang@sia.cn
[5] Institutes for Robotics and Intelligent Manufacturing, Chinese Academy of Sciences,
Shenyang 110169, China

Abstract. At present, various target detection algorithms are used in detection and classification. It is a problem to improve the accuracy and speed of target detection by using deep learning and neural network model to train the parameters of the data set. Aiming at the problems of insufficient detection, inaccurate detection of complex background and loss of small target feature information in yolov5 target detection, an improved method combining hole convolution, moving exponential average and retraining of positive samples is proposed in this paper. This method effectively solves the problems of complex background, occlusion and incomplete feature extraction, reduces the loss rate of target features, and improves the detection accuracy of each category. The improved yolov5 target detection algorithm in this paper can not only use multi feature fusion and accelerate positive sample matching, but also use cavity convolution to increase receptive field, improve the detection effect of small targets and reduce the occurrence of missed detection and false detection. Experiments show that the improved yolov5 algorithm has better detection rate and lower loss rate than many existing target detection algorithms.

Keywords: Target detection · YOLOv5 · Void convolution · Feature fusion

1 Introduction

1.1 A Subsection Sample Basic Introduction of Algorithm

In recent years, with the continuous development of computer vision technology, there are more and more application fields, in which target detection plays an important role and is widely used in face detection, intelligent medical treatment, pedestrian detection,

X. Zhao et al. (Eds.): WISA 2022, LNCS 13579, pp. 453–461, 2022.
https://doi.org/10.1007/978-3-031-20309-1_40

activity recognition and so on [1]. At present, the commonly used target detection algorithms have insufficient feature extraction, missed detection and false detection for the detection of some complex background, occlusion, light intensity change and some small targets. Guo Lei, Wang Qiulong and others [1] used mosaic-8 method to enhance data, add shallow feature map and adjust loss function to enhance the network's perception of small targets. Therefore, this paper proposes a method of hole convolution to increase the range of receptive field, improve the feature extraction ability of small targets and occluded targets, and solve the gradient disappearance problem through EMA. Compared with other target detection algorithms, the improved yolov5 algorithm has better classification effect and higher accuracy.

2   Target Detection Algorithm

2.1   Basic Introduction of Algorithm

At present, the algorithms for target detection are mainly divided into two categories. One is the two stage algorithm, which represents fast RCNN, and the other is the one stage algorithm, which represents SSD and Yolo series. The two stage algorithm has complex structure and slow speed, which has the advantage of high detection accuracy. The one stage algorithm is fast. It can also improve the accuracy of detection and achieve the effect of target detection through data enhancement, non maximum suppression, enhanced feature extraction, multiple matching and other methods. Basic flow of target detection algorithm (Fig. 1):

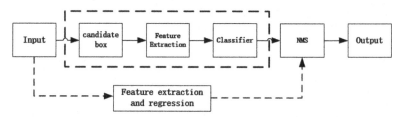

Fig. 1. Algorithm flow

2.2   YOLOv5 Algorithm

Yolov5 is a single-stage target detection algorithm, which improves the idea of yolov4 and greatly improves its speed and accuracy. Yolov5 is mainly divided into three parts: backbone, FPN and Yolo head. Backbone feature extraction network is suitable for extracting the features of input pictures and constructing effective feature layer; FPN mainly carries out feature fusion on the effective feature layer and continuously extracts the depth feature information, in which panet structure will also carry out up sampling and down sampling to realize feature fusion; Yolo head is the classifier and regressor of Yolo V5; Judge whether the feature points in the feature layer correspond to the detected

Fig. 2. Network flow

targets, and realize the classification and regression of targets. Yolov5 network workflow (Fig. 2):

The structure and main features of yolov5 backbone extraction network csparknet are as follows:

Residual Network Structure

The residual structure of the backbone network of yolov5 is composed of 1×1 and 3×3. The residual part does not need to be processed, and directly combines the input and output of the backbone network. This residual network can improve the accuracy of target detection by increasing the depth of the network, which is easy to optimize, and also conveniently solves the problem of gradient disappearance in the network. As shown in the figure below (Fig. 3).

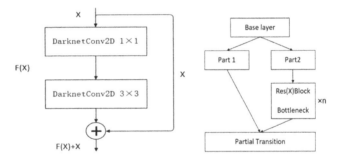

Fig. 3. Residual network structure

Cspnet Network Structure

In cspnet, the residual blocks are divided into two parts: one part is the stacking of the original residual blocks; The other part is directly connected to the end through a small amount of processing, forming a large residual edge.

Spp Structure and Silu Activation Function

Spp structure can extract network features from the maximum pool of different pool core sizes and improve the receptive field of the network. Silu activation function is better than relu activation function in depth network model. It has no upper bound and lower bound, smooth and non monotonic.

3 Improved YOLOv5-X Algorithm

3.1 Convolution with Holes

After the depth feature extraction, the feature layer extraction of small targets with occlusion and light intensity change is not obvious. In order to better extract the feature information of these affected targets and increase the receptive field of convolution kernel, we use void convolution to replace the original convolution kernel to enhance the feature extraction ability of targets. Left figure shows the 3 × 3 convolution with a receptive field of 3 × 3. Right figure shows the empty convolution with a receptive range of 5 × 5 and only 4 additional weighted points. However, it can feel more information of feature layer, expand perception field and reduce the missed detection ability of occluded targets (Fig. 4).

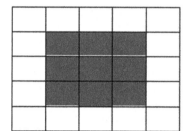 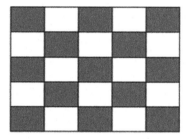

Fig. 4. Empty convolution

In the process of target detection, the feature extraction network of YOLOV5-X is improved as follows: Void convolution is used to replace the convolution of three scales in FPN feature pyramid with empty convolution to improve multi-feature extraction ability. The improved NETWORK structure of SPDarknet is shown in Fig. 5:

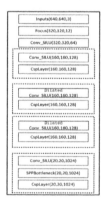

Fig. 5. Network structure

3.2 Exponential Moving Average (EMA)

The gradient of the training process based on the weights of ordinary parameters of YOLOV5-X will be unstable. Therefore, EMA is added to take the weighted average of the training gradient, so as to allocate weights and gradients more reasonably, thus improving the detection indicators and increasing the robustness of the model.

$$\text{Formula}: Vt = \beta \cdot Vt - 1 + (1 - \beta) \cdot \theta t$$

θt: The weights of all parameters obtained after the t-th update
Vt: Moving average of all parameters updated for the t time
β: Weight parameter

Finally, on the basis of YOLOv5 algorithm, the matching process of positive samples is retrained in order to strengthen the training and the matching speed of positive samples.

4 Experiment and Result Analysis

4.1 Data Source and Processing

The experimental data set adopts the first 2008 pictures of voc2007 data set, and then makes the corresponding label file. The processing of the experimental data set is to divide the training set and verification set, with a ratio of 9:1, generate the required target information, and the format of the label file is XML, which should correspond to the detected image one by one. As shown in Figs. 6 and 7.

000199   000200   000201

Fig. 6. Test picture

000199	2007/4/8 0:37	XML 文档	1 KB
000200	2007/4/8 0:37	XML 文档	2 KB
000201	2007/4/8 0:37	XML 文档	2 KB

Fig. 7. Lable file

The data set is divided into 20 categories, including the number and classification of statistical targets. As shown in Table 1.

Table 1. Date ste categories.

Category	Quantity
Bottle	120
Chair	284
Bird	150
Person	1737
…	…

Table 2. Environment configuration.

Development environment	Environment configuration
CPU	Intel Corei7-10750H CPU 2.60 GHz
OS	Windows 10
RAM	16 GB
IDE	PyCharm
Development language	Python
Third-party libraries	Numpy, Opencv, Torch, Pandas…

4.2 Experimental Environment and Parameter Setting

The experimental environment is shown in Table 2.

In this experiment, the detection results will be obtained through three main parts. First, the picture does not need to be fixed in size, and will automatically resize after inputting the picture, and then feature extraction to obtain the effective feature layer. Stack the obtained feature layers in the backbone network, expand the input channel, and enhance the mosaic data to enrich the background of the detection target. Secondly, use the FPN feature pyramid to strengthen the feature extraction, and fuse the feature layers of different sizes after hole convolution to extract better target features. Finally, use Yolo head to judge whether the feature points correspond to the target, A non maximum suppression operation is performed to obtain the detection result.

In this paper, the experimental parameter setting: to ensure the stable operation of model training, this paper uses multi-threaded data reading, trains 100 epochs in total, the maximum learning rate is 0.01, the learning rate decline mode is cos, uses SGD optimizer to set it to 1e-2, and uses weight decay weight attenuation to prevent over fitting.

4.3 Evaluation Criteria

In the detection and classification task of deep learning or machine learning, the general evaluation criteria generally include precision, recall, F1 value and map value. According to these three indicators, a confusion matrix is formed to evaluate the performance of target detection algorithm, as shown in Table 3.

Table 3. Confusion matrix.

	Positive samples	Negative samples	Aggregate
Positive samples (prediction)	TP	FP	TP + FP
Negative samples (prediction)	FN	TN	FN + TN
Aggregate	TP + FN	FP + TN	N

Where, T is true, F is false, P is positive and N is negative, which means: T or F represents whether the sample is correctly classified, and P or N represents whether the sample is predicted to be a positive sample or a negative sample.

TP (True positives) indicates that the detection is a positive sample, and the number of positive samples is detected;

TN (True negatives) indicates that the detection is a negative sample, and the number of negative samples is detected;

FP (False positives) indicates that the detection is a positive sample, but the number of negative samples is detected;

FN (False negatives) indicates that the detection is negative, but the number of positive samples is detected.

The specific evaluation methods in deep learning are:

Precision represents the proportion of correctly predicted positive samples in the classified positive samples, as shown in the formula.

$$precision = \frac{TP}{TP + FP} \tag{1}$$

Recall represents the proportion of correctly predicted positive samples in all classified positive samples, as shown in the formula (Table 4).

$$recall = \frac{TP}{TP + FN} \tag{2}$$

The F1 value represents the harmonic average of accuracy and recall, as shown in the formula.

$$F1 = 2\frac{precision \times recall}{precision + recall} \tag{3}$$

AP and mAP values. AP refers to the area corresponding to the curve formed by different positions and recall points, as shown in the figure. Then mAP is the average of AP values of all categories.

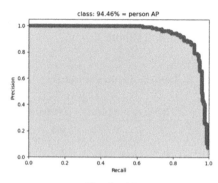

Fig. 8. AP

4.4 Effect Criteria

In order to test the effect of the improved yolov5 algorithm, the Fast R-CNN, yolox and yolov5 detection algorithms commonly used in the field of target detection are compared. The comparison indicators are mAP value, recall, loss rate, accuracy and testing time. The following table is the comparison detection results.

Table 4. Experimental results of various algorithms.

Algorithm	Precision	mAP	Recall	Loss	Time/h
Fast-RCNN	0.878	0.776	0.891	0.157	3
YOLOX	0.893	0.814	0.859	0.150	4.5
YOLOv5	0.914	0.862	0.876	0.127	4.2
Modified YOLOv5	0.933	0.90	0.896	0.094	4.4

It can be seen from the above table that the improved yolov5 has the best detection effect on the voc2007 dataset. In terms of accuracy, it is 1.9% higher than the unmodified yolov5, 4% higher than yolox and 5.5% higher than Fast-Rcnn. At the same time, the map of the improved algorithm is the highest, indicating that the detection and classification of each category is the best and its loss rate is the lowest, indicating that the difference between the target with correct classification and the predicted target is the smallest. However, in terms of detection time, the detection time of improved YOLOv5 is relatively long, which can be improved in the future.

Therefore, in terms of the effect of target detection, the improved yolov5 algorithm has better detection effect.

5 Conclusion

Through the improved yolov5 algorithm, this paper proposes a method to improve the detection of targets with complex background by using hole convolution, adding

EMA moving index average and retraining of positive samples, which not only obtains higher classification accuracy, but also reduces the loss rate. Experiments show that the improved algorithm in this paper performs well in target detection, which is significantly better than the commonly used Fast-RCNN, YOLOX and other algorithms. In addition, for different data sets, various parameters can be flexibly set to achieve the optimal performance index. In terms of data processing, multi feature fusion is used to screen the confidence of the predicted results, so as to make the model detection effect better. There are still some areas that need to be improved in this experiment. For example, the adjustment of training time is not done well under the data set of the same size. Therefore, the next step is to adopt the improved training model to improve the detection speed and ensure the accuracy of target detection at the same time.

Acknowledgments. This work was partially supported by National Key Research and Development Program of China under Grant(NO. 2018YFE0205803);

This work was partially supported by National Natural Science Foundation of China under Grant(NO.91648204);

Scientific research project of Liaoning Provincial Department of education in 2020(LJ2020024).

References

1. Guo, L., Wang, Q., Xue, W., Guo, J.: Small target detection algorithm based on improved YOLOv5. J. Univ. Electron. Sci. Technol. **51**(02), 251–258 (2022)
2. Li, A.: Improvement of YOLOv5 algorithm and its practical application. Zhongbei University (2021). https://doi.org/10.27470/d.cnki.ghbgc.2021.000232
3. Jin, R., Niu, Q., Spagnolo, P.: Automatic fabric defect detection based on an improved YOLOv5. Math. Probl. Eng. **2021** (2021)
4. Yuan, Z., Fang, W., Zhao, Y., Sheng, V.S.: Research of insect recognition based on improved YOLOv5. J. Artif. Intell. **3**(4), 145 (2021)
5. Lv, Z., Li, H., Liu, Y.: Garbage detection and classification method based on YoloV5 algorithm. Tianjin University of Technology and Education (China), Institut für Technische Optik (Germany), Institute for Information Transmission Problems (Kharkevich Institute) (Russian Federation), Sichuan University (China) (2022)
6. Jiang, L., Cui, Y.: Small target detection based on yolov5. Comput. Knowl. Technol. **17**(26), 131–133 (2021). https://doi.org/10.14004/j.cnki.ckt.2021.2620
7. Ge, L.: YOLOv5 multistage improvement target detection algorithm. Lanzhou University (2021). https://doi.org/10.27204/d.cnki.Glzhu.2021.000295
8. Xiaoling, Y., Weixin, J., Haoran, Y.: Traffic sign recognition and detection based on YOLOv5. Inf. Technol. Informatization **04**, 28–30 (2021)
9. Cheng, L.: Research on small target detection algorithm based on improved YOLOv5. Yangtze River Inf. Commun. **34**(09), 30–33 (2021)
10. Wang, K., Teng, Z., Zou, T.: Metal defect detection based on YOLOv5. J. Phys. Conf. Ser. **2218**(1), 012050 (2022)
11. Li, Z.: Road aerial object detection based on improved YOLOv5. J. Phys. Conf. Ser. **2171**(1), 012039 (2022)
12. Kim, H. (ed.): WISA 2021. LNCS, vol. 13009. Springer, Cham (2021). https://doi.org/10.1007/978-3-030-89432-0

Diagnostic Prediction for Cervical Spondylotic Myelopathy Based on Multi-source Data in Electronic Medical Records

Shuhao Zheng[1,3], Guoyan Liang[2], Junying Chen[1,3(✉)], Yongyu Ye[2],
Yunbing Chang[2], Yi Cai[1,3], and Shaowu Peng[1]

[1] School of Software Engineering, South China University of Technology,
Guangzhou 510006, China
jychense@scut.edu.cn

[2] Department of Spine, Guangdong Provincial People's Hospital, Guangdong
Academy of Medical Sciences, Guangzhou 510080, China

[3] Key Laboratory of Big Data and Intelligent Robot (SCUT), Ministry of Education,
Guangzhou 510006, China

Abstract. For a long time, cervical spondylotic myelopathy has a high
incidence in middle-aged and elderly people. In reality, the diagnosis of
cervical spondylotic myelopathy by spine surgeons is a comprehensive
process of aggregating information from multiple clinical data sources,
which requires a comprehensive consideration based on the multi-source
data. This process requires extensive experience and years of study by
spine surgeons. The artificial intelligence method proposed in this work
can greatly speed up this learning process. The proposed method comprehensively analyzes multi-source data in the patients' electronic medical
records, and provides diagnostic predictions to assist spine surgeons in
efficient diagnosis. More importantly, the impact of different data sources
on the diagnostic results is analyzed in depth.

Keywords: Cervical spondylotic myelopathy · Disease prediction ·
Electronic medical record · Multi-source data analysis

1 Introduction

Spine-related disorders are among the most common diseases in the middle-aged
and elderly population. In addition, with changes in lifestyle, these conditions

Supported in part by the National Natural Science Foundation of China under Grant
61802130 and Grant 81802217, in part by the Guangdong Natural Science Foundation
under Grant 2021A1515012651, Grant 2019A1515012152 and Grant 2019A1515010754,
and in part by the Guangzhou Science and Technology Program Key Projects under
Grant 2021053000053. (Corresponding author: Junying Chen.)

X. Zhao et al. (Eds.): WISA 2022, LNCS 13579, pp. 462–470, 2022.
https://doi.org/10.1007/978-3-031-20309-1_41

are also common among adolescents in recent years, and the topic of increasing youthfulness of spine-related disorders is gaining more attention. Cervical spondylotic myelopathy (CSM) is the most prevalent type of spine-related disorders, and the symptoms include sensory disturbances of extremities, manual inflexibility, gait stiffness, and urinary dysfunction [7]. If CSM is diagnosed and treated at the early stage, the impact on the patient's life and work can be greatly reduced.

The diagnosis of CSM needs to consider a combination of information from many aspects of the patient, including the patient's personal health information, physical examination, and radiology reports, coming from different sources. Hence, the accurate diagnosis of CSM requires extensive clinical experience to handle the combination of multi-source medical information. To solve such challenge, it is important to introduce the artificial intelligence technology as an auxiliary tool to provide diagnostic predictions to assist spine surgeons in efficient diagnosis.

Electronic medical records (EMRs) collected in this work are a collection of data from multiple sources, including personal information, past medical history, laboratory results, radiology reports, diagnostic reports, and clinical treatment records. In the early stages, research on EMRs mainly used traditional machine learning methods that required hand-crafted feature designs, which was time-consuming and laborious [2]. The quality of the feature design directly determines how effective the machine learning approach can be, and it is difficult to transfer the research base from a specific disease to other diseases. With the rapid development of deep learning techniques, analysis and research on EMRs [10] are gradually expanding to deep neural networks [8]. The deep learning methods no longer rely on hand-crafted feature design and can fuse multi-source data for deeper research.

Existing studies of EMRs have focused on the field of internal diseases [4] and have been analyzed mainly on the basis of quantitative laboratory results. For surgical diseases such as CSM, multi-source clinical data are needed for a comprehensive diagnosis, and the impact of different data sources on the diagnostic results should be analyzed in depth.

2 CSM Diagnostic Prediction Based on EMRs

Multi-source Data in EMR Dataset. The EMR dataset studied in this work includes multi-source information of 855 patients hospitalized with cervical spine disorders. In the dataset, 627 patients were diagnosed with CSM, and the remaining 228 patients were diagnosed with other types of spine diseases. Table 1 lists the data items contained in our EMR dataset. The bold data items in Table 1 are the main data items of interest in this work, including physical examination (gender, age, muscle atrophy, grip strength, feeling, muscle tension, knee reflex, Hoffman sign, Babinski sign), medical history (chief complaint, present medical history), clinical indicators (JOACMEQ scores), 10-s grip and release (G&R) test (left/right-hand G&R cycles, left/right-hand average/max/min G&R speed, left/right-hand max/min G&R cycles per minute), radiology reports (preoperative X-ray/CT/MRI reports), and diagnostic reports (primary diagnosis). The hand motion data of the 10-s G&R test is obtained by a hand motion analysis model developed by ourselves [13].

Table 1. Data items in our EMR dataset.

Type	Items	Structure	Source
Physical examination	**gender, age,** height, weight, **muscle atrophy, grip strength, feeling, muscle tension, knee reflex, Hoffman sign, Babinski sign**	structured	spine surgeon
Medical history	**chief complaint, present medical history,** past medical history	unstructured	patient self-reporting
Blood sampling	random blood sugar, white blood cell count, C-reactive protein, erythrocyte sedimentation rate	structured	laboratory physician
Clinical indicators	**JOACMEQ scores,** mJOA scores	structured	spine surgeon
10-second G&R test	**left/right-hand G&R cycles, left/right-hand average/max/min G&R speed, left/right-hand max/min G&R cycles per minute**	structured	hand motion analysis model
Radiology reports	**preoperative X-ray/CT/MRI reports,** postoperative X-ray/CT/MRI reports, Electromyography reports	unstructured	imaging physician
Diagnostic reports	**primary diagnosis,** secondary diagnosis, surgical methods, surgical segments	structured	spine surgeon

As shown in Table 1, the multi-source data in EMR dataset can be classified into two categories: structured data and unstructured data. The structured data is quantitative data, including the patient's physical examination, blood sampling, clinical indicators, and hand motion data in 10-s G&R test. The unstructured data is free text data, including the patients' medical history and radiology reports.

EMR Analysis Model Design. The architecture of the proposed EMR analysis model is shown in Fig. 1. As shown in Fig. 1, the unstructured data is first pre-processed, and then goes through the named entity recognition (NER) module and the word embedding process (in the Word2Vec module) to generate a vectorized representation, which is then used as the input for text feature extractor to obtain the features of the unstructured data. On the other side, the structured data is first normalized using the Z-score method to obtain a vectorized representation, which is then fused with the representation of the unstructured data in feature fusion module to obtain the complete representation (features) of each EMR.

Since unstructured data is a serialized form of free text, which naturally contains logical sequential relationships between contexts, therefore models such as recurrent neural networks (RNNs) are well suited for the feature extraction. In contrast, the structured data comes from various clinical examinations, which do not have relevant sequential relationships with each other and whose features cannot be effectively extracted using models like RNNs. But the structured data already has obvious feature representation after normalization, and can be directly used as feature vectors to fuse with features of unstructured data. The feature fusion method used in this model is concatenation of feature vectors. As there is no direct correspondence between unstructured and structured data, using adding or averaging fusion methods between feature vectors without direct correspondence does not bring performance improvement, but rather makes the feature vectors interfere with each other.

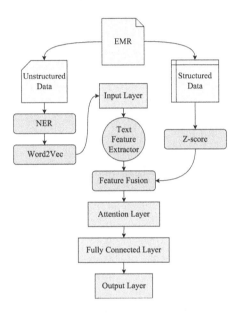

Fig. 1. Architecture of the proposed EMR analysis model.

The fused features are sent to the attention layer to reinforce the feature representation. The attention layer is adaptively adjusted to assign different weights for each dimension of the features according to the input feature representation, making the model focus more on strong features, *i.e.*, the features that have stronger correlation with the labels. The enhanced feature representation using the attention layer has stronger representational capability and also enables the model to obtain higher diagnostic prediction accuracy. Finally, the diagnostic prediction results are obtained through the fully connected layer.

3 Experiments

Experimental Settings. In the dataset, the patients diagnosed with CSM are considered as positive cases, and the patients diagnosed with other types of spine diseases are considered as negative cases. The dataset is evenly divided into training set, validation set and test set in the ratio of 3:1:1 according to the ratio of positive and negative cases. In addition, because the completeness of each patient's EMR is not uniform, for example, some patients may not have undergone a 10-s G&R test or have not taken an X-ray examination, we set the patient's missing unstructured data as a 0 vector and the missing structured data as 0 for pre-processing.

Model Performance Using Different Feature Extractors. In the proposed EMR analysis model, feature extraction of unstructured text data is a key step and has a direct impact on the overall classification performance of the model. In this work, different neural networks (with or without attention mechanism) are used as feature extractors, including RNN [12], LSTM [9], Bi-LSTM [3], DPCNN [5], Text-GCN [11], BERT [1] and BERTGCN [6] models. Related experiments are set up to compare the model performance with different feature extractors, as shown in Table 2.

Table 2. Model performance using different text feature extractors.

Feature extractor	Accuracy	Precision	Recall	F1-score
RNN	0.57	0.77	0.59	0.67
RNN(attn.)	0.62	0.81	0.63	0.71
DPCNN	0.64	0.83	0.65	0.73
LSTM	0.66	0.83	0.68	0.75
LSTM(attn.)	0.68	0.84	0.70	0.76
Bi-LSTM	0.73	0.87	0.74	0.80
Bi-LSTM(attn.)	0.79	0.89	0.81	0.85
Text-GCN	0.72	0.83	0.77	0.80
BERTGCN	0.77	0.88	0.80	0.84
BERT	**0.91**	**0.94**	**0.94**	**0.94**

As seen from Table 2, the model using BERT as the feature extractor for unstructured text data achieves the best performance, especially in the metric of recall, and far outperforms the model using other feature extractors in terms of the overall F1-score. Compared with the RNN, its variants (LSTM and Bi-LSTM) perform better due to the gating mechanism introduced in the variants which can effectively improve the long-term memory of the network and are more suitable for processing longer text sequences. The relatively poor performance of Text-GCN and BERTGCN may be due to the characteristics of GCN. In the graph constructed by the GCN-based models, the words in the text and the text itself are used as nodes. Since the text used in this work is more concise and the number of corresponding nodes in the graph is less, the results achieved are relatively poor. In addition, the introduction of the attention mechanism in RNN and its variants also provides a slight performance improvement compared to the original network structures.

Impact of Multi-source Data on Model Performance. The EMR-based CSM prediction model studied in this work emulates the real-world diagnostic process of an spine surgeon, in which all medical information about the patient is considered and analyzed before making a diagnosis, rather than considering only a single source of information. The patient's multi-source data used in this paper is divided into five main parts: (1) medical history, (2) physical examination, (3) hand motion data in 10-s G&R test, (4) clinical indicators, (5) preoperative radiology reports. In order to verify the effect of multi-source data on the prediction accuracy, experiments are designed to test the model performance using different combinations of multi-source data as input. The experimental results are shown in Table 3, and the analysis model in the experiments uses BERT as the feature extractor for unstructured text data.

As seen in Table 3, the best performance is achieved using all multi-source data as input. The comparison of combinations (b) and (c) shows that the hand motion data in G&R test has a higher accuracy than physical examination, probably because the hand motion data of patients with CSM in the G&R test differs significantly from those of patients with other types of spine diseases. The addition of this data source makes it easier for the model to distinguish patients with different types of spine diseases, thus showing that the hand motion data of G&R test has an important improvement in aiding the diagnosis. The long text of the radiology reports contains a lot of disease-related information, so this data source has the most significant improvement on the performance of the model.

Table 3. Model performance with different combinations of multi-source data.

	Multi-source data combination	Accuracy	Precision	Recall	F1-score
(a)	Medical history	0.62	0.79	0.66	0.72
(b)	Medical history physicial examination	0.64	0.80	0.67	0.73
(c)	Medical history 10-second G&R test	0.67	0.84	0.68	0.75
(d)	Medical history clinical indicators	0.70	0.84	0.74	0.78
(e)	Medical history radiology reports	0.74	0.84	0.79	0.81
(f)	Medical history radiology reports physicial examination	0.76	0.85	0.82	0.83
(g)	Medical history radiology reports 10-second G&R test	0.78	0.88	0.81	0.84
(h)	Medical history radiology reports clinical indicators	0.83	0.89	0.87	0.88
(i)	All	**0.91**	**0.94**	**0.94**	**0.94**

Table 4. Ablation experiment of the attention layer in the model with BERT as feature extractor.

	Accuracy	Precision	Recall	F1-score
w/ attention layer	0.88	0.93	0.90	0.92
w/o attention layer	**0.91**	**0.94**	**0.94**	**0.94**

Effectiveness of Attention Layer. As shown in the model structure in Fig. 1, this work uses an attention layer to reinforce the features of EMRs with adaptively adjusted weights to enhance the characterization. In order to verify the influence of the attention layer on the model, the ablation experiment of the attention layer is conducted, and the experimental results are shown in Table 4. It can be seen that the introduction of the attention layer improves the performance of the model's diagnostic prediction, especially the improvement in the recall rate is more obvious.

Interpretability of Proposed Model. The adaptive weights of the attention layer can be used to find the strong features among all the features, and each dimension of the features corresponds to a data item in the EMR, which means that the adaptive weights can be used to determine the main basis of the analysis process, so as to provide some interpretability for the EMR analysis model.

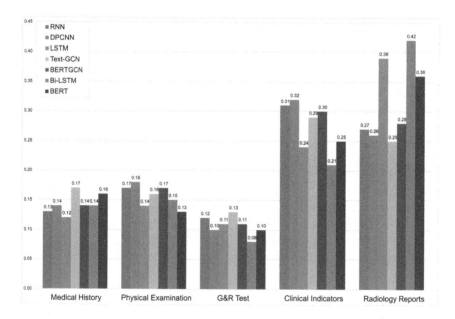

Fig. 2. Visualization of attention layer weights for EMR analysis model with different feature extractors.

The visualization of the weights can be obtained after using Softmax to normalize the average weights of the attention layers during the model testing, as shown in Fig. 2. Figure 2 also shows weight ratio in models with different feature extractors.

As seen from the weight ratios in Fig. 2, the multi-source data used in this work all contributes to the analysis process of the model. Especially, the radiology reports reach the highest average weight ratio in different feature extractors. This is probably due to the long length of the text in the radiology reports, which contain a large amount of information. While the lower ratio of medical history is due to the short length of the text and the relatively limited information contained in the medical history. The Text-GCN and BERTGCN have a lower weight ratio on radiology reports, which may be caused by the poor performance of GCN in text classification with a small vocabulary. Compared with them, LSTM and Bi-LSTM have a much higher weight ratio on radiology reports, and they are better at capturing the dependencies of words in long texts.

Overall, physical examination, clinical indicators and radiology reports are three most important data sources, which are also in line with the reality of spine surgeons in analyzing patients' diseases.

4   Conclusion

In this work, we propose a diagnostic prediction method for CSM based on EMRs, which integrates unstructured text data and structured numerical data. In our experiments, the diagnostic accuracy for CSM reached 91%. In addition, we also study the weighting ratio of each data item in the analysis process by visualizing the attention layer weights in the model, which provides interpretability for the diagnostic results of the model, helping to reveal the influence of different data sources on the diagnosis results. In future research, this method can be further improved according to the characteristics of EMR data, including the way of fusion of unstructured and structured data and more refined disease diagnosis.

References

1. Devlin, J., Chang, M.W., Lee, K., Toutanova, K.: Bert: pre-training of deep bidirectional transformers for language understanding. arXiv:1810.04805 (2018)
2. El-Sappagh, S., Elmogy, M., Riad, A., Zaghlol, H., Badria, F.A.: Ehr data preparation for case based reasoning construction. In: Proceedings of International Conference on Advanced Machine Learning Technologies and Applications, pp. 483–497 (2014)
3. Huang, Z., Xu, W., Yu, K.: Bidirectional LSTM-CRF models for sequence tagging. arXiv:1508.01991 (2015)
4. Jia, Z., Zeng, X., Duan, H., Lu, X., Li, H.: A patient-similarity-based model for diagnostic prediction. Int. J. Med. Inform. **135**, 104073 (2020)

5. Johnson, R., Zhang, T.: Deep pyramid convolutional neural networks for text categorization. In: Proceedings of Annual Meeting of the Association for Computational Linguistics, pp. 562–570 (2017)

6. Lin, Y., et al.: Bertgcn: transductive text classification by combining gcn and bert. In: Findings of the Association for Computational Linguistics: ACL-IJCNLP, pp. 1456–1462 (2021)

7. Nakashima, H., et al.: Validity of the 10-s step test: prospective study comparing it with the 10-s grip and release test and the 30-m walking test. Eur. Spine J. **20**(8), 1318–1322 (2011)

8. Rajkomar, A., et al.: Scalable and accurate deep learning with electronic health records. NPJ Digit. Med. **1**(1), 1–10 (2018)

9. Shi, X., Chen, Z., Wang, H., Yeung, D.Y., Wong, W.K., Woo, W.C.: Convolutional LSTM network: a machine learning approach for precipitation nowcasting. Adv. Neural Inf. Process. Syst. **28**, 802–810 (2015)

10. Wu, Y., Zhang, Y., Wu, J.: Configurable in-database similarity search of electronic medical records. In: Proceedings of International Conference on Web Information Systems and Applications, pp. 62–73 (2021)

11. Yao, L., Mao, C., Luo, Y.: Graph convolutional networks for text classification. In: Proceedings of AAAI conference on Artificial Intelligence, vol. 33, pp. 7370–7377 (2019)

12. Zaremba, W., Sutskever, I., Vinyals, O.: Recurrent neural network regularization. arXiv:1409.2329 (2014)

13. Zheng, S., Liang, G., Chen, J., Duan, Q., Chang, Y.: Severity assessment of cervical spondylotic myelopathy based on intelligent video analysis. IEEE J. Biomed. Health Inform. **26**(9), 4486–4496 (2022)

Query Processing and Algorithm

Weighted Cost Model for Optimized Query Processing

Xiaorui Qi[1], Minhui Wang[2], Yanlong Wen[1(✉)], Haiwei Zhang[1],
and Xiaojie Yuan[1]

[1] College of Computer Science, Nankai University, Tianjin, China
qixiaorui@mail.nankai.edu.cn, {wenyl,zhhaiwei,yuanxj}@nankai.edu.cn
[2] China Mobile Communication Group Tianjin Co., Ltd., Tianjin, China
wangmh_nk@126.com

Abstract. Query processing is one of the most commonly used database procedures as well as a significant criterion for evaluating database performance. A key study direction in the database field is how to optimize queries and enhance the efficiency of database queries. The cost-based optimizer (CBO) is the current industry standard. Traditional cost models just sum all the cost factors together to get the overall cost, which makes it impossible to thoroughly assess queries and match customers' expectations. This paper proposes a **W**eighted **C**ost **M**odel (WCM) that emphasizes the balance of different cost factors. Constants and operators that fit WCM are also rewritten. We discuss the correlation between transmission cost and other factors by introducing it as a new factor. Then, we simulate the transmission cost using the correlation we discovered. Finally, we integrate WCM into Apache Calcite's Cascade-style query optimizer framework. In a variety of systems, we test WCM using both real data sets and a virtual TPC-H environment. Our experimental results show that WCM is more stable and performs better than (or equal to) Calcite in 90% of systems, with an 18-fold optimization on a single query and a 2.3-time improvement in the virtual TPC-H environment.

Keywords: Weighted cost model · Query optimization · Apache Calcite

1 Introduction

Query processing is the most important aspect of database technology. The query processing performance will have a direct impact on the user's appraisal of the entire database system, and the query technique used will directly affect the query time and execution cost. In the actual use procedure, complex queries are unavoidable. Improper processing will cause a significant drop in database performance, resulting in increased demand for database query optimization technology research.

CBO, which analyzes query costs to find individual execution plans with the lowest cost weights, is currently the most widely used technology. For a classic centralized database, the cost model primarily considers CPU and I/O costs.

Though this idea of "*ONE SIZE FITS ALL*" features of a single model [11] has strong generalization ability and coverage of query conditions, it comes at the expense of query accuracy, resulting in poor performance of the database's prediction results in increasingly complicated query contexts. To avoid costly physical join schemes, [13] introduces partitioning budgeting strategies that guarantee monotonous behavior and integrate boundaries into the optimizer framework. [14] focuses on multi-objective parametric query optimization and proposes the piecewise linear plan cost function and algorithm. In the exponential planning space, schemes are limited by high enumeration costs. Those based on dynamic programming necessitate a lot of computing power, whereas those based on heuristics are ineffective due to accuracy and precision issues.

As a result of the key issues with query optimization technology, this paper proposes WCM that incorporates models from current database solutions. To match WCM, we recast the majority of the constants and operators. We also introduce transmission cost as a new factor, simulating it with constants and operators based on the correlations we discovered. With both optimizations on a single query and in the virtual environment, WCM demonstrates a significant increase in experimental outcomes.

The main contributions of this paper are:

(1) We propose WCM, a weighted cost model that pays more attention to the balance between different cost factors and can evaluate query performance comprehensively.
(2) We implement the rewrite of cost constants and operators for WCM, which makes the cost evaluation faster and more accurate.
(3) We introduce transmission cost as a new factor and simulate its generation. It achieves effective improvement on Apache Calcite middleware eventually.

2   Related Work

Optimizer. In each stage of database development, the query optimizer has stayed largely constant, and two model design methods have reigned. One is the Cascades [1]/Volcano [2] framework-based top-down design style. The other is a bottom-up approach based on dynamic programming [3]. More and more teams and businesses are utilizing or will soon be using the latter, adapting the framework to specific conditions. Most domestic database solutions, such as TiDB [4] and OceanBase [5], are migrating to this architectural paradigm, while open source optimizers like Apache Calcite [6] and Orca [7], being Cascades' follow-up research, remain active in academic and industrial circles.

Cost Model. Traditional cost models calculate the overall cost of query processing to be the sum of CPU and I/O costs. To forecast query performance, cost evaluation often uses formula modeling based on cardinality estimation and projected space statistics. TiDB keeps track of statistics to help with index selection. OceanBase is similar except it concentrates on the table (row) and column information, whereas TiDB focuses on the column and index. Furthermore, [8,9] produce specific physical costs using a standard histogram fusion genetic technique, while [10] uses particle swarm optimization to find proper weights. [12] performs adaptive sorting and optimization using the TiDB star model, sorting with a filter missing rate. [15] is based on machine learning, paying more attention to aggregate queries with user defined functions. [16] focuses on the setting of reward value $r(s_i, a_i)$, producing improved optimization strategies by querying a_i parameters under the updated condition s_i, according to reinforcement learning techniques. [17] investigates the integration of different operators across heterogeneous hardware platforms and develops a more precise cost expression.

3   Weighted Cost Model

In this section, we go through WCM in depth, including its definition, constant and operator rewriting rules, and how we may use it to improve queries.

3.1   Definition

Consider the classic cost model before moving on to WCM. Existing cost models mostly concentrate on three costs: CPU, I/O, and simple cardinality estimation (e.g., row count). Equation 1 shows the default cost evaluation formula:

$$C_{old} = c_{CPU} + c_{I/O} + c_{card} \tag{1}$$

where c_{card} represents cost of cardinality estimation. The formula demonstrates that the three basic costs are of equal relevance, i.e., the total cost is just a simple sum. This approach to cost evaluation is used in many publications, including Apache Calcite, but its interpretation is questionable. From a subjective standpoint, users have varied preferences for basic costs in different query contexts, so it should not be assumed that all basic costs are equally important. From an objective standpoint, the hardware environment will have a particular inclination to the fundamental cost across time and query circumstances. Intensive computer jobs, for example, will trend toward CPU optimization, while massive data tasks will tend toward I/O optimization, and so on.

We keep the three basic costs in WCM while adding new cost factors and adjusting the weight of the basic cost to respond to different query conditions. The traditional cost model is favorably correlated with WCM, as follows:

$$C_{old} = \alpha \cdot c_{CPU} + \beta \cdot c_{I/O} + \chi \cdot c_{card} \tag{2}$$

$$C_{new} \propto C_{old} + \delta \cdot c_{others} \tag{3}$$

where $\alpha, \beta, \chi, \delta$ are all parameters. c_{others} refers to all other cost factors except for the three basic costs. We will go over it in Sect. 4 later. The weight of the CPU cost is ω_{CPU}, normalized as:

$$\omega_{CPU} = \frac{\alpha}{\alpha + \beta + \chi + \delta} \times 100\% \tag{4}$$

The other three weights, $\omega_{I/O}, \omega_{card}, \omega_{others}$ can be obtained in the same way. To adapt to varied query conditions, weights are employed to balance the importance of each basic cost. Using WCM with N basic costs as an example, the total query cost is calculated as the product sum of all basic costs and their respective weights. The following is a summary of the formula:

$$C_{new} = \Sigma_i^N \omega_i \cdot c_i \tag{5}$$

3.2   Constants

Cost comparison is a part of cost evaluation. As for cost evaluation indicators, a well-designed cost model should provide preset cost constants. We define rules to rewrite several types of constants, such as zero, unit, small, medium, large, infinite (unreachable), and so on. The cost evaluation error ε, in particular, remains fixed at 1.0×10^{-5} by default.

The zero constant sets all basic costs to 0, including information like cardinality that should never be 0. We can get the smallest value when the cost is set to the zero constant, since the cost has no negative value.

The unit constant Δ_c sets the unit span of cost evaluation in the model. The value can be adjusted according to the cost granularity, for example, ε, 1, and so on.

As indicated in the equation below, the small constant established all of the basic costs, which are measured in unit constant as the basic unit. Where f is a variable that can be changed to fit various conditions. For example, if the unit constant is 1 ms and the delay acceptable to users is 1 s, the cost reaches a small constant when $f = 1.0 \times 10^3$, which is within the acceptable range of users.

$$C_{tiny} = f \cdot \Delta_c \tag{6}$$

The large constant is determined in the same way, and all basic costs are calculated using the formula above. To provide a consistent spread between the cost constants, the unit constant in the formula can be replaced by the small constant. Otherwise, the parameter should be changed to f' to calculate using the unit constant.

$$C_{huge} = g \cdot C_{tiny} = f' \cdot \Delta_c \tag{7}$$

The medium constant lies between the small and large constants, calculated from both. In general, averages and medians based on statistical information are two possible solutions.

The infinite constant ∞ is an endless constant that sets different fundamental costs to infinity. The query cost has reached its limit, and it cannot be further optimized. After marking, the query will be removed from the optimization list.

3.3   Operators

Calcite has four types of arithmetic operators and five types of comparison operators, for which rewriting rules are provided below. The original definitions of addition, subtraction, and multiplication remain unchanged in the arithmetic operator section, which calculates each base cost independently. The division method has been changed from geometric average to traditional division.

The comparison operator section is the same, with Formula 5 changing all the right-hand sides of each operator equation. In the situation of approaching equality, the following is the definition:

$$c_A \simeq c_B \rightarrow \left| \Sigma_{i=1}^{n_A} \omega_{A_i} \cdot c_{A_i} - \Sigma_{i=1}^{n_B} \omega_{B_i} \cdot c_{B_i} \right| < \varepsilon \tag{8}$$

Other operator definitions are similar. Because of the peculiarity of infinity, the operator remains unaltered when the cost fits the criterion of tending to infinity. Otherwise, use Formula 5 to see if the cost increases to infinity after the sum is calculated.

3.4 Evaluation Framework

This section will show the overall framework for cost evaluation (see Fig. 1 below). The basic cost information (CPU, I/O, etc.) and other cost information (network, etc.) are counted from the query and syntax tree, respectively.

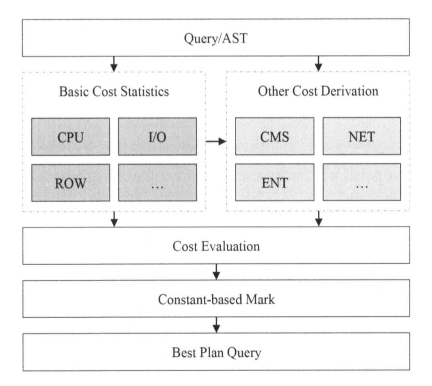

Fig. 1. Overall framework of cost evaluation. In the figure, CMS, NET and ENT represent **C**ount-**M**in **S**ketch, a conflict adjustment data structure of TiDB, **NET**work transmission quality and **E**ffective **N**umber of **T**ransmissions.

Other cost information is influenced by external elements like data and hardware, and is mapped from basic cost information to subsets of other cost sets using equations such as Formula 9 (Sect. 4.1).

4 Transmission Cost Factor

The transmission cost factor has now been introduced. We begin by looking at the correlation between this factor and others. Then, using the correlation we discovered, we illustrate how to simulate the transmission cost. Finally, we'll go over how to include this cost factor into WCM and Apache Calcite.

4.1 Correlation

In the previous section, c_{others} is described as cost factors other than the three fundamental costs in traditional cost models. Specifically defined as, given the cost set $C = \{c_1, c_2, ..., c_N\}$ and its corresponding weight set $W = \{w_1, w_2, ..., w_N\}$, (c_{others}, C, W) satisfies Eq. 5 to calculate the cost.

Transmission costs are a class of cost in the cost set c_{others} that have a significant impact on database query speed, especially in distributed setups. The existing transmission scenarios are constrained by a number of factors that make it impossible to objectively reflect the prevalence of problems. We'll talk about more generic transmission scenarios. Many factors influence the cost of transmission between nodes. In general, the transmission start time is determined by a single node's processing efficiency, which includes CPU performance and I/O ratio, while the transmission cost is proportional to data quantity and network quality. The following equation describes the cost of transmission:

$$c_{trans} \propto c_{CPU}, c_{I/O}, c_{card}, c_{net}, ... \tag{9}$$

where c_{net} represents the network transmission quality, other factors have been described previously. In practice, we look for different mapping functions for the above factors individually or in groups to simulate the transmission cost. (e.g., $F(c_i), G(c_i, c_j, ...)$, where F, G are functions selected and c_i, c_j for different factors) The selection of mapping functions has a significant impact on the simulation results.

4.2 Simulation

Now we'll use Calcite's predefined cost factors to simulate the transmission costs. The network quality is reduced to a constant when the transmission boundary condition is considered. The transmission is then calculated by mapping different costs together. Formula 9 can be summarized as follows:

$$c_{trans} \rightarrow F(c_{CPU}, c_{I/O}, c_{card}) \tag{10}$$

where F denotes a mapping function to calculate transmission cost from three basic factors above. Considering the nonlinear relationship among all kinds of costs affecting transmission cost, we select geometric average as the mapping function of WCM to simulate transmission cost, which goes like this: $\sqrt[n]{\Pi_{i=1}^n c_i}$

4.3   Integration

This part demonstrates how to put it all together and integrate it into Calcite. The constant adjustments are as follows: the small constant is calculated using $f = 1.0 \times 10^3, \Delta_c = \varepsilon = 1.0 \times 10^{-5}$, whereas the huge constant is calculated using $f = 1.0 \times 10^5, \Delta_c = 1.0$. If users are willing to accept a query latency of $2\,\text{min}$, the latter can better represent the large cost value than the initial cost constant setting in Calcite, rather than being transformed to the infinity constant due to the large value.

WCM is directly integrated into the Volcano Planner of Calcite. According to Formula 5, the cost evaluation is instantiated into the following:

$$\omega_{CPU} \cdot c_{CPU} + \omega_{I/O} \cdot c_{I/O} + \omega_{card} \cdot c_{card} + \omega_{trans} \cdot c_{trans} \qquad (11)$$

To simplify the model, WCM defaults the base cost to other cost ratio to 1, and the parameters $\omega_{CPU}, \omega_{I/O}, \omega_{card}$, and ω_{trans} to 0.3, 0.1, 0.1, and 0.5 roughly, according to the available statistics [11].

5   Experiment

This section is divided into three parts to show the evaluation of WCM. To begin, we show how to integrate multiple systems, such as Babel and Cassandra, with the Apache Calcite query optimization framework, with a great majority (90%) of the systems experiencing reduced runtimes. Then we look at individual query optimization examples from the H2, MySQL, and PostgreSQL database systems, achieving an 18-fold increase in query optimization. Finally, we describe the performance of WCM in the virtual TPC-H environment. TPC-H query performance is enhanced by 2.3 times.

5.1   Experimental Settings

Dataset. Foodmart contains 164, 558 instances of 1998 sales data facts and 86, 837 of 1997 generated by Foodmart Corporation between 1997 and 1998. In the virtual TPC-H environment, we use data on a certain scale, including 11, 745, 000 inventory data and 2, 880, 404 store-sales data and so on.

Approach. The experiment is carried out in the HOST+VM environment, with 16GB and 4GB DIMM memory and 256GB and 40GB SSD allocated respectively. On the basis of the correct query results, we judge WCM by the length of execution time. For test points with similar query time before and after optimization, we will look at specific query plan generated.

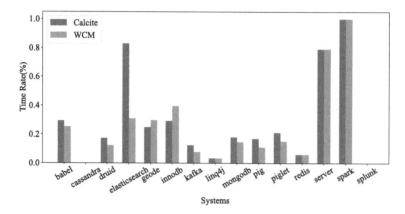

Fig. 2. Tests of weighted cost model under systems integration. Please note that Cassandra and Splunk are not empty, as they are too short to show up in the diagram. The former is 0.003/0.002 and the latter 0.001/0.001.

5.2 Integration Tests

Fig. 2 depicts the results of tests on over 10 systems integrating the Apache Calcite query optimization framework, including Babel, Cassandra, Druid, etc. According to the findings of the experiments, over 90% of the systems using WCM perform better than (or equal to) the original Calcite framework.

Under the original Calcite framework, Elasticsearch, Kafka, and Pig run for 4 min 39.55 s, 41.79 s, and 56.706 s, respectively, while WCM runs for 1 min 44.84 s, 25.956 s, and 36.94 s. The performance increased by 166.64%, 61.00%, and 53.51%. Ling4j, Redis, and Spark execute under WCM at the same speed as under Calcite.

Further, we find that there are certain features between systems of similar performance. The better performing systems all support distributed processing and are mostly based on Hadoop framework. Meanwhile, those performing less well are mostly memory based in the key technologies.

Experiments have shown that WCM can be well adapted to most data types. Whether it is JSON data for Elasticsearch, streaming data for Kafka or query languages such as CQL and SQL, the integration of WCM can achieve a faster execution time.

It's worth noting that even after integrating WCM, a tiny fraction of systems still scores poorly. Runtimes of Geode and InnoDB are 1 min 23.94 s and 1 min 38.5 s with the old Calcite framework, respectively, while 1 min 40.42 s and 2 min 13.41 s under WCM, indicating a performance drop. Taking the Geode system as an example, the specific test results are shown in Table 1:

Table 1. Geode test results

Tests	Size	Calcite(s)	WCM(s)
Geode.Bookstore	36	43.837	**27.399**
Geode.Zips	14	**20.216**	32.895

In the Bookstore test, WCM takes less time to run, but it performs poorly in the Zips test. The given result demonstrates that WCM has limits and does not cover all data types completely.

5.3   Specific Query Optimization Instances

Table 2 stores the results of H2, MySQL, and PostgreSQL in the same JDBC test environment. There are 332 test points in total. Results show that WCM outperforms the original Calcite framework in all three systems. PostgreSQL, in particular, reduces runtime by approximately 6 s and improves performance by 10.40%.

Table 2. H2, MySQL & PostgreSQL test results

Database	Calcite(s)	WCM(s)	Boost Rate(%)
H2	83.510	**82.000**	1.841
MySQL	52.836	**52.390**	0.851
PostgreSQL	57.097	**51.719**	10.398

We now separate the queries into simple (e.g., ordered, single-value selection, etc.) and complex (e.g., join, etc.). Instances of the three database systems are depicted in Fig. 3:

Part (a) (c) (e) of Fig. 3 describes the optimization of 66 simple query instances in total in three database systems. WCM performs better in all three, with a maximum 18-fold increase in H2, 9-fold in MySQL, and 15-fold in PostgreSQL.

The rest of Fig. 3 describes the performance of 9 complex query instances. WCM supports common complicated queries such as join (e.g., joinManyWay), long text (e.g., text), and complex conditions (e.g., havingNot). Data show that WCM has a lower intensity in the complex query optimization than in the simple condition.

Let us see the same test point in different database systems. Table 3 selects one query instance from each of the three systems and compares it with the other two. The results show that WCM can optimize with the same intensity in different systems, while the original Calcite framework fluctuates greatly.

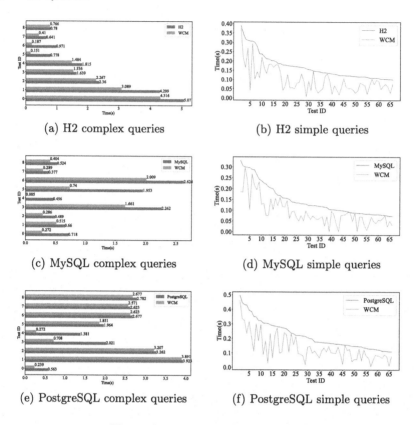

Fig. 3. Query optimization instances

5.4   TPC-H Analysis

Table 4 above shows the results of WCM in the virtual TPC-H environment. The results show that WCM takes 1 min 47.49 s in TPC-H queries, which is 4 min less than and improves performance by 2.3 times over the original Calcite framework. The other two tests, Enumerable and Basic, also perform well on the Foodmart dataset, with effective time reductions of 1 s and 3 min, and performance improvements of 3.57% and 44.29%, respectively.

Table 3. The same test point in different database systems

Tests	H2		MySQL		PostgreSQL	
	Calcite(s)	WCM(s)	Calcite(s)	WCM(s)	Calcite(s)	WCM(s)
modelWithModifableView	0.390	**0.376**	1.953	**0.740**	2.021	**0.708**
orderUnionStarByExpr	0.092	**0.054**	0.287	**0.215**	1.381	**0.272**
winRowNumber	0.224	**0.162**	0.524	**0.404**	0.446	**0.335**

Table 4. Results of virtual TPC-H environment

Tests	Size	Calcite(s)	WCM(s)	Boost Rate(%)
TPC-H	29	353.97	**107.49**	229.31
Enumerable	111	8.12	**7.84**	3.57
Basic	4390	560.10	**388.18**	44.29

6 Conclusion and Future Work

This paper underlines query optimization technology's central role in the database field, as well as the fact that the cost model is a vital component of query optimization technology. Our major task is to design and implement a new **W**eighted **C**ost **M**odel (WCM) based on the Apache Calcite open-source database query optimizer framework middleware. We model transmission costs as a new cost factor and integrate them into Calcite on this basis. In a large majority of systems, WCM performs better than Calcite. However, there are still many problems worth studying and exploring in database query optimization technology based on cost evaluation. Dynamic design for richer cost factors includes two tasks. One is to explore and design new cost factors. The other one is to rely on new algorithms to form a dynamic cost model. With the new hardware platforms changing the design concept of databases and the fierce development of machine learning, we believe that optimization based on new hardware and learning-based methods for estimation and evaluation are also two questions worth studying in the future.

Acknowledgments. This research is supported by Chinese Scientific and Technical Innovation Project 2030 (No. 2018AAA0102100), National Natural Science Foundation of China (No. 62077031). We thank the reviewers for their constructive comments.

References

1. Graefe, G.: The Cascades framework for query optimization. Data Eng. Bull. IEEE **18**(3), 19–28 (1995)
2. Graefe, G., Mckenna, W.J.: The Volcano optimizer generator: extensibility and efficient search. In: International Conference on Data Engineering, pp. 209–218. IEEE (1993)
3. Selinger, P., Astrahan, M., Chamberlain, D., Lorie, R., Price, T.: Access path selection in a relational database management system. In: Readings in Artificial Intelligence and Databases, pp. 511–522 (1979)
4. PingCAP. TiDB. https://pingcap.com/zh/product. Accessed 9 Mar 2022
5. Alibaba. OceanBase. https://www.oceanbase.com. Accessed 9 Mar 2022
6. Begoli, E., Rodriguez, J.C., Hyde, J., Mior, M.J., Lemire, D.: Apache calcite: a foundational framework for optimized query processing over heterogeneous data sources. In: Proceedings of the 2018 International Conference on Management of Data, pp. 221–230 (2018)

7. Soliman, M.A., et al.: Orca: a modular query optimizer architecture for big data. In: Proceeding of the 2014 ACM SIGMOD International Conference on Management of Data, pp. 337–348 (2014)
8. Nie, X.X.: Optimization design and implementation of ecommerce platform server performance. A Master Thesis Submitted to University of Electronic Science and Technology of China (2020)
9. Liao, X.: Design and implementation of query optimization module for distributed column database based on memory. A Master Thesis Submitted to University of Electronic Science and Technology of China (2021)
10. Xu, Q., Wu, S.: Improving medical record search performance by particle swarm optimization based data fusion techniques. In: Xing, C., Fu, X., Zhang, Y., Zhang, G., Borjigin, C. (eds.) WISA 2021. LNCS, vol. 12999, pp. 87–98. Springer, Cham (2021). https://doi.org/10.1007/978-3-030-87571-8_8
11. Siddiqui, T., Jindal, A., Qiao, S., Patel, H., Le, W.C.: Cost models for big data query processing: learning, retrofitting, and our findings. In: Proceedings of the 2020 ACM SIGMOD International Conference on Management of Data, pp. 99–113 (2020)
12. Fan, M.: Optimization of query algorithm for distributed relational database. A Master Thesis Submitted to University of Electronic Science and Technology of China (2020)
13. Cai, W., Balazinska, M., Suciu, D.: Pessimistic cardinality estimation: tighter upper bounds for intermediate join cardinalities. In: Proceedings of the 2019 International Conference on Manage of Data. SIGMOD, pp. 18–35 (2019)
14. Trummer, I., Koch, C.: Multi-objective parametric query optimization. In: Proceedings of the 2014 International Conference on Very Large Data Bases, vol. 8, no. 3, pp. 221–232 (2014)
15. Duan, Y., Zhang, Y., Wu, J.: Database native approximate query processing based on machine-learning. In: Xing, C., Fu, X., Zhang, Y., Zhang, G., Borjigin, C. (eds.) WISA 2021. LNCS, vol. 12999, pp. 74–86. Springer, Cham (2021). https://doi.org/10.1007/978-3-030-87571-8_7
16. Yu, X., Chai, C.L., Zhang, X.N., Tang, N., Sun, J., Li, G.L.: AlphaQO: robust learned query optimizer. Ruan Jian Xue Bao/J. Softw. **33**(3), 814–831 (2022). (in Chinese). http://www.jos.org.cn/1000-9825/6452.htm
17. Tu, Y.F., Chen, X.Q., Zhou, S.J., Bian, F.S., Wu, F., Chen, B.: Geno: cost-based heterogeneous fusion query optimizer. Ruan Jian Xue Bao/J. Softw. **33**(3), 774–796 (2022). (in Chinese). http://www.jos.org.cn/1000-9825/6441.htm

A Data Dimensionality Reduction Method Based on mRMR and Genetic Algorithm for High-Dimensional Small Sample Data

Yong Ji[1], Jun Li[1], Zhigang Huang[1], Weidong Xie[2], and Dazhe Zhao[2(✉)]

[1] Neusoft Corporation, Shenyang, China
[2] Northeastern University, Shenyang 110000, Liao Ning, China
zhaodazhe@mail.neu.edu.cn

Abstract. With the development of microarray sequencing technology, researchers can obtain expression data of a large number of genes or proteins from patients at one time for analysis of biomarkers that cause disease. However, limited by the number of patient cohorts, the number of samples is usually small, so this type of data is often referred to as high-dimensional small sample data, also known as microarray data. In order to effectively select valid biomarkers, effective dimensionality reduction of the data is essential for further analysis. This paper proposes a two-stage feature selection method for data dimensionality reduction. The proposed method first improves two quantization functions of Max-Relevance and Min-Redundancy (mRMR) to make it applicable to microarray data for initial dimensionality reduction of the data. Subsequently, the improved genetic algorithm is used for further dimensionality reduction of the data. The proposed method combines the growth tree clustering algorithm with the genetic algorithm's selection and crossover process to improve the crossover efficiency. In addition, we combine the feature recursive elimination module with the genetic algorithm iteration process for further dimensionality reduction of the data. The proposed method is demonstrated to be effective and advanced by conducting comparative experiments on four publicly available data.

Keywords: Feature selection · Genetic algorithm · Microarray · mRMR

1 Introduction

In the field of bioinformatics, with the development of DNA microarray technology, researchers can obtain a large amount of gene or protein expression data of patients at one time. Further analysis of these data can effectively identify biomarkers that may cause a specific disease and then achieve downstream tasks such as disease diagnosis and drug development [1]. However, this type of data is

Supported by Software foundation of the Ministry of industry and information technology of China (Grant No. 2105-370171-07-02-860873).

usually characterized by high feature dimensionality and a small sample size due to sample size limitations. Direct analysis of such data can lead to low accuracy and poor robustness of machine learning models due to feature noise. Therefore, feature selection before data analysis and feature dimensionality reduction of data is essential for the subsequent model building task [2].

Generally speaking, feature selection techniques can effectively reduce data dimensionality, remove redundant features, and ensure that the acquired low-dimensional data still have feature interpretability [3]. According to different processing processes, standard feature selection methods can be divided into Filter, Wrapper, and Embedded methods [4]. The Filter method is based on a statistical information test or information entropy, etc. The technique is simple to implement, fast to process, and usually used for the initial filtering of features [5]. The wrapper method can find the best combination of features based on a given classifier by a heuristic optimization algorithm, achieving high model classification accuracy [6]. The embedded method outputs feature weights by analyzing the degree of feature contribution in the machine learning model construction process with high classification accuracy and moderate processing speed [7].

Researchers currently widely adopt hybrid methods because they combine the advantages of standard feature selection [8]. Hybrid methods usually employ a combination of the processing efficiency advantages of the Filter method and the accuracy advantages of the Wrapper method, using the Filter method for the initial filtering of features, followed by the Wrapper method for further feature selection. The features selected by the hybrid approach have been proven to have high classification accuracy and robustness. The algorithm has low complexity, so it is widely used as a dimensionality reduction method for high-dimensional data.

However, there are few papers on filter processing methods that consider the continuous characteristics of microarray data attributes in the current research. In contrast, the efficient search strategy and the effective feature number constraint method in the search process are easy to ignore in wrapper methods [9]. Considering the continuity of microarray data attributes and borrowing the idea of Peng et al. [10] to embed a feature recursive elimination module in a genetic algorithm, this paper proposes a two-stage hybrid feature selection algorithm for data dimensionality reduction. In the first stage of the proposed method, we improve the two quantization functions used in the mRMR method for quantizing between features and between features and labels to make it more applicable to microarray data with continuous attributes. In the second stage, the improved genetic algorithm is used for further feature selection. The proposed method uses a growing tree clustering algorithm to improve the selection and crossover process in the original genetic algorithm to improve search accuracy and efficiency. It combines the recursive feature elimination (RFE) process into the algorithm iteration process for eliminating poorly adapted chromosomes. The main innovations of the proposed approach are as follows.

1. Improved the two quantization functions used by the mRMR method for quantifying inter-feature and feature-label relationships to make it applicable to attribute-continuous microarray data for coarse-scale feature selection.

2. The growth tree clustering method is used in the selection and crossover process of the genetic algorithm to improve the accuracy and efficiency of the algorithm, and the RFE process is embedded in the iteration of the genetic algorithm to achieve feature selection.

3. A two-stage hybrid feature selection method is proposed. Experiments are conducted on four publicly available datasets, comparing traditional machine learning algorithms with advanced algorithms, and demonstrating the effectiveness and advancement of the proposed method.

2   Related Work

Researchers have widely adopted hybrid feature selection methods due to low algorithm complexity, low redundancy of selected features, and the ability to build accurate classification models effectively [4]. Gunavathi proposed a hybrid feature selection algorithm based on genetic algorithm, which first uses t-test, signal-to-noise ratio, and F-test to select a subset of features, and then applies KNN and SVM as genetic algorithm of the fitness function for further feature selection, which finally achieves an accurate tumor classification task [11].

Pardo et al. proposed an integrated feature selection strategy for feature selection of microarray data, which fuses the results of different feature selection methods for feature importance assessment and produces a unified ranking that is finally input to an SVM model for classification accuracy prediction. The results on seven publicly available datasets show that the proposed method has a high classification accuracy [12].

Wang et al. combined the Markov blanket with an improved sequential forward selection method (SFS). The Markov blanket was used as the feature selection method in the first stage, which fully considers feature dependencics and can effectively reduce the number of features. The SFS method was used in the second stage for further feature selection. The method's effectiveness was demonstrated on ten publicly available microarray data, and the technique reduced the algorithm's time complexity [13].

Theera et al. proposed a hybrid feature selection method based on neuro-fuzzy and firefly algorithms, where neuro-fuzzy can select good feature sets and generate rule sets as classifiers. Firefly algorithm was used for further feature selection, which proved effective and provide more comprehensible rule sets [14].

Musheer et al. effectively combined the independent component analysis (ICA) algorithm with the artificial bee colony (ABC) algorithm to propose a hybrid feature selection algorithm. ICA can extract the same number of features as the number of samples to provide a subset of features for further feature selection for the ABC algorithm, which used a Naive Bayesian (NB) as a classifier and was used in six benchmark datasets was proved to be effective [15].

Ram et al. proposed a quantum-inspired genetic algorithm (QIGA) for feature selection of high-dimensional small sample data, which combined with SVM can effectively reduce the data dimensionality and obtain high classification accuracy. The authors present a detailed comparative analysis of the proposed method with traditional genetic algorithms to demonstrate the effectiveness of the QIGA algorithm [16].

3   Proposed Method

In this section, we describe the proposed data dimensionality reduction algorithm, hereafter collectively referred to as the feature selection algorithm, whose overall framework is shown in Fig. 1.

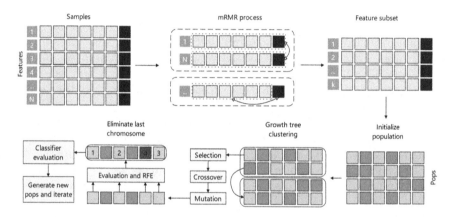

Fig. 1. The overall framework of the proposed method. Filtering features is first achieved using the mRMR method to quantify feature-to-feature and feature-to-label relationships. The subset of features after initial filtering is input to the genetic algorithm for further feature selection. In the improved genetic algorithm, the individuals are first clustered using the growth tree algorithm and undergo selection, crossover, and variation processes. Finally, the optimal individuals are further evaluated, and the lowest-scoring chromosomes are eliminated using the RFE mechanism and enter into a new iteration process, which finally realizes the dimensionality reduction process of the original data.

In the proposed method, the raw data is first quantified using the improved mRMR algorithm to quantify the relationship between features and the relationship between features and labels. Redundant features are filtered based on the relationship between features. Features that are not relevant to the labels are filtered based on the quantified relationship between features and labels to obtain the initial filtered feature subset, which is the first step of the proposed method. In the second stage, an improved genetic algorithm is used for a further selection of features by first initializing the population based on the feature subset, clustering individuals according to the growth tree algorithm, and performing selection, crossover, and mutation processes. Subsequently, the chromosomes (features) corresponding to the best individuals are evaluated, and the lowest-scoring chromosomes are eliminated. Then the classification accuracy of the current individuals is evaluated, and a new population is generated for the next cycle. We describe each improvement method in detail below.

The basic principle of the traditional mRMR algorithm for feature selection is to minimize the redundancy between features and maximize the correlation between features and labels, and the specific quantization functions are shown in Eq. 1 and Eq. 2, respectively.

$$\max A(S, C) = \frac{1}{n} \sum_{f_i \in S} I(f_i, C) \tag{1}$$

$$\min R(S) = \frac{1}{n^2} \sum_{f_i, f_j \in S} I(f_i, f_j) \tag{2}$$

where, S is the selected feature subset, C is the data label, f_i and f_i are the two features in S. $A(S, C)$ denotes the correlation between the selected feature subset and the label, $R(S)$ denotes the redundancy between the selected feature subset, and $I(*)$ denotes the mutual information.

Therefore, in the proposed method, we propose a new quantification function, as shown in Eq. 3 and Eq. 4. Specifically, the proposed method uses Pearson correlation to quantify the redundancy between features and uses t-test to quantify the correlation between features and labels.

$$t(f_i) = \frac{\left| \bar{f}_{i_{pos}} - \bar{f}_{i_{ne.g.}} \right|}{\sqrt{S^2_{i_{pos}}/n_{pos} + S^2_{i_{ne.g.}}/n_{ne.g.}}} \tag{3}$$

$$\rho(f_i, f_j) = \frac{\sum (f_i - \bar{f}_i)(f_j - \bar{f}_j)}{\sqrt{\sum (f_i - \bar{f}_i)^2 \sum (f_j - \bar{f}_j)^2}} \tag{4}$$

where $\bar{f}_{i_{pos}}$ denotes the mean value of feature f_i in the positive sample, $\bar{f}_{i_{neg}}$ denotes the mean value of feature f_i in the negative sample; $S^2_{i_{pos}}$ and $S^2_{i_{neg}}$ denote the variance of feature f_i in the positive and negative samples, respectively; n_{pos} and n_{neg} denote the number of samples in the positive and negative samples, respectively.

In traditional genetic algorithms, the selection and crossover processes are usually performed with the best individuals, which limits the diversity of the population, is not conducive to the co-production of divergent individuals, and reduces the efficiency of the algorithm and the merit-seeking process. Therefore, we optimized the selection and crossover process of the genetic algorithm based on the idea of growing tree clustering. Specifically, we first used the growth tree clustering algorithm to divide the population into k families. The optimal individuals were evaluated among different families separately, and subsequently, the crossover process was performed among the individuals with higher fitness values from other families. In addition, we used the elite retention strategy as the selection operator. The proposed method can retain the best individuals and ensure the population diversity effectively, and the specific flow of the algorithm is shown in Algorithm 1.

Algorithm 1. Improved mRMR algorithm

Input: Dataset $X \in R^{n \times m}$, labels Y, feature sets F
Output: Filtered feature subset F_s
1: Randomly generate the initial population $Pop = p1, p2, ..., pn$, where n is the number of features
2: Cluster the population into k clusters using a growth-tree based clustering algorithm, each cluster as a family
3: Calculate the fitness of individuals in each family
4: Select individuals with high fitness from different families for crossover, and the newly generated individuals are used as parent individuals
5: Select a number of individuals from the same family for crossover, and the newly generated individuals will be the parents
6: Perform mutation on the population, compare the fitness of the mutated individuals with that of their parents, and keep the individuals with higher fitness
7: If the termination condition is satisfied, the algorithm ends, otherwise go to step2

In addition, considering the number of features that the original genetic algorithm can retain is closely related to the random initialization population. It is difficult to effectively reduce the number of features during the algorithm's selection, crossover, and mutation process. We introduced the RFE module embedded in the iterative process of the genetic algorithm. Specifically, the proposed method further evaluates the optimal individuals after mutation. The employed machine learning model performs a feature evaluation operation, where for each possible feature selected, its importance is evaluated using a classifier and ranked. For the chromosome at the end of the ranking, we consider it has a low contribution to the classifier. Therefore, the chromosome is eliminated in the next operation.

4   Experiment Result

4.1   Datasets

This subsection describes the datasets and evaluation metrics used for the experiments. Four publicly available microarray data were used in our experiment, and the details of these data are shown in Table 1, which are typical of high-dimensional small sample data. Colon is a colon cancer dataset containing 40 tumor samples and 22 normal samples, each containing 2000 genetic information. DLBCL is a lymphoma dataset containing 59 diffuse large B-cell lymphoma

(DLBCL) samples and 59 follicular lymphoma (FL) samples. DLBCL is a lymphoma dataset containing 59 Diffuse Large B-Cell Lymphoma (DLBCL) samples and 19 Follicular Lymphoma (FL) samples, each containing 7070 genetic information. It contains 25 samples for Acute Myeloid Leukemia (AML) and 47 samples for Acute Lymphocytic Leukemia (ALL), each containing 7129 genetic information. Lymphoma is a lymphoma dataset with 22 positive and 23 negative samples, each containing 4026 genetic information.

Table 1. Datasets used in this paper. Ur means unbalanced rate.

Datasets	Samples	Pos	Neg	Features	Ur
Colon	62	40	22	2000	1.82
DLBCL	77	58	19	7129	3.05
Leukemia	72	47	25	7129	1.88
Lymphoma	45	22	23	4026	0.96

4.2 Analysis of Experimental Results

In this subsection, we conducted experiments on the dataset in Table 1 using the proposed method, and the experimental parameters were set as shown in Table 2. The proposed method first uses mRMR for feature filtering and retains 500 features for further feature selection. We then used the proposed improved genetic algorithm to implement further filtering of features until the algorithm stopped after all features were eliminated. Then we select the better adapted corresponding individual as the final feature selection result based on the fitness curve.

Table 2. Experimental parameters.

Parameters	Values
Iterations	250
Pop size	10
Chromosome	Features
Crossover rate	0.6
Mutation rate	0.01
Cluster rate	0.3
RFE rate	0.8

In our experiments, we use a decision tree as a classification model, which has been shown to be effective for classification tasks and in providing feature weights. The proposed method uses the classification accuracy of the dimensionality-reduced data in the decision tree model to judge the effect of dimensionality reduction, i.e., the feature selection effect. The average classification accuracy of the five-fold cross-validation on the four datasets is shown in Fig. 2, and we also counted the ROC curves in Fig. 3. The results show that the proposed method has good classification performance on the four publicly available microarray data, and the model is stable and robust.

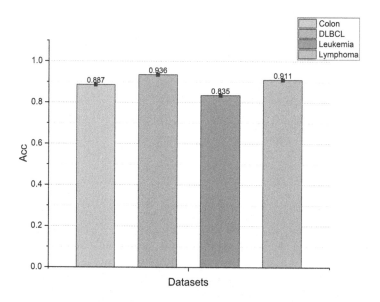

Fig. 2. The classification accuracy of the proposed method in different data sets

We also plotted the change curve of the number of feature iterations versus the fitness function, and the experimental results are shown in Fig. 4.

The horizontal axis represents the number of iterations and the vertical axis represents the average classification accuracy of five-fold cross-validation. In the proposed method, the number of features keeps decreasing with iterations. We selected individuals with high fitness as the final feature selection results. The percentage decrease in the number of features and the percentage increase in classification accuracy from the beginning of the iteration nodes are labeled in the figure, which shows that the proposed method can reduce the data dimensionality by 88% to 99% and also improve the classification accuracy by 3.7% to 7.9% on the four publicly available microarray datasets. The result shows that the proposed method effectively reduces the data dimensionality, performs effective feature selection, removes redundant features, and improves the model classification accuracy.

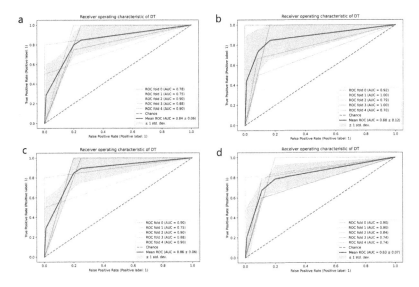

Fig. 3. The ROC curves of the proposed method in different data sets, figure a represents Colon data set, figure b represents DLBCL data set, figure c represents Leukemia data set, and figure d represents Lymphoma data set.

4.3   Comparison with Traditional Methods

In this subsection, we compare the feature selection results of the proposed method with traditional machine learning algorithms. The compared methods include Lasso regression, logistic regression (LR), ridge regression (Ridge), correlation coefficient (Corr), and recursive feature elimination (RFE) algorithms, and the experiments use decision trees as the classification models. The average classification accuracy of five-fold cross-validation is used as the evaluation index. The detailed results are shown in Table 3, which shows that the proposed method outperforms the traditional feature selection methods on all four data sets, proving the effectiveness of the proposed method.

Table 3. Comparison between the proposed method and the traditional method.

Methods	Colon	DLBCL	Leukemia	Lymphoma
Lasso	0.696	0.819	0.833	0.777
LR	0.661	0.597	0.652	0.400
Ridge	0.821	0.702	0.807	0.822
Corr	0.819	0.819	0.821	0.866
RFE	0.581	0.623	0.502	0.577
Proposed	0.887	0.936	0.835	0.911

4.4   Comparison with Advanced Methods

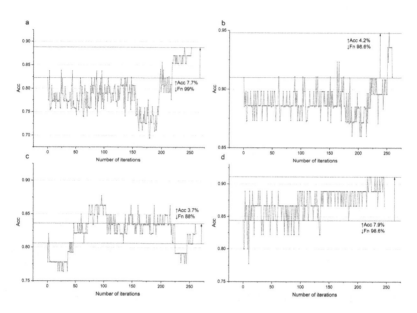

Fig. 4. The classification accuracy of the proposed method versus the number of iterations, with the marked points in the figure indicating the percentage improvement in classification accuracy and the percentage decrease in the number of features (Fn) compared to the initial node.

To demonstrate the advancedness of the algorithm, we compared the proposed method with the advanced hybrid feature selection algorithm. And these compared methods and results are shown in Table 3, and all the compared methods are described in detail in the related work section. From the results in Table 4, it is easy to find that the proposed method outperforms all the compared methods on the Colon dataset and can obtain higher classification accuracy on fewer features. On the DLBCL dataset, the proposed method achieves the highest classification accuracy. However, it is slightly higher than Pardo and Ram's method regarding the number of features. Overall, the proposed method effectively reduces the data dimensionality and ensures high model classification accuracy.

Table 4. Comparison between the proposed method and the advanced method.

Literatures	Method	Colon		DLBCL	
Gunavathi et al. [11]	GA+KNN	10	0.750	10	0.800
Pardo et al. [12]	mRMR+ABC	5	0.850	5	0.933
Wang et al. [13]	MB	11.1	0.857	10.6	0.900
Theera et al. [14]	FA+FC	11	0.769	10.6	0.833
Musheer et al. [15]	ICA+ABC	4	0.821	/	/
Ram et al. [16]	QIGA	/	/	5	0.920
Proposed	mRMR+GA	4	0.887	8	0.936

5 Conclusion

In this paper, we propose a two-stage hybrid feature selection algorithm for data dimensionality reduction for high-dimensional small sample data. The proposed method first uses the improved mRMR method for the initial filtering of features, i.e., the relationship between features and between features and labels is analyzed by two improved quantization functions, making it applicable to attribute continuous data. The improved genetic algorithm is then used for further filtering the features. The proposed method combines the growth tree clustering and RFE methods into an iterative process of the genetic algorithm to increase population diversity, improve algorithm efficiency, and reduce the number of features. Multiple comparative experiments on publicly available datasets demonstrate the advancement and effectiveness of the proposed method, which is essential for the analysis of high-dimensional small sample data.

References

1. Tong, D.L., Schierz, A.C.: Hybrid genetic algorithm-neural network: feature extraction for unpreprocessed microarray data. Artif. Intell. Med. **53**(1), 47–56 (2011)
2. Wang, X., Wang, Y., Wong, K.C., et al.: A self-adaptive weighted differential evolution approach for large-scale feature selection. Knowl.-Based Syst. **235**, 107633 (2022)
3. Yu, K., Xie, W., Wang, L., et al.: Determination of biomarkers from microarray data using graph neural network and spectral clustering. Sci. Rep. **11**(1), 1–11 (2021)
4. Alhenawi, E., Al-Sayyed, R., Hudaib, A., et al.: Feature selection methods on gene expression microarray data for cancer classification: a systematic review. Comput. Biol. Med. **140**, 105051 (2022)
5. Xie, W., Fang, Y., Yu, K., et al.: MFRAG: Multi-Fitness RankAggreg Genetic Algorithm for biomarker selection from microarray data. Chem. Intell. Lab. Syst., 104573 (2022)
6. Amini, F., Hu, G.: A two-layer feature selection method using genetic algorithm and elastic net. Expert Syst. Appl. **166**, 114072 (2021)

7. Hira, Z.M., Gillies, D.F.: A review of feature selection and feature extraction methods applied on microarray data. Adv. Bioinform. **2015** (2015)

8. Yu, K., Xie, W., Wang, L., et al.: ILRC: a hybrid biomarker discovery algorithm based on improved L1 regularization and clustering in microarray data. BMC Bioinform. **22**(1), 1–19 (2021)

9. Xie, W., Chi, Y., Wang, L., et al.: MMBDE: a two-stage hybrid feature selection method from microarray data. In: 2021 IEEE International Conference on Bioinformatics and Biomedicine (BIBM), pp. 2346–2351. IEEE (2021)

10. Peng, C., Wu, X., Yuan, W., et al.: MGRFE: multilayer recursive feature elimination based on an embedded genetic algorithm for cancer classification. IEEE/ACM Trans. Comput. Biol. Bioinf. **18**(2), 621–632 (2019)

11. Gunavathi, C., Premalatha, K.: Performance analysis of genetic algorithm with kNN and SVM for feature selection in tumor classification. Int. J. Comput. Electr. Autom. Control Inf. Eng. **8**(8), 1490–1497 (2014)

12. Seijo-Pardo, B., Bolón-Canedo, V., Alonso-Betanzos, A.: Using a feature selection ensemble on DNA microarray datasets. In: ESANN (2016)

13. Wang, A., An, N., Yang, J., et al.: Wrapper-based gene selection with Markov blanket. Comput. Biol. Med. **81**, 11–23 (2017)

14. Jinthanasatian, P., Auephanwiriyakul, S., Theera-Umpon, N.: Microarray data classification using neuro-fuzzy classifier with firefly algorithm. In: 2017 IEEE Symposium Series on Computational Intelligence (SSCI), pp. 1–6. IEEE (2017)

15. Musheer, R.A., Verma, C.K., Srivastava, N.: Novel machine learning approach for classification of high-dimensional microarray data. Soft. Comput. **23**(24), 13409–13421 (2019)

16. Ram, P.K., Bhui, N., Kuila, P.: Gene selection from high dimensionality of data based on quantum inspired genetic algorithm. In: 2020 11th International Conference on Computing, Communication and Networking Technologies (ICCCNT), pp. 1–5. IEEE (2020)

Efficient Subhypergraph Containment Queries on Hypergraph Databases

Yuhang Su[1(✉)], Yang Song[2], Xiaohua Li[1], Fangfang Li[1], and Yu Gu[1]

[1] College of Computer Science and Engineering, Northeastern University,
Shenyang 110819, Liaoning, China
suyuhang_neu@163.com, {lixiaohua,lifangfang,guyu}@mail.neu.edu.cn
[2] College of Information Science and Engineering, Northeastern University,
Shenyang 110819, Liaoning, China

Abstract. In the real world, many complex systems consist of a large number of interacting groups of entities. A hypergraph consists of vertices and hyperedges that can connect multiple vertices. Since hypergraphs can effectively simulate complex intergroup relationships among entities, they have a wide range of applications such as computer vision and bioinformatics. In this paper, we study the subhypergraph containment query problem which is one of the most basic problems in the processing of hypergraphs. Existing methods on the subgraph query are designed for ordinary graphs and do not consider hypergraph features. If they are directly applied to subhypergraph containment query, they will suffer from hyperedge semantic incompleteness and label diversity sensitivity issues, resulting in inefficient algorithm performance. This motivates us to improve the performance by exploiting hyperedge features. In our work, we propose a novel framework for subhypergraph containment query called hyperedge filtering vertex testing. Based on the features of hypergraph, we propose an efficient filtering algorithm that can reduce the cost of the traditional filtering stage. In addition, we further propose efficient isomorphism testing techniques based on hyperedge vertex candidates to improve the performance. Extensive experiments on real datasets validate the superiority of our algorithm compared to existing methods.

Keywords: Hypergraph · Subhypergraph · Subhypergraph containment query · Hyperedge filtering vertex testing

1 Introduction

Benefiting from the powerful expressiveness of ordinary graphs, graph structures are prevalently used for modeling complex structures such as social networks, molecular structures, and proteins [2]. However, in the real world, interactions in many complex systems are grouped rather than paired, such as joint interactions of proteins and co-purchases of items. These group interactions cannot be represented by edges in a graph. On the other hand, a hypergraph consists of vertices and hyperedges that can connect multiple vertices [1]. As a generalized form of ordinary graphs, hypergraphs can naturally simulate more complex inter-group relationships between entities by hyperedges. Therefore, in smart-grid technologies [9], Bioinformatics [11], pattern recognition [5], computer vision [19] and other fields, the hypergraphs are widely applied.

In this paper, we focus on subhypergraph containment query, which is a fundamental and important searching problem on hypergraph data. Given a hypergraph database $D = \{g_1, g_2, ..., g_n\}$ and a query hypergraph q, subhypergraph containment query aims to retrieve all hypergraphs $g_i \in D$ that contain q. Subhypergraph containment query has a wide range of applications such as hypergraph-based 2D object search [6] and complex pattern search in collaborative networks and social networks [14]. The following example illustrates the application scenarios of subhypergraph containment query.

Example 1. **Finding a protein complex network set containing a specific protein complex group with the specific function:** This is a typical representative for computing property generalization in bioinformatics [11,15]. A protein complex network is naturally modeled as a hypergraph. Here, a protein represents a vertex and a protein complex represents a hyperedge. Multiple protein complexes form a protein complex group containing certain specific functions. Biologists may need to find a protein complex network set on the protein complex network database. In the set, each protein complex network contains the specific protein complex group. It can be naturally regarded as a subhypergraph containment query problem.

Existing solutions mainly focus on subgraph query and most of their frameworks follow the **IFV** paradigm, namely indexing-filtering-verification. However, the latest research [18] shows that the slow verification method in existing **IFV** algorithms can lead us to over-estimate the gain of filtering. Specifically, complex indexing and filtering algorithms occupy a large proportion of the overall time, but cannot bring significant speedup to verification. The advanced framework **vcFV** is proposed by Sun et al. [18]. The **GFQL** [18] method in the **vcFV** framework does not require indexing. It utilizes the preprocessing techniques in the **CFL** [3] for filtering and uses enumeration the techniques in the **GraphQL** [10] to speed up isomorphism testing. Although the **GFQL** methods can effectively solve the subgraph query problem, applying them directly to subhypergraph containment query has the following two problems [17]. 1) **Hyperedge semantic incompleteness**. Subgraph query methods cannot guarantee the integrity of hyperedge semantics. We need to modify its pruning rules for different vertices to ensure accuracy of the results. 2) **Sensitive to label diversity**. The pruning methods in subgraph query rely on the diversity of labels. Hypergraphs have only two types of labels (hyperedge vertices and normal vertices). It will result in processing many invalid neighbors in hyperedges and generating a large amount of intermediate data without the help of label information. This motivates us to fully explore hypergraph features to efficiently solve subhypergraph containment query problem. The major contributions are concluded as follows:

- We propose a novel framework for subhypergraph containment query called hyperedge filtering vertex testing. Based on the framework, we only need to compute the candidate set of hyperedges, which greatly shortens the time of the filtering stage.
- Based on the features of hyperedges, we propose a pruning method for hyperedge candidates. This method not only ensures a high filtering precision but also improves the filtering efficiency.
- We propose an isomorphism testing algorithm based on hyperedge candidates which is competitive in time performance with existing methods.
- We conduct extensive experiments to verify that our algorithm outperforms existing algorithms.

The rest of this paper is organized as follows: Sect. 2 introduces the existing works related to ours; Sect. 3 describes the basic concepts related to subhypergraph containment query and the framework of our algorithm; Sect. 4 mainly expounds our pruning method for hyperedge candidates. How to test for subhypergraph isomorphism by hyperedge candidates is explained in Sect. 5; Sect. 6 finally concludes this paper.

2   Related Work

A subgraph query is also called a subgraph containment query. Given a graph database D, which contains a collection of multiple medium or small data graphs $D = \{g_1, ..., g_n\}$, the goal of subgraph query is to find all data graphs $A \subseteq D$ that contain a given query graph q. Obviously, the subgraph query is different from the traditional subgraph matching [21]. Most of the existing subgraph query methods follow an indexing-filtering-verification paradigm (**IFV**), which is divided into online and offline stages. In the offline phase, the features of all data graphs in the graph database are extracted to build an index on the entire graph database. In the online phase, the query graph features are extracted and combined with the index created by the graph database for filtering, leaving the data graphs that meet the conditions for the final verification. Due to the differences in the structure of index features and the methods of feature extraction, the existing subgraph query methods can be roughly divided into two categories. **Mining-Based Approaches.** Mining-based Approaches give a trade-off between "frequent" features and "discrimination" features, only mining and extracting part of the features, such as **FG-Index** [7], **SwiftIndex** [16], **Lindex** [20]. **Enumeration-Based Approaches.** Enumeration-based Approaches are the methods which completely enumerate all specified features and store them in indexes such as **CT-Index** [13], **GraphGrepSX** [4], **Grapes** [8]. Previous subgraph query algorithms focus on how to efficiently build indexes and how to extract features with stronger filtering capabilities, but ignored the gains that advanced verification methods may bring to the overall performance of the algorithm. Therefore, the recent research is focused on integrating the subgraph query algorithm and the advanced verification matching algorithm in the subgraph isomorphism search to achieve better performance [12,18]. It is worth noting that the **GFQL** method [18] in **vcFV** framework utilizes advanced subgraph matching techniques to make it overall outperform existing methods.

Hypersubgraph query is also called the subhypergraph containment query. In fact, as a general form of ordinary graphs, the characteristics of hypergraphs are not the same as ordinary graphs. Therefore, the subgraph query method cannot be directly applied to subhypergraph containment query and there is no algorithm specifically for subhypergraph containment query. Subhypergraph matching [9,17] is a topic closely related to subhypergraph containment query. Although both the subhypergraph containment query and subhypergraph matching involve subhyperbgraph isomorphism, they are inherently different. The subhypergraph matching requires enumerating all embeddings.

3 The Subhypergraph Containment Query

3.1 Basic Definition

Definition 1 (Hypergraph and Subhypergraph). *A hypergraph is represented by $g = (V, E)$, where V is a finite set of vertices, $E = \bigcup_{i=1}^{|E|} e_i$ is a finite set of hyperedges. Each hyperedge $e_i \in E$ is a non-empty subset of V. A subhypergraph g' of g is a hypergraph $g' = (V', E')$, where $V' \subseteq V$ and $E' = \{e_j \mid e_j \subseteq V'\} \subseteq E$.*

Definition 2 (Hypergraph Isomorphism). *A hypergraph isomorphism between two hypergraphs g' and g'' is a bijective mapping $f : V' \rightarrow V''$, and for any hyperedge $e_j = (x_1, ..., x_i) \in E'$ there exists a hyperedge $e_k = (f(x_1), ..., f(x_i)) \in E''$; also for any hyperedge $e_k = (f(x_1), ..., f(x_i)) \in E''$ there exists a hyperedge $e_j = (x_1, ..., x_i) \in E'$. If g' is a subhypergraph of g we say that g contains g''.*

Definition 3 (Subhypergraph Containment Query). *Given a hypergraph database D, which contains a collection of multiple medium or small data hypergraphs $D = \{g_1, ..., g_n\}$, the goal of subhypergraph containment query is to find the set $A \subseteq D$ of all data hypergraphs that contain a given query hypergraph q.*

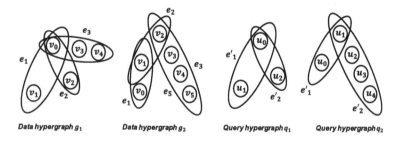

Fig. 1. Data and query hypergraphs

Example 2. The most commonly used form for representing hypergraphs is shown in Fig. 1. A circle is used to represent a vertice (such as v_1), and the hyperedge is simply expressed as a closed curve surrounding the corresponding vertices (such as e_1). As shown in Fig. 1, data hypergraph g_1 contains q_1. This is because q_1 is isomorphic to the subhypergraph of g_1, and the mapping relationship is $\{u_0 \rightarrow v_0, u_1 \rightarrow v_1, u_2 \rightarrow v_2, e'_1 \rightarrow e_1, e'_2 \rightarrow e_2\}$. It is worth mentioning that between g_1 and q_1, e_3 and e'_1 cannot establish a mapping relationship because they contain different numbers of vertices. This is what we call the hyperedge integrity constraint and it is an important feature that distinguishes hypergraphs from ordinary graphs. In Fig. 1, hypergraph database D is $\{g_1, g_2\}$ and the query hypergraphs are q_1 and q_2. The subhypergraph containment query result of q_1 is $A_1 = \{g_1, g_2\}$ and the subhypergraph containment query result of q_2 is $A_2 = \{g_2\}$.

3.2   The Framework of Hyperedge Filtering Vertex Testing

Our framework is different from traditional **IFV** frameworks with the index building and complex index-based filtering processes. The filtering precision and algorithm scalability are greatly improved by processing each hypergraph sequentially. This not only improves the efficiency of the algorithm but also eliminates the need to think about complex index update problems. Our framework is also different from the **vcFV** framework. The **vcFV** framework needs to find candidates for all vertices. However, our framework only requires hyperedge candidates by exploring the features of the hypergraph, which not only ensures stable filtering precision but also reduces the time cost of filtering. The details of our framework are given in Algorithm 1. After that, we give Theorem 1.

Algorithm 1: The Framework of HFVT

Input   :   A hypergraph database D and a query hypergraph q
Output: A set $A(q)$ keeping all data hypergraphs in D that contain q

1 **initialize**: Set $A(q) := \emptyset$;
2 **foreach** $g \in D$ **do**
3      $HC \leftarrow HyperedgeFiltering(q, g)$;
4   **if** $\forall e \in E(q), HC(e) \neq \emptyset$ **then**
5   **if** $VertexTesting(q, g, HC)$ *is ture* **then**
6              $A(q) \leftarrow A(q) \cup \{g\}$

7 *return* $A(q)$;

Theorem 1. *Given q and g in the subhypergraph containment query task, if there exists a hyperedge e'_i such that HC(e'_i) is empty, then g does not contain q.*

Proof. If g contains q, according to Definition 2, we can know that there is a subhypergraph g' of g that is isomorphic to q. Therefore, there must be a hyperedge in g' that has a bijective relationship with e'_i, which contradicts that $HC(e'_i)$ is empty. This is because $HC(e'_i)$ contains all candidates that can establish a mapping relationship with e'_i. Thus, the proposition is proved by contradiction.

As shown in Algorithm 1, it first initializes the hyperedge candidate set HC (*Line* 1) and then processes each data hypergraph sequentially (*Line* 2). The processing of each data graph can be divided into two parts **Hyperedge Filtering Stage** (*Lines* 3–4) and **Vertex Testing Stage** (*Lines* 5–6). It is worth mentioning that this framework does not test all data hypergraphs. When we are calculating hyperedge candidates, if there is a hyperedge candidate set that is empty $(HC(e) = \emptyset)$, then according to Theorem 1, we know that the data hypergraph g does not contain the query graph q. In this way, redundant testing procedures are discarded (*Line* 4). In fact, by exploring the features of the hypergraph, we can get a more accurate set of hyperedge candidates to reduce the test hypergraph candidates. This will improve the filtering precision and reduce the time cost of the algorithm.

4   Computing Hyperedge Candidates Based on Hyperedge Features

Definition 4 (Hyperedge-Projected Graph). *A hyperedge-projected graph of $g = (V, E)$ is an ordinary graph $PG = (E, H)$, where $H = \{(e_i, e_j) \mid e_i \cap e_j \neq \emptyset\}$.*

Definition 5 (Hyperedge-Projected Neighbors and Degree). *For a hyperedge $e \in E$ in a hyperedge projection graph $PG = (E, H)$, its associated hyperedges are called its projected neighbors, denoted by $N(e) = \{e_i \in E \mid e \cap e_i \neq \emptyset\}$. The number of its projected neighbors is called the projected degree denoted as $degree(e) = |N(e)|$.*

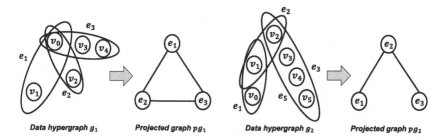

Fig. 2. Data hypergraph and hyperedge-projected graph

Example 3. As shown in Fig. 2, we can transform the hypergraph into a hyperedge projected graph to describe the binary relationship between hyperedges. For g_1 and pg_1, there are $N(e_1) = \{e_2, e_3\}$ and $degree(e_1) = 2$. For g_2 and pg_2, there are $N(e_1) = \{e_2\}$ and $degree(e_1) = 1$. In fact, the transformation of the hypergraph to the hyperedge projected graph loses the relevant information of the vertices. However, since projected graphs can better represent the binary relationship between hyperedges, it is often used for hypergraph studies.

Theorem 2. *Suppose the query hypergraph is q and the data hypergraph is g. There is a hyperedge e_i' and its projected neighbor e'' in q, and hyperedge $e_i \in HC(e')$ in g. If e' and e_i can establish isomorphic mapping, a hyperedge $e_j \in HC(e'')$ must be found, satisfying $|e' \cap e''| = |e_i \cap e_j|$.*

Proof. We use contradiction to prove the theorem. For any $e_j \in HC(e'')$, there is $|e' \cap e''| \neq |e_i \cap e_j|$. If $|e' \cap e''| > |e_i \cap e_j|$, there must be a vertex $u \in |e' \cap e''|$ and we cannot find the corresponding vertex $v \in |e_i \cap e_j|$, which violates Definition 2. Similarly, the case of $|e' \cap e''| < |e_i \cap e_j|$ can be proved. Obviously, e' and e_i can not establish isomorphic mapping. This contradicts the given conditions and the original proposition is proved.

Algorithm 2: HyperedgeFiltering

Input : A data hypergraph g and a query hypergraph q
Output: The hyperedge candidate set HC

1 **foreach** $e' \in E(q)$ **do**
2   **foreach** $e \in E(g)$ **do**
3   **if** $degree(e') = degree(e)$ && $|N(e')| \leq |N(e)|$ **then**
4           $HC(u) \leftarrow e$;

5 **foreach** $e' \in E(q)$ **do**
6   **foreach** $e'' \in N(e')$ **do**
7   **foreach** $e_i \in HC(e'')$ **do**
8   **foreach** $e_j \in HC(e')$ **do**
9               $count := 0$;
10   **if** $|e'' \cap e'| = |e_i \cap e_j|$ **then**
11                  $count ++$; $break$;

12   **if** $count = 0$ **then**
13               $remove$ e_i $from$ $HC(e'')$;

14 *return HC* ;

As shown in Algorithm 2, the processing of hyperedge filtering can be divided into two parts, **Hyperedge Degree Filtering** (*Lines* 1–4) and **Hyperedge Neighbors Pruning** (*Lines* 5–13). It first filters the unqualified hyperedges of data hypergraphs (*Line* 3) according to the hypergraph isomorphism Definition 2, and builds a hyperedge candidate set for each hyperedge of the query hypergraph (*Line* 4). Then it sequentially uses each hyperedge candidate set to prune the corresponding neighbor hyperedge candidates (*Lines* 5–6). It traverses each candidate hyperedge in $HC(e')$ to verify whether the corresponding hyperedge e_i conforms to Theorem 2 (*Lines* 7–11). If a qualified hyperedge is found, we jump out of the current loop (*Line* 11). If the counter is still 0 when the traversal ends, it means that a qualified hyperedge cannot be found, and e_i is removed (*Lines* 12–13). Since each hyperedge is only refined once by its neighbors, the worst time complexity of the algorithm is $O(Maxdegree(E(q)) \times |E(q)| \times |Max(HC(e))|^2)$.

5   Vertex-Based Isomorphism Testing for Hypergraphs

When we get the candidate sets of all hyperedges of the query hypergraph, we can enumerate all possible hyperedge mapping combinations through the hyperedge projection graph. If any combination of hyperedge mapping passes the vertex-based isomorphism testing, it means that the subhypergraph containment relationship exists. The specific details are given by Algorithm 3.

Algorithm 3: VertexTesting

Input   : A data hypergraph g, a query hypergraph q and HC
Output: True or false

1   **foreach** $M \in ProjectedEnumeration(q, HC)$ **do**
2   │ $V_{rc} \leftarrow ReconstructVertex(M, g)$
3   │   **if** $|V(q)| \neq |V_{rc}|$ **then**
4   │   │ $break$;
5   │   **foreach** $u \in V(q)$ **do**
6   │   │   **foreach** $v \in V_{rc}$ && v *is unmarked* **do**
7   │   │   │   **if** $E_q(u) = M(E_g(v))$ **then**
8   │   │   │   │ $Mark\ u$;
9   │   │   │   │ $Mark\ v$
10 │   │   **if** u *is unmarked* **then**
11 │   │   │ $break$;
12 │   **if** $\forall v \in V_{rc}$ *is marked* **then**
13 │   │ $return\ true$
14 $return\ false$;

As shown in Algorithm 3, the processing of vertex testing can be divided into three parts.

Projected Graph Enumeration. The hyperedge projection graph can be regarded as an ordinary graph. When we get all the hyperedge candidates, we can use the subgraph enumeration method to calculate the embedding of the query hypergraph projection on the data hypergraph projection (*Line* 1). In the implementation, we use the enumeration method of **GraphQL** [10]. It is worth noting that the conditions for connecting vertices are not only the existence of edges but also the satisfaction of Theorem 2.

Reconstruct Vertex. When we get a combination of hyperedge mapping M, we use the strategy of reconstructing vertices (*Lines* 2–4). Specifically, we calculate the number of vertices contained in this set of hyperedges and what are the hyperedges associated with each vertex (*Line* 2). If the number of vertices is different from the number of vertices in the query hypergraph, the mapping combination fails the vertex testing (*Lines* 3–4).

Vertex Testing. Vertex testing is to find the corresponding vertices of all query hypergraph vertices (*Lines* 5–13). For each vertex of the query hypergraph, it traverses the unmarked vertices in the reconstructed vertex set V_{rc} (*Lines* 5–6). If all hyperedge mappings of a reconstructed vertex $M(E_g(v))$ are identical to the hyperedges of this query graph vertex $E_q(u)$ (*Line* 7), it marks both vertices (*Lines* 8 − 9). If a query hypergraph vertex cannot find a corresponding vertex, it breaks out of the loop (*Lines* 10–11). If all query hypergraph vertices can find their corresponding vertices, the vertex testing is passed (*Lines* 12–13).

6 Experimental Settings and Results Analysis

6.1 Experimental Settings

Competitive Algorithms. In fact, there is no method specifically for the subhypergraph query problem. We can handle the subhypergraph query problem by modifying the existing subgraph query methods. Our competitive algorithm is **MCFQL**, which is achieved by modifying state-of-the-art subgraph query algorithms **CFQL** [18]. We specifically modify the *FilterCandidates* function related to the vertex degree in these algorithms so that they can conform to Definition 2.

Experiment Environment. We obtained the source code of **CFQL** from the authors of [18]. All the algorithms are implemented in $C + +$. The compiler for compiling source code is $g + + 4.9.3 − O3 \ flag$. We conduct all experiments on a PC machine with equipment of $Intel$ $i5$ 3.20 GHz and 16 GB RAM.

Metrics. We measure the execution time in milliseconds (ms). For a query, the total processing time can be divided into two parts, one of which is the filtering time (that is, the time spent on the constructing candidate set), and the other is the verification time (that is, the time spent on isomorphism testing). Another important metric is the filtering precision given by the formula below. Q denotes the query hypergraph set and $C(q)$ denotes the data hypergraph set for isomorphism testing when the query hypergraph is q.

$$FilteringPrecision = \frac{1}{|Q|} \sum_{q \in Q} \frac{|A(q)|}{|C(q)|} \tag{1}$$

Datasets. Due to commercial intellectual property protection, there is no publicly available hypergraph database in related fields. Therefore, we adopt the method of extracting data hypergraphs on real-world large hypergraphs. The *Github* and *Flickr* datasets are from the *KONECTProject* (http://konect. cc/). The *Github* can be viewed as a sparse hypergraph with 56,591 vertices and 120,867 hyperedges. The average hyperedge and vertex degrees are 3.64 and 7.79, respectively. The *Flickr* can be viewed as a no-sparse hypergraph with 395,979 vertices and 103,631 hyperedges. The average hyperedge and vertex degrees are 21.58 and 82.46, respectively. We extract 200 hypergraphs, each of which contains 200 vertices from two large hypergraphs to form sparse and non-sparse hypergraph databases. The extraction method is the same as the breadth-first extraction method in [17].

Query Sets. We also extract subhypergraphs from the hypergraph databases as the query hypergraph. We have established 6 query sets for each hypergraph database, which are divided into 3 sparse query sets and 3 non-sparse query sets. There are 100 queries in each query set. Q_{iS} and Q_{iN} represent the sparse query set and the non-sparse query set containing i vertices, respectively. According to the value of i, 6 query sets are Q_{8S}, Q_{8N}, Q_{16S}, Q_{16N}, Q_{32S} and Q_{32N}. The extraction method is also the same as the extraction method in [17].

6.2   Experimental Results Analysis

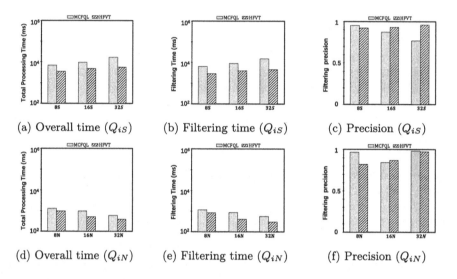

Fig. 3. Performance on sparse real hypergraph datasets.

Performance on the Sparse Hypergraph Database. Figures 3(a)–(f) show the performance on sparse hypergraph database. Based on the experimental results, we can obtain the following conclusions. 1) **Overall time:** As shown in Figs. 3(a) and (d), our algorithm always maintains advantages. Different query sets have a significant impact on the overall time. Non-sparse query sets can have an order of magnitude advantage over sparse query sets. More notably, as the number of query hypergraph vertices increases, the time cost of non-sparse query sets also decreases. This is because query hypergraphs that are not sparse or have more vertices can have more discriminative features. Therefore, in the early stage of the algorithm, more invalid candidates can be pruned to reduce the time cost. 2) **Filtering time:** As shown in Figs. 3(b) and (e), filtering time has a high proportion of our overall time (80 to 95% on average). The filtering time of our algorithm as a proportion of the overall time is slightly lower than that of **MCFQL**. This is because **MCFQL** computes the candidates for all vertices in the filtering stage, which brings a speedup to the subsequent isomorphism testing. 3) **Filtering Precision** Both algorithms have relatively high filtering precision and our algorithm performs slightly better than **MCFQL**. This shows

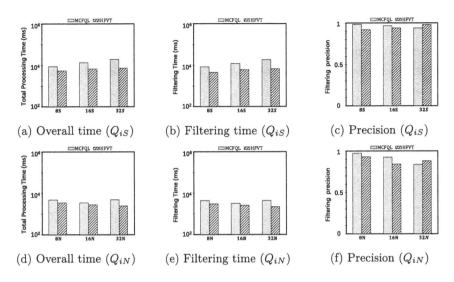

Fig. 4. Performance on non-sparse real hypergraph datasets.

that even if we only compute hyperedge candidates, fully exploring hyperedge features can lead to higher filtering precision.

Performance on Non-sparse Real Hypergraph Datasets. As shown in Figs. 4(a)–(f), we can see that 1) The relative performances of the two algorithms are similar on sparse graphs and non-sparse databases. 2) The time cost is higher on non-sparse datasets than on sparse graphs. This is because the closer the relationship between vertices is, the more candidate vertices there are. Therefore, the time cost of filtering and isomorphism testing increases.

7   Conclusion

In this paper, we propose effective techniques for subhypergraph containment query. In our work, we first propose a novel framework for subhypergraph containment query called hyperedge filtering vertex testing. Based on this framework, we fully explore hyperedge features, propose a low-cost hyperedge filtering algorithm and implement an efficient vertex testing method by reconstructing vertices. Extensive experiments on real datasets show that our method is superior to the existing solutions.

Acknowledgements. This work is supported by the National Nature Science Foundation of China (62072083) and the Fundamental Research Funds of the Central Universities (N2216017).

References

1. Berge, C.: Hypergraphs: Combinatorics of Finite Sets, vol. 45. Elsevier, Amsterdam (1984)
2. Berman, H.M., et al.: The protein data bank. Nucl. Acids Res. **28**(1), 235–242 (2000)
3. Bi, F., Chang, L., Lin, X., Qin, L., Zhang, W.: Efficient subgraph matching by postponing cartesian products. In: Proceedings of the 2016 ACM SIGMOD International Conference on Management of Data, pp. 1199–1214 (2016)
4. Bonnici, V., Ferro, A., Giugno, R., Pulvirenti, A., Shasha, D.: Enhancing graph database indexing by suffix tree structure. In: Dijkstra, T.M.H., Tsivtsivadze, E., Marchiori, E., Heskes, T. (eds.) PRIB 2010. LNCS, vol. 6282, pp. 195–203. Springer, Heidelberg (2010). https://doi.org/10.1007/978-3-642-16001-1_17
5. Bretto, A., Cherifi, H., Aboutajdine, D.: Hypergraph imaging: an overview. Pattern Recogn. **35**(3), 651–658 (2002)
6. Bunke, H., Dickinson, P., Kraetzl, M., Neuhaus, M., Stettler, M.: Matching of hypergraphs: algorithms, applications, and experiments. In: Bunke, H., Kandel, A., Last, M. (eds) Applied Pattern Recognition, vol. 91, pp. 131–154. Springer, Heidelberg (2008). https://doi.org/10.1007/978-3-540-76831-9_6
7. Cheng, J., Ke, Y., Ng, W., Lu, A.: FG-Index: towards verification-free query processing on graph databases. In: Proceedings of the 2007 ACM SIGMOD International Conference on Management of Data, pp. 857–872 (2007)
8. Giugno, R., Bonnici, V., Bombieri, N., Pulvirenti, A., Ferro, A., Shasha, D.: Grapes: a software for parallel searching on biological graphs targeting multi-core architectures. PLoS ONE **8**(10), e76911 (2013)
9. Ha, T.W., Seo, J.H., Kim, M.H.: Efficient searching of subhypergraph isomorphism in hypergraph databases. In: IEEE International Conference on Big Data and Smart Computing (2018)
10. He, H., Singh, A.K.: Graphs-at-a-time: query language and access methods for graph databases. In: Proceedings of the 2008 ACM SIGMOD International Conference on Management of Data, pp. 405–418 (2008)
11. Hwang, T.H., Tian, Z., Kuang, R., Kocher, J.P.: Learning on weighted hypergraphs to integrate protein interactions and gene expressions for cancer outcome prediction. In: Eighth IEEE International Conference on Data Mining (2008)
12. Katsarou, F., Ntarmos, N., Triantafillou, P.: Hybrid algorithms for subgraph pattern queries in graph databases. In: 2017 IEEE International Conference on Big Data (Big Data), pp. 656–665. IEEE (2017)
13. Klein, K., Kriege, N., Mutzel, P.: CT-Index: fingerprint-based graph indexing combining cycles and trees. In: 2011 IEEE 27th International Conference on Data Engineering, pp. 1115–1126. IEEE (2011)
14. Knoke, D., Yang, S.: Social Network Analysis. Sage Publications, Thousand Oaks (2019)
15. Ramadan, E., Tarafdar, A., Pothen, A.: A hypergraph model for the yeast protein complex network. In: 18th International Parallel and Distributed Processing Symposium. Proceedings, p. 189. IEEE (2004)
16. Shang, H., Zhang, Y., Lin, X., Yu, J.X.: Taming verification hardness: an efficient algorithm for testing subgraph isomorphism. Proc. VLDB Endow. **1**(1), 364–375 (2008)
17. Su, Y., Gu, Y., Wang, Z., Zhang, Y., Qin, J., Yu, G.: Efficient subhypergraph matching based on hyperedge features. IEEE Transactions on Knowledge and Data Engineering (2022)

18. Sun, S., Luo, Q.: Scaling up subgraph query processing with efficient subgraph matching. In: 2019 IEEE 35th International Conference on Data Engineering (ICDE), pp. 220–231. IEEE (2019)
19. Wong, A.K.C., Lu, S.W.: Recognition and shape synthesis of 3-d objects based on attributed hypergraphs. IEEE Trans. Pattern Anal. Mach. Intell. **11**(3), 279–290 (1989)
20. Yuan, D., Mitra, P.: Lindex: a lattice-based index for graph databases. VLDB J. **22**(2), 229–252 (2013)
21. Zhang, H., Xie, X., Wen, Y., Zhang, Y.: A twig-based algorithm for top-k subgraph matching in large-scale graph data. In: Wang, G., Lin, X., Hendler, J., Song, W., Xu, Z., Liu, G. (eds.) WISA 2020. LNCS, vol. 12432, pp. 475–487. Springer, Cham (2020). https://doi.org/10.1007/978-3-030-60029-7_43

Logistics Distribution Route Optimization Using Hybrid Ant Colony Optimization Algorithm

Chao Zhang[1], Yuhan Cai[2(✉)], Peng Hu[3], Peng Quan[3], and Wei Song[2]

[1] "Internet + Tobacco" Integration Innovation Laboratory of Hubei Provincial Tobacco Monopoly Administration (Company), Wuhan, China
[2] School of Computer Science, Wuhan University, Wuhan, China
{Caiyuhan,songwei}@whu.edu.cn
[3] Shiyan Branch of Hubei Tobacco Company, Shiyan, China

Abstract. Logistics route distribution optimization problem (LRDOP) belongs to traveling salesman problem (TSP), but it not only has higher requirements on the running efficiency of the path planning algorithm, but also is easier to fall into local optimum. Ant colony optimization (ACO) is one of the dominant algorithms for solving TSP. In ACO, α and β parameters are critical and specific. Symbiotic organisms search (SOS) is a non-parametric algorithm, so the α and β parameters of the ACO can be dynamically optimized by using SOS. In this paper, we introduce a hybrid ant colony optimization SOSACOp, which uses a mixture of ACO and SOS, and adjusts the results by using local optimization strategies and another pheromone updating rules. Experimental results show that SOSACOp has better comprehensive performance than ACO.

Keywords: Logistics route distribution optimization problem · Traveling salesman problem · Ant colony optimization · Symbiotic organisms search

1 Introduction

Traveling salesman problem (TSP) is an important issue in combinatorial optimization, and it is a NP-complete problem [1]. For specific tobacco logistics route distribution optimization problem (LRDOP), there are many new challenges:

1. Real-time results are emphasized in LRDOP, since new merchants with demands may appear every day, efficient algorithm is needed to calculate path planning results.
2. In the LRDOP, since the actual vehicle distance is used, the distance from place A to place B may be different from that from place B to place A, which will make algorithm iteration easier to fall into local optimization.

This work is supported by Key R&D project of China National Tobacco Corporation No. 110202102031 and Project of Science and Technology Project of Hubei Tobacco Company No. 027Y2021-046.

For these challenges, we propose an improved ant colony optimization named SOSACOp, which is a combination of symbiotic organisms search (SOS) and ant colony optimization (ACO). In addition, a local optimization strategy and a new pheromone updating rule are used in the algorithm to speed up convergence and improve algorithm efficiency. The main contributions of this paper are as follows:

1. Combine ACO with SOS, so that SOSACOp can adjust parameters adaptively to deal with specific problems.
2. A local optimization strategy is used in SOSACOp to improve the convergence speed of the algorithm.
3. A new pheromone updating rule is used in SOSACOp to further improve the convergence speed of the algorithm without making the algorithm result fall into local optimum.

According to the analysis of experimental results, each new strategy applied to our algorithm makes the algorithm better.

2   Related Work

As a classical combinatorial optimization problem, Tsp has produced many solutions since it was proposed [2]. In recent years, many swarm Intelligence algorithms have been produced, which are similar to ACO. B. A. S. Emambocus et al. [3] proposed a variant of dragonfly algorithm (DA). The DA is not originally suitable for solving TSP, but the variant adjusts its equations and updates the position of the artificial dragonflies in the search space by exchanging sequences, making it suitable for solving TSP. Li et al. [4] proposed an improved whale optimization algorithm for optimal welding path of welding robot. Zhang et al. [5] proposed an improved Artificial Bee Colony algorithm. In the improved Artificial Bee Colony, self-adaptive strategies for updating food source is introduced to ensure time and accuracy of algorithm.

According to the experimental results of Bi et al. [6], ACO performs better when the number of sites increases. In the LRDOP, a large number of merchants need to be distributed, so we choose ACO to solve the problem. Wang et al. [7] proposed a novel min-max ant colony system (MMACS) approach to solve the cooperative task allocation problem for multiple unmanned aerial vehicles (UAVs). Wang et al. [8] designed a novel automatic tuning parameter method based on MAX-MIN Ant System (MMAS) to make a dynamic balance between the effects and the efficiencies. D. M. Chitty [9] proposed a fully combined GA and ACO meta-heuristic solution, which took into account the operational variation of greedy mode and made it sustainable. W. Elleuch et al. [10] developed a system based on a time-dependent ACO algorithm to solve the problem of finding the fastest traffic path instead of the shortest path. For the newly proposed improved ACO, some of the algorithms solve different problems than ours, and some of them sacrifice one part of the performance in order to improve another aspect. However, some of the algorithms do put forward interesting ideas, which deserve our consideration and inspire our work.

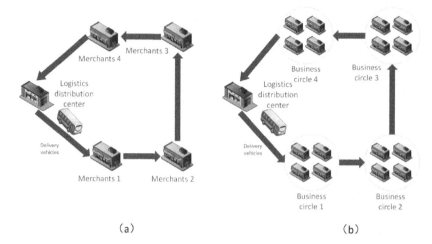

Fig. 1. Tobacco logistics route distribution optimization problem model before clustering (a) and after clustering (b)

3   System Model

In LRDOP, the distribution vehicle starts from the logistics and distribution center, sends the goods to all the merchants that need to be distributed without repeating, and finally returns to the logistics and distribution center, as shown in the Fig. 1(a). Different from the TSP, our problem fixes the starting point of a vehicle and uses the actual distance between two points instead of the Euclidean distance in the calculation of the optimal route, which is more realistic. For the former, the path we plan is a loop, which means that every point on the loop can be both the starting point and the ending point. Therefore, it has no impact on the execution effect of the algorithm, and only requires certain processing of the calculation results. The latter will make the distance from point A to point B and from point B to point A different, thus increasing the difficulty of path planning, which requires our algorithm to have higher efficiency. In addition, there are a large number of merchants in LRDOP, and it is obviously very inefficient and difficult to calculate the distribution path planning for each merchant as an independent individual. Therefore, we first cluster the merchants into business circles according to a certain rule, and carry out path planning in business circles as a unit, as shown in the Fig. 1(b).

4   Proposed Method

Our method is improved on the basis of ACO. For ACO, the α and β parameters have a very important influence on the performance of the algorithm, so we can optimize the algorithm by choosing the appropriate α and β values. However, how to choose α and β is also a problem. Dorigo et al. found that the algorithm

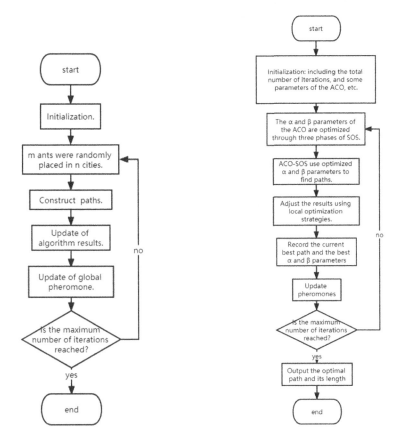

Fig. 2. Flow chart of ACO   **Fig. 3.** Flow chart of SOSACOp

works well when $\alpha = 1$ and $\beta = 5$ [11], but this can only be used as a general case, and the values of α and β suitable for different problems may be different. Therefore, we can use an algorithm to adjust the α and β parameters adaptively during the operation of ACO. From many algorithms, we choose SOS. From the above introduction of SO, it can be seen that the biggest characteristic of SOS is that it has no parameters, so it is more appropriate to use this algorithm to adjust the parameters of ACO. To further improve the efficiency of ACO, we use a local search algorithm to adjust the results after each iteration, and we also introduce another pheromone update rule.

4.1 Ant Colony Optimization

Ant Colony System (ACS) was first proposed by Dorigo and Gambardella [12], which made many modifications to the early ant colony algorithm and became the basic model of various ACO widely studied later [13]. The basic flow of ACO is shown in the Fig. 2.

In the process of constructing the path, the probability of each ant moving to the next city is calculated by Eq. (1) and Eq. (2):

$$p_k(i,j) = \begin{cases} \frac{\tau(i,j)^\alpha \eta(i,j)^\beta}{\sum_{u \in allowed_k} \tau(i,u)^\alpha \eta(i,u)^\beta} & j \in allowed_k \\ 0 & Otherwise \end{cases} \tag{1}$$

$$\eta(i,j) = \frac{1}{d(i,j)} \tag{2}$$

When all the paths have been computed, on every ant walking path pheromone update, pheromone update in addition to ants walked the path of the concentration increased, all path need a certain amount of pheromone evaporation. Pheromone updating rules are expressed by Eq. (3) and Eq. (4):

$$\tau(i,j) = (1-\rho)\tau(i,j) + \sum_{k=1}^{m} \Delta\tau_k(i,j) \tag{3}$$

$$\Delta\tau_k(i,j) = \begin{cases} \frac{Q}{L_k}, & if(i,j) \in tour \ done \ by \ ant \ k \\ 0, & Otherwise, \end{cases} \tag{4}$$

4.2 Symbiotic Organisms Search

Symbiotic organisms search algorithm is inspired by the symbiotic relationship between organisms in nature, and organisms use symbiotic relationship to achieve the purpose of adapting to nature [14]. This algorithm is developed to solve the numerical optimization problem of continuous search space. Different from other heuristic algorithms, this algorithm has no parameters.

At the beginning, SOS randomly generates a group of organisms as the initial population in the ecosystem. Each organism represents a solution to the problem. Each organism has a fitness value that represents how well it fits into the ecosystem. With each iteration, ancestral organisms evolve into their successors through three phases of mutualism, commensalism, and parasitism. During iteration, organisms with higher fitness values will be maintained to the next generation. The iteration process is repeated until the maximum number of iterations is reached. At this point, the organism with the highest fitness value is output [15].

4.3 The Local Optimization Strategy

After a path is generated using a heuristic algorithm, there are ways to adjust the resulting path to try to make the path shorter, these ways including local optimization strategies. This paper introduces a local optimization strategy for TSP as follows: given a path L, there are n points on it, which are respectively $v_1, v_2 \ldots \ldots, v_n$, denoted as $L(v_1, v_2 \ldots \ldots, v_n)$. Do the following for each point on the path:

1. find another point closest to the current point.
2. reverse the path between the two points so that the two points are adjacent in the path. If the length of the new path is shorter than that of the original path, the new path is used to replace the original path. Otherwise, the path remains unchanged.

4.4   The Hybrid Ant Colony Optimization SOSACOp

The flow chart of our hybrid ant colony optimization is shown in the Fig. 3. In the iteration part of the SOSACOp, we first adjust α and β using three phases of the SOS algorithm. After generating new α and β, each stage uses them to run the basic ACO process once to get the results, and compares the results with the previous optimal results, then keep better α and β. After the parameters adjustment in the three phases of SOS algorithm, the basic flow of ACO is run once with the optimized α and β and the results are obtained. Then the end of this iteration. Therefore, one iteration in SOSACOp is equivalent to four iterations of the ACO. In addition to adjusting parameters, the difference between SOSACOp and the ACO is that after each run of the SOSACOp, we will use local optimization strategy to adjust the results to improve the efficiency of the algorithm. Also, our pheromone update rule has changed. In our algorithm, only the top 20% of ants are updated according to a certain weight. If the length of the shortest path in this iteration is longer than the length of the shortest path recorded before, the original shortest path will be added to the result as a new ant and participate in the pheromone update. In this way, the convergence speed of the algorithm can be further improved, and the possibility of shorter paths will not be ignored, so as to avoid falling into local optimum.

5   Experiments and Analysis

In order to prove the superiority of SOSACOp and the improvement of every part of it is meaningful, we compared SOSACOp with SOSACOp (SOSACOp1) that removed the local optimization strategy, SOSACOp (SOSACOp2) that removed the SOS algorithm, SOSACOp (SOSACOp3) that did not change the pheromone update rule and ACO. The experiment was conducted in Python using AMD Ryzen 7 5800H with Radeon Graphics 3.20 GHz processor and 16G RAM. The experimental results are shown in the Fig. 4.

We use the number of sites of different sizes to conduct comparative experiments. n in the table is the number of sites, ranging from 30 to 80, to test the operation efficiency and results of different algorithms under different scale data. In this experiment, the parameters of the algorithm without SOS are set as follows: the number of ants is set as equal to the number of sites, pheromone volatility rate $\rho = 0.1$, completion rate $Q = 1$, initial $\alpha = 1$, initial $\beta = 5$, and the total number of iterations is set as 200. The parameters of the algorithm with SOS differ from the previous one in that the total number of iterations is set to 50. For each round of testing, each algorithm is run 10 times, producing

10 results, after which we take the average of the 10 results for comparison. From the analysis of the experimental results, it can be seen that the results of SOSACOp are the best. This shows that every improvement of SOSACOp is meaningful and effective. In addition, we also found that the iteration rounds of final results obtained by SOSACOp1 and SOSACOp3 are later than that of SOSACOp, indicating that the convergence speed of SOSACOp is significantly accelerated after the addition of local optimization strategy and new pheromone update rules, that is, the algorithm efficiency is significantly improved.

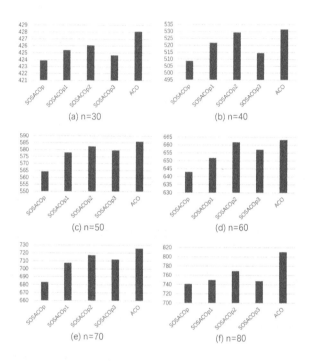

Fig. 4. Comparative experimental results

6   Conclusions and Prospect

By combining ACO with SOS, the parameters of ACO are optimized by using the parameterless characteristics of SOS, and the results are adjusted by using certain local optimization strategies, as well as a different pheromone updating rule. The SOSACOp formed by the above steps is not only more optimized, but also faster convergence than the basic ACO algorithm. It is more efficient to use SOSACOp to solve TSP. In the future, because the local search strategy used in SOSACOp is actually more suitable for the case of using Euclidean distance, a more efficient local optimization strategy can be used to replace the local optimization strategy in SOSACOp, so as to improve its performance for LRDOP. In addition, it can also be adapted to other problems suitable for the algorithm to improve its applicability.

References

1. Arora, S.: Polynomial time approximation schemes for Euclidean traveling salesman and other geometric problems. J. ACM (JACM) **45**(5), 753–782 (1998)
2. Jiang, L., Lai, Y., Chen, Q., Zeng, W., Yang, F., Yi, F.: Shortest path distance prediction based on CatBoost. In: Xing, C., Fu, X., Zhang, Y., Zhang, G., Borjigin, C. (eds.) WISA 2021. LNCS, vol. 12999, pp. 133–143. Springer, Cham (2021). https://doi.org/10.1007/978-3-030-87571-8_12
3. Emambocus, B.A.S., Jasser, M.B., Amphawan, A.: A discrete adapted dragonfly algorithm for solving the traveling salesman problem. In: Fifth International Conference on Intelligent Computing in Data Sciences (ICDS) 2021, pp. 1–6 (2021)
4. Li, D., Xiao, P., Zhai, R., Sun, Y., Wenbin, H., Ji, W.: Path planning of welding robots based on developed whale optimization algorithm. In: 2021 6th International Conference on Control, Robotics and Cybernetics (CRC), pp. 101–105 (2021)
5. Zhang, J., Zhang, Z., Lin, X.: An improved artificial bee colony with self-adaptive strategies and application. In: 2021 International Conference on Computer Network, Electronic and Automation (ICCNEA), pp. 101–104 (2021)
6. Bi, H., Yang, Z., Wang, M.: The performance of different algorithms to solve traveling salesman problem. In: 2021 2nd International Conference on Big Data & Artificial Intelligence & Software Engineering (ICBASE), pp. 153–156 (2021)
7. Wang, S., Liu, Y., Qiu, Y., Zhang, Q., Ma, J., Zhou, J.: Cooperative task allocation for multiple UAVs based on min-max ant colony system. In: 2021 5th Asian Conference on Artificial Intelligence Technology (ACAIT), pp. 283–286 (2021)
8. Wang, C., Li, W., Huang, Y.: An automatic heterogenous-based MAX-MIN ant system with pheromone reconstruction mechanism. In: 2021 17th International Conference on Computational Intelligence and Security (CIS), pp. 252–256 (2021)
9. Chitty, D.M.: A greedy approach to ant colony optimisation inspired mutation for permutation type problems. In: 2021 IEEE Symposium Series on Computational Intelligence (SSCI), pp. 1–8 (2021)
10. Elleuch, W., Wali, A., Alimi, A.M.: Time-dependent ant colony optimization algorithm for solving the fastest traffic path finding problem in a dynamic environment. In: 2021 IEEE International Conference on Systems, Man, and Cybernetics (SMC), pp. 1649–1654 (2021)
11. Dorigo, M., Caro, G.D., Gambardella, L.M.: Ant algorithms for discrete optimization. Artif. Life **5**(2), 137 C172 (1999)
12. Dorigo, M., Gambardella, L.M.: Ant colony system: a cooperative learning approach to the traveling salesman problem. IEEE Trans. Electron. Comput. **1**(1), 53–66 (1997)
13. Deng, X.Y., Yu, W.L., Zhang, L.M.: A new ant colony optimization with global exploring capability and rapid convergence. In: Proceedings of the 10th World Congress on Intelligent Control and Automation. IEEE (2012)
14. Cheng, M.Y., Prayogo, D.: Symbiotic Organisms Search: a new metaheuristic optimization algorithm. Comput. Struct. **139**, 98–112 (2014)
15. Wang, Y., Han, Z.: Ant colony optimization for traveling salesman problem based on parameters optimization. Appl. Soft Comput. **107**(2), 107439 (2021)

Recommendation

Multi-preference Book Recommendation Method Based on Graph Convolution Neural Network

Shuai Li, Xing Xing$^{(\boxtimes)}$, Yunshuo Liu, Zhongxuan Yang, Yong Niu,
and Zhichun Jia$^{(\boxtimes)}$

School of Information Science and Technology, Bohai University, Jinzhou 121013, China
{xingxing,jiazhichun}@qymail.bhu.edu.cn

Abstract. In the book recommendation system, the relationship between users and books can be regarded as a bipartite graph. The user's interest preferences are mined from the graph through Collaborative Filtering recommendation method, and then use the graph convolution neural network to effectively aggregate the characteristics of users and books, so as to form the book recommendation content that user interest. However, the mining of user interest in the existing book recommendation system is always based on a single user preference, ignoring the diversity of user preferences. We propose a multi-preference book recommendation method based on graph convolution neural network to observe the potential reading interest of users when interacting with books. By capturing these reading interests, we can get more information about users' preferences, so as to recommend books more in line with their preferences. We extract the recommendation dataset from the real scenario of Bohai University Library between May 2014 and May 2021, and evaluate our method on it. The experimental results show that our method effectively improves the performance of book recommendation.

Keywords: Book recommendation · Graph convolution neural network · Collaborative filtering · Recommendation system

1 Introduction

The library of each university is often a collection and distribution center of literature and information resources of the University. The rich resources of the library have brought indispensable help to the study, scientific research and life of teachers and students. With the continuous development of computer science, the storage form of documents in the library is no longer limited to the storage of paper documents. Meanwhile, more and more electronic documents are saved in the library system for teachers and students to consult. The scale and volume of information resources in the library also show an explosive growth. Facing the massive information documents in the library, it is often not easy for users to find the information and literature they are interested in in a short time, which leads to the phenomenon of 'information overload'. Although indexing and other methods can provide quick search direction for some users with clear goals, the

X. Zhao et al. (Eds.): WISA 2022, LNCS 13579, pp. 521–532, 2022.
https://doi.org/10.1007/978-3-031-20309-1_46

effect of these methods is not obvious for some users with unclear reading goals. These users need an intelligent system to screen and provide them with books and documents they may be interested in. Obviously, the recommendation system can obviously solve this problem perfectly.

In recent years, the research on recommendation system has never stopped. The earliest recommendation system generally has two methods: user-based recommendation and item-based recommendation. Both methods are calculated based on the similarity of the interaction between users and items. However, there are some problems in these two methods, that is, the ability to deal with sparse matrix is weak, and the calculated similarity is not accurate. Therefore, in order to make collaborative filtering [1] better deal with the sparse matrix problem and enhance the generalization ability, Koren Y [2] proposed the Latent Factor Model (LFM), which applies the matrix decomposition to the model-based recommendation system to recommend users by mining the latent relationship between users and items. However, in essence, LFM is a single-layer neural network without nonlinear process, and its generalization ability is poor. When encountering the complex behavior of users, it is easy to produce overfitting phenomenon. Faced with this problem, He X [3] proposed the Neural network-based Collaborative Filtering model, which uses a neural network to replace the original operation of inner product of users and items with matrix decomposition. Guo H [4] proposed a new neural network framework DeepFM, which combines the advantages of Factorization Machine [5] in recommendation and Deep Neural Networks (DNN) [6–8] in feature learning, so that it can interact between low-order features like FM and high-order features like DNN.

However, the recommendation system based on neural network can effectively deal with structured data, neural network cannot deal with unstructured data, such as social network, knowledge map and so on. Therefore, Graph Neural Network (GNN) [9–11] was proposed to solve these unstructured data, which further expanded the application scope of recommendation system [12]. Generally speaking, GNN has one more adjacency matrix than neural network. GNN can obtain more information through these adjacency matrices to optimize the recommendation performance. Recently, GNN are popular, including graph convolution neural network (GCN) [13, 14] and graph attention network(GAT) [15–17]. Among them, GCN has excellent performance, which can effectively aggregate the characteristics of users and item and alleviate the problem of data sparsity. GAT can assign different attention scores to each neighbor, so as to identify more important neighbors.

Recommender systems based on graph convolutional neural networks are undoubtedly successful, but most graph convolutional recommender systems usually only consider a single motivation. In fact, user-item interaction should not consider only one latent component. As far as book recommendation is concerned, a person's motivation for reading is diverse, not just a single one. For example, some people may only choose books with good titles, while others may be motivated only by professional needs. Therefore, our method considers using multiple latent components to distinguish different reading motivations, so as to capture more user preference information, deal with more complex interaction features can be processed, and more accurate recommendation effects can be achieved.

In this paper, we propose a multi-preference graph convolution book recommendation method (MGC). The user interaction method is divided into two parts. In the first part, the method first decomposes the user-item interaction model to obtain each reading motivation component of each user or item. In the second part, these components are combined according to the weight to obtain a final user or item embedding.

2 The Proposed Framework

2.1 Definitions and Notations

We set the user set as $U = \{u_1, u_2, \cdots, u_{N_u}\}$, and the project set as $I = \{i_1, i_2, \cdots, i_{N_i}\}$, where N_u and N_i are the number of user and item respectively. Suppose users and items have feature matrices Q and P, respectively, where $Q = \{q_1, q_2, \cdots, q_{N_u}\} \in \mathbb{R}^{L_u \times N_u}$, L_u is the dimension of user features; $P = \{p_1, p_2, \cdots, p_{N_i}\} \in \mathbb{R}^{L_i \times N_i}$, L_i is the dimension of item features (Fig. 1).

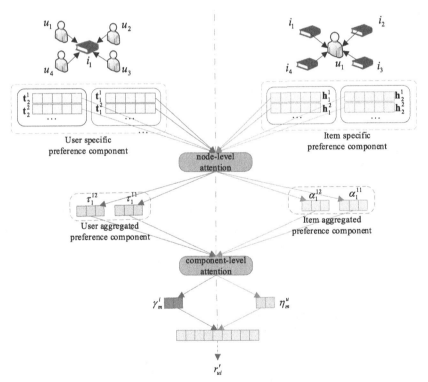

Fig. 1. The framework of multi-preference book recommendation method based on graph convolution neural network (MGC).

2.2 Preference Weight Decomposition

Let the user-item bipartite graph be G, assuming that it can be represented by M latent components. First, we need to extract the specific preference components of users and items from these M latent components. We set the user specific preference conversion matrix $W = \{W_1, W_2, \cdots, W_M\}$ and the item specific preference conversion matrix $V = \{V_1, V_2, \cdots, V_M\}$ to extract different features corresponding to feature preferences. Among the M latent components, we use the preference transformation matrix V_m of the m-th item i to capture the m-th user or item specific preference component h_m^i in the user-item bipartite graph:

$$h_m^i = V_m p_i \tag{1}$$

For user u, the specific preference component t_m^u of its m-th user is u as follow:

$$t_m^u = W_m q_u \tag{2}$$

User u and item i have M specific components $\{t_m^u\}_{m=1}^M$ and $\{h_m^i\}_{m=1}^M$, respectively. User u's motivation to read item i based on the m-th component does not need to be represented by aggregate all the components of item i, so when we infer the possibility of the user reading the item because of the component, we use the deep node-level attention neural network, and the formula is as follows:

$$e_m^{ui} = att_{node}(t_m^u, h_m^i, m) \tag{3}$$

Among them, att_{node} represents a deep neural network for node-level attention. For the m-th specific preference component, all nodes based on the specific preference component m share att_{node}, It is not difficult to see from Eq. (3) that the probability of user u actually reading the item due to the m-th component is determined by the characteristics of this component.

After obtaining the preference probability e_m^{ui} based on the preference components, we normalize them through the softmax function to obtain the weight α_m^{ui}, and then we can get the contribution of each item under preference M:

$$\alpha_m^{ui} = softmax(e_m^{ui}) = \frac{\exp(\sigma(a_m^T \cdot [t_m^u \oplus h_m^i]))}{\sum_{i \in P_u} \exp(\sigma(a_m^T \cdot [t_m^u \oplus h_m^i]))} \tag{4}$$

where σ denotes the activation function, and we use the ReLU activation function. We set the m-th component's vector is a_m. It can be seen from Eq. (4) that the weight of preference component m depends on its feature preference component.

Finally, we learn the user latent factor z_m^u for a specific preference component by aggregating all about the m-th specific component in the item set p_u, as follows:

$$z_m^u = \sigma(\sum_{i \in P_u} \alpha_m^{ui} \cdot h_m^i) \tag{5}$$

We call z_m^u the item aggregation component, and the corresponding reading motivation represented by the component can be obtained according to this component, Meanwhile, each user has correspondingly M item aggregation components, and then we need to combine these M item aggregation components to get the final user embedding.

2.3 Node Feature Combinations

After obtaining the aggregation components z_m^u, we need to give weights to these components respectively. According to the weight, we can select the components within a certain threshold for aggregation, and then we can get the embedded components we finally need.

Since each preference of a user is affected by different degrees of external influence, the contribution of each preference to the user is not unique. Here we use a component-level attention mechanism [18, 19] to obtain the weight of each item aggregated component as follows:

$$d_m^u = \sigma(C_m \cdot [z_m^u \oplus t_m^u] + b_m) \tag{6}$$

Equation (6) is to connect z_m^u and s_m^u to obtain a unified embedding item, which takes into account the preference of user u while considering item aggregation components, where C_m is the weight matrix and b_m is the bias vector.

$$w_m = \sigma(q^T \cdot d_m^u + b) \tag{7}$$

Equation (7) is to extract preference information fused with user embedded features through a deep neural network. Where q represents the component-level attention vector, b represents the bias vector, and q and b are shared parameters.

$$\eta_m^u = \frac{\exp(w_m)}{\sum_{k=1}^{M} \exp(w_k)} \tag{8}$$

Equation (8) is to normalize w_m to obtain the final weight η_m^u. It is not difficult to see that the higher the η_m^u, the stronger the m-th reading motivation.

Through the above steps, we get the weight of each item aggregation component $\{\eta_m^u\}_{m=1}^M$. According to the weight, we can aggregate all user preference information to obtain the final embedding of user u:

$$z_u = \sum_{m=1}^{M} \eta_m^u \cdot z_m^u \tag{9}$$

The above is the user modeling learning process of our method. Similarly, the process of item modeling learning is not repeated here. We set the final item embedding as k_i.

2.4 Rating Prediction

In this subsection, we apply our proposed method to the recommendation task of rating prediction, we first concatenate the user's final embedding z_u and the item's final embedding k_i, and then take them as input to make predictions through Multilayer Perceptron (MLP), and finally get the rating r'_{ui}:

$$g_1 = [z_u \oplus k_i], \tag{10}$$

$$g_2 = \sigma(W_2 \cdot g_1 + b_2), \tag{11}$$

$$g_{l-1} = \sigma(W_l \cdot g_{l-1} + b_l), \tag{12}$$

$$r'_{ui} = w^T \cdot g_{l-1}, \tag{13}$$

where l is the index of a hidden layer.

2.5 Method Training

We use the following objective function formula to optimize the parameters in our method to minimize the gap between the predicted rating and the realistic rating:

$$Loss = \frac{1}{2|O|} \sum_{(u,i) \in O} (r'_{ui} - r_{ui})^2, \tag{14}$$

where O is the set of observed ratings and r_{ui} is user u's base ground truth rating for item i.

In order to optimize our objective function, we adopt L0 regularization [20] to the objective function. This method is used to prune the neural network and sparse the multi-component extraction matrix W and V, which can effectively improve the training speed and model generalization ability and avoid the occurrence of over fitting problems. Meanwhile, for the item read by the user, we will weight and sort all neighborhoods in descending order, and add a threshold at the same time. we use the sampling method to select neighborhood for convolution operation when the number of neighborhoods exceeds the threshold we set. This can effectively speed up method optimization.

Algorithm 1. multi-preference graph convolution algorithm

Input: user set U, item set I, item set p_u interacting with user u. user specific preference conversion matrix W, user specific preference conversion matrix V

Output: User rating prediction r'_{ui}

1:   **for** i to n, $i \in U$ **do**
2:       Calculate the transformation matrix of users and items under each preference component by Eq.(1) and Eq.(2)
3:       Calculate the attention weight of items under each preference component m by Eq(3) and Eq.(4)
4:       Calculate the user characteristics under the preference component m by Eq.(5)
5:       Calculate attention weights for each preference component by Eq.(6) to Eq.(8).
6:       Calculate user feature representation by Eq.(10)
7:   **end for**
8:   Learning parameters by
9:   Calculate rating score r'_{ij}, by Eq. (11) to Eq.(14)
10: Estimate by MAE and RMSE

---

3 Experiments

We extract the recommendation dataset from the real scenario of Bohai University Library between May 2014 and May 2021, and used the dataset to evaluate our experiment. We compare our method with the current mainstream recommendation algorithm to evaluate our method. Meanwhile, we also explore the impact of different parameters on the method, find the optimal value of parameters and improve the effectiveness of the method.

3.1 Experimental Settings

Datasets. We extracted and sorted out the database information of Bohai University Library, obtained the recommendation dataset of teachers and students between May 2014 and May 2021, and ensured the authenticity and reliability of the dataset. The data set contains 772,371 books, 61,947 users and 1,271,964 borrowing records. In order to ensure the density of the data, we extracted 32217 books and the scoring records of 8543 users are from the database as the experimental dataset. The specific information of the data set is shown in Table 2. Here, we divide the score into {1,2,3,4,5} five levels.

Table 2. Information of Bohai University library dataset.

Dataset	Number
Number of users	32217
Number of items	8543
Number of ratings	28967
Number of timestamp	32217

Evaluation Metrics. In order to effectively evaluate the performance of the algorithm, we use two most commonly used evaluation metrics, namely mean absolute error (MAE) and root mean square error (RMSE) [21], to evaluate the prediction accuracy. When MAE and RMSE are smaller, the prediction accuracy is higher. It is worth noting that minor improvements in MAE and RMSE usually have a relatively large impact on the quality of the algorithm. MAE and RMSE are calculated as follows:

$$MAE = \frac{\sum_{i=1}^{n} |y_i - \hat{y}_i|}{n}, \tag{15}$$

$$RMSE = \sqrt{\frac{\sum_{i=1}^{n} (y_i - \hat{y})^2}{n}}, \tag{16}$$

where y_i is the true value of the test dataset, $\hat{y}_i$ is the predicted value of the test dataset, and n is the amount of data in the test set.

Baseline. In order to evaluate the performance of the algorithm, we selected three representative groups of algorithms, including two groups of traditional recommendation algorithms and one group of recommendation algorithms based on neural networks, and compared their performance with our proposed method on the same dataset. The following are their details

PMF [22]: Probabilistic Matrix Factorization adds a probability model to the traditional Matrix Factorization, which can better deal with user data with few rates and very large datasets.

AutoRec [23]: This method proposed a Collaborative Filtering method based on AutoEncoder, which combines Collaborative Filtering with AutoEncoder, and uses AutoEncoder to learn the compressed vector expression of row or column data in the rating matrix to accurately infer the user's score value of non-rated items.

SVD++ [24]: This method considers the implicit feedback of users on the basis of Singular Value Decomposition algorithm.

Implementation. Our experimental environment is Ubuntu 20.04.2LTS operating system, equipped with NVIDIA graphics card driver, version number 470.103.01 and CUDA version 11.4. The excellent experimental environment strongly supports the rapid operation and implementation of our algorithm. Our code is based on PyTorch machine learning library to complete the construction of our method.

We tested several parameters of the method for several rounds to achieve the best experimental effect. We set the number of components of the method as K and its range as $\{2, 3, 4, 5\}$. For the embedded latitude D, we set it to be changed within the range of $\{8, 16, 32, 64\}$. Batch size and learning rate are searched in $\{64, 128, 256, 512\}$ and $\{0.0005, 0.001, 0.002, 0.0025\}$ respectively. For all neural networks, the activation function of neural networks is ReLU function. We use gaussian distribution to randomly initialize the method parameters. Dropout is used in the method. The dropout rate is tested in the range of $\{0.1, 0.4, 0.5, 0.6\}$.

In terms of evaluation, we use two widely used evaluation protocols: RMSE and MAE as evaluation metrics. For each experiment, we will initialize the method to ensure the randomness of each dataset selection, and do the same test 5 times, and take the average value as the final result [25].

3.2 Performance Comparison

We used RMSE and MAE to compare the recommended performance of all methods. Figure 2 shows the overall rating prediction error. In the baseline, PMF is based on matrix decomposition, AutoRec is based on automatic encoder, and SVD++ is based on feature decomposition. Obviously, AutoRec is superior to SVD++ and PMF, which also reflects the powerful function of neural network model. Our method is obviously superior to all baselines, which reflects the effectiveness and superiority of our method. The comparison results show that our method is better than the representative baseline, it can better provide book recommendation. We can also find that the neural network model can improve recommendation performance.

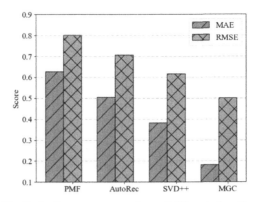

Fig. 2. Performance comparison of different algorithm

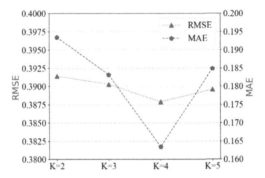

Fig. 3. Effect of the number of latent component K

Since our method needs to consider multiple preferences, we set the range of component K to be $\{2, 3, 4, 5\}$. On the premise that other parameters remain unchanged, we get the recommendation results shown in Fig. 3. It can be seen from the Fig. 3 that at the beginning, with the increase of the number of components, the performance of the method is also steadily improving. When $K = 4$, it reaches a minimum value, and then when $K = 5$, the performance begins to decline. It is concluded that the method has the best performance when $K = 4$.

After obtaining the optimal number of components K, we tested the impact of the embedded dimension size on the performance of the method with $K = 4$ unchanged. The experimental results are shown in Fig. 4. We set four groups of embedded dimension sizes $\{8, 16, 32, 64\}$. Initially, the performance will be gradually optimized with the increase of embedded dimension d, because embedded dimension d represents the representation ability of the method to a certain extent. The larger d is, the stronger the representation ability is. When $d = 16$, the embedded dimension is the best value, and increasing the embedded dimension d will degrade the performance of the method.

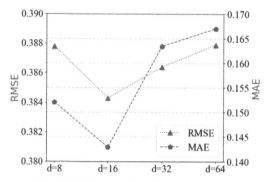

Fig. 4. Effect of embedding dimension d

4   Conclusion

In this paper, we propose multi-preference book recommendation method based on graph convolution neural network. Firstly, we decompose the user-item inter-action to obtain multi-preference components of each user or item, and each component is given a weight through the attention mechanism. Then, through the aggregation algorithm, these components are combined according to the weight to obtain the final embedding of users or item. Finally, the method is scored and predicted by multi-layer neural network and obtain the characteristic representation of users and items. We evaluate the algorithm on the loan information dataset of Bohai University Library between May 2014 and May 2021. The experimental results show that our method achieve good results in multiple evaluation metrics.

Acknowledgements. This paper is partially supported by the National Natural Science Foundation of China under Grant No.61972053 and No.62172057, The Project is sponsored by 'Liaoning BaiQianWan Talents Program' under Grant No.2021921024, Master's innovation fund project of Bohai University YJC2021–053.

References

1. Wang, X., He, X., et al.: Neural graph collaborative filtering. In: Proceedings of the 42nd International ACM SIGIR Conference on Research and Development in Information Retrieval, SIGIR 2019, July 2019, pp. 165–174 (2019)
2. Koren, Y., Bell, R., Volinsky, C.: Matrix factorization techniques for recommender systems. Computer **42**(8), 30–37 (2009)
3. He, X., Liao, L., et al.: Neural collaborative filtering. In: Proceedings of the 26th International Conference on World Wide Web, WWW 2017, April 2017, pp. 173–182 (2017)
4. Guo, H., Tang, R., et al.: DeepFM: a factorization-machine based neural network for CTR prediction. In: Proceedings of the 26th International Joint Conference on Artificial Intelligence, IJCAI 2017, August 2017, pp. 1725–1731 (2017)
5. Rendle, S.: Factorization machines. In: Proceedings of the 10th IEEE International Conference on Data Mining, ICDM 2010, December 2010, pp. 14–17 (2010)

6. Chen, J., Zhou, D., et al.: Closing the generalization gap of adaptive gradient methods in training deep neural networks. In: Twenty-Ninth International Joint Conference on Artificial Intelligence and Seventeenth Pacific Rim International Conference on Artificial Intelligence, IJCAI 2020, pp. 3267–3275 (2020)
7. Cheng, H.T., Koc, L, et al.: Wide & deep learning for recommender systems. In: Proceedings of the 1st Workshop on Deep Learning for Recommender Systems, DLRS 2016, September 2016, pp. 7–10 (2016)
8. Covington, P., Adams, J., Sargin, E.: Deep neural networks for YouTube recommendations. In: Proceedings of the 10th ACM Conference on Recommender Systems 2016, pp. 191–198 (2016)
9. Gao, C., Zheng Y., et al.: Graph neural networks for recommender systems: challenges, methods, and directions. In: Proceedings of the Fifteenth ACM International Conference on Web Search and Data Mining, WSDM 2022, February 2022, pp. 1623–1625 (2022)
10. Wu, Z., Pan, S., et al.: A comprehensive survey on graph neural networks. IEEE Trans. Neural Netw. Learn. Syst. **32**, 4–24 (2019)
11. Fan, W., Ma, Y., et al.: Graph neural networks for social recommendation. In: Proceedings of the World Wide Web Conference, WWW 2019, May 2019, pp. 417–426 (2019)
12. Shuman, D.I., Narang, S.K., et al.: The emerging field of signal processing on graphs: extending high-dimensional data analysis to networks and other irregular domains. IEEE Signal Process. Mag. **30**(3), 83–98 (2013)
13. Defferrard, M., Bresson, X., Vandergheynst, P.: Convolutional neural networks on graphs with fast localized spectral filtering. In: Proceedings of the 30th International Conference on Neural Information Processing Systems, NIPS 2016, December 2016, pp. 3844–3852 (2016)
14. Chiang, W.L., Liu, X., et al.: Cluster-GCN: an efficient algorithm for training deep and large graph convolutional networks. In: Proceedings of the 25th ACM SIGKDD International Conference on Knowledge Discovery and Data Mining, KDD 2019, July 2019, pp. 257–266 (2019)
15. Velikovi, P., Cucurull, G., et al.: Graph attention networks. In: Proceedings of the 6th International Conference on Learning Representations (2018)
16. Zhao, H., et al.: Multivariate time-series anomaly detection via graph attention network. In: IEEE International Conference on Data Mining 2020, pp. 269–279 (2020)
17. Song, W., Xiao, Z., et al.: Session-based social recommendation via dynamic graph attention networks. In: Proceedings of the Twelfth ACM International Conference on Web Search and Data Mining, WSDM 2019, January 2019, pp. 555–563 (2019)
18. Chen, C., Zhang, M., et al.: Neural attentional rating regression with review-level explanations. In: Proceedings of the 2018 World Wide Web Conference, WWW 2018, April 2018, pp.1583–1592 (2018)
19. Yang, Z., Yang, D., et al.: Hierarchical attention networks for document classification. In: Proceedings of the 2016 Conference of the North American Chapter of the Association for Computational Linguistics: Human Language Technologies, pp.1480–1489 (2016)
20. Louizos, C., Welling, M., Kingma, D.P.: Learning sparse neural networks through regularization. In: Proceedings of the 6th International Conference on Learning Representations (2019)
21. Wang, S., Tang, J., et al.: Exploring hierarchical structures for recommender systems. IEEE Trans. Knowl. Data Eng. **30**, 1 (2018)
22. Salakhutdinov, R.: Probabilistic matrix factorization. In: Proceedings of the 20th International Conference on Neural Information Processing Systems, NIPS 2007 December 2007, pp.1257–1264 (2007)
23. Sedhain, S., Menon AK, et al.: AutoRec: autoencoders meet collaborative filtering. In: Proceedings of the 24th International Conference on World Wide Web Companion, WWW 2015, May 2015, pp.111–112 (2015)

24. Koren, Y.: Factorization meets the neighborhood: a multifaceted collaborative filtering model. In: Proceedings of the 14th ACM SIGKDD International Conference on Knowledge Discovery and Data Mining, KDD 2008, August 2008, pp. 426–434 (2008)
25. Li, C., Zhai, R., Zuo, F., Yu, J., Zhang, L.: Mixed multi-channel graph convolution network on complex relation graph. In: Xing, C., Fu, X., Zhang, Y., Zhang, G., Borjigin, C. (eds.) WISA 2021. LNCS, vol. 12999, pp. 497–504. Springer, Cham (2021). https://doi.org/10.1007/978-3-030-87571-8_43

Sentiment-Aware Neural Recommendation with Opinion-Based Explanations

Lingyu Zhao[1], Yue Kou[1(✉)], Derong Shen[1], Tiezheng Nie[1], and Dong Li[2]

[1] Northeastern University, Shenyang 110004, China
{kouyue,shenderong,nietiezheng}@cse.neu.edu.cn
[2] Liaoning University, Shenyang 110036, China
dongli@lnu.edu.cn

Abstract. Explainable recommendation systems are crucial for complex decision making, which provide users with the recommendation results as well as the reasons why such items are recommended. However, most existing explainable recommendation methods only consider one aspect of user sentiment, such as ratings or reviews, which fails to capture the fine-grained user sentiment. In this paper, we propose a novel sentiment-aware neural recommendation model, named SNROE, which jointly performs a rating prediction task and an explanation generation task, to guarantee both the accuracy of recommendation and the personalization of explanations. For the rating prediction task, we adopt MLP to learn user/item representations. For the explanation generation task, we propose a sentiment-aware explanation generation method, which utilizes pretrained Transformer to generate opinion-based explanations by fusing users' rating-level sentiment, aspect-level sentiment and review-level sentiment. We also propose a joint training algorithm to jointly optimize the above two tasks. The experiments demonstrate the effectiveness and the efficiency of our proposed model compared to the baseline models.

Keywords: Explainable recommendation · Sentiment-aware neural recommendation · Opinion-based explanations · Rating prediction

1 Introduction

Recently, more and more attention has been paid to explainable recommendation, because it can provide the corresponding explanations while giving users the recommendation results. Undoubtedly, these provided explanations can improve the persuasiveness of recommender systems and help users make better decisions quickly [1,2]. The explanations can be classified into: user-based or item-based

This work was supported by the National Natural Science Foundation of China under Grant Nos. 62072084, 62172082 and 62072086, the Science Research Funds of Liaoning Province of China under Grant No. LJKZ0094, the Natural Science Foundation of Liaoning Province of China under Grant No. 2022-MS-171, the Science and Technology Program Major Project of Liaoning Province of China under Grant No. 2022JH1/10400009.

explanations, feature-based explanations, visual explanations, social explanations, and opinion-based explanations.

The opinion-based explanations are generated by extracting the user opinions from the user-generated content. Now more and more reviews written by users are available from the Internet. Figure 1 shows a review made by a user after purchasing a keyboard. The review contains the user's rich feelings and experiences. Then, users are more likely to accept recommendations if we provide explanations in similar sentence forms. However, most existing methods only consider a certain aspect of user sentiment. Obviously, such an explanation is difficult to gain user trust.

★★★★☆ **Great and fun keyboard**

Reviewed in the United States on March 6, 2022
Keyboard is nice and the couples are great but the keys have some dimness on them it's not as bright as how the pictures advertise them to be. The different modes for the lights are great depending on the mood you are in and show pretty light shows.

Fig. 1. An example of a user review.

Table 1 shows an example of different explanations, each of which is generated from an aspect of user sentiment. From Table 1, we can see that if only one aspect of the user sentiment is considered, the generated explanation is either over-generalized or opposite to the sentiment of real reviews. If we can generate an explanation by taking into account multiple aspects of user sentiment, it will greatly enhance the credibility of the explanation.

Table 1. An example of sentiment-aware explanation generation.

Ground-truth	Review: The buffet dinner and breakfast are top notch and have many western and chinese choices. Rating: 5.0
Rating-level	The food is good
Aspect-level	The buffet dinner and breakfast are not up to the standards
Review-level	The breakfast is top notch
Hybrid-level	The buffet dinner and breakfast are top notch and have many choices

In this paper, we propose a novel sentiment-aware neural recommendation model, named SNROE, which jointly performs a rating prediction task and an explanation generation task, to guarantee both the accuracy of recommendation and the personalization of explanations. The main contributions of this paper are as follows:

- We propose a multi-task learning framework based on the interactions between the rating prediction task and the explanation generation task.

Through interactive learning, the performance of two tasks is improved simultaneously.

- For the rating prediction task, we adopt MLP to learn user/item representations. For the explanation generation task, we propose a sentiment-aware explanation generation method, which utilizes pretrained Transformer to generate opinion-based explanations by fusing users' rating-level sentiment, aspect-level sentiment and review-level sentiment.
- Extensive experimental studies based upon two real-world datasets are conducted. The experiments demonstrate the effectiveness of our proposed method.

The rest of this paper is organized as follows. Section 2 introduces the related work. Section 3 introduces our model. Section 4 proposes a joint training algorithm based on SNROE. Section 5 shows the experimental results. Section 6 concludes the paper and presents the future work.

2   Related Work

Existing explainable recommendation methods mainly include matrix/tensor factorization approaches, topic modeling approaches, graph-based approaches [3], deep learning approaches, knowledge-based approaches, rule mining approaches, and post-hoc/model-agnostic approaches [1]. Here we mainly introduce the deep learning approaches.

In some existing works, they use highlighted review words as explanations. HANCI [4] uses an RNN text processor in review feature extraction and utilizes a hierarchical attention network based on multi-level review text analysis to extract more precise user preferences and item latent features. CARL [5] combines user-item interaction and review information in a unified framework. User reviews are summarized as content features, which are further integrated using user and item embeddings to predict final ratings. TARMF [6] co-learns user and item information from ratings and customer reviews, by optimizing matrix factorization and an attention-based GRU network.

Different from the above methods, some researchers generate textual sentences as explanations by RNN/LSTM. NETE [2] generates template-controlled sentences that comment about specific features. NRT [7] can simultaneously predict precise ratings and generate abstractive tips with gated recurrent neural networks. Att2Seq [8] uses multilayer perceptrons with one hidden layer to encode attribute information into a vector and the decoder is built upon multilayer recurrent neural networks (RNNs) with long short-term memory (LSTM) units. CAML [9] designs an encoder-selector-decoder architecture inspired by human's information-processing model in cognitive psychology. HSS [10] proposes a hierarchical sequence-to-sequence model for personalized explanation generation and further proposes an auto-denoising mechanism based on topical item feature words for sentence generation.

Recently, some researchers generate explanations by transformer [11–13]. They take advantage of the powerful language modeling capabilities of the transformer and use the comment information to generate corresponding explanations. Meanwhile, instead of generating explanations, some researchers adopt user reviews or image to explain [14,15].

However, most of the existing works only consider an aspect of the user's sentiment, which does not deeply mine the fine-grained sentiment, and they are not supported by large corpora. Compared to them, our model utilizes a multi-task learning framework to generate explanations that fuse three aspect of user sentiment.

3   The Proposed Model

Our proposed model (i.e. SNROE) consists of two tasks including rating prediction and explanation generation. Our aim is to predict a rating $\hat{r}_{u,v}$ and generate an explanation e for given a user and an item. At the training stage, users, items, ratings, features, and explanation text are used as inputs. At the testing stage, users, items and features are used as inputs for recommendation and explanation. In this section, we first introduce the overall framework of the model, then introduce the method for rating prediction, finally introduce the method for explanation generation.

3.1   Model Overview

The overall framework of SNROE is shown in Fig. 2. SNROE mainly consists of two parts: rating prediction and explanation generation.

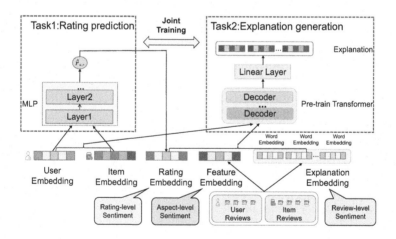

Fig. 2. Overall framework of SNROE.

- Rating prediction. It is to predict ratings, which are used as the users' rating-level sentiment input into the explanation generation task. Meanwhile, user and item representations can also be better learned, which in turn improves personalization and accuracy.

- Explanation generation. It is to generate sentiment-aware explanations, by fusing users' rating-level sentiment, aspect-level sentiment and review-level sentiment, making the generated explanations more specific and personalized.

3.2   Task 1: MLP-Based Rating Prediction

In most existing latent factor models, the rating is computed by a linear combination of user latent factors, item latent factors, and biases. The learned latent factors may fail to capture the complex structure implicit in the user's historical interactions. Therefore, we employ MLP (a nonlinear transformation model) with L hidden layers to capture interactions between users and items, as shown on the left side of Fig. 2. Specifically, we first get the embedding vectors of user and item ids, and then feed them into the MLP. The encoding of user ids and item ids is described in the explanation generation task. The rating score is calculated as follows:

$$\begin{cases} \boldsymbol{h}_0 = \sigma(\boldsymbol{W}_0[\boldsymbol{u}, \boldsymbol{v}] + \boldsymbol{b}_0) \\ \boldsymbol{h}_1 = \sigma(\boldsymbol{W}_1\boldsymbol{h}_0 + \boldsymbol{b}_1) \\ \quad\cdots\cdots \\ \boldsymbol{h}_L = \sigma(\boldsymbol{W}_L\boldsymbol{h}_L - 1 + \boldsymbol{b}_L) \end{cases} \tag{1}$$

where $\boldsymbol{W}_0 \in \mathbb{R}^{d_h \times 2d}$, $\boldsymbol{b}_0 \in \mathbb{R}^{d_h}$, $\boldsymbol{W}_x \in \mathbb{R}^{d_h \times d_h}$ and $\boldsymbol{b}_x \in \mathbb{R}^{d_h}$ are the mapping matrix and bias vectors in the hidden layers, L is the index of a hidden layer and $\sigma(\cdot)$ denotes the sigmoid function, $[\boldsymbol{u}, \boldsymbol{v}]$ is the concatenated vector of user and item embedding vector. The output layer transforms $\boldsymbol{h}_L$ into a real-valued rating $\hat{r}$:

$$\hat{r} = \boldsymbol{w}\boldsymbol{h}_L + \boldsymbol{b} \tag{2}$$

where $\boldsymbol{w} \in \mathbb{R}^{d_h}$, $\boldsymbol{b} \in \mathbb{R}$.

To minimize the difference between the true and predicted ratings, and optimize the parameters, we adopt the mean squared error loss as its objective function:

$$L_r = \frac{1}{|\Omega|} \sum_{(u,v)\in\Omega} (r_{u,v} - \hat{r}_{u,v})^2 \tag{3}$$

where Ω represents the training set. $r_{u,v}$ is the ground truth rating assigned by the user u to the item v.

3.3   Task 2: Sentiment-Aware Explanation Generation

The main disadvantage of existing RNN-based or GRU-based models is training language models that generate textual explanations from scratch, which makes training them computationally expensive and not cost-effective. Furthermore, the fluency of the generated explanations is questionable since their respective language models are not trained on the huge corpus. Therefore, we adopt a pretrained Transformer for text generation.

Multi-level Sentiment Representation. The input sequence X is consist of user id, item id, rating, features, explanation text. First, we take user id u and item id v as input of the explanation generation task, making the generated explanations more personalized. Then, the rating $\hat{r}$ in the rating prediction task is taken as the user's rating-level sentiment. Since the user id u, item id v, and user's rating-level sentiment $\hat{r}$ are all special word tokens relative to the vocabulary of the pretrained Transformer model. So, we separately encode representations of user, item, and rating-level sentiment. We prepare three sets of randomly initialized token embeddings U, V and R respectively. After performing embedding look-up, we can obtain their embedding representations. Then, we take the features f extracted from user and item reviews as the user's aspect-level sentiment. During the training phase, we take the explanation text t, which consists of a sequence of words, as the user's review-level sentiment. Because features and explanation text are sequences of words, their representations come from the vocabulary of the pretrained Transformer model.

Multi-level Sentiment Fusing. Consider that the explanation generation task is to predict the next word based on what is read in the text. So we adopted an autoregressive masking mechanism, which is used during the training stage so that the attention calculation can only see what is before the word, but not what is after the word. Next, we describe the process in detail.

Firstly, we add the token representation of the sequence X and its position representation to get the input representation of the sequence X:

$$E(X) = TE(X) + PE(X) \tag{4}$$

where $TE(X)$ denotes the token embedding representation of sequence X, which is concatenated from the embedding representations of user, item, rating, features, and explanation text. $PE(X)$ denotes the position embedding representation of sequence X.

Multi-head attention allows the model to jointly attend to information from different representation subspaces at different positions. Then, the multi-head self-attention layer is defined as follows:

$$A^{(1)} = MultiHead(E(X), E(X), E(X)) \tag{5}$$

$$MultiHead(E(X), E(X), E(X)) = Concat(head_1, ..., head_h)W^O$$
$$where \ head_i = Attention(E(X)W_i^Q, E(X)W_i^K, E(X)W_i^V) \tag{6}$$

where the projections are parameter matrices $W_i^Q, W_i^K, W_i^V \in \mathbb{R}^{d_{model} \times d}$, $W^O \in \mathbb{R}^{hd \times d_{model}}$.

Next, we input the results obtained above into the position-wise feed-forward networks as follows:

$$O^{(1)} = FFN(A^{(1)}) \tag{7}$$

$$FFN(x) = max(0, xW_1 + b_1)W_2 + b_2 \tag{8}$$

This consists of two linear transformations with a ReLU activation in between.

Similarly, for other layers:

$$A^{(n)} = MultiHead(O^{(n-1)}, O^{(n-1)}, O^{(n-1)}) \tag{9}$$

$$O^{(n)} = FFN(A^{(n)}) \tag{10}$$

The final sequence representation O is the output states from the last layer.

Objective Function. Then, O is passed through a linear layer, we can get the probability distribution of words $p(y)$. Finally, we adopt the Negative Log-Likelihood (NLL) as the explanation generation task's loss function, and compute the mean of user-item pairs in the training set:

$$L_e = \frac{1}{|\Omega|} \sum_{(u,v) \in \Omega} \frac{1}{|X_{u,v}|} \sum_{t=1}^{|X_{u,v}|} -\log p(y_{3+|f|+t}) \tag{11}$$

where $|f|$ is the length of features and $p(y_{3+|f|+t})$ is offset by $3+|f|+t$ positions because the explanation is placed at the end of the sequence.

4   Joint Training Algorithm

We integrate the rating prediction task and the explanation generation task into a multi-task learning framework. Its objective function is:

$$J = min(\lambda_r L_r + \lambda_e L_e + \lambda_n ||\Theta||^2) \tag{12}$$

where L_r is the rating regression loss from Eq. 3, L_e is the explanation generation loss from Eq. 11. Θ is the set of parameters. λ_r and λ_e are the weight proportion of each term. The whole framework can be efficiently trained using back-propagation in an end-to-end paradigm.

SNROE-based joint training algorithm includes the following steps (as shown in Algorithm 1):

Step1: Parameter initialization (Line 1). For user id and item id and rating, we encode them separately to get their embedding representations. The representation of features and explanation text is then obtained from the vocabulary in the pretrained Transformer.

Step2: Rating prediction (Line 3–4). Input the user's embedding vector and the item's embedding vector into the function f_r to get the predicted score and calculate the loss function.

Step3: Explanation generation (Line 5–6). Input the user's embedding vector and item's embedding vector, as well as the embedding vector of the predicted score and the embedding vector of the features into the function f_t to get the predicted explanation and calculate the loss function.

Step4: Joint training (Line 7–14). Putting two tasks into a multi-task learning framework and calculate the final loss function. Then, update the model parameters by gradient descent. If the stop condition is satisfied, and then stop the iteration.

Step5: Generate predicted rating and explanation (Line 15). By training to get the optimized model parameters, we can get the final prediction scores and generated explanations.

Algorithm1: Joint training algorithm

Input: user id u, item id v, feature f
Output: predicted rating, generated explanation
1   Initialize all trainable parameters
2   while step != total_step   //batch samples for training simultaneously
3       $\hat{r} = f_r(\boldsymbol{u},\boldsymbol{v})$ //user $\boldsymbol{u}$, item $\boldsymbol{v}$
4       $r\_loss = \text{rating\_criterion}(r,\hat{r})$ //the loss of rating prediction
5       $\boldsymbol{e} = f_t(\boldsymbol{u},\boldsymbol{v},\boldsymbol{f},\hat{r})$ // feature $\boldsymbol{f}$, rating $\hat{r}$, explanation $\boldsymbol{e}$
6       $t\_loss = loss(t,\boldsymbol{e})$ //the loss of explanation generation
7       $loss = rating\_reg * r\_loss + text\_reg * t\_loss$
8   update $loss$
9   Optimize the parameter set
10   step++
11   if step == total_step
12   break
13   end if
14   end while
15   return $\hat{r}$, $\boldsymbol{e}$

---

5   Experiments

5.1   Experimental Settings

Datasets. We choose two real-world datasets: TripAdvisor and Amazon. The statistics of the two datasets are shown in Table 2.

Table 2. Statistics of the experimental datasets.

Dataset	Users	Items	Reviews	Features
TripAdvisor	9765	6280	320023	5069
Amazon	7506	7360	441783	5399

Evaluation Metrics. For the rating prediction task, we use Root Mean Square Error (RMSE) and Mean Absolute Error (MAE) to measure the recommendation results. For the two metrics, a lower value indicates a better performance.

For the explanation generation task, we adopt two evaluation matrices: BLEU [16] and ROUGE [17]. Then, we use BLEU-1, BLEU-4 and Recall, Precision, and F-measure of ROUGE-1 (R1) and ROUGE-2 (R2) to evaluate the quality of the generated explanation.

Parameter Settings. We optimize our model with AdamW, and set batch size to 64. The learning rate is set to 0.001 and we set the maximum length of generated explanations to 20. For the optimization objective, we let the weight parameter $\lambda_r = \lambda_e = 1.0$. At each epoch, if the model reaches the minimum loss, we save the model on the validation set, but when the loss is not reduced by a factor of 5, we stop training and load the saved model for prediction.

5.2   Performance Comparison

Baselines. For the rating prediction task, we choose the following two methods for comparison:

- NETE [2]: This model generates template-controlled sentences that comment about specific features.
- NRT [7]: This model can simultaneously predict precise ratings and generate abstractive tips with gated recurrent neural networks.

For the explanation generation task, in addition to the above two methods, we also compared with the following two methods:

- Att2Seq [8]: This model uses multilayer perceptrons with one hidden layer to encode attribute information into a vector and the decoder is built upon multilayer recurrent neural networks (RNNs) with long short-term memory (LSTM) units.
- ACMLM [11]: An aspect-conditional masked language model which can generate diverse justifications based on templates extracted from justification histories.

Rating Prediction. For the rating prediction task, the performance comparison results of our model and other models are shown in Table 3. We can see that the performance of our model and other models is close. Considering that our model is a multi-task joint training model, our focus is on the explanation generation task, so the advantage of rating prediction performance may not be obvious. Another reason may be that our model is based on a pre-trained language model, which involves many parameters, so it may lead to overfitting problems.

Table 3. The experimental results for rating prediction

	TripAdvisor		Amazon	
	RMSE	MAE	RMSE	MAE
NETE	0.792	0.608	0.961	0.727
NRT	0.792	0.605	0.957	0.718
SNROE	0.804	0.608	0.959	0.730

Explanation Generation. For the explanation generation task, the performance comparison results of our model and other models are shown in Table 4-5. R1-P, R1-R, R1-F, R2-P, R2-R and R2-F denote Precision, Recall and F1 of ROUGE-1 and ROUGE-2. We can see that our experimental performance is always better than the baseline models. It also proves that the user's multiple sentiments can be interacted, which ensures that the generated explanations is personalized and accurate to a certain extent. Meanwhile, the advantage of explainable recommendation based on pretrained Transformer is that it is trained on a large corpus, so the fluency of generated explanations is guaranteed.

Table 4. The experimental results for explanation generation on TripAdvisor

	BLEU-1	BLEU-4	R1-P	R1-R	R1-F	R2-P	R2-R	R2-F
ACMLM	3.45	0.02	4.86	3.82	3.72	0.18	0.2	0.16
NRT	14.26	0.8	17.57	16.52	16.56	2.45	2.64	2.48
Att2Seq	14.76	1.01	19.26	14.45	15.83	2.43	1.96	2.06
NETE	22.39	3.66	35.68	24.86	27.71	10.2	6.98	7.66
SNROE	**23.43**	**4.26**	**39.13**	**29.20**	**30.34**	**12.18**	**8.56**	**9.02**

Table 5. The experimental results for explanation generation on Amazon

	BLEU-1	BLEU-4	R1-P	R1-R	R1-F	R2-P	R2-R	R2-F
ACMLM	9.52	0.22	11.65	10.39	9.69	0.71	0.81	0.64
NRT	14.02	0.57	23.57	14.24	16.87	2.53	1.7	1.92
Att2Seq	12.78	1.01	20.53	13.49	15.42	2.77	1.87	2.09
NETE	18.76	2.46	33.87	21.43	24.81	7.58	4.77	5.46
SNROE	**19.28**	**2.60**	**33.98**	**23.83**	**25.88**	**8.19**	**5.67**	**6.08**

Ablation Study. The results of our ablation experiments are shown in Fig. 3. We compared with three variant models: no_f, no_r and no_s. This no_f model does not consider aspect-level sentiment as input, we can see that it turns out to be the worst. This no_r model does not consider the rating prediction task, so the representation learning for users and items suffers greatly. This no_s model does not consider rating-level sentiment as input, and we can see that its effect is better than the no_r variant model, but it is still worse than our proposed

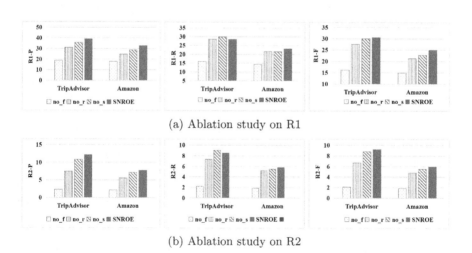

(a) Ablation study on R1

(b) Ablation study on R2

Fig. 3. The experimental results for ablation study

model. This is because our proposed model takes into account three aspects of sentiment, making our generated explanations more accurate.

Case Analysis. Examples of explanations generated by our proposed model, as well as variant models, are shown in Table 6. Since the no_r variant model does not consider the rating prediction task, the model naturally has no sentiment embedding as input. That is, the variant model also does not take into account rating-level sentiment. So here we only show the results of no_r variant model. From the table, we can see that the explanations generated by our model are consistent with the sentiment of the real reviews, and the content is relevant. But the reviews generated by the variant model are inaccurate, contrary to the real content, and even repeat words. In conclusion, the quality of explanations generated by our proposed model is satisfactory.

Table 6. Examples of the generated explanations.

Ground-truth	Review: The general manager was in the lobby each morning chatting to guests and the team was very helpful. Rating: 5.0
no_f	The bed was very comfortable and the bathroom was very clean and modern
no_r	The hotel is a bit far from the main road and the guests are not allowed to use the hotel
SNROE	The hotel was very well kept and the team was very friendly and helpful
Ground-truth	Review: I absolutely cannot believe how bad this movie is Rating: 1.0
no_f	This movie is so bad
no_r	This movie is not only a masterpiece but a masterpiece
SNROE	This movie is not worth watching

6   Conclusions

This paper proposes a model that jointly trains the rating prediction task and the explanation generation task, named SNROE. For the rating prediction task, we utilize MLP for rating prediction to better learn user and item representations. For the explanation generation task, we fuse three aspects of sentiment: rating-level sentiment, aspect-level sentiment, review-level sentiment to generate natural language explanations on a pretrained Transformer. Compared with other existing methods, our proposed model performs the best. In the future, we are ready to further improve the sentiment control of generated text.

References

1. Zhang, Y., Chen, X.: Explainable recommendation: a survey and new perspectives. Found. Trendső Inf. Retr. **14**, 1–101 (2020). https://doi.org/10.1561/1500000066
2. Li, L., Zhang, Y., Chen, L.: Generate neural template explanations for recommendation. In: Proceedings of the ACM International Conference on Information Knowledge Management, pp. 755–764. ACM (2020)
3. Wang, H., Kou, Y., Shen, D., Nie, T.: An explainable recommendation method based on multi-timeslice graph embedding. In: Wang, G., Lin, X., Hendler, J., Song, W., Xu, Z., Liu, G. (eds.) WISA 2020. LNCS, vol. 12432, pp. 84–95. Springer, Cham (2020). https://doi.org/10.1007/978-3-030-60029-7_8
4. Yang, C., Zhou, W., Wang, Z., Jiang, B., Li, D., Shen, H.: Accurate and explainable recommendation via hierarchical attention network oriented towards crowd intelligence. KBS **213**, 106687 (2021)
5. Wu, L., Quan, C., Li, C., Wang, Q., Zheng, B., Luo, X.: A context-aware user-item representation learning for item recommendation. ACM Trans. Inf. Syst. **37** (2019). https://doi.org/10.1145/3298988
6. Lu, Y., Dong, R., Smyth, B.: Coevolutionary recommendation model: mutual learning between ratings and reviews. In: Proceedings of the 2018 World Wide Web Conference, pp. 773–782. WWW (2018). https://doi.org/10.1145/3178876.3186158
7. Li, P., Wang, Z., Ren, Z., Bing, L., Lam, W.: Neural rating regression with abstractive tips generation for recommendation. In: Proceedings of the 40th International ACM SIGIR Conference on Research and Development in Information Retrieval, pp. 345–354. ACM (2017). https://doi.org/10.1145/3077136.3080822
8. Dong, L., Huang, S., Wei, F., Lapata, M., Zhou, M., Xu, K.: Learning to generate product reviews from attributes. In: Proceedings of the 15th Conference of the European Chapter of the Association for Computational Linguistics: Volume 1, Long Papers, pp. 623–632. ACL (2017). https://aclanthology.org/E17-1059
9. Chen, Z., et al.: Co-attentive multi-task learning for explainable recommendation. In: Proceedings of the Twenty-Eighth International Joint Conference on Artificial Intelligence, IJCAI 2019, pp. 2137–2143. IJCAI (2019). https://doi.org/10.24963/ijcai.2019/296
10. Chen, H., Chen, X., Shi, S., Zhang, Y.: Generate natural language explanations for recommendation. CoRR abs/2101.03392 (2021)
11. Ni, J., Li, J., McAuley, J.: Justifying recommendations using distantly-labeled reviews and fine-grained aspects. In: Proceedings of the 2019 Conference on Empirical Methods in Natural Language Processing and the 9th International Joint Conference on Natural Language Processing (EMNLP-IJCNLP), pp. 188–197. ACL (2019). https://aclanthology.org/D19-1018
12. Li, L., Zhang, Y., Chen, L.: Personalized prompt learning for explainable recommendation. arXiv preprint arXiv:1511.05644 (2022)
13. Li, L., Zhang, Y., Chen, L.: Personalized transformer for explainable recommendation. In: Proceedings of the 59th Annual Meeting of the Association for Computational Linguistics and the 11th International Joint Conference on Natural Language Processing (Volume 1: Long Papers), pp. 4947–4957. ACL (2021)
14. Chen, C., Zhang, M., Liu, Y., Ma, S.: Neural attentional rating regression with review-level explanations. In: Proceedings of the 2018 World Wide Web Conference, pp. 1583–1592. WWW (2018). https://doi.org/10.1145/3178876.3186070
15. Du, F., Plaisant, C., Spring, N., Crowley, K., Shneiderman, B.: EventAction: a visual analytics approach to explainable recommendation for event sequences. ACM Trans. Interact. Intell. Syst. **9** (2019). https://doi.org/10.1145/3301402

16. Papineni, K., Roukos, S., Ward, T., Zhu, W.J.: Bleu: a method for automatic evaluation of machine translation. In: Proceedings of the 40th Annual Meeting of the Association for Computational Linguistics, pp. 311–318. ACL (2002). https://aclanthology.org/P02-1040

17. Lin, C.Y.: ROUGE: a package for automatic evaluation of summaries. In: Text Summarization Branches Out, pp. 74–81. ACL (2004). https://aclanthology.org/W04-1013

Dual-level Hypergraph Representation Learning for Group Recommendation

Di Wu[1], Yue Kou[1(✉)], Derong Shen[1], Tiezheng Nie[1], and Dong Li[2]

[1] Northeastern University, Shenyang 110004, China
{kouyue,shenderong,nietiezheng}@cse.neu.edu.cn
[2] Liaoning University, Shenyang 110036, China
dongli@lnu.edu.cn

Abstract. Group Recommendation (GR) is the task of recommending items for a group of users. Most of existing studies adopt heuristic or attention-based preference aggregation strategies to learn group preferences, which ignores the composition of the group and suffers seriously from the problem of group-item interactions sparsity. In this paper, we propose a new group recommendation model based on Dual-level Hypergraph Representation Learning (called DHRL), which well models the group decision-making process by considering user-item interactions, group-item interactions and group-group interactions. Specifically, we design a member-level hypergraph convolutional network to learn group members' personal preferences from user-item interactions. We also design a group-level hypergraph convolutional network to capture group preferences with full consideration of both group-item interactions and group-group interactions. Finally, we propose a joint training strategy to ease data sparsity by combining the group recommendation task with the user recommendation task. The experiments demonstrate the effectiveness and the efficiency of our proposed method compared to several state-of-the-art methods in terms of HR and NDCG.

Keywords: Group recommendation · Hypergraph convolutional network · Group decision-making · Representation learning · Group-group interactions

1 Introduction

Due to the rapid development of social media in recent years, users with similar interests now have the opportunity to form various groups on the web. The traditional personalized recommendation system [10] can no longer satisfy the interests of groups. Therefore, it is urgent to develop a practical group recommendation system. The purpose of group recommendations is to reach a consensus among group members and generate recommendations that conform to the majority of group members. The final decision on group recommendation is not made by individuals but is the result of joint negotiation among group members. As shown in Fig. 1, when each member of the group goes to the cinema alone,

X. Zhao et al. (Eds.): WISA 2022, LNCS 13579, pp. 546–558, 2022.
https://doi.org/10.1007/978-3-031-20309-1_48

the mom usually chooses literary films, the dad chooses horror films, and the child prefers animated films. However, when couples watch movies together, this group will choose affectional films, while a family of three will choose educational films. In this case, group preference is not only dependent on the individual preference of each member, but also closely related to the composition of the group. Hence, group embeddings learned only by aggregating group members' personal preferences do not reflect group preferences well.

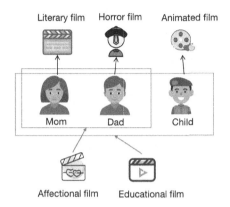

Fig. 1. An example of group recommendation.

However, most of existing studies adopt heuristic or attention-based preference aggregation strategies to learn group preferences, which ignores the composition of the group. Many of them adopt the predefined strategy, such as average [2], least misery [1] and so on, which are short of the ability to adjust decision weights of group members dynamically. In recent years, the attention mechanism has been applied to group recommendation, which endows group members with different attention weights when making group decisions [3,4,7]. Nevertheless, these models neglect inter- and intra-group high-order complex interactions. Besides, group-item interactions are sparse, as groups are mostly formed accidentally.

In this paper, we propose a new group recommendation model based on Dual-level Hypergraph Representation Learning (called DHRL), which well models the group decision-making process by considering user-item interactions, group-item interactions ,and group-group interactions. We summarise our contributions as follows.

- A new group recommendation model based on Dual-level Hypergraph Representation Learning (called DHRL) is proposed, which includes member-level representation learning and group-level group representation learning to capture interactions within and beyond groups respectively. Different from previous work, our model takes user-item interactions that reflect user preferences into account to learn fine-grained group representation and leverage group-group interactions to relieve group-item interaction sparsity.

- For member-level representation learning, we design a member-level hypergraph convolutional network to learn group members' personal preferences from user-item interactions. An attentional mechanism is used for learning the influence of each user in a group to better model the group decision-making process.
- For group-level representation learning, we design a group-level hypergraph convolutional network to capture group preferences with full consideration of both group-item interactions and group-group interactions.
- We propose a joint training strategy to ease data sparsity by combining the group recommendation task with the user recommendation task.
- We conduct extensive experiments on two real-world datasets and the experimental results demonstrate the effectiveness of our proposed DHRL model for group recommendation.

2   Related Work

Group recommendation has been widely researched and applied in various domains in recent years. In general, existing group recommendations can be divided into two categories, including memory-based and model-based methods. The memory-based methods can be further divided into the score aggregation methods and preference aggregation methods. The score aggregation strategy calculates candidates' scores for each user and then aggregates the scores of group members through some predefined score aggregation strategies such as average satisfaction [2] and least misery [1] as group preferences. The preference aggregation methods [13] aggregate the profile of each member to form a group profile at first, and then apply a personalized recommender system to recommend items to groups.

Unlike memory-based methods, model-based methods attempt to model the group decision-making process. COM [14] assumes that the individual influences are thematically relevant, both the group's topic preferences and individual preferences affect the final group decision. Attention-based aggregation strategies have appeared recently. AGREE [3] combines the attention network with the neural collaborative filtering method [8]. SIGR [11] takes the attention mechanism and the bipartite graph embedding model BGEM as a building block to learn the social influence of individuals in the group from both group-item interaction and social network. Cao et al. [4] exploited social followee information as user-user relationship to obtain member preferences by aggregating user's social friends' representations. Both models rely on the social network, but ignore interactions between groups. We capture the inter-group collaborative signals through HyperGCN and meanwhile employ user-item interactions to model user behavior.

Hypergraph [12] has emerged with enormous potential in modeling complex high-order connectivity and has been widely used to solve various problems. Zhang et al. [15] proposed a novel neural group recommendation with the hierarchical hypergraph convolutional network and self-supervised learning strategy to capture inter-and intra-group user interactions and alleviate the data sparsity. Guo et al. [6] proposed a hierarchical GNN-based group recommender Hyper-Group to learn the group preference via the hyperedge embedding technique. Jia et al. [9] propose a dual-channel Hypergraph Convolutional network for group Recommendation (HCR) which separately formulates group data as member-level hypergraph and group-level overlap graph tailored for GCN to learn group representation. However, HCR does not make full use of user-item interactions to generate fine-grained group representation.

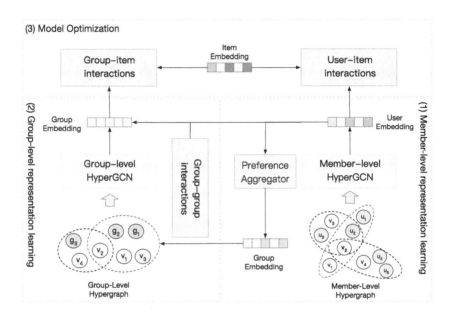

Fig. 2. Overall framework of DHRL.

3   Preliminaries

Suppose we have m users $\mathcal{U} = \{u_1, u_2, ..., u_m\}$, n items $\mathcal{V} = \{v_1, v_2, ..., v_n\}$, and k groups $\mathcal{G} = \{g_1, g_2, ..., g_k\}$. The t-th group $g_t \in \mathcal{G}$ consists of a set of users, and we use $g_t = \{u_1, u_2, ..., u_{|g_t|}\}$, where $u_j \in \mathcal{U}$, $|g_t|$ is the size of g_t. We use $\mathcal{R}^U = [r_{j,h}^U]_{m \times n}$ to denote the user-item interactions, and $\mathcal{R}^G = [r_{t,h}^U]_{k \times n}$ to denote the group-item interactions, where $r \in \{1, 0\}$ denotes whether a group/user has a direct interaction with the corresponding item. We use $\mathcal{N}^G = [n_{g,g'}^G]_{k \times k}$ to denote the group-group interactions, where $n_{g,g'}$ denotes the number of shared common members between g and g'. Then, give a target group g_t, the task of GR is to recommend a ranked list of items that g_t might be interested in.

Definition 1 Member-level Hypergraph. Let $\mathcal{G}^m = (\mathcal{V}^m, \varepsilon^m)$ denote a member-level hypergraph, where $\mathcal{V}^m$ is the vertex set containing M unique vertices, and ε^m is the hyperedge set containing N hyperedges. Each hyperedge $\epsilon \in \varepsilon^m$ represents a group and contains multiple vertices, which are group members and their interacted items. The hyperedge can be represented by an incidence matrix $\boldsymbol{H} \in \mathbb{R}^{M \times N}$, where $h_{v\epsilon} = 1$ if the hyperedge ϵ contains the vertex $v \in \mathcal{V}^m$, otherwise $h_{v\epsilon} = 0$. For each hypergraph, we use the diagonal matrix $\boldsymbol{D}$ to denote the degrees of vertex, where $d_v = \sum_{\epsilon=1}^{N} w_\epsilon h_{v\epsilon}$. $\boldsymbol{W} \in \mathbb{R}^{N \times N}$ is the diagonal matrix of the weight of the hyperedge. Let diagonal matrix $\boldsymbol{B}$ denote the degree of hyperedge, where $b_\epsilon = \sum_{v=1}^{M} h_{v\epsilon}$ represents the number of vertices connected by the hyperedge ϵ.

Definition 2 Group-level Hypergraph. Let $\mathcal{G}^g = (\mathcal{V}^g, \varepsilon^g)$ denote a group-level hypergraph, where $\mathcal{V}^g$ contains the hyperedges in the member-level hypergraph $\mathcal{G}^m$ and items in the group-item interactions. There are two cases where two nodes of $\mathcal{G}^g$ are connected. One is when the corresponding hyperedge in $\mathcal{G}^g$ shares at least one common node, and the other is when the group node and the item node interact directly. If two or more groups interact directly with the same item, these nodes are added to a hyperedge. The hyperedge can also be represented by an incidence matrix $\boldsymbol{H}_g$, $\boldsymbol{D}_g$ denoting the degrees of vertex, and $\boldsymbol{B}_g$ denoting the degrees of hyperedge.

4   DHRL: The Proposed Model

In this work, we propose the model DHRL for Group recommendation, which is designed to relieve the data sparsity problem and learn more effective group embedding to enhance the accuracy of GR with dual-level HyperGCN. The overall framework of the DHRL is shown in Fig. 2, which consists of three components: member-level representation learning, group-level representation learning, and model optimization.

4.1   Member-level Representation Learning

Member-level Representation Learning consists of two components: user behavior modeling based on member-level HyperGCN and preference aggregator.

User Behavior Modeling based on Member-level HyperGCN. In this subsection, we model user individuals' preferences captured from user-item interactions. Firstly, we construct a member-level hypergraph in which each hyperedge consists of the group members and the interacted items to capture the high-order relations between group members and items. Then, we learn user and item embeddings from the member-level hypergraph convolutional network. For the specific user node p_u and item node q_v, the member-level hypergraph convolutional network firstly aggregates the node representations of all its connective hyperedges e_j. Then the representation of node p_u and q_v would integrate the related hyperedge feature information, and the hypergraph convolution is defined as follows:

$$P_u^{(l+1)} = D_u^{-1} H_u W_u B_u^{-1} H_u^T P_u^{(l)} \qquad (1)$$

where D_u denotes the vertex degree matrix of the user behavior-view hypergraph, and B_u denotes the hyperedge degree matrix. H_u is regarded as the incidence matrix to describe the relationship between nodes and hyperedges, and W_u as the weight matrix of the hyperedges that each hyperedge is assigned the same weight initially. $P_u^{(l)}$ is the users' embeddings in the l-th member-level hypergraph convolutional network, where $P_u^{(0)}$ is randomly initialized and will be trained via back-propagation. Assume that the number of layers in the member-level HyperGCN is K, then the member-level user representation is P_u. Similarly, the item representation is Q_v.

Preference Aggregator. Our goal is to obtain the group embeddings to estimate their preferences for items. We endow each member of the group different weight, because group members have different influence in group decision-making process. Next, We sum the representations of group members obtained from the previous step according to the different weights.

$$G_u = \sum_{u \in g_t} \alpha_u P_u, \qquad (2)$$

$$\alpha_u = Softmax[h^T \sigma(W(P_u, P_v) + b)] \; for \; u \in g_t, v \in g_t \qquad (3)$$

where α_u is the weight of a user u in the group decision-making, P_v denotes other members embeddings in the same group g_t, and h, W, b are learnable parameters which are learned by an attention network, parameterized by a MLP layer.

4.2   Group-level Representation Learning

The group representation merely obtained from member-level representation learning module neglects the inherent group preferences that may be different from the preferences of group members. In addition, group-group interactions record the same users between any two groups and groups with common members should have similar group-level preferences, so we leverage the common users in the group-group interactions to model the fine-grained group representation further. The group-group interactions not only record the number of common users, but also who common users are, because different users have different effects on group decision-making. Hence, based on member-level group representation learning, we exploit group-group interactions to capture group-level group representation.

Firstly, we calculate the number of the shared common members $n_{g,g'}^{(t)}$, and obtain the embedding set of these members $emb_{g,g'}^{(t)}$, between any two groups from group-group interactions. Then, we employ the mean aggregator to compute $emb_{g,g'}^{(t)}$, as defined in Eq. 4:

$$emb_{g,g'}^{(t)} = Aggregator(P_u, \forall u \in \{\mathcal{G}(g_t) \cap \mathcal{G}(g_t')\}) \tag{4}$$

Finally, the fine-grained group representation is updated to the product of the number of common members and the set of embeddings in which groups have the common members, as defined in Eq. 5:

$$G_s = \sum_{\forall g' \in \mathcal{N}(g)} n_{g,g'}^{(t)} emb_{g,g'}^{(t)} \tag{5}$$

In addition to the group-group interactions, the collaborative signal beyond groups can also be captured from the group-item interactions. In this paper, groups and items are constructed as a group-level hypergraph, and then the group-level hypergraph convolution network is further employed to learn group representation from the group-level hypergraph directly. Each group-level hyperedge includes groups that have shared common users and the items they interact with.

$$G_i^{(l+1)} = D_g^{-1} H_g W_g B_g^{-1} H_g^T G_i^{(l)} \tag{6}$$

where $G_i^{(l)}$ is the groups' embeddings in the l-th group-level hypergraph convolutional network, $G_i^{(0)} = G_u$. Finally, we aggregate the group embeddings from two kinds of interaction datas to obtain the final group representation G_g, which is defined in Eq. 7

$$G_g = \omega G_s + (1 - \omega) G_i \tag{7}$$

where ω is a hyper-parameter controlling the contributions of the two aspects.

4.3   Model Optimization

For the group-item pair (g_t, v_h) and user-item pair (p_u, v_h), we feed the concatenation of them into a Multi-Layer Perception (MLP) for interaction learning and preference prediction. Firstly, we compute the hidden interaction vector c. Then, DHRL leverages a FCL with Sigmoid activation function to predict the preferences of group g_t (or user p_u) on the specific item v_h, as defined in Eq. 8 and Eq. 9 respectively.

$$c_g = MLP_{interaction}([g_t^v \odot v_h]), \quad \hat{y}_{gv} = Sigmoid(w^T c_g + b) \tag{8}$$

$$c_u = MLP_{interaction}([p_u^v \odot v_h]), \quad \hat{y}_{uv} = Sigmoid(w^T c_u + b) \tag{9}$$

where w and b are learnable parameters.

Finally, since we recommend top-K items for groups based on implicit interactions, we employ pairwise learning for optimizing parameters. Due to the group-item interaction sparsity, we introduce user-item interactions into a joint training strategy to optimize the group and user recommendation task simultaneously. In this strategy, we jointly train loss function $\mathcal{L}_G$ (as defined in Eq. 10) on the group-item interactions and $\mathcal{L}_U$ (as defined in Eq. 11) on the user-item interactions.

$$\mathcal{L}_G = \sum_{(g,v,v' \in \mathcal{O})} (\hat{y}_{gv} - \hat{y}_{gv'} - 1)^2 \tag{10}$$

$$\mathcal{L}_U = \sum_{(u,v,v' \in \mathcal{O}')} (\hat{y}_{uv} - \hat{y}_{uv'} - 1)^2 \tag{11}$$

where $\mathcal{O}$ is the set of group training instances, each instance (g, v, v') indicates that group g has interacted with item v but not interacted with item v', and $\mathcal{O}'$ is the same of the set of user training instances. The final loss function is defined as the following equation:

$$\mathcal{L} = \mathcal{L}_G + \mathcal{L}_U \tag{12}$$

5   Experiments

5.1   Datasets

We conduct experiments on two public datasets: Mafengwo [4] and Weeplaces. Mafengwo is a tourism website where users can record their traveled venues, create or join a group travel, which retained the groups which have at least 2 members and have traveled at least 3 venues, and collected their traveled venues as the group-item interactions. Weeplaces dataset includes the users' check-in history in a location-based social network. The information of datasets is summarized in Table 1.

Table 1. The Statistics of Datasets.

Dataset	#Users	#Items	#Groups	#U-I interactions	#G-I interactions	Avg. group size
Mafengwo	5275	1513	995	39761	3595	7.19
Weeplaces	8643	25081	22733	1,358,458	180,229	2.9

5.2   Baselines and Evaluation Metrics

Baseline Methods. We compare DHRL with four baseline methods:

- **Pop** [5]: This method is a non-personalized method that recommends items to users and groups based on the popularity of items. The popularity of an item is measured by its number of interactions in the training set.

Table 2. Top-N Performance of Both Recommendation Tasks for Users and Groups on Mafengwo.

Overall Performance Comparison (Mafengwo)								
	User		Group		User		Group	
	HR@5	NDCG@5	HR@5	NDCG@5	HR@10	NDCG@10	HR@10	NDCG@10
Pop	0.4047	0.2876	0.3115	0.2169	0.4971	0.3172	0.4251	0.2537
NCF	0.6363	0.5432	0.4701	0.3657	0.7417	0.5733	0.6269	0.4141
AGREE	0.6383	0.5502	0.4814	0.3747	0.7491	0.5775	0.6400	0.4244
SoAGREE	0.6510	0.5612	0.4898	0.3807	0.7610	0.5865	0.6481	0.4301
DHRL	**0.6713**	**0.5901**	**0.5201**	**0.4478**	**0.7954**	**0.6178**	**0.7094**	**0.6118**

Table 3. Top-N Performance of Group Recommendation Tasks on Weeplaces.

Overall Performance Comparison (Weeplaces)				
	HR@20	HR@50	NDCG@20	NDCG@50
Pop	0.126	0.176	0.063	0.074
NCF	0.271	0.295	0.193	0.244
AGREE	0.354	0.671	0.224	0.267
SoAGREE	0.382	0.739	0.240	0.281
DHRL	**0.394**	**0.776**	**0.323**	**0.357**

- **NCF** [8]: This method treats the group as a virtual user and ignores the member information of the group. This is to verify whether the traditional Collaborative Filtering (CF) method can be directly applied to the GR task.
- **AGREE** [3]: This is a state-of-the-art OGR method that models the group-item interactions under the Neural Collaborative Filtering (NCF) framework and adopts a vanilla attention mechanism to learn the weight of each user in a group.
- **SoAGREE** [4]: This is a social information enhanced method for group recommendation. It designs a hierarchical attention network learning the representation of groups and users in a hierarchical structure.

Evaluation Metrics. To measure the performance of all methods, we employ the metrics $HR@K$ and $NDCG@K$, with $K = \{5, 10\}$ in the Mafengwo dataset and $K = \{20, 50\}$ in the Weeplaces dataset. $HR@K$ is the fraction of relevant items that have been retrieved in the Top-K relevant items. $NDCG@K$ evaluates the ranking of true items in the recommendation list.

5.3 Parameter Setup

DHRL is implemented with the Pytorch and trained by minimizing the pairwise loss in Eq. 12 with 10 negative instances for each user or group. The embedding layer is initialized by the Glorot strategy and the other layers are randomly initialized by a Gaussian distribution of a mean of 0 and a standard deviation of 0.1. The embedding size is set as 64. We apply the Adam optimizer, where the batch size and learning rate are searched in [32, 64, 128, 256, 512] and [0.001, 0.005, 0.01, 0.05, 0.1], respectively. In the dual-level hypergraph GCN, we empirically set the size of the convolutional layer as 3.

5.4   Performance Evaluation

Overall Performance. We intuitively compare and analyze the experimental results of DHRL with the baselines, as shown in Table 2 and Table 3 on the two datasets respectively. For the group recommendation task, DHRL consistently outperforms all baseline methods. The above experimental results illustrate the effectiveness of DHRL in group representation learning. The performance of neural network-based methods (i.e. NCF, AGREE, SoAGREE, DHRL) is superior to that of the non-personalized approach (Pop). This demonstrates the superiority of neural networks. Among these methods, the attention-based group models (i.e. AGREE, SoAGREE, DHRL) outperform others due to their ability to dynamically model the user interactions within groups and learn different weights of different members in the group. In the models based on attention mechanism, DHRL is better than most models. We believe that the performance improvement of DHRL verifies the effectiveness of member-level and group-level hypergraph convolutional network modules to exploit high-order interactions. Besides, SoAGREE performs better than AGREE due to considering social followee information to represent user-user interactions.

Effect of Model Hyper-parameters. We evaluate the impact of the hyper-parameters on the performance of group recommendation, as shown in Fig. 3. We analyze hyper-parameters including the depth of hypergraph convolutional layers, the number of negative samples, the learning rate, and the batch size. We only show the experimental results on the Mafengwo dataset on account of limited space. As shown in Fig. 3(a), when the depth of HyperGCN layers is 3, the model is optimal. With the increase of convolutional layers, the multi-layer hypergraph convolutional network would lead to the problem of over-smoothing. As shown in Fig. 3(b), when the number of negative samples is 10, our model achieves the best performance. With the increase of the number of negative samples, the model performance decreases and tends to be stable. According to Fig. 3(c), we observe $5e-2$ is sufficient for the learning rate. As the rate increases, the performance rises first and then decreases. Figure 3(d) depicts the influence of different batch sizes from 32 to 512. As the batch size increases, the model performance is gradually enhanced.

Ablation Experiment. Finally, we conducted ablation experiments. Here we compare DHRL with two variants: (1) DHRL-noMRL without the group embeddings from the member-level representation learning module, we only input initialized vectors of all groups only into GRL. (2) DHRL-noGRL without the group-level representation learning module, we directly use the group representation obtained by the MRL module to complete group recommendations. The

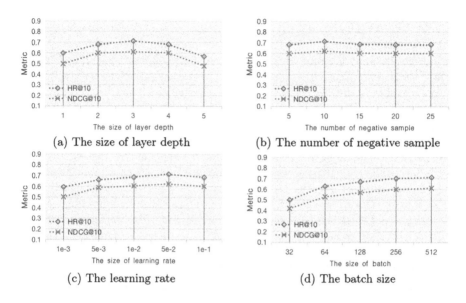

(a) The size of layer depth

(b) The number of negative sample

(c) The learning rate

(d) The batch size

Fig. 3. Parameter setting

experimental results are shown in Fig. 4. Compared with DHRL, in the absence of MRL or GRL, the NDCG on both datasets decreases significantly, indicating the effectiveness of dual-level hypergraph representation learning. In addition, compared to DHRL-noMRL, the performance of DHRL-noGRL is even worse, which indicates that mining correlation between groups from group-item and group-group interactions through group-level hyperGCN is indeed necessary.

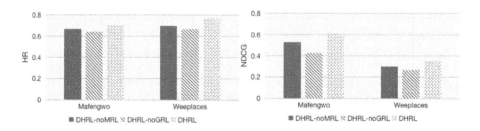

Fig. 4. Comparisons of variants on HR@{10,50} and NDCG@{10,50}.

6   Conclusions

In this paper, we propose a new group recommendation model based on Dual-level Hypergraph Representation Learning (called DHRL). In this way, our model not only model the group decision-making process well but also capture inter- and intra-group interactions, containing user-item interactions, group-item interactions, and group-group interactions. We firstly leverage the member-level HyperGCN module to generate fine-grained group representations with user-item interactions. Secondly, we design a group-level hypergraph convolutional network to capture group preferences with both group-item interactions and group-group interactions. Finally, We propose a joint training strategy to ease data sparsity by combining the group recommendation task with the user recommendation task. Next, we can continue to study the interpretability and social information of GR for boosting group recommendation performance.

Funding. This work was supported by the National Natural Science Foundation of China under Grant Nos. 62072084, 62172082 and 62072086, the Science Research Funds of Liaoning Province of China under Grant No. LJKZ0094, the Natural Science Foundation of Liaoning Province of China under Grant No. 2022-MS-171, the Science and Technology Program Major Project of Liaoning Province of China under Grant No. 2022JH1/10400009.

References

1. Amer-Yahia, S., Roy, S.B., Chawla, A., Das, G., Yu, C.: Group recommendation: semantics and efficiency. Proc. VLDB Endow. **2**(1), 754–765 (2009)
2. Baltrunas, L., Makcinskas, T., Ricci, F.: Group recommendations with rank aggregation and collaborative filtering. In: RecSys 2010, pp. 119–126. ACM (2010)
3. Cao, D., He, X., Miao, L., An, Y., Yang, C., Hong, R.: Attentive group recommendation. In: SIGIR 2018, pp. 645–654. ACM (2018)
4. Cao, D., He, X., Miao, L., Xiao, G., Chen, H., Xu, J.: Social-enhanced attentive group recommendation. IEEE Trans. Knowl. Data Eng. **33**(3), 1195–1209 (2021)
5. Cremonesi, P., Koren, Y., Turrin, R.: Performance of recommender algorithms on top-n recommendation tasks. In: RecSys 2010, pp. 39–46. ACM (2010)
6. Guo, L., Yin, H., Chen, T., Zhang, X., Zheng, K.: Hierarchical hyperedge embedding-based representation learning for group recommendation. ACM Trans. Inf. Syst. **40**(1), 3:1–3:27 (2022)
7. Guo, L., Yin, H., Wang, Q., Cui, B., Huang, Z., Cui, L.: Group recommendation with latent voting mechanism. In: ICDE 2020, pp. 121–132 (2020)
8. He, X., Liao, L., Zhang, H., Nie, L., Hu, X., Chua, T.: Neural collaborative filtering. In: WWW 2017, pp. 173–182 (2017)
9. Jia, R., Zhou, X., Dong, L., Pan, S.: Hypergraph convolutional network for group recommendation. In: ICDM 2021, pp. 260–269 (2021)
10. Wang, H., Kou, Y., Shen, D., Nie, T.: An explainable recommendation method based on multi-timeslice graph embedding. In: Wang, G., Lin, X., Hendler, J., Song, W., Xu, Z., Liu, G. (eds.) WISA 2020. LNCS, vol. 12432, pp. 84–95. Springer, Cham (2020). https://doi.org/10.1007/978-3-030-60029-7_8
11. Yin, H., Wang, Q., Zheng, K., Li, Z., Yang, J., Zhou, X.: Social influence-based group representation learning for group recommendation. In: ICDE 2019, pp. 566–577. IEEE (2019)
12. Yoon, S., Song, H., Shin, K., Yi, Y.: How much and when do we need higher-order information in hypergraphs? A case study on hyperedge prediction. In: WWW 2020, pp. 2627–2633 (2020)

13. Yu, Z., Zhou, X., Hao, Y., Gu, J.: TV program recommendation for multiple viewers based on user profile merging. User Model. User Adapt. Interact. **16**, 63–82 (2006)
14. Yuan, Q., Cong, G., Lin, C.: COM: a generative model for group recommendation. In: KDD 2014, pp. 163–172. ACM (2014)
15. Zhang, J., Gao, M., Yu, J., Guo, L., Li, J., Yin, H.: Double-scale self-supervised hypergraph learning for group recommendation. In: CIKM 2021, pp. 2557–2567. ACM (2021)

GeoGTI: Towards a General, Transferable and Interpretable Site Recommendation

Yunfan Gao[1], Dong Han[2], Haofen Wang[3(✉)], Maohong Zhang[2], Fangjie Hou[2], Dongqing Yu[2], and Yun Xiong[1(✉)]

[1] School of Computer Science, Fudan University, Shanghai, China
[2] Shanghai Wayz Information Technology, Shanghai, China
[3] College of Design and Innovation, Tongji University, Shanghai, China
carter.whfcarter@gmail.com

Abstract. Lack of data and weak interpretability are the main problems faced by store site recommendations. This paper presents a unified site recommendation system called GeoGTI (**G**eneral,**T**ransferable and **I**nterpretable), which applies to different brands in various industries. Different from the existing single-dimensional transfer methods, we adopt multi-layer knowledge transfer, leveraging knowledge from industries, competitive brands, upper administrative districts, and other cities to alleviate the problems of data scarcity and cold-start. Besides, to fill in the gap of weak interpretability, we score the candidate locations into a five-dimension radar chart from population, business, living, working, and transportation aspects, making the recommended result more convincing and instructive. Extensive experiments are conducted on real-world datasets from various industries, demonstrating GeoGTI's practicability and effectiveness on store site recommendations.

Keywords: Site recommendation · Knowledge transfer · Model interpretability

1 Introduction

Location is considered a key factor to the success of a store in the modern retail industry. Picking an optimal location to place a new store plays crucial roles in brick-and-mortar enterprises. However, with the rapid development of Internet technology and urbanization, the amount of store information from various sources has grown explosively, such as the flow of people, surrounding businesses and traffic conditions. Selecting locations for new stores has become more and more challenging. Therefore, an effective chain store recommendation system becomes necessary for enterprise managers.

Traditional site selection relies on professionals and detailed on-the-spot research or survey. Experts need to conduct demographic and geographic analyses on all candidate locations, which is time-consuming and labor-intensive. Recently, with the proliferation of spatial-temporal data and urban data, many data-driven methods have been proposed [1,8,16]. Researchers resort to addressing store site recommendations by leveraging data mining and machine learning

© The Author(s), under exclusive license to Springer Nature Switzerland AG 2022
X. Zhao et al. (Eds.): WISA 2022, LNCS 13579, pp. 559–571, 2022.
https://doi.org/10.1007/978-3-031-20309-1_49

techniques. Many studies utilize LBSN data, such as social-medium check-in numbers [2,9,13,18] and user-generated reviews [16,19], to establish various predictive indicators to help find the popular locations.

However, current works still face many challenges. 1) Poor universality. These works usually focus on a specific chain store and are challenging to use in various industries widely. 2) Limited data. Due to the uneven distribution of the number of stores in different cities, the data in some cities is incomplete and limited. Besides, for some niche industry brands, the problem of data scarcity is more obvious. Some recent works have noticed the transferable problem in the site selection task [5,12], but they are all limited to the knowledge transfer between similar cities. 3) Weak interpretability. Current model-based site selection approaches are incomprehensible to non-computer majors. Models cannot clearly explain why this location is better than any other candidate location to open a new store, resulting in poor guidance for follow-up works.

We propose our method GeoGTI, a general, transferable and interpretable general site recommendation system to address the above challenges. First, we set a general site selection module to incorporate different site selection requirements from various industries based on multi-source geographic and demographic data. Secondly, to solve the data scarcity problem, we design a multi-layer knowledge transfer framework by leveraging knowledge from industry, competitive brands, upper administrative district, and other cities. Finally, we propose an interpretable module, attributing the recommended location to a five-dimension radar chart. The whole framework is s proved effective on real-world datasets from the banking and retail industries.

In summary, we make the following contributions:

- We propose a unified and comprehensive store site recommendation system called GeoGTI, which recommends optimal locations for new stores based on multi-source data.
- To the best of our knowledge, this is the first work that studies how to transfer knowledge from multi-layer source data under the site selection task. We leverage data from industry, competitive brands, upper administrative districts, and other cities to alleviate the data scarcity problem in site recommendations.
- To solve the problem of weak interpretability, we score the candidate locations in a five-dimension radar chart from population, business, living, working, and transportation, which makes the recommended result more convincing.
- We conduct experiments on the banking and retail industries respectively, and the experimental results show that our methods can achieve F1 greater than 65 in various industries. At the same time, compared with non-transfer experiments, our transfer method can improve F1 by 2–3.

2   Related Work

In recent years, site recommendation has been a trend in the research field due to the rapidly increasing urban data that allows researchers to predict new locations

[10,20]. Karamshuk et al. [6] proposed the Geo-Spotting model to study the influence of geographical location factors and human mobility factors on the location of retail stores in a particular city. Yu et al. [19] adopt a Geographic Information System to mine location information in social networks, studying business site selection and business circle analysis. At the same time, a large amount of data is necessary to guarantee good results for the model. However, in many real-world applications, this condition is challenging to hold. A few studies leverage transfer learning to address the no-training-data issue in the computing field. Fan et al. [3] proposed CityCoupling, which builds an inter-city spatial mapping in one city as input and reproduces human mobility in another city. Wei et al. [17] proposed a flexible transfer learning method for air quality prediction, aiming to transfer knowledge from cities with multi-modal data and sufficient labels to cities with insufficient data and scarce labels. These studies are based on one-fold knowledge transfer and lack of knowledge transfer exploration from different dimensions and different sources.

3 Proposed Method: GeoGTI

The framework is illustrated in Fig. 1. It consists of four major components: Feature Extraction, General Site Selection, Transfer Site Selection, and Site Selection Interpretability.

3.1 Problem Definition

An enterprise e, which belongs to industry ind, is an enterprise that has already opened many stores in different locations and has the need to open new ones. A city is represented as a set of two-dimensional location grids, each of which has a size of $150\,\mathrm{m} \times 150\,\mathrm{m}$ and can be seen as a basic unit of area for evaluation. Suppose a city c has n grids, denoted as $G^c = \{g_1^c, g_2^c, ...g_n^c\}$. For each grid g_i^c, we mainly extract two kinds of features, observable geographic features $F_{geo}^c = \{f_{geo\_i}^c\}$ and user profile features $F_{user}^c = \{f_{user\_i}^c\}$. The objective of store site recommendation is to select the optimal grid to open a new store. In detail, based on multi-source data, we train a classification model to discriminate ground-truth locations from other store locations and then score each gird g_i^c to judge the suitability.

3.2 Feature Extraction

Geographic Features. Refer to features that describe the environment around the location. More specifically, we further measure geospatial characteristics from the following perspectives.

- ***Diversity.*** We calculate the number of POIs (Point of Interest) of different categories in each grid to reflect the diversification and heterogeneity of the

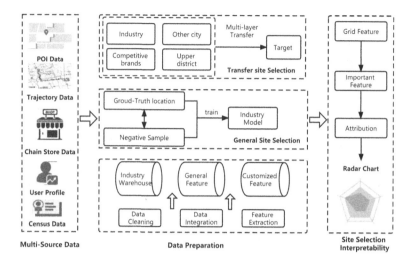

Fig. 1. The framework of GeoGTI

environment, as shown in Eq. 1. $N_c(i)$ is the number of POIs of category c in $grid_i$, and the Γ is the set of all the POIs categories.

$$f_i^{div} = -\sum_{c \in \Gamma} \frac{N_c(i)}{N(i)} \times log \frac{N_c(i)}{N(i)} \qquad (1)$$

- **_Traffic Convenience._** It reflects the accessibility to different means of transportation $trans$ (e.g. bus station, subway station, ferry station, etc.), as shown in Eq. 2. t represents a certain type of transportation, $num(t_i)$ is the number of all stations of the corresponding transportation type t in $grid_i$.

$$f_i^{traffic} = \sum_{t \in trans} num(t_i) \qquad (2)$$

- **_Competitiveness._** Stores of the same category in a region will form a competitive relationship and influence each other. We define the competitiveness feature in Eq. 3. N_c represents the total number of stores of the same category of the target stores in the area around the candidate location j, and $N_{c/tar}$ is the number of stores of the same category except for the target stores in $grid_i$.

$$f_i^{compet} = -\frac{N_{c/tar}(i)}{N_c(i)} \qquad (3)$$

User Profile Features. The possibility of consuming is highly related to the people living around the location. Therefore, we take some demographic features and mobility features of each grid into account.

- **Gender Distribution.** Since users of different genders have different consumption tendencies, we measure the gender distribution in Eq. 4. Among them, $N_m(i)(N_f(i))$ is the number of male(female) users in $grid_i$.

$$f_i^{gender} = \frac{N_m(i)}{N_f(i)} \tag{4}$$

- **Occupation Distribution.** We use the top 10 occupations of users living in $grid_i$ to denote the occupation distribution. We represent the occupation by label-encoder and record the number of people in the top 10 occupations, as shown in Eq. 5. le denotes the label-encoder and O represents the type of occupation. Finally, we get 20-dimensional features.

$$f_i^{occ} = \{le(O_1), num(O_1), ..., le(O_{10}), num(O_{10})\} \tag{5}$$

- **Popularity.** It reflects the popularity of the area. We use the total number of people living in $gird_i$, as defined in Eq. 6. Furthermore, we count the flow of people per hour t in the day defined in Eq. 7.

$$f_i^{flow} = num(user)(user \in grid_i) \tag{6}$$

$$f_i^{flow_t} = \{num(user_t)\}(t = 1, 2, ...24) \tag{7}$$

3.3   General Site Selection

In order to meet the site selection requirements of different industries, GeoGTI uses a general framework to model the features of each grid. This Module can be divided into data preparation, offline training, and real-time prediction. The architecture of general site selection is illustrated in Fig. 2.

Data Preparation. The target industry ind is mapped to a self-owned industry brand database, and the historical location information of all the brands in this industry is selected according to the administrative district. We take these locations as positive samples. For each positive sample, we select n locations with different industries within the radius of r as negative samples, r is an adjustable hyper-parameter. After collecting positive and negative samples, we map them into grids and then extract the corresponding grid features f_{grid_i}.

Offline Training. After collecting the positive and negative samples and features, we construct the training pipeline to offline train an industry-specified binary classification model. Through feature analysis, we choose important features to update the feature engineering and help construct interpretability.

Real-Time Prediction. The general site selection module provides a real-time prediction service for a specific industry. When the user inputs the locations and specifies the industry, the prediction service extracts the grid features and feeds them into the corresponding industry model to get the predicted results.

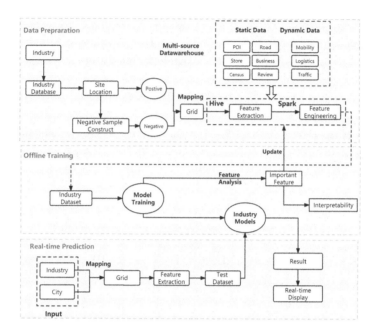

Fig. 2. General site selection module pipeline

3.4   Transfer Site Selection

Ideally, the data of various industries are large and sufficient, but in practice, due to the uneven distribution of the data, we often suffer from the problem of data scarcity and cold-start, especially for some remote areas or niche industries. Therefore, we design a multi-layer transfer framework to solve the above problems.

Industry Transfer. For niche brands, their data is often insufficient. To address this issue, we propose an industry transfer method. Given an enterprise e, we can train an enterprise instance model $M_e(\ )$ by transferring knowledge from the industry model $M_{ind}(\ )$, where $e \in ind$. For example, we can pre-train a bank industry model based on all kinds of banks, and fine-tune it on a specific bank brand (e.g., CCB bank), $M_{bank}(\ ) \mapsto M_{CCB}(\ )$.

Competitive Brands Transfer. Different from the top-down method of industry transfer, we also propose a method of using the same level of competitive brands transfer. We can train an enterprise instance model $M_e(\ )$ by transferring knowledge from the competitive brand model $M_{compet}(\ )$, where e, $compet$ are competing brands. For example, we can pre-train ABC bank model, and fine-tune it on CCB bank, $M_{ABC}(\ ) \mapsto M_{CCB}(\ )$.

Upper Administrative District Transfer. Due to the regional nature of some brands, the data is uneven in different cities. Therefore, we can train an enterprise instance model $M_e(city)$ in a city by transferring knowledge from the upper administrative district model $M_e(up\_admin)$, where $city \in up\_admin$. For example, we can fine-tune Shanghai_CCB bank model on Nation_CCB bank model. $M_{CCB}(Nation) \mapsto M_{CCB}(Shanghai)$.

City Transfer. Like the competitive brands transfer, we also propose a regional same-level transfer method, city transfer. We can train an enterprise instance model $M_e(city)$ in a city by transferring knowledge from the brand model in other cities $M_e(other\_city)$, where $city, other\_city$ are cities of the same level that belong to the same province. For example $M_{CCB}(Beijing) \mapsto M_{CCB}(Shanghai)$.

3.5 Site Selection Interpretability

The interpretability of the model has always been a blank in the site selection task. Common site selection models only give the model predicted score of this location, but it is difficult for business personnel to understand. We design an interpretability module to help understand why these locations are worth recommending, as shown in Fig. 3. In order to better conform to users' intuitive understanding, features are converted into BI attribution and classified into five categories: population, business, living, working, and transportation. The scores of each dimension are calculated based on the rules of feature importance and feature value and finally presented in the form of radar chart.

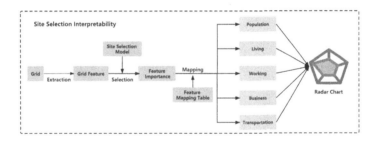

Fig. 3. Site selection interpretability pipeline

3.6 Summary

Further, we compare GeoGTI with other proposed systems regarding data, generality, transferability, and interpretability. Table 1 shows the comparison results between the various systems. In detail, We measure the Geographic Info level by judging whether the system uses POIs data, enterprise data, and surrounding commercial data, and the User Profile Info level by judging whether the system uses static user features (location, gender), user behavior features (income level, check-in) and dynamic user data (flow, trajectory). The Brand Scope measures whether the system is designed for only one chain enterprise or multiple similar enterprises in the same industry or various industries. Applicable Area refers to the region that the system covers, such as a city or cities in several provinces or most cities in the nation. Transferability measures how the system leverages knowledge from other sources. For example, whether it transfers knowledge from single similar data or utilizes more than one way to deal with source data or designs a multi-transfer architecture. Interpretability refers to whether the system considers the explanation of results and makes business attributions to the recommended results. As can be seen from Table 1, GeoGTI is a more complete and sufficient system than other systems.

Table 1. The comparison between store site recommendation systems

System	Geographic Info	User Profile Info	Brand Scope	Applicable Area	Transferability	Interpretability
Ge et al. [4]	★★★	★★★	★☆☆	★☆☆	☆☆☆	☆☆☆
CityTransfer [5]	★★★	★★★	★★☆	★★☆	★★☆	☆☆☆
DeepStore [11]	★★★	★★☆	★☆☆	★★★	☆☆☆	☆☆☆
TakingData-SmartLBS[a]	★★★	★★★	★★★	★★★	☆☆☆	☆☆☆
Baidu-HuiYan[b]	★★★	★★★	★★★	★★★	☆☆☆	☆☆☆
Shahriari-Mehr et al. [15]	★★☆	★★☆	★★★	★☆☆	☆☆☆	☆☆☆
Roy et al. [14]	★★★	★★★	★★★	★★★	☆☆☆	☆☆☆
WANT [12]	★★★	★★☆	★★☆	★★☆	★★☆	☆☆☆
GeoGTI (ours)	★★★	★★★	★★★	★★★	★★★	★★☆

[a] https://www.talkingdata.com/product-SmartLBS.jsp
[b] https://huiyan.baidu.com/

4 Experiments

4.1 Data Collection and Analysis

To demonstrate the effectiveness and practicality of our methods, we use massive real-world urban data from ten industries. In particular, we choose the banking and retail industries for further research, and the corresponding data scale is shown in Table 2. Multi-source data used in this work can be categorized into:
Store Data. It contains the store name, city, location (longitude and latitude), category, and the industry it belongs to.
User Data. It refers to the user profile data on the grid, including gender distribution, occupation distribution, and trajectories after aggregation[1].
POI Data. It contains the name, location, and category of POIs. There are 372 categories in total, such as shop, restaurant, bank, etc.

4.2 Experiment Settings

We choose LightGBM [7] as the classifier model, a GBDT framework that supports efficient parallel training, faster training speed and lower memory consumption. Each grid is represented as a vector of geographic features and user profile features through the feature extraction module. The radius r and the number

---

[1] Note that, considering user privacy, we only use aggregation or vague indicators.

of locations n used for constructing negative samples are the core parameters in our methods. We try different values for radius r (*e.g.*, 100 m, 300 m, 500 m) and sample ratio, and we find that the best performance can be achieved when the r is 500 m with 1 : 4 positive and negative sample ratio. In the experiment, the method of Bayesian Optimization is adopted to search for hyperparameters of the model. We split the training set and test set by 8:2, and in order to reduce the disturbance of the experiment, we perform a 5-fold cross-validation on the training set. Finally, we employ macro-precision, macro-recall, and macro-f1-score to evaluate the result.

4.3 Experiment Results and Analysis

General Site Selection. The results of national models for ten industries are illustrated in Fig. 4, from which we can see most of the models have achieved an F1 score greater than 65. The top 20 feature importance distribution is calculated by measuring how many industry models the feature ranks in the top 20 for feature importance. The x-coordinate is the number of features, 478 in total. The results exhibits that very few features play a key role (rank in the top 20 important features) in multiple industry models, showing the robustness of our method. Further, as a case study, we chose two nationally-operated banks (CCB and ABC) and two regionally-operated convenient stores (MYJ and JY). The results are demonstrated in Table 2. The above results demonstrate our methods' generality and effectiveness.

In particular, we notice that the results at the city level are significantly better than the ones at the national level, which is due to the geographical characteristics of the industry. In addition, we also observe that the results for the Shanghai all-bank brands are better than those of Shanghai ABC bank and lower than those of Shanghai CCB bank, which we infer is because the all-brand bank data includes both ABC data and CCB data.

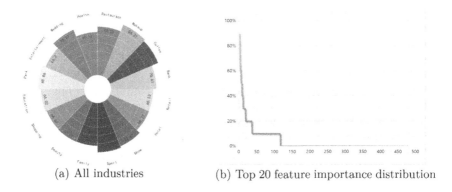

   (a) All industries   (b) Top 20 feature importance distribution

Fig. 4. General site selection results

Table 2. The results of general site selection models

Industry	Region	Brand	Training	Test	P	R	F1
Bank	Nation	All	465,198	116,300	73.67	69.75	70.61
		ABC	91,000	22,750	80.41	79.10	79.72
		CCB	111,058	27,765	88.36	89.13	88.73
	Shanghai	All	20,797	5,200	90.38	88.34	89.30
		ABC	1,616	404	83.67	78.70	80.70
		CCB	2,809	703	91.48	89.30	90.36
	Beijing	CCB	3,472	868	87.32	84.64	85.87
Retail	Nation	All	202,936	50,735	66.26	65.03	65.53
	Guangdong	All	39,831	9,958	75.71	74.46	75.04
		MYJ	17,563	4,391	77.90	77.43	77.66
		JY	10,711	2,678	76.05	80.57	77.95
	Guangzhou	MYJ	12,700	3,175	76.90	76.32	76.60
	Shenzhen	MYJ	17,380	4,345	80.94	77.17	78.77

Transfer Site Selection. Figure 5 presents the results of the proposed transfer methods: industry transfer (T_{ind}), competitive brands transfer (T_{compet}), upper administrative district transfer (T_{upper}) and city transfer (T_{city}). For the case study of banking and retail industries, we select Shanghai_CCB (Guangzhou_MYJ) as the baseline, Shanghai_All (Guangzhou_All) for T_{ind}, Shanghai_ABC (Guangzhou_JY) for T_{compet}, Nation_CCB (Guangdong_MYJ) for T_{upper}, and Beijing_CCB (Shenzhen_MYJ) for T_{city}. In this section, according to the results of transfer methods, we try to answer the following questions.

Question 1: Whether the proposed transfer methods can solve the problem of data scarcity and cold-start in the chain store site recommendation ?

Answer 1: We conduct experiments with cold-start and fine-tune methods, respectively. In detail, when the target brand has no data, we adopt the cold-start method and directly use the existing model (such as the industry model or the competitive brand model) to evaluate the target brand without adjusting parameters. When the target brand has only a small amount of data, we use the fine-tune method to initialize the model with the existing model and update the parameters on the target brand dataset[2]. From the experimental results, the cold start method achieves an F1 value greater than 63. Compared with the non-transfer method, the fine-tune method can optimally improve the F1 value by 2–3 on average.

Question 2: Since there are four different dimensional transfer methods, in practice, which method should we choose ?

---

[2] In order to prevent the problem of label leakage, when we initialize the model, we remove the target brand data and retrain the model.

Answer 2: In practice, the adequacy of source data is the first consideration. Often, we can hardly determine an appropriate competitive brand and get corresponding data. Second, we consider the effects of different transfer methods. Based on empirical experience, We have concluded a general order for different layer transfers. For common brands, industry transfer usually gives good results and thus should be selected first. For brands with obvious regional or local characteristics, we give priority to upper administrative district transfer. Due to the difficulty in finding similar cities and competitive brands, city transfer and competitive brands transfer come last.

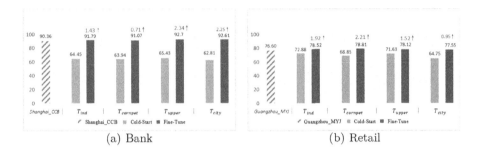

(a) Bank   (b) Retail

Fig. 5. Transfer site selection results

Interpretability Experiment. For the verification of the interpretability of the models, we choose the locations with discrimination in population, business, living, working, and transportation. We select Shanghai Railway Station (a densely-populated place) and Shanghai Xiangke Road (a sparsely-populated place) as examples. The Shanghai all-brands bank model gives a score of 98.56 at the Shanghai Railway Station while 38.43 at the Xiangke Road. Meanwhile, Fig. 6 shows the attribution results at these two locations respectively. It can be seen from the figure that the score of Shanghai Railway Station is higher than that of Xiangke Road in terms of population and transportation, which is basically consistent with the actual situation.

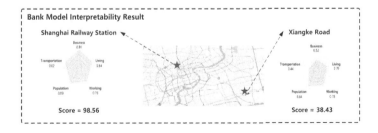

Fig. 6. Interpretable radar chart

5   Conclusion

This paper proposes a transferable and interpretable general site recommendation system called GeoGTI using multi-source urban data. It aims to set a unified framework for different industries and different regions. To solve the problem of data scarcity and cold-start, GeoGTI use a multi-layer knowledge transfer architecture, leveraging knowledge from other sources. For better interpretability, GeoGTI adopts the BI attribution method to map features to a five-dimensional radar chart. As for future works, we will explore more dimensions of data transfer issues, provide more intuitive analysis, and consider customized methods to improve the results of the site recommendation task.

Acknowledgements. Supported by the Shanghai Municipal Science and Technology Major Project (2021SHZDZX0100) and the Fundamental Research Funds for the Central Universities.

References

1. Chen, Q., Ma, K., Hou, M., Kong, X., Xia, F.: Decision behavior based private vehicle trajectory generation towards smart cities. In: Xing, C., Fu, X., Zhang, Y., Zhang, G., Borjigin, C. (eds.) WISA 2021. LNCS, vol. 12999, pp. 109–120. Springer, Cham (2021). https://doi.org/10.1007/978-3-030-87571-8_10
2. Chen, Y.M., Chen, T.Y., Chen, L.C.: On a method for location and mobility analytics using location-based services: a case study of retail store recommendation. Online Inf. Rev. **45**, 297–315 (2021)
3. Fan, Z., Song, X., Shibasaki, R., Li, T., Kaneda, H.: Citycoupling: bridging intercity human mobility. In: Proceedings of the 2016 ACM International Joint Conference on Pervasive and Ubiquitous Computing, UbiComp 2016, pp. 718–728. Association for Computing Machinery, New York (2016)
4. Ge, D., Hu, L., Jiang, B., Su, G., Wu, X.: Intelligent site selection for bricks-and-mortar stores. Modern Supply Chain Res. Appl. **1**, 88–102 (2019)
5. Guo, B., Li, J., Zheng, V.W., Wang, Z., Yu, Z.: CityTransfer. In: Proceedings of the ACM on Interactive, Mobile, Wearable and Ubiquitous Technologies, vol. 1, pp. 1–23, January 2018
6. Karamshuk, D., Noulas, A., Scellato, S., Nicosia, V., Mascolo, C.: Geo-spotting: mining online location-based services for optimal retail store placement, June 2013
7. Ke, G., et al.: LightGBM: a highly efficient gradient boosting decision tree. In: Guyon, I., Luxburg, U.V., Bengio, S., Wallach, H., Fergus, R., Vishwanathan, S., Garnett, R. (eds.) Advances in Neural Information Processing Systems, vol. 30. Curran Associates, Inc. (2017)
8. Li, J., Guo, B., Wang, Z., Li, M., Yu, Z.: Where to place the next outlet? Harnessing cross-space urban data for multi-scale chain store recommendation, pp. 149–152. Association for Computing Machinery, Inc, September 2016
9. Lian, D., Zheng, K., Ge, Y., Cao, L., Chen, E., Xie, X.: GeoMF++: scalable location recommendation via joint geographical modeling and matrix factorization. ACM Trans. Inf. Syst. **36** (2018)

10. Lin, J., Oentaryo, R., Lim, E.P., Vu, C., Vu, A., Kwee, A.: Where is the goldmine? Finding promising business locations through Facebook data analytics. In: Proceedings of the 27th ACM Conference on Hypertext and Social Media, pp. 93–102 (2016)

11. Liu, Y., et al.: DeepStore: an interaction-aware wide amp;deep model for store site recommendation with attentional spatial embeddings. IEEE Internet Things J. **6**(4), 7319–7333 (2019). https://doi.org/10.1109/JIOT.2019.2916143

12. Liu, Y., et al.: Knowledge transfer with weighted adversarial network for cold-start store site recommendation. ACM Trans. Knowl. Discov. Data (TKDD) **15**(3), 1–27 (2021)

13. Rahman, M.K., Nayeem, M.A.: Finding suitable places for live campaigns using location-based services, pp. 37–42. Association for Computing Machinery, Inc, May 2017

14. Roy, A.C., Arefin, M.S., Kayes, A., Hammoudeh, M., Ahmed, K.: An empirical recommendation framework to support location-based services. Future Internet **12**(9), 154 (2020)

15. Shahriari-Mehr, G., Delavar, M.R., Claramunt, C., Araabi, B.N., Dehaqani, M.R.A.: A store location-based recommender system using user's position and web searches. J. Locat. Based Serv. **15**(2), 118–141 (2021)

16. Wang, F., Chen, L., Pan, W.: Where to place your next restaurant? Optimal restaurant placement via leveraging user-generated reviews, 24–28 October 2016, pp. 2371–2376. Association for Computing Machinery, October 2016

17. Wei, Y., Zheng, Y., Yang, Q.: Transfer knowledge between cities, 13–17 August 2016, pp. 1905–1914. Association for Computing Machinery, August 2016

18. Xu, Y., Shen, Y., Zhu, Y., Yu, J.: AR2net: an attentive neural approach for business location selection with satellite data and urban data. ACM Trans. Knowl. Discov. Data **14** (2020)

19. Yu, Z., Tian, M., Wang, Z., Guo, B., Mei, T.: Shop-type recommendation leveraging the data from social media and location-based services. ACM Trans. Knowl. Discov. Data **11** (2016)

20. Zhao, G., et al.: Location recommendation for enterprises by multi-source urban big data analysis. IEEE Trans. Serv. Comput. **13**(6), 1115–1127 (2017)

LFM-C: A Friend Recommendation Algorithm for Campus Mutual Aid System

Lufeng Han[✉]

Nanjing University of Finance and Economics, Nanjing, China
njuejwc@163.com

Abstract. At present, the campus mutual aid system in colleges and universities is increasingly prevalent, but generally, such systems lack the function of customized friend recommendations. Based on the LFM algorithm and the cosine similarity algorithm, a friend recommendation algorithm, LFM-C, is proposed in this paper. Taking the current students and alumni as data sets, this algorithm establishes connections between current students and alumni and effectively recommends the alumni who graduated from the majors at the universities that are of interest to the current students. The algorithm gives full play to the role of alumni as a mentor and helps students who are preparing to pursue postgraduate studies. Experiments show that the LFM-C algorithm is more accurate and efficient than User-CF.

Keywords: Campus mutual aid · Friend recommendation · LFM-C

1 Introduction

The intra-campus mutual aid systems have been developed by multiple domestic colleges and universities, providing a good platform for students to communicate, share resources, and collaborate in learning. However, such systems usually only provide students with the function to follow others actively, but do not enable intelligent recommendations of friends to students through the system. Therefore, there is a gap in the current research: a customized friend recommendation system based on a friend recommendation algorithm.

With the development of the times, the number of applicants for postgraduate entrance examinations has been soaring in recent years. Pursuing postgraduate studies is one of their future development directions for many college graduates. On the one hand, when faced with a grim employment situation, a better education background means more substantial competence in the workplace. On the other hand, to seek positions in government agencies or the public sector, students will need postgraduate degrees as they are the "hard currency" in the job market [1]. There are also many domestic students who are considering pursuing postgraduate studies abroad. This trend is followed by the emergence of various postgraduate counseling companies and various agencies. To find the most suitable university quickly and effectively from those recommended by such a lot of agencies, listening to the advice of one's predecessors is an essential part. Every university has its student information management system. Based on this system, if the university's interpersonal resources can be reasonably used to build a mutual aid

© The Author(s), under exclusive license to Springer Nature Switzerland AG 2022
X. Zhao et al. (Eds.): WISA 2022, LNCS 13579, pp. 572–580, 2022.
https://doi.org/10.1007/978-3-031-20309-1_50

platform on campus, it will not only help students avoid the fraud of some of the illegal agencies in the society and reduce unnecessary expenses but also help them save time in finding a practical solution, thereby improving the success rate of students preparing for postgraduate program application.

To solve the above problems, we designed a hybrid recommendation algorithm, LFM-C. TBased on their historical behaviors,the LFM algorithm will forecast the postgraduate universities that may be interested to the current students.From now on we can referred to these universities as "target universities". And then used the weighted algorithm to calculate the active degree of alumni and added it to the alumni information attribute. After that, the cosine similarity algorithm was used to obtain the similarity matrix between the current students and the alumni of the target university and finally generate the TOP-N alumni sequences recommended to the current students. In the last step, the effectiveness of the LFM-C algorithm was verified by experiments. The experiment results indicated that the LFM-C algorithm proposed in this paper could effectively recommend TOP-N friend sequences.

2 Related Work

Since the recommendation system was proposed in the 1990s, it has gone through many years of development. Many outcomes have been achieved in both academic research and industrial applications. The four main purposes of the recommendation system are: first, to help users find the products they want and explore the long tail; second, to reduce information overload; third, to increase the click-through rate or conversion rate of the site; fourth, to deepen the understanding of the users and to provide customized services for users [2]. In the research on friend recommendation on social networks, there are mainly demographic-based recommendation algorithms, content-based recommendation algorithms, collaborative filtering-based recommendation algorithms, rule-based recommendations, and hybrid recommendations.

Most of the recommendation algorithms are used to recommend things to people. For example, a user-based collaborative filtering algorithm identifies the users' preference for products or contents through the users' historical behavioral data (such as product purchases, favorites, content comments, or sharing). These preferences are also measured and scored. Then the algorithm calculates the relationship between users based on different users' degrees of attitude and preference for the same products or content and recommends products among users with the same preferences. Item-based collaborative filtering algorithms are very similar to user-based collaborative filtering algorithms, only replacing users with items. The relationship between items is obtained by calculating the ratings of different users for different items. The algorithm then recommends similar items to users based on the relationship between items. The rating here represents the user's attitude and preference for the product. To put it simply, if user A purchases both commodity 1 and commodity 2, it means that commodity 1 and commodity 2 are highly correlated. When user B also buys commodity 1, it can be inferred that he also needs to purchase commodity 2 [3].

Both Facebook's "Friends You May Know" feature and Twitter's "People You May Know" feature can recommend friends that users may want to follow, that is, to recommend people to people. Facebook is mainly for people to connect with their family and friends; people use Facebook to share photos, videos, and regular updates; Twitter is mainly for sharing ideas, real-time information, and news sharing.

Based on the above two aspects, this article proposes a concept integrating recommending things to people and recommending people to people and applies this concept to the mutual aid system of the university campus. Using the university's student information system database, the idea of recommending related alumni for current students is realized. The function of recommending relevant alumni. For example, if student A wants to apply for a master's program in Finance at Nanjing University, we can recommend the alumni who were already admitted to Nanjing University's Finance major to come into contact and student A will gain more experience in the postgraduate entrance examination. For another example, if student B wants to apply for a job at IBM, we can recommend the university's alumni. They have already worked at IBM to establish contact with them and learn about their experience in written examinations and interviews, or even future job referrals. In this way, not only can the role of alumni be brought into full play, but also the sense of achievement of the alumni will be enhanced, and the friendship between the current students and alumni will also be strengthened. The actual realization of the mentoring role of the alumni has great practical implications for the development of students and universities.

3   Friend Recommendation Algorithm LFM-C

3.1   Overall Framework and Process

This article aims to accurately recommend alumni in the target universities that the designated students are most interested in. For this purpose, a hybrid recommendation algorithm based on the preference analysis of the current students and the calculation of alumni correlation, called the LFM-C algorithm, was built. The overall framework of the algorithm is shown in Fig. 1. The model in this paper is mainly divided into three modules: 1. The student preference analysis module, where the current student-university scoring matrix is formed based on the historical forum data that students have browsed or clicked on. The LFM algorithm is used to analyze the matrix to obtain the preference matrix of the current students. 2. The alumni attributes generation module calculates the active degree of the alumni according to the number of interactions and calculates the active degree of the alumni according to the number of interactions and the alumni's followers and adds the active degree to alumni information attributes. 3. The recommendation moduleadopts the cosine similarity algorithm to calculate and sort the similarity of the attributes of students and alumni to accurately recommend highly active alumni in the target universities.

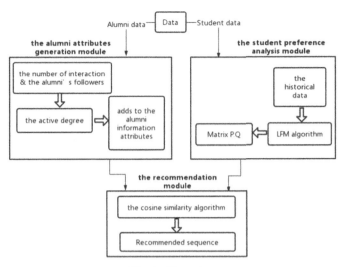

Fig. 1. Flow chart

3.2 The LFM Recommendation Algorithm

The implicit semantic model recommendation algorithm is a collaborative filtering recommendation algorithm that recommends relevant content based on the user's historical behavioral data. It is an extension of the collaborative filtering algorithm and expresses the similarity of users and the similarity of items through a method called implicit feature [14]. The user-based or recommended content-based recommendation algorithm needs to maintain a correlation matrix for the recommended content in collaborative filtering algorithms. When there are many users or recommended contents, the matrix has a large dimension and takes up huge storage space. However, the correlation matrix is actually sparse. The LFM algorithm introduces an implicit feature so that the sparse correlation matrix can be decomposed into two relatively dense matrices, which greatly reduces the space complexity [4–6].

Since the number of universities that students are interested in will be much lower than the number of existing universities in the database, the scoring matrix between the current students and the target universities will be sparse. Therefore, we use the LFM algorithm in the user preference analysis module, which effectively solves the problem of a sparse scoring matrix, and reduces the dimension, thereby improving the algorithm's efficiency.

The LFM calculation idea is shown in Fig. 2.

The formula for calculating user U's interest in item I by the implicit semantic model is:

$$R(U, I) = T_{UI} = P_U^T Q_I = \sum_{f=1}^{F} P_{U,K} Q_{I,K} \tag{1}$$

where Pu,k represents the relationship between user U's interest and the k-th implicit feature, while Qi,k represents the number of item i and the k-th implicit feature, and r

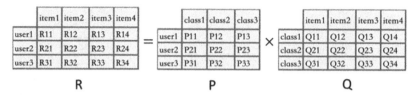

Fig. 2. LFM calculation idea

is the user's interest in the item [7]. We know that LFM finally seeks to find P and Q through the formula.

In this article, we applied the algorithm to study the degree of interest of current student U in alumni's university I in the current alumni database. In the formula, K is the implicit feature, the current student U's attention on the implicit feature K and the weight relationship between the implicit feature K and the alumni's university I. The P matrix we expect to get is the relationship matrix between the current student U and the Kth implicit feature. The Q matrix represents the number matrix of the university I of the alumni in the current alumni database and the Kth implicit feature. When solving P and Q, the optimized loss function is generally used, and the formula is as follows:

$$C = \sum_{(U,I) \in K} (R_{UI} - \hat{R}_{UI})^2 = \sum_{(U,I) \in K} (R_{UI} - \sum_{K=1}^{K} P_{U,K} Q_{K,I})^2 + \lambda ||P_U||^2 + \lambda ||Q_I||^2 \quad (2)$$

In Formula 2, the loss function is used to calculate the score between the true interest of the student U in the target university I and the error of the predicted interest. The model is optimal if the square of the error is the smallest. The regularization penalty term in the formula $\lambda ||P_U||^2 + \lambda ||Q_I||^2$ is used to prevent the model from overfitting.

The commonly used algorithm is the gradient descent algorithm to find the minimized loss function. Gradient descent algorithms include BGD, SGD, and MBGD. This paper adopts SGD, that is, the stochastic gradient descent algorithm. When using SGD, the partial derivation of the unknown parameters is first obtained. The formula is as follows:

$$\frac{\partial C}{\partial P_{Uk}} = -2 \left(R_{UI} - \sum_{k=1}^{K} P_{U,k} Q_{k,I} \right) Q_{kI} + 2\lambda P_{Uk} \quad (3)$$

$$\frac{\partial C}{\partial Q_{kI}} = -2 \left(R_{UI} - \sum_{k=1}^{K} P_{U,k} Q_{k,I} \right) P_{Uk} + 2\lambda Q_{kI} \quad (4)$$

The iterative calculation is also needed to continuously optimize the parameters until the parameters converge. The iterative formula is as follows:

$$P_{Uk} = P_{Uk} + \alpha \left(\left(R_{UI} - \sum_{k=1}^{K} P_{U,k} Q_{k,I} \right) Q_{kI} - \lambda P_{Uk} \right) \quad (5)$$

$$Q_{kI} = Q_{kI} + \alpha \left(\left(R_{UI} - \sum_{k=1}^{K} P_{U,k} Q_{k,I} \right) P_{Uk} - \lambda P_{kI} \right) \quad (6)$$

In the above formula, α is the learning rate. The larger the α is set, the faster the iterative descent will be [8].

In this way, two implicit factor matrices are obtained, which realizes the automatic clustering of the alumni universities based on the historical behavior of the current students.

3.3 Cosine Similarity Algorithm

Cosine similarity measurement, also known as cosine similarity, measures the similarity between two vectors by calculating the cosine value of the angle between them. The cosine of an angle of 0 degrees is 1, and the cosine of any other angle is not greater than 1; its minimum value is -1. The cosine of the angle between the two vectors thus determines whether the two vectors are pointing roughly in the same direction. When the two vectors have the same direction, the cosine similarity value is 1; when the angle between the two vectors is 90°, the cosine similarity value is 0; when the two vectors point in completely opposite directions, the cosine similarity value is −1. This result is independent of the length of the vector and is only related to the direction of the vector [9–11].

Applying cosine similarity to the friend recommendation algorithm is to calculate the similarity between users through user attributes. The variables for calculating cosine similarity are two vectors that are the user's attributes. Specifically, the preference of the current students obtained by the LFM algorithm is set as a vector in the n-dimensional space, and the similarity analysis is carried out with the relevant data in the alumni database to get the sequence of alumni in the target universities of the current student based on the forecast scoring. The result can be calculated with the following Formula 7:

$$Sim(i,j) = \cos(I, J) = \frac{I * J}{\|I\| \bullet \|J\|} = \frac{\sum\limits_{c=1}^{n} R_{i,c} R_{j,c}}{\sqrt{\sum\limits_{c=1}^{n} R_{i,c}^2} \bullet \sqrt{\sum\limits_{c=1}^{n} R_{j,c}^2}} \qquad (7)$$

In Formula 7, i is the target user, and j is the neighbor user, indicating recommending friends to the current student i.

This article adds a dimension of "active degree" to the alumni attributes to recommend alumni with a high active degree to current students. The weighted value can obtain this value according to the number of interactions between alumni and current students and the number of the alumni's followers. The calculation formula is as follows:

$$Active\ degree = 1 + \alpha \cdot M + \beta \cdot F \qquad (8)$$

In this formula, the active degree represents the active level of the alumni. Suppose the alumni do not interact with the current students and are not followed. In that case, the default activity degree is 1. α and β are coefficients that represent the weights of M and F. In this experiment, the value of α is 0.6, and the value of β is 0.4. That is to say, and we pay more attention to the number of interactions between alumni and current students. The alumni recommended according to this rule usually have a shorter response time when answering the current student's questions. M represents the number of alumni and current students interactions, and F represents the number of alumni followers. In the calculation, if the alumni reply to a message once, then $M = 1$, and if they reply to two messages, then $M = 2$. The value of F follows suit.

4   Experiment and Analysis

4.1   Introduction to the Experiment

The experimental data in this paper adopted the data of the class of 2022 students in the calibration system of the author's university and the 2021 alumni data on 91job. The School of Finance student data was extracted from the two databases as the experimental data set. The alumni data included the student number, major, employment category names, company name, type of company, industry category, etc. There are 500 pieces of data in the student data set and 516 pieces of data in the alumni data set. The experiment in this paper is simulated with MATLAB. The algorithm LFM-C of this article and the user-based collaborative filtering algorithm (User-CF) as the similarity measure of the friend recommendation algorithm were compared and analyzed in the accuracy rate, recall rate, and F-1 measure standard.

Precision, recall, and F1-measure are three evaluation indicators commonly used in recommendation systems [12]. The following diagram illustrates them (Fig. 3):

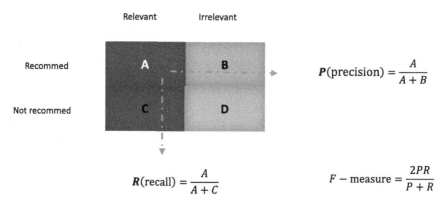

Fig. 3.   Precision, recall and F1-measure

In the figure, A represents what is recommended and wanted; B represents recommended but useless; C represents not recommended, but actually wanted; D represents not recommended and useless.

The accuracy rate refers to the proportion of the user's friends in the recommended list given by the system in the total number of recommended users in the list. The higher the index is, the more useful the data in the recommended list is to the user. The recall rate refers to the proportion of the number of recommended friends in the number of friends in the test data. The higher the index, the better the effect of the algorithm. F1-measure is a comprehensive measurement index, a combination of the two. The higher the precision rate and the recall rate, the better, but there could be contradictions between the two in some cases. In this case, the F1-measure indicator is needed for measurement. When the F1-measure indicator has a higher value, the method is effective [13].

4.2 Experimental Results and Analysis

The experiments in this paper compared the accuracy of the LFM-C algorithm and the user-based collaborative filtering algorithm. The experimental results are shown in Figs. 4, 5 and 6:

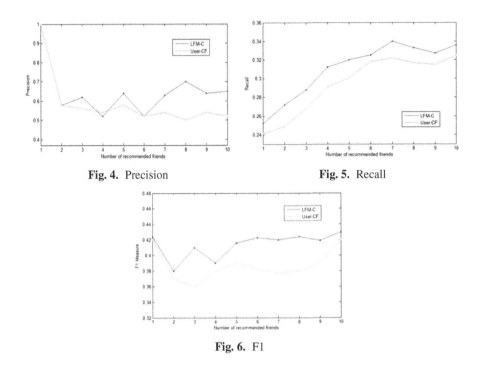

Fig. 4. Precision

Fig. 5. Recall

Fig. 6. F1

Based on the analysis of the experimental results, with the increase in the number of friends, the accuracy rate of all algorithms gradually decreases, the recall rate gradually increases, and the F1 gradually increases as a whole. It can be seen from the experimental results that the evaluation index of the LFM-C algorithm is generally better than that of the User-CF algorithm, because we took into consideration the problem of a sparse scoring matrix and simultaneously added the active degree attribute to the alumni attributes, which makes the clustering more accurate and recommendations more optimized. Overall, the experimental results are in line with the expectations.

5 Conclusion

At present, although the informatization of universities has become quite a prevalent phenomenon, the systems are not yet fully intelligent. This article mainly aims to provide solutions to recommend alumni more accurately for current students preparing for the postgraduate entrance examinations through the campus mutual aid system, rather than unilaterally searching and following alumni. To solve the problem of a sparse scoring matrix, a hybrid algorithm based on LFM and cosine similarity algorithm was proposed, which improved the accuracy of recommendation. At the same time, the active degree attribute is added to the information attributes of the alumni, which improves the optimality of recommendation. The experimental results show that the LFM-C algorithm can effectively recommend the alumni who have studied in the relevant majors of the relevant universities. This recommendation idea can also be applied to the employment recommendation in the campus mutual aid system.

References

1. Bradley, P.S., Fayyad, U.M.: Refining initial points for k-means clustering. In: Proceedings of the Fifteenth International Conference on Machine Learning ICML, pp. 91–99 (1998)
2. Joo, Y.J., So, H., Kim, N.H.: Examination of relationships among students' self-determination, technology acceptance, satisfaction, and continuance intention to use K-MOOCs. Comput. Educ. **122**, 260–272 (2018)
3. Tang, F., Zhang, B., Zheng, J., Gu, Y.: Friend recommendation based on the similarity of micro-blog user model. In: Proceedings of the 2013 IEEE International Conference on Green Computing and Communications and IEEE Internet of Things and IEEE Cyber, Physical and Social Computing, pp. 2200–2204 (2013)
4. Li, B., et al.: Link prediction friends recommendation algorithm for online social networks named JAFLink. J. Chin. Comput. Syst. **38**(08), 1741–1745 (2017)
5. Lu, L., Zhou, T.: Link prediction in complex networks: a survey. Physica A Stat. Mech. Appl. **390**(6), 1150–1170 (2011)
6. Tang, F., Zhang, B., Zheng, J., Gu, Y.: Friend recommendation based on the similarity of micro-blog user model. In: Proceedings of the 2013 IEEE International Conference on Green Computing and Communications and IEEE Internet of Things and IEEECyber, Physical and Social Computing, pp. 2200–2204 (2013)
7. Xie, F., Chen, Z., Shang, J., Feng, X., Li, J.: A link prediction approach for item recommendation with complex number. Knowl. Based Syst. **81**, 148–158 (2015)
8. Kang, J., Zhang, J., Song, W., Yang, X.: Friend relationships recommendation algorithm in online education platform. In: Xing, C., Fu, X., Zhang, Y., Zhang, G., Borjigin, C. (eds.) WISA 2021. LNCS, vol. 12999, pp. 592–604. Springer, Cham (2021). https://doi.org/10.1007/978-3-030-87571-8_51
9. Zhao, H., Chen, J., Xu, L.: Semantic web service discovery based on LDA clustering. In: Ni, W., Wang, X., Song, W., Li, Y. (eds.) WISA 2019. LNCS, vol. 11817, pp. 239–250. Springer, Cham (2019). https://doi.org/10.1007/978-3-030-30952-7_25
10. Zou, W., Hu, X., Pan, Z., Li, C., Cai, Y., Liu, M.: Exploring the relationship between social presence and learners' prestige in MOOC discussion forums using automated content analysis and social network analysis. Comput. Hum. Behav. **115**, 106582 (2021)
11. Li, X., Wang, M., Liang, T.P.: A multi-theoretical kernel-based approach to social network-based recommendation. Decis. Support Syst. **65**(5), 95–104 (2014)
12. Quijano-sánchez, L., Díaz-agudo, B., Recio-garcía, J.A.: Development of a group recommender application in a social network. Knowl. Based Syst. **71**(S1), 72–85 (2014)
13. Li, Y.M., Hsiao, H.W., Lee, Y.L.: Recommending social network applications via social filtering mechanisms. Inf. Sci. **239**(4), 18–30 (2013)
14. Chaturved, A.: An efficient modified common neighbor approach for link prediction in social networks. IOSR J. Comput. Eng. **12**(3), 25–34 (2013)

Recommending Online Course Resources Based on Knowledge Graph

Xin Chen[1]([✉]) [iD], Yuhong Sun[1] [iD], Tong Zhou[2] [iD], Yan Wen[1] [iD], Feng Zhang[1] [iD], and Qingtian Zeng[1] [iD]

[1] College of Computer Science and Engineering,
Shandong University of Science and Technology, Qingdao 266590, People's Republic of China
xxwar@163.com
[2] College of Civil Engineering and Architecture,
Shandong University of Science and Technology, Qingdao 266590, People's Republic of China

Abstract. Nowadays, it is challenging for college students or lifelong education learners to choose the courses they need under the constant growth of massive online course resources. Therefore the recommendation systems are used to meet their personalized interests. In the scenario of course recommendation, traditional collaborative filtering (CF) is not applicable because of the sparsity of user-item interactions and the cold start problem. Learned from MKR, MKCR is proposed to enhance online courses sources recommendation when the interaction between students and courses is extremely sparse. MKCR is an end-to-end framework that utilizes a knowledge graph embedding task to assist recommendation tasks. The experiment data partially come from the MOOC platform of Chinese universities. The results show MKCR is better performance than other methods in the experiments.

Keywords: Recommendation System · Knowledge Graph · Collaborative Filtering · Online Course

1 Introduction

The popularization and development of the MOOC platform have brought tremendous changes to the education field, especially during COVID-19, students use the Internet for online learning, and online learning is considered a new type of education mode. To this end, colleges and universities across the country have launched many learning resources in various forms such as text, audio, and video on major education platforms, i.e., China University MOOC, XuetangX, CNMOOC, etc. For the regular full-time undergraduates at school and those lifelong or adult education learners, online course learning is a common way to provide them a large number of open learning opportunities. However, these online course platforms do not organize online courses systematically and progressively. It is difficult to find courses that suit their level. As a result, students are unsure which courses they need to learn to be more effective and may lose interest in learning. Finding a solution for course search or recommendation is a challenge for online

© The Author(s), under exclusive license to Springer Nature Switzerland AG 2022
X. Zhao et al. (Eds.): WISA 2022, LNCS 13579, pp. 581–588, 2022.
https://doi.org/10.1007/978-3-031-20309-1_51

education platforms. Only in this way can the value of online courses be maximized. Recommendation systems have been widely used in movies, books, social networks, and shopping platforms. In order to reduce the difficulty for students to find courses and arouse their learning interest, the MOOC platform recommends courses to students by using the recommendation system. Learned from MKR [1], MKCR is proposed to enhance online courses sources recommendation when the interaction between students and courses is extremely sparse.

2   Related Work

The existing course recommendation method usually uses the collaborative filtering (CF) method [2], which uses the historical interaction of the students-courses to realize recommendations based on the shared preferences of students. However, CF ignores the relationship between instances and courses. The recommendation method considers it will integrate edge information (such as social networks [3], user-item attributes [4], context [5], Knowledge Graph(KG) [1], etc.) into CF. There are also problems of data sparseness and cold start. The neural network methods, such as a neural network-based collaborative filtering recommendation NCF [3], a neural factorization machine NFM [2], and the wide&deep model [2], are introduced to solve the problem of a highly sparse interaction matrix. KG for recommendation is now used in movies, books, music, news, and other fields. These methods mainly include RippleNet [7], MKR, KGPolicy [8], and others [9, 10]. RippleNet is a model similar to a memory network. It spreads potential user preferences in KG and explores their hierarchical interests, but the importance of relationships in RippleNet is weakly characterized. MKR is an end-to-end deep recommendation framework that utilizes a KG embedding task to assist recommendation. Knowledge Graph Policy Network (KGPolicy) leverages knowledge graphs to provide rich relationships between items and entities. It develops negative sampling models through reinforcement learning to infer ground-truth negative samples from unobserved interactions.

3   Models and Methods

The recommendation question is formalized and the course Knowledge Graph is constructed based on the data of China University MOOC. The MKCR model is proposed to enhance the course recommendation task based on MKR. MKCR uses the Knowledge Graph as auxiliary information to capture rich semantic information between different types of entities, enhancing the recommendation's performance and increasing the recommendation method's interpretability and accuracy.

3.1   Problem Formulation

The online course recommendation problem is to recommend courses (items) to students (users). In the course recommendation scenario, set $U = \{u_1, u_2, \ldots, u_M\}$ contains M students and set $V = \{v_1, v_2, \ldots, v_N\}$ contains N courses. The interaction matrix that is expressed as $Y = \{y_{uv}|u \in U, v \in V\}$ is defined according to the implicit feedback of students. The student interacts with the course in the form of learning, browsing and other activities. As shown in Eq. (1).

$$y_{uv} = \begin{cases} 1, & if interaction(u, v) is observed \\ 0, & otherwise \end{cases} \quad (1)$$

The course KG is composed of a large number of entity-relation-entity triples (h, r, t), where $h \in E, r \in R, t \in E$ represent the head entity, relation, and tail entity in the KG respectively. E and R respectively represent the entity set and relation set in the KG. The purpose of the course recommendation problem is to predict whether student u is potentially interested in courses v that have not been interacted before, given the interaction matrix Y and the KG $\mathcal{G}$. The goal of the model is to learn $\hat{y}_{uv} = F(u, v; \Theta)$, $\hat{y}_{uv}$ represents the possibility that student u clicks on the course v. And Θ represents the model parameters of the function F.

3.2 Construction of KG for Online Courses

The construction of the course knowledge graph (KG) adopts the bottom-up construction method. The construction method of the course KG is shown in Fig. 1.

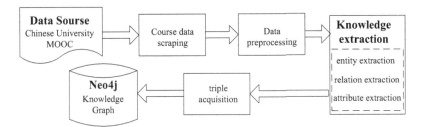

Fig. 1. Course KG construction method.

Table 1. Snippets of course data

ID	Course Name	Teacher	School	Course Category
1	Programming Introduction - C Language	Kai Weng	ZJU	Program design and development
2	C Language Programming Advanced	Kai Weng	ZJU	Program design and development
3	Data Structure	Chunbao Li	WHU	Hardware and software systems and principles
	......	......	......	......

The course attributes of computer and music&dance are crawled from the MOOC platform of Chinese universities. the crawled data has a certain relational structure, and this kind has a certain relationship. The knowledge elements of the structure are summarized and organized into the course KG. The course data snippet is shown in Table 1. Knowledge graph triples are shown in Table 2.

Table 2. Course KG triples examples table

Head Entity	Relation	Tail Entity
Programming Introduction - C Language	Teacher	Kai Weng
Data Structure	Teacher	Chunbao Li
Data Structure	School	WHU
......	......	......

3.3 Models and Methods

The MKCR framework for online course recommendation is shown in Fig. 2. And consists of three modules: recommendation module, Knowledge Graph embedding module, and cross-compression unit. The recommendation module on the left takes students and courses as input. It uses multi-layer perceptron (MLP) and cross-compression units to extract short and dense features for students and courses, respectively. The extracted student and course features are then fed into another MLP to output the predicted probabilities. Similar to the left side, the Knowledge Graph embedding module on the right also uses a multi-layer perceptron (MLP) and a cross-compression unit to extract features from the relation and head entities in the Knowledge Graph. MKCR outputs the representation of the predicted tail entity under the supervision of the score function and the real tail entity. The recommended and Knowledge Graph embedded modules are connected through a specially designed cross-compression unit. Among them, the cross-compression unit can automatically learn the high-level feature interaction between the courses in the recommendation system and the entities in the Knowledge Graph.

The learning algorithm of MKCR is proposed in Algorithm 1, where a training epoch consists of two stages: recommendation task (lines 3–7) and knowledge graph embedding task (lines 8–10). In each iteration, the experiment repeats the training for the recommendation task 4 times before training the knowledge graph embedding task once in each epoch to ensure the accuracy of the experiment.

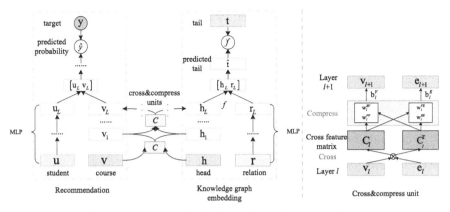

Fig. 2. Framework of MKCR for Course Recommendation

Algorithm 1:MKCR algorithm

Input: Interaction matrix Y, Knowledge Graph $\mathcal{G}$

Output: predicted probability $\hat{y}_{uv}$

1: Initialize all parameters

2: **for** the number of training iteration **do**

3:   **for** t steps **do**

4:     Sample minibatch of positive and negative interactions from Y;

5:     Sample $e \sim S(c)$ for each course c in the minibatch;

6:     Update parameters of F by gradient descent on Eq. (1)-(6),(9);

---

4 Experiment

4.1 Data Sets

Two data sets are used to verify whether MKCR can solve the problem of data sparsity in course recommendation. The data set Mooc-Music is the learning data in the MOOC music and dance category of Chinese universities. Mooc-Music contains 109 courses and 15,914 students' ratings of 109 courses. Interactive data of students in music courses is sparse because most students only choose one course for learning and the data sparsity of Mooc-Music dataset is 1.04%. The number of computer interactions in the MOOC platform of Chinese universities is more than that of other course categories.Mooc-Computer contains 297 courses and 113,630 students' ratings on 297 courses. 9022 students with more than 4 interactions with Mooc-Computer are filtered as Mooc-Computer dataset and the data sparsity of Mooc-Computer dataset is 1.04%, as shown as Table 3.

Table 3. Statistics of the datasets

Dataset	#Students	#Courses	#Interactions	#Entities	#Relations	#Triples
Mooc-Music	15914	109	18135	271	4	436
Mooc-Computer	9022	297	44600	667	5	1417

4.2 Evaluation Baseline

The course recommendation method proposed in this article will be compared with the following benchmarks. SVD decomposes the high-dimensional student course scoring matrix into low-dimensional student feature vector matrix, course feature matrix and diagonal matrix of singular values. The parameters can be updated according to the existing scoring data, once the student eigenvector matrix and the course eigenvector matrix is acquired. SVD takes the student's course scoring matrix as input. LibFM is a widely used feature-based decomposition model that takes the original features of students and courses as the input of LibFM. The dimension is {1, 1, 8}, and the number of training epochs is 50. Wide&Deep combines a deep recommendation model with linear channels and nonlinear channels. The input of Wide&Deep is the same as in LibFM. The dimension of the course and students is 64, using two layers of depth channels (dimensions 100 and 50 respectively) and breadth channels. RippleNet is a method similar to a memory network, which can spread student's preferences on the KG for a recommendation. The hyperparameters are set to $d=8$, $H=2$, $\lambda_1 = 10^{-6}$, $\lambda_2=0.01$, $\eta = 0.02$.

4.3 Experimental Results and Analysis

The method in two experiment scenarios is evaluated as the following: (1) In click-through rate (CTR) prediction, we apply the trained model to each piece of interactions in the test set and output the predicted click probability. We use AUC and Accuracy(ACC) to evaluate the performance of CTR prediction. (2) In top-K recommendation, we use the trained model to select K courses with highest predicted click probability for each student in the testset, and choose Precision@K, Recall@K, and F1@K to evaluate the

Table 4. The results of AUC and ACC in CTR prediction

Model	Mooc-Music				Mooc-Computer		
	AUC		ACC		AUC		ACC
SVD	0.5782		0.5345		0.6641		0.6375
LibFM	0.5890		0.5542		0.7005		0.6533
Wide &Wide	0.6332		0.6241		0.7142		0.7074
RippleNet	0.7165		0.7003		0.8042		**0.7473**
MKCR	**0.7802**		**0.7131**		**0.8224**		0.7422

recommended set. The results of CTR prediction and top-K recommendation of all methods are presented in Table 4 and the Fig. 3.

We can observe from the experimental results:

(1) The model experiment effect of Knowledge-Graph-added course recommendation is better than traditional recommendation methods such as SVD, LibFM, Wide&Deep. It can show good performance even under the sparse interaction of student and the course.

(2) RippleNet performs best in all baselines, especially Mooc-Computer recommendation accuracy is better than MKCR. That shows that RippleNet can accurately capture student interest, especially in the case ofstudent-courseintensive interaction. However, RippleNet in Mooc-Music performs worse than MKCR. Hence MKCR is more suitable for sparse scenes.

(3) Generally speaking, the recommendation performance of the model method on Mooc-Computer is better than that of Mooc-Music. However, it can be found that the improvement effect of the method of using KGs on Mooc-Music is higher than that of Mooc-Computer. The recommended algorithm model using KGs can solve problem in the sparse scenario effectively because the Mooc-Music interactive data is sparser than the Mooc-Computer.

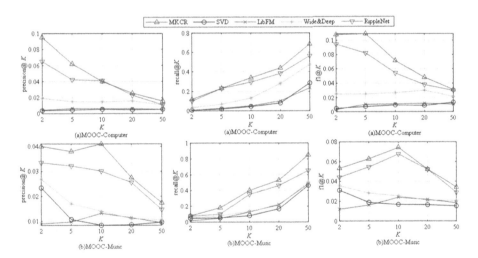

Fig. 3. The results of Precision@K, Recall@K, F1@K in top-K recommendation.

5 Conclusions and Future Work

A course KG is established to capture the rich semantic information between different types of entities and incorporate it into the representation learning process. A multi-task knowledge graph is proposed for the course recommendation method MKCR, which consists of a recommendation module and a knowledge graph embedding module. Experiments are carried out on the MOOC platform course scene in Chinese universities.

The results show that the course recommendation method combined with the knowledge graph is more satisfactory than the traditional course recommendation method. For future work, because each course is composed of multiple videos on the MOOC platform, and each video covers some specific knowledge points, fine-grained knowledge recommendation and the order of learning of knowledge points are challenges for online education platforms.

Acknowledgement. This work was supported in part by the Distinguished Teachers Training Plan Program of Shandong University of Science and Technology (MS20211105), in part by the Teaching Reform Research Project of the Teaching Steering Committee of Electronic Information Specialty in Higher Education and Universities of the Ministry of Education, in part by the Special Project of China Association of Higher Education, in part by the Education and Teaching Research Project of Shandong Province, in part by the Taishan Scholar Program of Shandong Province, in part by the University-Industry Collaborative Education Program (201902316015, 202102402001), and in part by the Open Fund of the National Virtual Simulation Experimental Teaching Center for Coal Mine Safety Mining (SDUST 2019).

References

1. Wang, H., Zhang, F., Zhao, M., Li, W., Xie, X., Guo, M.: Multi-task feature learning for knowledge graph enhanced recommendation. In: Proceedings of the World Wide Web Conference, WWW **2019**, 2000–2010 (2019)
2. He, X., Liao, L., Zhang, H., Nie, L., Hu, X., Chua, T.S.: Neural collaborative filtering. In: International World Wide Web Conferences Steering Committee, pp. 173–182 (2017)
3. Liu, J., Fu, L., Wang, X., Tang, F., Chen, G.: Joint recommendations in multilayer mobile social networks. IEEE Trans. Mob. Comput. **19**(10), 2358–2373 (2020)
4. Wang, H., Zhang, F., Hou, M., Xie, X., Guo, M., Liu, Q.: SHINE: signed heterogeneous information network embedding for sentiment link prediction. In: Proceedings of the 11th ACM International Conference on Web Search and Data Mining, pp. 592–600 (2018)
5. Sun, Y., Yuan, N.J., Xie, X., McDonald, K., Zhang, R.: Collaborative intent prediction with real-time contextual data. In: ACM Transactions on Information Systems (2017)
6. Cheng, H.T., Koc, L., Harmsen, J., et al.: Wide & Deep learning for recommender systems. In: Proceedings of the 1st Workshop on Deep Learning for Recommender Systems. ACM (2016)
7. Wang, H., Zhang, F., Wang, J., Zhao, M., Li, W., Xie, X., Guo, M.: RippleNet: propagating user preferences on the knowledge graph for recommender systems. In: International Conference on Information and Knowledge Management Proceedings, pp. 417–426 (2018)
8. Wang, H., et al.: RippleNet: propagating user preferences on the knowledge graph for recommender systems, pp. 417–426. ACM (2018)
9. Wen, Y., Kang, S., Zeng, Q., Duan, H., Chen, X., Li, W.: Session based recommendation with GNN and time-aware memory network, mobile information systems, vol. 2022, Article ID 1879367, 12 pages (2022). doi: https://doi.org/10.1155/2022/1879367
10. Li, J., Xu, Z., Tang, Y., Zhao, B., Tian, H.: Deep Hybrid Knowledge Graph Embedding for Top-N Recommendation. In: Wang, G., Lin, X., Hendler, J., Song, W., Xu, Z., Liu, G. (eds.) WISA 2020. LNCS, vol. 12432, pp. 59–70. Springer, Cham (2020). https://doi.org/10.1007/978-3-030-60029-7_6

Data Privacy and Security

Multi-party Privacy-Preserving Record Linkage Method Based on Trusted Execution Environment

Xuefei He[1], Haiping Wei[1], Shumin Han[2(✉)], and Derong Shen[3]

[1] School of Information and Control Engineering, Liaoning Petrochemical University, Fushun 113001, China
[2] School of Artificial Intelligence and Software, Liaoning Petrochemical University, Fushun 113001, China
hanshumin_summer@yeah.net
[3] School of Computer Science and Engineering, Northeastern University, Shenyang 110167, China
shenderong@ise.neu.edu.cn

Abstract. As the amount of data in the real world explodes, linking data and making decisions about it is critical. The multi-party privacy-preserving record linkage (PPRL) technology is proposed to find all the record information corresponding to the same entity from multiple data sources, and the sensitive information of the data source should not be disclosed during the process. Existing multi-party PPRL methods often use homomorphic encryption to ensure data security, but there are still some shortcomings. For example, malicious collusion among participants will lead to the disclosure of private keys, and the calculation process is complicated, which challenges the scalability of the multi-party PPRL method. Based on the shortcomings of the current research status, to improve the security and shorten the matching time to make it more suitable for the real big data environment, we propose a multi-party PPRL method based on Trusted Execution Environment (TEE), which avoids the possibility of malicious collusion and reduces the loss of data. It can better resist privacy attacks in the process of linking while shortening the runtime, showing better performance and scalability.

Keywords: Record linkage · Privacy-preserving · Security · Trusted Execution Environment

1 Introduction

In the era of the wide variety of data and high-speed circulation, the first step of data analysis is to integrate it from different sources. Record linkage is one of the core technologies of data integration. However, privacy attacks such as frequency attacks and differential attacks appear in the linking process. This is a great threat to privacy protection in the process of multi-party record linkage. Therefore, considering the privacy security of sensitive data information, it is necessary to use secure privacy-preserving

X. Zhao et al. (Eds.): WISA 2022, LNCS 13579, pp. 591–602, 2022.
https://doi.org/10.1007/978-3-031-20309-1_52

methods to link information from multiple data sources. To solve this problem, Privacy-Preserving Record Linkage (PPRL) is designed to identify matching records of the same entity in different databases during linking. Without compromising the privacy and confidentiality of other unmatched entities, only the final matching result is shared among the data sources, and other unmatched results are not leaked.

The existing PPRL methods have some disadvantages. To protect the privacy of the record usually adopt some kinds of encryption methods. However, there is a certain degree of information loss in the process, and the linkage quality would be reduced. And most of the existing multi-party PPRL methods rely on a trusted third party. But in real applications, there is no guarantee that there will always be a fully trusted third party. Therefore, a homomorphic encryption algorithm that generates pairs of public and private keys is often used in the encryption process of multi-party PPRL, which has higher security [1]. Nevertheless, if a party holding a private key, maliciously colludes with another party, the key would be leaked, which could lead to a total leak of sensitive data. In addition, multiple parties use the same set of public and private keys, so it is impossible to determine which party will generate them. Therefore, the application of homomorphic encryption in PPRL is limited.

Due to the shortcomings of existing methods, we propose a multi-party PPRL method based on Trusted Execution Environment (TEE). The main contributions of this paper are as follows:

1. TEE is introduced during linking, it is a secure area on the main processor of the device. It ensures the security, confidentiality, and completeness of the code in which data is loaded inside it. It also improves privacy without relying on trusted third parties.
2. Let the PPRL take place inside the TEE, taking advantage of the unique properties of TEE, it can more effectively defend against privacy attacks from outside parties or other participants, this improves the security of the method and conducts PPRL directly on TEE also improves efficiency.
3. Experiments using real data prove that our method can perform multi-party PPRL more efficiently, and compared with the existing methods, our method has higher security and better scalability.

The rest of the paper follows: Sect. 2 briefly describes the work related to this method; Sect. 3 describes definitions and preparations; Sect. 4 describes the process and algorithm of multi-party PPRL based on TEE; Sect. 5 describes the experimental results and analysis; Sect. 6 summarizes the content of the paper and the direction of future work.

2 Related Work

First-generation PPRL often employs Secure Multi-Party Computation (SMC) [2], and sensitive information is secured through performing federated computing. Paillier homomorphic encryption is often used, although the SMC protocol is effective and reliable, it is computationally expensive, so the running time is too long.

The second generation of PPRL relies on the method of Bloom filter codings [3], but Bloom filter codings have been shown to be vulnerable to cryptanalysis under certain circumstances [4, 5]. Furthermore, Bloom filter encoding provides no guarantee of accuracy in preserving the original distance of the string, it can harm the quality of linkage results. Lawati et al. [6] used secure TF-IDF [7] to generate weight vectors. Using secure hash signatures for each record, these signatures are used as keys in the chunking mechanism. This method increases privacy but reduces efficiency. Churches and Christen [8] proposed to extract q-grams from string QID values, these values are sent as hash codes to third parties for comparison. But the communication overhead of the method is higher and vulnerable to frequency attacks by third parties, after that, the frequency of certain hash character sequences can be calculated, thereby breaking privacy.

A breakthrough method in the third generation of PPRL is the classified neighborhood method [9] and its privacy-preserving variant. Although the method proposed in [10] eliminates the risk of collusion between participants, it would generate higher computing and communication cost. Kerschbaum [11] proposed an anonymous scheme by setting isometric grid points and calculating the distance of anonymous values from these points. But it reveals a lot to a third party by showing them the exact initial distance between the two values. Randall et al. [12] extended the Bloom filter encoding protocol. They combined the homomorphic encryption method, send encrypted data information to an imaginary trusted the third party, and complete the similarity calculation. To solve the problem that the method is vulnerable to frequency attack. Due to the complexity of the homomorphic encryption process and higher computational cost, and the reliance on an imaginary trusted third party, the application of the method, in reality, has great limitations.

Given the problems of the existing methods, such as longer running time, lower information matching accuracy, and poor security. The method we proposed in this paper combines TEE with homomorphic encryption to reduce the number of decryption in the process and shorten the running time of PPRL so that the method has better scalability. The unique features of TEE ensure data is completed, thus improving the quality of matching while improving the security of PPRL.

3   Definition and Preparation

3.1   Trusted Execution Environment

The Trusted Execution Environment protects computing and operations involving private data utilizing hardware isolation. Tee-enabled CPUs have a specific area called Enclave that is isolated from the outside environment, providing a more secure space for data and executing code, ensuring the confidentiality and completeness of applications. By using both hardware and software to protect data and code, even the server cannot retrieve the user's execution logic and user data in this area. TEE is a physically isolated key storage space. It avoids repeated encryption and decryption during the linking process, avoiding possible data loss and shortening the running time. TEE can directly obtain information about the external environment. But an attacker cannot read or interfere with memory regions, without cracking the hardware, and an attacker can't directly access private data and keys. TEE is more secure than the operating system. Trusted applications running in the TEE have access to the full functionality of the device's main processor and memory, while hardware isolation protects these components from user-installed applications running in the main operating system. Only the processor can decrypt and execute the applications in this area to ensure the confidentiality and completeness of the information. TEE is a hardware extension that ensures the authenticity and confidentiality of the code running on it. It also ensures the completeness of data in the runnable state. The contents of TEE are dynamic and can be updated during execution. An ideal TEE can resist all software and hardware attacks. Based on the above features, the solution combined with TEE shows higher security and efficiency compared with existing privacy protection solutions.

3.2 Homomorphic Encryption Algorithm

Homomorphic encryption technology is an effective encryption method. It is characterized that it does not require direct access to the plaintext. The basic idea is to encrypt the data first. Then the encrypted ciphertext is calculated to obtain the encryption result, and the result obtained after decryption is consistent with the result of the direct calculation of the original data (plaintext) [13]. This method not only achieves the purpose of data protection but also does not affect the calculation of data. The types can be divided into addition homomorphic, multiplication homomorphic, and full homomorphic (support both addition and multiplication). Taking additive homomorphic encryption as an example, there are:

$$\text{Enc}_{pk}(m_1) = c_1 \, \text{Enc}_{pk}(m_2) = c_2 \tag{1}$$

$$\text{Dec}_{sk}(c_1 \odot c_2) = m_1 + m_2 \tag{2}$$

Among them, pk is the public key, sk is the private key, the ciphertext c_1, and c_2 are the encryption results of the plaintext m_1, m_2, and $\odot$ is some kind of multiplication or addition operation.

3.3 Privacy Attacks

PPRL technology is vulnerable to privacy attacks or has a privacy gap. The main privacy attacks of PPRL are dictionary attacks, frequency attacks, cryptanalysis attacks, combination attacks, and collusion.

A dictionary attack is possible with masking functions, where an adversary masks a large list of known values using various existing masking functions until a matching masked value is identified. Frequency attack is still possible even with a keyed masking approach, where the frequency distribution of a set of masked values is matched with the distribution of known unmasked values in order to infer the original values of the masked values. A cryptanalysis attack is a special category of frequency attack that applies to Bloom filter-based data masking techniques. Depending upon certain parameters of Bloom filter masking, such as the number of hash functions employed and the number of bits in a Bloom filter. Using a constrained satisfaction solver allows the iterative mapping of individual masked values back to their original values. A composition attack can be successful by combining knowledge from more than one independent masked datasets to learn sensitive values of certain records. An attack on distance-preserving perturbation techniques, allows the original values to be re-identified with a high level of confidence if knowledge about mutual distances between values is available. Collusion is another vulnerability associated with multi-party or three-party PPRL techniques, where some of the parties involved in the protocol work together to find out about another database owner's data.

4 A Multi-party PPRL Based on TEE

The method proposed in this paper is mainly divided into the following three parts: data preprocessing, calling TEE, and building Paillier homomorphic encryption. The flow chart of the method is shown in Fig. 1.

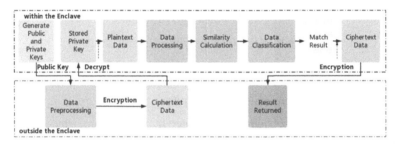

Fig. 1. The flow of PPRL based on Trusted execution environment.

4.1 Data Preprocessing

At this stage, before data entry, each part ensures that the parameters for all participants are consistent. To ensure the security of data, homomorphic encryption is adopted. Our method generates the corresponding encrypted public and private keys in the Enclave, the public key is transferred outside the Enclave and the private key is stored inside it to prevent leakage of the private key. Before encrypting the data, each party needs to receive the public key distributed from the Enclave, the public key is used to encrypt the data into ciphertext data and transmit the encrypted data to the Enclave. Decrypting the ciphertext data into plaintext data using the private key stored in the Enclave, then performing corresponding calculations on the data in the Enclave. In our method, TEE is a physically isolated key store space, the private key exists only in the Enclave in plain text and cannot be read or intercepted by outsiders. It avoids the disclosure of the private key caused by malicious collusion.

4.2 Call TEE

TEE needs to be called before PPRL can be performed, to make it create a new and uninitialized Enclave. In the TEE architecture, user programs are encapsulated and obtain valid signature authentication from the application provider. After the authentication is passed, the required parameters are called using the public key for encryption and sending the security service request and data to the TEE driver. The driver transfer ciphertext data and call parameters to Enclave's isolated runtime environment. The private key is used in the Enclave to decrypt the ciphertext data into plaintext and complete the agreed calculation to return the plaintext result. Finally, the plaintext result is encrypted and transmitted outside the Enclave. The processing flow is shown in Fig. 2.

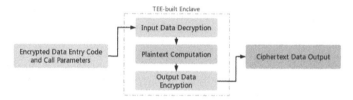

Fig. 2. The flow of data processing in TEE.

4.3 Building Paillier Homomorphic Encryption

In this paper, we use Paillier homomorphic encryption algorithm, which is a difficult problem based on the composite residual class. Paillier homomorphic encryption is a public-key encryption algorithm and an asymmetric encryption algorithm with additive homomorphism, i.e., the ciphertext multiplication is equal to the plaintext addition. The Paillier homomorphic encryption algorithm consists of three parts: key generation, encryption process, and decryption process.

Paillier Key Generation. Generate two large random prime numbers p and q, ensure that the lengths of p and q are the same, and calculate $n = p \cdot q$, $\lambda = \text{lcm}(p - 1, q - 1)$, where lcm(,) is the function of calculating the least common multiple of two numbers. Choose a random integer g, let $g \in Z_{n^2}{}^*$ ($Z_{n^2}{}^*$ denotes a set of coprime integers in Z_{n^2}), where $n^2 \cdot \mu = (L(g^{\lambda} \bmod n^2))^{-1}$, and $\bmod n$ must exist, L is a fun- ction of μ, denoted by $L(\mu) = \frac{\mu-1}{n}$. The public key of the participant is represented as $pk = \,<n,g>$, and the private key of the participant is represented as $sk = \,<\lambda, \mu>$.

Paillier Encryption Function. Given a plaintext message m and public key $pk = \,<n,g>$, where m satisfies $m \in Z_{n^2}$, each message can be represented as a unique integer. Select a random number $r_i, r_i \in Z_{n^2}{}^*$, can calculate this. The ciphertext of the message c, where $c = E(m_i) = g^m \cdot r_i{}^n \bmod n^2$ [14].

The calculation method of verifying the ciphertext information to satisfy the Paillier additive homomorphism is shown in the following formula [14]. Where r_i and r_j are random numbers.

$$E(m_i)E(m_j) = c_i \cdot c_j = g^{m_i} \cdot r_i{}^n \cdot g^{m_j} \cdot r_j{}^n \bmod n^2$$

$$= g^{m_i} \cdot r_i^n \cdot g^{m_j} \cdot r_j^n \mathrm{mod} n^2 = E(m_i + m_j) \tag{3}$$

According to this formula, *Corollary* 1 can be obtained.

Corollary 1: If $m_i + m_j = 0$ is true, then also $E(m_i)E(m_j) = E(0)$ is true.

Paillier Decryption Function. Given the ciphertext c of the message and the private key $sk = \ <\lambda, \mu>$, the plaintext m can be recovered, and the Paillier decryption function is as follows [14]:

$$m = D(c)L(c^\lambda \mathrm{mod} n^2)\mu \mathrm{mod} n = \frac{L(c^\lambda \mathrm{mod} n^2)}{L(g^\lambda \mathrm{mod} n^2)} \mathrm{mod} n \tag{4}$$

4.4 Implementation of Multi-party PPRL in TEE

Algorithm 1 Multi-party PPRL method based on TEE

Input: Participant data

Output: Match the result

1: Each participant receives their own encryption key and uses the public key to encrypt plaintext data into ciphertext data
2: Call TEE
3:     $TEE.create() \to E(\emptyset, \emptyset)$
4:     $TEE.add(E(\emptyset, \emptyset), data, code) \to E(data, code)$
5:     $TEE.extend(E(data, code)) \to M(E(data, code))$
6:     $TEE.init(E(data, code)) \to HMAC(M(E(data, code)))$
7:     Building Paillier homomorphic encryption
8:     $TEE.KeyDerive(E(data, code)) \to key, Enc_{pk}$
9:     Verify whether the product of the decrypted ciphertext is 0, and determine whether the plaintext segment m matches the feature segment f
10: **if** $L(c^\lambda mod \ n^2) mod \ n = 0$ **then**
11:     Match
12: **else**
13:     Mismatch
14: **end if**
15:     Output matching result
16: $TEE.remove(E(data, code)) \to E(\emptyset, \emptyset)$

As shown in Algorithm 1, the first line preprocesses the participant data to obtain the corresponding ciphertext data. Lines 2–6 create a new and uninitialized Enclave, which we denote as E(data,code). The Enclave measurement is expressed as M(E(data,code)), it is a hash of the data and code placed inside the Enclave, including sorting and positioning. First of all, the user deplores the required programs in the Enclave. The internal data generated by running in the Enclave cannot be easily read by programs outside it. Even if there is a participant who wants to conduct malicious collusion. It cannot obtain the corresponding decryption private key to obtain the private data, so privacy is guaranteed. The data of each party is encrypted and transmitted to the Enclave, and then the measured values of the Enclave are updated. Its value (M(E(.,.)) is used by the remote party for proof purposes to establish trust, which then sets the initialization of the Enclave as true. Open the way for the loaded code. Finally determines the hashing of the Enclave metrics. Lines 7–14 are homomorphic encryption in Enclave. By L $(c^\lambda \bmod n^2) \bmod n = 0$ can judge whether the plaintext fragment m matches the feature fragment f, and find out all the record information corresponding to the same entity from the data source of multiple participants. Lines 15–16 print the result of the match and clear all allocated memory, removing the initialization processing power allocated to the Enclave.

In the isolated operating environment of Enclave, in addition to returning the result in the plain text directly. The user can also deploy the return value encryption key attached to the corresponding program call parameter to encrypt the result during the program deployment phase. The result is returned in ciphertext, ensuring that only the consumer can decrypt it. When the sensor of the module storing data in TEE detects external malicious attacks, the data will be cleared and protected. TEE features effectively improve the security of the PPRL process.

5   Experiment and Analysis

5.1   Experimental Setup

The experimental host is an Intel(R) Core(TM) i7-4710MQ, 2.50GHz quad-core processor with 8GB of memory and a 64-bit Windows 7 operating system. Using PyCharm (2020.1) to implement the method proposed in this paper. Our experiment uses the North Carolina voter registration list (NCVR) dataset, and the data records in this dataset are all publicly available real information.

Evaluation Criteria. For the method proposed in this paper, experiments were evaluated from four aspects: Runtime, Recall, Precision, and F-measure [15]. Runtime is the main measure used to evaluate the scalability of our method. Recall, Precision and F-measure are used to evaluate the linkage quality of the proposed method.

Recall: The ratio between the number of truly matched record groups in the candidate record group and the number of truly matched record groups in the data set. The higher the ratio is, the more comprehensive the truly matched records found by this method are.

Precision: The ratio between the number of truly matched record groups and the number of candidate record groups in the candidate record group, the higher the ratio, the more accurate the results of this method are.

F-measure: The value of F is derived from the harmonic mean of recall and precision, usually expressed as: $F = 2 \times \frac{Recall \times Precision}{Recall + Precision}$.

Comparison Method and Parameter Determination. In order to evaluate the performance of all aspects of our method, from the data source size, the number of participants, and different degrees of disturbance of the data set to evaluate the changes of three variables to evaluate it. We extracted 5K, 10K, 50K, 100K, 500K, and 1000K records from NCVR and distributed them to each participant. At the same time, to ensure the quality of the linkage, it is necessary to ensure that at least half of the records of each participant in data sources of different sizes are common. The number of parties selected is 3, 5, 7, 9. In this paper, three scrambled data sets are generated on the original data set, so that each record in each data source has at most one spelling error (Mod-1), at most two spelling errors (Mod-2), and at most three spelling errors (Mod-3). That includes common spelling errors such as character insertion, malicious deletion, and random replacement of characters.

5.2 Experimental Results and Analysis

This part will compare our method with the current four mainstream methods. They are the PPRL method based on homomorphic encryption (HE-PPRL) proposed by Randall et al. [16], the method is based on the composite Bloom filter (RBF-PPRL) proposed by Durham et al. [17], the method is based on the FEDERAL framework (F-PPRL) proposed by Karaapiperis et al. [18], and the method (MD-PPRL) proposed by Vatsalan et al., which combines Bloom filter, secure summation, and Dice coefficient similarity calculation protocol [19].

Scalability Assessment. Firstly, the scalability of our method is evaluated, how the running time of our method changes as the data source size increases. When the number of participants is 5, it can be seen from Fig. 3 that the running time of our method is significantly lower than that of HE-PPRL but slightly higher than the other three methods. Since the method in this paper also uses homomorphic encryption, the required data needs to be encrypted, so it increases the running time compared with other methods. But running in Enclave can effectively reduce the number of encryption and decryption operations, so the running time of our method is significantly lower than that of HE-PPRL. Therefore, our method has better scalability with high security.

Secondly, test our method of running time. With the number of participants, the situation of data set size is $|D_i|$=5K. As our method eliminates frequent encryption and decryption processes, it can be seen from Fig. 4 that as the number of participants increases. The running time of other methods increases exponentially, while our method increases slowly, so our method has good scalability.

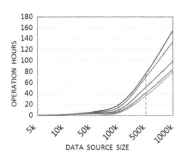 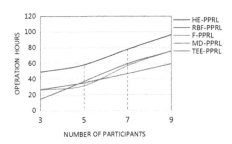

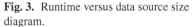

Fig. 3. Runtime versus data source size diagram.   **Fig. 4.** Runtime and number of parties diagram.

Method Performance Evaluation. The performance of our method is evaluated from Recall, Precision and F-measure respectively. Select three different degree of disturbance data sets Mod-1, Mod-2 and Mod-3.

In the case of Mod-1, the data set the size of $|D_i|=5K$. The recall rate evaluates the relationship of recall rate, precision rate, and F-measure between our method and the other four methods as the number of participants increases. As can be seen from Fig. 5, the Recall of our method is superior to the other four methods in terms of disturbing data set Mod-1. When the number of participants is 9, the Recall of our method is still higher than that of the other four methods, and always remains at a high level above 0.6. Because TEE can guarantee the integrity of data, there is no data loss due to multiple encryption and decryption of privacy protection methods, so the method in this paper has a high rate of detection in the experiment.

As can be seen from Fig. 6, in terms of disturbing data set Mod-1. The Precision of our method is better than the other four methods. Also because TEE ensures data is completed. When the number of participants is 3, the Precision is greater than 0.9, and as the number of participants increases to 9, the rate is close to the high level of 0.5.

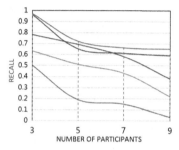

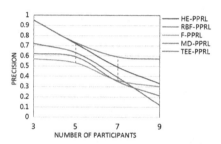

Fig. 5. Recall and number of parties diagram.

Fig. 6. Precision and number of parties diagram.

It can be seen from Fig. 7 that the F-measure of our method is higher than the other four methods. When the number of participants is 3, the F-measure of HE-PPRL is slightly higher than our method. However, as the number of participants increases, the F-measure of HE-PPRL decreases, and the value of its F-measure is lower than our

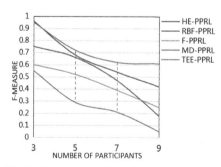

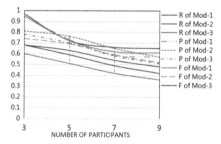

Fig. 7. F-measure and number of parties diagram.

Fig. 8. Evaluation measure and the number of parties in different degrees.

method. When the number of participants increases to 9, the F-measure of our method is still above 0.6, so our method still shows better linkage quality.

When in the different disturb degrees with the sizes of data sets $|D_i|=5K$, the changes of the three evaluation indexes of our method as the number of participants increases are shown in Fig. 8. With the increase of the degree of disturbance, the Recall, Precision, and F-measure of our method all decrease. Because some truly matched records are more likely to be lost as the level of perturbation increases. However, the F-measure shows that our method still maintains a better effect.

Through the experiments, it can be seen that our method has a significant improvement in operation efficiency, lower time cost, and better scalability. TEE ensures the completeness of data and makes our method have higher linkage quality.

Privacy Performance Analysis. To ensure the data security of all participants in the PPRL and get rid of the dependence on an imaginary trusted the third party, in this paper, we propose the method transfers the whole PPRL process to TEE. At first, public and private keys are generated in the Enclave, and the public keys are then used to encrypt the data. The decryption private key is stored in the Enclave. Therefore, malicious attackers cannot obtain sensitive data by obtaining private keys, thus preventing privacy disclosure caused by malicious collusion. TEE also ensures that data running in Enclave cannot be easily read by programs outside the Enclave, which is better protection against possible privacy attacks such as dictionary attacks, frequency attacks, cryptanalysis attacks, and combination attacks, enhanced security of the linking process. Our multi-party PPRL method combined with TEE significantly improves privacy performance.

6   Conclusion

At present, PPRL still has some problems, such as longer running time, lower matching quality, and poor security. Therefore, we propose a method combined with TEE, which can effectively solve the problem of privacy disclosure caused by privacy attacks in the process of multi-party PPRL and enhance the security of the linkage process. At the same time, TEE is combined with a homomorphic encryption algorithm to reduce the time cost and data loss caused by repeated encryption and decryption. In the case of large data sets or more parties, our method has a significant improvement in efficiency, lower time cost, and better scalability. TEE ensures the completeness of data and makes our method has higher linkage quality. Our method has a profound influence on the development of information society. In future work, since TEE still has some limitations, we will further investigate better methods to improve the scalability of the method proposed in this paper.

References

1. Li, T., Liu, Q., Huang, R.: Multi-user fully homomorphic encryption scheme based on policy for cloud computing. In: Xing, C., Fu, X., Zhang, Y., Zhang, G., Borjigin, C. (eds.) WISA 2021. LNCS, vol. 12999, pp. 274–286. Springer, Cham (2021). https://doi.org/10.1007/978-3-030-87571-8_24

2. Goldreich, O., Micali, S., Wigderson, A.: How to play ANY mental game. ACM (1987)
3. Bloom, Burton, H.: Space/time trade-offs in hash coding with allowable errors. Commun. ACM .**13**(7), 422–426 (1970)
4. Kuzu, M., Kantarcioglu, M., Durham, E., Malin, B.: A constraint satisfaction cryptanalysis of bloom filters in private record linkage. In: Fischer-Hübner, S., Hopper, N. (eds.) PETS 2011. LNCS, vol. 6794, pp. 226–245. Springer, Heidelberg (2011). https://doi.org/10.1007/978-3-642-22263-4_13
5. Ali, I., Murat, K., Elisa, B., Monica, S.: A hybrid approach to private record linkage. In: 2008 IEEE 24th International Conference on Data Engineering (ICDE 2008), pp. 496–505. IEEE Computer Society, USA (2008)
6. Ali, A., Dongwon, L., Patrick, M.D.: Blocking-aware private record linkage. In: Proceedings of the 2nd International Workshop on Information Quality in Information Systems (IQIS 2005), pp. 59–68. Association for Computing Machinery, New York, USA (2005)
7. Ravikumar, P., Cohen, W.W., Fienberg, S.E.: A secure protocol for computing string distance metrics. In: IEEE International Conference on Data Mining (ICDM 2004), Workshop on Privacy and Security Aspects of Data Mining (2004)
8. Churches, T., Christen, P.: Some methods for blindfolded record linkage. BMC Med Inform Decis Mak **4**(9), 1–17 (2004)
9. Hernández, M.A., Stolfo, S.J.: Real-world data is dirty: data cleansing and the Merge/Purge problem. Data Min. Knowl. Disc. **2**, 9–37 (1998)
10. Inan, A., Kantarcioglu, M., Bertino, E., Scannapieco, M.: A hybrid approach to private record linkage. In: 2008 IEEE 24th International Conference on Data Engineering, pp. 496–505. Cancun, Mexico (2008)
11. Florian, K.: Distance-preserving pseudonymization for timestamps and spatial data. In: Proceedings of the 2007 ACM Workshop on Privacy in Electronic Society (WPES 2007), pp. 68–71. Association for Computing Machinery, New York, USA (2007)
12. Herranz, J., Nin, J., Rodríguez, P., Tassa, T.: Revisiting distance-based record linkage for privacy-preserving release of statistical datasets. Data Knowl. Eng. **100**, 78–93 (2015)
13. Zhang, C., Li, S., Xia, J., et al.: Batchcrypt: efficient homomorphic encryption for cross-silo federated learning. In: 2020 USENIX Annual Technical Conference (USENIX ATC 20), pp. 493–506 (2020)
14. Pascal, P.: Public-key cryptosystems based on composite degree residuosity classes. In: Proceedings of the 17th International Conference on Theory and Application of Cryptographic Techniques (EUROCRYPT 2004), pp. 223–238. Springer, Heidelberg (2004)
15. Peter, C.: A survey of indexing techniques for scalable record linkage and deduplication. IEEE Trans. Knowl. Data Eng. **24**(9), 1537–1555 (2012)
16. Randall, S.M., Brown, A.P., Ferrante, A.M., Boyd, J.H., Semmens, J.B.: Privacy preserving record linkage using homomorphic encryption. In: First International Workshop on Population Informatics for Big Data (PopInfo 2015) (2015)
17. Durham, E.A., Kantarcioglu, M., Xue, Y., Toth, C., Kuzu, M., Malin, B.: Composite bloom filters for secure record linkage. In: IEEE Transactions on Knowledge and Data Engineering, pp. 2956–2968 (2014)
18. Karapiperis, D., Gkoulala, A., Verykios, V.S.: FEDERAL: a framework for distance aware privacy-preserving record linkage. In: IEEE Transactions on Knowledge and Data Engineering, vol. 30, no. 2, p. 292–304 (2018)
19. Vatsalan, D., Peter, C.: Scalable privacy-preserving record linkage for multiple databases. In: Proceedings of the 23rd ACM International Conference on Conference on Information and Knowledge Management (CIKM 214), pp. 1795–1798. Association for Computing Machinery, New York, USA (2014)

Efficient Differential Privacy Federated Learning Mechanism for Intelligent Selection of Optimal Privacy Protection Levels

Mingyuan Gao[1,2], Fang Zuo[1,2(✉)], and Guanghui Wang[1,3(✉)]

[1] Henan International Joint Laboratory of Intelligent Network Theory and Key Technology, Henan University, Kaifeng 475000, China
zuofang@henu.edu.cn, gwang@vip.henu.edu.cn
[2] Subject Innovation and Intelligence Introduction Base of Henan Higher Educational Institution-Software Engineering Intelligent Information Processing Innovation and Intelligence Introduction Base of Henan University, Kaifeng 475000, China
[3] School of Software, Henan University, Kaifeng 475000, China

Abstract. Differential privacy (DP) is considered as an effective privacy-preserving method in federation learning to defend against privacy attacks. However, recent studies have shown that it can be exploited to perform security attacks (e.g., false data injection attacks), leading to degraded Federated Learning (FL) performance. In this paper, we systematically study poisoning attacks using the DP mechanism in order to perform them from an adversarial perspective. We demonstrate that although the DP mechanism provides a certain degree of privacy assurance, it can also be a vector for poisoning attacks by adversaries. As a countermeasure, we propose FedEDP, a concise and effective differential privacy federation learning (DPFL) algorithm that uses the parameters and losses of differential privacy to intelligently generate an optimal privacy level for edge nodes (clients) to defend against possible poisoning attacks. We conducted experiments on the datasets MNIST and CIFAR10, respectively, and the experimental results show that FedEDP significantly improves the privacy-utility trade-off over the state-of-the-art in DPFL.

Keywords: Federated learning · Differential privacy · Noise interference · Edge computing

This work was supported by Graduate Education Innovation and Quality Improvement Action Plan project of Henan University (No. SYLJD2022008 and No. SYLKC2022028), 2022 Discipline Innovation Introduction Base cultivation project of Henan University and Key Technology Research and Development Project of Henan Province under Grant 222102210055, in part by the Henan Provincial Major Public Welfare Project (201300210400), the China Postdoctoral Science Foundation (2020M672211 and 2020M672217), the Key Scientific Research Projects of Henan Provincial Colleges and Universities (21A520003), and the Key Technology Research and Development Program of Henan (182102210106, 212102210090, 212102210094, and 222102210133).

1   Introduction

With the development and popularity of cloud services and smart devices, edge computing [1] has penetrated into all levels and aspects of our lives, improving the efficiency of our lives and work. In living and industrial environments, there are various smart devices that connect to servers without human intervention, such as medical facilities, smart grids, smart home systems, vehicles and roadside unit networks [2]. The efficiency of cloud computing cannot support the large amount of data generated by these devices, nor can it meet the fast response requirements in certain scenarios. Therefore, edge computing with low latency and heterogeneity is the trend in the development of remote data services. As service providers are deployed close to end devices, request latency may be lower while throughput may be higher. At the same time, the use of edge computing networks can avoid potential security and privacy risks in centralized systems. For example, data processing can occur on local servers rather than centralized servers, thus reducing the security risk of data leakage.

The concept of federated learning was first introduced by Google in 2017 [3]. Federated learning is a high-performance training paradigm that addresses the security and privacy issues of decentralized data by sharing the model rather than the data itself. That is, each client trains locally based on local private data only, and by using a central aggregation server to accumulate the learning gradients of the local model, data holders are able to collaboratively participate in the training of the global model without exposing their private data. With recent developments, federation learning has been successfully applied to a variety of academic research and industrial tasks. The privacy-preserving and distributed nature of federation learning makes it a natural fit for edge computing.

Edge computing combined with federated learning techniques holds good research promise, but there are still many challenges to be addressed in this area. For example, there is a possible threat of data privacy leakage during the training and testing phases of the models [4,16,17]. A great deal of research has been conducted to alleviate these problems, among which differential privacy (DP) [5] has become a standard privacy specification in the last few years due to its effectiveness in protecting users' confidential information. It utilizes a random noise addition mechanism that follows well known statistical distributions (e.g., Laplace, Gaussian, exponential, etc.). Due to its privacy guarantees and low computational cost, DP has been widely used in numerous machine learning (ML) and Fl applications [6].

However, recent studies on privacy and security challenges in edge computing have pointed out that DP mechanisms can be used to perform disruptive poisoning attacks [7,8]. In particular, in FL-based edge computing, where models are trained on decentralized networks of edge devices using local data, attackers can use new attack vectors to perform covert poisoning attacks via DP [9,10]. Thus, while DP can be effective against certain privacy-invasive attacks in edge computing (e.g., membership inference attacks (MIAs)) [6,8], it can be exploited to falsify sensor data (data poisoning) or manipulate local model parameters through covert poisoning (model poisoning). Figure 1 illustrates such an attack

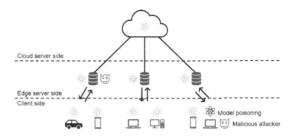

Fig. 1. Scenarios for attacks on Differential Privacy Federation Learning (DPFL) processes.

scenario during federation learning (DPFL) using differential privacy mechanisms. The adversary can extract some spurious noise from some adversarial distributions that hold similar properties to benign statistical distributions (e.g., Gaussian, Laplacian, etc.). The adversary can then inject this malicious data as a form of DP noise into the benign local model to degrade the performance of the global model. Now, due to the differential structure of the malicious noise, the anomaly detector will consider the noise as DP noise (unless the noise level is too high to be easily detected).

In this paper, we address the opportunity for DP to be exploited for poisoning attacks. We first analyze the adversarial impact of the DP mechanism in FL. In addition, we consider poisoning attacks in the model sharing phase of the FL process. Our evaluation shows the potential adversarial damage that can occur under the DP mechanism. To address the above issues, we propose FedEDP, an algorithm to help users adaptively choose the level of privacy protection. We summarize the main contributions of our paper as follows:

- We investigate how attackers exploit the DP mechanism in DPFL in the context of edge computing in order to perform model poisoning attacks. The results of the adversarial impact analysis show a degradation in FL performance in the presence of adversarial.
- We propose a concise and effective algorithm, called FedEDP, which is based on reinforcement learning (RL) and can intelligently generate an optimal privacy level for a client by using differential privacy parameters (privacy loss, information leakage probability, etc.) and losses. It enables FL in adversarial environments to prevent potential attacks by ensuring the best privacy protection level for the clients involved in the training while guaranteeing the learning performance.
- In addition to performing theoretical analysis, we also evaluated the performance of the FedEDP algorithm on the MNIST and CIFAR10 datasets. The evaluation shows that our proposed FedEDP algorithm performs better than the current optimal DPFL algorithm.

The remainder of this paper is organized as follows. Section 2 provides an overview of related work. Section 3 models the problem of possible poisoning

Table 1. Important notations

Symbol	Definition
N	The total number of clients participating in training
w_0	The initial Model State of the DPFL
w_{t+1}^a	The models with adversarial
w_{t+1}^f	The models with benign
γ	The threshold of detection
ϵ	The loss of privacy or budget
l_a	The loss of attacker
l_f	The loss of Federated
α	The attacker's tolerance to attack detection

attacks in DPFL, as well as describing the FedEDP algorithm in detail. Section 4 shows the simulation results and analysis. Finally, Sect. 5 concludes the paper.

2   Related Work

McMahan et al. proposed the FL process [3], where the model is trained on a decentralized network of edge nodes using their local data. However, there are excessive communication overheads and long delays that cause the training process to become inaccurate [14]. [15] showed that placing the parameter servers close to the edge (e.g., base stations), the latency of computation is comparable to the communication latency to the edge parameter servers, potentially pursuing a better trade-off between computation and communication.

Most of the existing defense strategies against DP-exploited poisoning attacks [8,12,13] focus on the post-attack defense (exfiltration phase), by modeling the bad data detection algorithms. For example, [8] developed optimal attack detection defenses following a game-theoretic approach to minimize the impact of attacks exploited by DPs. Similarly, to address poisoning attacks on FL, [12] introduced a two-stage defense algorithm, called Local Malicious Factor (LoMar). However, their poisoning attacks are not performed by exploiting DP noise. Instead, in our work, we consider that the malicious noise used for poisoning attacks comes from an adversarial distribution with properties similar to any benign Laplace or Gaussian distribution. In addition, to design an effective defense strategy, it is necessary not only to detect the attack but also to limit the attack surface. Work closer to ours [6] discusses non-targeted local model poisoning attacks on Byzantine clients in FL. They also proposed anomaly detection schemes based on error rate rejection (ERR) and loss function rejection (LFR) to defend against such attacks. However, they do not consider attacks based on DP itself, but aim to limit the attack surface.

3   Threat and Problem Modeling

In this section, we first introduce the application of differential privacy in federation learning (DPFL), then analyze how attackers can use the DP mechanism to perform poisoning attacks on FL, and finally, we model possible attack behaviors in DPFL, to simulate an untargeted, stealthy local model poisoning attack on the system by a potential attacker.

3.1   DPFL

We consider a set of clients K that hold n_k samples of data, respectively, and the loss function of the k-th client is $F_k(w) = \frac{1}{n_i} \sum_{j=1}^{n_k} l(w; x_{k,j})$, where w is the model parameter and $l(w; x_{k,j})$ denotes the loss function of the model parameter w at the sample j of client k. The main notations are shown in Table 1. The optimal solution for w is w^*.

$$w^* = \min_w F(w) \tag{1}$$

$$F(w) = \sum_{k=1}^{N} p_k F_k(w) \tag{2}$$

where $\sum_{k=1}^{N} p_k = 1$, p_k are the weights of the clients.

In the FL of LDP, the t round of training, the client uses its local data for t_n rounds of local training $w_{t_n+1}^n \leftarrow w_{t_n}^n \xi_n \nabla f_n(w_{t_n}^n)$ to update the model with weights of $\triangle w_t^n \leftarrow w_{T_n+1}^n - w_t$. Then the gradient cropping technique C is used to crop $\triangle w_t^n$ and update the model to

$$\triangle w_t^n \leftarrow w_t + \triangle w_t^n / max(1, \frac{||\triangle w_t^n||_2}{C}) \tag{3}$$

Finally the noise is added to the updated model $\triangle w_{t+1}^n \leftarrow \triangle w_t^n + N(0, \sigma^2 I)$ and the Gaussian noise $N \sim (0, \sigma^2 I)$ added by the local model should satisfy $\sigma \geq \sqrt{2rT \ln \frac{1}{\delta}} \times \frac{2C}{\epsilon}$. Finally, the server aggregates the received model with noise, i.e.

$$w_{t+1} \leftarrow w_t + \frac{1}{N} \sum_{n=1}^{N} \triangle w_{t+1}^n \tag{4}$$

3.2   The Basic Mechanism of Poisoning Attacks Using DP

We assume that the anomaly detector (on the server side) expects an update weight of $\triangle w_t^h$ for a particular model during communication round t and the actual received weight of $\triangle w_t^r$ for that particular model. If the actual weight $\triangle w_t^r$ exceeds the pre-defined detection range $S = [\triangle w_t^h \pm \gamma]$, where γ is the detection threshold, the detector raises an alarm.

If we consider a differential privacy with a noise level of $\pm\beta$ added to the learning process, then the modified value of the weights received by the server will be $\triangle w_t^r \pm \beta$. To accommodate this modified weight, the detector also needs to adjust its detection range to, e.g. $S' = [\triangle w_t^h \pm \gamma']$, where the new detection threshold is $\gamma' = (\gamma \pm \beta)$. This adjustment mechanism of the detection range opens an additional window of (false) noise injection for the attacker, which can be expressed as follows:

$$[\gamma + \delta, \gamma - \delta] = [\{(\triangle w_t^r - \beta) - (\triangle w_t^h - \gamma')\}, \{(\triangle w_t^h + \gamma') - (\triangle w_t^r - \beta)\}] \quad (5)$$

where $\delta = \triangle w_t^r - \triangle w_t^h$ is the weight bias. An attacker can use this fake noise injection window (i.e., $\gamma \pm \delta$) to make fake noise data $\beta_a \leftarrow N_a(\mu_a, \sigma = \frac{\triangle f}{\epsilon})$. In particular, a potential attacker can extract the adversarial noise β_a from an adversarial distribution f_a similar to a Gaussian distribution, and then inject this noise into the fully corrupted local model, i.e.,

$$\triangle w^a = \sum_{i=1}^{a} [\triangle w^i + \beta_a^i] \forall \{a \in n | 0 < an\} \quad (6)$$

Such optimal attack distribution f_a^* and the optimal attack impact β_a^* are derived and presented in [8] by solving a multi-criteria optimization problem that addresses two conflicting adversarial goals: (1) maximum damage, and (2) minimum disclosure. From the attacker's perspective, these goals are inherently contradictory, as maximum damage leads to easier detection of the attack, while minimum disclosure limits the damage. The optimal adversarial distribution and the optimal attack impact are denoted as

$$f_a^*(x) = \frac{1}{\sqrt{2\pi}\sigma_x} e^{-\frac{(x-\theta-\sqrt{2\alpha}\sigma_x)^2}{2\sigma_x^2}} \quad (7)$$

$$\mu_a^* = \theta + \sqrt{2\alpha}\sigma_x \quad (8)$$

where α is the concealment parameter, i.e., the attacker's tolerance to attack detection. The server aggregates both the adversarial model $\triangle w^a$ and the benign model $\triangle w^f$, and the global model is updated to

$$w_{t+1}^a = w_t + \frac{1}{N}(\sum_{i=1}^{a} w_{t+1}^a + \sum_{i=1}^{f} w_{t+1}^f) \quad (9)$$

3.3   Modeling Covert Model Poisoning Attacks in the DPFL Process

Following Sects. 3.1 and 3.2, we model the possible covert model poisoning attacks during DPFL. The specific process is shown in Algorithm 1. In this case,

---

Algorithm 1: Differential privacy federated learning process with covert model poisoning attacks

---

Input: Initial model w_0, training set D_r, test set D_e, loss threshold γ

Output: Global Model $\triangle w_T$

1 Initialize model w_0;

2 **for** *each round* $t = 0, 1, ..., T$ **do**

3   Delivery of w_t to all clients $n \in N$ participating in the training;

4   **for** $n \in N$ **do**

5   Receive the global model w_t;

6   **if** *Attackers* **then**

7              $\triangle w_{t+1}^a \leftarrow \triangle w^a + N_a(\mu_a, \sigma = \frac{\triangle f}{\epsilon})$;

8   **else**

9              $w_0^f \leftarrow w_t$;

10   **for** *each round* $t_n = 0, 1, ..., T_n$ **do**

11                 $w_{t_n+1}^f \leftarrow w_{t_n}^f \xi_n \nabla f_n(w_{t_n}^f)$

12   **end**

13             $\triangle w_t^f \leftarrow w_{T_n+1}^f - w_t$;

14             $\triangle w_t^f \leftarrow w_t + \triangle w_t^f / max(1, \frac{\|\triangle w_t^n\|_2}{C})$;

15             $\triangle w_{t+1}^f \leftarrow \triangle w_t^f + N(0, \sigma^2 I)$;

16   **end**

17   Send $\triangle w_{t+1}$ to the server;

18   **end**

19   Update global model $w_{t+1} \leftarrow w_t + \frac{1}{N}(\sum_{i=1}^a w_{t+1}^a + \sum_{i=1}^f w_{t+1}^f)$;

20   **if** $Test(w_{t+1}, D_e) < \gamma_t$ **then**

21   **repeat**

22   round t;

23   **until** $Test(w_{t+1}, D_e) \geq \gamma_t$;

24   **end**

25 **end**

---

the client $n \in N$ involved in training performs T_n rounds of SGD locally, trains a local update model $w_{T_n+1}^n$ using local data, adds noise $N(0, \sigma^2 I)$ conforming to the Gaussian mechanism to the updated model parameters after a gradient cropping has been performed, and finally returns the updated model $\triangle w_{t+1}^n$ to the server. The server aggregates all the received update models according to the weights and updates the global model. Finally, the loss of the current global model is tested on the test dataset D_e, and if the validation loss is below the expected loss threshold γ_t, the training continues, otherwise, the current global model is rolled back to the previous global model.

We assume that the attacker masquerades as a benign client. The attacker adds adversarial noise β_a to the local model and uploads the malicious local model w^a to the server disguised as a benign model. Since the server cannot effectively identify whether the accepted model is malicious or not, it aggregates the malicious model together with the benign model to update the global model, resulting in a degradation of the system learning performance.

3.4 Optimal Difference Level Selection Algorithm in Differential Privacy Federated Learning (FedEDP)

As described in Sects. 3.2 and 3.3, the use of DP mechanisms in FL, while securing data privacy through noise, may also create a new avenue of poisoning attacks for attackers. One way to prevent this is to reduce the privacy protection level (i.e., reduce the added noise). However, a low privacy protection level can again lead to data privacy attacks. Therefore, it is crucial to design an optimal privacy level for DPs in FL. In this paper, we adjust the privacy loss ϵ based on the Q-learning learning method in reinforcement learning (RL) to intelligently choose the best privacy protection level. The detailed process is described in Algorithm 2.

We were inspired in the Q-learning learning approach. Q-learning follows an action value function that gives the expected utility of taking a particular action in a particular state. The objective for FedEDP is to minimize the maximum attack accuracy as well as maximize the federated accuracy. FedEDP use the event-driven manner approach where the defensive agent makes a decision when a new event occurs. When learning starts, the states are initialized. We define the state space as $S = (l_a, l_f, \epsilon)$, where l_a represents the set of losses of the poisoning attack, l_f represents the set of federal losses, and ϵ is the set of privacy losses. The set of losses of poisoning attacks l_a and the set of federal losses l_f are predefined before the training starts and need to be obtained by prior adversarial and non-adversarial training respectively. The attacker's tolerance α at different values should be considered in the adversarial training. On the other hand, to ensure the integrity and accuracy of the FedEDP process, the same privacy loss ϵ needs to be used for the adversarial and non-adversarial training. Then, we define an action space $A = \{increase, decrease, static\}$ where first the agent observes the current system state $s \in S$, then takes an action $i \in A$ in this state s and obtains the next state s_{t+1}. The future rewards of s_{t+1} are updated using the feedback rewards under state s and the future rewards of s_{t+1}. The reward function is defined by the following equation

$$r = \lambda_1 \frac{l_a^{max}}{l_a} + \lambda_2 \frac{l_f^{max}}{l_f} + \lambda \frac{1}{\epsilon} \tag{10}$$

where l_a^{max} and l_f^{max} denote the maximum value of the poisoning attack loss and the federal loss, while λ_1, λ_2 and λ_3 denote the balancing parameters to mitigate the exploration the exploration and exploitation dilemmaWe set the initial exploration probability at 1.0, and gradually reduce the exploration probability over episodes until it matches with the minimum exploration probability (which we assume 0.05 in this paper). Moreover, for simplicity, we select the maximum number of episodes as the stopping criterion or terminating condition. Finally we solve for the optimal value of Q, i.e. Q^*, and from the Bellman equation we get

$$Q^*(s, i) = r(s, i) + \zeta \max_i \sum_{s_{t+1}} P(s_{t+1}|s, i) Q^*(s_{t+1}, i) \tag{11}$$

Algorithm 2: FedEDP

    Input: Attacker's loss l_a, Federated loss l_f, Epsilon set E, State set S, Reward
           function r

    Output: Optimal privacy loss ϵ^*

1 **for** ϵ *in* E **do**

2      $s_t \leftarrow (l_a, l_f, \epsilon)$;

3   Choose $i \in A$ from s_t using policy derived from $Q(e.g., \epsilon - greedy)$;

4   Take action i, observe r_t, s_{t+1};

5   Update $Q(s_t, i_t) \leftarrow Q(s_t, i_t) + \tau[r_t + \zeta \max_i Q(s_{t+1}, i) - Q(s_t, i_t)]$;

6 **end**

---

where $\sum_{s_{t+1}} P(s_{t+1}|s, i)$ is the state transition probability and taking action i will result in the highest expected reward of r. Without loss of generality, we define $Q*$ as

$$Q^*(s, i) = r(s, i) + \zeta \max_i Q^*(s_{t+1}, i) \tag{12}$$

According to Q^* we can get the optimal state policy ϕ^* and use it to get the optimal privacy loss ϵ^*.

4 Performance Evaluation

4.1 Simulation Setup

We perform image classification tasks on the MNIST and CIFAR10 datasets to verify the performance of the FedEDP algorithm. Before training the neural network we normalize the image data to speed up the model convergence. For MNIST, there are 60,000 training samples and 10,000 testing samples. Suppose the system has 100 clients, where each client is assigned 600 training samples and 100 test samples. We adopt the same convolutional neural network (CNN) model. For the CIFAR10 dataset, there are 50,000 training samples and 10,000 test samples. For a maximum of K= 100 clients, each is given 500 training and 100 testing samples, we also use the CNN model. Finally we use a softmax layer. We tested separately with different paradigm constraints C, privacy protection levels ϵ, the probability of violating strict differential privacy δ. Calculate training loss by cross entropy and evaluate model performance by test accuracy. We conducted simulation experiments using MATLAB software running on a Dell desktop computer with Inter core i5-9500 CPU @ 3.00-GHz processor and 8.00 GB (7.81 GB available) RAM. The results of all simulation experiments are an average of 10,000 independent experiments. The aim is to reduce the effect of the randomness of random variables on the results.

4.2 Evaluation Results

Results of Resistance Impact Analysis. As shown in Fig. 2, using DP in FL can secure data privacy, but the loss increases when we increase the privacy

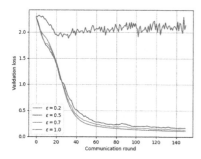
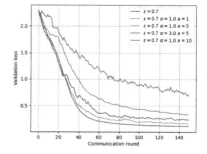

Fig. 2. The loss of DPFL at different privacy loss ϵ.

Fig. 3. The loss of DPFL at different client-attacker ratio n/a and different attacker's tolerance α

level (i.e., reduce the value of privacy loss ϵ). That is, if we add more noise to achieve a higher level of privacy protection, the loss increases. And, there is a safety threshold when using differential privacy for privacy preservation in federal edge learning. That is, it will only affect the learning performance until reaches the threshold, but when is less than the threshold, the FL system will crash. In addition, the loss increases if there is an attacker that maliciously attacks the model, i.e., the attacker can use the DP mechanism in the federal learning process to make the global model suboptimal, as shown in Fig. 3. For the same number of clients and the same level of privacy protection, the global model containing the malicious local model ($a = 1$) has a greater loss. When the client-attacker (n/a) ratio is low, then the global model is still closer to the optimal global model. If this ratio increases, the global model starts to deviate from the optimal model. However, attacking a large number of local models may lead to easy attack detection, while a small number of malicious ones may not have much impact on the global model. Therefore, for a successful attack, an attacker needs to carefully choose the number of models to attack. The attacker's tolerance α is also an important determinant of the loss level during adversarial joint learning. If the attacker maintains a high level of stealth, then the impact of the attack will remain small. As the attacker keeps increasing α, the loss keeps increasing and the model eventually becomes a suboptimal model.

Performance Analysis of FedEDP Algorithm. To defend against model poisoning attacks by attackers using the DP mechanism, the differential privacy level of the client needs to be chosen carefully. Too much noise leads to poorer model performance and a larger attack window, while less noise does not guarantee user privacy security. As shown in Fig. 4, the RL agent learns optimal policy as the episodes increase. After sufficient episodes are executed, it converges to an optimal policy that ensures the desired privacy, utility, and security. The policy converges around episode 60 and stays high for the rest of the period. That means the agent is making optimal actions at this stage. The FedEDP algorithm not only limits the attack surface by adaptively selecting the privacy protection

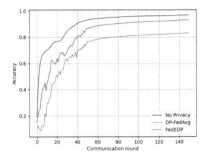

Fig. 4. Performance analysis of FedEDP algorithm.

level, but also facilitates attack detection techniques. In particular, privacy level selection by FedEDP algorithm does not need to be performed in real time; it can be performed offline during the design phase of DPFL through some test experiments. This ensures its applicability on edge devices.

5   Conclusion

Adversarial attacks on edge learning and its federated learning (FL)-based applications can have a catastrophic impact on national economic growth and public safety. In this paper, we analyze one of such adversarial attacks - a poisoning attack on DPFL in an edge computing scenario. Our adversarial analysis shows that the attack can make the FL model suboptimal or even, cause the system to crash. To defend against this attack, we propose a privacy-preserving algorithm based on reinforcement learning that can adaptively select the privacy level. Experimental evaluation shows that the algorithm can limit the attack surface. In the future, we will consider optimizing the communication efficiency of the system while protecting privacy, as well as the robustness of FedEDP in the Non-IID case.

References

1. Shi, W., Cao, J., Zhang, Q., Li, Y., Xu, L.: Edge computing: vision and challenges. IEEE Internet Things J. **3**(5), 637–646 (2016)
2. Hu, X., Wang, G., Jiang, L., Ding, S., He, X.: Towards efficient learning using double-layered federation based on traffic density for internet of vehicles. In: Xing, C., Fu, X., Zhang, Y., Zhang, G., Borjigin, C. (eds.) WISA 2021. LNCS, vol. 12999, pp. 287–298. Springer, Cham (2021). https://doi.org/10.1007/978-3-030-87571-8_25
3. McMahan, B., Moore, E., Ramage, D., Hampson, S., Arcas, B.A.: Communication-efficient learning of deep networks fromdecentralized data. In: Artificial intelligence and statistics. PMLR, pp. 1273–1282 (2017)
4. Su, Z., et al.: Secure and effificient federated learning for smart grid with edgecloud collaboration. IEEE Trans. Industr. Inform. **18**(2), 1333–1344 (2021)

5. Dwork, C., McSherry, F., Nissim, K., Smith, A.: Calibrating noise to sensitivity in private data analysis. In: Halevi, S., Rabin, T. (eds.) TCC 2006. LNCS, vol. 3876, pp. 265–284. Springer, Heidelberg (2006). https://doi.org/10.1007/11681878_14

6. Geyer, R.C., Klein, T., Nabi, M.: Differentially private federated learning: a client level perspective. arXiv preprint arXiv:1712.07557 (2017)

7. Hossain, M.T., Badsha, S., Shen, H.: Privacy, security, and utility analysis of differentially private cpes data. arXiv preprint arXiv:2109.09963 (2021)

8. Giraldo, J., Cardenas, A., Kantarcioglu, M., Katz, J.: Adversarial classification under differential privacy. In: Network and Distributed Systems Security (NDSS) Symposium (2020)

9. Hossain, M.T., Islam, S., Badsha, S., Shen, H.: Desmp: differential privacy-exploited stealthy model poisoning attacks in federated learning. arXiv preprint arXiv:2109.09955 (2021)

10. Fang, M., Cao, X., Jia, J., Gong, N.: Local model poisoning attacks to byzantine-robust federated learning. In: 29th USENIX Security Symposium (USENIX Security 20), pp. 1605–1622 (2020)

11. Mothukuri, V., Parizi, R.M., Pouriyeh, S., Huang, Y., Dehghantanha, A., Srivastava, G.: A survey on security and privacy of federated learning. Futur. Gener. Comput. Syst. **115**, 619–640 (2021)

12. Li, X., Qu, Z., Zhao, S., Tang, B., Lu, Z., Liu, Y.: Lomar: a local defense against poisoning attack on federated learning. IEEE Trans. Dependable Secure Comput. (2021)

13. Gao, J., et al.: Secure aggregation is insecure: category inference attack on federated learning. IEEE Trans. Dependable Secure Comput. (2021)

14. Bonawitz, K., et al.: Towards federated learning at scale: system design. In: Proceedings of the 2nd SysML Conference, Palo Alto, CA, USA (2019)

15. Tran, N.H., Bao, W., Zomaya, A., Minh, N.H., Hong, C.S.: Federated learning over wireless networks: optimization model design and analysis. In: IEEE INFOCOM 2019, pp. 1387–1395 (2019)

16. Wang, G., Xu, Y., He, J., Pan, J., Zuo, F., He, X.: Resilient participant selection under vulnerability-induced colluding attacks for crowdsourcing. IEEE Trans. Veh. Technol. **71**(7), 7904–7918 (2022)

17. Wang, G., He, J., Shi, X., Pan, J., Shen, S.: Analyzing and evaluating efficient privacy-preserving localization for pervasive computing. IEEE Internet Things J. **5**(4), 2993–3007 (2018)

Patient-Friendly Medical Data Security Sharing Scheme Based on Blockchain and Proxy Re-encryption

Xinhao Xu[1], Xiaomei Dong[1(✉)], Xin Li[1], Guangyu He[2,3], and Shicheng Xu[2,3]

[1] School of Computer Science and Engineering, Northeastern University,
Shenyang 110169, China
xmdong@mail.neu.edu.cn
[2] Research Center of Liaoning Promotion for Block Chain Engineering Technology,
Shenyang 110179, China
[3] Neusoft Corporation, Shenyang 110179, China

Abstract. In recent years, with the wide application of blockchain technology, blockchain-based medical data sharing has become a major hot spot in the current research on medical data sharing. Using the decentralized, tamper-proof and traceable features of blockchain, data sharing of electronic medical data in various medical institutions has been realized. With the help of privacy protection technology, the security and privacy of medical data sharing links have been improved. However, there are still problems. For example, patients cannot manage their medical data independently and cannot guarantee the confidentiality of their private keys. And the problems arising from the actual situation of patients are often ignored. In this paper, we propose a secure medical data sharing scheme based on blockchain and proxy re-encryption. The scheme enables patients to manage their own medical data directly and uses improved distributed key management technology to ensure the confidentiality of patients' private key. In addition, we use a combination of on-chain and off-chain for medical data storage to ensure high efficiency and the independence of hospital medical data. We also design a blockchain-based off-chain database security verification algorithm.

Keywords: Electronic medical data · Proxy re-encryption · Blockchain · Distributed key management

1 Introduction

With the rise of the era of big data, medical information systems play a key role in the sharing of medical data as the infrastructure of modern hospitals. For the electronic smart medical system, the medical industry has been shared and intelligent [1]. Since the outbreak of COVID-19 in 2020, timely sharing of medical data and resources has played an increasingly important role [2]. However, there are some problems along with the convenience. Traditional centralized storage solutions are unable to meet the demands of exponential growth of medical data. There are problems, such as lack of trustworthiness

© The Author(s), under exclusive license to Springer Nature Switzerland AG 2022
X. Zhao et al. (Eds.): WISA 2022, LNCS 13579, pp. 615–626, 2022.
https://doi.org/10.1007/978-3-031-20309-1_54

of data and data loss during storage and sharing, which make medical data sharing difficult [3]. In addition, centralized storage also faces serious security challenges, the risk of malicious attacks, data leakage and tampering of medical systems will also increase. And based on the cloud server storage, the massive amount of data will make the cloud storage space consumption is huge [4]. With the rise of blockchain technology in recent years, the application of blockchain technology in medical data sharing has become more and more widespread. Its decentralized, transparent and tamper-proof characteristics are good for solving some current problems [5]. The blockchain-based medical information system has become one of the key points of current research. Even so, there are still three major challenges that need to be over-come:

- The true data owner, the patient, cannot directly manage his or her own data, that reduces the patient's autonomy over his or her own medical data.
- The confidentiality of the user's private key and the confidentiality of the patient's medical data are challenged in the process of data sharing.
- A large amount of medical data cannot be stored on the blockchain. Also, the independence of hospital medical data needs to be ensured in order to facilitate device performance evaluation by different medical device manufacturers. How to store medical data safely and efficiently is an urgent problem to be solved.

Based on the above problems, we propose a secure medical data sharing scheme based on blockchain and proxy re-encryption, which meets all the above practical requirements with reliable security. The main contributions of this paper can be summarized as:

- We propose an improved distributed key management scheme. We increase the pre-processing operation, reduce the overhead generated by key generation and adopt a threshold secret sharing scheme to distribute the user private key for storage on nodes without hospitals. It ensures the confidentiality of the patient private key and at the same time can effectively solve the single point of failure problem.
- We use a proxy re-encryption technique to ensure the autonomy of patients to their own medical data. Patient authorization is required in the process of medical data sharing.
- We propose a security verification scheme for off-chain database storage. In order to ensure the independence of each hospital's data, hospitals store their own medical data. Meanwhile, obtain specific communication tokens from blockchain and communicate with the other database in a P2P manner.

The rest of the paper is organized as follows: In Sect. 2, we discuss the work related to secure blockchain-based healthcare data sharing. In Sect. 3, we detail our proposed scheme design. System and scheme analysis and evaluation report in Sect. 4, and we summarize the full paper in Sect. 5.

2   Related Work

With the wide application of blockchain technology, blockchain-based medical information sharing has become a major hot spot in the current research on medical data sharing. In related studies, researchers have made various improvements and enhancements, and the system proposed by Cao et al. [6] used blockchain for secure storage of medical data. They focused on architectural design, but lacked system implementation details and data sharing process. Nguyen et al. [7] designed a specific system workflow in detail to achieve sharing or access control of medical data. However, the proposed system is implemented based on Ethereum, which is inefficient and not applicable to hospital information systems.

Due to the wide range of uses and privacy of medical data, the privacy protection needs cannot be met by using blockchain technology alone. In blockchain systems, the transparency of transaction records greatly increases the risk of privacy leakage. For example, patient data can be easily leaked and the analysis of transaction records can obtain the transaction patterns of users, so it needs to be coupled with more prominent privacy protection technologies. Kassab et al. [8] and Abu-Elezz et al. [9] studied common data protection mechanisms and analyzed the advantages of blockchain in medical data security protection and sharing. Xu et al. [10] designed a blockchain technology-based privacy protection scheme for large-scale health data with fine-grained access control of medical data. Patients can revoke or add authorization to doctors autonomously through key management.

For privacy protection techniques, the proxy re-encryption proposed by BLAZE is widely used [11], which is gradually applied to different scenarios. Sun et al. [12] proposed a patient-friendly blockchain electronic medical data sharing system based on main-sub structured blockchain, which enables patients to manage their personal medical data directly. Chen et al. [13] proposed a secure medical data sharing scheme based on agent heavy encryption algorithm to achieve secure storage, management and access control of medical data. However, neither of these solutions above addresses solves the key escrow problem. The storage of private keys completely relied on third parties to store them, which could not guarantee their confidentiality and had the risk of leakage. Tang et al. [14] proposed an electronic prescription sharing scheme based on blockchain and conditional proxy re-encryption, which achieves access control and has fine-grained decryption authority assignment by proxy re-encryption with conditional restrictions, and uses distributed key generation technology to solve the key escrow problem, but cannot prevent single point of failure.

In this paper, we propose a secure medical data sharing scheme based on blockchain and proxy re-encryption. This scheme includes three major parts. The first part is the improved distributed key generation and management phase, which uses Shamir(t, n) threshold secret sharing scheme [15]. The second part is the proxy re-encryption phase, which can ensure the autonomy of patients to their own medical data. The last part is the off-chain data query phase, which uses a security verification mechanism.

3 Scheme Design

According to our proposed scheme, we design the constructure of a corresponding system, in which the three parts of our scheme are integrated. Firstly, we describe the composition of the system (see Sect. 3.1). Then we design the detailed process of three parts of our scheme, including on improved distributed key generation and management phase, proxy re-encryption phase and off-chain data query phase (see Sect. 3.2).

3.1 System Model

In order to both conform to the traditional hospital organization and management and solve the existing problems, the system consists of three major parts: a blockchain system jointly maintained by multiple hospitals, an improved distributed key management system and off-chain medical databases. Its system model is shown in Fig. 1.

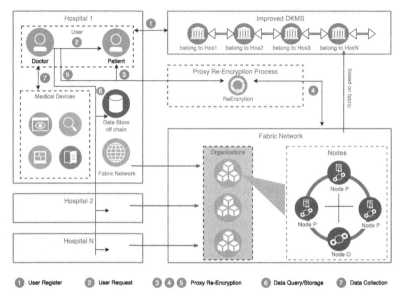

Fig. 1. System model

And the characteristics and functions of each entity are described as follows:

1. Blockchain System (BS): Hospitals work together to maintain the federated chain system. Each hospital acts as an organization in the system network. Information stored in the chain, such as indexes of data stored in the off-chain medical database.
2. Distributed Key Management System (DKMS): DKMS consists of peer nodes for each hospital. Its major work can be divided into two steps. Firstly, performing pre-processing operations to ensure the trustworthiness of CA nodes in the key generation and choosing a leader node to generate the user's key pair. Secondly, using shamir(t, n) threshold secret sharing scheme to store the user's private key for improving the ability to resist malicious attacks. Compared with the traditional distributed key generation scheme, it is able to recover keys under no more than n-t node failures.
3. Off-Chain Medical Database (OCMD): Hospitals manage their own offline databases, that can ensure the independence of each hospital's data. A P2P approach is used for communication to improve efficiency. In addition, we design a blockchain-based three-party security verification scheme to ensure the security of off-chain data queries.

4. Data Owner (DO): As the actual owner of medical data, patients have autonomy in their own medical data. Their rights include encrypting their medical data and generating proxy re-encryption keys to decide who receives the medical data during the sharing process.
5. Data Requester (DR): A consumer of medical data with the doctor as the subject. The doctor diagnoses and treats the patient and sends the generated electronic medical record (EMR) to the patient for confirmation and storage. After being shared by the patient's consent, the doctor can decrypt the returned ciphertext or re-ciphertext, thus enabling data sharing and effective monitoring and review.

In addition, we should also focus on the design of the data structure stored on the blockchain. After a patient completes treatment, the patient will perform a series of processes on the data. The data will eventually be uploaded to the chain, including on doctor's ID, patient's ID, storage index, key of the patient's medical data cipher and ID of the off-chain database. See Fig. 2.

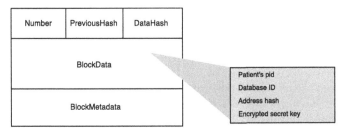

Fig. 2. Block data structures

3.2 Description of Proposed Scheme

This scheme includes three major parts. The first part is the improved distributed key generation and management phase. The second part is the proxy re-encryption phase and the last part is the off-chain data query phase. As described below (see Table 1).

Table 1. Symbols used in this paper.

Symbols	Meaning of the symbols
PK_i, SK_i	The public key and the secret key of user i
M_i, C_i	M_i refers to plaintext of data i and C_i refers to ciphertext of data i
$RK_{i \rightarrow j}$	The proxy re-encryption conversion key, which is generated by user i and will be used to convert the for user j
I_y	Database identification of database y

Improved Distributed Key Generation and Management Phase. Its major work can be divided into two steps.

1. $Setup(1^\lambda) \to params$. This is a randomized algorithm that takes no input other than the implicit security parameter λ. The system negotiates to generate a large prime q and uses q to generate the elliptic curve group G and the generating element P. Define two collision-resistant hash functions H_1: $H_1(G, G) \to Z_q^*$, H_2: $H_2(\{0, 1\}^*, G) \to Z_q^*$. Select the security parameter to generate the common parameter $params = \{G, g, q, P, H_1, H_2\}$. Set up a federated chain node $P_i(i = 1, 2, \cdots, n)$. Each node P_i randomly selects a polynomial $F_i(x)$ of order $n-1$ based on the common parameters. It then calculates and broadcasts $P_{ik} = g^{p_{ik}}(mod q)(k = 0, 1, \cdots, n-1)$. P_i sends secret values $s_{ij} = F_i(j)$ to other nodes $P_j(j = 1, 2, \cdots, n, j \neq i)$. Finally, each node P_i is verified whether (2) holds. If it does not hold, then the secret is invalid and needs to be checked and processed. Conversely, the secret value is valid and P_j can be proved to be honest and trustworthy.

$$F_i(x) = p_{i0} + p_{i1}x + \cdots + p_{i(n-1)}x^{n-1} \tag{1}$$

$$g^{s_{ji}} = \prod_{k=0}^{n-1}(P_{jk})^{i^k}(mod q) \tag{2}$$

P_i selects $SK_i \in Z_q^*$ and calculates $PK_i = SK_i \cdot P$. Then P_i updates the parameter $params = \{G, g, q, P, H_1, H_2, PK_1, PK_2, \cdots PK_n\}$ and publishes it to the Genesis block, while deploying the contract to support querying the CA public key list. If a CA is added in the future, it is done according to the rules specified in the Genesis block, which is to synchronize the existing blockchain locally to participate in the subsequent authentication service.

2. $Generate(Pid_i) \to s_r$. Hospital performs the anonymization process for id of user i. Firstly, it generates a random number w and calculates $Pid_i = hash(w + id_i)$. It establishes a local mapping. Next, user i sends pid_i to the elected CA node in the organization for registration. CA returns $SK_i = H_2(Pid_i, G)$ to user i and stores the $PK_i = SK_i \cdot P$ on the blockchain. Finally, user i stores PK_i on peers with Shamir$(t, , n)$ and selects $\{x_1, x_2, \cdots, x_n\}$ from $GF(p)$. $x_r(r = 1, 2, \cdots, n)$ corresponds to each shadow secret holding node $P_r = \{P_1, P_2, \cdots, P_n\}$. Then user i generates shadow secret s_r for P_r. In the above formulas, p is a large prime number and $p > SK_i$. When used again, it is recovered by no less than t peer nodes.

$$s_r = f(x_r) = \sum_{j=1}^{t-1} a_0 + a_j x_r^j(mod p)a_j \in GF(p) \tag{3}$$

Proxy Re-Encryption Phase. There are five parts in the process of uploading and sharing medical data with doctors at the time of patient's visit.

1. $Upload(Pid_i, I_y, C_{dek}, Hash_{address}) \to PatientDataEntity$. After the patient finishes treatment, the doctor generates EMR data M_{EMR} for the patient based on disease

information or historical medical data, and sends it to the patient. Then the patient selects a random symmetric key $dek \in Z_q^*$. The patient applies the AES algorithm and dek to generate the ciphertext $C_{EMR} = En_{dek}(M_{EMR})$. Next, the patient uploads C_{EMR} to the off-chain database to obtain $Hash_{address}$ and I_y. The patient encrypts the dek to generate the ciphertext $C_{dek} = En_{PK_p}(dek)$. Finally, the patient uploads $\{Pid_i, I_y, C_{dek}, Hash_{address}\}$ to blockchain system (see **Table 2**).

Table 2. Patient data entity

Key	Value
1	Pid_i
2	I_y
3	C_{dek}
4	$Hash_{address}$

2. $KeyGen(s_1, s_2, \cdots, s_n) \rightarrow SK_i$. This function is performed by the patient. The user i recovers private key SK_i through the shadow secret $\{s_1, s_2, \cdots, s_n\}$ held by any t peer nodes $\{P_1, P_2, \cdots, P_n\}$.

$$SK_i = f(0) = \sum_{j=1}^{t} f(x_l) \prod_{v=1,v\neq l}^{t} \frac{-x_v}{x_l - x_v}(mod\,p) \tag{4}$$

3. $ReKeyGen(SK_i, PK_j) \rightarrow RK_{i\rightarrow j}$. Re-encryption key generation algorithm. Enter the patient's private key PK_i and the authorized doctor's public key PK_j to generate the re-encryption key $RK_{i\rightarrow j}$ by the local server.
4. $ReEncrypt(C_{dek}, RK_{i\rightarrow j}) \rightarrow C'$. Proxy re-encryption algorithm is performed by a local proxy server. Firstly, a query is performed from the blockchain to get the patient data entity. Secondly, C_{dek} and $RK_{i\rightarrow j}$ are input as parameters to calculate the re-encrypted ciphertext C'. Finally, CT' is sent to the doctor j.

$$CT' = (C' = ReEncrypt(C_{dek}, RK_{i\rightarrow j}), I_y, Hash_{address}) \tag{5}$$

5. $Decrypt(C', SK_j) \rightarrow dek$. After receiving CT', the doctor j calculates $C_{dek} = De_{SK_j}(C')$ with the private key SK_j. The doctor applies the AES algorithm to generate dek. Then the doctor will get patient medical data from off-chain database.

Off-Chain Data Query Phase. The process consists of two parts.

1. $GenTransKey(Pid_j) \rightarrow Token$. Communication token generation algorithm. The doctor j sends the request data including his Pid_j to CA. Then CA generates $Token$ and sends $En_{PK_j}(Token)$ to this doctor ($\varphi \in Z_q^*$).

$$Token = (PK_{trans} = \varphi \cdot P, SK_{trans} = \varphi + SK_{ca} \cdot H_1(PK_{trans}, PK_j)) \tag{6}$$

2. $Request(Token, Hash_{address}) \rightarrow C_{EMR}$. The doctor j sends the request data including $Token$ and $Hash_{address}$ to off-chain database according to I_y. After verifying (7) to determine the legitimacy of this request, database returns C_{EMR} at the $Hash_{address}$. The doctor receives the C_{EMR} and applies the AES algorithm to generate $M_{EMR} = De_{dek}(C_{EMR})$.

$$PK_{trans} + PK_{ca} \cdot H_1(PK_{trans}, PK_j) = SK_{trans} \cdot P \tag{7}$$

4   Performance Analysis and Evaluation

IN This Section, We Will Analyze and Evaluate the System Solution of This Paper in Terms of Correctness, Security and Performance.

4.1   Correctness Proof

There are two places to prove in the design of the mechanism in Sect. 3.2.

1. Honesty and trustworthiness verification of peer nodes at initialization.

$$g^{S_{ji}} = g^{f_j(i)} = g^{\sum_{k=0}^{n-1} p_{jk} i^k} = \prod_{k=0}^{n-1} g^{p_{jk} i^k} = \prod_{k=0}^{n-1} (P_{jk})^{i^k} \tag{8}$$

2. Legitimacy verification when launching query requests to off-chain databases.

$$\begin{aligned} SK_{trans} \cdot P &= (\varphi + SK_{ca} \cdot H_1(PK_{trans}, PK_j)) \cdot P \\ &= \varphi \cdot P + SK_{ca} \cdot P \cdot H_1(PK_{trans}, PK_j) \\ &= PK_{trans} + PK_{ca} \cdot H_1(PK_{trans}, PK_j) \end{aligned} \tag{9}$$

4.2   Security Analysis

The scheme in this paper has the following security properties.

1. Security of Patient Medical Data.

   a. Our scheme uses symmetric encryption algorithms to store patient medical data in plaintext encryption in the off-chain database. To obtain patient medical data, attackers send a request to database. However, they will get $Token$ beforehand. It is effective to against unauthorized access attacks.

   b. If attackers obtain the ciphertext of medical data, they need dek to decrypt it. However, dek is encrypted using the patient's private key. ECC is based on the principle of discrete logarithm over the elliptic curve domain, with exponential level of encryption strength. Even if they crack the smallest elliptic curve $prime\,192v1$, they would take about $17 * \left(\sqrt{2^{192}}/\sqrt{2^{109}}\right) \approx 5 * 10^{13}$ months and therefore cannot decrypt the data without the patient's private key. It is effective to against brute-force cracking attacks.

2. Security of Patient Private Keys.

   a. Our scheme uses an improved distributed key generation technique to implement user key management. It avoids the key escrow problem in the traditional identity-based cryptosystem and ensures the confidentiality of the patient's private key.
   b. If an attacker wants to obtain the patient's private key, then he needs to obtain no less than t shadow secrets. With t/n constant, the difficulty of a successful attack increases linearly as n increases. It is effective to increase the cost to the attacker.

4.3 Scheme Comparison

We will compare the functionality and security of this paper's scheme with similar schemes or systems (see Table 3). Xu [10] proposed a blockchain technology-based privacy protection scheme for large-scale health data, but without anonymizing the users. Chen [13] proposed an anonymous medical data sharing scheme based on cloud servers and proxy re-encryption algorithms. Both of the above schemes use a single key generation center, which suffers from the problem of key escrow. Once the private key generation center is attacked, the data encrypted by all users' private keys may be leaked. Tang [14] proposes an electronic prescription sharing scheme based on blockchain and proxy re-encryption, and the distributed key generation scheme used does not solve the single point of failure problem. Based on this, we improve and optimize it to propose a blockchain and agent re-encryption based patient-friendly medical data sharing scheme. The security and fault tolerance are improved while ensuring the performance within acceptable limits.

Table 3. Comparison with other schemes

	Xu [10]	Chen [13]	Tang [14]	Our scheme
User anonymity	✗	✓	✗	✓
Distributed key management	✗	✗	✓	✓
Single point of failure prevention	✗	✗	✗	✓
Decentralized storage	✓	✓	✓	✓

4.4 Cost Overhead Comparison

We evaluate the efficiency of the patient-friendly medical data sharing scheme based on blockchain and proxy re-encryption proposed in this paper. The experimental environment is an Intel Core i9-9880H CPU@ 2.40 GHz × 4, with an operating system of Ubuntu 20.04.2 LTS, 64-bit, and 4 GB of RAM.

As shown in Fig. 3, we conduct experiments on 100 kb of medical data. Because three schemes use the method of storing indexes in the blockchain and the small data stored on the chain, 100 KB of data is used to test the efficiency of encryption. We will judge the superiority of each encryption algorithm based on the trend of time spent as the data increases. The cost overhead of each link of the schemes proposed by Chen [13] and Tang [14] is compared with that of our proposed scheme. Because the consensus time of different schemes varies with the number of nodes and the approach and architecture of different schemes are different, the overhead incurred by consensus is not considered in this experiment.

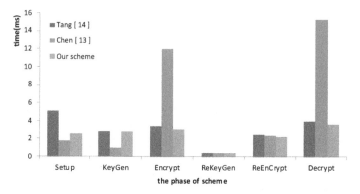

Fig. 3. Cost overhead comparison with other schemes

We can visualize the cost and overhead of each of these three schemes. Both Tang [14] and our scheme use distributed key generation techniques, and therefore consume more time in the initialization and key generation sessions. However, we have optimized it to improve the system performance and fault tolerance while ensuring security. In the encryption and decryption phase, the scheme of Chen [13] uses RSA encryption of medical data, which significantly increases the cost overhead. Our scheme, on the other hand, uses symmetric encryption algorithms to encrypt medical data and asymmetric encryption algorithms to encrypt symmetric keys, which reduces the cost overhead. A query request validation link is added to further improve the security.

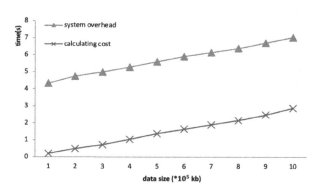

Fig. 4. System overhead and calculating cost

As shown in Fig. 4, we conduct another set of experiments for the size of the data volume to test the impact of the data size on the overall system overhead and the calculating cost of our scheme, respectively. The test environment simulates three hospitals consisting of three peer nodes and one orderer node, and the orderer node uses the draft consensus algorithm. The calculating cost of our solution shows a linear growth trend as the data grows, but the calculating cost is a relatively small percentage of the overall system overhead.

It is intuitive to see that the calculating cost is negligible for data volumes less than 10^5 KB. At greater than 10^6 KB, the calculating cost will reach more than 40% of the total system overhead. With the enlargement of data, the calculating cost has displaced the communication cost as a major cost. However, the growth trend is slower.

5 Conclusion

In this paper, we focus on the practical issues of privacy protection and management of personal electronic medical data in electronic medical data sharing systems. We further discuss the practical needs of patients for this problem. To be able to meet these needs, we propose a patient-friendly medical data security sharing scheme based on blockchain and proxy re-encryption. In our scheme, we use distributed key management technology to keep the private keys for patients and enhance the confidentiality of patients' private keys, as well as proxy re-encryption technology to ensure patients' autonomy over their medical data in order to solve the practical problems of different groups of patients. In addition, we store medical data in a distributed way and design a blockchain-based three-party security verification scheme to ensure the security of off-chain data queries. It also can ensure the independence of each hospital's data and facilitate device performance evaluation by different medical device manufacturers. Finally, we conducted several experiments to prove the effectiveness of this scheme. However, in our experiments, we found that when the data volume is larger than 10^6 KB, the encryption overhead will become the main overhead. It greatly reduces the rate of the whole system. In general, our scheme is applicable to blockchain-based electronic medical data sharing systems.

Acknowledgement. This work is supported by the Major projects of National Social Science Foundation of China (No. 21 & ZD124) and the Open Program of Neusoft Corporation (NCBETOP2001). We also thank anonymous reviewers for the helpful reports.

References

1. Hathaliya, J.J., Tanwar, S.: An exhaustive survey on security and privacy issues in Healthcare 4.0. Comput. Commun. 153, 311–335 (2020)
2. Butpheng, C., Yeh, K.-H., Xiong, H.: Security and privacy in IoT-cloud-based e-Health systems—a comprehensive Review. Symmetry. 12, 1191 (2020). https://doi.org/10.3390/sym120 71191
3. Li, C.T., Shih, D.H., Wang, C.C., Chen, C.L., Lee, C.C.: A Blockchain based data aggregation and group authentication scheme for electronic medical system. IEEE Access 8, 173904–173917 (2020)

4. Shahid, F., Ashraf, H., Ghani, A., Ghayyur, S.A.K., Shamshirband, S., Salwana, E.: PSDS–proficient security over distributed storage: a method for data transmission in cloud. IEEE Access **8**, 118285–118298 (2020)

5. Cao, B., et al.: When internet of things meets Blockchain: challenges in distributed consensus. IEEE Netw. **33**(6), 133–139 (2019)

6. Cao, S., Zhang, X., Xu, R.: Toward secure storage in cloud-based eHealth systems: a Blockchain-assisted approach. IEEE Netw. **34**(2), 64–70 (2020)

7. Nguyen, D.C., Pathirana, P.N., Ding, M., Seneviratne, A.: Blockchain for secure EHRs sharing of mobile cloud based E-Health systems. IEEE Access **7**, 66792–66806 (2019)

8. Kassab, M., DeFranco, J.F., Malas, T., Laplante, P.A., Destefanis, G., Neto, V.V.G.: Exploring research in Blockchain for healthcare and a roadmap for the future. IEEE Trans. Emerg. Top. Comput. (TETC) **9**(4), 1835–1852 (2021)

9. Abu-elezz, I., Hassan, A., Nazeemudeen, A., Househ, M.S., Abd-alrazaq, A.A.: The benefits and threats of Blockchain technology in healthcare: a scoping review. Int. J. Med. Inform. **142**, 104246 (2020)

10. Xu, J., et al.: Healthchain: a Blockchain-Based privacy preserving scheme for large-scale health data. IEEE Int. Things J. (IoT) **6**(5), 8770–8781 (2019)

11. Blaze, M., Bleumer, G., Strauss, M.: Divertible protocols and atomic proxy cryptography. In: Nyberg, K. (eds) Advances in Cryptology — EUROCRYPT'98. EUROCRYPT 1998. Lecture Notes in Computer Science, vol. 1403. Springer, Berlin, Heidelberg (1998). https://doi.org/10.1007/BFb0054122

12. Sun, Y., Song, W., Shen, Y.: Efficient Patient-Friendly Medical Blockchain System Based on Attribute-Based Encryption. In: Wang, G., Lin, X., Hendler, J., Song, W., Xu, Z., Liu, G. (eds) Web Information Systems and Applications. WISA 2020. Lecture Notes in Computer Science, vol. 12432. Springer, Cham (2020). https://doi.org/10.1007/978-3-030-60029-7_57

13. Chen, Z., Xu, W., Wang, B., Yu, h.: A blockchain-based preserving and sharing system for medical data privacy. Future Gener. Comput. Syst. (FGCS) 124, 338–350 (2021)

14. Tang, F., Chen, Y., Feng, Z.: Electronic prescription sharing scheme based on Blockchain and proxy Re-encryption. Comput. Sci. **48**, 498–503 (2021)

15. Shamir, A.: How to share a secret. Commun. ACM **22**(11), 612–613 (1979)

Spatial Data Publication Under Local Differential Privacy

Jian Zhuang[1], Ning Wang[1,2(✉)], Zhigang Wang[1], Xiaodong Wang[1], Haipeng Qu[1], and Zhiqiang Wei[1]

[1] Faculty of Information Science and Engineering,
Ocean University of China, Qingdao, China
zhuangjian@stu.ouc.edu.cn,
{wangning8687,wangzhigang,wangxiaodong,quhaipeng,weizhiqiang}@ouc.edu.cn
[2] Key Lab of Cryptologic Technology and Information Security,
Ministry of Education, Shandong University, Qingdao, China

Abstract. Local differential privacy (LDP), which has been applied in Google Chrome and Apple iOS, provides strong privacy assurance to users when collecting data from users. We focus on the sensitive spatial data collection, with the goal of obtaining high result utility while satisfying LDP. The existing methods for this problem mostly target at the task of range queries. They combine the frequency estimation technology and spatial decomposition method to publish the number of users located in some sub-spaces. However, these methods cannot well support distance-related applications such as k-means clustering, since they treat the sub-spaces not containing the user equally and do not consider the distances between the sub-space and user data. Motivated by this, we propose dimension-correlated piecewise mechanism (DCPM), a novel LDP perturbation mechanism with a well-designed probability density, in which the distance between the published value and the true one is considered. Extensive experiments on real-world data and synthetic data demonstrate that DCPM achieves significantly higher result utility compared to previous solutions.

Keywords: Local differential privacy · Spatial data · k-means · Range query

1 Introduction

Spatial data is of great significance for the development of smart city and the improvement of government governance capabilities. What's more, it can be used to improve the service quality of enterprises. For example, service providers can use spatial data to optimize the scheduling of their resources. However, since spatial data often contains sensitive personal information, such as the location of one user, direct publication would reveal sensitive information. Therefore, we need to fully consider the risk of privacy leakage when collecting and using spatial data. On the other hand, local differential privacy (LDP), as an emerging

X. Zhao et al. (Eds.): WISA 2022, LNCS 13579, pp. 627–637, 2022.
https://doi.org/10.1007/978-3-031-20309-1_55

privacy protection method, has been used by many companies such as Google [1], Microsoft [2]. It works under the setting that there is no trusted third-party data collector. To protect the privacy, the user locally perturbs data and sends the perturbed data to the collector. Compared to CDP [3], LDP provides a more rigorous privacy protection for individual users, as the true values of users never leave their local devices.

In this paper, we focus on the spatial data publication under LDP by considering the superiority of LDP. Intuitively, spatial data consisting of latitudes and longitudes can be regarded as two-dimensional numerical data. For the perturbation of multi-dimensional data under LDP, an intuitive method is to split privacy budgets into several parts, each of which is used to publish the value from one dimension. Applying the above technique for the spatial data would destroy the correlation between the latitude and latitude, further lead to inferior published results. Besides, the other line for spatial data publication is the hierarchical tree decomposition technique [4], which iteratively splits the spatial space into four sub-spaces and publishes the count of users located in each sub-space. In this way, the data distribution can be derived based on the tree structure. However, the LDP perturbation component for publishing the sub-space one user locates in treats the sub-spaces without this user equally without discrimination. Namely, the sub-spaces without the user are published with the same probability. However, it is undoubted that the sub-spaces closer to the one containing the user can reflect the true location information better. Unfortunately, the hierarchical tree decomposition ignores the above consensus, which leads to inferior performance.

This paper addresses the above challenges and makes several major contributions. First, we propose a novel mechanism, namely Dimension-correlated Piecewise Mechanism (DCPM), for collecting the spatial data under LDP, which obtains higher result accuracy compared to existing methods. In particular, a well-designed probability density function involved in the LDP perturbation component is proposed, which takes the distance between the spatial coordinate point and the actual coordinate as a parameter. And this mechanism makes the points closer to the true be reported with higher probability, which satisfies the intuitive design requirement of perturbation mechanisms. Besides, we conduct extensive experiments on both synthetic and real-world datasets to validate and compare the accuracy of our methods through k-means clustering and range queries. The experimental results show that our method has obvious advantages over the basic method, with better data utility.

2   Related Work

Local differential privacy(LDP) [5,6] mainly focused on frequency estimation in the early stage. Up to now, a large number of studies on frequency estimation have been proposed, like RAPPOR [1], RR [7], OUE, OLH [8], CM [9] and so on. Compared with other methods, OUE and OLH are the two most popular methods for frequency estimation, which can introduce less noise and easier to implement [10,11].

For spatial data, most existing mechanisms combine frequency estimation techniques with spatial decomposition method [12,13] with hierarchical tree structure [14]. HDG [15] introduced the idea of one-dimensional grid and user grouping. PCEP [16] combines SH [17] and personalized differential privacy [18]. However, these methods are optimized for counting the number of points located in region, i.e.range queries. They may not support well the distance-related applications like k-means clustering. In addition, the user needs to interact with the server in advance to discretize the data, and noisy data needs to be post-processed, which dramatically increases the communication cost and computing cost.

As a two-dimensional data, spatial data can apply multi-dimensional perturbation methods to perturb data. The whole processing method and sampling method are commonly used for multi-dimensional data in LDP. The whole processing means that each dimension invokes single-attribute perturbation algorithm separately. According to the combination theorem [19], it is proved that satisfies local differential privacy. But it destroys the correlation between dimensions. Sampling method means that only select one dimension randomly from multiple dimensions to perturb. Due to the particularity of spatial data, this method is not applicable here.

For the perturbation of numerical data, a classic method is Laplace mechanism [20], which adds a Laplace noise that satisfies local differential privacy on the actual value. Duchi et al. [21] first proposed a mean estimation method MeanEst for numerical properties under LDP by using random response. Building on the MeanEst method, Nguyen and Xiao et al. [22] proposed the Harmony-mean method by adopting the sampling technique. Geng and Kairouz et al. [23] proposed Staircase mechanism, which use piece-wise constant probability density distribution to replace Laplace distribution. To further improve the performance, Wang and Xiao et al. [24] proposed the Piecewise Mechanism(PM), which combines the advantages of the Laplace mechanism and random response.

3 Preliminary

3.1 Local Differential Privacy

LDP model fully considers the risk of privacy leakage from an untrusted server or data collector. It requires users to perturb actual data locally, and then send noisy data to server. The collector executes statistical analysis based on the collected data. The LDP algorithm should satisfy the following property.

Definition 1 (ϵ-Local Differential Privacy). For any two input values v and v' in dataset Z, M will satisfy ϵ-Local Differential Privacy if and only if it satisfies Eq. 1.

$$P[M(v) \in O] \leq e^{\epsilon} P[M(v') \in O] \qquad (1)$$

ϵ is privacy budget, which measures the degree of privacy protection.

3.2 Problem Definition

In this paper, we assume there are n users and an untrusted third-party data collector. Each user u_i has a spatial coordinate (x_i, y_i) in two-dimensional coordinate system and a total privacy budget ϵ. In the real world, the coordinate range is often irregular. In this paper, we use the circumcircle of the data range to cover the original data range and take this circumcircle as the input domain. The radius of the circumcircle is R. For the convenience of representation, we default R to 1 in the following.

Each user u_i privatizes the coordinates by invoking the LDP perturbation mechanism M and sends (x_i^*, y_i^*) to the collector. Our goal is proposing a non-interactive LDP protocol to collect and analyze user coordinate data while reducing computational and communication cost. Furthermore, we exploit the relationship between coordinate dimensions to improve the accuracy of distribution estimation.

4 DCPM: Dimension-Correlated Piecewise Mechanism

In this section, we propose dimension-correlated piecewise mechanism (DCPM), which is used to collect spatial data from users under LDP. This mechanism will be introduced in Sect. 4.1. In Sect. 4.2, we will demonstrate its privacy guarantee and other properties.

4.1 DCPM Algorithm

Inspired by the Piecewise Mechanism(PM) [24], we design DCPM algorithm for spatial data based on the following three principles:

   I. The output values x^*, y^* are unbiased, i.e. $E(x^*) = x$, $E(y^*) = y$.
  II. For the entire output domain$\mathcal{O}$, the sum of probabilities is 1.
 III. The points located in the nearby regions of the true data are reported with higher probabilities, compared with the ones in the regions far away.

Our proposal DCPM satisfies the above principles. In particular, DCPM takes as input a value (x, y), where $x^2 + y^2 \leq 1$, and outputs a perturbed value (x^*, y^*), where $x^{*2} + y^{*2} \leq (1 + 2r)^2$. r is the radius of $\mathcal{A}$, which satisfies Eq. 2.

$$(e^\epsilon - 1) r^3 - 4r^2 - 4r - 1 = 0 \tag{2}$$

The joint probability density function of x^*, y^* is a piecewise constant function as follows:

$$P\left(x^* = \hat{x}, y^* = \hat{y} \mid x, y\right) = \begin{cases} p, & \text{if } (\hat{x} - a)^2 + (\hat{y} - b)^2 \leq r^2 \\ \frac{p}{e^\epsilon}, & \text{otherwise} \end{cases} \tag{3}$$

where

$$p = \frac{e^\epsilon}{(1 + 4r + (3 + e^\epsilon) r^2) \pi}$$

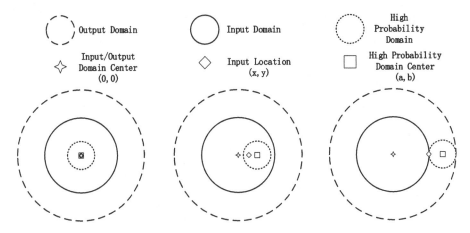

Fig. 1. The noisy output (x^*, y^*)'s probability density function when $(x, y) = (0, 0)$, $(0.5, 0)$ and $(1, 0)$.

a and b satisfy Eq. 4.

We use $P(x^*, y^* \mid x, y)$ as a shorthand for $P(x^* = \hat{x}, y^* = \hat{y} \mid x, y)$. Figure 1 represents the joint probability density function when $(x, y) = (0, 0)$, $(0.5, 0)$ and $(1, 0)$. As shown in Fig. 1, DCPM algorithm's output domain $\mathcal{O}$ consists of two parts. One part is the high probability area $\mathcal{A}$ which close to the input value (x, y). The other part is the low probability area $(\mathcal{O} - \mathcal{A})$. $\mathcal{A}$ is a circle whose center is (a, b) and radius is r. The sampling probability value of $\mathcal{A}$ is p, and $(\mathcal{O} - \mathcal{A})$ is $\frac{p}{e^{\epsilon}}$. Furthermore, we can see that the user's input value (x, y) and the center (a, b) of $\mathcal{A}$ are not necessarily the same point, which we will introduce later.

Algorithm 1 is the pseudo-code of DCPM. We first calculate the center of the circle $\mathcal{A}$ (a_i, b_i)(Line 1), then randomly sample from the entire output domain $\mathcal{O}$ with probability $\frac{p\pi(1+2r)^2}{e^{\epsilon}}$(Line 3–7), or randomly sample from high probability area $\mathcal{A}$ with probability $(1 - \frac{p\pi(1+2r)^2}{e^{\epsilon}})$ (Line 8–12). For convenience, we randomly sample the polar angle θ and polar diameter ρ in the form of polar coordinates to determine the coordinates of the two-dimensional coordinate system (x_i^*, y_i^*).

Choice of High Probability Area $\mathcal{A}$. For the selection of the high probability area $\mathcal{A}$ for each user, we are based on a precondition that the distance $s = \sqrt{(x_i - a_i)^2 + (y_i - b_i)^2}$ from (a_i, b_i) to (x_i, y_i) is proportional to the distance $l = \sqrt{x_i^2 + y_i^2}$ from (x_i, y_i) to $(0, 0)$ and three points lie on the same line. So we can get the follow Equation:

$$
\begin{aligned}
a_i &= x_i(1 + r) \\
b_i &= y_i(1 + r)
\end{aligned}
\tag{4}
$$

We can know from Eq. 4 that when (x_i, y_i) is closer to $(0, 0)$, (a_i, b_i) is closer to (x_i, y_i).

Algorithm 1. DCPM algorithm

Input: actual coordinate data (x_i, y_i), privacy budget ϵ
Output: noisy coordinate data (x_i^*, y_i^*)
1: $(a_i, b_i) = (x_i(1 + r), y_i(1 + r))$
2: $z = random(0, 1)$
3: **if** $z < \frac{p\pi(1+2r)^2}{e^\epsilon}$ **then**
4:    $\theta = 2\pi \cdot random(0, 1)$
5:    $\rho = (1 + 2r) \cdot \sqrt{random(0, 1)}$
6:    $x_i^* = \rho \cdot \cos\theta$
7:    $y_i^* = \rho \cdot \sin\theta$
8: **else**
9:    $\theta = 2\pi \cdot random(0, 1)$
10:    $\rho = r \cdot \sqrt{random(0, 1)}$
11:    $x_i^* = a_i + \rho \cdot \cos\theta$
12:    $y_i^* = b_i + \rho \cdot \sin\theta$
13: **end if**
14: **return** (x_i^*, y_i^*)

4.2 Theoretical Guarantees

Theorem 1. *DCPM algorithm satisfies ϵ-local differential privacy.*

Proof. Given two input tuples (x_i, y_i) and (x_i', y_i') satisfying $x_i^2 + y_i^2 \leq 1$, $x_i'^2 + y_i'^2 \leq 1$, output (x_i^*, y_i^*) satisfying $x_i^{*2} + y_i^{*2} \leq (1 + 2r)^2$. According Eq. 3, we have

$$\frac{P\{x^*, y^* \mid x_i, y_i\}}{P\{x^*, y^* \mid x_i', y_i'\}} \leq e^\epsilon \tag{5}$$

Theorem 2. *For the output (x_i^*, y_i^*) of Algorithm 1, $E(x_i^*) = x_i$, $E(y_i^*) = y_i$.*

Proof. Here we take (x_i, y_i) being in the first quadrant as an example.

$$
\begin{aligned}
E(x_i^*) &= \int_{-1-2r}^{1+2r} \int_{-\sqrt{1-x^2}}^{\sqrt{1+x^2}} \frac{p}{e^\epsilon} x \, dy \, dx + \int_{a_i-r}^{a_i+r} \int_{b_i-\sqrt{r^2-(x-a_i)^2}}^{b_i+\sqrt{r^2-(x-a_i)^2}} \left(p - \frac{p}{e^\epsilon}\right) x \, dy \, dx \\
&= \int_{-a_i-r}^{-a_i+r} \int_{b_i-\sqrt{r^2-(x-a_i)^2}}^{b_i+\sqrt{r^2-(x-a_i)^2}} \frac{p}{e^\epsilon} x \, dy \, dx + \int_{a_i-r}^{a_i+r} \int_{b_i-\sqrt{r^2-(x-a_i)^2}}^{b_i+\sqrt{r^2-(x-a_i)^2}} p x \, dy \, dx \\
&= 2\left(p - \frac{p}{e^\epsilon}\right) a_i \int_{-r}^{r} \sqrt{r^2 - t^2} \, dt = x_i
\end{aligned}
\tag{6}
$$

$$E\left(y_{i}^{*}\right)=\int_{-1-2r}^{1+2r}\int_{-\sqrt{1-y^{2}}}^{\sqrt{1+y^{2}}}\frac{p}{e^{\epsilon}}ydxdy+\int_{b_{i}-r}^{b_{i}+r}\int_{a_{i}-\sqrt{r^{2}-(y-b_{i})^{2}}}^{a_{i}+\sqrt{r^{2}-(y-b_{i})^{2}}}\left(p-\frac{p}{e^{\epsilon}}\right)ydxdy$$

$$=\int_{-b_{i}-r}^{-b_{i}+r}\int_{a_{i}-\sqrt{r^{2}-(y-b_{i})^{2}}}^{a_{i}+\sqrt{r^{2}-(y-b_{i})^{2}}}\frac{p}{e^{\epsilon}}ydxdy+\int_{b_{i}-r}^{b_{i}+r}\int_{a_{i}-\sqrt{r^{2}-(y-b_{i})^{2}}}^{a_{i}+\sqrt{r^{2}-(y-b_{i})^{2}}}pxdydx$$

$$=2\left(p-\frac{p}{e^{\epsilon}}\right)b_{i}\int_{-r}^{r}\sqrt{r^{2}-t^{2}}dt=y_{i}$$

$$(7)$$

Theorem 3. *The sum of the probabilities of all output domain equal 1.*

Proof.

$$\iint P\left(x_{i}^{*},y_{i}^{*}\right)dydx=\frac{p\pi(1+2r)^{2}}{e^{\epsilon}}+\left(p-\frac{p}{e^{\epsilon}}\right)\pi r^{2}=1 \qquad (8)$$

5   Experiments

In this section, we assess the performance of our method and the basic method on real-world and synthetic data. We first conduct $|Q|=100$ range queries and measure the usability of the method by the mean-square error(MSE).

$$MSE=\frac{\sum_{i=1}^{|Q|}\left(F_{i}-F_{i}'\right)^{2}}{|Q|} \qquad (9)$$

where F_{i} and F_{i}' is frequency of the range query Q_{i} on actual and noisy data. $|Q|$ is the number of queries.

Then we apply the noisy data to the k-means clustering algorithm to get noisy cluster centers. We use the average of the distances L from each user's actual coordinate (x_{i}, y_{i}) to the nearest cluster noisy center to assess performance of method.

$$L=\frac{\sum_{i=1}^{n}|(x_{i},y_{i}),C_{i}^{*}|_{2}}{n} \qquad (10)$$

where C_{i}^{*} is the nearest noisy cluster center to u_{i}.

Datasets and Parameters. We normalize the domain of real-world and synthetic data in advance. For the fairness of the experiment, we normalize the data coordinates to a $[-1,1] \times [-1,1]$ rectangle.

- **MixGauss:** For the x-axis data x, we mix $\mathcal{N}_{1}(-0.8, 0.1)$, $\mathcal{N}_{2}(-0.4, 0.1)$, $\mathcal{N}_{3}(0, 0.1)$, $\mathcal{N}_{4}(0.4, 0.1)$ and $\mathcal{N}_{5}(0.8, 0.1)$ with equal weights. For the y-axis data y, we mix $\mathcal{N}_{1}(-0.5, 0.1)$ and $\mathcal{N}_{2}(0.5, 0.1)$ with equal weights. This dataset contains 30000 coordinate data.
- **BeiJing** [25]: This is a T-Drive trajectory dataset, which contains pickup and drop-off locations of Beijing taxis. Here, we only use pickup location data. This dataset contains 30000 coordinate data.

Mechanisms. We compare our algorithm with the whole processing methods applying two numerical data perturbation algorithms.

- **Laplace:** A basic method used in LDP for numerical data. We use the Laplace mechanism to process the x and y respectively
- **PM:** A numerical data publishing method under LDP which have lower noisy bound.
- **DCPM:** Our method in Algorithm 1. In order to meet the requirements of the algorithm on the input domain and the fairness of the experiment, we process the data into a circle with radius 1 and perturb it locally. And restore it when send to server.
- **NP:** Running on non-private data.

Results and Analysis. All experimental results are repeated 10 times and averaged, thus can reflect the authenticity and universality of the results.

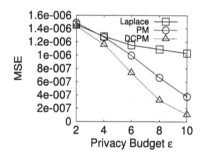

Fig. 2. Vary Privacy Budget(MixGauss).

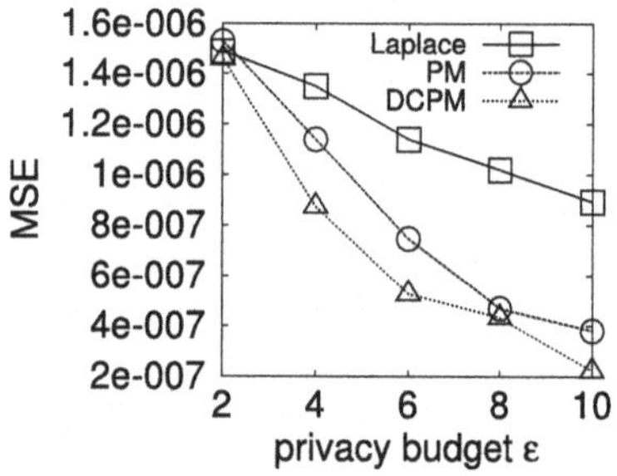

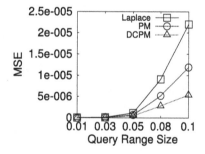

Fig. 3. Vary Privacy Budget(BeiJing).

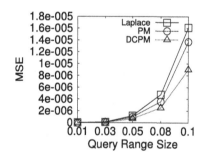

Fig. 4. Vary Range Size(MixGauss).

Fig. 5. Vary Range Size(BeiJing).

Figure 2 and Fig. 3 show the average MSE for 100 queries corresponding to different privacy budgets when the query range size is 0.05×0.05. We can see that our method significantly outperforms PM and Laplace mechanism under

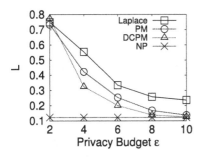

Fig. 6. Vary Privacy Budget(MixGauss).

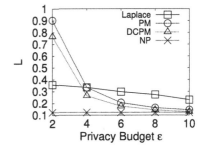

Fig. 7. Vary Privacy Budget(BeiJing).

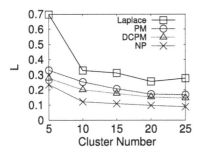

Fig. 8. Vary Cluster Number(MixGauss).

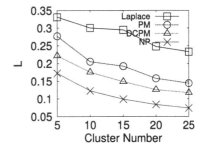

Fig. 9. Vary Cluster Number(BeiJing).

different privacy budgets. One of the main reasons is that we publish data uniformly instead of perturbing dimensions independently, which reduces the noise caused by privacy budget segmentation. Moreover, DCPM enables noisy data points to be closer to actual data points.

Figure 4 and Fig. 5 test the accuracy under different query range size $\{0.01 \times 0.01, 0.03 \times 0.03, 0.05 \times 0.05, 0.08 \times 0.08, 0.1 \times 0.1\}$ when privacy budget is 6. As the query size changes, DCPM can still perform better than PM and Laplace.

Figure 6 and Fig. 7 show the average distance L with the change of privacy budget when clusters number is set to 10. We can see that DCPM outperforms PM and Laplace in most cases, but it performs worse than Laplace when the privacy budget is small, because when the privacy budget is small, the radius r will increases sharply, causing DCPM performs poorly. PM has the same problem.

Figure 8 and Fig. 9 show that with the change of the number of cluster centers, the average distance L changes, here we set the ϵ to 6. We can see that for different number of clusters, DCPM can still show good performance.

6   Conclusion

In this paper, we focus on the perturbation and collection strategy of two-dimensional coordinate data under LDP. We first analyze the shortcomings of the traditional multi-dimensional processing scheme on spatial data perturbation. Then we introduce the idea of dimension-correlation and proximity principle, proposing a two-dimensional density distribution estimation method DCPM. This algorithm enables user to output perturbed data close to the actual data with high probability, and preserve the characteristics of numerical data as possible. Experiments show that our proposed mechanism outperforms traditional multi-dimensional data processing schemes for spatial data applications. In future work, we plan to extend DCPM to more cases like 3D situation.

Acknowledgements. This work was supported by the National Natural Science Foundation of China (61902365 and 61902366), Open Project Program from Key Lab of Cryptologic Technology and Information Security, Ministry of Education, Shandong University, and the Fundamental Research Funds for the Central Universities (202042008).

References

1. Erlingsson, Ú., Pihur, V., Korolova, A.: RAPPOR: randomized aggregatable privacy-preserving ordinal response. In: CCS, pp. 1054–1067. ACM (2014)
2. Ding, B., Kulkarni, J., Yekhanin, S.: Collecting telemetry data privately. In: NIPS, pp. 3571–3580 (2017)
3. Liu, Y., et al.: Differentially private linear regression analysis via truncating technique. In: Xing, C., Fu, X., Zhang, Y., Zhang, G., Borjigin, C. (eds.) WISA 2021. LNCS, vol. 12999, pp. 249–260. Springer, Cham (2021). https://doi.org/10.1007/978-3-030-87571-8_22
4. Zhang, J., Xiao, X., Xie, X.: Privtree: a differentially private algorithm for hierarchical decompositions. In: SIGMOD Conference, pp. 155–170. ACM (2016)
5. Kasiviswanathan, S.P., Lee, H.K., Nissim, K., Raskhodnikova, S., Smith, A.: What can we learn privately? SIAM J. Comput. **40**(3), 793–826 (2011)
6. Duchi, J.C., Jordan, M.I., Wainwright, M.J.: Local privacy and statistical minimax rates. In: FOCS, pp. 429–438. IEEE Computer Society (2013)
7. Warner, S.L.: Randomized response: a survey technique for eliminating evasive answer bias. J. Am. Stat. Assoc. **60**, 63–66 (1965)
8. Wang, T., Blocki, J., Li, N., Jha, S.: Locally differentially private protocols for frequency estimation. In: USENIX Security Symposium. pp. 729–745. USENIX Association (2017)
9. Wang, S., Li, J., Qian, Y., Du, J., Lin, W., Yang, W.: Hiding numerical vectors in local private and shuffled messages. In: IJCAI, pp. 3706–3712. ijcai.org (2021)
10. Ye, Q., Hu, H.: Local differential privacy: tools, challenges, and opportunities. In: U, L.H., Yang, J., Cai, Y., Karlapalem, K., Liu, A., Huang, X. (eds.) WISE 2020. CCIS, vol. 1155, pp. 13–23. Springer, Singapore (2020). https://doi.org/10.1007/978-981-15-3281-8_2
11. Yang, M., Lyu, L., Zhao, J., Zhu, T., Lam, K.: Local differential privacy and its applications: a comprehensive survey. CoRR abs/2008.03686 (2020)

12. De Berg, M.T., Van Kreveld, M., Overmars, M., Schwarzkopf, O.: Computational Geometry: Algorithms and Applications. Springer, Heidelberg (2000). https://doi.org/10.1007/978-3-662-04245-8
13. Samet, H.: Foundations of Multidimensional and Metric Data Structures. Morgan Kaufmann, San Francisco (2006)
14. Cormode, G., Kulkarni, T., Srivastava, D.: Answering range queries under local differential privacy. Proc. VLDB Endow. **12**(10), 1126–1138 (2019)
15. Yang, J., Wang, T., Li, N., Cheng, X., Su, S.: Answering multi-dimensional range queries under local differential privacy. Proc. VLDB Endow. **14**(3), 378–390 (2020)
16. Chen, R., Li, H., Qin, A.K., Kasiviswanathan, S.P., Jin, H.: Private spatial data aggregation in the local setting. In: ICDE, pp. 289–300. IEEE Computer Society (2016)
17. Bassily, R., Smith, A.D.: Local, private, efficient protocols for succinct histograms. In: STOC, pp. 127–135. ACM (2015)
18. Jorgensen, Z., Yu, T., Cormode, G.: Conservative or liberal? personalized differential privacy. In: ICDE, pp. 1023–1034. IEEE Computer Society (2015)
19. Dwork, C., Roth, A.: The algorithmic foundations of differential privacy. Found. Trends Theor. Comput. Sci. **9**(3–4), 211–407 (2014)
20. Dwork, C., McSherry, F., Nissim, K., Smith, A.: Calibrating noise to sensitivity in private data analysis. In: Halevi, S., Rabin, T. (eds.) TCC 2006. LNCS, vol. 3876, pp. 265–284. Springer, Heidelberg (2006). https://doi.org/10.1007/11681878_14
21. Duchi, J.C., Wainwright, M.J., Jordan, M.I.: Minimax optimal procedures for locally private estimation. CoRR abs/1604.02390 (2016)
22. Nguyên, T.T., Xiao, X., Yang, Y., Hui, S.C., Shin, H., Shin, J.: Collecting and analyzing data from smart device users with local differential privacy. CoRR abs/1606.05053 (2016)
23. Geng, Q., Kairouz, P., Oh, S., Viswanath, P.: The staircase mechanism in differential privacy. IEEE J. Sel. Top. Signal Process. **9**(7), 1176–1184 (2015)
24. Wang, N., et al..: Collecting and analyzing multidimensional data with local differential privacy. In: ICDE, pp. 638–649. IEEE (2019)
25. Yuan, J., Zheng, Y., Xie, X., Sun, G.: Driving with knowledge from the physical world. In: KDD, pp. 316–324. ACM (2011)

Big Data Analysis for Anti-Money Laundering: A Case of Open Source Greenplum Application

Chaochen Hu[1], Ran Li[2], Chao Li[1(✉)], Hengshuo Miao[1], Zongyou Yang[3], and Tengda Zhang[4]

[1] Tsinghua University, Beijing, China
{hcc20,miaohs}@mails.tsinghua.edu.cn, li-chao@tsinghua.edu.cn
[2] Communication University of China, Beijing, China
ranlee@cuc.edu.cn
[3] Beijing University of Posts and Telecommunications, Beijing, China
dryang0624@bupt.edu.cn
[4] Beijing AgileCentury Information Technology Co., Ltd., Beijing, China
zhangtengda@agilecentury.com

Abstract. For the anti-money laundering regulations of financial institutions, it is important to do the timely big data analysis to continuously optimize the comprehensiveness, correctness and effectiveness of transaction monitoring models. And at the same time, how to reduce the cost investment for the solution is also important. The traditional solutions are usually based on Hadoop family, which is costly both in technical stack building and in code programming. And sometimes, although these costly solutions can be borne, the response time cannot be borne. This paper is an application case to build the solution on open source distributed database Greenplum. By utilizing the share-nothing MPP structure and the software ecosystem of Greenplum, a solution of high efficient big data analysis through large scale parallel computing is designed. After ETL processing by the tools such as gpfdist, we use advanced SQL features to support the calculation of several key transaction monitoring statistical indicators for anti-money laundering. The test cases show that, through this way, we can not only reduce the cost in technical stack building and in code programming, but also get the analysis results in just few seconds. Thus, this is a feasible solution for the similar big data analysis tasks.

Keywords: Big data · Data analysis · Anti-Money Laundering · Open source · Distributed database · Case study · Ecosystem of software · SQL · Advanced queries

1 Introduction

In the anti-money laundering regulatory of financial institutions, it is necessary to continuously optimize the transaction monitoring mode's comprehensiveness,

Supported by Beijing AgileCentury Information Technology Co., Ltd.

correctness, and effectiveness and consider the low-cost investment. This research project was started in 2021 to realize high-efficiency analysis through large-scale parallel computing and meet big data analysis needs. By integrating the customer information, account information, contract information, transaction information and other data of multiple business line information systems of financial institutions, enter the big data platform from the original system according to a certain format, and then store the original data. At the same time, integrate the main body according to the certificate number and name to form a unified customer view, and calculate the transaction statistical indicators of various dimensions on a daily, weekly and monthly basis, Form a new customer profile. And complete the query of a certain period and the current status of a certain type of customer within two seconds, and query and display the profile of a customer within two seconds.

2   Related Work

2.1   Traditional Data Processing Based on Hadoop

Hadoop distributed file system (HDFS) provides a certain method that can be adopted in such realistic situations [5]. A typical solution based on HDFS involves Hive, a data warehouse tool built on top of Hadoop. Hive consists of HDFS used for data storage as the bottom level of the overall architecture, and MapReduce engine lying on the top level, which is a software framework aiming to process big data in a parallel manner. With the assistance of HiveQL (HQL) queries, SQL querying language will be converted into an executable MapReduce program, eliminating the need of using Java APIs directly [3]. In terms of the application considered in the previous part, Hadoop method can be applied with elastic search [4] storing the general profile of clients, and redis [8] storing the profile corresponding with the status of a certain client during a certain period.

However, there are several drawbacks that should not be ignored when applying HDFS. One prominent point is that the architecture of this method is relatively complicated, containing multiple kinds of tools and frameworks in the final solution.

2.2   Data Processing Based on Greenplum

Greenplum is a open source massive parallel processing (MPP) database based on PostgreSQL for transactional and analytical workloads [7]. Greenplum meets the need of SQL standards (including SQL-1992, SQL-1999, SQL-2003, SQL-2011 and SQL-2016) perfectly, and can be integrated to analyze different sorts of data, no matter it is structured, semi-structured (e.g. JSON and XML) or unstructured (text). Graph data and data from Geographic Information System (GIS) are accepted as well [2,9].

Greenplum database has plenty of advantages worth listing out below [6].

1. Greenplum supports efficient distributed query and analysis because data is distributed evenly between nodes according to distribution strategy.

2. Greenplum is highly available because the standby master node takes over the master node's job when the master node is no longer active and each segment generates a duplicate in other segments.
3. Greenplum supports SQL standard and standard application programming interface such as Java database connection (JDBC) and Open database connection (ODBC).
4. Greenplum is highly extensible with a one-stop technology stack and a complete ecosystem. It provides data loading tool (gpfdist), text analysis tool (gptext), native machine learning library (madlib) [1] and other tools to meet various needs. In addition, it supports procedural languages such as python, R, Java, Perl, etc., and allows users to develop extensions as needed, such as PostGIS. It is even possible to develop blockchain applications on Greenplum based on the design of [10].

The above advantages of Greenplum make its application more simple and convenient. One reason to use Greenplum instead of Hadoop is also revealed, namely that the one-stop technology stack and standardized APIs build a more flexible and efficient system that better focuses on specific goals.

3   Solution Overview

The overall demand can be decomposed into four parts: data collection from various sources, data integration, statistical analysis and advanced analysis.

3.1   Data Collection From Various Sources

Demand. In the case of the study, data is from three independent sources, which are respectively economy subsystem, invest bank subsystem and asset management subsystem, and hence data should be acquired automatically from diverse platforms.

Solution. The Platform Extension Framework (PXF) of Greenplum provides data accessing that is parallel and allows extensive data, as well as federal query function. With PXF, Greenplum and SQL can be utilized to access these data sources such as Hadoop, Apache Ignite, MySQL, ORACLE, Microsoft SQL Server, DB2, PostgreSQL (Via JDBC). Therefore, PXF is capable of acquiring daily updated data from each finance subsystem without any need of loading it into the disk, making it convenient to update the data every day.

3.2   Data Integration

Demand. Considering simplicity and efficiency of further analysis, data for processing needs to be integrated. Moreover, and data along with its form and number may conflict among subsystems, thus data cleaning is definitely essential, which can be included in the procedure of data integration.

Solution. Greenplum loads local data with a large quantity by gpfdist and gpload, which are a set of parallel data loading tools aiming to increase loading speed of huge amounts of data. By making use of these tools, the goal of data integration will be easily achieved.

3.3   Fundamental Statistical Analysis

Demand. The solution scheme should be able to support fundamental statistical analysis to meet the need of financial services. For instance, it ought to provide information like date, range, client type, product or service type, channel type etc. Indexes like the amount of clients, service branches and transactions also need to be computed under given conditions.

Solution. Greenplum database is qualified to satisfy this demand, as SQL instructions can accomplish all queries and computation that is required above. It accepts data analysis using SQL in the database directly, namely the database native analysis.

3.4   Advanced Analysis

Demand. There might be some relatively complicated requirements in the field of finance business, which fundamental statistical function of database cannot satisfy. So machine learning algorithms are in need.

Solution. MADlib was originally the SQL for Analytics tool set developed by UCB. It has implemented over 50 categories of algorithms for statistical analysis, graph computation and machine learning. Adopting MADlib collaborated with Greenplum will help to solve the problem.

4   Case Study

Since anti-money laundering data loading is done on T+1, 'current day' refers to the day before the actual date, e.g., April 1 is processed for March 31 data. Traditional database calculation considers performance, so it will store the current day data to the temporary table mdl_cur_stif, and the table structure is the same as t_stan_stif which is used to store information about standard transactions.

Description of some fields of t_stan_stif:

- CTIF_ID: The identification number corresponding to each customer.
  Data Type: varchar(64)
- TSTM: The time when the transaction took place.
  Data Type: varchar(14)
- CRAT: Transaction amounts denominated in the actual currency used in the transaction.
  Data Type:decimal(20,2)

Next, take the calculation of three indicators as an example.

4.1 Daily Cumulative Amount

The table ind_amt is used to store the cumulative amount of transactions for each customer for the day. This is achieved by taking the column CRAT of the transaction table mdl_stan_stif and grouping the sums by column CTIF_ID.

Meaning of this indicator: Provides underlying indicator data for calculating multi-day dimensional metrics and is used to monitor sudden transactions on dormant accounts and identify money laundering scenarios where there is a significant increase in short-term account transactions.

Q1(Take the example of processing the data of date 21):

```
01 |   SELECT ctif_id,
02 |          sum(crat)
03 |   FROM
04 |     (SELECT *,
05 |             to_timestamp(tstm, 'yyyyMMddHH24MISS')::
           TIMESTAMP AS tstmp
06 |       FROM stif) AS stiftemp
07 |   WHERE date_part('day', tstmp)=21
08 |   GROUP BY ctif_id;
```

4.2 Daily Accumulation-Whole Hundred-Number of Transactions

The table ind_100_num is used to store the number of transactions where the transaction amount is an integer multiple of 100 for each customer on that day. The number of transactions is calculated by taking the cloumn CRAT of the mdl_stan_stif transaction table, and the transaction records where the crat remainder of 100 is 0, grouped by customer ID CTIF_ID.

Meaning of this indicator: This type of transaction is usually mixed in multiple transactions, and the number of whole hundred transactions accounted for a large proportion, commonly used to identify drug transaction money laundering scenarios.

Q2(Take the example of processing the data of date 21):

```
01 |   SELECT ctif_id,
02 |          count(crat)
03 |   FROM
04 |     (SELECT *,
05 |             to_timestamp(tstm, 'yyyyMMddHH24MISS')::
           TIMESTAMP AS tstmp
06 |       FROM stif) AS stiftemp
07 |   WHERE date_part('day', tstmp)=21
08 |     AND mod(crat, 100)=0
09 |   GROUP BY ctif_id;
```

4.3   Daily Accumulation Amount Between Time 23:00 to 6:00

Table ind_2306_amt is used to store the amount of transactions that occurred between 23:00 to 6:00 for each customer, then take the column TSTM of the transaction table t_stan_stif and determine the records that match the time range, and take the column CRAT of the table and sum them grouping by the customer number CTIF_ID.

Meaning of this indicator: Abnormal transactions at sensitive times, commonly used to identify drug transactions, online gambling money laundering scenarios.

Q3(Take the example of processing the data of date 21):

```
01 |  SELECT ctif_id,
02 |         sum(crat)
03 |  FROM
04 |    (SELECT *,
05 |           to_timestamp(tstm, 'yyyyMMddHH24MISS')::
           TIMESTAMP AS tstmp
06 |      FROM stif) AS stiftemp
07 |  WHERE date_part('day', tstmp)=21
08 |    AND (date_part('hour', tstmp)>=23
09 |         OR date_part('hour', tstmp)<6)
10 |  GROUP BY ctif_id;
```

5   Experiment

5.1   Experiment Setup

A total of 6 machines are allocated, including 1 master node and 5 segment nodes. CPU type is Intel(R) Xeon(R) Gold 5220 CPU@2.20 GHz, each machine is allocated 8 CPU cores, 32 GB RAM and 500 GB mechanical hard disk storage. The master node is responsible for command distribution and result integration only, while the five segment nodes are responsible for the actual storage and computation.

5.2   Data Loading and Querying

gpfdist is a Greenplum database parallel file distribution program. Use gpfdist to serve external table files to all Greenplum database Segment in parallel. Each time load 5 million data 100,000 users per day for a total of ten days to 50 million 1 million users.

The query process uses dbeaver to query the data in the database remotely. At the same time, in order to ensure that the query is executed on all data and the number of rows of the final result is fixed so that the data transfer does not interfere with the query time statistics, we use "limit" to limit the number of rows of the query displayed by dbeaver, thus ensuring that the time tested is the time used to query the data in the database.

5.3 Experiment Results

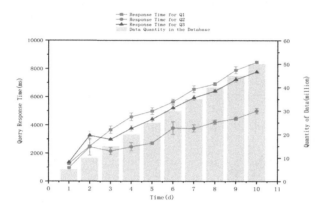

Fig. 1. Query response time/Time

The increase in the daily data volume is an increase of 100,000 customers (50,000 for individual type customers and 50,000 for institutional type customers) per day for a total of 5 million transactions.

According to Fig. 1, it can be seen that the data query response time is linearly related to the increase of data volume. The slope of the growth curve of response time using greenplum database is low compared with the traditional type of database, which is basically consistent with the expected effect.

6 Conclusion

This paper is an application case to build the big data analysis solution for anti-money laundering on open source distributed database Greenplum. By utilizing the share-nothing MPP structure and the software ecosystem of Greenplum, a solution of high-efficiency big data analysis through large-scale parallel computing is designed. Several key transaction monitoring statistical indicators for anti-money laundering is calculated in the test cases. The Experiment results show that, compared with traditional costly solutions based on Hadoop or mySQL, this way can not only reduce the cost in technical stack building and in code programming, but also get the analysis results in just few seconds. Besides, the response time increases linearly with the amount of data, not exponentially as other traditional solutions. All these benefits make it a feasible solution for the similar big data analysis tasks.

Due to the limited time and available experimental resources, the deficiency of the existing work is that several but not all representative indicators are selected for practical application verification. And the load test data set is simulates a representative scene of 10 consecutive days instead of a longer time range. In the future, details of the solution can be optimized on the basis of increasing experiments and application verification, in order to improve the amount of loaded data along with the universality and ease of use in a variety of application scenarios, as well as enhance its practical application value.

Acknowledgements. This work is supported by Open Research Project of CCF (China Computer Federation): Design of distributed data architecture for financial big data analysis (NO. CCFIS2020-04-01).

References

1. Albertini, O., et al.: Image classification in greenplum database using deep learning. In: Proceedings of the ACM SIGMOD Rec, pp. 1–4 (2020)
2. greenplum db: Greenplum github. [EB/OL]. https://github.com/greenplum-db/gpdb. Accessed 16 May 2022
3. Dean, J., Ghemawat, S.: Mapreduce: simplified data processing on large clusters. In: Brewer, E.A., Chen, P. (eds.) 6th Symposium on Operating System Design and Implementation (OSDI 2004), San Francisco, California, USA, 6–8 December 2004, pp. 137–150. USENIX Association (2004). http://www.usenix.org/events/osdi04/tech/dean.html
4. Elastic: what-is elasticsearch. [EB/OL]. https://www.elastic.co/cn/what-is/elasticsearch. Accessed 16 May 2022
5. Foundation, A.S.: Hdfs architecture. [EB/OL]. https://hadoop.apache.org/docs/current/hadoop-project-dist/hadoop-hdfs/HdfsDesign.html. Accessed 16 May 2022
6. Gollapudi, S.: Getting started with Greenplum for big data analytics. Packt Publishing Ltd. (2013)
7. Lyu, Z., et al.: Greenplum: a hybrid database for transactional and analytical workloads. In: Li, G., Li, Z., Idreos, S., Srivastava, D. (eds.) SIGMOD 2021: International Conference on Management of Data, Virtual Event, China, 20–25 June 2021, pp. 2530–2542. ACM (2021). https://doi.org/10.1145/3448016.3457562
8. Redis: Introduction to redis. [EB/OL]. https://redis.io/docs/about/ Accessed 16 May 2022
9. VMware: Greenplum community. [EB/OL]. https://cn.greenplum.org/. Accessed 16 May 2022
10. Zhao, X., Lei, Z., Zhang, G., Zhang, Y., Xing, C.: Blockchain and distributed system. In: Wang, G., Lin, X., Hendler, J., Song, W., Xu, Z., Liu, G. (eds.) WISA 2020. LNCS, vol. 12432, pp. 629–641. Springer, Cham (2020). https://doi.org/10.1007/978-3-030-60029-7_56

Blockchain

Efficient Multi-party Privacy-Preserving Record Linkage Based on Blockchain

Haoshan Yao[1], Haiping Wei[1], Shumin Han[2(✉)], and Derong Shen[3]

[1] School of Information and Control Engineering, Liaoning Petrochemical University, Fushun 113001, China
[2] School of Artificial Intelligence and Software, Liaoning Petrochemical University, Fushun 113001, China
hanshumin_summer@yeah.net
[3] School of Computer Science and Engineering, Northeastern University, Shenyang 110167, China
shenderong@ise.neu.edu.cn

Abstract. With the explosive growth of data, it is increasingly important to integrate data. Privacy-preserving record linkage (PPRL) refers to linking multiple data sources, matching the same entity to be shared by all parties, without disclosing other data. However, most existing PPRL methods rely on an untrusted party to generate matching records, which may lead to privacy leakage and is difficult to ensure the security of linkage. Therefore, an efficient multi-party PPRL method based on Blockchain is proposed. First of all, the data is encoded into Bloom Filters and then split to reduce the amount of information shared during the comparison step of PPRL. Then, homomorphic encryption technology is adopted to further protect data privacy. To improve the efficiency, we construct optimized binary storage trees, which store the records to calculate the similarity, to reduce the number of comparisons between records. In our method, an auditable protocol deployed on the Blockchain is introduced, to detect malicious attacks by untrusted parties. Experimental results show that the proposed method has high linkage quality and efficiency, with strong security of linkage.

Keywords: Record linkage · Privacy-preserving · Bloom filter · Blockchain · Homomorphic encryption

1 Introduction

With the high popularity of Internet applications, the data generated by organizations and individuals has exploded. The primary task in the era of big data is to reduce data redundancy and achieve data sharing. Record linkage, also known as entity matching, record matching [1], refers to the technique of matching records from multiple databases that refer to the same entities in the real world. Record linkage is widely used in financial, administrative, medical, government, and other fields. However, when data sets from multiple data sources are linked, they are extremely vulnerable to frequency attacks and differential attacks, which pose a great threat to personal privacy or sensitive information.

© The Author(s), under exclusive license to Springer Nature Switzerland AG 2022
X. Zhao et al. (Eds.): WISA 2022, LNCS 13579, pp. 649–660, 2022.
https://doi.org/10.1007/978-3-031-20309-1_57

Therefore, the privacy protection issues involved in record linkage have promoted the development of privacy-preserving record linkage technology (PPRL) [2, 3]. The PPRL technology means that in the process of linking data from multiple data sources, only matching entities are shared by each data source, and other unmatched data are not leaked. For example, in terms of medical treatment, the same patient can realize the mutual identification of diagnosis and treatment information in different hospitals, which is conducive to a more accurate analysis of the disease.

At present, the existing PPRL methods can guarantee data privacy and security issues in the record linkage to a certain extent. On the one hand, most existing PPRL methods use the Bloom Filter technique to protect data, which is efficient but vulnerable to frequency and cryptanalysis attacks. On the other hand, most existing PPRL methods use a semi-trusted third party (STTP) to calculate similarity. However, it is not completely reliable and secure [4–6]. At present, Blockchain technology is introduced to resolve STTP's existing problems. The disadvantage is that the matching time is expensive, the efficiency is low in the big data environment, and there is still the danger of malicious parties stealing private data. Therefore, it is an urgent problem to develop a more secure and efficient PPRL scheme that can resist third-party malicious attacks [7].

In view of the problems of the above methods, this paper proposes an efficient multi-party PPRL (MP_PPRL) method based on Blockchain. The method proposed in this paper ensures that while strengthening privacy protection, the linkage efficiency is improved, and it has more practical application value in the era of big data.

The contributions of this paper are:

- The use of homomorphic encryption technology to encrypt the Bloom Filter can ensure that the process of PPRL is more secure and reliable, and the introduction of Blockchain technology to verify the credibility of the third party can improve the security of the linkage.
- An optimized binary storage tree is used to store the data records, and the Jaccard similarity function is used to calculate the distance value between two records, which effectively reduces the number of comparisons between records after the introduction of the Blockchain, and improves the linkage efficiency.
- Experimental evaluation conducted on a real-world dataset shows our method in this paper has the advantage of keeping high efficiency. We compare our method with other techniques and demonstrate the increases in security and accuracy.

2 Related Work

At present, PPRL methods at home and abroad are mostly aimed at improving the linkage quality and scalability, and the research is still in the preliminary stages. Most of the early PPRL methods used the embedding space technology to embed the data to be linked into the progress space [8–10] to protect the similarity value between data attributes, but this technology can only effectively solve the problem of linking numerical data. In the next generation literature [11], Durham et al. proposed to use Bloom Filter to encode the data, which improves the linkage efficiency, but cannot resist the frequency attack in the linkage process. After that, the methods in [12, 13] are all based on the

PPRL technology of Bloom Filter coding. In [12], Han et al. proposed a combination of Bloom Filter coding technology, inspection mechanism, security summation, and Dice similarity method, to the data records of multiple data sources are securely and reliably linked, but this scheme relies on STTP for similarity calculation, and there is a problem of STTP malicious tampering and leaking data privacy. In [13], Vatsalan et al. proposed a multi-party PPRL method using Bloom Filter for exact matching, distributing the computation among the participants, but only for exact matching.

Most of the current PPRL methods assume an honest but curious (HBC) security model, but there is a privacy problem of STTP malicious tampering with data leakage. In [13], Thiago et al. proposed a privacy protection protocol, in which each participant splits the Bloom Filter into an equal amount of Splitting Bloom Filters, and converts STTP into intelligent data hosted on the Blockchain. The contract, in which each participant and STTP only use Splitting Bloom Filter for similarity calculation, thereby reducing the amount of data shared in the process of PPRL similarity calculation, and reducing the possibility of cryptanalysis attacks, but the disadvantage of this scheme is that it is repeatedly Iterative calculation of similarity has high matching time cost and low linkage efficiency, and there is still a hidden danger of malicious participants stealing information according to the Splitting Bloom Filter, which cannot guarantee the privacy and security of data. Aiming at the problems of high computational cost and privacy leakage in this scheme, this paper designs and builds a storage scheme for optimized binary storage trees, which improves the linkage efficiency by reducing the number of Jaccard similarity function calculations and comparisons; and uses homomorphic encryption technology to encrypt the Splitting Bloom Filter. Each bit of the encoded bit array is encrypted, and then split and similarity calculation is performed, so that the PPRL process has stronger security.

3   Problem Definition

3.1   Problem Formulation

Definition 1 (MP-PPRL) Suppose there are p participants $(P_1, ..., P_p)(p \geq 3)$, whose data sources are $(D_1, ..., D_p)$ respectively, each data source is composed of a set of different entities $D_p = \{e_1, ..., e_n\}_{\neq}$, identify the record groups representing the same entity in different data sources. The private data has not been disclosed.

Definition 2 (Bloom Filter [15], BF) is a random data structure that consumes less time and space, consisting of binary vectors and random mapping functions. In the initial state, BF is a bit array of m length bits, and each bit is set to 0. For a set $S = \{x_1, x_2, ..., x_n\}$ containing n elements, BF uses k independent hash functions $h_1(x_i), h_2(x_i), ..., h_k(x_i)$, maps each element x_i to the range $\{1, ..., m\}$, and sets the corresponding position of BF to 1.

Definition 3 (Jaccard similarity function [12]) is a set similarity measure function, which is usually used to calculate the similarity between two BFs. The larger the Jaccard coefficient value, the higher the similarity. Assuming the similarity of P similar bit arrays $B_1, B_2, ..., B_p$, formula (1) is as follows:

$$Jaccard(B_1, ..., B_p) = \frac{B_1 \cap B_2 ... \cap B_p}{B_1 \cup B_2 ... \cup B_p} \tag{1}$$

where $|B_i|$ represents the number of 1 in the i-th bit array.

Definition 4 (Splitting Bloom Filter [14], SBF) the SBF intends to split the original BF in splits $[\phi^0, ..., \phi^{s-1}]$, where each split will have a fraction of the original BF length. SBF's basic idea is to perform iterative similarity computation using only a small piece (split) of the original BF to reduce the amount of information shared during the comparison step of PPRL. Assuming the BF of the entity e is $BF(e) = e^\tau$, the SBF is $SBF(e^\tau, s) = [\phi^0, ..., \phi^{s-1}]$, where $\phi^i = \left[b_j, ..., b_{j+\left(\frac{l}{s}-1\right)}\right]$, $j = i \times \frac{l}{s}, \forall(i)|0 \le i \le s - 1$.

Based on SBF, the similarity $Jaccard\ (e_a^\tau, e_b^\tau)$ is calculated between two different entities, and their split similarity can be calculated separately, as shown in formula (2):

$$Jaccard\_SBF(e_a^\tau, e_b^\tau, s) = \frac{1}{s} \sum_{i=0}^{s} \frac{|\phi_a^i \cap \phi_b^i|}{|\phi_a^i \cup \phi_b^i|} \tag{2}$$

where s represents the number of splits, ϕ_a^i and ϕ_b^i represents the i-th split in $SBF(\phi_a^i, s)$ and $SBF(\phi_b^i, s)$, respectively.

On the basis of acknowledging that the segmentation of SBF is different, we also assume that the similarity of the segmentation is slightly different from the BF similarity, that is, there is an error value $error\ (\varepsilon)$, $0 \le \varepsilon \le 1$. The calculation is shown in formula (3):

$$Jaccard\ (e_a^\tau, e_b^\tau) = Jaccard\ (\phi_a^i, \phi_b^i) + \varepsilon \tag{3}$$

To illustrate the similarity difference between SBF and BF, it is shown in Fig. 1. The similarity of the two BF is 0.55.

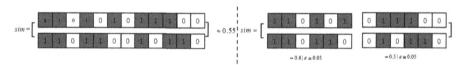

Fig. 1. Similarity and error of BF and SBF.

Definition 5 (Blockchain [16]) Blockchain originated from Bitcoin and was first proposed by Satoshi Nakamoto in 2008. Blockchain is a technology that maintains the states and the historical transactions, using a peer-to-peer network, without any central node to enforce compliance with the rules. Blockchain provides immutable storage of transactions in a chain of blocks, by storing data records in blocks that are linked using cryptography tools, i.e., the previous block hash, the transaction owner signature, and the identifier (id) of the miner that executed the transaction. In the Blockchain, the block content is difficult to tamper with [17, 18].

4 An Efficient MP_PPRL Method Based on Blockchain

The MP-PPRL method designed in this paper can be divided into a data preparation module, an approximate matching module, and an auditable module.

The main function of the data preparation module is to set various parameters, use BF to encode the dataset, and split it into SBF, which reduces the amount of data shared by the participants. The SBF is encrypted with homomorphic encryption technology, which further reduces the possibility of cryptanalysis. The flow chart of the data preparation module is shown in Fig. 2.

At first, each participant presets the number of splits s, thresholds α and error value ε, and encodes the data into BF. Secondly, the participants split BF into SBF, and then use the homomorphic encryption technology to encrypt the SBF, to obtain the ciphertext for homomorphic calculation. Each participant only sends a segment of SBF into the STTP of the Blockchain to calculate the similarity, and stores the possibly similar entity ID and similarity value in the table and sends it to each participant.

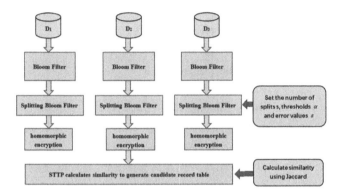

Fig. 2. Flow chart of data preparation module.

The main function of the approximate matching module is: based on the table ζ obtained above, each participant alternately exchanges the remaining SBFs of the entities in the table, stores them in the optimized binary storage tree, calculates the similarity, and judges whether they are from the same entity, and finally filters out the matching entities.

The main function of the auditable module is to convert STTP into a piece of code executed in the Blockchain environment. And we use the homomorphically encrypted SBF to calculate the similarity, which reduces the amount of data shared by each participant.

4.1 Data Preparation Module

At first, each participant P needs to agree on the input parameters and ensure that the parameters are consistent. The parameters used in this method and their meanings are shown in Table 1. Each participant anonymizes the entity to obtain e^τ, and randomly generates a unique ID for each entity. We use k hash functions to map the dataset D_p^τ to N_i bit array of length l. And then split BF into s SBF ϕ, and SBF is encrypted by homomorphic encryption technology, to obtain the ciphertext for homomorphic calculation $\phi' = E(\phi)$.

Each participant agrees to send the homomorphically encrypted SBF ϕ_i' to STTP deployed in the Blockchain environment, STTP calculates the similarity according to formula (1), and stores entities' ID and similarity values with similarity value $> a$ in a table ζ. Thereby, the data records that are approximately matched are initially screened out, the number of similarity calculations and comparisons in the subsequent matching is reduced, and the linkage efficiency of the MP_PPRL method is improved.

Table 1. MP_PPRL approach parameter table.

Parameter	Description
P	Participant of PPRL
e	Entity
D_p	Dataset of participant p
D_p^τ	Anonymized dataset of participant p
l	Bloom filter length
k	Bloom filter hash functions $h_1, h_2, ..., h_k$
s	Number of splits
ϕ	Splitting bloom filter SBF
ϕ'	Homomorphic encryption value of SBF
ζ	List of entities (ids) pairs with their similarity values
φ	A set of SBF (ϕ')
ε	Error
i	The i-th participant

4.2 Approximate Matching Module

In the approximate matching module, according to the ID values of the entities stored in the table ζ, the participants alternately exchange the corresponding splits ϕ', and each time a segment of SBF is exchanged, each participant receives a set of splits φ, where $|\varphi| = \frac{s-1}{|P|}$. Since the data records sent by all participants are SBFs after homomorphic encryption, data privacy is fully guaranteed in the approximate matching module.

To improve the efficiency of similarity calculation, in this paper, we use an optimized binary storage tree to reduce the number of similarity calculations. The left subtree of the optimized binary storage tree stores its own homomorphically encrypted SBF, while the right subtree stores the homomorphically encrypted value of the SBF sent by other participants, as shown in Fig. 3.

Based on the optimized binary storage tree, each participant performs the similarity calculation, according to Algorithm 1.

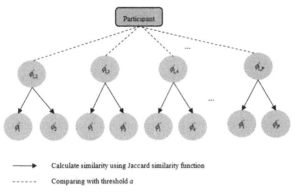

Fig. 3. Optimized binary storage tree.

Algorithm 1 Approximate matching algorithm

Input: Part of the split proliferating filter code encrypted by the P participants

Output: The table ξ of closely matched candidate record groups

 1: Build an optimized binary tree to store the split bloom filter codes of each participant

 2: **for** each fixed encrypted value ϕ_i in $D_i, 1 \leq i \leq P$ **do**

 3:     **for** others fixed encrypted value ϕ_j in $D_j, j \neq i, 1 \leq j \leq P$ **do**

 4:        $LeftChildNode \leftarrow \phi_i, 1 \leq i \leq P$

 5:        $RightChildNode \leftarrow \phi_j, 1 \leq j \leq P$

 6:        Calculate the Jaccard similarity value between left and right subtree split bloom filter codes

 7:        $Jaccard(\phi_i, \phi_j) = \frac{\phi_i \cap \phi_j}{\phi_i \cup \phi_j}$

 8:        Compared with the similarity γ in the table ξ, ϵ is the error value error

 9:        **if** $|Jaccard(\phi_i, \phi_j) - \gamma| < \epsilon$ **then**

10:           Indicates that ϕ_i, ϕ_j two split bloom filter codes match

11:        **end if**

12:     **end for**

13: **end for**

Lines 1 to 5 are to build an optimized binary storage tree to store the encrypted values of the sensitive information of each participant. Lines 6 to 10 calculate the similarity of the left and right subtrees through the Jaccard similarity function. According to the transitive characteristics of the optimized binary storage tree, if ϕ'_1 and ϕ'_2 are matched, ϕ'_1 and ϕ'_8 are matched, it can be inferred that ϕ'_2 and ϕ'_8 are matched. Therefore, the similarity of different SBFs can be inferred without calculating the record pairs composed of all the right subtrees. This mechanism can significantly reduce the number of Jaccard similarity calculations and can be better applied to the big data environment.

4.3 Auditable Module

Since the previous studies have assumed an STTP to perform similarity calculation, it poses a huge threat to data privacy. Therefore, all parties involved in this article will audit the STTP. To make the similarity calculation performed by STTP auditable, the Blockchain converts STTP into a piece of code that is executed in the Blockchain environment. By considering the STTP as a Smart Contract, we offer three important characteristics inherited from the Blockchain.

The first is the tamper-evident characteristic of the STTP. In other words, once the STTP Smart Contract is deployed to the Blockchain, it cannot be modified; in other terms, it is impossible for the malicious party to change the Smart Contract code. Therefore, STTP can be deployed on the Blockchain.

The second characteristic is the decentralized execution model. This characteristic enables the STTP to be executed in different machines eliminating the need for a centralized computing environment. In other words, the PPRL parties could create a private network with their machines, and the computations will be executed in all machines of the network, without a central server.

The final characteristic is the audibility (transparency) of the Blockchain. Once that STTP is stored in the Blockchain, the STTP code, inputs, and outputs can be read (audited) by the Blockchain members. This advantage provides the MP_PPRL audibility capability while posing as a privacy disadvantage by making available private data to every Blockchain member.

As shown in Algorithm 2, in order to protect entity privacy from STTP attacks of MP_PPRL participants, this paper uses SBF to calculate the similarity to reduce the amount of shared information. At the same time, each participant will check the similarity calculation value, to judge whether the other party is credible.

Algorithm 2 Auditable Module

Input: Encrypted value ϕ_i', φ_p, threshold α, a table ξ consisting of approximately matching candidate record groups

Output: The set M composed of the true matching candidate record groups

1: The input of each participant is split bloom filter code ϕ_i', STTP verifies the input of each participant

2: STTP performs similarity calculation and stores entity pairs with high similarity probability in table ξ

3: Send the table ξ to all participants

4: The participants exchange the rest ϕ_i' of the entities in the table

5: $Jaccard(\phi_i, \phi_j) = \frac{\phi_i \bigcap \phi_j}{\phi_i \bigcup \phi_j}$

6: Compared with the similarity γ in the table ξ , ϵ is the error value error

7: **if** $|Jaccard(\phi_i, \phi_j) - \gamma| < \epsilon$ **then**

8:    Indicates that STTP is trusted

9: **end if**

10: The participants exchange the similarity calculated in the previous step and audited whether the participants were trustworthy

11: $Jaccard(e_\alpha^\tau, e_\beta^\tau, s) = \frac{1}{s} \sum_{i=0}^{s} \frac{|\phi_a^i \bigcap \phi_b^i|}{|\phi_a^i \bigcup \phi_b^i|}$

12: **if** $Jaccard(e_\alpha^\tau, e_\beta^\tau, s) > \alpha$ **then**

13:    Add it to M

14: **end if**

Lines 1 to 3 are STTP using formula (1) to calculate the similarity of the data, store the entities with similarity $> a$ in the table and send it to each participant. Lines 4 to 9 show that each participant uses the formula (1) to calculate the similarity between SBFs according to the entity ID value in the table, and compares it with the similarity in the table. If the difference is less than the error value ε, it means that STTP is acceptable. Lines 10 to 13 are the exchange of similarity values calculated in the previous step between the participants to check whether each participant is credible. Afterward, the similarity value of the BF is calculated by using the formula (2), and the entities with the similarity value > a are stored in the matching entity set.

5 Experiments and Analysis

5.1 Experiment Preparation

Experimental Dataset. The dataset used in this experiment is the North Carolina voter registration list (NCVR), which can be downloaded from ftp://alt.ncsbe.gov/data. The dataset used is real public voter records information. In this paper, we use Python (version 3.7) to implement this method.

All tests were conducted on a computer server with a 64-bit, 8.0G of RAM Intel Core (3.30 GHz) CPU. The programs and test datasets are available from the authors.

Evaluation Criteria. Three well-known quality measures (*Precision, Recall, F-measure*) from Information Retrieval are employed to evaluate the effectiveness of our approach. *Runtime* is used to evaluate the scalability of our method. *Precision* refers to the ratio of the actually matched record groups to the candidate record groups. *Recall* refers to the ratio of the number of actually matched record groups to the actually matched record groups in the candidate record groups. *F*-measure is used to synthesize the results of the evaluation method, as shown in Eq. 4:

$$F = 2 \times \frac{Recall \times Precision}{Recall + Precision} \tag{4}$$

Comparison Method and Parameter Determination. In this paper, we compare our approach (MP_PPRL) with the existing methods such as ABEL [13], Darham [10], Vatsalan [12], and Karaplperis [9]. The independent variables of the experiment are dataset size, number of participants, and disturbance ratio. Evaluate the changes in the above three indicators. The experimental data disturbance ratio is 30%, and the dataset sizes are selected as 5000, 10000, 50000, 100000, 500000, and 1000000 respectively. The number of participants is selected from 3, 5, 7, and 9 respectively.

5.2 Experimental Results and Analysis

Scalability Assessment. We first evaluate how the running time of our method varies with the size of the data source. Among them, the number of participants is 3, the disturbance ratio is 30%, and the number of splits is 3. Figure 4 (a) below shows that when the number of participants is 3, the *Runtime* of this method is significantly lower than that of Darham, and Vatsalan, but slightly higher than that of Karaplperis, Overall, although our method uses homomorphic encryption technology to protect data privacy,

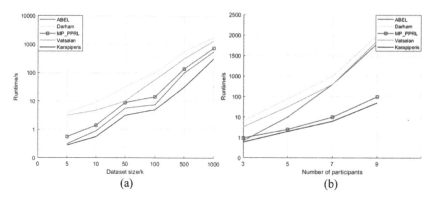

Fig. 4. *Runtime* with different values for (a) dataset sizes and (b) number of participants *P*.

due to the use of an optimized binary storage tree, *Runtime* of our method and the ABEL method are almost equal. Therefore, among the methods, our method in this paper has good scalability in the era of big data.

In the experiment in Fig. 4 (b), the data disturbance ratio is 30%. When the number of participants is 3 and the dataset size is 5000, the *Runtime* of our method and ABEL method are almost the same. With the increase in the number of participants, our method is significantly lower than that of ABEL, Darham, and Vatsalan, and slightly higher than that of Karaplperis. The *Runtime* of these matching algorithms all increase, but in terms of *Runtime* growth rate, the growth rate of our method is smaller than that of other methods, and the growth rate is slower. It shows that using the optimized binary storage tree to calculate the similarity can reduce the number of calculations and comparisons of the similarity. Therefore, our method is efficient in the Big data.

Method Performance Evaluation. In this paper, we evaluate the performance of MP_PPRL in terms of *Precision, Recall*, and *F*-measure, and compare it with the method ABEL to judge the relationship between performance and threshold *a*. The number of participants is 3, the disturbance ratio is 30%, and the number of splits is 3. As shown in Fig. 5, the *Precision* of both methods increases with increasing *a*, and the results are similar. Because both methods compare the SBF similarity with a preset threshold *a*, when the threshold is larger, the *Precision* is higher.

As shown in Fig. 6, the *Recall* of both methods decreases with the increase of *a*. When *a* > 0.4, the *Recall* of the ABEL method is slightly better than that of the MP_PPRL method. This is because the MP_SPPRL method uses an optimized binary storage tree, and does not need to calculate the similarity of the record pairs composed of all the right subtrees, which reduces the number of similarity comparisons.

The variation trend of the *F*-measure of the two methods is shown in Fig. 7. When *a* < 0.4, both methods increase with the increase of *a*, and when *a* > 0.4, both methods increase with *a*. However, the reduction speed of the MP_PPRL method is slightly larger than that of ABEL. But the *F*-measure is still above 0.6, so the MP_PRRL method proposed in this paper has better *Recall* and *Precision*.

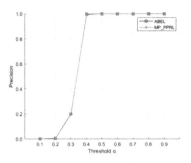
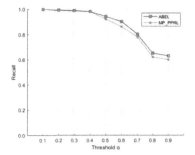

Fig. 5. *Precision* with different *a.*   **Fig. 6.** *Recall* with different *a.*

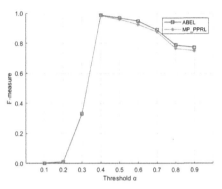

Fig. 7. *F*-measure with different *a.*

Privacy Performance Analysis. In order to ensure the security of PPRL, all similarity calculations in this paper use SBF and homomorphic encryption technique. Therefore, the amount of information shared by the participants is reduced, and the possibility of cryptanalysis is reduced. At the same time, in order to solve the problem of the STTP, in this paper we introduce an auditable module to convert STTP into a smart contract deployed on the Blockchain so that each participant can audit the behavior of STTP, which ensures privacy and security during the linkage process.

6   Conclusion

In this paper, we propose an auditable PPRL method. In the Blockchain environment, SBF and homomorphic encryption techniques are used, which reduces the amount of information shared by all participants. At the same time, an optimized binary storage tree is constructed, which effectively reduces the number of similarity calculations and comparisons, and improves efficiency. Blockchain technology is used to implement the similarity calculation for STTP to provide auditability for the PPRL. Therefore, in this paper, our method can safely and efficiently complete the task of entity matching among multiple data sources. Our method in this paper has good scalability and good linkage quality. In future work, further research will be done to improve linkage quality by filling missing values efficiently, so that it can be better applied in the real world.

References

1. Elmagarmid, A.K., Ipeirotis, P.G., Verykios, V.S.: Duplicate record detection: a survey. IEEE Trans. Knowl. Data Eng. **19**(1), 1–16 (2007)
2. Lai, P., Yiu, S., Chow, K., Chong, C., Hui, L.: An efficient Bloom filter based solution for multi-party private matching. In: Proceedings of the Conference on SAM (2006)
3. Sehili, Z., Rohde, F., Franke, M., et al.: Multi-party privacy preserving record linkage in dynamic metric space. BTW 2021 (2021)
4. Tong, D.N., Shen, D.R., Han, S.M., Nie, T.Z., Kou, Y., Yu, G.: Multi-party strong-privacy-preserving record linkage method. J. Front. Comput. Sci. Technol. **13**(03), 394–407 (2019)
5. Kuzu, M., Kantarcioglu, M., Durham, E., Malin, B.: A constraint satisfaction cryptanalysis of Bloom filters in private record linkage. In: Fischer-, S., Hopper, N. (eds.) PETS 2011. LNCS, vol. 6794, pp. 226–245. Springer, Heidelberg (2011). https://doi.org/10.1007/978-3-642-22263-4_13
6. Zhao, L., Zhang, E., Qin, L.Y., Li, G.L.: Multi-party privacy preserving k-means clustering scheme based on block-chain. J. Comput. Appl. 1−10. http://kns.cn.ki.net/kcms/dtail/51.1307.TP.20220331.1656.004.html
7. Schnell, R., Bachteler, T., Reiher, J.: Privacy preserving record linkage using Bloom filters. MIBM **9**(1), 41 (2009)
8. Wang, Y.Y., Huang, D.Y., Xu, D.X.: Record linking protocol based on privacy-preserving in vetor space. Mod. Electr. Technol. **32**(14), 138–141 (2009)
9. Karapiperis, D., Gkoulalas-Divanis, A., Verykios, V.S., et al.: FEDERAL: a framework for distance-aware privacy-preserving record linkage. IEEE Trans. Knowl. Data Eng. **30**(2), 292–304 (2018)
10. Karapiperis, D., Gkoulalas-Divanis, A., Verykios, V.S.: Distance-aware encoding of numerical values for privacy-preserving record linkage. In: 2017 33th IEEE International Conference on Data Engineering, pp. 135−138 (2017)
11. Darham, E.A., Kantarcioglu, M., Xue, Y.: Composite bloom filters for secure record linkage. IEEE Trans. Knowl. Data Eng. **26**(12), 2956–2968 (2014)
12. Han, S.M., Shen, D.R., Nie, T.Z., Kou, Y., Yu, G.: Multi-party privacy-preserving record linkage approach. J. Softw. **28**(9), 2281–2292 (2017)
13. Vatsalan, D., Christen, P.: Multi-party privacy-preserving record linkage using bloom filters. In: 2014 In: Proceedings of ACM Confernece in Information and Knowledge Management, pp. 1795−1798 (2014)
14. Nóbrega, T., Pires, C., Nascimento, D.C.: Blockchain-based privacy-preserving record linkage enhancing data privacy in an untrusted environment. Inf. Syst. **102**, 101826 (2021)
15. Vatsalan, D., Christen, P., Rahm, E.: Scalable multi-database privacy preserving record linkage using counting bloom filters. In: Proceedings of the 23th International Conference on Information and Knowledge Management, pp. 1795−1798. ACM, New York Press (2014)
16. Zhu, L.H., et al.: Survey on privacy preserving techniques for blockchain technology. J. Comput. Res. Dev. **54**(10), 2170–2186 (2017)
17. Mao, X., Li, X., Guo, S.: A blockchain architecture design that takes into account privacy protection and regulation. In: Xing, C., Fu, X., Zhang, Y., Zhang, G., Borjigin, C. (eds.) WISA 2021. LNCS, vol. 12999, pp. 311–319. Springer, Cham (2021). https://doi.org/10.1007/978-3-030-87571-8_27
18. Hasan, H.R., Salah, K.: Combating deepfake videos using blockchain and smart contracts. IEEE Access **7**, 41596–41606 (2019)

A Blockchain-Based Scheme for Efficient Medical Data Sharing with Attribute-Based Hierarchical Encryption

Xin Li[1], Xiaomei Dong[1]([✉]), Xinhao Xu[1], Guangyu He[2,3], and Shicheng Xu[2,3]

[1] School of Computer Science and Engineering, Northeastern University, Shenyang 110169, China
xmdong@mail.neu.edu.cn
[2] Research Center of Liaoning Promotion for Block Chain Engineering Technology, Shenyang 110179, China
[3] Neusoft Corporation, Shenyang 110179, China

Abstract. Most of the traditional medical information management system adopt centralized storage and management. So, there are some problems such as single point of failure and data sharing difficulty. Considering the advantages of decentralization, non-tampering and traceability, the sharing of medical data can be realized based on the blockchain system. But the risk of privacy leakage will still be faced. In order to protect the privacy of data in sharing and guarantee the medical data controlled by the data owner, we propose an attribute-based hierarchical encryption scheme to realize the access control of medical data. In order to improve the query efficiency of users, we propose a blockchain query index with a combination of skip list and Bloom filter. However, users still may encounter the privacy leakage during the query process. Therefore, supervision institutions are introduced to supervise the historical inquiry records of user groups to prevent them from doing evil things. Finally, the scheme is proved to be effective by relevant experiments.

Keywords: Electronic medical record · Blockchain · Attribute-based hierarchical encryption · Medical data sharing

1 Introduction

Traditionally, most medical data is generated by medical institutions and stored in centralized databases, which brings the risk of suffering single point of failure and vulnerable to attacks. Meanwhile, the real owners of these medical data, i.e. patients, cannot control their medical data. Because their medical data is stored in the hospital's local database, when they go to another hospital, they are unable to provide their previous medical data to the doctors and get adjunctive treatment.

Distributed storage for medical big data was proposed to solve the problems of centralized data storage, but the scheme did not achieve medical data sharing [1]. Then

© The Author(s), under exclusive license to Springer Nature Switzerland AG 2022
X. Zhao et al. (Eds.): WISA 2022, LNCS 13579, pp. 661–673, 2022.
https://doi.org/10.1007/978-3-031-20309-1_58

medical data sharing through cloud storage technology was proposed [2, 3], but they may have the risk of medical data privacy leakage due to the lack of supervision.

Blockchain is essentially a distributed shared database with features such as decentralization, tamper-evident and traceability. In the medical field, storing and managing medical data through blockchain offers greater security compared to cloud storage. However, the storage capacity of each block is limited, and storing the complete Electronic Medical Records (EMRs) on the blockchain will undoubtedly increases the storage overhead. Hybrid storage was applied in some research works [4–6]. The complete data were stored on the storage server, and the blockchain stores the digest of these data, which contains the storage location, hash value, and other relevant information of the complete data. According to these studies, the complete medical data is encrypted and stored on the storage server or blockchain. If someone else can access and decrypt it, he will also get the complete medical data, which may increase the network overhead in the query process and the risk of privacy leakage.

Overall, there are still three major challenges that need to be over-come:

- EMRs are encrypted completely. But some of the contents are not necessary to visitors, which may cause privacy leakage.
- With the increasing number of blocks in the blockchain, it would be inefficient to query them by smart contracts.
- Medical data is sensitive, and there is a lack of regulatory measures for the blockchain that store EMRs. It is a challenge to protect the users' privacy while the relevant agencies are regulating it.

Based on the above issues, we propose an attribute-based hierarchical encryption scheme to solve the above problems. The main contributions can be summarized as follows:

- First, we adopt hybrid storage to reduce the storage overhead of the blockchain. Then we introduce ciphertext-policy attribute-based encryption (CP-ABE) [7] to achieve hierarchical encryption of EMRs, and divide the complete EMR into three sub-medical records according to the actual demand, so as to achieve fine-grained privacy protection of EMRs. We also add the update of the key to the CP-ABE.
- Second, we design a query index that combines skip list with Bloom filter to improve the query efficiency for users.
- Finally, we introduce anonymous IDs for users who queries the block data to protect privacy under the regulating.

The rest of the paper is organized as follows. In Sect. 2, we discuss related works on existing medical data sharing. In Sect. 3, we describe our scheme in detail. In Sect. 4, we analyze the scheme and perform an experimental evaluation, and the conclusion is presented in Sect. 5.

2 Related Work

In some blockchain-based medical data sharing schemes, smart contracts were adopted for access control instead of cryptographic algorithms [8–10]. However, smart contracts may have vulnerabilities.

Zeng Chen et al. [6] used proxy re-encryption to ensure the secure sharing of medical data, which increases the cost of patient participation. Lu XiaoFeng et al. [10] applied CP-ABE to blockchain to achieve access control. But it's not efficient enough. H. Guo et al. [11] proposed an attribute-based multiple signatures (ABMS) based on attribute-based encryption. Because of the encryption of the whole medical data, it has the risk of privacy leakage. Besides using CP-ABE to achieve access control, Sun. Y et al. [12] also designed an index structure to improve query efficiency. However, the complete data is encrypted and stored directly on the blockchain, which also undoubtedly increases the storage burden. J. Tao et al. [5] implemented a distributed identity management scheme based on attribute encryption, which ensures a more realistic membership management. But its overhead time is large. J. Liu et al. [13] proposed an attribute-based searchable encryption scheme. But with the increasing number of attributes, it is less efficient in encryption.

As far as we know, there is no such scheme that combines the idea of hierarchical permission with CP-ABE. In this paper, we propose a blockchain-based hybrid storage scheme for implementing fine-grained storage and access control of EMRs. We also design an efficient query index. To enhance the security supervision of data queries, we extend another blockchain for storing users' query records and anonymize the queries to protect the privacy of data users.

3 Description of the Proposed Scheme

In our scheme, we adopt a hybrid storage method. The cipher of medical data is stored in off-chain databases and the key information such as the location of these ciphers is stored on the blockchain. The CP-ABE is used to protect the privacy of patients' EMRs. A query index structure for the user's quick query is designed. This scheme allows the supervision institution to supervise the query history.

3.1 System Model

In our scheme, an EMR sharing system includes seven entities: medical institution, key generation center, authoritative supervision institution, data owner, data user, main-blockchain and sub-blockchain. The system model is illustrated in Fig. 1.

The characteristics and functions of each entity are described as follows:

1. Key Generation Center (KGC): responsible for generating system public parameters and creating master system key MK. At the same time, the KGC generates key pairs for patients and data users.

2. Medical Institutions (MI): They consist of hospitals with medical capacity. They provide medical services to patients and generate EMRs of patients. After the registration of a hospital, KGC generates public-private key pairs for the hospital. Each hospital generates a set of attributes for each doctor to describe their identity, and then the hospital generates a private key for each doctor to use for decryption based on their identity attributes. Meanwhile the hospital provides storage services for the patient's cipher.

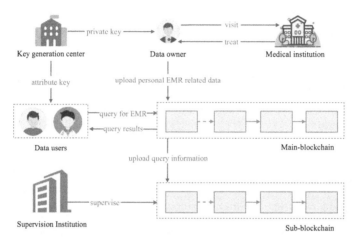

Fig. 1. System model

3. Supervision Institution (SI): responsible for maintaining the sub-blockchain, which stores the query information of the data user to the main-blockchain. The supervision institution uses the public key of the user to encrypt its identity ID, generates the anonymous identity ID′, and maintains the mapping table of ID and ID′ locally.
4. Data owner (DO): mainly composed of patient groups. The complete EMR is divided into three parts, and patients use three different symmetric keys to encrypt each part of data, store the cipher into the local database of the hospital. After getting the storage location, patients use CP-ABE to encrypt these data, and set the access policies and put them into the cipher. Then they are uploaded to the main-blockchain.
5. Data user (DU): mainly composed of medical staff of each hospital. They all have their own identity attributes generated by their hospitals and their keys.
6. Main-Blockchain (MB): used for storing the information related to the EMRs and returning the results of the query according to the query request. Then MB sends the query information to the sub-blockchain.
7. Sub-blockchain (SB): The sub-blockchain stores the history information of data users querying the MB. It is mainly monitored and maintained by the SI.

3.2 Hierarchical Encryption

We divide the complete EMR data into three parts, namely, the patient's personal identity, medical examination, diagnosis and treatment results. The personal identity involves the patient's sensitive privacy information, including name, home address, telephone number, etc. The medical examination includes gender, date of birth, medical history, symptoms, etc. The diagnosis and treatment results include the doctor's diagnosis and treatment plan, as well as the use of drugs.

We set different access policies for different types of medical data through CP-ABE to achieve hierarchical access control. The corresponding storage structure is also designed on the blockchain to apply this hierarchical access control. In fact, it is easy to know that the most sensitive personal information is no-very meaningful to assist treatment, and the personal data also involves very sensitive privacy. Therefore, we do not store personal data on the blockchain.

When a patient goes to a hospital, the hospital will concatenate the hospital number with the patient's ID number as the patient's ID (i.e. PID) for the hospital visit. We stipulate that EID = Hash (PID | PSW), Hash () represents the hash function and PSW is the patient's self-set password string. So, only the patient knows his or her own EID, which guarantees that the patient has complete ownership of his or her own EMR. When a requestor wants to access the patient's EMR, he or she must obtain the patient's EID, i.e. he or she must get the patient's authorization.

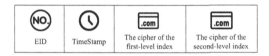

Fig. 2. The structure of the record in main-blockchain

With the above analysis, we design the storage structure of each record in MB as shown in Fig. 2. The first level of index cipher is the cipher of the index of medical examination, and the second level is the cipher of the index of treatment results.

We support fine-grained access control which consists of the following five algorithms, and the symbols involved in the algorithms are shown in Table 1.

Let G_1 be a multiplicative cyclic group of prime order p, and g is a generator of G_1. Let G_2 be a multiplicative cyclic group of prime order q. Let $e: G_1 \times G_1 \rightarrow G_2$ denotes the bilinear map. The security parameter λ determines the size of the group G_2. We also define the Lagrange coefficient: $\Delta_{i,S}(x) = \prod_{j \in S, j \neq i} \frac{x-j}{i-j}, i \in Z_p$. S is a set of members belonging to Z_p We additionally employ a hash function $H: H : \{0, 1\}^* \rightarrow G_1$. The specific construction steps are as follows:

1. Setup $(1^\lambda) \rightarrow (MK, PK)$. KGC outputs the master key MK and the public key PK by operating this Setup algorithm and entering a random security parameter λ.

    KGC sets two random numbers $\alpha, \beta \in Z_p$ and then computes: $PK = (G_1, g, h = g^\beta f = g^{1/\beta}, e(g, g)^\alpha), MK = (\beta, g^\alpha)$.

2. KeyGen $(MK, Y) \rightarrow SK$. KGC inputs the user's attribute set Y and the master key MK, outputs the private key SK bound to that attribute set Y, and sends it to the user. The function is to choose a random number $r \in Z_p$ for each attribute $j \in Y$, in other

Table 1. Symbols used in this scheme.

Symbols	Meaning of the symbols
PK	The public key
MK	The system master key
SK	The decryption key
AK	The AES symmetric key
m	The plaintext data
C	The AES encrypted cipher
M	The Data to be attribute encrypted
Y	The authorized attribute set Y
T	The access strategy
E_T	The attribute encrypted data

word, to choose a random number $r_j \in Z_p$, and compute the private key:

$$SK = \left(D' = g^{(\alpha+r)/\beta}, \forall j \in Y : D_j = g^r \cdot H(j)^{r_j}, D'_j = g^{r_j} \right) \quad (1)$$

3. Enc (PK, m, AK, T) → E_T. The user inputs the PK, the plaintext m, the symmetric key AK and the access strategy T, the output is encrypted medical data E_T.

   The algorithm uses AES algorithm to generate the cipher C. After the cipher C is stored to the database and the storage location information L is obtained. Then the algorithm generates M by concatenating the L and AK.

   The algorithm then automatically selects a polynomial q_x for each node x in the access control tree T. The polynomial is selected in a top-down manner starting from the root node R. For each node in the access control tree, the order d_x of the polynomial q_x and the threshold value k_x for this node satisfy $d_x = k_x - 1$. Then a random number $s \in Z_p$ is chosen starting from the root node. For the root node R, let $q_R(0) = s$, while the values of the polynomial q_R at the other d_R nodes are defined randomly. Down to the other nodes x, let $q_x(0) = q_{parent(x)}(index(x))$, while the values of the other d_x points are defined randomly. Finally, let Y denote the set of all leaf nodes in the access control tree T.

$$E_T = \left(T, C' = M \cdot e(g, g)^{\alpha \cdot s}, C = h^s, \forall y \in Y : C_y = g^{q_y(0)}, C'_y = H(att(y))^{q_y(0)} \right) \quad (2)$$

4. Dec (E_T, SK) → m/⊥. The user inputs SK and cipher E. If SK satisfies the access policy of cipher E_T, it can be decrypted successfully and output plaintext M. Otherwise, decryption fails and the algorithm outputs ⊥. For the specific calculation, we need to define a recursive algorithm DecryptNode (E_T, D, x). Let $i = att(x)$. If x is a leaf node and $i \in y$, then DecryptNode (E_T, D, x):

$$DecryptNode(E_T, D, x) = \frac{e(D_i, C_x)}{e(D'_i, C'_x)} = \frac{e\left(g^r \cdot H(i)^{r_i}, h^{q_x(0)}\right)}{e\left(g^{r_i}, H(i)^{q_x(0)}\right)} = e(g, g)^{r \cdot q_x(0)} \quad (3)$$

If x is not a leaf node, then the recursive case should be considered. For each child z of node x, let $F_z = \text{DecryptNode}(E, D, z)$. There exists a set S_x and the number of elements in the set is a random value k_x. The nodes in the set are all children of x. Let $i = \text{index}(x)$, $S'_x = \{\text{index}(z) : z \in S_x\}$.

$$F_x = \prod_{z \in S_x} F_z^{\Delta_{i,s'_x}(0)} = \prod_{z \in S_x} \left(e(g,g)^{r \cdot q_{parent(z)}(index(z))} \right)^{\Delta_{i,j'_x}(0)}$$
$$= \prod_{z \in S_x} e(g,g)^{r \cdot q_x(i) \cdot \Delta_{i,s'_x}(0)} = e(g,g)^{r \cdot q_x(0)} \tag{4}$$

Now we have the complete definition of the function DecryptNode, which decryption algorithm should call for the value of the function at the root node. Let $A = \text{DecryptoNode}(E_T, SK, r) = e(g,g)^{r \cdot q_R(0)} = e(g,g)^{r \cdot SS}$, if and only if the cipher properties satisfy the access control tree structure. The specific calculation steps are as follows.

$$\frac{C'}{\frac{e(C,D')}{A}} = M \cdot \frac{e(g,g)^{\alpha \cdot s}}{\frac{e\left((g^\beta)^s, g^{\frac{\alpha+r}{\beta}}\right)}{e(g,g)^{r \cdot s}}} = M \tag{5}$$

Thus, we have completed the decryption: $M = (L, AK)$. The algorithm then automatically performs the decryption of AES after accessing the location information L.

5. Delegate $(SK, \tilde{Y}) \rightarrow \widetilde{SK}$. The algorithm takes a private key SK based on the set of attributes Y and a new set of attributes $\tilde{Y}$, where $\tilde{Y} \subseteq Y$, and outputs a new key $\widetilde{SK}$, which is based on the set of attributes $\tilde{Y}$. The private key is of the form $SK = \left(D, \forall j \in Y, D_j, D'_j\right)$, then the algorithm selects random numbers $\tilde{r}$ and $\tilde{r}_k$, where $\forall k \subseteq \tilde{Y}$. Thus, we can then construct a new private key.

$$\widetilde{SK} = \left(D' = Df^{\tilde{r}}, \forall k \in \tilde{Y} : \widetilde{D}_k = D_k \cdot g^{\tilde{r}} \cdot H(k)^{\tilde{r}_k}, \widetilde{D}'_k = D'_k g^{\tilde{r}_k}\right) \tag{6}$$

3.3 Structure of the Index

In order to improve the efficiency of data query, we design an index structure. Inspired by [14], we use the timestamp in the block header to build a skip list index and construct a Bloom filter based on the EID recorded in the block. The Bloom filter can quickly verify whether an element is in a set. So, after getting the patient's EID and the approximate time of uploading EMR, the data requester can quickly locate the block where the patient's EMR is roughly located based on the skip list. Therefore, we design the block structure of the MB as shown in Fig. 3. Let Num indicate the height of the current block, PreBkHash indicate the hash value of the previous block and DataHash indicate the hash value of the current block.

The query of blockchain in most schemes is to traverse the data in each block one by one, and the query overhead time will become larger and larger as the number of blocks increases. The time complexity of traversing the block head is $O(n)$ in the worst case, where n denotes the number of blocks. The time complexity is $O(\log n)$ when using the skip list. Meanwhile, the query index can quickly verify whether the record is in the block by introducing Bloom filter. It's more efficient than traversing the block. Therefore, the query efficiency is greatly improved.

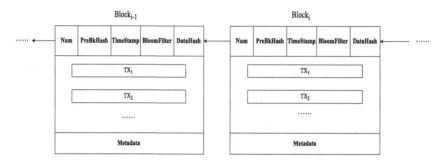

Fig. 3. Blocks in main-Blockchain

3.4 Design of the Sub-blockchain

The SB stores the query history information of the MB and is supervised by a SI. If a user wants to do data query, he first goes to the MB, which triggers a smart contract to respond to his query, while the MB transmits the information related to the user's query to the SB for storage. And then the SI maintains a mapping table of the user's identity. The block structure of the sub-blockchain is shown in Fig. 4.

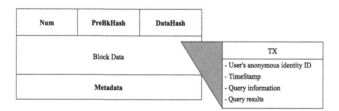

Fig. 4. The block structure of the sub-blockchain

The user's anonymous ID is the cipher of the user's identity encrypted using the user's public key. The mapping table between the user's real identity and the anonymous ID is monitored by the SI. In case of query information leakage, the SI can find out which user's query has caused the information leakage based on the mapping table, and do further investigation and tracing.

4   Security Analysis and Performance Evaluation

4.1   Security Analysis

Analysis of the Security of Medical Data
Collusion Attack
According to our scheme, the secret information is embedded in the cipher instead of the user's private key. The attacker must recover $e(g, g)^{\alpha \cdot s}$ if he wants to crack it. Then the adversary must make the part of the cipher match with the part of the user's private key. In this way, although the value of $e(g, g)^{\alpha \cdot s}$ can be obtained, it is blinded by $e(g, g)^{r \cdot s}$. The value of $e(g, g)^{r \cdot s}$ can be derived if and only if the user's private key satisfies the access structure of the cipher, at which time the collusion attack cannot work because the blind factor of each user's private key blinded by the user is random. So, this scheme is chosen for plaintext security.

Dos Attack
Dos attacks are prone to occur in centralized schemes where an attacker can launch attacks on that centralized system to block normal requests from other users. In this paper, we use a distributed storage scheme of blockchain to solve this problem. The system is guaranteed to operate normally under Dos attacks as long as the attacker cannot disable most of the nodes in the blockchain.

Anonymity of Data Requesters
To protect the identity of the data user, the identity of the user is anonymized. The identity is encrypted with the user's own public key, which can be decrypted only by the user. The supervision institution manages the mapping table between the real identity and the anonymized ID, so that in case of privacy leakage, it can be traced and accountable in time.

4.2   Evaluation of Performance

In this subsection, we experimentally validate our scheme and compare it with several other schemes. Our experimental environment is under Ubuntu 21.04 with Intel(R) Core (TM) i7-9700 CPU @ 3.00 GHz and 16 GB RAM, mainly using the Hyperledger Fabric framework and writing SDK and smart contracts using Golang.

We compare the security and performance of our scheme with other schemes. In this paper, we combine AES and hierarchical encryption to implement a hierarchical CP-ABE scheme on this basis, while supporting the update of private keys. The results of scheme comparison are shown in Table 2.

First, we measure the calculating time. Here we compare two schemes. Our Scheme is to divide the complete data and then encrypt part of it in two times using our encryption algorithm, which is also the scheme proposed in this paper. The second one is the scheme in the paper [7]. Since the running time of AES is different for files with different sizes, we use files with size from $2 * 10^3$ KB to $2 * 10^4$ KB according to the actual storage size of the EMR. The number of attributes is set to 10.

Table 2. Comparison between proposed scheme and other schemes

Scheme	Paper [5]	Paper [7]	Paper [13]	Our scheme
Attribute-based encryption	Y	Y	Y	Y
Hierarchical encryption	N	N	N	Y
Update of the key	N	N	N	Y
Query index	N	N	Y	Y
Anonymous identity	N	N	Y	Y

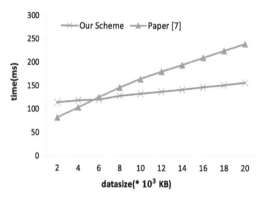

Fig. 5. Cost of calculations for different file size

The experimental results are shown in Fig. 5. We can see that the overall growth rate of the computational overhead of Our Scheme in file encryption and decryption is smaller than that in [7] from Fig. 5. This is because that encryption and decryption combined with AES is more efficient than CP-ABE as the file size gets larger.

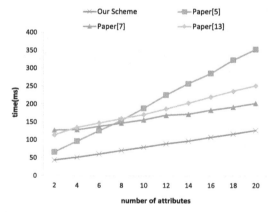

Fig. 6. Cost of calculations for different number of attributes

Second, we set the file size to 10^4 KB and measure the total time for encryption and decryption by changing the number of attributes in the access policy. The experimental results of our scheme comparing with the schemes in [5, 7] and [13] are shown in Fig. 6. Our scheme performs better than the scheme in [7], because in our scheme location information L and an AES key are encrypted in the CP-ABE, which is much smaller than a file with size of 10^4 KB. The scheme in [5] has relatively large overhead because it involves numerous membership negotiations. The calculating cost of the scheme in [13] is large because it additionally computes the index used in searchable encryption in the CP-ABE. All the overhead time grows linearly as the number of encryption attributes increases. But the growth rate of our encryption scheme is smaller.

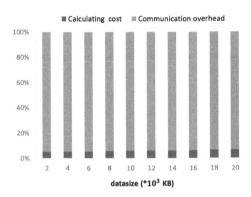

Fig. 7. Percentage of calculating cost

Finally, we also compute the proportion of the encryption and decryption time in the whole operation when running in combination with the Fabric blockchain system. In the Fabric blockchain, three MI organizations and one orderer organization are used. Each MI organization contains three peer nodes, and the orderer organization uses the raft consensus algorithm. From Fig. 7, we can see that the total time for encryption and decryption accounts for less than 10% of the total system running time. It should be noted that in reality the complete EMR may exceed $2 * 10^4$ KB. However, according to the experimental results in Fig. 7. Percentage of calculating cost, the percentage of calculating cost grows slowly. When the file does not exceed $5 * 10^4$ KB, the proportion does not exceed 15%. Therefore, the total cost of encryption and decryption to achieve hierarchical encryption used in our scheme is not obvious to the whole system, thus ensuring the feasibility of this scheme.

5   Conclusion

In this paper, we propose a blockchain-based attribute hierarchical encryption for medical data security sharing. In order to achieve fine-grained access control of the content, the file is divided and hierarchically encrypted to protect the privacy of patients' medical data. In the sharing, we design a data structure combining skip list and Bloom filter to improve the efficiency of data sharing. To enhance the access control of data, we introduce a supervision institution to supervise the query operation of users. We have conducted experiments based on the actual EMR file size. And the experimental results proved that our scheme is effective and applicable to the blockchain-based medical data sharing system under the premise of protecting user privacy. However, this scheme also has some limitations. For example, the system also needs more symmetric keys of AES after hierarchical encryption, which increases the cost of key management for users. At the same time, it is difficult for our scheme to support batch queries for a certain type of medical records. In future research work, we will optimize and improve the above scheme limitations to reduce the cost of key management while implementing hierarchical encryption. As well, we will support more efficient and richer query methods.

Acknowledgement. This work is supported by the Major projects of National Social Science Foundation of China (No. 21&ZD124) and the Open Program of Neusoft Corporation (NCBETOP2001). We also thank anonymous reviewers for the helpful reports.

References

1. De Macedo, D.D.J., Von Wangenheim, A., Dantas, M.A.R.: A data storage approach for large-scale distributed medical systems. In: 2015 Ninth International Conference on Complex, Intelligent, and Software Intensive Systems (CISIS), pp. 486–490 (2015)
2. Wu, Y., et al.: Adaptive authorization access method for medical cloud data based on attribute encryption. In: Ni, W., Wang, X., Song, W., Li, Y. (eds.) WISA 2019. LNCS, vol. 11817, pp. 361–367. Springer, Cham (2019). https://doi.org/10.1007/978-3-030-30952-7_36
3. Marwan, M., Kartit, A., et al.: A cloud based solution for collaborative and secure sharing of medical data. Int. J. Enterp. Inf. Syst. (IJEIS) **14**, 128–145 (2018)
4. Wang, B., Li, Z.: Healthchain: a privacy protection system for medical data based on blockchain. Future Internet **13**(10), 247 (2021)
5. Tao, J., Ling, L.: Practical medical files sharing scheme based on blockchain and decentralized attribute-based encryption. IEEE Access **9**, 118771–118781 (2021)
6. Chen, Z., Xu, W.D., et al.: A blockchain-based preserving and sharing system for medical data privacy. Future Gener. Comput. Syst. **124**, 338–350 (2021)
7. Bethencourt, J., Sahai, A., Waters, B.: Ciphertext-policy attribute-based encryption. In: IEEE Symposium on Security and Privacy (S&P), pp. 321–334 (2007)
8. Xu, L.: Design of a security model for healthcare data based on consortium blockchain. Microcomput. Appl. **37**(9), 143–145+154 (2021). (in Chinese)
9. Alnssayan, A.A., Hassan, M.M., Alsuhibany, S.A.: VacChain: a blockchain-based EMR system to manage child vaccination records. Comput. Syst. Sci. Eng. **40**(3), 927–945 (2022)
10. Lu, X.F., Fu, S.B.: Access control scheme for trusted data based on a combination of attribute-based encryption and blockchain. Netinfo Secur. **21**(3), 7–14 (2020). (in Chinese)
11. Guo, H., Li, W., Meamari, E., et al.: Attribute-based multi-signature and encryption for EHR management: a blockchain-based solution. In: IEEE International Conference on Blockchain and Cryptocurrency (ICBC), pp. 1–5 (2020)
12. Sun, Y., Song, W., Shen, Y.: Efficient patient-friendly medical blockchain system based on attribute-based encryption. In: Wang, G., Lin, X., Hendler, J., Song, W., Xu, Z., Liu, G. (eds.) WISA 2020. LNCS, vol. 12432, pp. 642–653. Springer, Cham (2020). https://doi.org/10.1007/978-3-030-60029-7_57

13. Liu, J., Wu, M., Sun, R., et al.: BMDS: a blockchain-based medical data sharing scheme with attribute-based searchable encryption. In: IEEE International Conference on Communications (ICC), pp. 1–6 (2021)
14. Xu, C., Zhang, C., Xu, J.J.: vChain: enabling verifiable boolean range queries over blockchain databases. In: Proceedings of the 2019 International Conference on Management of Data (SIGMOD 2019), pp. 141–158 (2019)

B-store, a General Block Storage and Retrieval System for Blockchain

Xiaofei Gao[1], Tiezheng Nie[1(✉)], Derong Shen[1], Yue Kou[1], Guangyu He[2,3],
and Shicheng Xu[2,3]

[1] School of Computer Science and Engineering, Northeastern University,
Shenyang 110169, China
`nietiezheng@cse.neu.edu.cn`
[2] Neusoft Corporation, Shenyang 110179, China
[3] Liaoning Blockchain Engineering Technology Research Center, Shenyang 110179, China

Abstract. The emergence of Hyperledger provides a variety of feasible and feature-rich solutions for smart contracts on the blockchain, promoting the popularity and spread of blockchain systems. Similar with Bitcoin and Ethereum, Hyperledger has accelerated the dramatic increasement in the scale of blockchain data. Applications based on blockchain systems need to quickly retrieve transaction data in blocks while processing and analyzing data. Currently, Hyperledger Fabric only provides limited retrieval functions, such as searching for blocks based on single fields like block height, block hash, and transaction hash. Such retrieval functions cannot meet the needs of current blockchain applications. In this paper, we propose a general blockchain data storage and retrieval system B-Store for the blockchain platform. B-Store divides the blockchain data into data segment and index segment. The index segment uses B+ tree as the underlying data structure, and provides two retrieval options: block retrieval and transaction retrieval. B-Store supports single-field or multi-field equivalent retrieval, range retrieval and top-k retrieval on block data and transaction records. We compare two blockchain storage solutions, Hyperledger Fabric and B-Store. The final experimental results show that B-Store achieves low additional storage and performance overhead when adding new blocks, but provides rich and efficient retrieval functions for blocks and transactions.

Keywords: Blockchain · Hyperledger fabric · B+ Tree

1 Introduction

Blockchain is a decentralized distributed database maintained by multiple parties [1]. It is a new application model of computer technologies such as distributed data storage, point-to-point transmission protocols, consensus mechanisms, and encryption algorithms [2]. Blockchain is a chain data structure that combines data blocks in a sequential manner according to time sequence. It is a cryptographically guaranteed immutable and unforgeable distributed ledger [3]. Blockchain has received more and more attention and application due to its characteristics of decentralization, information immutability, and autonomy.

In Hyperledger Fabric, the ledger is one of the core concepts, which stores important information about business objects in smart contracts [4]. The ledger consists of two parts: the world state and the transaction log. The world state describes the state of the ledger at a given time. By default, the world state is implemented by the LevelDB key-value store database, which can also be configured as CouchDB. The state database allows applications to quickly access the current value of the ledger state without traversing the entire transaction log to calculate the latest state value. Compared with LevelDB database, CouchDB supports creating indexes for data values and provides rich query functions. Another part of the ledger is an ordered transaction log, which records the history of all transactions that changed the world state. They are organized into interconnected ordered blocks, and the header of each block contains the hash of all transactions in the current block and the hash of the previous block. In this way, all transactions on the ledger are ordered and cryptographically linked together. The transaction log is stored differently than the world state. It is stored in multiple files in ascending order of block height by appending blocks. Once they are written to disk, they cannot be changed. To speed up the retrieval of blocks and transactions, Fabric uses LevelDB as an index database. LevelDB stores the mapping of several key fields (such as block hash, block height, transaction number) to block addresses or transaction addresses. However, the retrieval functions provided by Fabric are limited, and only the following retrieval methods are provided: 1. Retrieve blocks by block hash; 2. Retrieve blocks by block height; 3. Retrieve blocks by transaction number; 4. Retrieve transactions by transaction number; 5. Retrieve transactions by block height and transaction sequence number. However, LevelDB is implemented based on LSM-tree, and this structure does not support complex query.

Because the retrieval function provided by Fabric is limited and cannot meet the needs of current blockchain applications, we propose a general block storage and retrieval system B-Store. It provides rich and efficient retrieval functions for the blockchain platform and supports custom retrieval fields. At the same time, B-Store fully considers the uncertainty of block fields and transaction fields in different blockchain systems, so it is not only applicable to the Hyperledger Fabric, but also to other blockchain platforms. The main contributions of this paper are as follows:

1. We analyze the storage model of Hyperledger Fabric and describe its shortcomings.
2. We propose a general block storage and retrieval system B-Store based on the B + tree, which provides a block storage solution for the current popular blockchain platforms. B-Store provides two retrieval options: block retrieval and transaction retrieval. It supports single-field or multi-field equivalent search, range search and top-k search on block data and transaction data.
3. We compare B-Store with Fabric's existing storage solutions through experiments. The experimental results show that B-Store provides rich retrieval functions, and the retrieval efficiency of B-Store is also close to Fabric's existing solutions. B-Store can meet the needs of current blockchain applications.

We introduce related work in Sect. 2, and then introduce the system architecture and storage model of B-Store in Sects. 3 and 4, respectively. Section 5 provides related experiments and experimental results. Finally, Sect. 6 describes the conclusion of this paper.

2 Related Work

In order to realize the storage, retrieval and analysis of blockchain data, researchers in this field have proposed many solutions for different needs.

ForkBase [5] is a storage engine designed for blockchain and forkable applications. It proposes a new index structure POS-Tree, which can store and manage multi-version data. ForkBase supports efficient query and can detect duplicate data to remove redundancy. According to the design and characteristics of ForkBase, it is obviously more suitable for storing the world state of Hyperledger Fabric, because the world state is multi-version data. The transaction log is stored in an append mode, and once written, it will not be changed. So ForkBase is not suitable for the storage of the transaction log.

EBTree [6] is an index structure for Ethereum data, which provides complex retrieval functions for blockchain data. It has the following problems: first, its retrieval granularity is the block, and it does not further realize the retrieval function of transactions; secondly, it uses B + tree as the index structure, but still uses LevelDB to store the nodes of B + tree, which will cause performance loss; finally, it only provides indexes for the Ethereum blockchain platform, so it lacks generality and cannot provide retrieval to other blockchain platforms. EtherQL [7], an efficient query layer designed for Ethereum, provides efficient primitives for analyzing blockchain data. EtherQL, like EBTree, is only designed for Ethereum and not applicable to other blockchain platforms. Moreover, EtherQL repeatedly stores the data of the blockchain in MongoDB, which will generate huge storage overhead.

BlockSci [8] mainly focuses on the analysis of block and transaction data, while systems such as vChain and VQL pay more attention to the verification of query [9–11]. BlockchainDB [12], adds a database layer on the blockchain to realize data sharing, but only supports simple key-value query, not complex query. However, the limited query function is one of the shortcomings of the blockchain system [13].

The B-Store proposed in this paper is a general block storage and retrieval system. As a storage layer for popular blockchain platforms, it can provide efficient complex query services.

3 System Architecture

Hyperledger Fabric is one of the most popular blockchain platforms. Its blocks are serialized and stored directly in the block file. The default maximum size of the block file is 64MB. When a new block is generated, it will judge whether the remaining space of the latest block file can store the latest block. If the remaining space is sufficient, the latest block will be directly appended to the end of the file; if it cannot be stored, a new block file will be generated, and the latest block will be stored in the new block file. After a new block is written to the file, Fabric will index the block in the LevelDB database. The keys of the index are block height, block hash, transaction number, and the combined field of block height and transaction sequence number. The value of the index is the address of the block or transaction in the block file. When querying a block, Fabric needs to query the address of the block in LevelDB, and then read the corresponding block in the block file according to the address.

Due to the limited retrieval function provided by Fabric, we propose B-Store based on the storage model of Fabric. B-Store is a general block storage and retrieval system, which can provide the storage service of historical transaction logs for the blockchain platform, and generate customized index data as needed. B-Store provides richer and more efficient block retrieval and transaction retrieval.

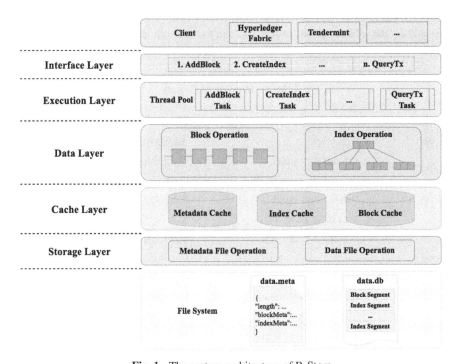

Fig. 1. The system architecture of B-Store

As shown in Fig. 1, B-Store is divided into five layers, namely interface layer, execution layer, data layer, cache layer and storage layer. The interface layer provides a total of 11 methods shown in Table 1. Clients (such as Hyperledger Fabric, Tendermint) can create block indexes or transaction indexes through the interface. After adding a block, the client can query the block or transaction through the index field. The query interface is divided into two categories, one is for querying blocks, and the other is for querying transactions. Both types of query interfaces correspond to 4 retrieval methods, namely equivalent query, range query, maximum top-k query and minimum top-k query. B-Store adopts the remote procedure call (RPC) method to provide interfaces, which reduces the coupling between the block storage system and the blockchain platform.

The execution layer maintains a thread pool and creates corresponding tasks (such as adding block tasks, querying transaction tasks, etc.) to process requests from the interface layer. For example, after the interface layer receives a new block request, the execution layer will create a new block task and execute it asynchronously to complete block storage and index construction. Asynchronous execution can greatly improve the response efficiency of client requests.

The data layer mainly maintains block operations and index operations. Block operations include block read operations, block write operations, and transaction read operations. The index operation maintains a B+ tree for each index, which is responsible for the construction, search and persistence of the B+ tree. The keys of leaf nodes and inner nodes of the B+ tree are index fields. The index field can be a single field or multiple fields. The value of the leaf node is the storage address of the block or transaction in the file (such as the address offset of the block in the file, the length of the block, and the sequence number of the transaction in the block).

Table 1. Interface of B-Store

Category	Description	Method
Block	Add block	AddBlock(block)
	Create block index	CreateBlockIndex(fields)
	Equivalent query block	QueryBlock(condition)
	Range query block	RangeQueryBlock(begin, end)
	Max top-k query block	MaxTopKQueryBlock(fields, k)
	Min top-k query block	MinTopKQueryBlock(fields, k)
Transaction	Create transaction index	CreateTxIndex(fields)
	Equivalent query transaction	QueryTx (condition)
	Range query transaction	RangeQueryTx (begin, end)
	Max top-k query transaction	MaxTopKQueryTx (fields, k)
	Min top-k query transaction	MinTopKQueryTx (fields, k)

The data layer relies on the storage layer to achieve data persistence. The storage layer realizes the transparent transmission of data, that is, it does not distinguish whether the data is a block or an index. It is only responsible for writing data to or reading from a file as a stream of bytes. In order to avoid frequently reading or writing data from the disk and improve data access efficiency, a cache layer is added between the data layer and the storage layer. Both block data and index data processed by the data layer need to flow into or out of the storage layer through the cache layer. A background thread in the cache layer periodically writes changed cache data back to disk. The cache layer contains three data cache pools, namely metadata cache pool, block cache pool and index cache pool. The metadata cache pool will call the metadata file operation of the storage layer to read and write data. The block cache pool and the index buffer cache uniformly call the data file operations of the storage layer.

4   Storage Model

The storage model of B-Store is divided into two parts: storage structure and index structure. The storage structure is responsible for the allocation of storage space, as well as the reading and writing of block data and index data. The index structure is responsible for index building and searching. The building of the index involves the creation, rotation and splitting of nodes in the B+ tree.

4.1   Storage Structure

The storage model of B-Store has two files: data files and metadata files. As shown in Fig. 2, both block data and index data are stored in data files. The file space consists of segments, regions, and pages. According to different storage contents, segments can be divided into block segments and index segments. The index segments include inner node segments and leaf node segments, which are used to store two different types of nodes in the B+ tree. The region is the allocation unit of segments. A segment is made up of multiple regions. The default size of each region is 4 MB, and a region is a space composed of contiguous pages. The page is the smallest unit of the B-Store storage model, and it is also the data read and write unit of the B-Store storage model. Like segments, pages are divided into block pages, inner node pages, and leaf node pages. Each block page stores one block. Since the block fields stored by different blockchain platforms are different, and the number of transactions contained in the block is not fixed, the length of the block page is not fixed. When a block page is allocated, its length is determined by the block length. And each inner node page and leaf node page respectively correspond to the inner node and leaf node in the B+ tree, and their content is determined, so their lengths are fixed. The default length of the node page is 16KB, and each region can store 256 inner node pages or leaf node pages.

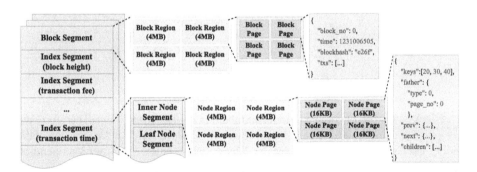

Fig. 2. The storage structure of B-Store

The metadata file records the allocation and usage of the storage space of the data file, such as the current length of the data file, the size of the region, and the allocated region space. In addition, the metadata file also records the information of the block index and transaction index in the B-Store, such as the name and type of the index key, and the address of the root node of the index.

4.2 Index Structure

The index structure of the B-Store is shown in Fig. 3. It is organized by B+ trees, and each index corresponds to a B+ tree. There are two types of nodes in a B+ tree, inner nodes and leaf nodes. Inner nodes consist of a list of index keys, parent nodes, left siblings, right siblings, and child nodes, while leaf nodes consist of a list of index keys, parent nodes, left siblings, right siblings, and index values. Since all nodes need to be serialized and persisted to disk, the parent node, left sibling node, right sibling node and child node list attributes store the address of the node in the file. If the type of the index is a block index, the index value list of the leaf node stores the block address. If the type of the index is a transaction index, the index value list of the leaf node stores the block address and the sequence number of the transaction in the block. Inner nodes or leaf nodes of the same level have left sibling nodes and right sibling nodes, which form a doubly linked list. At the same time, the parent node pointer in the node will facilitate the rotation and splitting of the node.

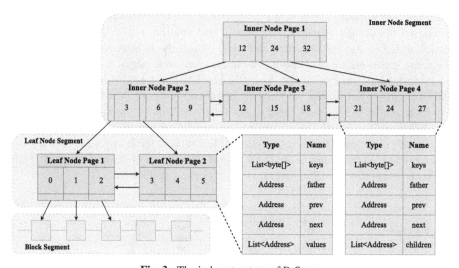

Fig. 3. The index structure of B-Store

Blocks are stored in append mode, and there is no delete operation. Therefore, the B+ tree of the block index or transaction index only needs to consider the insertion of the index, and does not need to consider the deletion of the index. The insertion operation of B + tree mainly has the following steps:

1. Read the root node address from the metadata file and load the root node.
2. If the root node is null, create a leaf node as the root node. Insert the key and value of the index into it, and the insert operation is over.
3. If the root node is an inner node, traverse the key list of the node to find the target child node. If the child node is still an inner node, repeat this step until a leaf node to be inserted is found.

4. No matter whether the root node is a leaf node or an inner node, after step 3, the leaf node to be inserted has been found, and then the key and value of the index can be inserted into the leaf node.

The steps for inserting leaf nodes are as follows:

1. Insert the key and value of the index into the leaf nodes.
2. If the leaf node does not overflow, the insertion is over.
3. If the leaf node overflows, the leaf node needs to be split. First create a new leaf node, and then take out the middle key of the original leaf node. Keep all keys before the middle key in the original leaf node, and move the rest (including the middle key) to the new leaf node. Finally, insert the middle key into the parent node, that is, insert the middle key into the inner node.

The steps of inserting an inner node are as follows:

1. The insertion operation of inner nodes is bottom-up, so the inner node to be inserted may be null. In this case, create a new inner node and insert the key and two child nodes into it.
2. If the inner node is not null, insert the key and child nodes into the inner node. If the inner node does not overflow, the insert operation is over.
3. If the inner node overflows, the inner node needs to be split, and the steps are the same as the leaf node. The only difference is that the intermediate key is not moved into the new inner node.

Since the split operation of nodes will not only generate new nodes, but also change the content of child nodes and parent nodes, which will lead to many disk read and write operations. Therefore, the split operation of nodes should be avoided as much as possible. Before splitting, try to use the rotation operation to move the key to the sibling node. In addition, if the index is added in ascending order of keys, once the node is split, the subsequently added index will not be inserted into the previous node. And the node page is pre-allocated, the split operation will waste half the space of the node page. The rotation operation will solve this problem. Moving spilled indexes to under-full siblings can save a lot of disk space. The rotation operation can be divided into left rotation and right rotation: left rotation refers to moving the key to the left sibling node; right rotation refers to moving the key to the right sibling node. The rotation of inner nodes is slightly different from the rotation of leaf nodes. The following is the left rotation step of the leaf node:

1. If the left sibling is empty, it cannot rotate left.
2. If the left sibling is full, it also cannot rotate left.
3. When the key of the left sibling node is not full, find the ancestor nodes of the current node and the left sibling node. Move the smallest key of the current node to the left sibling node, and update the key of the ancestor node to the new smallest key of the current node.

The right rotation operation of a leaf node is to move the maximum key of the current node to the right sibling node. Then update the key of the ancestor node to the new smallest key of the right sibling node. The left rotation operation of the inner node is to insert the key of the ancestor node and the first child of the current node into the left sibling node. Then update the key of the ancestor node to the smallest key of the current node. Finally remove the smallest key and first child of the current node. The right rotation operation of an inner node is to insert the key of the ancestor node and the last child of the current node into the right sibling node. And update the key of the ancestor node to the largest key of the current node. Finally remove the largest key and last child from the current node.

5   Experiments

In this section, we conduct a comprehensive evaluation of the design of the B-Store. First, we evaluate the throughput of block storage and block query. Then we conducted a set of comparative experiments to compare the storage space, storage time and query time of blocks under the three solutions. Finally, we tested the performance of each query function provided by B-Store. Our experiments use two datasets, Bitcoin and Ethereum, see Table 2 for details.

Table 2. The datasets of experiments

Datasets	Number of blocks	Block height
Bitcoin	600,000	0 –599,999
Ethereum	2,500,000	10,000,000 –12,499,999

5.1 Throughput

To verify the generality of B-Store, in the throughput experiments, we use two datasets: Bitcoin and Ethereum. Figure 4(a) is the throughput of block storage. When the number of blocks is between 150,000 and 600,000, B-Store can store 818 to 1026 blocks per second. Figure 4(b) is the throughput of block query. In the experiment, we will generate a random block height, and then query the block by the block height. Experimental results show that B-Store can query 55 to 129 times per second. The experimental results of the two datasets are roughly equal, which shows that B-Store has good generality.

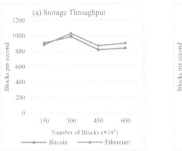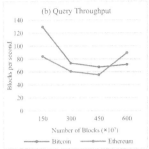

Fig. 4. Throughput

5.2 Comparative Experiment

In addition to B-Store, two other solutions were selected in the comparative experiment: (1) LevelDB: The block data of the blockchain is directly stored in LevelDB. The stored key is the block height and the value is the block content. (2) Fabric: Hyperledger Fabric stores blocks in files in ascending order of block height, and then stores block addresses in LevelDB. The key stored in LevelDB is the block height, and the value is the storage address of the block in the file.

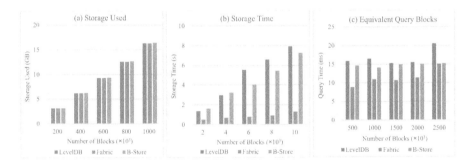

Fig. 5. Comparison results of the three solutions

Figure 5 (a) shows the storage space usage of the three solutions. When the number of blocks ranges from 200,000 to 1,000,000, the disk space used by the three solutions for storing blocks is roughly equal, and all of them increase linearly with the number of blocks.

The storage efficiencies of the three solutions are shown in Fig. 5(b). When 10,000 blocks are inserted, the first solution takes about 7.91 s. Since it directly stores a large amount of block data in LevelDB, it takes a lot of time. The second solution takes the least time, about 1.29 s, because it only needs to store the block address in LevelDB. The third solution is B-Store. When it adds new blocks, it will generate a lot of disk read and write overhead due to index creation. We calculated the usage of B-Store's internal storage space, and the results are shown in Table 3. Since the number of transactions far exceeds the number of blocks, the transaction index requires far more disk space than the block index. When the number of blocks is 250,000, the total storage space used by B-Store is about 4.44 GB, of which block data accounts for 87.61%, and the block index and transaction index account for 0.68% and 11.71%, respectively. It turns out that indexes use relatively little disk space in the B-Store.

As shown in Fig. 5(c), when the number of blocks is between 500,000 and 2,500,000, the query time of a single block of B-Store is about 15ms, and its performance is close to the solution provided by Fabric. Moreover, the query performance of B-Store is relatively stable, which is determined by the B+ tree structure used in the index of B-Store. When the number of ways of the B+ tree is 100 and the height is 4, the B+ tree can store 100 million indexes. Therefore, when B-Store retrieves a block, it only needs to read the disk 4 times to traverse 4 nodes to obtain the block address. Then it reads the block data according to the block address. This process only needs to read the disk 5 times in total.

Table 3. Storage usage of B-Store

Number of block	Total storage (GB)	Block data		Block index		Transaction index	
		Storage (GB)	Percentage	Storage (GB)	Percentage	Storage (GB)	Percentage
50,000	1.00	0.89	89.00%	0.01	1.00%	0.10	10.00%
150,000	2.73	2.40	87.91%	0.02	0.73%	0.31	11.36%
250,000	4.44	3.89	87.61%	0.03	0.68%	0.52	11.71%

5.3 Performance Analysis

We also conduct performance experiments on each query function provided by B-Store. Figure 6 is the experimental results of the performance of the three methods for querying blocks.

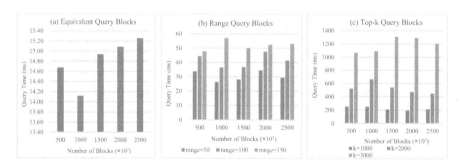

Fig. 6. The performance result of query blocks

Figure 6 (a) shows the response time of the equivalent query block, whose average query time is around 15 ms. Figure 6(b) is the response time of querying blocks by range. When the query range is 150, the average query time is around 50 ms. Figure 6(c) is the experimental result of the top-k query block. The response time for querying the top-1000 blocks is around 200 ms and grows linearly with k.

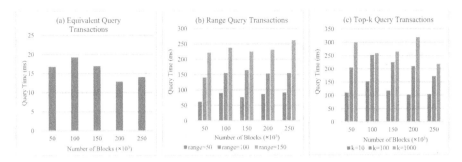

Fig. 7. The performance result of query transactions

As shown in Fig. 7(a), the average response time of equivalent query transactions is within 20 ms. Figure 7(b) is the experimental result of the range query transaction. The time of the range query transaction grows with the expansion of the range. When the number of blocks is 250,000 and the range of block heights is 150, the range query transaction time is about 250 ms. This far exceeds the time required to range query blocks. This is because the number of transactions is much larger than the number of blocks. Figure 7(c) is the experimental result of top-k query transactions, and the query time increases as k increases.

6   Conclusion

In this paper, we propose a general block storage and retrieval system B-Store for blockchain platforms. B-Store provides two retrieval options: block retrieval and transaction retrieval. It supports equivalence search, range search and top-k search on single or multiple fields. We compare the storage solutions of B-Store and Fabric through experiments. B-Store provides rich retrieval functions, and the retrieval efficiency of B-Store is close to Fabric's existing solutions. B-Store can meet the needs of current blockchain applications.

Acknowledgment. This work is supported by the National Natural Science Foundation of China (62072086, 62172082, 62072084), the Fundamental Research Funds for the central Universities (N2116008), the Open Project Fund of Neusoft Corporation (NCBETOP2002).

References

1. Zhao, X., et al.: Blockchain and distributed system. In: International Conference on Web Information Systems and Applications, pp. 629–641 (2020)
2. Yu, G., Nie, T., Li, X., et al.: Distributed data management technology in blockchain system-challenges and prospects. Chin. J. Comput. 1–27 (2019). (in Chinese)
3. Fan, J., Li, X., Nie, T., Yu, G.: Overview of smart contract technology in blockchain system. Comput. Sci. **46**(11), 1–10 (2019). (in Chinese)
4. Cachin, C.: Architecture of the hyperledger blockchain fabric. In: Workshop on Distributed Cryptocurrencies and Consensus Ledgers, vol. 310, no. 4, pp. 1–4 (2016)
5. Wang, S., et al.: Forkbase: An efficient storage engine for blockchain and forkable applications. arXiv preprint arXiv:1802.04949 (2018)
6. XiaoJu, H., et al.: Ebtree: a b-plus tree based index for ethereum blockchain data. In: Proceedings of the 2020 Asia Service Sciences and Software Engineering Conference, pp. 83–90 (2020)
7. Li, Y., Zheng, K., Yan, Y., Liu, Q., Zhou, X.: EtherQL: a query layer for blockchain system. In: Candan, S., Chen, L., Pedersen, T.B., Chang, L., Hua, W. (eds.) DASFAA 2017. LNCS, vol. 10178, pp. 556–567. Springer, Cham (2017). https://doi.org/10.1007/978-3-319-55699-4_34
8. Kalodner, H., et al.: Blocksci: design and applications of a blockchain analysis platform. In: 29th USENIX Security Symposium (USENIX Security 20), pp. 2721–2738 (2020)
9. Wang, H., et al.: Vchain: a blockchain system ensuring query integrity. In: Proceedings of the 2020 ACM SIGMOD International Conference on Management of Data, pp. 2693–2696 (2020)
10. Wu, H., et al.: VQL: efficient and verifiable cloud query services for blockchain systems. IEEE Trans. Parallel Distrib. Syst. **33**(6), 1393–1406 (2021)
11. Zhang, C., et al.: Authenticated keyword search in scalable hybrid-storage blockchains. In: 2021 IEEE 37th International Conference on Data Engineering (ICDE), pp. 996–1007. IEEE (2021)
12. El-Hindi, M., et al.: BlockchainDB: a shared database on blockchains. Proc. VLDB Endow. **12**(11), 1597–1609 (2019)
13. Wang, Q., He, P., Nie, T., Shen, D., Yu, G.: Overview of data storage and query technology of blockchain system. Comput. Sci. **45**(12), 12–18 (2018). (in Chinese)

Enabling Verifiable Single-Attribute Range Queries on Erasure-Coded Sharding-Based Blockchain Systems

Dongyang Pan[1], Derong Shen[1(✉)], Tiezheng Nie[1], Yue Kou[1], Guangyu He[2,3], and Shicheng Xu[2,3]

[1] Northeastern University, Shenyang 110004, China
pandy1715@foxmail.com, {shenderong,nietiezheng}@cse.neu.edu.cn
[2] Neusoft Corporation, Shenyang 110179, China
{hegy,xushicheng}@neusoft.com
[3] Liaoning Blockchain Engineering Technology Research Center, Shenyang 110179, China

Abstract. As the amount of data grows, blockchain has an increasing need for data storage and how to query data efficiently and securely. The combination of erasure coding technology and permissioned chain can reduce the storage consumption of each block from O(n) to O(1), but the query strategy it adopts limits the query function, reduces the query efficiency, and increases the network load. In this paper, we propose a query method for blockchains based on erasure code sharding, which combines the idea of pushing queries and accumulator to make up for the shortcomings of the original query strategy and ensure the integrity of the query results. We design an accumulator-based authentication data structure (ADS), which supports verifiable single-attribute range queries. Second, in order to solve the problem of amplifying the effect of node mischief brought by distributed computing, we propose an error correction strategy that uses the erasure code recovery mechanism to locally recover the fault interval and correct the erroneous results. Finally, it is implemented on the open source blockchain system Tendermint. It has been proved through sufficient experiments that the system query efficiency has been improved while ensuring the availability of the system.

Keywords: Blockchain · Erasure coding · Query processing · Authenticated data structure

1 Introduction

Blockchain technology has come a long way with the advent of Bitcoin [1], Ethereum [2] and HyperLedger Fabric [3]. But whether the blockchain uses PoW or PBFT [4] consensus protocols, each node of the blockchain uses huge storage space to keep a complete copy of the block data. Methods to reduce blockchain storage, such as data compression [5], full node combined with light nodes [1] and per-group slices [6], do not fundamentally solve the problem of full-copy storage. The erasure code (EC) sharding method can provide lower storage overhead than the multi-copy method at the same fault tolerance level [7]. Qi et al. [8] implemented a blockchain system based on erasure codes to solve the Byzantine fault problem, but it currently only supports single-block query.

The unstructured data storage system based on the Key-Value model commonly used in blockchain does not support complex queries. In order to manage blockchain data more conveniently, such as BigchainDB [9], EtherQL [10] try to combine blockchain and database. They process the query in the database after transferring the transaction into the database, but this method cannot verify the integrity of the query result. Some scholars have applied some researches on cryptography and outsourced databases to the blockchain, such as SEBDB [11], vChain [12], which enriches blockchain queries while ensuring integrity. However, its query still relies on full nodes to return query results and proofs, which cannot be directly applied to the blockchain system based on erasure code sharding.

In this paper, we focus on how to provide a single-attribute range query in an erasure code sharding-based blockchain system, such as querying that an attribute is equal to a certain value or greater than or less than a certain value. We use the relational database as the query engine, and each node queries the block data it holds. Two problems need to be solved for this. (i) We need to design an authentication data structure (ADS) that is suitable for an erasure code sharding-based blockchain system, contains relational semantics, and can provide proof of non-existence. Because in the case of storage shards, each node can only use the block header data to verify the proofs provided by other nodes. Also, we need an ADS that can handle relational data and is not limited to specific tables or attributes. Finally, the ADS should be able to provide integrity verification when a malicious node returns little or no data. (ii) We need to design an error correction strategy. When a node distributes query requests to other nodes for distributed queries, if the malicious nodes involved in the query do not return the correct data, the final query result is necessarily incorrect and incomplete.

To solve the above problems, firstly, we propose an ADS based on accumulator [13] and MB-Tree [14], i.e., Single Attribute Range Query Tree (SA-Tree). The main idea is to build the accumulator hierarchically, from top to bottom, the database layer, the table layer, the attribute layer, and the value layer. Only the accumulator value of the top layer is included in the block header, and the connection relationship between each layer is represented by the proof of accumulator values. Secondly, we propose an error correction strategy based on erasure code. The seed node being queried can verify that the query results are collected. For incorrect results, the seed node uses the erasure code to recover the involved blocks to correct the incorrect query results. Our contributions are summarized as follows:

- We propose a universal accumulator-based ADS SA-Tree, which supports verifiable single-attribute range queries.
- We propose an error correction strategy based on erasure codes, which solves the problem that the evil effect of malicious nodes is amplified.

- We conducted relevant experiments to verify that in the blockchain system based on erasure code sharding, this query verification strategy improves query efficiency while ensuring validity.

The rest of this paper is organized as follows. Section 2 introduces related work. Section 3 introduces the system model. Section 4 introduces the design of the SA-Tree. Section 5 introduces the query result error correction strategy. Section 6 gives the security analysis. Section 7 contains a series of experiments. Section 8 concludes the paper.

2   Related Work

In this section, we briefly introduce the related work on the blockchain's certifiable query processing. Most of the underlying data storage systems of blockchain systems use key-value storage based on log-structured merge tree (LSM-tree). This unstructured data storage system based on the Key-Value model does not support complex queries. Due to the existence of Byzantine nodes, the well-defined data index structure for various queries in traditional databases is not applicable in blockchain systems. Blockchain needs to provide Soundness and Completeness guarantee for query results, which has been extensively studied in outsourced databases [15,16]. There are three solutions that can be applied to the blockchain: one is the policy-based solution adopted by Fabric [3], and each chaincode needs to specify an endorsement policy. Fabric assumes that enough endorsing nodes are honest, and this scheme is a compromise between security and performance. One is the secure-hardware-based solution adopted by xu et al. [17,18], which is also very efficient, but depends on a specific hardware environment. The third cryptography-based solution can be divided into two types: tree-based ADS and cryptographic accumulator-based ADS. Most tree-based ADSs are based on Merkle Tree [19] and its variants to support specific queries, such as ForkBase [20] using POStree to support data comparison, data deduplication and other functions. $GEM^2 - Tree$ implements range queries under mixed storage. Zhu et al. [11] leveraged hierarchical authentication indexes to support verifiable SPJ queries. For ADS based on cryptographic accumulators, vChain [12] supports verifiable Boolean range queries on the blockchain based on Merkle trees and cryptographic accumulators. Zhu et al. [21] implemented aggregation functions. However, in the blockchain system based on erasure code sharding, there is currently no verifiable single-attribute range queries.

3   System Model

Figure 1 shows the system model, there are two types of nodes in the system: (i) storage nodes based on erasure code sharding; (ii) query user. Each time the storage node collects a certain number of blocks, it will be encoded by the encoder to form an erasure code encoded block. Each node stores the corresponding encoded block according to its position in the consensus list, and discards the remaining

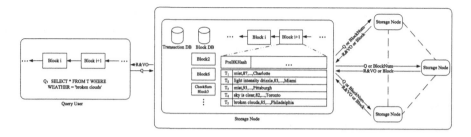

Fig. 1. System model

blocks. The transaction database only retains data within locally stored blocks. The storage node can use the data of the block to construct the SA-Tree and provide query proof i.e. verification objects (VO) based on it (detailed in Sect. 4). During the distributed query process, the accessed storage nodes can send query requests to other storage nodes and verify the results. At the same time, the node can also request block data from other storage nodes to correct erroneous query results (detailed in Sect. 5). The query user is a light node that keeps track of the block headers only. User can verify the integrity of query results (R) based on VO and block header data.

4   Single Attribute Range Query Tree

In order to implement verifiable queries for a specific attribute in a blockchain system based on erasure code storage shards. We propose a general tree structure that can provide integrity verification for single-attribute range queries. In this section, we first introduce the structure and construction method of SA-Tree. Then we introduce how to utilize SA-Tree to generate proofs of query results. Finally, we introduce how to use SA-Tree to verify query results. For the sake of demonstration, we first consider how to query in the case of full replica storage. For the case of storage sharding, it will be described in detail in Sect. 5.

4.1   SA-Tree Structure

In the data model we designed, a transaction is actually a tuple of the database. The database can be divided into database, table, attribute, value according to the level, and we also build the accumulator according to the level. As shown in Fig. 2, the top layer of SA-Tree is the accumulator value of a database, which contains the accumulator value of each table, and the same is true for table-level accumulators and attribute-level accumulators. It should be noted that the bottom accumulator is the attribute accumulator, and its value is composed of the attribute name, the different attribute values of the attribute, and the hash of the sum of all transactions corresponding to the attribute value, in the current block. If it is a numeric attribute, an MB-Tree root of the attribute is also added here. MB-Tree is a structure that combines B+Tree and Merkle Tree

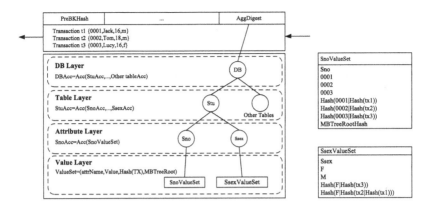

Fig. 2. The SA-Tree structure on a block

to implement verifiable range queries. For transactions involving attribute values we use $Hash(AttrValue|Hash(tn|Hash(tm|...)))$ to calculate.

For simplicity, we assume that only one table is declared in the block chain and that the SA-Tree has only a three-level structure of table, attributes and values without constructing a database layer. The Algorithm 1 shows how to construct a SA-Tree for a block based on the transactions of that block. First, an empty table-level accumulator is initialized, which contains the attribute-level accumulators corresponding to the number of attributes. The table-level accumulator, the table structure, and the set of transactions for the block are taken as input. The accumulator values of the table-level accumulators are used as outputs. In line 1, a map array is first initialized, with each attribute corresponding to a map. Each map represents a mapping of attribute values to the hash of the set of transactions containing the attribute values. In lines 2–6, the algorithm iterates through each tuple of each transaction in the set of input transactions. Line 3 parses a tuple into the format of a collection of attributes. The algorithm uses the input table information to iterate through each attribute value in the tuples and maintain the map corresponding to each attribute value. Lines 7–17 maintain the accumulator for each attribute level. Line 11 adds the attribute value and line 13 adds the set of transactions corresponding to that attribute value. If it is a numeric attribute, the Hash root of the MB-Tree is added again on line 16, and the attribute name of the attribute is added on line 17.

4.2   Query Processing

We've already covered how to build a SA-Tree based on a block of data in the case of a table. Now we'll introduce how to handle queries. The queries we support include equality and range queries that can be performed on a single property. Verifiability for range queries is provided by MB-Tree.

Algorithm 1: ADS Construction

Input: $TableAcc, Table, Tuples$
Output: Acc

1 $attrSet \leftarrow make([\ ] map [string] [\ ] byte, Table.attributeCount)$
2 **for** $tuple\ in\ Tuples$ **do**
3 | $strs \leftarrow TupleToStringBytes\ (tuple)$
4 | **for** $index, value := range\ strs$ **do**
5 | | $left \leftarrow attrSet [index] [value]$
6 | | $attrSet [index] [value] = Hash.Sum\ (left\|tuple)$

7 **for** $index, value := range\ attrSet$ **do**
8 | $AttrName \leftarrow Table.GetName(index)$
9 | $AttributeAcc \leftarrow TableAcc.AttributeAcc[AttrName]$
10 | **for** $key, hashvalue := range\ value$ **do**
11 | | AttributeAccAdd(key)
12 | | $hashAndKey := Hash.Sum(key\|hashvalue)$
13 | | AttributeAcc.Add(hashAndKey)

14 | **if** $Table.GetType(index)==int$ **then**
15 | | $MBTreeRootHash \leftarrow BuildMBTree(tuple)$
16 | | AttributeAcc.Add(MBTreeRootHash)

17 | AttributeAccAdd(AttrName)

18 **for** $v := range TableAcc.AttributeAcc$ **do**
19 | $TableAcc.Add(v.GetA())$
20 **return** $TableAcc.GetA();$

Query Processing on Single Block. The Algorithm 2 describes the query process, and in line 1 we call the local database to get the query result. Lines 3–12 describe the equivalence query process. If query results exist, lines 4–10 reconstruct the hash of the set of transactions added at the value level from the result set in the SA-Tree. If no query results exist, line 12 proves that there is no attribute value to be queried in the accumulator of this attribute. Lines 13–18 describe the processing of range queries. Range queries do not distinguish between existence and non-existence, since the proof of non-existence only needs to return one side of the MB-Tree. Line 16 returns the root of the MB-Tree in the proof. Line 17 returns the range query proof of the MB-Tree. Line 18 proves that the root of the MB-Tree exists in the attribute-level accumulator. Line 19 proves that the attribute name of the query is in the accumulator of this attribute, in other words, this attribute accumulator is the accumulator of the attribute to be queried. Lines 20 and 21 return the value of the attribute-level accumulator and the proof of its existence in the table-level accumulator.

Query Processing on Multiple Blocks. The query on multiple blocks is not special compared to the query on single block, but just adds the integration of result sets to the single block query. The query results and proofs of all blocks can be regarded as a two-dimensional array. Due to the limitation of space not to do code demonstration of multi-block query.

Algorithm 2: Query Processing on Single Block

   Input: Req : query request, H : the block height you want to query
   Output: R : result, P : proof
1 $tuples \leftarrow table.db.Select(Req, H)$
2 $AttrAcc \leftarrow TableAcc.AttributrACC[attrName]$
3 **if** Req is not range query **then**
4   **if** $tuples$ not null **then**
5        $tempResult \leftarrow [\,]$
6   **for** $tuple$ in $tuples$ **do**
7           $left \leftarrow tempResult$
8           $tempResult = Hash.Sum(left\|tuple)$
9        $value \leftarrow Hash.Sum(attrvalue\|tempResult)$
10       $P.ValueProof \leftarrow AttrAcc.ProofValueinAcc(value)$
11   **else**
12       $P.ValueNotExistProof \leftarrow AttrAcc.ProveNonMembership(attrvalue)$
13 **else**
14     $value \leftarrow MBTreeGetRoot(attrname)$
15     $rangeProof \leftarrow MBTreeMakeProof(Req, tuples)$
16     $P.MBRoot \leftarrow value$
17     $P.rangeProof \leftarrow rangeProof$
18     $P.ValueProof \leftarrow AttrAcc.ProofValueinAcc(MBRoot)$
19 $P.AttrNameProof \leftarrow AttrAcc.ProofValueinAcc(AttrName)$
20 $P.AttrAccA \leftarrow AttrAcc.GetA()$
21 $P.AttrProof \leftarrow TablrAcc.ProofValueinAcc(P.AttrAccA)$
22 **return** tuples,P

4.3 Validate Query Results

The verification process is similar to the proof generation process. The light node first has to recover the value at the value level based on the query request and the query result, and then use the proof ValueProof and AttrAcc of the query result to prove that the value is in the accumulator of this attribute. Then based on AttrName, AttrNameProof and AttrAccA prove that the attribute accumulator is attribute specific. Finally the light node uses the table-level accumulator value stored in the block header, and AttrProof to prove that the attribute-level accumulator value exists in it. Note that it makes sense to add attribute names to the value-level accumulator to prove that the attribute accumulator is the accumulator of the attribute to be queried. For validating nonexistent queries: If a certain age of the query does not exist, malicious nodes can use the gender accumulator to generate a proof of nonexistence for the specified age, and the gender accumulator value is verifiable by the table-level accumulator. Due to space limitations, no pseudo-code display will be made here.

5 Query Result Error Correction Strategy

In Sect. 4 we describe how to apply our proposed SA-Tree for single-attribute verifiable queries in the full copy case. In this section, we focus on how to query in the blockchain based on erasure code shard storage. We adopt the idea of pushing query to the data. We distribute the query to each block-holding node, and each node returns the query result and proof for the local block. This approach reduces the amount of data transmitted by the network and allows for parallel execution of queries to improve efficiency. But the consequent problem is that the query security issue is magnified. To ensure the correctness of the query, we propose an error correction strategy based on the erasure code recovery mechanism. The seed node executes the query locally on the problem block and obtains the query result.

Algorithm 3: Query Processing Based on EC Fragment Storage

Input: Req : query request
Output: R : The query results include the results and proofs of all blocks

1 $queryHeight \leftarrow app.Height$
2 $queryResult \leftarrow make([]acctable.ResultSet, QueryHeight + 1)$
3 $problemBlock \leftarrow [\ ]$
4 $queryMap \leftarrow QueryAllocation(QueryHeight)$
5 $problemBlock \leftarrow [\ ]$
6 $wg \leftarrow sync.WaitGroup$
7 **for** $k, v := range\ queryMap$ **do**
8 $\quad$ go RPCQuery(Req, k, v, queryResult, wg)

9 wg.Wait()
10 **for** $i = 1$ to $queryHeight$ **do**
11 $\quad$ **if** $!Verifiy(Acc(i), queryResult[i].T, queryResult[i].P, Req$ **then**
12 $\quad\quad$ problemBlock.add(i)

13 $recoveredBlock \leftarrow ec.Recovery(problemBlock)$
14 **for** $k, v := range\ recoveredBlock$ **do**
15 $\quad tempAcc \leftarrow ReBuildAcc(v)$
16 $\quad elem, p \leftarrow app.QueryInBlockI(reqQuery, k)$
17 $\quad QueryResult[k] \leftarrow acctable.ResultSetelem, p$

18 **return** queryHeight[1:]

The Algorithm 3 describes this query process. Lines 1–3 initialize the query height, the query result set and the set of problem blocks. Line 4 implements that the map of the node's corresponding block set is obtained by QueryAllocation, which represents the set of blocks for which the corresponding node needs to execute the query. For our current stage implementation, it is getting the blocks held by each node, but this function can be optimized into a better allocation scheme based on multiple copies. Lines 5–9 implement the query distribution and collection by sending the queries to each node. Each node queries in parallel and the main thread blocks until the results are all collected. Lines 10–12 implement verifying the query results and collecting the faulty blocks. Line 13 recovers the faulty block using erasure code. Lines 14–17 reconstruct the SA-Tree and correct the faulty query results. The last 18 lines return the results and proofs from the 1st block to the specified query height.

6   Security Analysis

In this section, we perform a security analysis on verifiable range queries based on SA-Tree.

Definition 1. *Our verifiable query algorithm is safe if, for all PPT adversaries, the probability of success is negligible in the following experiments:*

- *Create an SA-Tree tree on the dataset D according to the SA-Tree creation algorithm, and send D to the attacker Adv.*
- *The attacker Adv outputs a verifiable multidimensional aggregate query q, a fake result R' and a fake proof VO', where R' is not equal to the correct query result, and passes the verification.*

Theorem 1. *Our proposed SA-Tree based verifiable query range algorithm is safe if the hash function is collision-resistant and the accumulator is safe.*

Proof. We prove Theorem 1 by proof by contradiction. For a false attribute-level accumulator value, if it contains the specified attribute name and passes the existence proof of the table-level accumulator. This means that other attributes in the fake attribute-level accumulator collide with other attributes of the real attribute-set accumulator, which contradicts the accumulator being safe. For a fake numeric-level hash value $Vt=Vf$, that means (i) there are two MB-trees with different objects but the same root hash, or (ii) There is another object whose value and hash of the transaction are the same as the real value and hash of the transaction. In either case, it means a successful collision of the underlying hash function, which contradicts our assumption.

7   Experimental Evaluation

All experiments are performed on a node with 2.9 GHz, Intel Core i7-10700 CPU and 16 GB RAM. The code is based on the Tendermint blockchain written in Go and tested on a local area network.

WorkLoad. Since MB-Tree is only used in range queries, range queries have less process of building a value layer than equivalent queries. The calculation of the range query is extremely dependent on the locally constructed MB-Tree, and the generation and verification of the proof have little to do with our proposed SA-Tree. So in the experiments we do not test for range queries. We use the query statement: WHERE SEX = F; WHERE WEATHER = 'broken clouds'.

Dataset. Our experiments use the Student Alcohol Consumption[1] dataset and the Weather[2] dataset. The student data were obtained in a survey of students math and portuguese language courses in secondary school. It contains a lot of interesting social, gender and study information about students. The Weather dataset contains 1.5M hourly weather records for 36 cities in US, Canada, and Israeli during 2012–2017.2 For each record, it contains seven numerical attributes (such as humidity and temperature) and one weather description attribute with 2 keywords on average. The transactions we constructed contain 8 attributes, each transaction is about 100bytes, each A block can contain up to 50 transactions.

---

[1] https://www.kaggle.com/datasets/uciml/student-alcohol-consumption.
[2] https://www.kaggle.com/datasets/selfishgene/historical-hourly-weather-data.

Metrics. We evaluate the performance of our proposed SA-Tree by its build time, query time, and validation time. The performance of the two query methods in the case of storage sharding is evaluated by query response time.

7.1   Performance of SA-Tree

We first test the performance of the proposed SA-Tree, which is built for each block. Based on this structure, query validation within a single block is supported. We tested SA-Tree construction time, query time, verification time and proof size using two datasets.

Figure 3 shows the total time required to build a block SA-Tree as transactions within the block increase from 1–50. The attribute value types and the number of different values of the same attribute in the two datasets are different, so even if the number of transactions is the same, the construction time is different. The more complex the property value type, and the more distinct property values a property has, the longer it takes to build.

Figures 4, 5 and 6 show the query and verification time and verification object (VO) size on a block as the number of transactions in the block grows. Thanks to the SA-Tree we designed, all three metrics do not fluctuate significantly with the number of transactions within a block. The SA-Tree is data independent, and both proof generation and verification are only related to the number of layers in the SA-Tree.

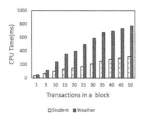
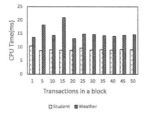
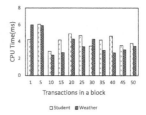

Fig. 3. SA-Tree build time   **Fig. 4.** Query performance   **Fig. 5.** Verify performance

7.2   Performance of the Query Method

Next, we test the query time of our proposed distributed query method and local recovery block method in a blockchain system based on erasure code shard storage. We start with 50 blocks until the query involves 200 blocks. Among them, Fig. 7 contains 25 transactions per block, and Fig. 8 contains 50 transactions per block.

It can be seen that the overall trend of the query time of the two methods increases with the number of blocks involved in the query, which is almost linear. The method of recovering blocks is about 4 times longer than the method of

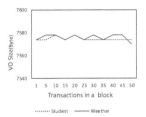 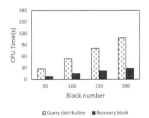 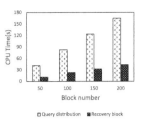

Fig. 6. VO size

Fig. 7. Multi-block query performance with 25 transactions per block

Fig. 8. Multi-block query performance with 50 transactions per block

distributed query, which is enough to explain the time-effectiveness of the method of distributed query. Because the most time-consuming part of the whole process is to build the SA-Tree, and the method of restoring the block needs to build 3 times more SA-Tree. At the same time, the query speed of distributed query and local query is also greatly improved. Our proposed method can improve query efficiency while ensuring security.

8 Conclusion

In this paper, we introduce a method that supports verifiable single-attribute range queries in blockchains stored in erasure-coded shards. We propose a generic accumulator-based ADS SA-Tree that supports verifiable single-attribute range queries. In order to solve the problem of malicious node amplification caused by storage sharding, we propose an error correction strategy. We conduct relevant experimental verifications, and the empirical results demonstrate the effectiveness and efficiency of our proposed method.

Acknowledgements. This work was supported by the National Natural Science Foundation of China (62172082, 62072084, 62072086), the Fundamental Research Funds for the central Universities (N2116008) and the Open Project Fund of Neusoft Corporation (NCBETOP2002).

References

1. Nakamoto, S.: Bitcoin: A peer-to-peer electronic cash system. Decentralized Bus. Rev. 21260 (2008)
2. Wood, G., et al.: Ethereum: a secure decentralised generalised transaction ledger. Ethereum Project Yellow Paper **151**(2014), 1–32 (2014)
3. Androulaki, E., et al.: Hyperledger fabric: a distributed operating system for permissioned blockchains. In: Proceedings of the Thirteenth EuroSys Conference, pp. 1–15 (2018)
4. Castro, M., Liskov, B., et al.: Practical byzantine fault tolerance. In: OsDI, vol. 99, pp. 173–186 (1999)

5. Xu, Y.: Section-blockchain: a storage reduced blockchain protocol, the foundation of an autotrophic decentralized storage architecture. In: 2018 23rd International Conference on Engineering of Complex Computer Systems (ICECCS), pp. 115–125. IEEE (2018)
6. Al-Bassam, M., Sonnino, A., Bano, S., Hrycyszyn, D., Danezis, G.: Chainspace: a sharded smart contracts platform. arXiv preprint arXiv:1708.03778 (2017)
7. Weatherspoon, H., Kubiatowicz, J.D.: Erasure coding vs. replication: a quantitative comparison. In: Druschel, P., Kaashoek, F., Rowstron, A. (eds.) IPTPS 2002. LNCS, vol. 2429, pp. 328–337. Springer, Heidelberg (2002). https://doi.org/10.1007/3-540-45748-8_31
8. Qi, X., Zhang, Z., Jin, C., Zhou, A.: A reliable storage partition for permissioned blockchain. IEEE Trans. Knowl. Data Eng. **33**(1), 14–27 (2020)
9. McConaghy, T., et al.: BigChainDB: a scalable blockchain database. In: White Paper, BigChainDB (2016)
10. Li, Y., Zheng, K., Yan, Y., Liu, Q., Zhou, X.: EtherQL: a query layer for blockchain system. In: Candan, S., Chen, L., Pedersen, T.B., Chang, L., Hua, W. (eds.) DASFAA 2017. LNCS, vol. 10178, pp. 556–567. Springer, Cham (2017). https://doi.org/10.1007/978-3-319-55699-4_34
11. Zhu, Y., Zhang, Z., Jin, C., Zhou, A., Yan, Y.: Sebdb: semantics empowered blockchain database. In: 2019 IEEE 35th international conference on data engineering (ICDE). pp. 1820–1831. IEEE (2019)
12. Xu, C., Zhang, C., Xu, J.: vchain: Enabling verifiable Boolean range queries over blockchain databases. In: Proceedings of the 2019 International Conference on Management of Data, pp. 141–158 (2019)
13. Boneh, D., Bünz, B., Fisch, B.: Batching techniques for accumulators with applications to IOPs and stateless blockchains. In: Boldyreva, A., Micciancio, D. (eds.) CRYPTO 2019. LNCS, vol. 11692, pp. 561–586. Springer, Cham (2019). https://doi.org/10.1007/978-3-030-26948-7_20
14. Li, F., Hadjieleftheriou, M., Kollios, G., Reyzin, L.: Authenticated index structures for aggregation queries. ACM Trans. Inf. Syst. Secur. (TISSEC) **13**(4), 1–35 (2010)
15. Pang, H., Tan, K.L.: Authenticating query results in edge computing. In: Proceedings of 20th International Conference on Data Engineering, pp. 560–571. IEEE (2004)
16. Narasimha, M., Tsudik, G.: Authentication of outsourced databases using signature aggregation and chaining. In: Li Lee, M., Tan, K.-L., Wuwongse, V. (eds.) DASFAA 2006. LNCS, vol. 3882, pp. 420–436. Springer, Heidelberg (2006). https://doi.org/10.1007/11733836_30
17. Xu, C., Zhang, C., Xu, J., Pei, J.: SlimChain: Scaling blockchain transactions through off-chain storage and parallel processing. Technical report (2021)
18. Jiang, Qin, An, Yanjun, Qi, Yong, Fang, Hai: Oblivious data structure for secure multiple-set membership testing. In: Xing, Chunxiao, Fu, Xiaoming, Zhang, Yong, Zhang, Guigang, Borjigin, Chaolemen (eds.) WISA 2021. LNCS, vol. 12999, pp. 299–310. Springer, Cham (2021). https://doi.org/10.1007/978-3-030-87571-8_26
19. Merkle, R.C.: Protocols for public key cryptosystems. In: Secure Communications and Asymmetric Cryptosystems, pp. 73–104. Routledge (2019)
20. Wang, S., et al.: ForkBase: an efficient storage engine for blockchain and forkable applications. arXiv preprint arXiv:1802.04949 (2018)
21. Zhu, Y., Zhang, Z., Jin, C., Zhou, A.: Enabling generic verifiable aggregate query on blockchain systems. In: 2020 IEEE 26th International Conference on Parallel and Distributed Systems (ICPADS), pp. 456–465. IEEE (2020)

An Efficient Query Architecture for Permissioned Blockchain

Xiabin Huang[1], Derong Shen[1(✉)], Tiezheng Nie[1], Yue Kou[1], Guangyu He[2,3],
and Shicheng Xu[2,3]

[1] Northeastern University, Shenyang 110004, China
`{shenderong,nietiezheng,kouyue}@cse.neu.edu.cn`
[2] Neusoft Corporation, Shenyang 110179, China
`{hegy,xushicheng}@neusoft.com`
[3] Liaoning Blockchain Engineering Technology Research Center,
Shenyang 110179, China

Abstract. For the existing permissioned blockchain system, consensus nodes process not only query requests but also write requests, which make themselves in a heavy workload that reduce the system's performance. In this paper, we propose an efficient architecture which separates part of the query tasks from the consensus nodes to the secondary nodes, and present EQblockchain, a permissioned blockchain system following this architecture, to solve the problem mentioned above. EQblockchain divides transactions into two types: one is read-transactions, the other is write-transactions. We describe how query requests, i.e., read-transactions, are processed by the consensus nodes and the secondary nodes and how write requests update the blockchain state. Because the read-transactions are processed by the consensus nodes and secondary nodes, the system throughput is significantly increased. Besides, EQblockchain constructs a few inverted indexes to support range query, which also enhances the query ability. Furthermore, we conduct several experiments to show that EQblockchain not only has efficient performance in query-heavy workload but also slightly improves the throughput of the write requests, and really has good range query ability in comparison to the baseline.

Keywords: Blockchain · Permissioned · Architecture · Query

1 Introduction

In Nakamoto's white paper [12], the original blockchain system Bitcoin was proposed as an electronic cash system. Nowadays, blockchain gains more and more attention [11] and is widely used in many fields, such as supply management, finance, healthcare and so on.

A blockchain consists of a set of blocks, which are sequentially linked by hash pointers as a chain, in an asynchronous distributed system. A block has two parts: one is the block header, the other is the block body. The block structure of different blockchains have little difference on the whole, but are different in

X. Zhao et al. (Eds.): WISA 2022, LNCS 13579, pp. 699–711, 2022.
https://doi.org/10.1007/978-3-031-20309-1_61

details, e.g., some blockchains' block body is a Merkle tree, the other is a directed acyclic graph [14].

A blockchain is a distributed architecture managed by a few nodes which do not trust each other, where nodes can reach agreement on transactions without a central node. In order to make all nodes meet consistency, i.e., all nodes process the same transactions and reach the same result, blockchain uses the state machine replication (SMR) algorithm to handle transactions. According to SMR, nodes agree on the ordering of incoming transactions and execute transactions in a completely consistent sequence, which can ensure the data of blockchain are identical.

At present, blockchain can be divided into two categories, i.e., permissionless blockchain and permissioned blockchain. For permissionless blockchain like bitcoin, the network is public, and anyone can anonymously participate in. In contrast, the network in permissioned blockchain is private and the nodes joining are restricted.

The existing permissioned blockchain's performance is not good enough when the transactions are too many. In the existing permissioned blockchain, the nodes need not only compose transactions into blocks, but also process the query requests from the clients. If the number of requests is not too large, there is no problem. However, if the amount exceeds the critical value, the node can not handle these tasks, resulting in the decrease of query efficiency.

In this paper, we propose an efficient query architecture with read-write separation for permissioned blockchain, which solves the mentioned problem above. It reduces the impact of the read tasks on the consensus nodes and improve query efficiency. The main contributions of this paper are as follows:

(1) We propose an architecture which has good performance in query-heavy workload and implement a permissioned blockchain system, Efficient Query blockchain (EQblockchain), following the architecture.
(2) We construct an inverted index to expand our blockchain supporting range query.
(3) We conduct several groups of experiments to prove that EQblockchain is good at processing transactions in query-heavy workload, and has good range query ability in comparison to the baseline.

The rest of this paper is organized as follows. Section 2 reviews related work. Section 3 introduces our system's architecture. Section 4 presents EQblockchain and explains how it works. Section 5 shows the performance evaluation. Section 6 concludes the paper.

2   Related Work

Sharding. Sharding is a traditional-distributed technology used in database and also used in many permissioned blockchains, e.g., Fabric [4], SharPer [1,3], AHL [6]. Sharding technology is to divide nodes into different partitions according to certain rules, and then allocate transactions to different partitions, so that different transactions can be processed in parallel. Compared with a single partition, multiple partitions can handle more transactions at the same time, which improves the throughput of the blockchain system.

Fabric implements sharding on blockchain through multi-channels [5]. A node can run one or more channels, and for each channel, the data in the channel is independent. So, different channels can be seen as different shards of the blockchain. Even if nodes run different channels, different channels still can not access each other directly.

SharPer is another sharding permissioned blockchain system, that supports not only consensus with Crash-only nodes, but also consensus with Byzantine nodes. SharPer describes the specific details of introducing Pbft and Raft into the partitioned blockchain system. It processes cross-shard transactions in a decentralized manner among the involved partitions without requiring a reference committee (mentioned in AHL).

AHL is similar to permissionless blockchain, e.g., Elastico [10], OmniLedger [9], Rapidchain [17], where nodes are randomly assigned to different partitions. In contrast to SharPer, AHL processes cross-shard transactions in a centralized way. It uses two-phase commit to process cross-shard transactions, where a node cluster (called reference committee) plays as the coordinator role and each partition acts as the participant role.

Ordering Architecture. From the ordering point of view, there are two architectures for permissioned blockchain systems, i.e., order-execute (OX) and execute-order-validate (XOV).

OX architecture, that traditional blockchain uses, is a pessimistic approach, that the blockchain system orders transactions before executing. The order-parallel execute (OXII) architecture [2] is an improvement on OX, which enhances the ability to handle transactions in parallel.

XOV architecture switches the sequence of the execution and ordering phases, which comes from optimistic concurrency technology. Fabric uses XOV architecture (which was first introduced by Eve [8]) to increase the number of transactions processed in parallel. Because the execution phase is first, the transactions may conflict, resulting in high abort rate of the transactions. To solve it, other XOV architecture blockchains, such as Fabric++ [16], FabricSharp [15] and XOX Fabric [7], use different methods to reduce abort rate. The first two, present their re-ordering techniques to eliminate unnecessary aborts and the latter adds a post-order after the validation step to re-execute unnecessary abort transactions.

The former (Sharding technology) divides nodes into different shards that means it has different blockchains, resulting in low throughput when the transactions are involved in multiple shards. The latter focuses on the sequence of ordering and executing of the transactions. Our work has only one blockchain, and we use order-execute architecture, but we can also use Sharding technology to expand our system or use different ordering architectures to replace the current one. Besides, EQblockchain uses secondary nodes to lighten the workload of consensus nodes.

3   System Architecture

In this section, we introduce our blockchain system architecture, which is a permissioned blockchain and also an order-execute paradigm. Figure 1 presents the architecture of our system. Our system uses a lot of secondary nodes to increase query ability, which improves the system's throughput. We mainly design a flexible secondary storage architecture, which can be compatible with the current blockchain performance optimization technology.

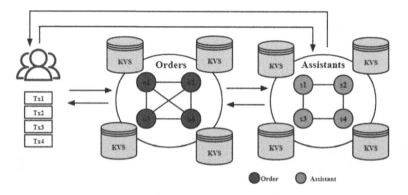

Fig. 1. System architecture

The system consists of a lot of nodes in an asynchronous distributed network and each node has its own role, i.e., Client, Order or Assistant. The nodes use p2p protocol to exchange messages signed by asymmetric encryption, e.g., RSA, which guarantees the messages can not be forged. Our blockchain is maintained by several organizations and each organization has its Clients, Orders and Assistants in the system.

- **Clients** are the initiator of the transaction and send transactions to the blockchain system for execution. Client can be seen as a light node, i.e., the node that only stores the block header of the block, and verify whether a transaction exists in the blockchain.
- **Orders** are consensus nodes and the key component of the blockchain, who agree on the order of the incoming transactions.
- **Assistants** are secondary nodes that support Orders to process query transactions and store data.

3.1   Orders

For a blockchain system, Orders are the key nodes as it should run the state machine replication algorithm according to the consensus protocol to make all of the Orders execute transactions in the same sequence, ensuring the Orders' consistency of data.

Like other blockchain systems, our system also uses a pluggable consensus protocol for ordering. Depending on the types of the Orders existing in the blockchain, i.e., the Orders are Byzantine or Crash-only, the consensus protocol may use Byzantine fault tolerance (BFT) protocol or crash fault tolerance (CFT) protocol.

It is important to choose a kind of consensus protocol because it decides not only the performance but also the safety of blockchain. For example, PBFT, one of BFT protocols, needs at least more than two of third Orders to be reliable and two rounds of broadcasting to provide safety for the blockchain system, where Orders may do malicious behaviours, e.g., Orders broadcast transactions in incorrect sequence. In contrast, if the Orders of the blockchain are crash-only, the system can choose Raft, one of CFT protocols, which only needs at least more than half Orders to be reliable and just one round of broadcasting.

After the primary of the Orders deciding the sequence of transactions and batching them into block, i.e., ordering, the primary of the Orders sends the block to the other Orders. Then, all Orders should execute all transactions in the block.

3.2   Assistants

For the existing blockchain systems, read-only transactions may consume computer performance which belongs to other tasks, e.g., process write transactions and produce blocks. Considering it, we can use Assistants to process read-only transactions to alleviate the workload of the Orders. Besides that, we also need to think over the system's scalability. The storage capacity of the Orders is limited, so we can expand the whole storage capacity of the system by extending Assistants.

In our system, the Assistants can be seen as the supplements to the Orders and there are mainly three functions for Assistants.

Firstly, every Assistant can become the Order, which can add the security for the blockchain. Because no matter what consensus protocol is chosen in the blockchain, the number of the Orders increases, and the system can tolerate more malicious Orders, which makes system more safer.

For example, Pbft is chosen as the consensus protocol, and at first, the number of Orders is four, i.e., the blockchain at most tolerates one Byzantine Order. Then, we should enhance the blockchain's system security, so we can choose three Assistants and let them be added into the Orders. At present, there are seven Orders, and it means the system can tolerate two Byzantine Orders.

Secondly, the Assistants can also process some read-only transactions for the Orders and enhance the query ability for the blockchain, e.g., fast range query while the existing blockchain may not support.

Thirdly, the blocks' data will be bigger and bigger, and the Orders' storage capacity is limited. To solve this problem, the Assistants can store the data for the Orders that means the Orders can delete it for saving storage space, and recover from the Assistants when necessary. Furthermore, the Assistants can use erasure code to reduce their own storing data by themselves [13].

4   EQblockchain

In this section, we present EQblockchain following the architecture on the above. Next, we will explain how our system works and how to ensure efficient query in the following subsections.

4.1   Transaction Process

First, we describe a normal request (transaction) event flow as follows. The Clients send requests to the Orders and the Orders agree on the sequence of the transactions. Then, the Orders construct a block with transactions and return results to the Clients after executing transactions.

We notice in this case that the read-only transactions will wait for themselves being batched into a block. Adding the read-only transactions into a block will increase the cost of the consensus. Because the number of the transactions in one block is limited, read-only transactions will take up the space of the block, which means the Orders need more time to send the block. Besides, read-only transactions will squeeze write transactions into the next block, resulting in the update of the blockchain being delayed. Furthermore, if the workload of the Orders is heavy, it is wasteful to query data by the Orders. For these reasons, we design our transaction process, and it has two parts. Figure 2 shows the two transaction process.

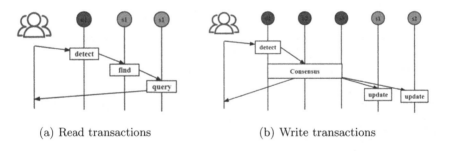

(a) Read transactions   (b) Write transactions

Fig. 2. The flow of processing transactions

Figure 2a shows how EQblockchain processes the query transactions. At first, Client sends a request to an Order. Then, the Order detects whether the Client belongs to its organization and the request is valid or not. If the request is valid, the Order detects whether the request is a read transaction. It is simple to detect it because we can get the read-write set of the transaction directly. If it is a query request, the Order sends it to the organization's Assistants. Assistants will find the location of the data according to data storage strategy, e.g., erasure code or hash, and execute the transaction to query the database. Finally, the Assistants will return the result to the Client.

Figure 2b shows how EQblockchian processes the write-transactions. Similar to query transactions, Client also sends a request to an Order. Then, the Order does the same thing to the request as above, but this time it will be detected as a write transaction. If the request is valid, the Order adds it to the set of waiting transactions. Next, the Order multicasts the waiting transactions to other Orders. Depending on the consensus protocol, the Orders establish a total order on transactions. After that, those transactions are batched into a block and sent to other Orders for executing. Once the consensus phase ends, the Orders send the results to the Client and the new state to the Assistants, updating the database.

4.2 Range Query

For the existing blockchain system, the range query may not be implemented or may be inefficient because it need to traverse all data in the key-value database. For enhancing the query ability of the EQblockchain, we use an inverted index to implement efficient range query in the Assistants.

In order to simplify the discussion, the data involved is integer. We build inverted indexes for the data, on which we want to support range queries, and there are two ways to build an inverted index: one is to construct a B+ tree and the other is to use a SkipList. We choose the SkipList as the actual implementation of the inverted index and put the state of the blockchain in memory without disk IO. Because SkipList has higher performance than B+ tree when adding or deleting records.

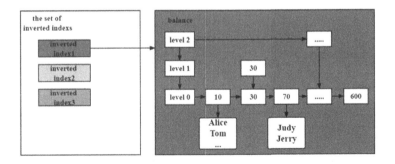

Fig. 3. The structure of inverted index

Figure 3 shows our inverted index implemented by SkipList. There are three inverted indexes and each one is a key-value pair, i.e., the key is the name of the inverted index, and the value is the specific SkipList. For example, in Fig. 3, the green inverted index is about accounts' balance and we can notice that the

SkipList is also the set of key-value pairs. In the green SkipList, we can find that the balances of "Alice" and "Tom" are 10 and the balances of "Judy" and "Jerry" are 70.

Algorithm 1 presents how our range query is performed. When we perform a range query, we should input the left value (LV) and right value (RV) of the query, and the name of the inverted index. Then, a set of Nodes SN is initiated to store the nodes involved. Next, the set of inverted indexes will be traversed to find the inverted index that we want. And then, we try to find the left node (LN) closest to the LV and the right node (RN) closest to RV. we sequentially traverse the nodes between the LN and the RN, and add them to SN. Finally, the set SN is returned. All data we want to query is in these nodes.

Algorithm 1. Range Query

1: **Input:** the **LV** and the **RV** of the range, the name of queried data **N**, the set of the inverted indexed **I**;
2: **Initialize:** set the set of Nodes **SN** to be empty;
3: **for** inverted index in I **do**
4:     **if** inverted index's name == N **then**
5:         LNode = findLeftNodeClosestToLeft(**LV**)
6:         RNode = findRightNodeClosestToRight(**RV**)
7:         **while** LNode ≠ RNode **do**
8:             Add LNode to SN
9:             LNode = LNode.next
10:         **end while**
11:         Add LNode to SN;
12:     **end if**
13:     break;
14: **end for**
15: **Output: SN**;

---

5   Experiment

5.1   Experimental Setup

In this section, we conduct several experiments to evaluate EQblockchain's query performance against baseline. We first demonstrate the throughput of the blockchain. Next, we report the efficiency of our range query. Our experiments are conducted by Tendermint on the Vmware Workstation and the computer is equipped with Ryzen-5600U CPU and 16GB RAM.

We implement a simple accounting application where each client has several accounts. Each account is a key-value pair of (Address, Balance), and the client sends requests to transfer assets from one account to another. For example, some client wants to transfer 1000 from Address1 to Address2 and produces two new pairs: (Address1, Balance1-1000) and (Address2, Balance2+1000).

For the first evaluation, the baseline is a normal blockchain, in which the Assistants do not exist. To show the effect of our query architecture, We adjust the number of Assistants from 1 to 3. For the second evaluation, the baseline approach is to traverse all data, and we build an inverted index to implement the range query. We change the number of the records and percentage of range queries to test how they impact on the query.

When reporting throughput measurements, we let each client connect to only one Order or one Assistant to send read-request, and five additional clients to send write-request. All the clients run on a single process, and the total number of the transactions in each group are fixed, i.e., 180000 transactions, while the read-write ratios of transactions in different group are not the same. Throughput is reported as the total number of the transactions divided by the processing time. When reporting range query measurements, we test many times and report the average time of these queries.

5.2   Throughput with Different Assistants

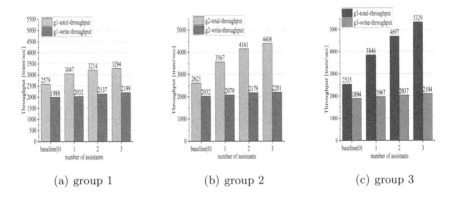

(a) group 1   (b) group 2   (c) group 3

Fig. 4. Throughput with different assistants

In this subsection, we report the impact of read-write ratio of transactions and the number of Assistants on system throughput. The read-write ratio of 1:2, 1:1, 2:1 are used in the experiment and the number of Assistants is changed from 0 to 3. As can be seen in Fig. 4, there are three groups of experiments. As mentioned before, the group 1(g1) has 120000 write-transactions and 60000 read-transactions, the group 2(g2) has 90000 write-transactions and 90000 read-transactions and group 3(g3) has 60000 write-transactions and 120000 read-transactions.

In these experiments, there is zero Assistant in baseline, and we change the number of Assistants to display the impact of Assistants on throughput. By increasing the number of Assistants, each group's throughput also increases as we distribute the query tasks to different nodes for parallel execution and the highest throughput of g1, g2 and g3 are 3294, 4408 and 5329 transactions per second respectively.

Here, it is worth noting that the increasing throughput in g2 and g3 almost linearly with the growth of the number of Assistants, while it does not happen in g1 due to the different read-write ratios of transactions.

In the EQblockchain, the Assistants only share the query tasks for the Orders. Thus, in the write-heavy workload like g1, the Assistants only have little effect against baseline. However, we should also note that by adding the amount of Assistants, the write-throughput of the system increases as well. Because the Assistants alleviate the Orders' workload, which makes the Orders have more CPU resource to process write-transactions, the total throughput is improved.

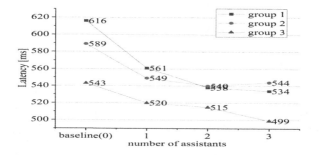

Fig. 5. Average latency with different Assistants

In Fig. 5, We can also notice that the average latency of transactions decreases when the number of Assistants increases. This is because our system is good at working with read-heavy workload, and therefore the average latency decreases as the read-write ratio of transactions increases and as the number of Assistants increases.

5.3   Performance About Range Query

We build inverted indexes to help our system improve the performance of the range query. In this set of experiments, we report the performance of range query with different number of records and query coverage in Figs. 6 and 7.

The baseline is to iterate over each item in database, as it does not know the location of the needed data. Our method builds an inverted index for the data, which may support range query, e.g., users' balance. There are three groups of experiments in Fig. 6, which represents the range query from one million, five million and ten million of records respectively and each group has three ratios of query coverage: 20

It is worth noting that the consuming time in SkipList and B+ tree is different, as the SkipList, we implement, uses linked list to store data but B+ tree uses array, i.e., range query is a kind of sequential access, where accessing

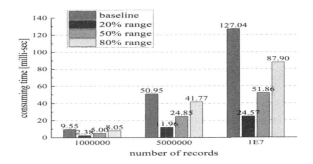

Fig. 6. Consuming time about range query using SkipList

array is faster than accessing linked list. For fair comparison, we choose the time accessing all data in the index as the baseline. Experiments show that the ratios of SkipList's consuming time to baseline are close to B+ tree's. While adding or deleting an item, SkipList is efficient than B+ tree, because the latter may re-balance while adding or deleting.

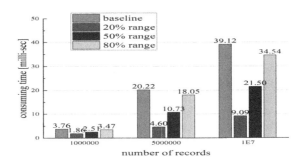

Fig. 7. Consuming time about range query using B+ Tree

6 Conclusion

In this paper, we propose our architecture and implement our permissioned blockchain, EQblockchain. EQblockchain effectively improves the query performance and slightly enhances the write performance, by introducing the nodes called Assistants to reduce the query pressure of the Orders. We explain the transaction process of our system and the support for range query. Our experimental evaluations show that EQblockchain has efficient performance in query-heavy workload, and good range query ability in comparison to the baseline.

Acknowledgements. This work was supported by the National Natural Science Foundation of China (62172082, 62072084, 62072086), the Fundamental Research

Funds for the central Universities (N2116008) and the Open Project Fund of Neusoft Corporation (NCBETOP2002).

References

1. Amiri, M.J., Agrawal, D., El Abbadi, A.: On sharding permissioned blockchains. In: 2019 IEEE International Conference on Blockchain (Blockchain), pp. 282–285. IEEE (2019)
2. Amiri, M.J., Agrawal, D., El Abbadi, A.: Parblockchain: Leveraging transaction parallelism in permissioned blockchain systems. In: 2019 IEEE 39th International Conference on Distributed Computing Systems (ICDCS), pp. 1337–1347. IEEE (2019)
3. Amiri, M.J., Agrawal, D., El Abbadi, A.: Sharper: sharding permissioned blockchains over network clusters. In: Proceedings of the 2021 International Conference on Management of Data, pp. 76–88 (2021)
4. Androulaki, E., et al.: Hyperledger fabric: a distributed operating system for permissioned blockchains. In: Proceedings of the Thirteenth EuroSys Conference, pp. 1–15 (2018)
5. Androulaki, E., Cachin, C., De Caro, A., Kokoris-Kogias, E.: Channels: horizontal scaling and confidentiality on permissioned blockchains. In: Lopez, J., Zhou, J., Soriano, M. (eds.) ESORICS 2018. LNCS, vol. 11098, pp. 111–131. Springer, Cham (2018). https://doi.org/10.1007/978-3-319-99073-6_6
6. Dang, H., Dinh, T.T.A., Loghin, D., Chang, E.C., Lin, Q., Ooi, B.C.: Towards scaling blockchain systems via sharding. In: Proceedings of the 2019 International Conference on Management of Data, pp. 123–140 (2019)
7. Gorenflo, C., Golab, L., Keshav, S.: XOX fabric: a hybrid approach to blockchain transaction execution. In: 2020 IEEE International Conference on Blockchain and Cryptocurrency (ICBC), pp. 1–9. IEEE (2020)
8. Kapritsos, M., Wang, Y., Quema, V., Clement, A., Alvisi, L., Dahlin, M.: All about eve:{Execute-Verify} replication for {Multi-Core} servers. In: 10th USENIX Symposium on Operating Systems Design and Implementation (OSDI 12), pp. 237–250 (2012)
9. Kokoris-Kogias, E., Jovanovic, P., Gasser, L., Gailly, N., Syta, E., Ford, B.: OmniLedger: a secure, scale-out, decentralized ledger via sharding. In: 2018 IEEE Symposium on Security and Privacy (SP), pp. 583–598. IEEE (2018)
10. Luu, L., Narayanan, V., Zheng, C., Baweja, K., Gilbert, S., Saxena, P.: A secure sharding protocol for open blockchains. In: Proceedings of the 2016 ACM SIGSAC Conference on Computer and Communications Security, pp. 17–30 (2016)
11. Mao, X., Li, X., Guo, S.: A blockchain architecture design that takes into account privacy protection and regulation. In: Xing, C., Fu, X., Zhang, Y., Zhang, G., Borjigin, C. (eds.) WISA 2021. LNCS, vol. 12999, pp. 311–319. Springer, Cham (2021). https://doi.org/10.1007/978-3-030-87571-8_27
12. Nakamoto, S., Bitcoin, A.: A peer-to-peer electronic cash system. Bitcoin. **4**, 2 (2008). https://bitcoin.org/bitcoin.pdf
13. Qi, X., Zhang, Z., Jin, C., Zhou, A.: A reliable storage partition for permissioned blockchain. IEEE Trans. Knowl. Data Eng. **33**(1), 14–27 (2020)
14. Ruan, P., Dinh, T.T.A., Lin, Q., Zhang, M., Chen, G., Ooi, B.C.: LineageChain: a fine-grained, secure and efficient data provenance system for blockchains. VLDB J. **30**(1), 3–24 (2021)

15. Ruan, P., Loghin, D., Ta, Q.T., Zhang, M., Chen, G., Ooi, B.C.: A transactional perspective on execute-order-validate blockchains. In: Proceedings of the 2020 ACM SIGMOD International Conference on Management of Data, pp. 543–557 (2020)
16. Sharma, A., Schuhknecht, F.M., Agrawal, D., Dittrich, J.: Blurring the lines between blockchains and database systems: the case of hyperledger fabric. In: Proceedings of the 2019 International Conference on Management of Data, pp. 105–122 (2019)
17. Zamani, M., Movahedi, M., Raykova, M.: RapidChain: scaling blockchain via full sharding. In: Proceedings of the 2018 ACM SIGSAC Conference on Computer and Communications Security, pp. 931–948 (2018)

MCQL: A Multi-node Consortium Blockchain Query Method Based on Node Dynamic Adjustment

Zhibo Zhou, Tiezheng Nie[(⊠)], Derong Shen, and Yue Kou

School of Computer Science and Engineering, Northeastern University, Shenyang 110004, China
{nietiezheng,shenderong,kouyue}@cse.neu.edu.cn

Abstract. Blockchain plays an important role in the secure storage and efficient query of application data, and blockchain query performance is also a challenging research task. The system may send a large number of query requests to a node with strong responsiveness, which will cause the node to be paralyzed and cause the system to crash. Due to the high time complexity of consensus reached by single-layer nodes in the consortium blockchain system, it affects the generation of new blocks, which indirectly affects the system query performance and reduces query efficiency. And for a multi-node system, using the same access priority for nodes with different corresponding rates will undoubtedly affect the query efficiency of the system. This paper proposes a query method in the case of multiple nodes in a consortium blockchain network. This method divides nodes into different levels according to their response capabilities, gives incentive values to nodes that successfully respond to query requests, and dynamically adjusts the level of nodes according to the incentive values. The higher the level, the higher the access priority, which fully considers the security and query performance. The experimental results show that the performance of the method proposed in this paper is better than the existing multi-node query methods of the consortium blockchain, which can improve the query performance of the consortium blockchain system and ensure the security of the system.

Keywords: Consortium blockchain query · Node hierarchy division · Dynamic adjustment · Node excitation

1 Introduction

With the continuous development of Internet technology, a large number of emerging technologies have been derived, such as block chain, cloud computing, big data, 5G and other technologies. With the maturity of the above emerging technologies, a large number of related applications and projects have been implemented. Blockchain technology since 2008 scholar Satoshi Nakamoto proposed Bitcoin white paper Bitcoin: A Peer-to-peer Electronic Cash System [1] has attracted continuous attention from A large number of researchers in the field of computer. Its underlying technologies include distributed ledger [2], consensus mechanism, smart contract and cryptography [3] algorithm. It has

© The Author(s), under exclusive license to Springer Nature Switzerland AG 2022
X. Zhao et al. (Eds.): WISA 2022, LNCS 13579, pp. 712–723, 2022.
https://doi.org/10.1007/978-3-031-20309-1_62

the characteristics of decentralization, immutability, transparency, traceability and final consistency.

Blockchain technology as a combination of a variety of basic technology of distributed Shared general ledger, all participants can be understood as a kind of network use and maintenance of the distributed database together, but different from the original database is that the parties have obtained all the data on the chain, improves the query efficiency of participants and system data security at the same time, It reduces the maintenance cost of the original database maintainer. Public blockchain uses a decentralized approach, such as blockchain nodes participating in transactions through the PoW consensus mechanism, jointly maintaining the integrity of the public blockchain and the persistence of transactions, ensuring the credibility and traceability of data. Thus, blockchain is considered an effective solution for the secure storage and query of data in many decentralized applications, such as the Internet of Things, healthcare, supply chain, and legal document management [4]. The architectural features of blockchain can protect account privacy [5].

The query of the blockchain is based on the structure of the blockchain. There are many blocks on the blockchain, and each block is linked by an encrypted hash pointer. Each block consists of a block header and a block body. The block header mainly includes the version number, the hash value of the previous block, the timestamp, the random number, and the Merkle root hash value. The structure of the Merkle tree is stored in the block body, and the leaf nodes of the Merkle tree store the target data to be queried and the hash value of the transaction data. The query needs to verify the security of the data. The security verification process is the Merkle proof formed by the search path for verification. The verifier can reconstruct the root hash according to the Merkle proof and the hash value of the object to be searched, and compare the constructed root hash with the root hash in the block header is compared. The matching results are consistent, which proves that the data object to be searched exists in the MHT and has not been tampered with, so the security of the data can be verified.

Consortium blockchain is a blockchain whose consensus mechanism is controlled by a group of pre-selected nodes. The pre-selected nodes have high credibility, and the permission control on the joining nodes takes effect when the number of signed nodes reaches two-thirds of the total in the PBFT consensus process, and the information may not be public, such as the root hash value of the public block. Consortium blockchain is a weakly centralized blockchain. Although sacrificing part of the decentralized nature may have an impact on security, consortium blockchain reduces the verification burden and transaction cost to improve the block rate and obtain a more flexible blockchain structure. At present, a large number of consortium blockchain systems are based on full copy backup, which will increase the storage burden of the system and reduce the query efficiency of target data blocks with the continuous writing of blocks. Using the same access priority for nodes with different response rates will undoubtedly reduce the query performance of the system. The single-layer PBFT consensus protocol of consortium blockchain can only tolerate no more than 1/3 of the total number of faulty nodes, and the communication complexity of consensus node group is relatively high.

Inspired by the above content, this paper proposes a block chain query method based on dynamic adjustment of node level, which solves the shortcomings of the previous work by dividing nodes into levels, motivating nodes and dynamically adjusting the level according to the excitation value of nodes. The main contributions of this paper are as follows:

- The nodes in the system are divided into different levels according to their response capability. The higher the level of the node is, the stronger the response capability of the node is. Moreover, multi-level PBFT consensus can reduce the communication complexity between nodes.
- Propose the method of node excitation and dynamic adjustment. The nodes in the system that respond successfully are rewarded with excitation values, and the nodes in the system can dynamically adjust their levels according to the excitation values.
- Realize the ability to distinguish the response of nodes for different numbers of nodes. The higher the level of the node, the faster the response rate to the system, thereby improving the query performance of the system.

The remainder of this paper is structured as follows: Sect. 2 describes the related work, term definitions and descriptions are given in Sect. 3, the query processing method is described in detail in Sect. 4, and related experiments are given in Sect. 5. Finally, the full text is summarized in Sect. 6.

2 Related Work

The query of blockchain data is based on the data storage method of blockchain, and different storage methods adopt different query methods. The data storage method of the permissionless blockchain is based on the full text mode, that is, all nodes in the permissionless blockchain system store a complete distributed ledger. The Merkle tree is constructed bottom-up. The basic transaction data is stored in the leaf nodes of the Merkle tree. The non-leaf nodes of the Merkle tree are composed of the hashes of its child nodes. Each search needs to query most of the blocks, and the verification of the data in the block requires the Merkle proof.Representative of the permissionless blockchain system is Bitcoin, Ethereum and so on.

Li et al. [6] designed EtherQL system based on the characteristics of blockchain and the high query efficiency of traditional database. This paper mainly used external database, through the block chain data copy to MongoDB, with MongoDB efficient query interface management data, so as to achieve a variety of queries, with high flexibility. Blockchain.com [7] was able to provide address information because it pre-stored historical transactions in a database.BlockSci [8] integrated an in-memory database to facilitate data queries for blockchain analytics. However, these systems assumed that the server always return the correct result based on the blockchain data. In fact, the server may return false results that conflict with real blockchain data due to some interest or security holes [9]. Inspired by version control system, Dinh et al. [10] proposed UStore, a data storage system with rich semantics, which maintained a directed acyclic graph for each key to facilitate the historical version query of key, and proposed UStore as the underlying storage of blockchain system. UStore ran efficiently, supported multiple query semantics, and effectively ensured data imtambility, shareability, and security. Wu

H et al. [11] pointed out the problem of direct query of blockchain. Searching each block took a lot of time, and indirect query of blockchain greatly reduced the authenticity of query results. A verifiable query layer (VQL), deployed in the cloud, was proposed to provide efficient and verifiable query services for blockchain systems, and to calculate password fingerprints for each constructed database to ensure the authenticity of query results. Cheng Xu et al. [12] proposed vChain, a verifiable framework, to solve the problem of how to verify data to ensure the integrity of results when lightweight nodes use full nodes for data query. vChain designed a kind of authentication data structure based on accumulator, which can support the dynamic aggregation of any query attributes, so as to realize the data verification of any attributes. Two data indexes and one index based on prefix tree were designed to improve query efficiency.

Himanshu et al. [13] summarized Fabric[16] 's approach to querying historical data, pointed out Fabric's shortcomings in handling temporal historical data, and proposed two new models to overcome temporal query limitations in Fabric by creating replicas and inserting data into metadata, respectively. Experiments compared the two models and the original methods of Fabric[17][16] in many aspects, and verified that the model in this paper greatly improved the efficiency of Fabric processing historical data. Wenyu Li et al. [14] pointed out that due to frequent communication between nodes, the node scalability of PBFT mechanism was poor, and it could only be applied to small networks with low query efficiency. In order to support PBFT in large-scale systems such as IOT ecosystem and blockchain, a scalable consensus mechanism based on multi-layer PBFT was proposed by hierarchical grouping of nodes and limiting intra-group communication, and the communication complexity of the algorithm was significantly reduced to improve query efficiency. Xiaodong Qi et al. [15] proposed a new storage engine called BFT-Store, which enhanced the scalability of storage by combining erase coding with Byzantine fault tolerance (BFT) consistency protocol. The first feature of BFT-Store is that the storage consumption of each block can be reduced to $O(1)$ for the first time. When more nodes join the blockchain, the overall storage capacity can be expanded. Although this scheme can reduce the redundant data of each node, the query request may need to be recovered by redundant blocks and data processing before the query result can be returned. Nodes need to access the index database of the target block file code block for query, which may reduce the query efficiency.

Based on the above research work, in order to solve the security and efficiency problems of blockchain query, this paper proposes a blockchain query method based on dynamic adjustment of node level, considering the communication cost of nodes and priority access of nodes in the query process.

3 Terminology Description and Definition

Blockchain is used to manage data, which is stored in the leaf nodes of a Merkle tree. In mathematical statistics, variables can be divided into two types, discrete type and continuous type. Before introducing the method in this paper, we first introduce the concepts related to data and node level, and make detailed formal definitions for data objects, data sets and nodes as follows:

Definition 1. Data Objects. Let X_i be a data object, C_i be a continuous variable of X_i, D_i be a discrete variable of X_i, $hash_i$ is the hash value of the data object X_i and used as the primary key of the data object X_i, the definition of the data object in this paper is as follow:

$$X_i = \langle C_i, D_i, hash_i \rangle \tag{1}$$

Definition 2. Data Collection. X is the data set of all data objects stored in a block in the blockchain system, is the data object X_i, and the data set X is defined as follows:

$$X = \langle X_1, X_2, X_3 \ldots X_n \rangle \tag{2}$$

Definition 3. Blockchain Node. Let r_i^m be the i-th node of the m-th layer, where the superscript m indicates the layer number of the blockchain node, and the subscript i indicates that the node is the index of the layer. is the leader of the m + 1th layer node r_i^{m+1}.

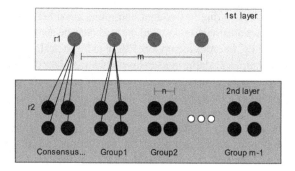

Fig. 1. Node Hierarchy Model diagram.

4 Query Processing

4.1 Node Layering

The blockchain system has the characteristics of decentralization, each node has the same status and communicates between nodes in a p2p manner. This paper first divides the nodes in the system into multiple levels according to their responsiveness. The structure of the levels in the system is shown in Fig. 1. The level of the node can be dynamically adjusted according to the excitation value of the node. The detailed process is involved in the following Sect. 4.3. Each level can accommodate a certain proportion of nodes, and the proportion increases as the level increases.

4.2 Node Query and Incentive

The overall framework of the query in this paper is shown in Fig. 2. The client node sends a query request to the blockchain network, and the blockchain system broadcasts the request message. After receiving the query request message, each layer node in the system calls the RPC interface of the local smart contract to find leveldb and query the area from the blockchain. The Merkle tree in the block obtains the data set. The search process in the Merkle tree is given by the query algorithm in 4.3. The obtained Merkle proof and the hash of the search data are connected to obtain the final hash value, and the final hash value is combined with the block header. The stored root hash values are compared. If they are consistent, the data has not been tampered with, and if they are inconsistent, the data is insecure. The node returns the found data to the client node. After the client node receives the response data, it gives the response node an incentive value. After the nodes are layered, the first query object of the client node is the node at a relatively high level. They have a high response rate and can well ensure the query performance of the system.

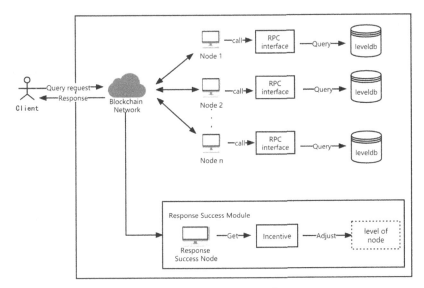

Fig. 2. The overall framework diagram

4.3 Data Query

This paper provides a Data query algorithm in the Merkle tree. When the user enters the query value d, the system will traverse from the root node of the Merkle tree in the last block, and return the result if the corresponding data of the query value d can be found; then Sequentially compare the previous block until the genesis block. The data query process is shown in Algorithm 1.

Algorithm 1: *Data query*

Input: Query Attribute d, Block Set BS

Output: The Result Set R

1. while !BS. isempty()
2.   node=BS. first(). tree. root;
3.   queue.add(node) ;
4.   while !queue.isempty()
5.   node=queue.pop();
6.   if node ∈ leafnode&&node. Dij ==d
7.   hash=node.hash,hash.Xi→R;
8.   else if node∉leafnode
9.   queue.add(node.left),queue.add(node.right) ;
10.   end if
11. BS.remove();
12. end while
13. return R;

In Algorithm 1, line 1 calls each block cyclically. Lines 2–3 put the root node of the block into the queue. Lines 4–12 traverse the Merkle tree hierarchically, and lines 6–7 judge the node Whether it is a leaf node, if so, compare whether its attribute is the queried d, if so, find the data hash value, and put its corresponding data into the result set R; Lines 8–9 are to determine whether the node is a non-leaf node, if its left and right child nodes are put into the queue. Line 13 returns the final result.

4.4   Dynamic Adjustment of Node Levels

The blockchain system is usually accompanied by the joining and exiting of nodes, or even sudden node downtime and network isolation. Under the condition of a large number of query transactions, it is very likely to cause network congestion. If it is not handled in time, the system will crash. Another example is that a new node is added to the system, the node can respond to the query request of the client node with a faster response speed, thereby obtaining the conditions for node upgrade and realizing the level rise in the hierarchical structure. The hierarchical dynamic adjustment process is shown in Algorithm 2.

In Algorithm 2, line 1 calls the consensus group in the system cyclically, and lines 4–17 of the algorithm traverse the nodes in the consensus group. Lines 6–8 judge the status of the 0-layer node, and lines 7–8 judge whether the node is down. If true, the node level drops. Lines 9–14 judge the status of non-level 0 nodes, line 12 judges whether the upper-level nodes in this group are down or exited, and lines 13–14 judge whether the current node incentive value is the largest in the group and the node at the same level, if it is true the node level goes up.

Algorithm 2: *level dynamic adjustment*

Input: Consensus Group Set GS;
Output: NULL
1. while ! GS.isempty()
2.   group＝BS.first();
3.   Init(index);
4.   while group.size()>0
5.   node＝group.get(index);
6.   if node.level=0
7.   if node.status==breakdown
8.   node.level-=1;
9.   else if node.level>0
10.   if node.status==breakdown
11.   node.level-=1;
12.   else if LastLevel_inGroupNode(node.level-1).status==breakdown or exit
13.   if NowLevelGroup_IsMax(node.incentive)==true
14.   node.level+=1;
15.   end if
16.   index++;
17.   continue;
18. end while

5   Experiments

5.1   Experimental Setup

Based on tendermint v0.1, this paper rewrites the node structure and transaction structure, and realizes the writing of the experimental code in this paper.

Experimental Environment: Intel(R) Core(TM) i7–8700 CPU; RAM is 16 GB; Operating system is 64-bit Windows 10; the underlying database uses leveldb. The virtual machine is Ubuntu 20.04LTS.

5.2   Comparative Experiment

This paper compares the original structure of tendermint. Since the original structure of tendermint does not involve query processing for multi-node cases, this paper rewrites the original node structure and transaction structure of tendermint, adds attribute values, and uses it as the baseline method. In the baseline, this paper adopts the random access method for the query in the case of multiple nodes. The client randomly accesses the nodes in the network system. The node to be searched will traverse all the data in the block, and return the result if it exists. Proposal represents the method proposed in this paper.

5.3 Experiment Results and Analysis

5.3.1 Time-Consuming Comparison of Querying Multiple Transactions

There are 4 nodes in the system and each node stores multiple transactions. This paper research the relationship between the number of query transactions and query time. Figure 3 compares the query time required for the number of query transactions to be 100,000, 200,000, 300,000, 400,000, and 500,000. It can be seen that the system query time increases with the number of query transactions, showing a linear growth trend (Fig. 3).

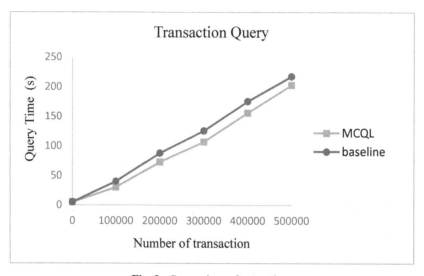

Fig. 3. Comparison of query time

5.3.2 Comparison of System Throughput

Based on the existence of four nodes in the system to store multiple transactions, this paper research the ratio of the number of query transactions to the required time to calculate the throughput of the system. Figure 4 compares the relationship between the system throughput and the number of query transactions. It can be seen from the figure that the throughput changes little with the increase in the number of query transactions, and is almost a constant.

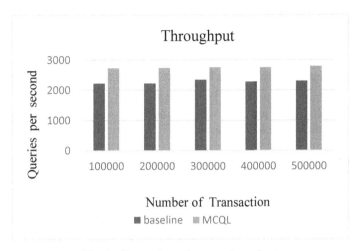

Fig. 4. Comparison of system throughput

5.3.3 Comparison of Query Time Under Multiple Blocks

Based on the existence of four nodes in the system, this paper research the relationship between block height and query time. Figure 5 compares the query time of the system under different block heights. It can be seen that the query time increases with the increase of the block height, showing a linear growth trend. This is because more transactions are committed, more blocks are generated. The query process needs to traverse the blocks,

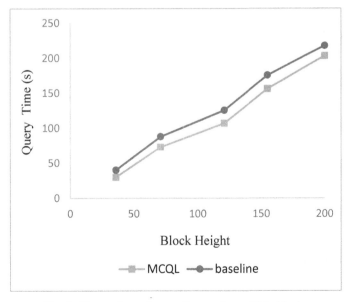

Fig. 5. Comparison of query time under multiple blocks

and as the number of blocks increases, the time required for the query will also increase. The experimental results are in line with expectations (Fig. 5).

5.3.4   Comparison of Query Time Under Multiple Nodes

This section studies the effect of the number of nodes on the query time. In this section, the experiment compares the query time of the two systems with different numbers of nodes under the condition of querying 500,000 transactions. As can be seen from Fig. 6, as the number of nodes increases, the query time required by the system also increases. When the number of nodes is 1, the time required for both is equal, because both are single-node systems at this time. This is as expected. When the number of nodes increases, the query time required by MCQL is less than baseline, which is in line with the expected results. Because as the number of query transactions increases, the level of nodes with faster response rates changes, and nodes with faster response rates have higher levels of nodes and are more preferentially accessed by clients (Fig. 6).

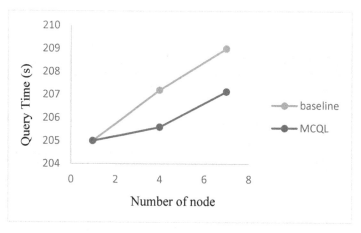

Fig. 6. Comparison of query time under multiple nodes.

6   Conclusion and Outlook

This paper proposes a new multi-node query method based on the dynamic adjustment of consortium blockchain nodes. By dividing the nodes in the consortium blockchain into layers, the client node sends a data query request to the consortium blockchain network, and successfully responds to the data to the client. The node gets the incentive value, and dynamically adjusts the level of the node through the incentive value. In this paper, the proposed method is compared with the original multi-node query method through experiments. The experimental results show that the method in this paper is effective in multi-node query. Next, I will do a more in-depth study of cluster and blockchain multi-node query optimization.

Acknowledgment. This work was supported by the National Social Science Foundation(21&ZD124) .

References

1. Nakamoto, S.: Bitcoin: A Peer-to-Peer Electronic Cash System[EB/OL].(2008–11–1) metz-dowd.com
2. Benčić, F.M., Podnar Žarko, I.: Distributed ledger technology: blockchain com-pared to directed acyclic graph. In: 2018 IEEE 38th International Conference on Distributed Computing Systems (ICDCS), pp. 1569–1570 (2018)
3. Lone, A.H., Naaz, R.: Demystifying cryptography behind blockchains and a vision for post-quantum blockchains. In: 2020 IEEE International Conference for Innovation in Technology (INOCON), pp. 1–6 (2020)
4. Zhang, C., Xu, C., Wang, H., et al.: Authenticated keyword search in scalable hybrid-storage blockchains. In: 2021 IEEE 37th International Conference on Data Engineering (ICDE), pp. 996–1007. IEEE (2021)
5. Xing, C., Fu, X., Zhang, Y., Zhang, G., Borjigin, C. (eds.): WISA 2021. LNCS, vol. 12999. Springer, Cham (2021). https://doi.org/10.1007/978-3-030-87571-8
6. Li, Y., Zheng, K., Yan, Y., Liu, Q., Zhou, X.: EtherQL: a query layer for blockchain system. In: Candan, S., Chen, L., Pedersen, T.B., Chang, L., Hua, W. (eds.) DASFAA 2017. LNCS, vol. 10178, pp. 556–567. Springer, Cham (2017). https://doi.org/10.1007/978-3-319-55699-4_34
7. Blockchain.com (2019). https://www.blockchain.com/explorer
8. Kalodner, H., et al.: BlockSci: Design and applications of a blockchain analysis plat-form. In: 29th USENIX Security Symposium (USENIX Security 20), pp. 2721–2738 (2020)
9. Ren, K., Wang, C., Wang, Q.: Security challenges for the public cloud. IEEE Internet Comput. **16**(1), 69–73 (2012)
10. Dinh, A., Wang, J., Wang, S., et al.: UStore: A distributed storage with rich semantics [EB/OL]. https://arxiv.org/abs/1702.02799, 2017–7–9
11. Wu, H., Peng, Z., Guo, S., et al.: VQL: efficient and verifiable cloud query services for blockchain systems. IEEE Trans. Parallel Distrib. Syst. **33**(6), 1393–1406 (2021)
12. Xu, C., Zhang, C., Xu, J.: vchain: enabling verifiable boolean range queries over block-chain databases. In: Proceedings of the 2019 International Conference on Management of Data, pp. 141–158 (2019)
13. Gupta, H., Hans, S., Aggarwal, K., et al.: Efficiently processing temporal queries on hyper-ledger fabric. In: 2018 IEEE 34th International Conference on Data Engineering (ICDE), pp. 1489–1494. IEEE (2018)
14. Li, W., Feng, C., Zhang, L., et al.: A scalable multi-layer PBFT consensus for blockchain. IEEE Trans. Parallel Distrib. Syst. **32**(5), 1146–1160 (2020)
15. Qi, X., Zhang, Z., Jin, C., Zhou, A.: A reliable storage partition for permissioned blockchain. IEEE TKDE **33**(1), 14–27 (2020)
16. Androulaki, E., Barger, A., Bortnikov, V., et al.: Hyperledger fabric: a distributed operating system for permissioned blockchains. In: Proceedings of the Thirteenth EuroSys Conference, pp. 1–15. ACM (2018)
17. Chacko, J.A., Mayer, R., Jacobsen, H.A.: Why do my blockchain transactions fail? a study of hyperledger fabric. In: Proceedings of the 2021 International Conference on Management of Data, pp. 221–234 (2021)

An Ethereum-Based Image Copyright Authentication Scheme

Xueqing Zhao[1]($\boxtimes$), Hao Liu[1], Shuning Hou[1], Xin Shi[1], Yun Wang[2], and Guigang Zhang[2]

[1] Shaanxi Key Laboratory of Clothing Intelligence, School of Computer Science, Xi'an Polytechnic University, Xi'an 710048, China
`zhaoxueqing@xpu.edu.cn`
[2] Institute of Automation, Chinese Academy of Sciences, Beijing 100190, China

Abstract. Decentralization and on-chain data Immutability, two of the primary characteristics of blockchain, drive the further technological development as well as effectively reduce the cooperation cost in present era of artificial intelligence. In this paper, an image copyright authentication scheme is proposed on the basis of Ethereum to avoid the watermarking embedding or third-party organizations' certification in traditional image copyright authentication mechanism. Firstly, image summary descriptions are calculated through the Secure Hash Algorithm 3(SHA-3) and stored subsequently in blockchain by means of smart contracts. Then, the ownership of the image is certified by timestamp on the blockchain. Besides, with the consideration of multiple and repeated transactions of image copyright, a new image copyright authentication traceable model is proposed to enable the image authentication and traceability by identifying every user-held timestamp from the very beginning when the image is generated to the last deal. Finally, our proposed scheme has been realized on Ganache, a fast and convenient Ethereum construction platform where the smart contracts have been tested and modified for many times. Simulation results show that the Ethereum based image copyright authentication scheme proposed in this paper is capable of effectively protecting the copyright of the image and tracking the source on the blockchain.

Keywords: Blockchain · Ethereum · Smart contract · Image copyright authentication

1 Introduction

With the fast development of internet and multimedia industry, the protection of copyright is becoming more and more important [8]. Usually, most medium information on the internet is organised in the form of image and has been traded as commodities, which means that image data could be owned individually or collectively as an asset. Therefore, an effective protection of image copyright has become an imperative requirement. Currently, two main methods of image

X. Zhao et al. (Eds.): WISA 2022, LNCS 13579, pp. 724–731, 2022.
https://doi.org/10.1007/978-3-031-20309-1_63

copyright protection have been proposed: third-party copyright certification and copyright-traceable image watermarks' retention. The former is complicated and costly to process the image copyright, and it is impractical to certificate image copyright when a large number of image works are produced. The latter commonly uses the Watermark to be embedded in the protected images, but the watermarks are easily corrupted.

Blockchain is a unique distributed ledger technology where every computer in each account node can share the blockchain's ledger. The blockchain's transaction records are timestamped and tamper-proof [9], which provides the underlying security for image copyright authentication, transactions and even traceability. From the perspective of copyright authentication, blockchain can be used to solve the problem of data trust. The successful operation of smart contracts on the Turing-complete Ethereum Virtual Machine (EVM) enables blockchain to play its due role in the field of decentralized applications. There are two key technologies in blockchain: cryptography and consensus mechanism. In cryptography, the irreversibility of hash function is used to increase the anti-modification and anti-collision of blockchain; asymmetric encryption is used to realize public key open and private key secretive. Consensus mechanism is the core technology to realize decentralization of blockchain. Generally, each node reaches a consensus through negotiation, and common consensus algorithms include pow, pos, etc.

The existing blockchain can be divided into public blockchain, private blockchain and consortium blockchain according to usage scenarios. In order to facilitate the deployment and debugging of smart contracts and reduce unnecessary communication time and consensus cost, the image copyright authentication scheme proposed in this paper will be deployed and realized based on the private blockchain. A smart contract can be regarded as an electronic contract managed by a computer. After the contents, conditions and specifications of the contract have been agreed upon by many parties, the contract will be handed over to the computer for storage and processing, and the computer will arbitrate when necessary, so as to avoid unnecessary fussiness and errors caused by human factors. Therefore, the smart contract is absolutely fair. From the other point of view, the smart contract is a program stored in the blockchain, which is usually triggered by a transaction. Once all the execution conditions of the smart contract are satisfied, the smart contract program will automatically execute without any intervention from a third party. Though the smart contract leads to longer execution time and higher cost, it guarantees the decentralized operation of the Ethereum blockchain [10].

As a new blockchain platform, Ethereum generates smart contract based on the original features of the blockchain, which can effectively track the execution of all contract-related situations. This paper is a practical application of blockchain technology on image copyright authentication by deploying the smart contracts on the ethereum blockchain.

2   Algorithms, Architecture and Implementation

The scheme of image copyright authentication proposed in this paper mainly focuses on two aspects: image encryption hash and image copyright authentication. Due to the requirement of images' ownership authentication and the limitation of blockchain storage, we use SHA3-256 to generate image summary description and upload it onto blockchain; besides, the timestamp of the copyright for each image transaction is also stored in blockchain. Furthermore, we use image upload timestamp and copyright trading timestamp to authenticate the image ownership.

2.1   Algorithm Introduction

Image Encryption and Hash. Since it is too costly to store the whole image in the blockchain, the scheme proposed in this paper firstly stores the original image in the database, then the image is encoded through base64 encoding method to transform the image into a String document. Afterwards, SHA3-256 operation is applied on the String document to get the hash value of the image which is interacted with the smart contract and packed into block to be stored in the blockchain. Base64 is a 64-base position counting method, which was first used to encrypt binary data transmission [2], after the image is converted into text by Base64, it is convenient for the next SHA-3 encryption operation. SHA-3 is a very secure hash algorithm, any size of data input can get a fixed-length hash summary description through SHA-3. SHA-3 functions are primarily used to provide attack resistance properties such as resistance to collision, forward image and second forward image attacks [4]. The hash function is a fundamental component in various information security applications and is used in this paper for the primary purpose of encrypting and compressing images [5].

Image Copyright Authentication. The scheme proposed in this paper has two user identities: the image creator and the image user. The structural model is shown in Fig. 1, where the smart contract is written in solidity, compiled by solc compiler and deployed to the ethereum blockchain through transactions. The interaction deployment process is as follows.

(1) Smart contract developers write smart contracts and deploy them to the ethereum blockchain by triggering transactions.
(2) The copyright owner deposits the image into the database via the client, and transaction triggers a smart contract to deposit the image hash into the blockchain, finally the client returns the image ID to the image creator.
(3) Image users find the images they need through the client.
(4) Image users can buy the copyright of the image creator through a transaction and trigger a smart contract for the transfer of the copyright.

The two core functions of this system are image tracking and image copyright trading. The image creator firstly triggers the smart contract to upload the image summary descriptions to the blockchain through trading, the smart contract

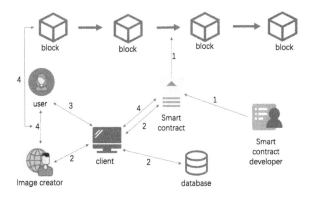

Fig. 1. The interaction process between image creator and image user.

will return the timestamp and image ID of the client's image uploading. After purchasing an image, the image ID is transferred to the buyer and returned the image ID and transfer timestamp to them, which are used to prove the user's ownership of the image. The process of copyright authentication and trading is explained in detail below in conjunction with the smart contract.

Smart Contract Model. In order to realize image copyright authentication and copyright transaction, two smart contracts are designed, image creation contract and image copyright transfer contract. Here, image creation contract is used to describe image of the current account to upload the blockchain. and image copyright transfer contract contains three private methods, namely, transferring the ownership of the creator to the target user, deleting the image ID from the creator's address and transferring the image ID to the target address. Through the interaction between these contracts, the images can not only circulate on the Ethereum blockchain, but also be traced back to the former holders of the images and the transaction records of the images, thus providing effective proof for copyright authentication when disputes arise. image copyright transfer contract is the most critical part of the whole scheme, aim to remove the image ID from the holder address to the target user address, after the transaction triggers the smart contract, the smart contract will run automatically without any third party interference. The interaction between these functions not only prevents illegal node transactions but also effectively prevents the double-spending problem similar to that of Bitcoin [1]. The key outputs after a transaction triggers the ownership transfer contract and completes the copyright transfer are shown in Table 1, including the seller, buyer, image ID, transaction hash and timestamp.

Table 1. Output information for image copyright transactions.

Seller	Buyer	Image ID	Transaction hash	Timestamp
0 × 5B38...ddC4	0 × Ab84...5cb2	946152261	0 × e629...14b3	May-1 16:28:57
0 × 5B38...ddC4	0 × 4B20...02db	636613330	0 × 7fc3...dbfc	May-1 16:31:22
0 × 5B38...ddC4	0 × 7873...abaB	749313250	0 × 7968...b32a	May-1 16:28:57

2.2   Architecture Design

The architecture of this scheme consists of three main layers, the storage layer, the consensus layer and the application layer. The main architecture of the proposed scheme is shown in Fig. 2.

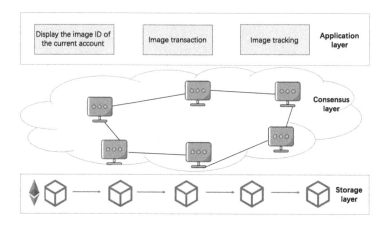

Fig. 2. The main architecture of the proposed scheme.

Storage layer: The storage layer is mainly composed of blocks, which can be regarded as a distributed storage system. Users store image summary descriptions and perform image copyright transactions, which will change the state variables in the smart contract, so they will trigger transactions. Once a transaction is triggered The parameters and other transaction information entered by the user during the transaction will be packaged by miners into blocks and become non-tamperable data for later image copyright authentication and image tracking. Therefore, the storage layer is the foundation of the entire system.

Consensus layer: The consensus mechanism is the core part of the decentralized application of blockchain. It is the basis for the decentralization of blockchain and the immutability of data on blockchain. It can be regarded as a kind of rule, which is used to specify a node from all the mined nodes and use it to calculate new block, thereby ensuring that all nodes in the entire blockchain system reach an effective consensus on each data transaction [6].

Application layer: The application layer is the bridge between the user and the system. After the above functions are implemented through smart contracts and deployed to the blockchain, the user can interact with it through the application layer, which also deploys the image summary description calculation module.

2.3   Implementation of Dapp

Ganache. Ganache is a personal blockchain platform for the rapid development of decentralized applications based on Ethereum. The development, testing and deployment of Dapps (short for decentralized application) can be carried out based on ganache. We use ganache to create an Ethereum private blockchain, and ganache could automatically generate several accounts to facilitate simulation experiments. At the same time, it provides each account with 100 virtual ETH for simulation of smart contract and transaction sending. Also, there is some key information contained in ganache, such as NETWORK ID, RPC SERVER address, etc.

Writing and Deploying Contracts. The smart contract of this paper is written and deployed to the blockchain through remix. Remix is an online integrated development environment officially provided by Ethereum for writing smart contracts based on solidity. It can facilitate the development, debugging and deployment of smart contracts. It can also provide bytecode and ABI for Web3 to interact with smart contracts [7]. Before the contract is deployed, connect ganache to the metamask wallet to facilitate the management of the Ethereum account. Metamask is a lightweight Ethereum wallet that supports the official Ethereum network and various Ethereum test networks, and can also be connected to private blockchain through custom rpc [3].

3   Experiments and Analysis

The analysis of the experimental results is shown in Fig. 3, where the image summary description is "f093023b07bf278faada85521af8cd8601059d189e45d 306163e05c14bb42263", the account 1 address is "0 × 6cc6419522aDD0147D6 Eb0a78e8E1a7D 8c6682e5", and the account 2 address is "0 × 8b46b20396723 Bf57dB98Cd9e9a2860e78 F1584D". The analysis of the experimental test results is as follows.

(1) **Image upload** Account1 is used as the current account to upload the image summary description to the blockchain, and we can see the image ID304956055 returned by the system, which can be used as an image identification for copyright transaction and copyright authentication.

(2) **Image copyright transfer** Account1 initiates an image copyright transaction to account2, and the image ID is transferred from the address of account1 to the address of account2.

(3) **Image transfer succeeded** The current account is switched to account2, which has successfully owned the image ID.

(4) **Image ID removed** Switching back to account1, we can see that the image ID has been successfully removed.

(5) **Tracking** Tracing the image ID, we can see that the image was created by account1 and successfully authorized to account2 through transaction.

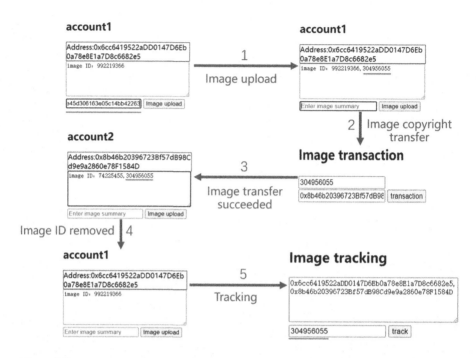

Fig. 3. The experimental result of proposed image copyright authentication scheme.

4   Conclusion

In this paper, an image copyright authentication scheme is proposed on the basis of Ethereum to avoid the watermarking embedding or third-party organizations' certification in traditional image copyright authentication mechanism. Image authenticated timestamp and copyright trading timestamp are used to prove the ownership of the image; at the same time, the traceability of the image can be realized by capturing the timestamp that each user purchases the image. As the two sides of the transaction can easily complete the image copyright transaction by consensus, the combination of blockchain and image copyright authentication not only solves the security problems and efficiency problems of the traditional image copyright authentication mechanism, but also effectively avoids possible disputes. Compared with the traditional third-party certification mechanisms, the image copyright authentication scheme based on Ethereum proposed in this paper gets twice the result with half the effort, it can not only provide a channel for image creators to defend their copyrights, but also integrate the image copyright transactions function. Our future works will concentrate on the copyright authentication of short videos and some other multimedia data formats and will try to attack smart contracts to find and fix the vulnerabilities of smart contracts.

References

1. Kang, K.Y.: Cryptocurrency and double spending history: transactions with zero confirmation. Econ. Theory. 1–39 (2022). https://doi.org/10.1007/s00199-021-01411-3
2. Luo, J.: Hybrid encryption algorithm based on md5 and base64. Comput. App. **32**, 47–49 (2012)
3. Pramulia, D., Anggorojati, B.: Implementation and evaluation of blockchain based e-voting system with Ethereum and metamask. In: 2020 International Conference on Informatics, Multimedia, Cyber and Information System (ICIMCIS, pp. 18–23. IEEE (2020). https://doi.org/10.1109/ICIMCIS51567.2020.9354310
4. Pritzker, P., Gallagher, P.D.: Sha-3 standard: permutation-based hash and extendable-output functions (2014). https://doi.org/10.6028/NIST.FIPS.202
5. Rachmawati, D., Tarigan, J., Ginting, A.: A comparative study of message digest 5 (md5) and sha256 algorithm. J. Phys. Conf. Ser. **978**, 012116 (2018). https://doi.org/10.1088/1742-6596/978/1/012116
6. Song, H., Zhu, N., Xue, R., He, J., Zhang, K., Wang, J.: Proof-of-contribution consensus mechanism for blockchain and its application in intellectual property protection. Inf. Process. Manage. **58**(3), 102507 (2021). https://doi.org/10.1016/j.ipm.2021.102507
7. Teo, E.: Introduction to blockchain smart contracts and programming with solidity for Ethereum. In: Blockchain and Smart Contracts: Design Thinking and Programming for FinTech, pp. 189–216. World Scientific (2021)
8. Wei, M., Gao, y., Zhang, X.: Research on infringement of we-media works and copyright protection. Trade Exhib. Econ. **4**, 66–68 (2022). https://doi.org/10.19995/j.cnki.CN10-1617/F7.2022.04.066
9. Zhao, X., Lei, Z., Zhang, G., Zhang, Y., Xing, C.: Blockchain and distributed system. In: Wang, G., Lin, X., Hendler, J., Song, W., Xu, Z., Liu, G. (eds.) WISA 2020. LNCS, vol. 12432, pp. 629–641. Springer, Cham (2020). https://doi.org/10.1007/978-3-030-60029-7_56
10. Zheng, Z., et al.: An overview on smart contracts: challenges, advances and platforms. Futur. Gener. Comput. Syst. **105**, 475–491 (2020). https://doi.org/10.1016/j.future.2019.12.019

Modifiable Blockchain Based on Chebyshev Polynomial and Chameleon Hash Function

Guizhong Xu and Haojun Sun[✉]

Shantou University, Shantou, China
{19gzxu,haojunsun}@stu.edu.cn

Abstract. Blockchain has become one of the most important technologies in the financial sector. The blockchain based on cryptographic hash algorithm has a strong anti-tampering function. But it also makes it nearly impossible for the blockchain to modify any bad transactions that could be detrimental to the economy as a whole. In order to solve the problem of difficulty in blockchain modification, this paper proposes a modifiable blockchain based on chameleon hash and Chebyshev polynomials (MBCC). We use Chebyshev polynomials to concatenate blocks and use the chameleon hash function to ensure transaction modifiability. The blockchain we designed inherits the advantages of the classic blockchain well, adopts a new proof-of-work mode, and supports transaction-level modification. And MBCC maintains strong consistency in the verification process of all blocks (modified and unmodified). Finally, we conducted an experimental analysis and believed that the structure of this blockchain can meet the actual requirements and have a wide range of application scenarios on the premise of ensuring security.

Keywords: Modifiable blockchain · Chameleon hash · Chebyshev polynomial

1 Introduction

Blockchain technology has experienced the development from distributed ledgers to smart contracts, and has been widely used around the world. Bitcoin created the blockchain in 2008, and the emergence of Ethereum in 2013 brought blockchain technology into the 2.0 era [1]. Since the birth of the blockchain, the technological development has been advancing by leaps and bounds. At present, the blockchain technology has gradually entered the practical stage, and this is what we often call the era of blockchain 3.0, and the blockchain is deeply applied to the fields of politics, economy and culture [2,3].

However, blockchain applications have some security issues that traditional centralized databases do not have. For example, The DAO (Distributed Autonomous Organization) event [4], the crowdfunding amount of this project

© The Author(s), under exclusive license to Springer Nature Switzerland AG 2022
X. Zhao et al. (Eds.): WISA 2022, LNCS 13579, pp. 732–739, 2022.
https://doi.org/10.1007/978-3-031-20309-1_64

is hundreds of millions of dollars. Hackers exploited two loopholes in the DAO to separate funds from the pool and avoid the funds being destroyed. In just a few hours, more than 30% of the ETH in the DAO contract was transferred. Generally speaking, we cannot avoid the vast majority of losses just by withdrawing the transaction like on other centralized systems [5]. This is because most blockchains today are based on linked lists of hashes. Since hash collisions are nearly impossible, the data recorded on the blockchain cannot be modified. Therefore, people began to try to abandon immutability in limited ways and design some modifiable blockchain structure. Existing modifiable blockchains have their own advantages and disadvantages. We propose an improved scheme: a modifiable blockchain based on chameleon hash and Chebyshev polynomials. We have improved the ledger structure of the existing modifiable blockchain and maintain as many excellent features as possible: support for fine-grained modification at the transaction level; sufficiently reliable verification methods; guarantee security and robustness. Our work starts with the definition of the underlying data structures, refines our design with minimal granularity around the entire lifecycle of the transaction, and ends with experimental simulations and summaries.

This article is divided into six Sections. Section 1: Introduction, which mainly introduces the background and significance of modifiable blockchain. The Sect. 2 mainly introduces the related work and achievements in the field of modifiable blockchain. Section 3 will describe our proposed modifiable blockchain structure. Section 4 describes the operation of modifiable blockchains. The Sect. 5 presents the experimental results, and the Sect. 6 is the summary and outlook of the work.

2   Related Work

Modification of the blockchain has always been a hot topic in the blockchain field. Data can be easily modified in other systems, but in blockchain systems this is often not so easy to achieve. The most famous work is Giuseppe Ateniese et al. in 2017, who proposed a blockchain data modification scheme based on the chameleon hash function. The chameleon hash is a hash function that contains trapdoors, knowing that this trapdoor can effectively create collisions, any edit to the blockchain by a participant holding a chameleon hash of a block is possible, including deletion, modification, and insertion of any number of blocks. The advantage is that it is almost compatible with most mainstream blockchain architectures. However, such a structure cannot provide transaction-level modification rights [6]. In another study, Kuhn DR addressed the problem of preserving hash-based integrity when deleting transactions from the blockchain, which delved into a core blockchain element, its data structure. The authors describe a data structure, block matrix, and algorithm that allow the safe deletion of arbitrary records while preserving hash-based integrity guarantees that other blocks remain unchanged. Although their scheme does not discuss the decentralized structure, it is still a very pioneering approach [7]. In addition, some schemes adopt a compromise method, such as I Puddu, A Dmitrienko et al.

After the transaction is released, the user can select a transaction in the queue as the active transaction by change strategy, so as to realize the modification with the transaction as the granularity. The security guarantees provided by the deformed blockchain [8], while the disadvantage is that more storage space is required, that only limited and predetermined modifications can be achieved because the transaction group is limited. There is also a common solution to have the blockchain only be used to store time information and hashes that point to actual information held off-chain [9,10]. In fact, off-chain technologies are a widely used technology because of their significant benefits, enhancing scalability and reducing blockchain data storage requirements. However, this workaround just bypasses and does not solve our problem [11]. More seriously, these techniques shift the responsibility of distributed data storage to other protocols, introducing complexity and delay, increasing the risk of more attacks and degrading security [12]. In the following we will show our work.

3   A MBCC Blockchain Structure

In this chapter, we will use the classic blockchain as a template to define the block structure and chain structure of MBCC.

3.1   Block Structure

A series of transaction data is recorded in each block. Combining the work of predecessors and the actual needs of MBCC, we define Block as shown in Table 1. A block contains a block header and a transaction set, a block header is mainly used to record important information of the block, and the transaction set stores the packaged transactions. MBCC is based on transaction granularity, so the height of the block cannot determine the longest chain. AllTransactionCounter can help us to easily determine the longest chain. Miners will keep processing new transactions, adding new blocks to the end of the chain.

Table 1. Definition of MBCC block.

Values	Explanations
BlockHeader	Several important data of the block
Transactions	The set of all transactions in the block
AllTransactionCounter	The total number of transactions on the current blockchain

BlockHeader is an important part of the block, as shown in Table 2. It consists of three data fields: PrevBlockCheby, Coordinate and Target. PrevBlockCheby is the Chebyshev polynomial calculated from the transaction set of the previous block, and is used as the pointer of MBCC. It uniquely points to the previous block, that is to say, it assumes part of the function of the hash encryption function. Coordinates are the coordinate values of the current hash value mapping of all transactions, and are parameters that ensure security on MBCC. We use it to check the consistency of this block. It ensures security through encrypted hash function and Chebyshev polynomial. Target is the difficulty factor, which represents the difficulty of adding a transaction to a block.

Table 2. Definition of MBCC BlockHeader.

Values	Explanations
PrevBlockCheby	The Chebyshev polynomial obtained by the interpolation of the previous block
Coordinate	The set of coordinates calculated by the transaction set of this block
Target	The transaction generation difficulty of this block

In order to meet the modification of transaction granularity, we define transaction as shown in Table 3. Among them, TX contains the specific details of a transaction; ChamPubKey is the chameleon hash public key of the transaction, and ChamNonce is the random number of the transaction. By calculating the chameleon hash operation **ChameHash**(TX $\|$ChamPubKey $\|$ChamNonce), we can get the chameleon hash value of the transaction, which is mapped to a real number in MBCC.

Table 3. Definition of MBCC Transaction.

Values	Explanations
TX	The content of the transaction
ChamPubKey	The Chameleon hash public key of this transaction
ChamNonce	The random value of this transaction

3.2 Chain Structure

MBCC blocks use Chebyshev polynomial polynomials as addresses. Use PrevBlockCheby as a pointer to point to the previous block, thus linking all blocks into a MBCC chain. Each block contains n transactions, and each transaction is mapped to a real number, which is matched with Chebyshev nodes in turn to obtain a sequence of coordinates $(x_1, y_1) \cdots (x_n, y_n)$, where the x and y coordinates are the Chebyshev points and the corresponding transaction hash. We know that these coordinate sequences can fit a unique Chebyshev polynomial of degree $n - 1$. In this way, a polynomial function can uniquely determine a block due to the non-collisability of the hash function. Generating two identical curves requires n hash collisions, which is nearly impossible. We also use Coordinate to ensure that the block's data is not maliciously tampered with. After hashing all the chameleon hashes of the current block, we substitute the resulting real numbers into the Chebyshev polynomial to get a series of coordinate values, which can be used to verify the validity of transaction information in MBCC. And AllTransactionCounter can be used to determine the longest chain, because in the structure of MBCC, the more transactions in the block, the more work required.

4   Operations

In this chapter, we will introduce how MBCC participants (e.g., Alice) publish transactions, how miners (e.g., Bob) synthesize blocks, and how to modify transactions. Finally, we analyze some problems that may arise in the blockchain system.

4.1   Publish Transaction

In this section we describe how to publish a transaction, which is a fairly complex task, and we need to consider more things in MBCC than in classic blockchains. Assuming such a situation, Alice is going to send 100 dollars to Carol, let's take a look at the work Alice needs. First, Alice needs to generate a new pair of Chameleon public and private keys for this transaction in addition to the public and private keys of her own account. The chameleon public key is public, while the chameleon private key needs to be kept secret. Then Alice generates the transaction content of TX and signs it. As shown in Table 3, Alice fills in TX and ChamPubKey. Finally, all the content is packaged and distributed through the P2P network.

4.2   Create Block Using MBCC POW

Miner Bob has received a series of transactions, and we introduce how he generates a block in this section. Bob first needs to obtain the difficulty coefficient through the nearest Target. For each transaction, Bob needs to generate a Cham-Nonce so that the generated chameleon hash matches the difficulty coefficient. Then as we introduced earlier, Bob fits all transactions into Chebyshev polynomials, fills in other data items and signs them. Finally publishes to the blockchain community.

Every time we generate a coordinate here is equivalent to completing a proof of work, and the work of entire block is divided into several non-interfering tasks. Therefore, the amount of calculation required by MBCC to generate a block is equal to the number of transactions is positively correlated. This can save computing resources compared to the traditional POW algorithm.

4.3   Block Modification

Going back to Alice's example, in this section we will detail how Alice modifies the transaction she publishes. If Alice regrets it, she just wants to send 10 dollars to Carlo, that is, replace the original message "Alice sends Carol 100 dollars" with "Alice sends Carol 10 dollars". Before the actual work starts, we first make the following agreement: Carol did not spend the 100 yuan; Alice has complete modification rights; and the community has a complete modification agreement. Alice first modifies the content of TX and reads the chameleon private key of the transaction. Then he needs to calculate the chameleon hash collision algorithm **ChamCol**(TX$_{new}$ ||ChamPubKey ||ChamHash) to calculate the new ChamNonce. After that, Alice re-signs the transaction and publishes it to the community. Although Alice's transaction content has changed, the chameleon hash has not changed. Therefore, the modified block address has not changed, and MBCC ensures the consistency of the verification method before and after the block modification. Moreover, Alice's modification of the transaction will not have any effect on other transactions or data on the same block.

4.4 Some Discussion

In this section we discuss several issues.

First, we believe that MBCC can be a good replacement for Merkle trees. In MBCC, we can quickly verify the legitimacy of transactions through PrevBlockCheby. And we set a brand new parameter coordinate, which can also quickly check whether all transactions match the BlockHeader. Therefore, MBCC does not need to generate a Merkle tree, and can still guarantee security and practicality.

Secondly, we discuss why MBCC uses Chebyshev polynomials. The advantage of Chebyshev interpolation over Lagrangian interpolation is that it avoids the Runge phenomenon. On the other hand, Chebyshev points are very easy to calculate and unique, and we hope to use such a string of numbers as anchor points to facilitate MBCC verification.

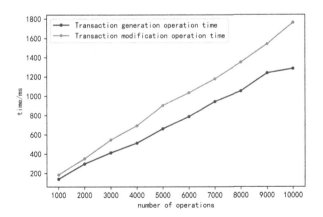

Fig. 1. The time consumption of the MBCC operation

In addition, our structure can reduce the consumption of computing resources. The maximum number of transactions in our block is 2000. Let's assume this situation: Bob packs 2000 transactions, but finds that the longest chain already contains 300 of them. In Bitcoin, we know that all of Bob's work is invalid at this time, but in MBCC, Bob still has 1700 transactions that are valid. This brings another benefit: the earlier the transaction is generated, the easier it is to be packaged and released.

5   Experiments

In this section we will perform performance analysis of MBCC. We developed MBCC with Golang and used Intel Xeon E3-1231 v3 CPU to simulate real usage environment. We first used a 1000-byte transaction to perform the chameleon hash operation and the chameleon hash collision operation. Figure 1 shows the time consumption results of MBCC performing 1000 to 10000 times. The first conclusion we can draw is that the modification operation of MBCC does not require a high amount of computation, and it only takes 1.761 s to modify a 1000-byte transaction 10,000 times. And we found that a Chameleon hash is more computationally expensive than SHA256. In the same environment, 10,000 times SHA256 takes about 38 ms, while CHash takes about 1284 ms. Of course this is not a disadvantage, the cryptographic hash operation itself is part of the proof of work. Second, we tested the effect of transaction size on the time consumption of modification operations. As shown in Fig. 2, we found that the transaction size is generally proportional to the time consumed by the modification operation. It takes only about 0.159 ms to modify a bitcoin-sized transaction. Through our experimental analysis, MBCC can meet the needs of actual use. And it should be noted that we did not call GPU resources for operation, and the hash operation capability of GPU is much higher than that of CPU.

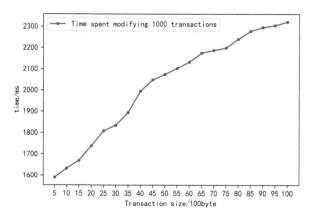

Fig. 2. The relationship between the time consumption of the modification operation and the Transaction size

6   Conclusion

As an emerging research direction, editable blockchain has a variety of different solutions. The use of editable blockchain to develop practical applications has broad prospects and has important practical significance and research value. This paper proposes MBCC to achieve modifiability of blockchain. Compared with the polynomial-based blockchain structures proposed by other researchers, the structure proposed in this paper adopts a novel proof-of-work method and supports transaction-level modification, which has certain advantages. The disadvantage is that there is no detailed scheme for assigning modification rights. In future work, we can study the scheme for assigning modification rights suitable for MBCC. In terms of security, this paper does not conduct a strict cryptographic analysis of MBCC. If it is subsequently applied to reality, a strict theoretical analysis is necessary.

References

1. von Haller, G.M.: Blockchain 2.0 smart, contracts and challenges. Comput. Law, SCL. Mag. **2016**, 1–5 (2016)
2. Maesa, D.D.F., Mori, P.: Blockchain 30 applications survey. J. Parallel Distrib. Comput. **138**, 99–114 (2020)
3. Mao, X., Li, X., Guo, S.: A blockchain architecture design that takes into account privacy protection and regulation. In: Xing, C., Fu, X., Zhang, Y., Zhang, G., Borjigin, C. (eds.) WISA 2021. LNCS, vol. 12999, pp. 311–319. Springer, Cham (2021). https://doi.org/10.1007/978-3-030-87571-8_27
4. Luu, L., Chu, D.H., Olickel, H., et al.: Making smart contracts smarter. In: 2016 Proceedings of the ACM SIGSAC Conference on Computer and Communications Security, pp. 254–269 (2016)
5. Dhillon, V., Metcalf, D., Hooper, M.: The DAO Hacked. Blockchain Enabled Applications, pp. 67–78. Apress, Berkeley (2017)
6. Politou, E., Casino, F., Alepis, E., et al.: Blockchain mutability: challenges and proposed solutions. IEEE Trans. Emerg. Top. Comput. **9**, 1972–1986 (2019)
7. Kuhn, D.R.: A data structure for integrity protection with erasure capability. NIST Cybersecur. Whitepaper (2018)
8. Puddu, I., Dmitrienko, A., Capkun, S.: μ chain: how to forget without hard forks. Cryptology ePrint Archive (2017)
9. Eberhardt, J., Tai, S.: On or off the blockchain? Insights on off-chaining computation and data. In: De Paoli, F., Schulte, S., Broch Johnsen, E. (eds.) ESOCC 2017. LNCS, vol. 10465, pp. 3–15. Springer, Cham (2017). https://doi.org/10.1007/978-3-319-67262-5_1
10. García-Barriocanal, E., Sánchez-Alonso, S., Sicilia, M.-A.: Deploying metadata on blockchain technologies. In: Garoufallou, E., Virkus, S., Siatri, R., Koutsomiha, D. (eds.) MTSR 2017. CCIS, vol. 755, pp. 38–49. Springer, Cham (2017). https://doi.org/10.1007/978-3-319-70863-8_4
11. Dorri, A., Kanhere, S.S., Jurdak, R.: MOF-BC: a memory optimized and flexible blockchain for large scale networks. Futur. Gener. Comput. Syst. **92**, 357–373 (2019)
12. Humbeeck, A.V.: The blockchain-GDPR paradox. J. Data Protect. Priv. (2019)

Author Index

Printed in the United States
by Baker & Taylor Publisher Services